SELECTED SOLUTIONS MANUAL

Mary Beth Kramer
University of Delaware

Kathleen Thrush Shaginaw
Particular Solutions, Inc.

CHEMISTRY

A MOLECULAR APPROACH

Nivaldo J. Tro

PEARSON
Prentice Hall

Upper Saddle River, NJ 07458

Editor-in-Chief, Science: Nicole Folchetti
Editor: Jeff Howard
Assistant Editor: Carol G. DuPont
Senior Managing Editor, Science: Kathleen Schiaparelli
Assistant Managing Editor, Science: Gina M. Cheselka
Project Manager: Ed Thomas
Supplement Cover Manager: Paul Gourhan
Supplement Cover Designer: Victoria Colotta
Senior Operations Supervisor: Alan Fischer
Director of Operations: Barbara Kittle

Pearson Prentice Hall
Pearson Education, Inc.
Upper Saddle River, NJ 07458

Printed in the United States of America

10 9 8 7 6 5 4 3 2 1

ISBN 13: 978-0-13-615116-6

ISBN 10: 0-13-615116-7

Pearson Education Ltd., *London*
Pearson Education Australia Pty. Ltd., *Sydney*
Pearson Education Singapore, Pte. Ltd.
Pearson Education North Asia Ltd., *Hong Kong*
Pearson Education Canada, Inc., *Toronto*
Pearson Educación de Mexico, S.A. de C.V.
Pearson Education—Japan, *Tokyo*
Pearson Education Malaysia, Pte. Ltd.

Table of Contents

Chapter 1
Matter, Measurement, and Problem Solving

1. "The properties of the substances around us depend on the atoms, ions, or molecules that compose them" means that the specific types of atoms and molecules that compose something tell us a great deal about the properties to expect. A material composed of only sodium and chlorine ions will have the properties of table salt. A material composed of molecules with one carbon atom and two oxygen atoms will have the properties of the gas carbon dioxide. If the atoms and molecules change, so do the properties that we expect the material to have.

3. The scientific approach to knowledge is based on observation and experiment. Scientists observe and perform experiments on the physical world to learn about it. Observations often lead scientists to formulate a hypothesis, a tentative interpretation or explanation of the observations. Hypotheses are tested by experiments, highly controlled procedures designed to generate such observations. The results of an experiment may support a hypothesis or prove it wrong—in which case the hypothesis must be modified or discarded. A series of similar observations can lead to the development of scientific law, a brief statement that summarizes past observations and predicts future ones. One or more well-established hypotheses may form the basis for a scientific theory. A scientific theory is a model for the way nature is and tries to explain not merely what nature does, but why.

   The Greek philosopher Plato (427 - 347 B.C.) took an opposite approach. He thought that the best way to learn about reality was not through the senses, but through reason. He believed that the physical world was an imperfect representation of a perfect and transcendent world (a world beyond space and time). For him, true knowledge came, not through observing the real physical world, but through reasoning and thinking about the ideal one.

5. Antoine Lavoisier studied combustion and made careful measurements of the mass of objects before and after burning them in closed containers. He noticed that there was no change in the total mass of material within the container during combustion. Lavoisier summarized his observations on combustion with the law of conservation of mass, which states that, "In a chemical reaction, matter is neither created nor destroyed."

7. The statement "that is just a theory" is generally taken to mean that there is no scientific proof behind the statement. This statement is the opposite of the meaning in the context of the scientific theory, where theories are tested again and again.

9. In solid matter, atoms or molecules pack close to each other in fixed locations. Although the atoms and molecules in a solid vibrate, they do not move around or past each other. Consequently, a solid has a fixed volume and rigid shape.

   In liquid matter, atoms or molecules pack about as closely as they do in solid matter, but they are free to move relative to each other, giving liquids a fixed volume but not a fixed shape. Liquids assume the shape of their container.

   In gaseous matter, atoms or molecules have a lot of space between them and are free to move relative to one another, making gases compressible. Gases always assume the shape and volume of their container.

11. A pure substance is composed of only one type of atom or molecule. In contrast, a mixture is a substance composed of two or more different types of atoms or molecules that can be combined in variable proportions.

13. A homogeneous mixture has the same composition throughout, while a heterogeneous mixture has different compositions in different regions.

15. Mixtures of miscible liquids (substances that easily mix) can usually be separated by distillation, a process in which the mixture is heated to boil off the more volatile (easily vaporizable) liquid. The volatile liquid is then recondensed in a condenser and collected in a separate flask.

17. Changes that alter only state or appearance, but not composition, are called physical changes. The atoms or molecules that compose a substance *do not change* their identity during a physical change. For example, when water boils, it changes its state from a liquid to a gas, but the gas remains composed of water

molecules, so this a physical change. When sugar dissolves in water, the sugar molecules are separated from each other, but the molecules of sugar and water remain intact.

In contrast, changes that alter the composition of matter are called chemical changes. During a chemical change, atoms rearrange, transforming the original substances into different substances. For example, the rusting of iron, the combustion of natural gas to form carbon dioxide and water, and the denaturing of proteins when an egg is cooked are examples of chemical changes.

19. Chemical energy is a potential energy. It is the energy that is contained in the bonds that hold the molecules together. This energy arises primarily from electrostatic forces between the electrically charged particles (protons and electrons) that compose atoms and molecules. Some of these arrangements—such as the one within the molecules that compose gasoline—have a much higher potential energy than others. When gasoline undergoes combustion the arrangement of these particles changes, creating molecules with much lower potential energy and transferring a great deal of energy (mostly in the form of heat) to the surroundings. A raised weight has a certain amount of potential energy (dependent on the height the weight is raised) that can be converted to kinetic energy when the weight is released.

21. The three different temperature scales are Kelvin (K), Celsius (°C), and Fahrenheit (°F). The size of the degree is the same in the Kelvin and the Celsius scales, and they are 1.8 times larger than the degree size for the Fahrenheit scale.

23. A derived unit is a combination of other units. Examples of derived units are: speed in meters per second (m/s), volume in meters cubed (m^3), and density in grams per cubic centimeter (g/cm^3).

25. An intensive property is a property that is independent of the amount of the substance. An extensive property is a property that depends on the amount of the substance.

27. In multiplication or division, the result carries the same number of significant figures as the factor with the fewest significant figures.

29. When rounding to the correct number of significant figures, round down if the last (or left-most) digit dropped is four or less; and round up if the last (or left-most) digit dropped is five or more.

31. Random error is error that has equal probability of being too high or too low. Almost all measurements have some degree of random error. Random error can, with enough trials, average itself out. Systematic error is error that tends towards being either too high or too low. Systematic error does not average out with repeated trials.

33. a) This statement is a theory because it attempts to explain why. It is not possible to observe individual atoms.

    b) This statement is an observation.

    c) This statement is a law, because it summarizes many observations and can explain future behavior.

    d) This statement is an observation.

35. a) If we divide the mass of the oxygen by the mass of the carbon the result is always 4/3.

    b) If we divide the mass of the oxygen by the mass of the hydrogen the result is always 16.

    c) These observations suggest that the masses of elements in molecules are ratios of whole numbers (4 and 3; and 16 and 1, respectively).

    d) Atoms combine in small whole number ratios and not as random weight ratios.

37. a) Sweat is a homogeneous mixture of water, sodium chloride, and other components.

    b) Carbon dioxide is a pure substance that is a compound (two or more elements bonded together).

c) Aluminum is a pure substance that is an element (element #13 in the periodic table).

d) Vegetable soup is a heterogeneous mixture of broth, chunks of vegetables, and extracts from the vegetables.

39.

substance	pure or mixture	Type
aluminum	pure	element
apple juice	mixture	homogeneous
hydrogen peroxide	pure	compound
chicken soup	mixture	heterogeneous

41. a) Pure substance that is a compound (one type of molecule that contains two different elements)

b) Heterogeneous mixture (two different molecules that are segregated into regions)

c) Homogeneous mixture (two different molecules that are randomly mixed)

d) Pure substance is an element (individual atoms of one type)

43. a) Physical property (color can be observed without making or breaking chemical bonds)

b) Chemical property (must observe by making or breaking chemical bonds)

c) Physical property (the phase can be observed without making or breaking chemical bonds)

d) Physical property (density can be observed without making or breaking chemical bonds)

e) Physical property (mixing does not involve making or breaking chemical bonds, so this can be observed without making or breaking chemical bonds)

45. a) Chemical property (burning involves breaking and making bonds, so bonds must be broken and made to observe this property)

b) Physical property (sublimation is a phase change and so can be observed without making or breaking chemical bonds)

c) Physical property (odor can be observed without making or breaking chemical bonds)

d) Chemical property (burning involves breaking and making bonds, so bonds must be broken and made to observe this property)

47. a) Chemical change (new compounds are formed as methane and oxygen react to form carbon dioxide and water)

b) Physical change (vaporization is a phase change and does not involve the making or breaking of chemical bonds)

c) Chemical change (new compounds are formed as propane and oxygen react to form carbon dioxide and water)

d) Chemical change (new compounds are formed as the metal in the frame is converted to oxides)

49. a) Physical change (vaporization is a phase change and does not involve the making or breaking of chemical bonds)

b) Chemical change (new compounds are formed)

c) Physical change (vaporization is a phase change and does not involve the making or breaking of chemical bonds)

51. a) To convert from °F to °C, first find the equation that relates these two quantities. $°C = \frac{°F - 32}{1.8}$ Now substitute °F into the equation and compute the answer. Note: The number of digits reported in this answer follow significant figure conventions, covered in section 1.6. $°C = \frac{°F - 32}{1.8} = \frac{0.}{1.8} = 0.\ °C$

b) To convert from K to °F, first find the equations that relates these two quantities. $K = °C + 273.15$ and $°C = \frac{°F - 32}{1.8}$ Since these equations do not directly express K in terms of °F, you must combine the equations and then solve the equation for °F. Substituting for °C: $K = \frac{°F - 32}{1.8} + 273.15$ rearrange $K - 273.15 = \frac{°F - 32}{1.8}$ rearrange $1.8\,(K - 273.15) = (°F - 32)$ finally $°F = 1.8\,(K - 273.15) + 32$ Now substitute K into the equation and compute the answer.
°F = 1.8 (77 K – 273.15) + 32 = 1.8 (-196 K) + 32 = -353 + 32 = -321 °F

c) To convert from °F to °C, first find the equation that relates these two quantities. $°C = \frac{°F - 32}{1.8}$ Now substitute °F into the equation and compute the answer.
$°C = \frac{-109°F - 32°F}{1.8} = \frac{-141}{1.8} = -78.3\ °C$

d) To convert from °F to K, first find the equations that relates these two quantities. $K = °C + 273.15$ and $°C = \frac{°F - 32}{1.8}$ Since these equations do not directly express K in terms of °F, you must combine the equations and then solve the equation for K. Substituting for °C $K = \frac{°F - 32}{1.8} + 273.15$
Now substitute °F into the equation and compute the answer.
$K = \frac{(98.6 - 32)}{1.8} + 273.15 = \frac{66.6}{1.8} + 273.15 = 37.0 + 273.15 = 310.2\ K$

53. To convert from °F to °C, first find the equation that relates these two quantities. $°C = \frac{°F - 32}{1.8}$ Now substitute °F into the equation and compute the answer. Note: The number of digits reported in this answer follow significant figure conventions, covered in section 1.6. $°C = \frac{-80.°F - 32°F}{1.8} = \frac{-112}{1.8} = -62.2\ °C$
Begin by finding the equation that relates the quantity that is given (°C) and the quantity you are trying to find (K). $K = °C + 273.15$ Since this equation gives the temperature in K directly, simply substitute in the correct value for the temperature in °C and compute the answer. K = -62.2 °C + 273.15 = 210.9 K .

55. Use Table 1.2 to determine the appropriate prefix multiplier and substitute the meaning into the expressions.
a) 10^{-9} implies "nano" so 1.2×10^{-9} m = 1.2 nanometers = 1.2 nm

b) 10^{-15} implies "femto" so 22×10^{-15} s = 22 femtoseconds = 22 fs

c) 10^{9} implies "giga" so 1.5×10^{9} g = 1.5 gigagrams = 1.5 Gg

d) 10^{6} implies "mega" so 3.5×10^{6} L = 3.5 megaliters = 3.5 ML

57. b) **Given:** 515 km **Find:** dm
Conceptual plan: **km → m → dm**

$\frac{1000\ m}{1km}$ $\frac{10\ dm}{1m}$

Solution: $515 \; \cancel{km} \times \dfrac{1000 \; \cancel{m}}{1 \; \cancel{km}} \times \dfrac{10 \; dm}{1 \; \cancel{m}} = 5.15 \times 10^6 \; dm$

Check: The units (dm) are correct. The magnitude of the answer (10^6) makes physical sense because a decimeter is a much smaller unit than a kilometer.

Given: 515 km   **Find:** cm
Conceptual plan: km $\rightarrow$ m $\rightarrow$ cm

$$\dfrac{1000 \; m}{1 km} \qquad \dfrac{100 \; cm}{1m}$$

Solution: $515 \; \cancel{km} \times \dfrac{1000 \; \cancel{m}}{1 \; \cancel{km}} \times \dfrac{100 \; cm}{1 \; \cancel{m}} = 5.15 \times 10^7 \; cm$

Check: The units (cm) are correct. The magnitude of the answer (10^7) makes physical sense because a centimeter is a much smaller unit than a kilometer and a decimeter.

c) **Given:** 122.355 s   **Find:** ms
   Conceptual plan: s $\rightarrow$ ms

$$\dfrac{1000 \; ms}{1s}$$

   Solution: $122.355 \; \cancel{s} \times \dfrac{1000 \; ms}{1 \; \cancel{s}} = 1.22355 \times 10^5 \; ms$

   Check: The units (ms) are correct. The magnitude of the answer (10^5) makes physical sense because a millisecond is a much smaller unit than a second.
   Given: 122.355 s   **Find:** ks
   Conceptual plan: s $\rightarrow$ ks

$$\dfrac{1 \; ks}{1000 \; s}$$

   Solution: $122.355 \; \cancel{s} \times \dfrac{1 \; ks}{1000 \; \cancel{s}} = 1.22355 \times 10^{-1} \; ks = 0.122355 \; ks$

   Check: The units (ks) are correct. The magnitude of the answer (10^{-1}) makes physical sense because a kilosecond is a much larger unit than a second.

d) **Given:** 3.345 kJ   **Find:** J
   Conceptual plan: kJ $\rightarrow$ J

$$\dfrac{1000 \; J}{1kJ}$$

   Solution: $3.345 \; \cancel{kJ} \times \dfrac{1000 \; J}{1 \; \cancel{kJ}} = 3.345 \times 10^3 \; J$

   Check: The units (J) are correct. The magnitude of the answer (10^3) makes physical sense because a joule is a much smaller unit than a kilojoule.
   Given: 3.345×10^3 J (from above)   **Find:** mJ
   Conceptual plan: J $\rightarrow$ mJ

$$\dfrac{1000 \; mJ}{1 \; J}$$

   Solution: $3.345 \times 10^3 \; \cancel{J} \times \dfrac{1000 \; mJ}{1 \; \cancel{J}} = 3.345 \times 10^6 \; mJ$

   Check: The units (mJ) are correct. The magnitude of the answer (10^6) makes physical sense because a millijoule is a much smaller unit than a joule.

59. **Given:** 1 m square 1 m^2 **Find:** cm^2
    Conceptual plan: 1 m^2 $\rightarrow$ cm^2

$$\dfrac{100 \; cm}{1 \; m}$$ Notice that for squared units, the conversion factors must be squared.

Solution: $1\ \text{m}^2 \times \frac{(100\ \text{cm})^2}{(1\ \text{m})^2} = 1 \times 10^4\ \text{cm}^2$

Check: The units of the answer are correct and the magnitude makes sense. The unit centimeter is smaller than a meter, so the value in square centimeters should be larger than in square meters.

61. **Given:** $m = 2.49$ g, $V = 0.349\ \text{cm}^3$ **Find:** d in g/cm^3 and compare to pure copper.

Conceptual plan: **m, V → d**

$d = m/V$

Compare to the published value. d (pure copper) = 8.96 g/cm^3 (This value is in Table 1.4.)

Solution: $d = \frac{2.49\ \text{g}}{0.349\ \text{cm}^3} = 7.13\ \frac{\text{g}}{\text{cm}^3}$

The density of the penny is much smaller than the density of pure copper (= 7.13 g/cm^3 < 8.96 g/cm^3) so the penny is not pure copper.

Check: The units (g/cm^3) are correct. The magnitude of the answer seems correct. Many coins are layers of metals, so it is not surprising that the penny is not pure copper.

63. **Given:** $m = 4.10 \times 10^3$ g, V = 3.25 L **Find:** d in g/cm^3
Conceptual plan: **m, V → d then L → cm^3**

$d = m/V$ $\qquad \frac{1000\ \text{cm}^3}{1\ \text{L}}$

Solution: $d = \frac{4.10 \times 10^3\ \text{g}}{3.25\ \text{L}} \times \frac{1\ \text{L}}{1000\ \text{cm}^3} = 1.26\ \frac{\text{g}}{\text{cm}^3}$

Check: The units (g/cm^3) are correct. The magnitude of the answer seems correct.

65. a) **Given:** d = 1.11 g/cm^3, V = 417 mL **Find:** m
Conceptual plan: **d, V → m then cm^3 → mL**

$d = m/V$ $\qquad \frac{1\ \text{mL}}{1\ \text{cm}^3}$

Solution: $d = m/V$ Rearrange by multiplying both sides of equation by V. $m = d \times V$

$m = 1.11\ \frac{\text{g}}{\text{cm}^3} \times \frac{1\ \text{cm}^3}{1\ \text{mL}} \times 417\ \text{mL} = 4.63 \times 10^2\ \text{g}$

Check: The units (g) are correct. The magnitude of the answer seems correct considering the value of the density is about 1 g/cm^3.

b) **Given:** d = 1.11 g/cm^3, m = 4.1 kg **Find:** V in L
Conceptual plan: **d, V → m then kg → g and cm^3 → L**

$d = m/V$ $\qquad \frac{1000\ \text{g}}{1\ \text{kg}}$ $\qquad \frac{1\ \text{L}}{1000\ \text{cm}^3}$

Solution: $d = m/V$ Rearrange by multiplying both sides of equation by V and dividing both sides of the equation by d.

$V = \frac{m}{d} = \frac{4.1\ \text{kg}}{1.11\ \frac{\text{g}}{\text{cm}^3}} \times \frac{1000\ \text{g}}{1\ \text{kg}} = 3.7 \times 10^3\ \text{cm}^3 \times \frac{1\ \text{L}}{1000\ \text{cm}^3} = 3.7\ \text{L}$

Check: The units (L) are correct. The magnitude of the answer seems correct considering the value of the density is about 1 g/cm^3.

67. In order to obtain the readings look to see where the bottom of the meniscus lies. Estimate the distance between two markings on the device.
a) 73.5 mL – the meniscus appears to be about half way between the 73 mL and the 74 mL marks.

b) 88.2 °C – the mercury is between the 84 °C mark and the 85 °C mark, but it is closer to the lower number.

c) 645 mL - the meniscus appears to be just above the 640m mL mark.

69. Remember that

1. Interior zeroes (zeroes between two numbers) are significant.
2. Leading zeroes (zeroes to the left of the of the first non-zero number) are not significant. They only serve to locate the decimal point.
3. Trailing zeroes (zeroes at the end of a number) are categorized as follows:
    - o trailing zeroes after a decimal point are always significant.
    - o trailing zeroes before an implied decimal point are ambiguous and should be avoided by using scientific notation or by inserting a decimal point at the end of the number.

a) 1,050,501 km

b) 0.0020 m

c) 0.000000000000002 s

d) 0.001090 cm

71. Remember all of the rules from section 1.7

a) Three significant figures. The 3, 1, and the 2 are significant (rule 1). The leading zeroes only mark the decimal place and are therefore not significant (rule 3).

b) Ambiguous. The 3, 1, and the 2 are significant (rule 1). The trailing zeroes occur before an implied decimal point and are therefore ambiguous (rule 4). Without more information, we would assume 3 significant figures. It is better to write this as 3.12×10^5 to indicate three significant figures or as 3.12000×10^5 to indicate six (rule 4).

c) Three significant figures. The 3, 1, and the 2 are significant (rule 1).

d) Five significant figures. The 1's, 3, 2, and 7 are significant (rule 1).

e) One significant figure. The 2 is significant (rule 1). The 0's are not significant because of there is no decimal point (rule 4).

73. a) This is not exact because π is an irrational number. 3.14 only shows three of the infinite number of significant figures that π has.

b) This is an exact conversion, because it comes from a definition of the units, and so has an unlimited number of significant figures.

c) This is a measured number and so it is not an exact number. There are 2 significant figures.

d) This is an exact conversion, because it comes from a definition of the units, and so it has an unlimited number of significant figures.

75. a) 156.9 – the 8 is rounded up since the next digit is a 5.

b) 156.8 – the last two digits are dropped since 4 is less than 5.

c) 156.8 – the last two digits are dropped since 4 is less than 5.

d) 156.9 – the 8 is rounded up since the next digit is a 9 which is greater than 5.

77. a) $9.15 \div 4.970 = 1.84$ – Three significant figures are allowed to reflect the three significant figures in the least precisely known quantity (9.15).

b) $1.54 \times 0.03060 \times 0.69 = 0.033$ – Two significant figures are allowed to reflect the two significant figures in the least precisely known quantity (0.69). The intermediate answer (0.03251556) is rounded up since the first non-significant digit is a 5.

c) $27.5 \times 1.82 \div 100.04 = 0.500$ – Three significant figures are allowed to reflect the three significant figures in the least precisely known quantity (27.5 and 1.82). The intermediate answer (0.50029988) is truncated since the first non-significant digit is a 2, which is less than 5.

d) $(2.290 \times 10^6) \div (6.7 \times 10^4) = 34$ – Two significant figures are allowed to reflect the two significant figures in the least precisely known quantity (6.7×10^4). The intermediate answer (34.17910448) is truncated since the first non-significant digit is a 1, which is less than 5.

79. a)

$$\begin{array}{r} 43.7 \\ -\ \ 2.341 \\ \hline 41.359 \end{array} = 41.4$$

Round the intermediate answer to one decimal place to reflect the quantity with the fewest decimal places (43.7). Round the last digit up since the first non-significant digit is 5.

b)

$$\begin{array}{r} 17.6 \\ +\ \ 2.838 \\ +\ \ 2.3 \\ +\ 110.77 \\ \hline 133.508 \end{array} = 133.5$$

Round the intermediate answer to one decimal place to reflect the quantity with the fewest decimal places (2.3). Truncate non-significant digits since the first non-significant digit is 0.

c)

$$\begin{array}{r} 19.6 \\ +\ 58.33 \\ -\ \ 4.974 \\ \hline 72.956 \end{array} = 73.0$$

Round the intermediate answer to one decimal place to reflect the quantity with the fewest decimal places (19.6). Round the last digit up since the first non-significant digit is 5.

d)

$$\begin{array}{r} 5.99 \\ -\ 5.572 \\ \hline 0.418 \end{array} = 0.42$$

Round the intermediate answer to two decimal places to reflect the quantity with the fewest decimal places (5.99). Round the last digit up since the first non-significant digit is 8.

81. Perform operations in parentheses first. Keep track of significant figures in each step, by noting which is the last significant digit in an intermediate result.

a) $(24.6681 \times 2.38) + 332.58 =$

$$\begin{array}{r} 58.\underline{7}10078 \\ +\ 332.58 \\ \hline 391.290078 \end{array} = 391.3$$

The first intermediate answer has one significant digit to the right of the decimal, because it is allowed three significant figures (reflecting the quantity with the fewest significant figures (2.38).) Underline the most significant digit in this answer. Round the next intermediate answer to one decimal place to reflect the quantity with the fewest decimal places (58.7). Round the last digit up since the first non-significant digit is 9.

b) $$\frac{(85.3 - 21.489)}{0.0059} = \frac{63.\underline{8}11}{0.0059} = 1.\underline{0}81542 \times 10^4 = 1.1 \times 10^4$$

The first intermediate answer has one significant digit to the right of the decimal, to reflect the quantity with the fewest decimal places (85.3). Underline the most significant digit in this answer. Round the next intermediate answer to two significant figures to reflect the quantity with the fewest significant figures (0.0059). Round the last digit up since the first non-significant digit is 8.

c) $(512 \div 986.7) + 5.44 =$

$$\begin{array}{r} 0.51\underline{8}9014 \\ +\ \ \ \ \ 5.44 \\ \hline 5.9589014 \end{array} = 5.96$$

The first intermediate answer has three significant figures and three significant digits to the right of the decimal, reflecting the quantity with the fewest significant figures (512). Underline the most significant digit in this answer. Round the next intermediate answer to two decimal places to reflect the quantity with the fewest decimal places (5.44). Round the last digit up since the first non-significant digit is 8.

d) $[(28.7 \times 10^5) \div 48.533] + 144.99 =$

$$\begin{array}{r} 59\underline{1}35.01 \\ +\quad 144.99 \\ \hline 59280.01 = 59300 \end{array} = 5.93 \times 10^4$$

The first intermediate answer has three significant figures, reflecting the quantity with the fewest significant figures (28.7×10^5). Underline the most significant digit in this answer. Since the number is so large this means that when the addition is performed, the most significant digit is the 100's place. Round the next intermediate answer to the 100's places and put in scientific notation to remove any ambiguity. Note that the last digit is rounded up since the first non-significant digit is 8.

83. a) **Given:** 154 cm **Find:** in

Conceptual plan: **cm → in**

$$\frac{1 \text{ in}}{2.54 \text{ cm}}$$

Solution: $154 \cancel{\text{cm}} \times \dfrac{1 \text{ in}}{2.54 \cancel{\text{cm}}} = 60.\underline{6}2992 \text{ in} = 60.6 \text{ in}$

Check: The units (in) are correct. The magnitude of the answer (60.6) makes physical sense because an inch is a larger unit than a cm. Three significant figures are allowed because 154 cm has three significant figures.

b) **Given:** 3.14 kg **Find:** g

Conceptual plan: **kg → g**

$$\frac{1000 \text{ g}}{1 \text{ kg}}$$

Solution: $3.14 \cancel{\text{kg}} \times \dfrac{1000 \text{ g}}{1 \cancel{\text{kg}}} = 3.14 \times 10^3 \text{ g}$

Check: The units (g) are correct. The magnitude of the answer (10^3) makes physical sense because a kg is a much larger unit than a gram. Three significant figures are allowed because 3.14 kg has three significant figures.

c) **Given:** 3.5 L **Find:** qt

Conceptual plan: **L → qt**

$$\frac{1.057 \text{ qt}}{1 \text{ L}}$$

Solution: $3.5 \cancel{\text{L}} \times \dfrac{1.057 \text{ qt}}{1 \cancel{\text{L}}} = 3.6995 \text{ qt} = 3.7 \text{ qt}$

Check: The units (qt) are correct. The magnitude of the answer (3.7) makes physical sense because a L is a smaller unit than a qt. Two significant figures are allowed because 3.5 L has two significant figures. Round the last digit up because the first non-significant digit is a 9.

d) **Given:** 109 mm **Find:** in

Conceptual plan: **mm → m → in**

$$\frac{1 \text{ m}}{1000 \text{ mm}} \qquad \frac{39.37 \text{ in}}{1 \text{ m}}$$

Solution: $109 \cancel{\text{mm}} \times \dfrac{1 \cancel{\text{m}}}{1000 \cancel{\text{mm}}} \times \dfrac{39.37 \text{ in}}{1 \cancel{\text{m}}} = 4.2\underline{9}133 \text{ in} = 4.29 \text{ in}$

Check: The units (in) are correct. The magnitude of the answer (4) makes physical sense because a mm is a much smaller unit than an in. Three significant figures are allowed because 109 mm has three significant figures.

85. **Given:** 10.0 km **Find:** minutes **Other:** running pace = 7.5 miles per hour

Conceptual plan: **km → mi → hr → min**

$$\frac{0.6214 \text{ mi}}{1 \text{ km}} \qquad \frac{1 \text{hr}}{7.5 \text{ mi}} \qquad \frac{60 \text{ min}}{1 \text{ hr}}$$

Solution: $10.0 \cancel{\text{km}} \times \dfrac{0.6214 \cancel{\text{mi}}}{1 \cancel{\text{km}}} \times \dfrac{1 \cancel{\text{hr}}}{7.5 \cancel{\text{mi}}} \times \dfrac{60 \text{ min}}{1 \cancel{\text{hr}}} = 4\underline{9}.712 \text{ min} = 50. \text{ min} = 5.0 \times 10^1 \text{ min}$

Check: The units (min) are correct. The magnitude of the answer (50) makes physical sense because she is running almost 7.5 miles (which would take her 60 min = 1 hr). Two significant figures are allowed because of the limitation of 7.5 mi/hr (two significant figures). Round the last digit up because the first non-significant digit is a 7.

87. **Given:** 14 km/L **Find:** miles per gallon

Conceptual plan: $\frac{\text{km}}{\text{L}} \rightarrow \frac{\text{mi}}{\text{L}} \rightarrow \frac{\text{mi}}{\text{gal}}$

$$\frac{0.6214 \text{ mi}}{1 \text{ km}} \quad \frac{3.785 \text{ L}}{1 \text{ gallon}}$$

Solution: $\frac{14 \text{ km}}{1 \text{ L}} \times \frac{0.6214 \text{ mi}}{1 \text{ km}} \times \frac{3.785 \text{ L}}{1 \text{ gallon}} = 32.927986 \frac{\text{miles}}{\text{gallon}} = 33 \frac{\text{miles}}{\text{gallon}}$

Check: The units (mi/gal) are correct. The magnitude of the answer (33) makes physical sense because the dominating factor is that a L is much smaller than a gallon, so the answer should go up. Two significant figures are allowed because of the limitation of 14 km/L (two significant figures). Round up the last digit up because the first non-significant digit is a 9.

89. a) **Given:** 195 m^2 **Find:** km^2

Conceptual plan: $m^2 \rightarrow km^2$

$$\frac{(1 \text{ km})^2}{(1000 \text{ m})^2}$$

Notice that for squared units, the conversion factors must be squared.

Solution: $195 \text{ m}^2 \times \frac{(1 \text{ km})^2}{(1000 \text{ m})^2} = 1.95 \times 10^{-4} \text{ km}^2$

Check: The units (km^2) are correct. The magnitude of the answer (10^{-4}) makes physical sense because a kilometer is a much larger unit than a meter.

b) **Given:** 195 m^2 **Find:** dm^2

Conceptual plan: $m^2 \rightarrow dm^2$

$$\frac{(10 \text{ dm})^2}{(1 \text{ m})^2}$$

Notice that for squared units, the conversion factors must be squared.

Solution: $195 \text{ m}^2 \times \frac{(10 \text{ dm})^2}{(1 \text{ m})^2} = 1.95 \times 10^{4} \text{ dm}^2$

Check: The units (dm^2) are correct. The magnitude of the answer (10^4) makes physical sense because a decimeter is a much smaller unit than a meter.

c) **Given:** 195 m^2 **Find:** cm^2

Conceptual plan: $m^2 \rightarrow cm^2$

$$\frac{(100 \text{ cm})^2}{(1 \text{ m})^2}$$

Notice that for squared units, the conversion factors must be squared.

Solution: $195 \text{ m}^2 \times \frac{(100 \text{ cm})^2}{(1 \text{ m})^2} = = 1.95 \times 10^{6} \text{ cm}^2$

Check: The units (cm^2) are correct. The magnitude of the answer (10^6) makes physical sense because a centimeter is a much smaller unit than a meter.

91. **Given:** 435 acres **Find:** square miles **Other:** 1 acre = 43,560 ft^2, 1 mile = 5280 ft

Conceptual plan: acres $\rightarrow ft^2 \rightarrow mi^2$

$$\frac{43560 \text{ ft}^2}{1 \text{ acre}} \quad \frac{(1 \text{ mi})^2}{(5280 \text{ ft})^2}$$

Notice that for squared units, the conversion factors must be squared.

Solution: $435 \text{ acres} \times \frac{43560 \text{ ft}^2}{1 \text{ acre}} \times \frac{(1 \text{ mi})^2}{(5280 \text{ ft})^2} = 0.6796875 \text{ mi}^2 = 0.680 \text{ mi}^2$

Check: The units (mi^2) are correct. The magnitude of the answer (0.7) makes physical sense because an acre is much smaller than a mi^2, so the answer should go down several orders of magnitude. Three significant figures are allowed because of the limitation of 435 acres (three significant figures). Round up the last digit up because the first non-significant digit is a 7.

93. **Given:** 14 lbs **Find:** mL **Other:** 80 mg/0.80 mL & 15 mg/kg body
Conceptual plan: **lb → kg body → mg → mL**

$$\frac{1 \text{ kg body}}{2.205 \text{ lb}} \quad \frac{15 \text{ mg}}{1 \text{ kg body}} \quad \frac{0.80 \text{ mL}}{80 \text{ mg}}$$

Solution: $$14 \text{ lb} \times \frac{1 \text{ kg body}}{2.205 \text{ lb}} \times \frac{15 \text{ mg}}{1 \text{ kg body}} \times \frac{0.80 \text{ mL}}{80 \text{ mg}} = 0.9\underline{5}23809524 \text{ mL} = 0.95 \text{ mL}$$

Check: The units (cm^3) are correct. The magnitude of the answer (1 mL) makes physical sense because it is reasonable amount of liquid to give to a baby. Two significant figures are allowed because of the statement in the problem. Truncate the last digit because the first non-significant digit is a 2.

95. **Given:** solar year **Find:** seconds
Other: 60 seconds/minute; 60 minutes/ hour; 24 hours/solar day; and 365.24 solar days/solar year
Conceptual plan: **yr → day → hr → min → sec**

$$\frac{365.24 \text{ day}}{1 \text{ solar yr}} \quad \frac{24 \text{ hr}}{1 \text{day}} \quad \frac{60 \text{ min}}{1 \text{ hr}} \quad \frac{60 \text{ sec}}{1 \text{ min}}$$

Solution:

$$1 \text{ solar yr} \times \frac{365.24 \text{ day}}{1 \text{ solar yr}} \times \frac{24 \text{ hr}}{1 \text{day}} \times \frac{60 \text{ min}}{1 \text{ hr}} \times \frac{60 \text{ sec}}{1 \text{ min}} = 3.155\underline{6}736 \times 10^7 \text{ sec} = 3.1557 \times 10^7 \text{ sec}$$

Check: The units (seconds) are correct. The magnitude of the answer (10^7) makes physical sense because each conversion factor increases the value of the answer – a second is many orders of magnitude smaller than a year. Five significant figures are allowed because all conversion factors are assumed to be exact, except for the 365.24 days/ solar year (five significant figures). Round up the last digit because the first non-significant digit is a 7.

97. a) Extensive - the volume of a material depends on how much there is present.

b) Intensive – the boiling point of a material is independent of how much material you have, so these values can be published in reference tables.

c) Intensive - the temperature of a material depends on how much there is present.

d) Intensive – the electrical; conductivity of a material is independent of how much material you have, so these values can be published in reference tables.

e) Extensive - the energy contained in material depends on how much there is present. Many times energy is expressed in terms of Joules/mole, which then turns this quantity into an intensive property.

99. **Given:** 130 °X = 212 °F and 10 °X = 32 °F **Find:** temperature where °X = °F.
Conceptual plan: Use data to derive an equation relating °X and °F. Then set °F = °X = z and solve for z.
Solution: Assume a linear relationship between the two temperatures (y = mx + b).
Let y = °F and let x = °X.
The slope of the line (m) is the relative change in the two temperature scales:

$$m = \frac{\Delta\ °F}{\Delta\ °X} = \frac{212\ °F - 32\ °F}{130\ °X - 10\ °X} = \frac{180\ °F}{120\ °X} = 1.5$$

Solve for intercept (b) by plugging one set of temperatures into the equation:
y = 1.5 x + b → 32 = (1.5)(130) + b → 32 = 15 + b → b = 17 → °F = (1.5) °X + 17
Set °F = °X = z and solve for z.
z = 1.5 z + 17 → -17 = 1.5 z – z → -17 = 0.5 z → z = - 34 → - 34°F = - 34 °X

Check: The units (°F & °X) are correct. Plugging the result back into the equation confirms that the calculations were done correctly. The magnitude of the answer seems correct, since it is known that the result is not between 32°F and 212 °F. The numbers are getting closer together as the temperature is dropped.

101. 1G. F = ma = kg(m/s^2). Let's call it N for Newton. Ten tons = 20,000 lb = 2.2 kg/lb x 20,000 lb = 4.4 x 10 x 10^4 kg, deceleration = 55 mi x .6 km/mi x 10^3 m/km x 1/3.6 x 10^3 s^2 Exponents = 10^4 x10^3 x 10^{-3} = 10^4. So the kN is convenient.

For one molecule, the mass is 10^{-20} kg and deceleration is 3 x 10^8 m/s^2. So Exponents = 10^{-20} x 10^8 = 10^{-12}. So the pN is convenient.

103.a) 1.76 x 10^{-3}/8.0 x 10^2 = 2.2 x 10^{-6} Two significant figures are allowed to reflect the quantity with the fewest significant figures (8.0 x 10^2).

b) Write all figures so that the decimal points can be aligned:

$$\begin{array}{rr} & 0.0187 \\ + & 0.0002 \\ - & \underline{0.0030} \\ & 0.0159 \end{array}$$

All quantities are known to four places to the right of the decimal place, so the answer should be reported to four places to the right of the decimal place or three significant figures.

c) [(136000)(0.000322)/0.082)](129.2) = 6.899910244 x 10^4 = 6.9 x 10^4 Round the intermediate answer to two significant figures to reflect the quantity with the fewest significant figures (0.082). Round up the last digit since the first non-significant digit is 9.

105.a) **Given:** cylinder dimensions - length = 22 cm, radius = 3.8 cm, d(gold) = 19.3 g/cm^3 & *d(sand)* = 3.00 g/cm^3
Find: m(gold) and m(sand)
Conceptual plan: **L, r → V then d, V → m**

$V = l\pi r^2$ $d = m/V$

Solution: V(gold) = V(sand) = (22 cm)(π)(3.8 cm)^2 = 99.80212 cm^3 $d = m/V$ Rearrange by multiplying both sides of equation by V. → $m = d \times V$

$$\text{m(gold)} = \left(19.3\ \frac{\text{g}}{\cancel{\text{cm}^3}}\right) \times (99.80212\ \cancel{\text{cm}^3}) = 1.926181 \times 10^4\ \text{g} = 1.9 \times 10^4\ \text{g}$$

Check: The units (g) are correct. The magnitude of the answer seems correct considering the value of the density is ~20 g/cm^3. Two significant figures are allowed to reflect the significant figures in 22 cm and 3.8 cm. Truncate the non-significant digits because the first non-significant digit is a 2.

$$\text{m(sand)} = \left(3.00\ \frac{\text{g}}{\cancel{\text{cm}^3}}\right) \times (99.80212\ \cancel{\text{cm}^3}) = 2.99206 \times 10^3\ \text{g} = 3.0 \times 10^3\ \text{g}$$

Check: The units (g) are correct. The magnitude of the answer seems correct considering the value of the density is 3 g/cm^3. This number is much lower that the gold mass. Two significant figures are allowed to reflect the significant figures in 22 cm and 3.8 cm. Round the last digit up because the first non-significant digit is a 9.

b) Comparing the two values 1.9 x 10^4 g versus 3.0 x 10^3 g shows a difference in weight of almost a factor of 10. This difference should be enough to trip the alarm and alert the authorities to the presence of the thief.

107. **Given:** 3.5 lb of titanium **Find:** volume in in^3 **Other:** density of titanium is 4.5 g/cm^3
Conceptual plan: **lb → g then m, d → V then cm^3 → in^3**

$\frac{453.6\ \text{g}}{1\ \text{lb}}$ $d = m/V$ $\frac{(1\ \text{in})^3}{(2.54\ \text{cm})^3}$

Solution: $3.5 \text{ lb} \times \frac{453.6 \text{ g}}{1 \text{ lb}} = 1.5876 \times 10^3 \text{ g}$. Since $d = m/V$ Rearrange by multiplying both sides of the equation by V and dividing both sides of the equation by d. →

$$V = \frac{m}{d} = \frac{1.5876 \times 10^3 \text{ g}}{4.5 \frac{\text{g}}{\text{cm}^3}} = 3.520177384 \times 10^2 \text{ cm}^3 = 3.5 \times 10^2 \text{ cm}^3 \times \frac{(1 \text{ in})^3}{(2.54 \text{ cm})^3} = 21 \text{ in}^3$$

Check: The units (in^3) are correct. The magnitude of the answer seems correct considering many grams we have. Two significant figures are allowed to reflect the significant figures in 3.5 lb. Truncate the non-significant digits because the first non-significant digit is a 2.

109.**Given:** cylinder dimensions: length = 2.16 in, radius = 0.22 in, m= 41 g **Find:** density (g/cm^3)
Conceptual plan: **in → cm** **then** **l, r → V** **then m, V → d**

$\frac{2.54 \text{ cm}}{1 \text{ in}}$ $V = l\pi r^2$ $d = m/V$

Solution: $2.16 \text{ in} \times \frac{2.54 \text{ cm}}{1 \text{ in}} = 5.4864 \text{ cm} = l$

$0.22 \text{ in} \times \frac{2.54 \text{ cm}}{1 \text{ in}} = 0.5588 \text{ cm} = r$

$V = l \pi r^2 = (5.4864 \text{ cm})(\pi)(0.5588 \text{ cm})^2 = 5.3820798 \text{ cm}^3$

$$d = \frac{m}{V} = \frac{41 \text{ g}}{5.3820798 \text{ cm}^3} = 7.6178729 \frac{\text{g}}{\text{cm}^3} = 7.6 \frac{\text{g}}{\text{cm}^3}$$

Check: The units (g/cm^3) are correct. The magnitude of the answer seems correct considering the value of the density of iron (a major component in steel) is 7.86 g/cm^3. Two significant figures are allowed to reflect the significant figures in 0.22 in and 41 g. Truncate the non-significant digits because the first non-significant digit is a 2.

111.**Given:** 185 cubic yards (yd^3) of H_2O **Find:** mass of the H_2O (pounds) **Other:** $d(H_2O) = 1.00 \text{ g/cm}^3$ at 0°C
Conceptual plan: $\mathbf{yd^3 \rightarrow m^3 \rightarrow cm^3 \rightarrow g \rightarrow lb}$

$\frac{(1 \text{m})^3}{(1.094 \text{ yd})^3}$ $\frac{(100 \text{ cm})^3}{(1 \text{m})^3}$ $\frac{1.00 \text{ g}}{1.00 \text{ cm}^3}$ $\frac{1 \text{ lb}}{453.59 \text{ g}}$

Solution:

$$185 \text{ yd}^3 \times \frac{(1 \text{ m})^3}{(1.094 \text{ yd})^3} \times \frac{(100 \text{ cm})^3}{(1 \text{ m})^3} \times \frac{1.00 \text{ g}}{1.00 \text{ cm}^3} \times \frac{1 \text{ lb}}{453.59 \text{ g}} = 3.114987377 \times 10^5 \text{ lbs} = 3.11 \times 10^5 \text{ lbs}$$

Check: The units (lb) are correct. The magnitude of the answer (10^5) makes physical sense because a pool is not a small object. Three significant figures are allowed because the conversion factor with the least precision is the density (1.00 g/cm^3 - 3 significant figures) and the initial size has three significant figures. Truncate after the last digit because the first non-significant digit is a 4.

113.**Given:** 15 liters of gasoline **Find:** kilometers **Other:** 52 mi/gal in the city
Conceptual plan: **L → gal → mi → km**

$\frac{1 \text{ gallon}}{3.785 \text{ L}}$ $\frac{52 \text{ mi}}{1.0 \text{ gallon}}$ $\frac{1 \text{ km}}{0.6214 \text{ mi}}$

Solution: $15 \text{ L} \times \frac{1 \text{ gallon}}{3.785 \text{ L}} \times \frac{52 \text{ mi}}{1.0 \text{ gallon}} \times \frac{1 \text{ km}}{0.6214 \text{ mi}} = 3.316327941 \times 10^2 \text{ km} = 3.3 \times 10^2 \text{ km}$

Check: The units (km) are correct. The magnitude of the answer (10^2) makes physical sense because the dominating conversion factor is the mileage, which increases the answer. Two significant figures are allowed because the conversion factor with the least precision is 52 mi/gallon (2 significant figures) and the initial volume (15 L) has 2 significant figures. Truncate the last digit because the first non-significant digit is a 1. It is best to put the answer in scientific notation so that it is unambiguous how many significant figures are expressed.

115.**Given:** radius of nucleus of the hydrogen atom = 1.0×10^{-13} cm; radius of the hydrogen atom = 52.9 pm.

Find: percent of volume occupied by nucleus (%)

Conceptual plan:

cm → m then pm → m then r → V then V_{atom}, $V_{nucleus}$ → % $V_{nucleus}$

$\frac{1 \text{ m}}{100 \text{ cm}}$ $\frac{1 \text{ m}}{10^{12} \text{ pm}}$ $V = (4/3)\pi r^3$ $\% V_{nucleus} = \frac{V_{nucleus}}{V_{atom}} x\, 100\%$

Solution: $1.0 \times 10^{-13} \text{ cm} \times \frac{1 \text{ m}}{100 \text{ cm}} = 1.0 \times 10^{-15} \text{ m}$ and $52.9 \text{ pm} \times \frac{1 \text{ m}}{10^{12} \text{ pm}} = 5.29 \times 10^{-11} \text{ m}$

$V = (4/3)\pi r^3$ Substitute into %V equation.

$\% V_{nucleus} = \frac{V_{nucleus}}{V_{atom}} x\, 100\% \rightarrow \% V_{nucleus} = \frac{(4/3)\, \pi r^3_{nucleus}}{(4/3)\, \pi r^3_{atom}} x\, 100\%$ Simplify equation.

$\% V_{nucleus} = \frac{r^3_{nucleus}}{r^3_{atom}} x\, 100\%$ Substitute numbers and calculate result.

$$\% V_{nucleus} = \frac{(1.0 \times 10^{-15} \text{ m})^3}{(5.29 \times 10^{-11} \text{ m})^3} \times 100\% = (1.\underline{8}90359168 \times 10^{-5})^3 \times 100\% = 6.\underline{7}55118685 \times 10^{-13}\ \% = 6.8 \times 10^{-13}\ \%$$

Check: The units (%) are correct. The magnitude of the answer seems correct (10^{-13} %), since a proton is so small. Two significant figures are allowed to reflect the significant figures in 1.0×10^{-13} cm. Round up the last digits because the first non-significant digit is a 5.

117.**Given:** mass of black hole (BH) = 1×10^3 suns; radius of black hole = one-half the radius of our moon. **Find:** density (g/cm^3) **Other:** radius of our sun = 7.0×10^5 km; average density of our sun = 1.4×10^3 kg/m^3; diameter of the moon = 2.16×10^3 miles

Conceptual plan: $d_{BH} = m_{BH}/V_{BH}$

Calculate m_{BH} : r_{sun} → V_{sun} km^3_{sun} → m^3_{sun} V_{sun}' d_{sun} → m_{sun} m_{sun} → m_{BH} kg → g

$V = (4/3)\pi r^3$ $\frac{(1000 \text{ m})^3}{(1 \text{ km})^3}$ $d_{sun} = \frac{m_{sun}}{V_{sun}}$ $m_{BH} = (1 \times 10^3) \times m_{sun}$ $\frac{1000 \text{ g}}{1 \text{ kg}}$

Calculate V_{BH}: d_{moon} → r_{moon} → r_{BH} mi → km → m → cm r → V

$r_{moon} = ½\, d_{moon}$ $r_{BH} = ½\, r_{moon}$ $\frac{1 \text{ km}}{0.6214 \text{ mi}}$ $\frac{1000 \text{ m}}{1 \text{ km}}$ $\frac{100 \text{ cm}}{1 \text{ m}}$ $V = (4/3)\pi r^3$

Substitute into $d_{BH} = m_{BH}/ V_{BH}$

Solution: Calculate m_{BH} $V_{sun} = (4/3)\pi r_{sun}^3 = (4/3)\pi(7.0 \times 10^5 \text{ km})^3 = 1.\underline{4}3675504 \times 10^{18} \text{ km}^3$

$1.\underline{4}3675504 \times 10^{18} \text{ km}^3 \times \frac{(1000 \text{ m})^3}{(1 \text{ km})^3} = 1.\underline{4}3675504 \times 10^{27} \text{ m}^3$ $d_{sun} = m_{sun}/V_{sun}$

Solve for m by multiplying both sides of the equation by V_{sun}.

$m_{sun} = V_{sun} \times d_{sun}$ $m_{sun} = (1.\underline{4}3675504 \times 10^{27} \text{ m}^3)(1.4 \times 10^3 \text{ kg/m}^3) = 2.\underline{0}11457056 \times 10^{30} \text{ kg}$ m_{BH} = $= (1 \times 10^3) \times m_{sun} = (1 \times 10^3) \times (2.\underline{0}11457056 \times 10^{30} \text{ kg}) = 2.\underline{0}11457056 \times 10^{33} \text{ kg}$

$2.\underline{0}11457056 \times 10^{33} \text{ kg} \times \frac{1000 \text{ g}}{1 \text{ kg}} = 2.\underline{0}11457056 \times 10^{36} \text{ g}$

Calculate V_{BH} $r_{moon} = ½\, d_{moon} = ½\,(2.16 \times 10^3 \text{ miles}) = 1.08 \times 10^3$ miles

$r_{BH} = ½\, r_{moon} = ½\,(1.08 \times 10^3 \text{ miles}) = 540.$ miles

$540. \text{ miles} \times \frac{1 \text{ km}}{0.6214 \text{ mi}} \times \frac{1000 \text{ m}}{1 \text{ km}} \times \frac{100 \text{ cm}}{1 \text{ m}} = 8.6\underline{9}00547 \times 10^7 \text{ cm}$

$V = (4/3)\pi r^3 = (4/3)\pi(8.6\underline{9}00547 \times 10^7 \text{ cm})^3 = 2.7\underline{4}888228 \times 10^{24} \text{ cm}^3$

Substitute into $d_{BH} = \frac{m_{BH}}{V_{BH}} = \frac{2.\underline{0}11457056 \times 10^{36} \text{ g}}{2.7\underline{4}888228 \times 10^{24} \text{ cm}^3} = 7.\underline{3}1737339 \times 10^{11} \frac{\text{g}}{\text{cm}^3} = 7.3 \times 10^{11} \frac{\text{g}}{\text{cm}^3}$

Check: The units (g/cm^3) are correct. The magnitude of the answer seems correct (10^{12}), since we expect extremely high numbers for black holes. Two significant figures are allowed to reflect the significant figures in the radius of our sun (7.0×10^5 km) and the average density of sun (1.4×10^3 kg/m^3). Truncate the non-significant digits because the first non-significant digit is a 4.

119.**Given:** cubic nanocontainers with an edge length = 25 nanometers

Find: a) volume of one nanocontainer; b) grams of oxygen could be contained by each nanocontainer; c) grams of oxygen inhaled per hour; d) minimum number of nanocontainers per hour; and e) minimum volume of nanocontainers.

Other: (pressurized oxygen) = 85 g/L; 0.28 g of oxygen per liter; average human inhales about 0.50 L of air per breath and takes about 20 breaths per minute; and adult total blood volume = ~ 5 L.

Conceptual Plan:

a) **nm → m → cm then l → V then cm^3 → L**

$$\frac{1\text{ m}}{10^9\text{ nm}} \quad \frac{100\text{ cm}}{1\text{ m}} \qquad V = l^3 \qquad \frac{1\text{ L}}{1000\text{ cm}^3}$$

b) **L → g pressurized oxygen**

$$\frac{85\text{ g oxygen}}{1\text{ L nanocontainers}}$$

c) hr → min → breaths → L_{air} → g_{O2}

$$\frac{60\text{ min}}{1\text{ hr}} \quad \frac{20\text{ breath}}{1\text{ min}} \quad \frac{0.50\text{ L}_{air}}{1\text{ breath}} \quad \frac{0.28\text{ g}_{co}}{1\text{ L}_{air}}$$

d) **grams oxygen → number nanocontainers**

$$\frac{1\text{ nanocontainer}}{\text{part (b) grams of oxygen}}$$

e) **number nanocontainers → volume nanocontainers**

$$\frac{\text{part (a) volume}}{\text{of 1 nanocontainer}}$$

Solution:

a) $25\text{ nm} \times \dfrac{1\text{ m}}{10^9\text{ nm}} \times \dfrac{100\text{ cm}}{1\text{ m}} = 2.5 \times 10^{-6}\text{ cm}$

$V = l^3 = (2.5 \times 10^{-6}\text{ cm})^3 = 1.\underline{5}625 \times 10^{-17}\text{ cm}^3 \times \dfrac{1\text{ L}}{1000\text{ cm}^3} = 1.\underline{5}625 \times 10^{-20}\text{ L} = 1.6 \times 10^{-20}\text{ L}$

b)

$1.\underline{5}625 \times 10^{-20}\text{ L} \times \dfrac{85\text{ g oxygen}}{1\text{ L nanocontainers}} = 1.\underline{3}28125 \times 10^{-18}\ \dfrac{\text{g pressurized O}_2}{\text{nanocontainer}} = 1.3 \times 10^{-18}\ \dfrac{\text{g pressurized O}_2}{\text{nanocontainer}}$

c) $1\text{ hr} \times \dfrac{60\text{ min}}{1\text{ hr}} \times \dfrac{20\text{ breath}}{1\text{ min}} \times \dfrac{0.50\text{ L}_{air}}{1\text{ breath}} \times \dfrac{0.28\text{ gO}_2}{1\text{ L}_{air}} = 1.\underline{6}8 \times 10^2\text{ g oxygen} = 1.7 \times 10^2\text{ g oxygen}$

d)

$1.\underline{6}8 \times 10^2\text{ g oxygen} \times \dfrac{1\text{ nanocontainer}}{1.3 \times 10^{-18}\text{ g of oxygen}} = 1.\underline{2}92307692 \times 10^{20}\text{ nanocontainers} = 1.3 \times 10^{20}\text{ nanocontainers}$

e) $1.\underline{2}92307692 \times 10^{20}\text{ nanocontainers} \times \dfrac{1.5625 \times 10^{-20}\text{ L}}{\text{nanocontainer}} = 2.\underline{0}19230769\text{ L} = 2.0\text{ L}$

This volume is much too large to be feasible, since the volume of blood in the average human is 5 L.

Check:

a) The units (L) are correct. The magnitude of the answer (10^{-20}) makes physical sense because these are very, very tiny containers. Two significant figures are allowed reflecting the significant figures in the starting dimension (25 nm – 2 significant figures). Round up the last digit because the first non-significant digit is a 6.

b) The units (g) are correct. The magnitude of the answer (10^{-18}) makes physical sense because these are very, very tiny containers and very few molecules can fit inside. Two significant figures are allowed reflecting the significant figures in the starting dimension (25 nm) and the given concentration (85 g/L) – 2 significant figures in each. Truncate the non-significant digits because the first non-significant digit is a 2.

c) The units (g oxygen) are correct. The magnitude of the answer (10^{2}) makes physical sense because of the conversion factors involved and the fact that air is not very dense. Two significant figures are allowed because it is stated in the problem. Round up the last digit because the first non-significant digit is an 8.

d) The units (nanocontainers) are correct. The magnitude of the answer (10^{20}) makes physical sense because these are very, very tiny containers and we need a macroscopic quantity of oxygen in these containers. Two significant figures are allowed reflecting the significant figures in both of the quantities in the calculation – 2 significant figures. Round up the last digit because the first non-significant digit is a 9.

e) The units (L) are correct. The magnitude of the answer (2) makes physical sense because the magnitudes of the numbers in this step. Two significant figures are allowed reflecting the significant figures in both of the quantities in the calculation – 2 significant figures. Truncate the non-significant digits because the first non-significant digit is a 1.

121. c) s the best representation. When solid carbon dioxide (dry ice) sublimes it changes phase from a solid to a gas. Phase changes are physical changes, so no molecular bonds are broken. This diagram shows molecules with one carbon atom and two oxygen atoms bonded together in every molecule. The other diagrams have no carbon dioxide molecules.

123. In order to determine which number is large, the units need to be compared. There is a factor of 1000 between grams and kg, in the numerator. There is a factor of $(100)^3$ or 1,000,000 between cm^3 and m^3. This second factor more than compensates for the first factor. Thus Substance A with a density of 1.7 g/cm^3 is more dense than Substance B with a density of 1.7 kg/m^3.

125. Remember that: An observation is the information collected when studying phenomena. A law is a concise statement that summarizes observed behaviors and observations and predicts future observations. A theory attempts to explain why the observed behavior is happening.

a) This statement is most like a law, because it summarizes many observations and can explain future behavior – many places and many days.

b) This statement is a theory because it attempts to explain why – gravitational forces.

c) This statement is most like an observation, because it is information collected in order to understand tidal behavior.

d) This statement is most like a law, because it summarizes many observations and can explain future behavior – many places and many days and months.

Chapter 2
Atoms and Elements

1. Scanning tunneling microscopy is a technique that can image, and even move, individual atoms and molecules. A scanning tunneling microscope works by moving an extremely sharp electrode over a surface and measuring the resulting tunneling current, the electrical current that flows between the tip of the electrode and the surface even though the two are not in physical contact.

3. The first people to propose that matter was composed of small, indestructible particles were Leucippus and Democritus. These Greek philosophers theorized that matter was ultimately composed of small, indivisible particles called atomos. In the sixteenth century modern science began to emerge. A greater emphasis on observation brought rapid advancement as the scientific method became the established way to learn about the physical world. By the early 1800's certain observations led the English chemist John Dalton to offer convincing evidence that supported the early atomic ideas of Leucippus and Democritus. The theory that all matter is composed of atoms grew out of observations and laws. The three most important laws that led to the development and acceptance of the atomic theory were the law of conservation of mass, the law of definite proportions, and the law of multiple proportions. John Dalton explained the laws with his atomic theory.

5. The law of definite proportions states: All samples of a given compound, regardless of their source or how they were prepared, have the same proportions of their constituent elements. This means that elements composing a given compound always occur in fixed (or definite) proportions in all samples of the compound.

7. The law of definite proportions applies to two or more samples of the same compound and states that the ratio of one element to the other in the samples will always be the same. The law of multiple proportions applies to two different compounds containing the same two elements (A and B) and states that the masses of B that combine with 1 g of A are related as a small whole number ratio.

9. In the late 1800's, an English physicist named J.J. Thomson performed experiments to probe the properties of cathode rays. Thomson found that these rays were actually streams of particles with the following properties: they traveled in straight lines; they were independent of the composition of the material from which they originated; and they carried a negative electrical charge. He measured the charge to mass ratio of the particles and found that the cathode ray particle was about 2000 times lighter than hydrogen.

11. The plum – pudding model of the atom, proposed by J.J. Thomson said the negatively charged electrons were small particles electrostatically held within a positively charged sphere.

13. Rutherford's nuclear model of the atom has three basic parts: 1) Most of the atom's mass and all of its positive charge are contained in a small core called the **nucleus.** 2) Most of the volume of the atom is empty space, throughout which, time, negatively charged electrons are dispersed. 3) There are as many negatively charged electrons outside the nucleus as there are positively charged particles within the nucleus, so that the atom is electrically neutral. The revolutionary part of this theory is the idea that matter – at its core – is much less uniform than it appears.

15. The three subatomic particles that compose atoms are:
    Protons, which have a mass of 1.67262×10^{-27} kg or 100727 amu and a relative charge of +1.
    Neutrons, which have a mass of 1.67493×10^{-27} kg or 100866 amu and a relative charge of 0.
    Electrons, which have a mass of 0.00091×10^{-27} kg or 0.00055 amu and a relative charge of – 1.

17. The atomic number, Z, is the number of protons in an atom's nucleus. The atomic mass number (A) is the sum of the neutrons and protons in an atom.

19. Isotopes are atoms with the same number of protons but different numbers of neutrons. Natural abundance is the relative amount of each different isotope in a naturally occurring sample of a given element.

21. An ion is a charged particle. Positively charged ions are called cations. Negatively charged ions are called anions.

23. Metals are found on the left side and the middle of the periodic table. They are good conductors of heat and electricity, they can be pounded into flat sheets (malleability); they can be drawn into wires (ductility); they are often shiny; and they tend to lose electrons when they undergo chemical change.
Nonmetals are found on the upper right side of the periodic table. Their properties are more varied: some are solids at room temperature, others are liquids or gases. As a whole they tend to be poor conductors of heat and electricity and they all tend to gain electrons when they undergo chemical changes.
Metalloids lie along the zigzag diagonal line that divides metals and nonmetals. They show mixed properties. Several metalloids are also classified as semiconductors because of their intermediate and temperature dependent electrical conductivity.

25. Main group metals tend to lose electrons, forming a cation with the same number of electrons as the nearest noble gas. A main group nonmetal tends to gain electrons, forming an anion with the same number of electrons as the nearest, previous noble gas.

27. In a mass spectrometer, the sample is injected into the instrument and vaporized. The vaporized atoms are then ionized by an electron beam. The electrons in the beam collide with the vaporized atoms, removing electrons from the atoms and creating positively charged ions. Charged plates with slits in them accelerate the positively charged ions into a magnetic field, which deflects them. The amount of deflection depends on the mass of the ions, - lighter ions are deflected more than heavier ones. Finally, the ions strike a detector and produce an electrical signal that is recorded.

29. A mole is an amount of material. It is defined as the amount of material containing 6.0221421×10^{23} particles (Avogadro's number). The numerical value of the mole is defined as being equal to the number of atoms in exactly 12 grams of pure carbon – 12. It is useful for converting number of atoms to moles of atoms and moles of atoms to number of atoms.

31. **Given:** 1.50 g hydrogen; 12.0 g oxygen **Find:** grams water vapor
Conceptual Plan: total mass reactants = total mass products
Solution: Mass of reactants = 1.50 g hydrogen + 12.0 g oxygen = 13.5 grams
Mass of products = mass of reactants = 13.5 grams water vapor.
Check: According to the Law of Conservation of Mass, matter is not created or destroyed in a chemical reaction, so, since water vapor is the only product the masses of hydrogen and oxygen must combine to form the mass of water vapor.

33. **Given:** sample 1: 38.9 g carbon, 448 g chlorine; sample 2: 14.8 g carbon, 134 g chlorine
Find: consistent with definite proportions.
Conceptual Plan: determine mass ratio of sample 1 and 2 and compare.

$$\frac{mass\ of\ chlorine}{mass\ of\ carbon}$$

Solution: $Sample 1: \frac{448\ g\ chorine}{38.9\ g\ carbon} = 11.5 \quad \text{Sample 2:} \frac{134\ \text{g chlorine}}{14.8\ g\ carbon} = 9.05$

Results are not consistent with the law of definite proportions because the ratio of chlorine to carbon is not the same.
Check: According to the Law of Definite Proportions, the mass ratio of one element to another is the same for all samples of the compound.

35. **Given:** mass ratio sodium to fluorine = 1.21:1; sample = 28.8 g sodium **Find:** g fluorine
Conceptual Plan: g sodium → g fluorine

$$\frac{mass\ of\ fluorine}{mass\ of\ sodium}$$

Solution: $28.8\ g\ \cancel{sodium}\ x \frac{1\ g\ fluorine}{1.21\ g\ \cancel{sodium}} = 23.8\ g\ fluorine$

Check: The units of the answer, g fluorine are correct. The magnitude of the answer is reasonable since it is less than the grams of sodium.

37. **Given:** 1 gram osmium: sample 1 = 0.168 g oxygen; sample 2 = 0.3369 g oxygen
Find: consistent with multiple proportions.

Conceptual Plan: determine mass ratio of oxygen

$$\frac{mass\ of\ oxygen\ sample\ 2}{mass\ of\ oxygen\ sample\ 1}$$

Solution: $\frac{0.3369\ g\ oxygen}{0.168\ g\ oxygen}=2.00$ Ratio is a small whole number. Results are consistent with multiple proportions

Check: According to the Law of Multiple Proportions, when two elements form two different compounds, the masses of element B that combine with 1 g of element A can be expressed as a ratio of small whole numbers.

39. **Given:** sulfur dioxide = 3.49 g oxygen and 3.50 g sulfur, sulfur trioxide = 6.75 g oxygen and 4.50 g sulfur.
Find: mass oxygen per g S for each compound and then determine the mass ratio of oxygen

$$\frac{mass\ of\ oxygen\ in\ sulfur\ dioxide}{mass\ of\ sulfur\ in\ sulfur\ dioxide}\quad\frac{mass\ of\ oxygen\ in\ sulfur\ trioxide}{mass\ of\ sulfur\ in\ sulfur\ trioxide}$$

$$\frac{mass\ of\ oxyen\ in\ sulfur\ trioxide}{mass\ of\ oxyen\ in\ sulfur\ dioxide}$$

Solution: $sulfur\ dioxide=\frac{3.49\ g\ oxygen}{3.50\ g\ sulfur}=\frac{0.997\ g\ oxygen}{1\ g\ sulfur}$ $sulfur\ trioxide=\frac{6.75\ g\ oxygen}{4.50\ g\ sulfur}=\frac{1.50\ g\ oxygen}{1\ g\ sulfur}$

$\frac{1.50\ g\ oxygen\ in\ sulfur\ trioxide}{0.997\ g\ oxygen\ in\ sulfur\ dioxide}=\frac{1.50}{1}=\frac{3}{2}$ Ratio is in small whole numbers and is consistent with multiple proportions.

Check: According to the Law of Multiple Proportions, when two elements form two different compounds, the masses of element B that combine with 1 g of element A can be expressed as a ratio of small whole numbers.

41. a) Sulfur and oxygen atoms have the same mass. INCONSISTENT with Dalton's atomic theory because only atoms of the same element have the same mass.

b) All cobalt atoms are identical. CONSISTENT with Dalton's atomic theory because all atoms of a given element have the same mass and other properties that distinguish them from atoms of other elements.

c) Potassium and chlorine atoms combine in a 1:1 ratio to form potassium chloride. CONSISTENT with Dalton's atomic theory because atoms combine in simple whole number ratios to form compounds.

d) Lead atoms can be converted into gold. INCONSISTENT with Dalton's atomic theory because atoms of one element cannot change into atoms of another element.

43. a) The volume of an atom is mostly empty space. CONSISTENT with Rutherford's nuclear theory because most of the volume of the atom is empty space, throughout which tiny, negatively charged electrons are dispersed.

b) The nucleus of an atom is small compared to the size of the atom. CONSISTENT with Rutherford's nuclear theory because most of the atom's mass and all of its positive charge are contained in a small core called the nucleus.

c) Neutral lithium atoms contain more neutrons than protons. INCONSISTENT with Rutherford's nuclear theory because it did not distinguish where the mass of the nucleus came from other than from the protons.

d) Neutral lithium atoms contain more protons than electrons. INCONSISTENT with Rutherford's nuclear theory because there are as many negatively charged particles outside the nucleus as there are positively charged particles within the nucleus.

45. **Given:** drop A = -6.9×10^{-19} C; drop B = -9.2×10^{-19} C; drop C = -11.5×10^{-19} C;
drop D = -4.6×10^{-19} C
Find: The charge on a single electron

Conceptual Plan: determine the ratio of charge for each set of drops.

$$\frac{\text{charge on drop 1}}{\text{charge on drop 2}}$$

Solution: $\frac{-6.9 \times 10^{-19} C \; drop\, A}{-4.6 \times 10^{-19} \text{ C drop D}} = 1.5 \quad \frac{-9.2 \times 10^{-19} C \; drop\, B}{-4.6 \times 10^{-19} \text{ C drop D}} = 2 \quad \frac{-11.5 \times 10^{-19} C \; drop\, C}{-4.6 \times 10^{-19} \text{ C drop D}} = 2.5$

The ratios obtained are not whole numbers, but, can be converted to whole numbers by multiplying by 2. Therefore, the charge on the electron has to be 1/2 the smallest value experimentally obtained. The charge on the electron = -2.3×10^{-19} C.

Check: The units of the answer, Coulombs, are correct. The magnitude of the answer is reasonable since all the values experimentally obtained are integer multiples of -2.3×10^{-19}.

47. **Given:** charge on body = - 15 μC. **Find:** number of electrons, mass of the electrons

Conceptual Plan: μC → C → number of electrons → mass of electrons

$$\frac{1\ C}{10^6\ \mu C} \qquad \frac{1 \text{ electron}}{-1.60 \times 10^{-19} C} \qquad \frac{9.10 \times 10^{-28} \text{ g}}{1 \text{ electron}}$$

Solution: $-15\ \mu C \times \frac{1\ C}{10^6\ \mu C} \times \frac{1 \text{ electron}}{-1.60 \times 10^{-19}\ C} = 9.\underline{3}75 \times 10^{13} \text{ electrons} = 9.4 \times 10^{13} \text{ electrons}$

$$9.\underline{3}75 \times 10^{13} \text{ electrons} \times \frac{9.10 \times 10^{-28} \text{ g}}{1 \text{ electron}} = 8.5 \times 10^{-14} \text{ g}$$

Check: The units of the answers, number of electrons and grams, are correct. The magnitude of the answers are reasonable since the charge on an electron and the mass of an electron are very small.

49. a) True: protons and electrons have equal and opposite charges

b) True: protons and electrons have opposite charge so they will attract each other

c) True: the mass of the electron is much less than the mass of the neutron

d) False: the mass of the proton and the mass of the neutron are about the same

51. **Given:** mass of proton **Find:** number of electron in equal mass

Conceptual Plan: mass of protons → number of electrons

$$\frac{1 \text{ electron}}{0.00091 \times 10^{-27} \text{ kg}}$$

Solution: $1.67262 \times 10^{-27} \text{ kg} \times \frac{1 \text{ electron}}{0.00091 \times 10^{-27} \text{ kg}} = 1.8 \times 10^3 \text{ electrons}$

Check: The units of the answer, electrons, are correct. The magnitude of the answer is reasonable sin $^{16}_{8}O$ ce the mass of the electron is much less than the mass of the proton.

53. For each of the isotopes: determine Z(the number of protons) from the periodic table then determine A(protons + neutrons). Then, write the symbol in the form $^{A}_{Z}X$. $^{23}_{11}Na$

a) The sodium isotope with 12 neutrons: Z = 11; A = 11 + 12 = 23. $^{23}_{11}Na$

b) The oxygen isotope with 8 neutrons: Z = 8; A = 8 + 8 = 16.

c) The aluminum isotope with 14 neutrons: Z = 13; A = 13 + 14 = 27 $^{27}_{13}Al$

d) The iodine isotope with 74 neutrons: Z = 53; A = 53 + 74 = 127 $^{127}_{53}I$

55. a) $^{14}_{7}N$: Z = 7 ; A = 14: protons = Z = 7; neutrons = A – Z = 14 – 7 = 7

b) : Z = 11: A = 23: protons = Z = 11; neutrons = A – Z = 23 – 11 = 12

c) $^{222}_{86}Rn$: Z = 86: A = 222: protons = Z = 86; neutrons = A – Z = 222 – 86 = 136

d) ${}^{208}_{82}Pb$: Z = 82: A = 208: protons = Z = 82; neutrons = A – Z = 208 – 82 = 126

57. Carbon – 14: A = 14, Z = 6: ${}^{14}_{6}C$ # protons = Z = 6 # neutrons = A – Z = 14 – 6 = 8

59. In a neutral atom the number of protons = the number of electrons = Z. For an ion, electrons are lost (cations) or gained (anions)

a) Ni^{2+}: Z = 28 = protons; Z – 2 = 26 = electrons

b) S^{2-}: Z = 16 = protons; Z + 2 = 18 = electrons

c) Br^{-}: Z = 35 = protons; Z + 1 = 36 = electrons

d) Cr^{3+}: Z = 24 = protons; Z – 3 = 21 = electrons

61. Main group metal atoms will lose electrons to form a cation with the same number of electrons as the nearest, previous noble gas.
Nonmetal atoms will gain electrons to form an anion with the same number of electrons as the nearest noble gas.

a) O^{2-} O is a nonmetal and has 8 electrons. It will gain electrons to form an anion. The nearest noble gas is neon with 10 electrons, so O will gain 2 electrons.

b) K^{+} K is a main group metal and has 19 electrons. It will lose electrons to form a cation. The nearest noble gas is argon with 18 electrons, so K will lose 1 electron.

c) Al^{3+} Al is a main group metal and has 13 electrons. It will lose electrons to form a cation. The nearest noble gas is neon with 10 electrons, so Al will lose 3 electrons.

d) Rb^{+} Rb is a main group metal and has 37 electrons. It will lose electrons to form a cation. The nearest noble gas is krypton with 36 electrons, so Rb will lose 1 electron.

63. Main group metal atoms will lose electrons to form a cation with the same number of electrons as the nearest, previous noble gas.
Nonmetal atoms will gain electrons to form an anion with the same number of electrons as the nearest noble gas.

Symbol	Ion formed	Number of Electrons in Ion	Number of Protons in Ion
Ca	Ca^{2+}	**18**	**20**
Be	Be^{2+}	2	**4**
Se	**Se^{2-}**	**36**	34
In	**In^{3+}**	**46**	49

65. a) Na Sodium is a metal

b) Mg Magnesium is a metal

c) Br Bromine is a nonmetal

d) N Nitrogen is a nonmetal

e) As Arsenic is a metalloid

67. a) Tellurium Te is in group 6A and is a main group element

b) Potassium K is in group 1A and is a main group element

c) Vanadium — V is in group 5B and is a transition element

d) Manganese — Mn is in group 7B and is a transition element

69. a) Sodium — Na is in group 1A and is an alkali metal

b) Iodine — I is in group 7A and is a halogen

c) Calcium — Ca is in group 2A and is an alkaline earth metal

d) Barium — Ba is in group 2A and is an alkaline earth metal

e) Krypton — Kr is in group 8A and is a noble gas

71. a) N and Ni would not be similar. Nitrogen is a nonmetal, nickel is a metal

b) Mo and Sn would not be most similar. Although both are metals, molybdenum is a transition metal and tin is a main group metal.

c) Na and Mg would not be similar. Although both are main group metals, sodium is in group 1A and magnesium is in group 2A.

d) Cl and F would be most similar. Chlorine and fluorine are both in group 7A. Elements in the same group have similar chemical properties.

e) Si and P would not be most similar. Silicon is a metalloid and phosphorus is a nonmetal.

73. **Given:** Rb – 85; mass = 84.9118 amu; 72.15%: Rb – 87; mass = 86.9092 amu; 27.85 % **Find:** atomic mass Rb

Conceptual Plan: % abundance → fraction and then find atomic mass

$$\frac{\%\,abundance}{100} \qquad \text{Atomic mass} = \sum_n (\text{fraction of isotope n}) \times (\text{mass of isotope n})$$

Solution: $\text{Fraction Rb - 85} = \frac{72.15}{100} = 0.7215 \qquad \text{Fraction Rb - 87} = \frac{27.85}{100} = 0.2785$

$$\text{Atomic mass} = \sum_n (\text{fraction of isotope n}) \times (\text{mass of isotope n}) = 0.7215(84.9116\text{ amu}) + 0.2785(86.9092\text{ amu}) = 85.47$$

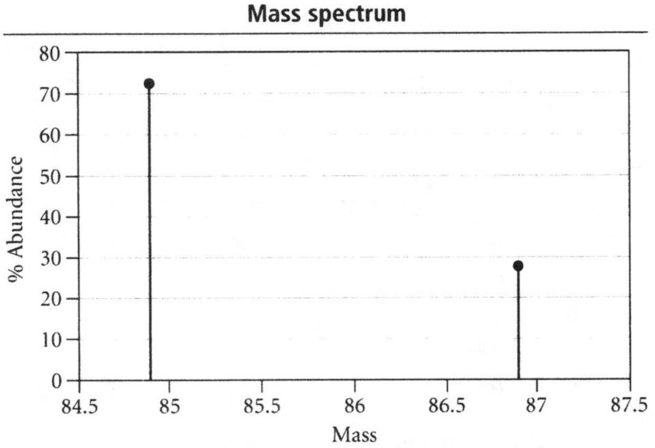

Check: Units of the answer, amu, are correct. The magnitude of the answer is reasonable because it lies between 84.9116 amu and 86.9092 amu and is closer to 84.9118, which has the higher % abundance. The mass spectrum is reasonable because it has two mass lines corresponding to the two isotopes and the line at 84.9116 is about 2.5 times larger than the line at 86.9092.

75. Fluorine has an isotope F –19 with a very large abundance so that the mass of fluorine is very close to the mass of the isotope and the line in the mass spectrum reflects the abundance of F – 19. Chlorine has 2 isotopes Cl – 35 and Cl – 37 and the mass of 35.45 amu is the weighted average of these two isotopes so there is no line at 35.45 amu.

77. **Given:** Isotope – 1; mass = 120.9038 amu; 57.4%: Isotope – 2; mass = 122.9042 amu; **Find:** atomic mass of the atom and identify the atom
Conceptual Plan:
% abundance isotope 2 → and then % abundance → fraction and then find atomic mass

$$100\% - \%\,abundance\,Isotope\,1 \qquad \frac{\%\,abundance}{100}$$

Atomic mass = $\sum_n$ (fraction of isotope n) x (mass of isotope n)

Solution: 100.0% - 57.4 % Isotope 1 = 42.6 % Isotope 2

$$\text{Fraction Isotope 1} = \frac{57.4}{100} = 0.574 \qquad \text{Fraction Isotope 2} = \frac{42.6}{100} = 0.426$$

Atomic mass = $\sum_n$ (fraction of isotope n) x (mass of isotope n) = 0.574(120.9038 amu) + 0.426(122.9042 amu) = 121.8

From the periodic table Sb has a mass of 121.757 amu, so it is the closest mass and the element is antimony.

Check: The units of the answer, amu, are correct. The magnitude of the answer is reasonable because it lies between 120.9038 and 122.9042 and is slightly less than halfway between the two values because the lower value has a slightly greater abundance.

79. **Given:** 3.8 mol sulfur **Find:** atoms of sulfur
Conceptual Plan: mol S → atoms S

$$\frac{6.022\,x10^{23}\,atoms}{mol}$$

Solution: $3.8\ \cancel{\text{mol S}} \times \frac{6.022 \times 10^{23}\ \text{atoms S}}{\cancel{\text{mol S}}} = 2.3 \times 10^{24}$ atoms S

Check: The units of the answer, atoms S, are correct. The magnitude of the answer is reasonable since there is more than 1 mole of material present.

81. a) **Given:** 11.8 g Ar **Find:** mol Ar
Conceptual Plan: g Ar → mol Ar

$$\frac{1\,mol\,Ar}{39.948\,g\,Ar}$$

Solution: $11.8\ \cancel{\text{g Ar}} \times \frac{1\ \text{mol Ar}}{39.948\ \cancel{\text{g Ar}}} = 0.295$ mol Ar

Check: The units of the answer, mol Ar, are correct. The magnitude of the answer is reasonable since there is less than the mass of 1 mol present.

b) **Given:** 3.55 g Zn **Find:** mol Zn
Conceptual Plan: g Zn → mol Zn

$$\frac{1\,mol\,Zn}{65.39\,g\,Zn}$$

Solution: $3.55\ \cancel{\text{g Zn}} \times \frac{1\ \text{mol Zn}}{65.39\ \cancel{\text{g Zn}}} = 0.0543$ mol Zn

Check: The units of the answer, mol Zn, are correct. The magnitude of the answer is reasonable since there is less than the mass of 1 mol present.

c) **Given:** 26.1 g Ta **Find:** mol Ta
Conceptual Plan: g Ta → mol Ta

$$\frac{1\,mol\,Ta}{180.948\,g\,Ta}$$

Solution: $26.1\ \cancel{\text{g Ta}} \times \frac{1\ \text{mol Ta}}{180.948\ \cancel{\text{g Ta}}} = 0.144$ mol Ta

Check: The units of the answer, mol Ta, are correct. The magnitude of the answer is reasonable since there is less than the mass of 1 mol present.

d) **Given:** 0.211 g Li **Find:** mol Li
Conceptual Plan: g Li → mol Li

$$\frac{1\,mol\,Li}{6.941\,g\,Li}$$

Solution: $0.211\ \cancel{g\,Li} \times \frac{1\text{ mol Li}}{6.941\ \cancel{g\,Li}} = 0.0304\text{ mol Li}$

Check: The units of the answer, mol Li, are correct. The magnitude of the answer is reasonable since there is less than the mass of 1 mol present.

83. **Given:** 3.78 g silver **Find:** atoms Ag
Conceptual Plan: g Ag → mol Ag → atoms Ag

$$\frac{1\text{ mol Ag}}{107.868\,g\,Ag} \quad \frac{6.022\,x10^{23}\,atoms}{mol}$$

Solution: $3.78\ \cancel{g\,Ag} \times \frac{1\ \cancel{mol\,Ag}}{107.868\ \cancel{g\,Ag}} \times \frac{6.022 \times 10^{23}\text{ atoms Ag}}{1\ \cancel{mol\,Ag}} = 2.11 \times 10^{22}\text{ atoms Ag}$

Check: The units of the answer, atoms Ag, are correct. The magnitude of the answer is reasonable since there is less than the mass of 1 mol of Ag present.

85. a) **Given:** 5.18 g P **Find:** atoms P
Conceptual Plan: g P → mol P → atoms P

$$\frac{1\text{ mol P}}{30.9738\,g\text{ P}} \quad \frac{6.022\,x10^{23}\,atoms}{mol}$$

Solution: $5.18\ \cancel{g\,P} \times \frac{1\ \cancel{mol\,P}}{30.9738\ \cancel{g\,P}} \times \frac{6.022 \times 10^{23}\text{ atoms P}}{1\ \cancel{mol\,P}} = 1.01 \times 10^{23}\text{ atoms P}$

Check: The units of the answer, atoms P, are correct. The magnitude of the answer is reasonable since there is slightly less than the mass of 1 mol of P present.

b) **Given:** 2.26 g Hg **Find:** atoms Hg
Conceptual Plan: g Hg **→ mol** Hg **→ atoms** Hg

$$\frac{1\text{ mol Hg}}{200.59\,g\text{ Hg}} \quad \frac{6.022\,x10^{23}\,atoms}{mol}$$

Solution: $2.26\ \cancel{g\,Hg} \times \frac{1\ \cancel{mol\,Hg}}{200.59\ \cancel{g\,Hg}} \times \frac{6.022 \times 10^{23}\text{ atoms Hg}}{1\ \cancel{mol\,Hg}} = 6.78 \times 10^{21}\text{ atoms Hg}$

Check: The units of the answer, atoms Hg, are correct. The magnitude of the answer is reasonable since there is less than the mass of 1 mol of Hg present.

c) **Given:** 1.87 g Bi **Find:** atoms Bi
Conceptual Plan: g Bi → mol Bi → atoms Bi

$$\frac{1\text{ mol Bi}}{208.98\,g\text{ Bi}} \quad \frac{6.022\,x10^{23}\,atoms}{mol}$$

Solution: $1.87\ \cancel{g\,Bi} \times \frac{1\ \cancel{mol\,Bi}}{208.98\ \cancel{g\,Bi}} \times \frac{6.022 \times 10^{23}\text{ atoms Bi}}{1\ \cancel{mol\,Bi}} = 5.39 \times 10^{21}\text{ atoms Bi}$

Check: The units of the answer, atoms Bi, are correct. The magnitude of the answer is reasonable since there is less than the mass of 1 mol of Bi present.

d) **Given:** 0.082 g Sr **Find:** atoms Sr

Conceptual Plan: g Sr → mol Sr → atoms Sr

$$\frac{1 \text{ mol Sr}}{87.62\, g \text{ Sr}} \quad \frac{6.022\, x 10^{23}\, atoms}{mol}$$

Solution: $0.082 \text{ g Sr} \times \frac{1 \text{ mol Sr}}{87.62 \text{ g Sr}} \times \frac{6.022 \times 10^{23} \text{ atoms Sr}}{1 \text{ mol Sr}} = 5.6 \times 10^{20} \text{ atoms Sr}$

Check: The units of the answer, atoms Sr, are correct. The magnitude of the answer is reasonable since there is less than the mass of 1 mol of Sr present.

87. **Given:** 52 mg diamond (carbon) **Find:** atoms C
Conceptual Plan: mg C → g C → mol C → atoms C

$$\frac{1 \text{ g C}}{1000\, mg\, C} \quad \frac{1 \text{ mol C}}{12.011\, g\, \text{C}} \quad \frac{6.022\, x 10^{23}\, atoms}{mol}$$

Solution: $52 \text{ mg C} \times \frac{1 \text{ g C}}{1000 \text{ mg C}} \times \frac{1 \text{ mol C}}{12.011 \text{ g C}} \times \frac{6.022 \times 10^{23} \text{ atoms C}}{1 \text{ mol C}} = 2.6 \times 10^{21} \text{ atoms C}$

Check: The units of the answer, atoms C, are correct. The magnitude of the answer is reasonable since there is less than the mass of 1 mol of C present.

89. **Given:** 1 atom platinum **Find:** g Pt
Conceptual Plan: atoms Pt → mol Pt → g Pt

$$\frac{1 mol}{6.022\, x 10^{23}\, atoms} \quad \frac{195.08 \text{ g Pt}}{1 \text{ mol Pt}}$$

Solution: $1 \text{ atom Pt} \times \frac{1 \text{ mol Pt}}{6.022 \times 10^{23} \text{ atoms Pt}} \times \frac{195.08 \text{ g Pt}}{1 \text{ mol Pt}} = 3.239 \times 10^{-22} \text{ g Pt}$

Check: The units of the answer, g Pt, are correct. The magnitude of the answer is reasonable since there is only 1 atom in the sample.

91. **Given:** 7.83 g HCN sample 1: 0.290 g H; 4.06 g N. 3.37 g HCN sample 2 **Find:** g C in sample 2
Conceptual Plan:
g HCN sample 1 → g C in HCN sample 1 → ratio g C to g HCN → g C in HCN sample 2

$$\text{g HCN - g H - g N} \qquad \frac{\text{g C}}{\text{g HCN}} \qquad \text{g HCN} \times \frac{\text{g C}}{\text{g HCN}}$$

Solution: 7.83 g HCN - 0.290 g H - 4.06 g N = 3.48 g C

$$3.37 \text{ g HCN} \times \frac{3.48 \text{ g C}}{7.83 \text{ g HCN}} = 1.50 \text{ g C}$$

Check: The units of the answer, g C, are correct. The magnitude of the answer is reasonable since the sample size is about half the original sample size, the g C are about half the original g C.

93. **Given:** In CO mass ratio O:C = 1.33:1; In compound X, mass ratio O:C = 2:1. **Find:** formula of X
Conceptual Plan: determine the mass ratio of O:O in the 2 compounds.
Solution:

For 1 gram of C: $\frac{2 \text{ g O in compound X}}{1.33 \text{ g O in CO}} = 1.5$ So, the ratio of O to C in compound X has to be 1.5:1 and the formula is C_2O_3.

Check: The answer is reasonable since it fulfills the criteria of multiple proportions and the mass ratio of O:C is 2:1.

95. **Given:** $^4He^{2+}$ = 4.00151 amu; **Find:** charge to mass ratio C/kg
Conceptual Plan: determine total charge on $^4He^{2+}$ and then amu $^4He^{2+}$ → g $^4He^{2+}$ → kg $^4He^{2+}$

$$\frac{+1.60218\, x 10^{-19}\, C}{proton} \quad \frac{1 g}{1.66054\, x 10^{-24}\, amu} \quad \frac{1 kg}{1000\, g}$$

Solution:

$$\frac{2 \text{ protons}}{1 atom\, ^4He^{2+}} \times \frac{+1.60218 \times 10^{-19} \text{ C}}{\text{proton}} = \frac{3.20436 \times 10^{-19} \text{ C}}{atom\, ^4He^{2+}}$$

$$\frac{4.00151 \text{ amu}}{1 atom\, ^4He^{2+}} \times \frac{1.66054 \times 10^{-24} \text{ g}}{1 \text{ amu}} \times \frac{1 \text{ kg}}{1000 \text{ g}} = \frac{6.64466742 x 10^{-27}\, kg}{1 atom\, ^4He^{2+}}$$

$$\frac{3.20436 \times 10^{-19} \text{ C}}{atom\, ^4He^{2+}} x \frac{1\, atom\, ^4He^{2+}}{6.64466742 x 10^{-27}\, kg} = 4.82245 \times 10^{7} \text{ C/kg}$$

Check: the units of the answer, C/kg, are correct. The magnitude of the answer is reasonable when compared to the charge to mass ratio of the electron.

97. $^{236}_{90}Th$ A – Z = number of neutrons. 236 – 90 = 146 neutrons. So, any isotope with 146 neutrons is an isotone of $^{236}_{90}Th$.

Some would be: $^{238}_{92}U$; $^{239}_{93}Np$; $^{241}_{95}Am$; $^{237}_{91}Pa$; $^{235}_{89}Ac$; $^{244}_{98}Cf$ etc.

99.

Symbol	Z	A	Number protons	Number electrons	Number neutrons	Charge
O^{2-}	8	16	8	10	8	2 –
Ca^{2+}	20	40	20	18	20	2+
Mg^{2+}	12	25	12	10	13	2+
N^{3-}	7	14	7	10	7	3 –

101. **Given:** r(nucleus) = 2.7 fm; r(atom) = 70 pm (assume 2 significant figures)
Find: vol(nucleus); vol(atom), % vol(nucleus)
Conceptual Plan:
r(nucleus)(fm) → r(nucleus)(pm) → vol(nucleus) and then r(atom) → vol(atom) and then % vol

$$\frac{10^{-15} m}{1 fm} \quad \frac{1 pm}{10^{-12} m} \quad V = \frac{4}{3}\pi r^3 \quad V = \frac{4}{3}\pi r^3 \quad \frac{vol(nucleus)}{vol(atom)} x100$$

Solution:

$$2.7 \text{ fm} \times \frac{10^{-15} \text{ m}}{\text{fm}} \times \frac{1 \text{ pm}}{10^{-12} \text{ m}} = 2.7 \times 10^{-3} pm \qquad V_{nucleus} = \frac{4}{3}\pi\,(2.7 \times 10^{-3}\text{pm})^3 = 8.2 \times 10^{-8} \text{ pm}^3$$

$$V_{atom} = \frac{4}{3}\pi\,(70 \text{ pm})^3 = 1.4 \times 10^{6} \text{ pm}^3 \qquad \frac{8.2 \times 10^{-8} \text{ pm}^3}{1.4 \times 10^{6} \text{ pm}^3} x100\% = 5.9 x 10^{-12}\%$$

Check: The units of the answer, % vol, are correct. The magnitude of the answer is reasonable because the nucleus only occupies a very small % of the vol of the atom.

103. **Given:** 6.022×10^{23} pennies **Find:** the amount in dollars; the dollars/person
Conceptual Plan: pennies → dollars → dollars/person

$$\frac{1 dollar}{100\, pennies} \quad 6.5 \text{ billion people}$$

Solution:

$$6.022 x 10^{23} \text{ pennies}\; x \frac{1 dollar}{100 \text{ pennies}} = 6.022 x 10^{21} dollars \qquad \frac{6.022 x 10^{21}\, dollars}{6.5 x 10^{9}\, people} = 9.3 x 10^{11} dollars\,/\,person$$

, billionaires

105. **Given:** O = 15.9994 amu when C = 12.011 amu **Find:** mass O when C = 12.00 amu
Conceptual Plan: determine ratio O:C for ^{12}C system then use the same ratio when C = 12.00

$$\frac{mass\, O}{mass\, C}$$

Solution: Based on ^{12}C = 12.00; O = 15.9994 and C = 12.011 so,

$$\frac{mass\,O}{mass\,C} = \frac{15.9994\,amu}{12.011\,amu} = \frac{1.33206\,amu\,O}{1\,amu\,C}$$

Based on C = 12.00, the ratio has to be the same,

$$12.000\ \text{amu C} \times \frac{1.33206\ \text{amu O}}{1\ \text{amu C}} = 15.985\ \text{amu O}$$

Check: The units of the answer, amu O, are correct. The magnitude of the answer is reasonable because the value for the new mass basis is smaller then the original mass basis, therefore, the mass of O should be less.

107. **Given:** Cu sphere: r = 0.935 in; d = 8.96 g/cm^3 **Find:** number of Cu atoms

Conceptual Plan: r in inch → r in cm → vol sphere → g Cu → mol Cu → atoms Cu

$$\frac{2.54\,cm}{1\,inch} \qquad V=\frac{4}{3}\pi r^3 \qquad \frac{8.96\,g}{cm^3} \qquad \frac{1\,mol\,\text{Cu}}{63.546\,g} \qquad \frac{6.022\,x10^{23}\,atoms}{mol}$$

Solution: $0.935\ in\ x\ \frac{2.54\,cm}{in} = 2.3749\,cm$

$$\frac{4}{3}\pi(2.3749\ \text{cm})^3 \times \frac{8.96\ \text{g}}{\text{cm}^3} \times \frac{1\ \text{mol Cu}}{63.546\ \text{g}} \times \frac{6.022 \times 10^{23}\ \text{atoms Cu}}{1\ \text{mol Cu}} = 4.76 \times 10^{24}\ \text{atoms Cu}$$

Check: The units of the answer, atoms Cu, are correct. The magnitude of the answer is reasonable because there are about 8 mol Cu present.

109. **Given:** Li – 6 = 6.01512 amu; Li – 7 = 7.01601 amu; B = 6.941 amu **Find:** % abundance Li – 6 and Li – 7

Conceptual Plan: Let x = fraction Li – 6 then 1 – x = fraction Li – 7 → abundances

$$\text{Atomic mass} = \sum_n (\text{fraction of isotope n}) \times (\text{mass of isotope n})$$

Solution:

$$\text{Atomic mass} = \sum_n (\text{fraction of isotope n}) \times (\text{mass of isotope n})$$

6.941 = (x)(6.01512 amu) + (1 - x)(7.01601 amu)

0.07501 = 1.00089 x

x = 0.07494 1 - x = 0.92506

Li – 6 = 0.07494 x 100 = 7.494 % and Li – 7 = 0.92506 x 100 = 92.506 %

Check: The units of the answer, %, abundance are correct. The relative abundances are reasonable because Li has an atomic mass closer to the mass of Li – 7 than to Li – 6.

111. **Given:** sun: d = 1.4 g/cm^3;r = 7 x 10^8 m; 100 billion stars/galaxy; 10 billion galaxies/universe

Find: number of atoms in the universe

Conceptual Plan: r (star) in m → r (star) in cm →vol (star) → g H/star → mol H star → atoms H/star

$$\frac{100\,cm}{m} \qquad V=\frac{4}{3}\pi r^3 \qquad \frac{1.4\,g\,H}{cm^3} \qquad \frac{1\,mol\,H}{1.008\,g} \qquad \frac{6.022\,x10^{23}\,atoms}{mol}$$

→ atoms H/galaxy → atoms H/universe

$$\frac{100\,x10^9\,stars}{galaxy} \qquad \frac{10\,x10^9\,galaxies}{universe}$$

Solution: $7 \times 10^8\ \text{m} \times \frac{100\ \text{cm}}{\text{m}} = 7 \times 10^{10}\ \text{cm}$

$$\frac{4}{3}\pi\frac{(7 \times 10^{10}\ \text{cm})^3}{\text{star}} \times \frac{1.4\ \text{g H}}{\text{cm}^3} \times \frac{1\ \text{mol H}}{1.008\ \text{g H}} \times \frac{6.022 \times 10^{23}\ \text{atoms H}}{\text{mol H}} \times \frac{100 \times 10^9\ \text{stars}}{\text{galaxy}} \times \frac{10 \times 10^9\ \text{galaxies}}{\text{universe}}$$

$= 1 \times 10^{78}$ atoms/universe

Check: The units of the answer, atoms/universe, are correct.

113.a) This is the Law of Definite Proportions: All samples of a given compound, regardless of their source or how they were prepared, have the same proportions of their constituent elements.

b) This is the Law of Conservation of Mass: In a chemical reaction, matter is neither created nor destroyed.

c) This is the Law of Multiple Proportions: When two elements form two different compounds, the masses of element B that combine with 1 g of element A can be expressed as a ratio of small whole numbers. In this example the ratio of O from hydrogen peroxide to O from water = 16:8 → 2:1, a small whole number ratio.

115. **Given:** a) Cr: 55.0 g; atomic mass = 52 g/mol b) Ti: 45.0 g; atomic mass = 48 g/mol and c) Zn: 60.0 g; atomic mass = 65 g/mol
Find: which has the greatest mol, and which has the greatest mass
Conceptual Plan: without calculation, compare grams of material to g/mol for each.
Solution: Cr would have the greatest mole amount of the elements. It is the only one whose mass is greater than the molar mass. Zn would be the greatest mass amount because it is the largest mass value.

Chapter 3
Molecules, Compounds, and Chemical Equations

1. The properties of compounds are generally very different from the properties of the elements that compose them. When two elements combine to form a compound, an entirely new substance results.

3. Chemical compounds can be represented by chemical formulas and molecular models. The type of formula or model you use depends on how much information you know about the compound and how much you want to communicate An empirical formula gives the relative number of atoms of each element in the compound. It contains the smallest whole number ratio of the elements in the compound. A molecular formula gives the actual number of atoms of each element in the compound. A structural formula shows how the atoms are connected. A ball and stick model shows the geometry of the compound. A space – filling model shows the relative sizes of the atoms and how they merge together.

5. Atomic elements are those that exist in nature with single atoms as their base units. Neon (Ne), gold (Au), potassium (K) are a few examples of atomic elements
Molecular elements do not normally exist in nature with single atoms as their base unit, rather they exist as molecules, two or more atoms of the same element bonded together. Most exist as diatomic molecules, for example hydrogen (H_2), nitrogen (N_2), oxygen (O_2). Some exist as polyatomic molecules; phosphorus (P_4) and sulfur (S_8).
Ionic compounds are generally composed of a one or more metal cations (usually one type of metal) and one or more nonmetal anions bound together by ionic bonds. Sodium chloride (NaCl), potassium sulfate (Na_2SO_4) would be examples of ionic compounds.
Molecular compounds are composed of two or more covalently bonded nonmetals. Examples would be water (H_2O), sulfur dioxide (SO_2), nitrogen dioxide (NO_2).

7. Binary ionic compounds are named by using the name of the cation (metal) and the base name of the anion (nonmetal) + the suffix – ide. Ionic compounds that contain a polyatomic anion are named by using the name of the cation (metal) and the name of the polyatomic anion.

9. To name a binary molecular inorganic compound: list the name of the first element with a prefix to indicate the number of atoms in the compound if there is more than one, followed by the base name of the second element with a prefix to indicate the number of atoms in the compound if there are more than one, followed by the suffix – ide.

11. Binary acids are composed of hydrogen and a nonmetal. The names for binary acids have the form: hydro plus the base name of the nonmetal + ic acid. Oxyacids contain hydrogen and an oxyanion. The names of oxyacids depend on the ending of the oxyanion and have the following forms: oxyanions ending with – ate: base name of the oxyanion + ic acid; oxyanions ending with – ite: base name of the oxyanion + ous acid.

13. The chemical formula indicates the elements present in the compound and the relative number of atoms of each.

15. Chemical formulas contain within them inherent relationships between atoms (or moles of atoms) and molecules (or moles of molecules). For example, the formula CCl_2F_2 tells us that one mole of CCl_2F_2 contains one mole of C atoms, two moles of Cl atoms and two moles of F atoms.

17. The molecular formula is a whole number multiple of the empirical formula. To find the molecular formula the molar mass of the compound must be known. The molar mass divided by the empirical formula molar mass gives the whole number multiple used to convert the empirical formula to the molecular formula.

19. To early chemist organic compounds came from living things, they were easily decomposed, and could not be synthesized in the laboratory. Inorganic compounds came from the earth, were more difficult to decompose and could be synthesized in the laboratory.

21. Functionalized hydrocarbons are hydrocarbons in which a functional group – a characteristic atom or group of atoms – has been incorporated into the hydrocarbon. The family or organic compounds known as alcohols have an – OH functional group.

23. The chemical formula gives you the kind of atom and the number of each atom in the compound.
   a) $Ca_3(PO_4)_2$ contains: 3 calcium atoms, 2 phosphorus atoms and 8 oxygen atoms
   b) $SrCl_2$ contains: 1 strontium atom and 2 chlorine atoms
   c) KNO_3 contains: 1 potassium atom, 1 nitrogen atom and 3 oxygen atoms
   d) $Mg(NO_2)_2$ contains: 1 magnesium atom, 2 nitrogen atoms and 4 oxygen atoms

25. a) 1 blue = nitrogen, 3 white = hydrogen: NH_3
   b) 2 black = carbon, 6 white = hydrogen: C_2H_6
   c) 1 yellow – green = sulfur, 3 red = oxygen: SO_3

27. a) Neon is an element and it is not one of the elements that exist as diatomic molecules therefore it is an atomic element

   b) Fluorine is one of the elements that exist as diatomic molecules therefore it is a molecular element.

   c) Potassium is not one of the elements that exist as diatomic molecules therefore it is an atomic element.

   d) Nitrogen is one of the elements that exist as diatomic molecules therefore it is a molecular element.

29. a) CO_2 is a compound composed of a nonmetal and a nonmetal therefore it is a molecular compound.
   b) $NiCl_2$ is a compound composed of a metal and a nonmetal therefore it is an ionic compound.
   c) NaI is a compound composed of a metal and a nonmetal therefore it is an ionic compound.
   d) PCl_3 is a compound composed of a nonmetal and a nonmetal therefore it is a molecular compound.

31. a) white – hydrogen: a molecule composed of two of the same element, therefore it is a molecular element.

   b) blue – nitrogen, white – hydrogen: a molecule composed of a nonmetal and a nonmetal therefore it is a molecular compound.

   c) purple – sodium: a substance composed of all the same atoms, therefore it is an atomic element.

33. To write the formula for an ionic compound: 1) Write the symbol for the metal cation and its charge and the symbol for the nonmetal anion and its charge. 2) Adjust the subscript on each cation and anion to balance the overall charge. 3) Check that the sum of the charges of the cations equals the sum of the charges of the anions.

a) magnesium and sulfur:	Mg^{2+}	S^{2-}	MgS	cations 2+; anions 2-
b) barium and oxygen:		Ba^{2+}	O^{2-}	BaO cations 2+; anions 2-
c) strontium and bromine:	Sr^{2+}	Br^-	$SrBr_2$	cation 2+; anions 2(1-) = 2-
d) beryllium and chlorine:	Be^{2+}	Cl^-	$BeCl_2$	cations 2+; anions 2(1-) = 2-

35. To write the formula for an ionic compound: 1) Write the symbol for the metal cation and its charge and the symbol for the polyatomic anion and its charge. 2) Adjust the subscript on each cation and anion to balance the overall charge. 3) Check that the sum of the charges of the cations equals the sum of the charges of the anions.
Cation = barium: Ba^{2+}

a) hydroxide:	OH^-	$Ba(OH)_2$	cation 2+, anion 2(1-) = 2-
b) chromate:	CrO_4^{2-}	$BaCrO_4$	cation 2+; anion 2-
c) phosphate:	PO_4^{3-}	$Ba_3(PO_4)_2$	cation 3(2+)=6+; anion 2(3-) = 6-
d) cyanide:	CN^-	$Ba(CN)_2$	cation 2+; anion 2(1-) = 2-

37. To name a binary ionic compound: name the metal cation followed by the base name of the anion + ide.
   a) Mg_3N_2: the cation is magnesium, the anion is from nitrogen which becomes nitride: magnesium nitride
   b) KF: the cation is potassium, the anion is from fluorine which becomes fluoride: potassium fluoride.

   c) Na_2O: the cation is sodium, the anion is from oxygen which becomes oxide: sodium oxide.

   d) Li_2S: the cation is lithium, the anion is from sulfur which becomes sulfide: lithium sulfide.

39. To name an ionic compound with a metal cation that can have more than one charge: name the metal cation followed by parentheses with the charge in roman numerals followed by the base name of the anion + ide.
   a) $SnCl_4$: the charge on Sn must be 4+ for the compound to be charge neutral: the cation is tin(IV), the anion is from chlorine which becomes chloride; tin(IV) chloride

   b) PbI_2: the charge on Pb must be 2+ for the compound to be charge neutral: the cation is lead(II), the anion is from iodine which becomes iodide; lead(II) iodide

   c) Fe_2O_3: the charge on Fe must be 3+ for the compound to be charge neutral; the cation is iron(III), the anion is from oxygen which becomes oxide: iron(III) oxide

   d) CuI_2: the charge on Cu must be 2+ for the compound to be charge neutral; the cation is copper(II), the anion is from iodine which becomes iodide: copper(II) iodide.

41. To name these compounds you must first decide if the metal cation is invariant or can have more than one charge. Then, name the metal cation followed by the base name of the anion + ide.
   a) SnO: Sn can have more than one charge. The charge on Sn must be 2+ for the compound to be charge neutral: the cation is tin(II), the anion is from oxygen which becomes oxide: tin(II) oxide.

   b) Cr_2S_3: Cr can have more than one charge. The charge on Cr must be 3+ for the compound to be charge neutral: the cation is chromium(III), the anion is from sulfur which become sulfide: chromium(III) sulfide.

   c) RbI: Rb is invariant. The cation is rubidium, the anion is from iodine which becomes iodide: rubidium iodide.

   d) $BaBr_2$: Ba is invariant. The cation is barium, the anion is from bromine which becomes bromide: barium bromide.

43. To name these compounds you must first decide is the metal cation is invariant or can have more than one charge. Then, name the metal cation followed by the name of the polyatomic anion.
   a) $CuNO_2$: Cu can have more than one charge. The charge on Cu must be 1+ for the compound to be charge neutral. The cation is copper(I), the anion is nitrite: copper(I) nitrite

   b) $Mg(C_2H_3O_2)_2$: Mg is invariant. The cation is magnesium, the anion is acetate: magnesium acetate.

   c) $Ba(NO_3)_2$: Ba is invariant. The cation is barium, the anion is nitrate: barium nitrate.

   d) $Pb(C_2H_3O_2)_2$: Pb can have more than one charge. The charge on Pb must be 2+ for the compound to be charge neutral. The cation is lead(II), the anion is acetate: lead(II) acetate.

   e) $KClO_3$: K is invariant. The cation is potassium, the anion is chlorate: potassium chlorate.

   f) $PbSO_4$: Pb can have more than one charge. The charge on Pb must be 2+ for the compound to be charge neutral. The cation is lead(II), the anion is sulfate: lead(II) sulfate.

45. To write the formula for an ionic compound: 1) Write the symbol for the metal cation and its charge and the symbol for the nonmetal anion or polyatomic anion and its charge. 2) Adjust the subscript on each cation and anion to balance the overall charge. 3) Check that the sum of the charges of the cations equals the sum of the charges of the anions.

a)	sodium hydrogen sulfite:	Na^+	HSO_3^-	$NaHSO_3$	cation 1+; anion 1-
b)	lithium permanganate:	Li^+	MnO_4^-	$LiMnO_4$	cation 1+; anion 1-
c)	silver nitrate:	Ag^+	NO_3^-	$AgNO_3$	cation 1+; anion 1-
d)	potassium sulfate:	K^+	SO_4^{2-}	K_2SO_4	cation 2(1+) = 2+; anion 2-
e)	rubidium hydrogen sulfate:	Rb^+	HSO_4^-	$RbHSO_4$	cation 1+; anion 1-
f)	potassium hydrogen carbonate:	K^+	HCO_3^-	$KHCO_3$	cation 1+; anion 1-

47. Hydrates are named the same way as other ionic compounds with the addition of the term *prefix*hydrate, where the prefix is the number of water molecules associated with each formula unit.

a) $CoSO_4 \bullet 7H_2O$ cobalt(II) sulfate heptahydrate
b) iridium(III) bromide tetrahydrate $IrBr_3 \bullet 4H_2O$
c) $Mg(BrO_3)_2 \bullet 6H_2O$ magnesium bromate hexahydrate
d) potassium carbonate dehydrate $K_2CO_3 \bullet 2H_2O$

49. a) CO The name of the compound is the name of the first element, *carbon*, following by the base name of the second element, *ox*, prefixed by *mono-* to indicate one and given the suffix – *ide*; carbonmonoxide

b) NI_3 The name of the compound is the name of the first element, *nitrogen*, followed by the base name of the second element, *iod*, prefixed by *tri-* to indicate three and given the suffix – *ide*.: nitrogen triiodide.

c) $SiCl_4$ The name of the compound is the name of the first element, *silicon*, followed by the base name of the second element, *chlor*, prefixed by *tetra-* to indicate four and given the suffix – *ide*.: silicon tetrachloride.

d) N_4Se_4 The name of the compound is the name of the first element, *nitrogen*, prefixed by *tetra-* to indicate four followed by the base name of the second element, *selen*, prefixed by *tetra-* to indicate four and given the suffix – *ide*.: tetranitrogen tetraselenide.

e) I_2O_5 The name of the compound is the name of the first element, *iodine*, prefixed by di to indicate two followed by the base name of the second element, *ox*, prefixed by *penta-* to indicate five and given the suffix – *ide*.: diiodine pentaoxide.

51. a) phosphorus trichloride: PCl_3
b) chlorine monoxide: ClO
c) disulfur tetrafluoride: S_2F_4
d) phosphorus pentafluoride: PF_5
e) diphosphorus pentasulfide:P_2S_5

53. a) HI: the base name of I is *iod* so the name is hydroiodic acid

b) HNO_3: the oxyanion is *nitrate,* which ends in *–ate;* therefore, the name of the acid is nitric acid.

c) H_2CO_3: the oxyanion is *carbonate,* which ends in *–ate;* therefore, the name of the acid is carbonic acid.

d) $HC_2H_3O_2$: the oxyanion is *acetate,* which ends in *–ate;* therefore, the name of the acid is acetic acid.

55. a) hydrofluoric acid: HF
b) hydrobromic acid: HBr
c) sulfurous acid: H_2SO_3

57. To find the formula mass, we sum the atomic masses of each atom in the chemical formula.

a) NO_2 formula mass = 1 x (atomic mass N) + 2 x (atomic mass O)
= 1 x (14.01 amu) + 2 x (16.00 amu)
= 46.01 amu

b) C_4H_{10} formula mass = 4 x (atomic mass C) + 10 x (atomic mass H)
= 4 x (12.01 amu) + 10 x (1.008 amu)
= 58.12 amu

c) $C_6H_{12}O_6$ formula mass = 6 x (atomic mass C) + 12 x (atomic mass H) + 6 x (atomic mass O)
= 6 x (12.01 amu) + 12 x (1.008 amu) + 6 x (16.00 amu)
= 180.16 amu

d) $Cr(NO_3)_3$ formula mass = 1 x (atomic mass Cr) + 3 x (atomic mass N) + 9 x (atomic mass O)
= 1 x (52.00 amu) + 3 x (14.01 amu) + 9 x (16.00 amu)
= 238.0 amu

59. a) **Given:** 6.5 g H_2O **Find:** number of molecules

Conceptual Plan: **g H_2O → mole H_2O → number H_2O molecules**

$$\frac{1 \text{ mol}}{18.02 \text{ g } H_2O} \qquad \frac{6.022 \times 10^{23} \ H_2O \text{ molecules}}{\text{mol } H_2O}$$

Solution:

$$6.5 \text{ g } H_2O \times \frac{1 \text{ mol } H_2O}{18.02 \text{ g } H_2O} \times \frac{6.022 \times 10^{23} \ H_2O \text{ molecules}}{\text{mol } H_2O} = 2.2 \times 10^{23} \ H_2O \text{ molecules}$$

Check: Units of the answer, H_2O molecules, are correct. The magnitude is a is appropriate because it is smaller than Avogadro's number as expected since we have less than 1 mole of H_2O.

b) **Given:** 389 g CBr_4 **Find:** number of molecules

Conceptual Plan: **g CBr_4 → mole CBr_4 → number CBr_4 molecules**

$$\frac{1 \text{ mol}}{331.6 \text{ g } CBr_4} \qquad \frac{6.022 \times 10^{23} \ CBr_4 \text{ molecules}}{\text{mol } CBr_4}$$

Solution:

$$389 \text{ g } CBr_4 \times \frac{1 \text{ mol } CBr_4}{331.6 \text{ g } CBr_4} \times \frac{6.022 \times 10^{23} \ CBr_4 \text{ molecules}}{\text{mol } CBr_4} = 7.06 \times 10^{23} \ CBr_4 \text{ molecules}$$

Check: Units of the answer, CBr_4 molecules, are correct. The magnitude is a is appropriate because it is large than Avogadro's number as expected since we have mare than 1 mole of CBr_4.

c) **Given:** 22.1 g O_2**Find:** number of molecules

Conceptual Plan: **g O_2 → mole O_2 → number O_2 molecules**

$$\frac{1 \text{ mol}}{32.00 \text{ g } O_2} \qquad \frac{6.022 \times 10^{23} \ O_2 \text{ molecules}}{\text{mol } O_2}$$

Solution:

$$22.1 \text{ g } O_2 \times \frac{1 \text{ mol } O_2}{32.00 \text{ g } O_2} \times \frac{6.022 \times 10^{23} \ O_2 \text{ molecules}}{\text{mol } O_2} = 4.16 \times 10^{23} \ O_2 \text{ molecules}$$

Check: Units of the answer, O_2 molecules, are correct. The magnitude is a is appropriate because it is smaller than Avogadro's number as expected since we have less than 1 mole of O_2.

d) **Given:** 19.3 g C_8H_{10} **Find:** number of molecules

Conceptual Plan: **g C_8H_{10} → mole C_8H_{10} → number C_8H_{10} molecules**

$$\frac{1 \text{ mol}}{106.16 \text{ g } C_8H_{10}} \qquad \frac{6.022 \times 10^{23} \ C_8H_{10} \text{ molecules}}{\text{mol } C_8H_{10}}$$

Solution:

$$19.3 \text{ g } C_8H_{10} \times \frac{1 \text{ mol } C_8H_{10}}{106.16 \text{ g } C_8H_{10}} \times \frac{6.022 \times 10^{23} \ C_8H_{10} \text{ molecules}}{\text{mol } C_8H_{10}} = 1.09 \times 10^{23} \ C_8H_{10} \text{ molecules}$$

Check: Units of the answer, C_8H_{10} molecules, are correct. The magnitude is a is appropriate because it is smaller than Avogadro's number as expected since we have less than 1 mole of C_8H_{10}.

61. **Given:** 1 H_2O molecule **Find:** mass in g

Conceptual Plan: **number H_2O molecules → mole H_2O → g H_2O**

$$\frac{1 \text{ mol } H_2O}{6.022 \times 10^{23} \ H_2O \text{ molecules}} \qquad \frac{18.02 \text{ g } H_2O}{1 \text{ mol } H_2O}$$

Solution:

$$1 \ H_2O \text{ molecules} \times \frac{1 \text{ mol } H_2O}{6.022 \times 10^{23} \ H_2O \text{ molecules}} \times \frac{18.02 \text{ g } H_2O}{1 \text{ mol } H_2O} = 2.992 \times 10^{-23} \text{ g } H_2O$$

Check: Units of the answer, grams H_2O, are correct. The magnitude is appropriate because there is much less than Avogadro's number of molecules so we have much less than 1 mole of H_2O.

63. **Given:** 1.8×10^{17} $C_{12}H_{22}O_{11}$ molecule **Find:** mass in mg

Conceptual Plan: number $C_{12}H_{22}O_{11}$ molecules → mole $C_{12}H_{22}O_{11}$ → g $C_{12}H_{22}O_{11}$ → mg $C_{12}H_{22}O$

$$\frac{1 \text{ mol } C_{12}H_{22}O_{11}}{6.022 \times 10^{23} \text{ } C_{12}H_{22}O_{11} \text{ molecules}} \quad \frac{342.3 \text{ g } C_{12}H_{22}O_{11}}{1 \text{ mol } C_{12}H_{22}O_{11}} \quad \frac{1 \times 10^{3} \text{ mg } C_{12}H_{22}O_{11}}{1 \text{ g } C_{12}H_{22}O_{11}}$$

Solution:

$$1.8 \times 10^{17} \text{ } C_{12}H_{22}O_{11} \text{ molecules} \times \frac{1 \text{ mol } C_{12}H_{22}O_{11}}{6.022 \times 10^{23} \text{ } C_{12}H_{22}O_{11} \text{ molecules}} \times \frac{342.3 \text{ g } C_{12}H_{22}O_{11}}{1 \text{ mol } C_{12}H_{22}O_{11}} \times \frac{1 \times 10^{3} \text{ mg } C_{12}H_{22}O_{11}}{\text{g } C_{12}H_{22}O_{11}} = 0.10 \text{ mg } C_{12}H_{22}O_{11}$$

Check: Units of the answer, milligrams $C_{12}H_{22}O_{11}$, are correct. The magnitude is appropriate because there is much less than Avogadro's number of molecules so we have much less than 1 mole of $C_{12}H_{22}O_{11}$.

65. a) **Given:** CH_4 **Find:** mass percent C

Conceptual Plan: $\text{mass } \%C = \frac{1 \text{ x molar mass C}}{\text{molar mass } CH_4} \text{ x } 100$

$1 \text{ x molar mass C} = 1(12.01\text{g/mol}) = 12.01 \text{ g C}$

$\text{molar mass } CH_4 = 1(12.01 \text{ g/mol}) + 4(1.008 \text{ g/mol}) = 16.04 \text{ g/mol}$

Solution: $\text{mass } \% \text{ C} = \frac{1 \text{ x molar mass C}}{\text{molar mass } CH_4} \text{ x } 100\%$

$= \frac{12.01 \text{ g/mol}}{16.04 \text{ g/mol}} \text{ x } 100\%$

$= 74.87\ \%$

Check: Units of the answer, %, are correct. The magnitude is reasonable because it is between 0 and 100% and carbon is the heaviest element.

b) **Given:** C_2H_6 **Find:** mass percent C

Conceptual Plan: $\text{mass } \%C = \frac{2 \text{ x molar mass C}}{\text{molar mass } C_2H_6} \text{ x } 100$

$2 \text{ x molar mass C} = 2(12.01\text{g/mol}) = 24.02 \text{ g C}$

$\text{molar mass } C_2H_6 = 2(12.01 \text{ g/mol}) + 6(1.008 \text{ g/mol}) = 30.07 \text{ g/mol}$

Solution: $\text{mass } \% \text{ C} = \frac{2 \text{ x molar mass C}}{\text{molar mass } C_2H_6} \text{ x } 100\%$

$= \frac{24.02 \text{ g/mol}}{30.07 \text{ g/mol}} \text{ x } 100\%$

$= 79.89\ \%$

Check: Units of the answer, %, are correct. The magnitude is reasonable because it is between 0 and 100% and carbon is the heaviest element.

c) **Given:** C_2H_2 **Find:** mass percent C

Conceptual Plan: $\text{mass } \%C = \frac{2 \text{ x molar mass C}}{\text{molar mass } C_2H_2} \text{ x } 100$

$2 \text{ x molar mass C} = 2(12.01\text{g/mol}) = 24.02 \text{ g C}$

$\text{molar mass } C_2H_2 = 2(12.01 \text{ g/mol}) + 2(1.008 \text{ g/mol}) = 26.04 \text{ g/mol}$

Solution: $\text{mass } \% \text{ C} = \frac{2 \text{ x molar mass C}}{\text{molar mass } C_2H_2} \text{ x } 100\%$

$= \frac{24.02 \text{ g/mol}}{26.04 \text{ g/mol}} \text{ x } 100\%$

$= 92.26\ \%$

Check: Units of the answer, %, are correct. The magnitude is reasonable because it is between 0 and 100% and carbon is the heaviest element.

d) **Given:** C_2H_5Cl **Find:** mass percent C

Conceptual Plan: $\text{mass \% C} = \dfrac{2 \text{ x molar mass C}}{\text{molar mass } C_2H_5Cl} \text{ x } 100$

Solution:

2 x molar mass C = 2(12.01g/mol) = 24.02 g C

molar mass C_2H_5Cl = 2(12.01 g/mol) + 5(1.008 g/mol) + 1(35.45 g/mol) = 64.51 g/mol

$$\text{mass \% C} = \frac{2 \text{ x molar mass C}}{\text{molar mass } C_2H_5Cl} \text{ x } 100\%$$

$$= \frac{24.02 \text{ g/mol}}{64.51 \text{ g/mol}} \text{ x } 100\%$$

$$= 37.23 \%$$

Check: Units of the answer, %, are correct. The magnitude is reasonable because it is between 0 and 100% and chlorine is heavier than carbon.

67. **Given:** NH_3 **Find:** mass percent N

Conceptual Plan: $\text{mass \% N} = \dfrac{1 \text{ x molar mass N}}{\text{molar mass } NH_3} \text{ x } 100$

1 x molar mass N = 1(14.01g/mol) = 14.01 g N

molar mass NH_3 = 3(1.008 g/mol) + (14.01 g/mol) = 17.03 g/mol

Solution: $\text{mass \% N} = \dfrac{1 \text{ x molar mass N}}{\text{molar mass } NH_3} \text{ x } 100\%$

$$= \frac{14.01 \text{ g/mol}}{17.03 \text{ g/mol}} \text{ x } 100\%$$

$$= 82.27 \%$$

Check: Units of the answer, %, are correct. The magnitude is reasonable because it is between 0 and 100% and nitrogen is the heaviest

Given: $CO(NH_2)_2$ **Find:** mass percent N

Conceptual Plan: $\text{mass \% N} = \dfrac{2 \text{ x molar mass N}}{\text{molar mass } CO(NH_2)_2} \text{ x } 100$

Solution:

2 x molar mass N = 1(14.01g/mol) = 28.02 g N

molar mass $CO(NH_2)_2$ = (12.01 g/mol) + (16.00 g/mol)+ 2(14.01 g/mol) + 4(1.008 g/mol) = 60.06 g/mol

$$\text{mass \% N} = \frac{1 \text{ x molar mass N}}{\text{molar mass } CO(NH_2)_2} \text{ x } 100\%$$

$$= \frac{28.02 \text{ g/mol}}{60.06 \text{ g/mol}} \text{ x } 100\%$$

$$= 46.65 \%$$

Check: Units of the answer, %, are correct. The magnitude is reasonable because it is between 0 and 100% there are two nitrogen

Given: NH_4NO_3 **Find:** mass percent N

Conceptual Plan: $\text{mass \% N} = \dfrac{2 \text{ x molar mass N}}{\text{molar mass } NH_4NO_3} \text{ x } 100$

$1 \text{ x molar mass N} = 2(14.01\text{g/mol}) = 28.02 \text{ g N}$

$\text{molar mass } NH_4NO_3 = +2\,(14.01 \text{ g/mol}) + 4(1.008 \text{ g/mol}) + 3(16.00 \text{ g/mol}) = 80.05 \text{ g/mol}$

Solution: $$\text{mass \% N} = \frac{2 \text{ x molar mass N}}{\text{molar mass } NH_4NO_3} \text{ x } 100\%$$

$$= \frac{28.02 \text{ g/mol}}{80.05 \text{ g/mol}} \text{ x } 100\%$$

$$= 35.00\ \%$$

Check: Units of the answer, %, are correct. The magnitude is reasonable because it is between 0 and 100% and the mass of nitrogen is less than the mass of oxygen and there are 3 oxygen.

Given: $(NH_4)_2SO_4$ **Find:** mass percent N

Conceptual Plan: $$\text{mass \% N} = \frac{2 \text{ x molar mass N}}{\text{molar mass } (NH_4)_2SO_4} \text{ x } 100$$

Solution:

$1 \text{ x molar mass N} = 2(14.01\text{g/mol}) = 28.02 \text{ g N}$

$\text{molar mass } (NH_4)_2SO_4 = 2(14.01 \text{ g/mol}) + 8(1.008 \text{ g/mol}) + (32.07 \text{ g/mol}) + 4(16.00 \text{ g/mol}) = 132.15 \text{ g/mol}$

$$\text{mass \% N} = \frac{2 \text{ x molar mass N}}{\text{molar mass } (NH_4)_2SO_4} \text{ x } 100\%$$

$$= \frac{28.02 \text{ g/mol}}{132.15 \text{ g/mol}} \text{ x } 100\%$$

$$= 21.20\ \%$$

Check: Units of the answer, %, are correct. The magnitude is reasonable because it is between 0 and 100% and the mass of nitrogen is less than the mass of oxygen and sulfur.
The fertilizer with the highest nitrogen content is NH_3 because is has the highest %N.

69. **Given:** 55.5 g CuF_2; 37.42 % F **Find:** g F in CuF_2
Conceptual Plan: **g CuF_2 → g F**

$$\frac{37.42 \text{ g F}}{100.0 \text{ g } CuF_2}$$

Solution: $$55.5 \text{ g } CuF_2 \text{ x } \frac{37.42 \text{ g F}}{100.0 \text{ g } CuF_2} = 20.77 = 20.8 \text{ g F}$$

Check: Units of the answer, g F, are correct. The magnitude is reasonable because it is less than the original mass.

71. **Given:** 150 μg I; 76.45% I in KI **Find:** μg KI
Conceptual Plan: **μg I → g I → g KI → μg KI**

$$\frac{1 \text{ g I}}{1 \text{ x } 10^6\ \mu\text{g I}} \quad \frac{100.0 \text{ g KI}}{76.45 \text{ g I}} \quad \frac{1 \text{ x } 10^6\ \mu\text{g KI}}{1 \text{ g KI}}$$

Solution: $$150\ \mu\text{g I x } \frac{1 \text{ g I}}{1\text{x}10^6\ \mu\text{g I}} \text{ x } \frac{100.0 \text{ g KI}}{76.45 \text{ g I}} \text{ x } \frac{1 \text{ x } 10^6\ \mu\text{g KI}}{1 \text{ g KI}} = 196\ \mu\text{g KI}$$

Check: Units of the answer, μg KI, are correct. The magnitude is reasonable because it is greater than the original mass.

73. a) red – oxygen, white – hydrogen: 2H:O H_2O
b) black – carbon, white – hydrogen: 2H:C CH_2
c) black – carbon, white – hydrogen, red – oxygen: 3C:O:H CH_3OH or CH_4O

75. a) **Given:** 0.0885 mol C_4H_{10} **Find:** mol H atoms
Conceptual Plan: **mol C_4H_{10} → mole H atom**

$$\frac{10 \text{ mol H}}{1 \text{ mol } C_4H_{10}}$$

Solution: $0.0885 \text{ mol } C_4H_{10} \times \frac{10 \text{ mol H}}{1 \text{ mol } C_4H_{10}} = 0.885 \text{ mol H atoms}$

Check: Units of the answer, mol H atoms, are correct. The magnitude is reasonable because it is greater than the original mol C_4H_{10}.

b) **Given:** 1.3 mol CH_4 **Find:** mol H atoms
Conceptual Plan: **mol CH_4 → mole H atom**

$$\frac{4 \text{ mol H}}{1 \text{ mol } CH_4}$$

Solution: $1.3 \text{ mol } CH_4 \times \frac{4 \text{ mol H}}{1 \text{ mol } CH_4} = 5.2 \text{ mol H atoms}$

Check: Units of the answer, mol H atoms, are correct. The magnitude is reasonable because it is greater than the original mol CH_4.

c) **Given:** 2.4 mol C_6H_{12} **Find:** mol H atoms
Conceptual Plan: **mol C_6H_{12} → mole H atom**

$$\frac{12 \text{ mol H}}{1 \text{ mol } C_6H_{12}}$$

Solution: $2.4 \text{ mol } C_6H_{12} \times \frac{12 \text{ mol H}}{1 \text{ mol } C_6H_{12}} = 29 \text{ mol H atoms}$

Check: Units of the answer, mol H atoms, are correct. The magnitude is reasonable because it is greater than the original mol C_6H_{12}.

d) **Given:** 1.87 mol C_8H_{18} **Find:** mol H atoms
Conceptual Plan: **mol C_8H_{18} → mole H atom**

$$\frac{18 \text{ mol H}}{1 \text{ mol } C_8H_{18}}$$

Solution: $1.87 \text{ mol } C_8H_{18} \times \frac{18 \text{ mol H}}{1 \text{ mol } C_8H_{18}} = 33.7 \text{ mol H atoms}$

Check: Units of the answer, mol H atoms, are correct. The magnitude is reasonable because it is greater than the original mol C_8H_{18}.

77. a) **Given:** 8.5 g NaCl **Find:** g Na
Conceptual Plan: **g NaCl → mole NaCl → mol Na → g Na**

$$\frac{1 \text{ mol NaCl}}{58.44 \text{ g NaCl}} \quad \frac{1 \text{ mol Na}}{1 \text{ mol NaCl}} \quad \frac{22.99 \text{ g Na}}{1 \text{ mol Na}}$$

Solution: $8.5 \text{ g NaCl} \times \frac{1 \text{ mol NaCl}}{58.44 \text{ g NaCl}} \times \frac{1 \text{ mol Na}}{1 \text{ mol NaCl}} \times \frac{22.99 \text{ g Na}}{1 \text{ mol Na}} = 3.3 \text{ g Na}$

Check: Units of the answer, g Na, are correct. The magnitude is reasonable because it is less than the original g NaCl.

b) **Given:** 8.5 g Na_3PO_4 **Find:** g Na
Conceptual Plan: **g Na_3PO_4 → mole Na_3PO_4 → mol Na → g Na**

$$\frac{1 \text{ mol } Na_3PO_4}{163.94 \text{ g } Na_3PO_4} \quad \frac{3 \text{ mol Na}}{1 \text{ mol } Na_3PO_4} \quad \frac{22.99 \text{ g Na}}{1 \text{ mol Na}}$$

Solution: $8.5 \text{ g } Na_3PO_4 \times \frac{1 \text{ mol } Na_3PO_4}{163.94 \text{ g } Na_3PO_4} \times \frac{3 \text{ mol Na}}{1 \text{ mol } Na_3PO_4} \times \frac{22.99 \text{ g Na}}{1 \text{ mol Na}} = 3.6 \text{ g Na}$

Check: Units of the answer, g Na, are correct. The magnitude is reasonable because it is less than the original g Na_3PO_4.

c) **Given:** 8.5 g $NaC_7H_5O_2$ **Find:** g Na
Conceptual Plan: **g $NaC_7H_5O_2$ → mole $NaC_7H_5O_2$ → mol Na → g Na**

$$\frac{1\ mol\ NaC_7H_5O_2}{144.10\ g\ NaC_7H_5O_2} \quad \frac{1\ mol\ Na}{1\ mol\ NaC_7H_5O_2} \quad \frac{22.99\ g\ Na}{1\ mol\ Na}$$

Solution: $$8.5\ g\ NaC_7H_5O_2 \times \frac{1\ mol\ NaC_7H_5O_2}{144.10\ g\ NaC_7H_5O_2} \times \frac{1\ mol\ Na}{1\ NaC_7H_5O_2} \times \frac{22.99\ g\ Na}{1\ mol\ Na} = 1.4\ g\ Na$$

Check: Units of the answer, g Na, are correct. The magnitude is reasonable because it is less than the original g $NaC_7H_5O_2$.

d) **Given:** 8.5 g $Na_2C_6H_6O_7$ **Find:** g Na
Conceptual Plan: **g $Na_2C_6H_6O_7$ → mole $Na_2C_6H_6O_7$ → mol Na → g Na**

$$\frac{1\ mol\ Na_2C_6H_6O_7}{236.1\ g\ Na_2C_6H_6O_7} \quad \frac{2\ mol\ Na}{1\ mol\ Na_2C_6H_6O_7}$$

$$\frac{22.99\ g\ Na}{1\ mol\ Na}$$

Solution:

$$8.5\ g\ Na_2C_6H_6O_7 \times \frac{1\ mol\ Na_2C_6H_6O_7}{236.1\ g\ Na_2C_6H_6O_7} \times \frac{2\ mol\ Na}{1\ mol\ Na_2C_6H_6O_7} \times \frac{22.99\ g\ Na}{1\ mol\ Na} = 1.7\ g\ Na_2C_6H_6O_7$$

Check: Units of the answer, g Na, are correct. The magnitude is reasonable because it is less than the original g $Na_2C_6H_6O_7$.

79. a) **Given:** 1.651 g Ag; 0.1224 g O **Find:** empirical formula

Conceptual Plan:
convert mass to mol of each element → write pseudoformula → write empirical formula

$$\frac{1\ mol\ Ag}{107.9\ g\ Ag} \quad \frac{1\ mol\ O}{16.00\ g\ O}$$ divide by smallest number

Solution: $$1.651\ g\ Ag \times \frac{1\ mol\ Ag}{107.9\ g\ Ag} = 0.01530\ mol\ Ag$$

$$0.1224\ g\ O \times \frac{1\ mol\ O}{16.00\ g\ O} = 0.007650\ mol\ O$$

$$Ag_{0.01530}O_{0.007650}$$

$$Ag_{\frac{0.01530}{0.007650}}O_{\frac{0.007650}{0.007650}} \rightarrow Ag_2O$$

The correct empirical formula is Ag_2O

b) **Given:** 0.672 g Co; 0.569 g As; 0.486 g O **Find:** empirical formula
Conceptual Plan: convert mass to mol of each element → write pseudoformula → write empirical formula

$$\frac{1\ mol\ Co}{58.93\ g\ Co} \quad \frac{1\ mol\ As}{74.92\ g\ As} \quad \frac{1\ mol\ O}{16.00\ g\ O}$$ divide by smallest number

Solution: $$0.672\ g\ Co \times \frac{1\ mol\ Co}{58.93\ g\ Co} = 0.0114\ mol\ Co$$

$$0.569\ g\ As \times \frac{1\ mol\ As}{74.92\ g\ As} = 0.00759\ mol\ O$$

$$0.486\ g\ O \times \frac{1\ mol\ O}{16.00\ g\ O} = 0.0304\ mol\ O$$

$$Co_{0.0114}As_{0.00759}O_{0.0304}$$

$$Co_{\frac{0.0114}{0.00759}}As_{\frac{0.00759}{0.00759}}O_{\frac{0.0304}{0.00759}} \rightarrow Co_{1.5}As_1O_4$$

$$Co_{1.5}As_1O_4 \times 2 \rightarrow Co_3As_2O_8$$

The correct empirical formula is $Co_3As_2O_8$

c) **Given:** 1.443 g Se; 5.841 g Br **Find:** empirical formula
Conceptual Plan:
convert mass to mol of each element → write pseudoformula → write empirical formula

$$\frac{1 \text{ mol Se}}{78.96 \text{ g Se}} \quad \frac{1 \text{ mol Br}}{79.90 \text{ g Br}} \qquad \text{divide by smallest number}$$

Solution:

$$1.443 \text{ g Se} \times \frac{1 \text{ mol Se}}{78.96 \text{ g Se}} = 0.01828 \text{ mol Se}$$

$$5.841 \text{ g Br} \times \frac{1 \text{ mol Br}}{79.90 \text{ g Br}} = 0.07310 \text{ mol Br}$$

$$Se_{0.01828}Br_{0.07310}$$

$$Se_{\frac{0.01828}{0.01828}}Br_{\frac{0.07310}{0.01828}} \rightarrow SeBr_4$$

The correct empirical formula is $SeBr_4$

81. a) **Given:** In a 100 g sample: 74.03 g C, 8.70 g H, 17.27 g N **Find:** empirical formula
Conceptual Plan:
convert mass to mol of each element → write pseudoformula → write empirical formula

$$\frac{1 \text{ mol C}}{12.01 \text{ g C}} \quad \frac{1 \text{ mol H}}{1.008 \text{ g H}} \quad \frac{1 \text{ mol N}}{14.01 \text{ g N}} \qquad \text{divide by smallest number}$$

Solution:

$$74.03 \text{ g C} \times \frac{1 \text{ mol C}}{12.01 \text{ g C}} = 6.164 \text{ mol C}$$

$$8.70 \text{ g H} \times \frac{1 \text{ mol H}}{1.008 \text{ g H}} = 8.63 \text{ mol H}$$

$$17.27 \text{ g N} \times \frac{1 \text{ mol N}}{14.01 \text{ g N}} = 1.233 \text{ mol N}$$

$$C_{6.164}H_{8.63}N_{1.233}$$

$$C_{\frac{6.164}{1.233}}H_{\frac{8.63}{1.233}}N_{\frac{1.233}{1.233}} \rightarrow C_5H_7N$$

The correct empirical formula is C_5H_7N

b) **Given:** In a 100 g sample: 49.48 g C, 5.19 g H, 28.85 g N, 16.48 g O **Find:** empirical formula
Conceptual Plan:
convert mass to mol of each element → write pseudoformula → write empirical formula

$$\frac{1 \text{ mol C}}{12.01 \text{ g C}} \quad \frac{1 \text{ mol H}}{1.008 \text{ g H}} \quad \frac{1 \text{ mol N}}{14.01 \text{ g N}} \quad \frac{1 \text{ mol O}}{16.00 \text{ g O}} \qquad \text{divide by smallest number}$$

Solution:

$$49.48 \text{ g C} \times \frac{1 \text{ mol C}}{12.01 \text{ g C}} = 4.120 \text{ mol C}$$

$$5.19 \text{ g H} \times \frac{1 \text{ mol H}}{1.008 \text{ g H}} = 5.15 \text{ mol H}$$

$$28.85 \text{ g N} \times \frac{1 \text{ mol N}}{14.01 \text{ g N}} = 2.059 \text{ mol N}$$

$$16.48 \text{ g O} \times \frac{1 \text{ mol O}}{16.00 \text{ g O}} = 1.030 \text{ mol O}$$

$$C_{4.120}H_{5.15}N_{2.059}O_{1.030}$$

$$C_{\frac{4.120}{1.030}}H_{\frac{5.15}{1.030}}N_{\frac{2.059}{1.030}}O_{\frac{1.030}{1.030}} \rightarrow C_4H_5N_2O$$

The correct empirical formula is $C_4H_5N_2O$

83. **Given:** 0.77 mg N, 6.61 mg N_xCl_y **Find:** empirical formula
Conceptual Plan:
Find mg Cl → convert mg to g for each element → convert mass to mol of each element →

$$mg\ N_xCl_y - mg\ N \quad \frac{1\ g}{1000\ mg} \quad \frac{1\ mol\ N}{14.01\ g\ N} \quad \frac{1\ mol\ Cl}{35.45\ g\ Cl}$$

write pseudoformula → write empirical formula
divide by smallest number

Solution: 6.61 mg N_xCl_y – 0.77 mg N = 5.94 mg Cl

$$0.77\ mg\ N \times \frac{1\ g\ N}{1000\ mg\ N} \times \frac{1\ mol\ N}{14.01\ g\ N} = 5.5 \times 10^{-5}\ mol\ N$$

$$5.94\ g\ Cl \times \frac{1\ g\ Cl}{1000\ mg\ Cl} \times \frac{1\ mol\ Cl}{35.45\ g\ Cl} = 1.7 \times 10^{-4}\ mol\ Cl$$

$$N_{5.5 \times 10^{-5}}\ Cl_{1.7 \times 10^{-4}}$$

$$N_{\frac{5.5x10^{-5}}{5.5x10^{-5}}}Cl_{\frac{1.7 \times 10^{-4}}{5.5x10^{-5}}} \rightarrow NCl_3$$

The correct empirical formula is NCl_3

85. a) **Given:** empirical formula = C_6H_7N, molar mass = 186.24 g/mol **Find:** molecular formula

Conceptual Plan: molecular formula = empirical formula x n $n = \dfrac{\text{molar mass}}{\text{empirical formula mass}}$

Solution: empirical formula mass = 6(12.01 g/mol) + 7(1.008 g/mol) + 1(14.01 g/mol) = 93.13 g/mol

$$n = \frac{\text{molar mass}}{\text{formula molar mass}} = \frac{186.24\ g/mol}{93.13\ g/mol} = 1.998 = 2$$

molecular formula = C_6H_7N x 2
= $C_{12}H_{14}N_2$

b) **Given:** empirical formula = C_2HCl, molar mass = 181.44 g/mol **Find:** molecular formula

Conceptual Plan: molecular formula = empirical formula x n $n = \dfrac{\text{molar mass}}{\text{empirical formula mass}}$

Solution: empirical formula mass = 2(12.01 g/mol) + 1(1.008 g/mol) + 1(35.45 g/mol) = 60.48 g/mol

$$n = \frac{\text{molar mass}}{\text{formula molar mass}} = \frac{181.44\ g/mol}{60.48\ g/mol} = 3$$

molecular formula = C_2HCl x 3
= $C_6H_3Cl_3$

c) **Given:** empirical formula = $C_5H_{10}NS_2$, molar mass = 296.54 g/mol **Find:** molecular formula

Conceptual Plan: molecular formula = empirical formula x n $n = \dfrac{\text{molar mass}}{\text{empirical formula mass}}$

Solution: empirical formula mass = 5(12.01 g/mol) + 10(1.008 g/mol) + 1(14.01 g/mol) + 2(32.07) = 148.28 g/mol

$$n = \frac{\text{molar mass}}{\text{formula molar mass}} = \frac{296.54\ g/mol}{148.28\ g/mol} = 2$$

molecular formula = $C_5H_{10}NS_2$ x 2
= $C_{10}H_{20}N_2S_4$

87. **Given:** 33.01 g CO_2, 13.51 g H_2O **Find:** empirical formula
Conceptual Plan: mass CO_2, H_2O → mol CO_2, H_2O → mol C, mol H → pseudoformula → empirical formula

$$\frac{1\text{ mol }CO_2}{44.01\text{ g }CO_2} \quad \frac{1\text{ mol }H_2O}{18.02\text{ g }H_2O} \quad \frac{1\text{ mol C}}{1\text{ mol }CO_2} \quad \frac{2\text{ mol H}}{1\text{ mol }H_2O}$$

divide by smallest number

Solution:

$$33.01\text{ g }CO_2 \times \frac{1\text{ mol }CO_2}{44.01\text{ g }CO_2} = 0.7500\text{ mol }CO_2$$

$$13.51\text{ g }H_2O \times \frac{1\text{ mol }H_2O}{18.02\text{ g }H_2O} = 0.7497\text{ mol }H_2O$$

$$0.7500\text{ mol }CO_2 \times \frac{1\text{ mol C}}{1\text{ mol }CO_2} = 0.7500\text{ mol C}$$

$$0.7497\text{ mol }H_2O \times \frac{2\text{ mol H}}{1\text{ mol }H_2O} = 1.499\text{ mol H}$$

$C_{0.7500}H_{1.499}$

$C_{\frac{0.7500}{0.7500}}H_{\frac{1.499}{0.7500}} \rightarrow CH_2$

The correct empirical formula is CH_2

89. **Given:** 4.30 g sample, 8.59 g CO_2, 3.52 g H_2O **Find:** empirical formula

Conceptual Plan:

mass CO_2, H_2O → mol CO_2, H_2O → mol C, mol H → mass C, mass H, mass O → mol O →

$$\frac{1\text{ mol }CO_2}{44.01\text{ g }CO_2} \quad \frac{1\text{ mol }H_2O}{18.02\text{ g }H_2O} \qquad \frac{1\text{ mol C}}{1\text{ mol }CO_2} \quad \frac{2\text{ mol H}}{1\text{ mol }H_2O} \qquad \frac{12.01\text{ g C}}{1\text{ mol C}} \quad \frac{1.008\text{ g H}}{1\text{ mol H}}$$

$$\text{g sample - gC - g H} \qquad \frac{1\text{ mol O}}{16.00\text{ g O}}$$

pseudoformula → empirical formula

divide by smallest number

Solution:

$$8.59\text{ g }CO_2 \times \frac{1\text{ mol }CO_2}{44.01\text{ g }CO_2} = 0.195\text{ mol }CO_2$$

$$3.52\text{ g }H_2O \times \frac{1\text{ mol }H_2O}{18.02\text{ g }H_2O} = 0.195\text{ mol }H_2O$$

$$0.195\text{ mol }CO_2 \times \frac{1\text{ mol C}}{1\text{ mol }CO_2} = 0.195\text{ mol C}$$

$$0.195\text{ mol }H_2O \times \frac{2\text{ mol H}}{1\text{ mol }H_2O} = 0.390\text{ mol H}$$

$$0.195\text{ mol C} \times \frac{12.01\text{ g C}}{1\text{ mol C}} = 2.34\text{ g C}$$

$$0.390\text{ mol }H_2O \times \frac{1.008\text{ g H}}{1\text{ mol H}} = 0.393\text{ g H}$$

4.30 g - 2.34 g - 0.393 g = 1.57 g O

$$1.57\text{ g O} \times \frac{1\text{ mol O}}{16.00\text{ g O}} = 0.0979\text{ mol O}$$

$C_{0.195}H_{0.390}O_{0.0979}$

$C_{\frac{0.195}{0.0979}}H_{\frac{0.390}{0.0979}}O_{\frac{0.0979}{0.0979}} \rightarrow C_2H_4O$

The correct empirical formula is C_2H_4O

91. **Conceptual Plan: write a skeletal reaction → balance atoms in more complex compounds → balance elements that occur as free elements → clear fractions**

Solution: Skeletal reaction: $SO_2(g) + O_2(g) + H_2O(l) \rightarrow H_2SO_4(aq)$

Balance O: $SO_2(g) + 1/2O_2(g) + H_2O(l) \rightarrow H_2SO_4(aq)$

Clear fraction: $2SO_2(g) + O_2(g) + 2H_2O(l) \rightarrow 2H_2SO_4(aq)$

Check:

left side	right side
2 S atoms	2 S atoms
8 O atoms	8 O atoms
4 H atoms	4 H atoms

93. **Conceptual Plan: write a skeletal reaction → balance atoms in more complex compounds → balance elements that occur as free elements → clear fractions**

Solution: Skeletal reaction: $Na(s) + H_2O(l) \rightarrow H_2(g) + NaOH(aq)$

Balance H: $Na(s) + H_2O(l) \rightarrow 1/2H_2(g) + NaOH(aq)$

Clear fraction: $2Na(s) + 2H_2O(l) \rightarrow H_2(g) + 2NaOH(aq)$

Check:

left side	right side
2 Na atoms	2 Na atoms
4 H atoms	4 H atoms
2 O atoms	2 O atoms

95. **Conceptual Plan: write a skeletal reaction → balance atoms in more complex compounds → balance elements that occur as free elements → clear fractions**

Solution: Skeletal reaction: $C_{12}H_{22}O_{11}(aq) + H_2O(l) \rightarrow C_2H_5OH(aq) + CO_2(g)$

Balance H: $C_{12}H_{22}O_{11}(aq) + H_2O(l) \rightarrow 4C_2H_5OH(aq) + CO_2(g)$

Balance C: $C_{12}H_{22}O_{11}(aq) + H_2O(l) \rightarrow 4C_2H_5OH(aq) + 4CO_2(g)$

Check:

left side	right side
12 C atoms	12 C atoms
24 H atoms	24 H atoms
12 O atoms	12 O atoms

97. a) **Conceptual Plan: write a skeletal reaction → balance atoms in more complex compounds → balance elements that occur as free elements → clear fractions**

Solution: Skeletal reaction: $PbS(s) + HBr(aq) \rightarrow PbBr_2(s) + H_2S(g)$

Balance Br: $PbS(s) + 2HBr(aq) \rightarrow PbBr_2(s) + H_2S(g)$

Check:

left side	right side
1 Pb atom	1 Pb atom
1 S atom	1 S atom
2 H atoms	2 H atoms
2 Br atoms	2 Br atoms

b) **Conceptual Plan: write a skeletal reaction → balance atoms in more complex compounds → balance elements that occur as free elements → clear fractions**

Solution: Skeletal reaction: $CO(g) + H_2(g) \rightarrow CH_4(g) + H_2O(l)$

Balance H: $CO(g) + 3H_2(g) \rightarrow CH_4(g) + H_2O(l)$

Check:

left side	right side
1 C atom	1 C atom
1 O atom	1 O atom
6 H atoms	6 H atoms

c) **Conceptual Plan: write a skeletal reaction → balance atoms in more complex compounds → balance elements that occur as free elements → clear fractions**

Solution: Skeletal reaction: $HCl(aq) + MnO_2(s) \rightarrow MnCl_2(aq) + H_2O(l) + Cl_2(g)$
Balance Cl $4HCl(aq) + MnO_2(s) \rightarrow MnCl_2(aq) + H_2O(l) + Cl_2(g)$
Balance O $4HCl(aq) + MnO_2(s) \rightarrow MnCl_2(aq) + 2H_2O(l) + Cl_2(g)$

Check:

left side	right side
4 H atoms	4 H atoms
4 Cl atoms	4 Cl atoms
1 Mn atom	1 Mn atom
2 O atoms	2 O atoms

d) **Conceptual Plan: write a skeletal reaction → balance atoms in more complex compounds → balance elements that occur as free elements → clear fractions**

Solution: Skeletal reaction: $C_5H_{12}(l) + O_2(g) \rightarrow CO_2(g) + H_2O(l)$
Balance C: $C_5H_{12}(l) + O_2(g) \rightarrow 5CO_2(g) + H_2O(l)$
Balance H: $C_5H_{12}(l) + O_2(g) \rightarrow 5CO_2(g) + 6H_2O(l)$
Balance O: $C_5H_{12}(l) + 8O_2(g) \rightarrow 5CO_2(g) + 6H_2O(l)$

Check:

left side	right side
5 C atoms	5 C atoms
12 H atoms	12 H atoms
16 O atoms	16 O atoms

99. a) **Conceptual Plan: balance atoms in more complex compounds → balance elements that occur as free elements → clear fractions**

Solution: Skeletal reaction: $CO_2(g) + CaSiO_3(s) + H_2O(l) \rightarrow SiO_2(s) + Ca(HCO_3)_2(aq)$
Balance C: $2CO_2(g) + CaSiO_3(s) + H_2O(l) \rightarrow SiO_2(s) + Ca(HCO_3)_2(aq)$

Check:

left side	right side
2 C atoms	2 C atoms
8 O atoms	8 O atoms
1 Ca atom	1 Ca atom
1 Si atom	1 Si atom
2 H atoms	2 H atoms

b) **Conceptual Plan: balance atoms in more complex compounds → balance elements that occur as free elements → clear fractions**

Solution: Skeletal reaction: $Co(NO_3)_3(aq) + (NH_4)_2S(aq) \rightarrow Co_2S_3(s) + NH_4NO_3(aq)$
Balance S: $Co(NO_3)_3(aq) + 3(NH_4)_2S(aq) \rightarrow Co_2S_3(s) + NH_4NO_3(aq)$
Balance Co: $2Co(NO_3)_3(aq) + 3(NH_4)_2S(aq) \rightarrow Co_2S_3(s) + NH_4NO_3(aq)$
Balance N: $2Co(NO_3)_3(aq) + 3(NH_4)_2S(aq) \rightarrow Co_2S_3(s) + 6NH_4NO_3(aq)$

Check:

left side	right side
2 Co atoms	2 Co atoms
12 N atoms	12 N atoms
18 O atoms	18 O atoms
24 H atoms	24 H atoms
3 S atoms	3 S atoms

c) **Conceptual Plan: balance atoms in more complex compounds → balance elements that occur as free elements → clear fractions**

Solution: Skeletal reaction: $Cu_2O(s) + C(s) \rightarrow Cu(s) + CO(g)$
Balance Cu: $Cu_2O(s) + C(s) \rightarrow 2Cu(s) + CO(g)$

Check:

left side	right side
2 Cu atoms	2 Cu atoms
1 O atom	1 O atom
1 C atom	1 C atom

d) **Conceptual Plan: balance atoms in more complex compounds → balance elements that occur as free elements → clear fractions**

Solution: Skeletal reaction: $H_2(g) + Cl_2(g) \rightarrow HCl(g)$
Balance Cl: $H_2(g) + Cl_2(g) \rightarrow 2HCl(g)$

Check: left side right side

2 H atom 2 H atom
2 Cl atom 2 Cl atom

101.a) Composed of metal cation and polyatomic anion – inorganic compound
b) Composed of carbon and hydrogen – organic compound
c) Composed of carbon, hydrogen and oxygen – organic compound
d) Composed of metal cation and nonmetal anion – inorganic compound

103.a) Contains a double bond – alkene
b) Contains only single bonds – alkane
c) Contains triple bond – alkyne
d) Contains only single bonds – alkane

105.a) prop = 3 C, ane = single bonds: $CH_3CH_2CH_3$
b) 3 C = prop, single bonds = ane: propane
c) oct = 8 C, and = single bonds: $CH_3CH_2\ CH_2\ CH_2\ CH_2\ CH_2\ CH_2\ CH_3$
d) 5 C = pent, single bonds = ane: pentane

107.a) Contains O: functionalized hydrocarbon: alcohol
b) Contains only C and H: hydrocarbon
c) Contains O: functionalized hydrocarbon: ketone
d) Contains N: functionalized hydrocarbon: amine

109.**Given:** 145 mL C_2H_5OH, d = 0.789g/cm^3 **Find:** number of molecules
Conceptual Plan: cm^3 → mL: mL C_2H_5OH → g C_2H_5OH → mol C_2H_5OH → molecules C_2H_5OH

$$\frac{1\ \text{cm}^3}{1\ \text{mL}} \qquad \frac{1\ \text{mL}\ C_2H_5OH}{0.789\ \text{g}\ C_2H_5OH} \qquad \frac{1\ \text{mol}\ C_2H_5OH}{46.07\ \text{g}\ C_2H_5OH}$$

$$\frac{6.022 \times 10^{23}\ \text{molecules}\ C_2H_5OH}{1\ \text{mol}\ C_2H_5OH}$$

Solution:

$$145\ \text{mL}\ C_2H_5OH \times \frac{0.789\ \text{g}}{\text{cm}^3} \times \frac{1\ \text{cm}^3}{1\ \text{mL}} \times \frac{1\ \text{mol}\ C_2H_5OH}{46.07\ \text{g}\ C_2H_5OH} \times \frac{6.022 \times 10^{23}\ \text{molecules}\ C_2H_5OH}{1\ \text{mol}\ C_2H_5OH} = 1.49 \times 10^{24}\ \text{molecules}$$

Check: Units of answer, molecules C_2H_5OH, are correct. The magnitude is reasonable because we had more than 2 moles of C_2H_5OH and we have more than 2 times Avogadro's number of molecules.

111.a) To write the formula for an ionic compound: 1) Write the symbol for the metal cation and its charge and the symbol for the nonmetal anion or polyatomic anion and its charge. 2) Adjust the subscript on each cation and anion to balance the overall charge. 3) Check that the sum of the charges of the cations equals the sum of the charges of the anions.
potassium chromate: K^+ CrO_4^{-2}; K_2CrO_4 cation 2(1+) = 2+; anion 2-

Given: K_2CrO_4 **Find:** mass percent of each element
Conceptual Plan: %K, then %Cr, then %O

$$\text{mass}\ \%\text{K} = \frac{2 \times \text{molar mass K}}{\text{molar mass}\ K_2CrO_4} \times 100$$

$$\text{mass}\ \%\text{Cr} = \frac{1 \times \text{molar mass Cr}}{\text{molar mass}\ K_2CrO_4} \times 100 \quad \text{mass}\ \%\text{O} = \frac{4 \times \text{molar mass O}}{\text{molar mass}\ K_2CrO_4} \times 100$$

molar mass of K = 39.10 g/mol, molar mass Cr = 52.00 g/mol, molar mass O = 16.00 g/mol
Solution: molar mass K_2CrO_4 = 2(39.10 g/mol) + 1(52.00 g/mol) + 4(16.00 g/mol) = 194.20 g/mol

2 x molar mass K = 2(39.10 g/mol) = 78.20 g K

$$\text{mass \% K} = \frac{\text{2 x molar mass K}}{\text{molar mass } K_2CrO_4} \times 100\%$$

$$= \frac{78.20 \text{ g/mol}}{194.20 \text{ g/mol}} \times 100\%$$

= 40.27 %

1 x molar mass Cr = 1(52.00 g/mol) = 52.00 g Cr

$$\text{mass \% Cr} = \frac{\text{1 x molar mass Cr}}{\text{molar mass } K_2CrO_4} \times 100\%$$

$$= \frac{52.00 \text{ g/mol}}{194.20 \text{ g/mol}} \times 100\%$$

= 26.78 %

4 x molar mass O = 4(16.00 g/mol) = 64.00 g O

$$\text{mass \% O} = \frac{\text{4 x molar mass O}}{\text{molar mass } K_2CrO_4} \times 100\%$$

$$= \frac{64.00 \text{ g/mol}}{194.20 \text{ g/mol}} \times 100\%$$

= 32.96 %

Check: Units of the answer, %, are correct. The magnitude is reasonable because each is between 0 and 100% and the total is 100%.

b) To write the formula for an ionic compound: 1) Write the symbol for the metal cation and its charge and the symbol for the nonmetal anion or polyatomic anion and its charge. 2) Adjust the subscript on each cation and anion to balance the overall charge. 3) Check that the sum of the charges of the cations equals the sum of the charges of the anions.

Lead(II)phosphate: Pb^{2+} PO_4^{3-} ; $Pb_3(PO_4)_2$ cation 3(2+) = 6+; anion 2(3-) = 6-

Given: $Pb_3(PO_4)_2$ **Find:** mass percent of each element

Conceptual Plan: %Pb, then % P, then %O

$$\text{mass \% Pb} = \frac{\text{3 x molar mass Pb}}{\text{molar mass } Pb_3(PO_4)_2} \times 100 \qquad \text{mass \% P} = \frac{\text{2 x molar mass P}}{\text{molar mass } Pb_3(PO_4)_2} \times 100$$

$$\text{mass \% O} = \frac{\text{8 x molar mass O}}{\text{molar mass } Pb_3(PO_4)_2} \times 100$$

Solution: molar mass $Pb_3(PO_4)_2$ = 3(207.2 g/mol) + 2(30.97 g/mol) + 8(16.00 g/mol) = 811.5 g/mol

3 x molar mass Pb = 3(207.2 g/mol) = 621.6 g Pb

$$\text{mass \% K} = \frac{\text{2 x molar mass K}}{\text{molar mass } Pb_3(PO_4)_2} \times 100\%$$

$$= \frac{78.20 \text{ g/mol}}{811.5 \text{ g/mol}} \times 100\%$$

= 76.60 %

2 x molar mass P = 2(30.97 g/mol) = 61.94 g P

$$\text{mass \% P} = \frac{\text{2 x molar mass P}}{\text{molar mass } Pb_3(PO_4)_2} \times 100\%$$

$$= \frac{61.94 \text{ g/mol}}{811.5 \text{ g/mol}} \times 100\%$$

= 7.632 %

4 x molar mass O = 8(16.00 g/mol) = 128.0 g O

$$\text{mass \% O} = \frac{\text{8 x molar mass O}}{\text{molar mass } Pb_3(PO_4)_2} \times 100\%$$

$$= \frac{128.0 \text{ g/mol}}{811.5 \text{ g/mol}} \times 100\%$$

= 15.77 %

Check: Units of the answer, %, are correct. The magnitude is reasonable because each is between 0 and 100% and the total is 100%.

c) sulfurous acid: H_2SO_3
Given: H_2SO_3 **Find:** mass percent of each element
Conceptual Plan: %H, then %S, then %O

$$\text{mass \% H} = \frac{2 \text{ x molar mass H}}{\text{molar mass HSO}_3} \text{ x } 100 \qquad \text{mass \% S} = \frac{1 \text{ x molar mass S}}{\text{molar mass HSO}_3} \text{ x } 100$$

$$\text{mass \% O} = \frac{3 \text{ x molar mass O}}{\text{molar mass HSO}_3} \text{ x } 100$$

Solution: molar mass H_2SO_3 = 2(1.008 g/mol) + 1(32.06 g/mol) + 3(16.00 g/mol) = 82.0$\underline{7}$6 g/mol
2 x molar mass H = 1(1.008 g/mol) = 2.016 g H

$$\text{mass \%H} = \frac{2 \text{ x molar mass H}}{\text{molar mass H}_2\text{SO}_3} \text{ x } 100\%$$

$$= \frac{2.016 \text{ g/mol}}{82.0\underline{7}6 \text{ g/mol}} \text{ x } 100\%$$

= 2.456 %

1 x molar mass S = 1(32.06 g/mol) = 32.06 g Cr

$$\text{mass \% S} = \frac{1 \text{ x molar mass S}}{\text{molar mass H}_2\text{SO}_3} \text{ x } 100\%$$

$$= \frac{32.06 \text{ g/mol}}{82.0\underline{7}6 \text{ g/mol}} \text{ x } 100\%$$

= 39.06 %

4 x molar mass O = 3(16.00 g/mol) = 48.00 g O

$$\text{mass \% O} = \frac{3 \text{ x molar mass O}}{\text{molar mass H}_2\text{SO}_3} \text{ x } 100\%$$

$$= \frac{48.00 \text{ g/mol}}{82.0\underline{7}6 \text{ g/mol}} \text{ x } 100\%$$

= 58.48 %

Check: Units of the answer, %, are correct. The magnitude is reasonable because each is between 0 and 100% and the total is 100%.

d) To write the formula for an ionic compound: 1) Write the symbol for the metal cation and its charge and the symbol for the nonmetal anion or polyatomic anion and its charge. 2) Adjust the subscript on each cation and anion to balance the overall charge. 3) Check that the sum of the charges of the cations equals the sum of the charges of the anions.
cobalt(II)bromide: Co^{2+} Br^-; $CoBr_2$ cation 2+ = 2+; anion 2(1-) = 2-
Given: $CoBr_2$ **Find:** mass percent of each element
Conceptual Plan: %Co, then %Br

$$\text{mass \% Co} = \frac{1 \text{ x molar mass Co}}{\text{molar mass CoBr}_2} \text{ x } 100 \qquad \text{mass \% Br} = \frac{2 \text{ x molar mass Br}}{\text{molar mass CoBr}_2} \text{ x } 100$$

Solution: molar mass $CoBr_2$ = (58.93 g/mol) + 2(79.90 g/mol) = 218.7 g/mol

2 x molar mass Co = 1(58.93 g/mol) = 58.93 g K

$$\text{mass \% Co} = \frac{1 \text{ x molar mass Co}}{\text{molar mass } CoBr_2} \text{ x } 100\%$$

$$= \frac{58.93 \text{ g/mol}}{218.7 \text{ g/mol}} \text{ x } 100\%$$

$= 26.94\ \%$

1 x molar mass Br = 2(79.90 g/mol) = 159.8 g Br

$$\text{mass \% Br} = \frac{2 \text{ x molar mass Br}}{\text{molar mass } CoBr_2} \text{ x } 100\%$$

$$= \frac{159.8 \text{ g/mol}}{218.7 \text{ g/mol}} \text{ x } 100\%$$

$= 73.07\ \%$

Check: Units of the answer, %, are correct. The magnitude is reasonable because each is between 0 and 100% and the total is 100%.

113.**Given:** 25 g CF_2Cl_2/mo. **Find:** g Cl /yr.

Conceptual Plan:g CF_2Cl_2/mo → g Cl/mo → g Cl/yr

$$\frac{70.09 \text{ g Cl}}{120.91 \text{ g } CF_2Cl_2} \qquad \frac{12 \text{ mo.}}{1 \text{ yr}}$$

Solution: $\frac{25 \text{ g } CF_2Cl_2}{\text{mo.}} \text{ x } \frac{70.90 \text{ g Cl}}{120.91 \text{ g } CF_2Cl_2} \text{ x } \frac{12 \text{ mo.}}{1 \text{ yr}} = 1.8 \text{ x } 10^2 \text{ g Cl/yr}$

Check: Units of answer, g Cl, is correct. Magnitude is reasonable because it is less than the total CF_2Cl_2 /yr.

115.**Given:** MCl_3, 65.57% Cl **Find:** identify M

Conceptual Plan: g Cl → mol Cl → mol M → atomic mass M

$$\frac{1 \text{ mol Cl}}{35.45 \text{ g Cl}} \qquad \frac{1 \text{ mol M}}{3 \text{ mol Cl}} \qquad \frac{\text{g M}}{\text{mol M}}$$

Solution: in 100 g sample: 65.57 g Cl, 34.43 g M

$$65.57 \text{ g Cl x } \frac{1 \text{ mol Cl}}{35.45 \text{ g}} \text{ x } \frac{1 \text{ mol M}}{3 \text{ mol Cl}} = 0.6165 \text{ mol M} \qquad \frac{34.43 \text{ g M}}{0.6165 \text{ mol M}} = 55.84 \text{ g/mol M}$$

molar mass of 55.84 = Fe

The identity of M = Fe

117.**Given:** In a 100 g sample: 79.37 g C, 8.88 g H, 11.75 g O, molar mass = 272.37g/mol

Find: molecular formula

Conceptual Plan:

convert mass to mol of each element → pseudoformula → empirical formula → molecular formula

$$\frac{1 \text{ mol C}}{12.01 \text{ g C}} \quad \frac{1 \text{ mol H}}{1.008 \text{ g H}} \quad \frac{1 \text{ mol O}}{16.00 \text{ g O}}$$

divide by smallest number

empirical formula x n

Solution:

$$79.37 \text{ g C x } \frac{1 \text{ mol C}}{12.01 \text{ g C}} = 6.609 \text{ mol C}$$

$$8.88 \text{ g H x } \frac{1 \text{ mol H}}{1.008 \text{ g H}} = 8.81 \text{ mol H}$$

$$11.75 \text{ g O x } \frac{1 \text{ mol O}}{16.00 \text{ g O}} = 0.7344 \text{ mol O}$$

$C_{6.609}H_{8.81}O_{0.7344}$

$$C_{\frac{6.609}{0.7344}}H_{\frac{8.81}{0.7344}}O_{\frac{0.7344}{0.7344}} \rightarrow C_9H_{12}O$$

The correct empirical formula is $C_9H_{12}O$

empirical formula mass = 9(12.01 g/mol) + 12(1.008 g/mol) + 1(16.00 g/mol) = 136.19 g/mol

$$n = \frac{\text{molar mass}}{\text{formula molar mass}} = \frac{272.37\text{ g/mol}}{136.19\text{ g/mol}} = 2$$

molecular formula $= C_9H_{12}O \times 2$
$= C_{18}H_{24}O_2$

119. **Given:** 13.42 g sample, 39.61 g CO_2, 9.01 g H_2O, molar mass = 268.34 g/mol
Find: molecular formula

Conceptual Plan:
mass CO_2, H_2O → mol CO_2, H_2O → mol C, mol H → mass C, mass H, mass O → mol O →

$$\frac{1\text{ mol }CO_2}{44.01\text{ g }CO_2}\quad \frac{1\text{ mol }H_2O}{18.02\text{ g }H_2O}\quad \frac{1\text{ mol C}}{1\text{ mol }CO_2}\quad \frac{2\text{ mol H}}{1\text{ mol }H_2O}\qquad \frac{12.01\text{ g C}}{1\text{ mol C}}\quad \frac{1.008\text{ g H}}{1\text{ mol H}}\quad \text{g sample - gC - g H}\quad \frac{1\text{ mol O}}{16.00\text{ g O}}$$

pseudoformula → empirical formula → molecular formula
divide by smallest number — empirical formula x n

Solution:

$$39.61\text{ g }CO_2 \times \frac{1\text{ mol }CO_2}{44.01\text{ g }CO_2} = 0.9000\text{ mol }CO_2$$

$$9.01\text{ g }H_2O \times \frac{1\text{ mol }H_2O}{18.02\text{ g }H_2O} = 0.5000\text{ mol }H_2O$$

$$0.9000\text{ mol }CO_2 \times \frac{1\text{ mol C}}{1\text{ mol }CO_2} = 0.9000\text{ mol C}$$

$$0.5000\text{ mol }H_2O \times \frac{2\text{ mol H}}{1\text{ mol }H_2O} = 1.000\text{ mol H}$$

$$0.9000\text{ mol C} \times \frac{12.01\text{ g C}}{1\text{ mol C}} = 10.81\text{ g C}$$

$$1.000\text{ mol }H_2O \times \frac{1.008\text{ g H}}{1\text{ mol H}} = 1.008\text{ g H}$$

13.42 g - 10.81 g - 1.008 g = 1.60 g O

$$1.60\text{ g O} \times \frac{1\text{ mol O}}{16.00\text{ g O}} = 0.100\text{ mol O}$$

$$C_{0.9000}H_{1.000}O_{0.100}$$

$$C_{\frac{0.9000}{0.100}}H_{\frac{1.000}{0.100}}O_{\frac{0.100}{0.100}} \rightarrow C_9H_{10}O$$

The correct empirical formula is $C_9H_{10}O$
empirical formula mass = 9(12.01 g/mol) + 10(1.008 g/mol) + 1(16.00 g/mol) = 134.2 g/mol

$$n = \frac{\text{molar mass}}{\text{formula molar mass}} = \frac{268.34\text{ g/mol}}{134.2\text{ g/mol}} = 2$$

molecular formula $= C_9H_{10}O \times 2$
$= C_{18}H_{20}O_2$

121. **Given:** 4.93 g $MgSO_4 \bullet xH_2O$, 2.41 g $MgSO_4$ **Find:** value of x

Conceptual Plan: g $MgSO_4$ → mol $MgSO_4$ $\frac{1 \text{ mol } MgSO_4}{120.38 \text{ g } MgSO_4}$ **g H_2O → mol H_2O** $\frac{1 \text{ mol } H_2O}{18.02 \text{ g } H_2O}$ **Determine mole ratio** $\frac{\text{mol } H_2O}{\text{mol } MgSO_4}$

Solution: $2.41 \text{ g } MgSO_4 \times \frac{1 \text{ mol } MgSO_4}{120.38 \text{ g } MgSO_4} = 0.0200 \text{ mol } MgSO_4$

$$2.52 \text{ g } H_2O \times \frac{1 \text{ mol } H_2O}{18.02 \text{ g } H_2O} = 0.140 \text{ mol } H_2O$$

$$\frac{0.140 \text{ mol } H_2O}{0.0200 \text{ mol } MgSO_4} = 7$$

x = 7

123. **Given:** molar mass = 177 g/mol, g C = 8(g H) **Find:** molecular formula

Conceptual Plan: C_xH_yBrO

Solution: in 1 mol compound, let x = mol C and y = mol H, assume mol Br = 1, mol O = 1

177 g/mol = x(12.01 g/mol) + y(1.008 g/mol) + 1(79.90 g/mol) + 1(16.00 g/mol)

x(12.01 g/mol) = 8 {y(1.008 g/mol)}

177 g/mol = 8y(1.008 g/mol) + y(1.008 g/mol) + 79.90 g/mol + 16.00 g/mol

81 = 9y(1.008)

y = 9 = mol H

x(12.01) = 9(1.008)

x = 6 = mol C

molecular formula = C_6H_9BrO

Check: molar mass = 6(12.01 g/mol) + 9(1.008 g/mol) + 1(79.90 g/mol) + 1(16.00 g/mol) = 177.0 g/mol

125. **Given:** 23.5 mg $C_{17}H_{22}ClNO_4$ **Find:** total number of atoms

Conceptual Plan: mg compound → g compound → mol compound → mol atoms → number of atoms

$$\frac{1000 \text{ mg}}{1 \text{ g}} \quad \frac{1 \text{ mol}}{339.8 \text{ g}} \quad \frac{45 \text{ mol atoms}}{1 \text{ mol compound}} \quad \frac{6.022 \times 10^{23} \text{ atoms}}{1 \text{ mol atoms}}$$

Solution: $\frac{23.5 \text{ mg}}{} \times \frac{1 \text{ g}}{1000 \text{ mg}} \times \frac{1 \text{ mol cpd}}{339.8 \text{ g}} \times \frac{45 \text{ mol atoms}}{1 \text{ mol cpd}} \times \frac{6.022 \times 10^{23} \text{ atoms}}{\text{mol}} = 1.87 \times 10^{21} \text{ atoms}$

Check: The units of the answer, number of atoms, is correct. The magnitude of the answer is reasonable since the molecule is so complex.

127. **Given:** MCl_3, 2.395 g sample, 3.606×10^{-2} mol Cl **Find:** atomic mass M

Conceptual Plan: mol Cl → g Cl → g X

$$\frac{35.45 \text{ g Cl}}{1 \text{ mol Cl}} \quad \text{g sample - g Cl}$$

mol Cl → mol M → atomic mass M

$$\frac{1 \text{ mol M}}{3 \text{ mol Cl}} \quad \frac{\text{g M}}{\text{mol M}}$$

Solution: $3.606 \times 10^{-2} \text{ mol Cl} \times \frac{35.45 \text{ g}}{1 \text{ mol Cl}} = 1.278 \text{ g Cl}$

2.395 g - 1.278 g = 1.117 g M

$$3.606 \times 10^{-2} \text{ mol Cl} \times \frac{1 \text{ mol M}}{3 \text{ mol Cl}} = 1.202 \times 10^{-2} \text{ mol M}$$

$$\frac{1.117 \text{ g M}}{0.01202 \text{ mol M}} = 92.93 \text{ g/mol M}$$

molar mass of M = 92.93 g/mol

129. **Given:** g NaCl + g NaBr = 2.00 g, g Na = 0.75 g **Find:** g NaBr

Conceptual Plan: Let x = mol NaCl, y = mol NaBr, then x(molar mass NaCl) = g NaCl, y(molar mass NaBr) = g NaBr

Solution: $x(58.4) + y(102.9) = 2.00$

$x(23.0) + y(23.0) = 0.75 \qquad y = 0.0326 - x$

$58.4x + 102.9(0.0326-x) = 2.00$

$58.4x + 3.354 - 102.9x = 2.00$

$44.5x = 1.354$

$x = 0.03043$ mol NaCl

$y = 0.0326 - 0.03043 = 0\ 00217$ mol NaBr

g NaBr = (0.00217)(102.9g/mol) = 0.224 g NaBr

Check: The units of the answer, g NaBr, are correct. The magnitude is reasonable since it is less than the total mass.

131. **Given:** Sample of $CaCO_3$ and $(NH_4)_2CO_3$ is 61.9% CO_3^{2-} **Find:** % $CaCO_3$

Conceptual Plan: Let x = $CaCO_3$, y = $(NH_4)_2CO_3$, then x(molar mass $CaCO_3$) = g $CaCO_3$, y(molar mass $(NH_4)_2CO_3$) = g $(NH_4)_2CO_3$ then, a 100.0 g sample contains: x(100.0) g $CaCO_3$; y(96.1) g $(NH_4)_2CO_3$; and 61.9 g CO_3^{2-}

Solution: $x(100.0) + y(96.1) = 100.0$

$x(60.0) + y(60.0) = 61.9 \qquad y = 1.032 - x$

$100.0x + 96.1(1.032-x) = 100$

$100.0x + 99.14 - 96.1x = 100$

$3.9x = 0.96$

$x = 0.22$ mol $CaCO_3$

$y = 1.032 - 0.22 = 0.81$ mol $(NH_4)_2CO_3$

g $CaCO_3$ = (0.22 mol)(100.0g/mol) = 22.0 g $CaCO_3$ in a 100 g sample:

mass % $CaCO_3$ = 22.0%

Check: The units of the answer, mass % $CaCO_3$, are correct. The magnitude is reasonable since it is between 0 and 100%

133. **Given:** 1.1 kg CF_2Cl_2/automobile, 25% leak/year, 100 x 10^6 automobiles **Find:** kg Cl/yr

Conceptual Plan: **kg CF_2Cl_2 /auto → kg CF_2Cl_2 leaked/yr → kg Cl/yr/auto → kg Cl**

$$\frac{25\text{ kg }CF_2Cl_2}{100\text{ kg }CF_2Cl_2} \qquad \frac{70.9\text{ g Cl}}{120.91\text{ g }CF_2Cl_2} \qquad \frac{100 \times 10^6\text{ auto}}{}$$

Solution:

$$\frac{1.1\text{ kg }CF_2Cl_2}{\text{auto}} \times \frac{25\text{ kg }CF_2Cl_2}{100\text{ kg }CF_2Cl_2} \times \frac{70.9\text{ kg Cl}}{120.91\text{ kg }CF_2Cl_2} \times \frac{100 \times 10^6\text{ auto}}{} = 1.6 \times 10^7\text{ kg Cl/yr}$$

Check: Units of the answer, kg Cl, are correct. The magnitude is reasonable because it is less than the kg CF_2Cl_2 leaked per year.

135. **Given:** rock contains: 38.0% PbS, 25.0% $PbCO_3$, 17.4% $PbSO_4$, **Find:** kg rock needed for 5.0 metric ton Pb

Conceptual Plan: determine kg Pb/ 100 kg rock then ton Pb→ kg Pb → kg rock

$$\frac{1000\text{ kg}}{\text{metric ton}} \qquad \frac{100\text{ kg rock}}{64.2\text{ kg rock}}$$

Solution: in 100 kg rock:

$$\left(\frac{38.0\text{ kg PbS}}{} \times \frac{207.2\text{ kg Pb}}{239.3\text{ kg PbS}}\right) + \left(\frac{25.0\text{ kg }PbCO_3}{} \times \frac{207.2\text{ kg Pb}}{267.2\text{ kg }PbCO_3}\right) + \left(\frac{17.4\text{ kg }PbSO_4}{} \times \frac{207.2\text{ kg Pb}}{303.2\text{ kg }PbSO_4}\right.$$

$$5.0\text{ metric ton Pb} \times \frac{1000\text{ kg Pb}}{\text{metric ton Pb}} \times \frac{100\text{ kg rock}}{64.2\text{ kg Pb}} = 7.8 \times 10^3\text{ kg rock}$$

Check: Units of answer, kg rock, are correct. Magnitude is reasonable since it is greater than the amount of Pb needed.

137. a) Atomic mass O > atomic mass C, % O would be higher.

b) Atomic mass N and O close, molecule contains 2N to 1 O, % N would be higher.

c) Atomic mass O > atomic mass C, same number of atoms, % O would be higher.

d) Atomic mass N much greater than atomic mass H, % N would be higher.

139. The statement is incorrect because equations are balanced based on the number and kind of atoms not molecules. The statement should read: "When a chemical equation is balanced, the number of atoms of each type on both sides of the equation will be equal."

Chapter 4
Chemical Quantities and Aqueous Reactions

1. Reaction stoichiometry is the numerical relationships between chemical amounts in a balanced chemical equation. The coefficients in a chemical reaction specify the relative amounts in moles of each of the substances involved in the reaction.

3. No, the percent yield would not be different if the actual yield and theoretical yield were calculated in moles. The relationship between grams and moles is the molar mass. This would be the same value for the actual yield and the theoretical yield.

5. Molarity is a concentration term. It is the amount of solute (in moles) divided by the volume of solution (in liters). The molarity of a solution can be used as a conversion factor between moles of the solute and liters of the solution.

7. Acids are molecular compounds that ionize – form ions – when they dissolve in water. A strong acid is one that completely ionizes in solution. A weak acid is one that does not completely ionize in water. A solution of a weak acid is composed mostly of the nonionized acid.

9. The solubility rules are a set of empirical rules that have been inferred from observations on many ionic compounds . the solubility rules allow us to predict if a compound is soluble or insoluble.

11. A precipitation reaction is one in which a solid or precipitate forms upon mixing two solutions. An example is: $2\ KI(aq) + Pb(NO_3)_2(aq) \rightarrow PbI_2(s) + 2\ KNO_3(aq)$

13. A molecular equation is an equation showing the complete neutral formulas for each compound in the reaction as if they existed as molecules. Equations which list individually all of the ions present as either reactants or precuts in a chemical reaction are called complete ionic equation. Equations which show only the species that actually change during the reaction are called net ionic equations.

15. When an acid and base are mixed, the $H^+(aq)$ from the acid combines with the OH^- from the base to form $H_2O(l)$. An example is: $HCl(aq) + NaOH(aq) \rightarrow H_2O(l) + NaCl(aq)$

17. Aqueous reactions that form a gas upon mixing two solutions are called gas – evolution reactions. An example is: $H_2SO_4(aq) + Li_2S(aq) \rightarrow H_2S(g) + Li_2SO_4(aq)$

19. Oxidation – reduction reactions or redox reactions are reactions in which electrons are transferred from one reactant to the other. An example is: $4\ Fe(s) + 3\ O_2(g) \rightarrow 2\ Fe_2O_3(s)$

21. To identify redox reactions by using oxidation states, begin by assigning oxidation states to each atom in the reaction. A change in oxidation state for the atoms indicates a redox reaction.

23. A substance that causes the oxidation of another substance is called an oxidizing agent. A substance that causes the reduction of another substance is called a reducing agent.

25. **Given:** 4.9 moles C_6H_{14} **Find:** balanced reaction, moles O_2 required
Conceptual Plan: balance the reaction then mol C_6H_{14} → mol O_2

$$2\ C_6H_{14}(g) + 19\ O_2(g) \rightarrow 12\ CO_2(g) + 14\ H_2O(g) \qquad \frac{19\ \text{mol}\ O_2}{2\ \text{mol}\ C_6H_{14}}$$

Solution: $4.9\ \cancel{\text{mol}\ C_6H_{14}} \times \dfrac{19\ \text{mol}\ O_2}{2\ \cancel{\text{mol}\ C_6H_{14}}} = 47\ \text{mol}\ O_2$

Check: The units, mol O_2, are correct. The magnitude is reasonable because much more O_2 is needed than C_6H_{14}.

27. a) **Given:** 1.3 mol N_2O_5 **Find:** mol NO_2

Conceptual Plan: mol N_2O_5 → mol NO_2

$$\frac{4\ NO_2}{2\ N_2O_5}$$

Solution: $1.3\ \cancel{mol\ N_2O_5} \times \frac{4\ mol\ NO_2}{2\ \cancel{mol\ N_2O_5}} = 2.6\ mol\ NO_2$

Check: The units of the answer, mol NO_2, are correct. The magnitude is reasonable since it is greater than mol N_2O_5.

b) **Given:** 5.8 mol N_2O_5 **Find:** mol NO_2
Conceptual Plan: mol N_2O_5 → mol NO_2

$$\frac{4\ NO_2}{2\ N_2O_5}$$

Solution: $5.8\ \cancel{mol\ N_2O_5} \times \frac{4\ mol\ NO_2}{2\ \cancel{mol\ N_2O_5}} = 11.6\ mol\ NO_2 = 12\ mol\ NO_2$

Check: The units of the answer, mol NO_2, are correct. The magnitude is reasonable since it is greater than mol N_2O_5.

c) **Given:** 10.5 g N_2O_5 **Find:** mol NO_2
Conceptual Plan: g N_2O_5 → mol N_2O_5 → molNO_2

$$\frac{1\ mol\ N_2O_5}{108.02\ g\ N_2O_5} \quad \frac{4\ NO_2}{2\ N_2O_5}$$

Solution: $10.5\ \cancel{g\ N_2O_5} \times \frac{1\ \cancel{mol\ N_2O_5}}{108.02\ \cancel{g\ N_2O_5}} \times \frac{4\ mol\ NO_2}{2\ \cancel{mol\ N_2O_5}} = 0.194\ mol\ NO_2$

Check: The units of the answer, mol NO_2, are correct. The magnitude is reasonable since 10 g is about 0.1 mol N_2O_5 and the answer is greater than mol N_2O_5.

d) **Given:** 1.55 kg N_2O_5 **Find:** mol NO_2
Conceptual Plan: kg N_2O_5 →g N_2O_5 → mol N_2O_5 → molNO_2

$$\frac{1000\ g\ N_2O_5}{kg\ N_2O_5} \quad \frac{1\ mol\ N_2O_5}{108.02\ g\ N_2O_5} \quad \frac{4\ NO_2}{2\ N_2O_5}$$

Solution: $1.55\ \cancel{kg\ N_2O_5} \times \frac{1000\ \cancel{g\ N_2O_5}}{\cancel{kg\ N_2O_5}} \times \frac{1\ \cancel{mol\ N_2O_5}}{108.02\ \cancel{g\ N_2O_5}} \times \frac{4\ mol\ NO_2}{2\ \cancel{mol\ N_2O_5}} = 28.7\ mol\ NO_2$

Check: The units of the answer, mol NO_2, are correct. The magnitude is reasonable since 1.5 kg is about 14 mol N_2O_5 and the answer is greater than mol N_2O_5.

29. **Given:** 3 mol SiO_2 **Find:** mol C, mol SiC, mol CO
Conceptual Plan: mol SiO_2 → mol C → mol SiC → mol CO

$$\frac{3\ C}{SiO_2} \quad \frac{SiC}{SiO_2} \quad \frac{2\ CO}{SiO_2}$$

Solution:

$3\ \cancel{mol\ SiO_2} \times \frac{3\ mol\ C}{\cancel{mol\ SiO_2}} = 9\ mol\ C$ $3\ \cancel{mol\ SiO_2} \times \frac{mol\ SiC}{\cancel{mol\ SiO_2}} = 3\ mol\ SiC$ $3\ \cancel{mol\ SiO_2} \times \frac{2\ mol\ CO}{\cancel{mol\ SiO_2}} = 6\ mol\ CO$

Given: 6 mol C **Find:** mol SiO_2, mol SiC, mol CO
Conceptual Plan: mol C → mol SiO_2 → mol SiC → mol CO

$$\frac{SiO_2}{3\ C} \quad \frac{SiC}{3\ C} \quad \frac{2\ CO}{3\ C}$$

Solution:

$6 \cancel{\text{mol C}} \times \frac{\text{mol } SiO_2}{3 \cancel{\text{mol C}}} = 2 \text{ mol } SiO_2$ $6 \cancel{\text{mol C}} \times \frac{\text{mol SiC}}{3 \cancel{\text{mol C}}} = 2 \text{ mol SiC}$ $6 \cancel{\text{mol C}} \times \frac{2 \text{ mol CO}}{3 \cancel{\text{mol C}}} = 4 \text{ mol CO}$

Given: 10 mol CO **Find:** mol SiO_2, mol C, mol SiC

Conceptual Plan: mol CO → mol SiO_2 → mol C → mol SiC

$\frac{SiO_2}{2\ CO}$ $\frac{3\ C}{2\ CO}$ $\frac{SiC}{2\ CO}$

Solution:

$10 \cancel{\text{mol CO}} \times \frac{\text{mol } SiO_2}{2 \cancel{\text{mol CO}}} = 5.0 \text{ mol } SiO_2$ $10 \cancel{\text{mol C}} \times \frac{3 \text{ mol C}}{2 \cancel{\text{mol CO}}} = 15 \text{ mol C}$ $10 \cancel{\text{mol CO}} \times \frac{\text{mol SiC}}{2 \cancel{\text{mol CO}}} = 5.0 \text{ mol SiC}$

Given: 2.8 mol SiO_2 **Find:** mol C, mol SiC, mol CO

Conceptual Plan: mol SiO_2 → mol C → mol SiC → mol CO

$\frac{3\ C}{SiO_2}$ $\frac{SiC}{SiO_2}$ $\frac{2\ CO}{SiO_2}$

Solution:

$2.8 \cancel{\text{mol } SiO_2} \times \frac{3 \text{ mol C}}{\cancel{\text{mol } SiO_2}} = 8.4 \text{ mol C}$ $2.8 \cancel{\text{mol } SiO_2} \times \frac{\text{mol SiC}}{\cancel{\text{mol } SiO_2}} = 2.8 \text{ mol SiC}$ $2.8 \cancel{\text{mol } SiO_2} \times \frac{2 \text{ mol CO}}{\cancel{\text{mol } SiO_2}} = 5.6 \text{ mol CO}$

Given: 1.55 mol C **Find:** mol SiO_2, mol SiC, mol CO

Conceptual Plan: mol C → mol SiO_2 → mol SiC → mol CO

$\frac{SiO_2}{3\ C}$ $\frac{SiC}{3\ C}$ $\frac{2\ CO}{3\ C}$

Solution:

$1.55 \cancel{\text{mol C}} \times \frac{3 \text{ mol } SiO_2}{3 \cancel{\text{mol C}}} = 0.517 \text{ mol } SiO_2$ $1.55 \cancel{\text{mol C}} \times \frac{\text{mol SiC}}{3 \cancel{\text{mol C}}} = 0.517 \text{ mol SiC}$ $1.55 \cancel{\text{mol C}} \times \frac{2 \text{ mol CO}}{3 \cancel{\text{mol C}}} = 1.03 \text{ n}$

SiO_2	C	SiC	CO
3	9	3	6
2	**6**	2	4
5.0	15	5.0	**10**
2.8	8.4	2.8	5.6
0.517	**1.55**	0.517	1.03

31. **Given:** 3.2 g Fe **Find:** g HBr; g H_2

Conceptual Plan: **g Fe → mol Fe → mol HBr → g HBr**

$\frac{\text{mol Fe}}{55.8 \text{ g Fe}}$ $\frac{2 \text{ mol HBr}}{\text{mol Fe}}$ $\frac{80.9 \text{ g HBr}}{\text{mol HBr}}$

g Fe → mol Fe → mol H_2 → g H_2

$\frac{\text{mol Fe}}{55.8 \text{ g Fe}}$ $\frac{1 \text{ mol } H_2}{\text{mol Fe}}$ $\frac{2.02 \text{ g } H_2}{\text{mol } H_2}$

Solution: $3.2 \text{ g} \times \frac{1 \text{ mol Fe}}{55.8 \text{ g Fe}} \times \frac{2 \text{ mol HBr}}{1 \text{ mol Fe}} \times \frac{80.9 \text{ g HBr}}{1 \text{ mol HBr}} = 9.3 \text{ g HBr}$

$3.2 \text{ g} \times \frac{1 \text{ mol Fe}}{55.8 \text{ g Fe}} \times \frac{1 \text{ mol } H_2}{1 \text{ mol Fe}} \times \frac{2.02 \text{ g } H_2}{1 \text{ mol } H_2} = 0.12 \text{ g } H_2$

Check: Units of answers, g HBr, g H_2, are correct. The magnitude of the answers is reasonable because molar mass HBr is greater than Fe and molar mass H_2 is much less than Fe.

33. a) **Given:** 2.5 g Ba **Find:** g $BaCl_2$

Conceptual Plan: **g Ba → mol Ba → mol $BaCl_2$ → g $BaCl_2$**

$$\frac{\text{mol Ba}}{137.33\text{ g Ba}} \quad \frac{1\text{ mol }BaCl_2}{1\text{ mol Ba}} \quad \frac{208.23\text{ g }BaCl_2}{1\text{ mol }BaCl_2}$$

Solution: $2.5\ \cancel{\text{g Ba}} \times \frac{1\ \cancel{\text{mol Ba}}}{137.33\ \cancel{\text{g Ba}}} \times \frac{1\ \cancel{\text{mol }BaCl_2}}{1\ \cancel{\text{mol Ba}}} \times \frac{208.23\text{ g }BaCl_2}{1\ \cancel{\text{mol }BaCl_2}} = 3.8\text{ g }BaCl_2$

Check: Units of answer, g $BaCl_2$, are correct. The magnitude of the answer is reasonable because it is larger than grams Ba.

b) **Given:** 2.5 g CaO **Find:** g $CaCO_3$

Conceptual Plan: **g CaO → mol CaO → mol $CaCO_3$ → g $CaCO_3$**

$$\frac{\text{mol CaO}}{56.08\text{ g CaO}} \quad \frac{\text{mol }CaCO_3}{1\text{ mol CaO}} \quad \frac{100.09\text{ g }CaCO_3}{\text{mol }CaCO_3}$$

Solution: $2.5\ \cancel{\text{g CaO}} \times \frac{1\ \cancel{\text{mol CaO}}}{56.08\ \cancel{\text{g CaO}}} \times \frac{1\ \cancel{\text{mol }CaCO_3}}{1\ \cancel{\text{mol CaO}}} \times \frac{100.09\text{ g }CaCO_3}{1\ \cancel{\text{mol }CaCO_3}} = 4.5\text{ g }CaCO_3$

Check: Units of answer, g $CaCO_3$, are correct. The magnitude of the answer is reasonable because it is larger than grams CaO.

c) **Given:** 2.5 g Mg **Find:** g MgO

Conceptual Plan: **g Mg → mol Mg → mol MgO → g MgO**

$$\frac{\text{mol Mg}}{24.30\text{ g Mg}} \quad \frac{\text{mol MgO}}{\text{mol Mg}} \quad \frac{40.30\text{ g MgO}}{\text{mol MgO}}$$

Solution: $2.5\ \cancel{\text{g Mg}} \times \frac{1\ \cancel{\text{mol Mg}}}{24.30\ \cancel{\text{g Mg}}} \times \frac{1\ \cancel{\text{mol MgO}}}{1\ \cancel{\text{mol Mg}}} \times \frac{40.30\text{ g MgO}}{1\ \cancel{\text{mol MgO}}} = 4.1\text{ g MgO}$

Check: Units of answer, g MgO, are correct. The magnitude of the answer is reasonable because it is larger than grams Mg.

d) **Given:** 2.5 g Al **Find:** g Al_2O_3

Conceptual Plan: **g Al → mol Al → mol Al_2O_3 → g Al_2O_3**

$$\frac{\text{mol Al}}{26.98\text{ g Al}} \quad \frac{2\text{ mol }Al_2O_3}{4\text{ mol Al}} \quad \frac{101.96\text{ g }Al_2O_3}{\text{mol }Al_2O_3}$$

Solution: $2.5\text{ g Al} \times \frac{1\ \cancel{\text{mol Al}}}{26.98\ \cancel{\text{g Al}}} \times \frac{1\ \cancel{\text{mol }Al_2O_3}}{1\ \cancel{\text{mol Al}}} \times \frac{101.96\text{ g }Al_2O_3}{1\ \cancel{\text{mol }Al_2O_3}} = 4.7\text{ g }Al_2O_3$

Check: Units of answer, g Al_2O_3, are correct. The magnitude of the answer is reasonable because it is larger than grams Al.

35. a) **Given:** 4.85 g NaOH **Find:** g HCl

Conceptual Plan: **g NaOH → mol NaOH → mol HCl → g HCl**

$$\frac{\text{mol NaOH}}{40.01\text{ g NaOH}} \quad \frac{1\text{ mol HCl}}{1\text{ mol NaOH}} \quad \frac{36.46\text{ g HCl}}{1\text{ mol HCl}}$$

Solution: $4.85\ \cancel{\text{g NaOH}} \times \frac{1\ \cancel{\text{mol NaOH}}}{40.01\ \cancel{\text{g NaOH}}} \times \frac{1\ \cancel{\text{mol HCl}}}{1\ \cancel{\text{mol NaOH}}} \times \frac{36.46\text{ g HCl}}{1\ \cancel{\text{mol HCl}}} = 4\ 42\text{ g HCl}$

Check: Units of answer, g HCl, are correct. The magnitude of the answer is reasonable since it is less than g NaOH.

b) **Given:** 4.85 g $Ca(OH)_2$ **Find:** g HNO_3

Conceptual Plan: **g $Ca(OH)_2$ → mol $Ca(OH)_2$ → mol HNO_3 → g HNO_3**

$$\frac{\text{mol }Ca(OH)_2}{74.10\text{ g }Ca(OH)_2} \quad \frac{2\text{ mol }HNO_3}{1\text{ mol }Ca(OH)_2} \quad \frac{63.02\text{ g }HNO_3}{1\text{ mol }HNO_3}$$

Solution:

$$4.85 \text{ g Ca(OH)}_2 \times \frac{1 \text{ mol Ca(OH)}_2}{74.10 \text{ g Ca(OH)}_2} \times \frac{2 \text{ mol HNO}_3}{1 \text{ mol Ca(OH)}_2} \times \frac{63.02 \text{ g HNO}_3}{1 \text{ mol HNO}_3} = 8.25 \text{ g HNO}_3$$

Check: Units of answer, g HNO_3, are correct. The magnitude of the answer is reasonable since it is more than g $Ca(OH)_2$.

c) **Given:** 4.85 g KOH **Find:** g H_2SO_4

Conceptual Plan: g KOH → mol KOH → mol H_2SO_4 → g H_2SO_4

$$\frac{\text{mol KOH}}{56.11 \text{ g KOH}} \qquad \frac{1 \text{ mol H}_2\text{SO}_4}{2 \text{ mol KOH}} \qquad \frac{98.09 \text{ g H}_2\text{SO}_4}{1 \text{ mol H}_2\text{SO}_4}$$

Solution: $$4.85 \text{ g NaOH} \times \frac{1 \text{ mol KOH}}{56.11 \text{ g KOH}} \times \frac{1 \text{ mol H}_2\text{SO}_4}{2 \text{ mol KOH}} \times \frac{98.09 \text{ g H}_2\text{SO}_4}{1 \text{ mol H}_2\text{SO}_4} = 4\ 24 \text{ g H}_2\text{SO}_4$$

Check: Units of answer, g H_2SO_4, are correct. The magnitude of the answer is reasonable since it is less than g KOH.

37\. a) **Given:** 2 mol Na; 2 mol Br_2 **Find:** Limiting Reactant

Conceptual Plan: mol Na → mol NaBr

$$\frac{2 \text{ mol NaBr}}{2 \text{ mol Na}}$$ **→ smallest mol amount determines limiting reactant**

mol Br_2 → mol NaBr

$$\frac{2 \text{ mol NaBr}}{1 \text{ mol Br}_2}$$

$$2 \text{ mol Na} \times \frac{2 \text{ mol NaBr}}{2 \text{ mol Na}} = 2 \text{ mol NaBr}$$

Solution: $$2 \text{ mol Br}_2 \times \frac{2 \text{ mol NaBr}}{1 \text{ mol Br}_2} = 4 \text{ mol NaBr}$$

Na is Limiting Reactant

Check: Answer is reasonable since Na produced smallest amount of product.

b) **Given:** 1.8 mol Na; 1.4 mol Br_2 **Find:** Limiting Reactant

Conceptual Plan: mol Na → mol NaBr

$$\frac{2 \text{ mol NaBr}}{2 \text{ mol Na}}$$ **→ smallest mol amount determines limiting reactant mol Br_2 → mol NaBr**

$$\frac{2 \text{ mol NaBr}}{1 \text{ mol Br}_2}$$

$$1.8 \text{ mol Na} \times \frac{2 \text{ mol NaBr}}{2 \text{ mol Na}} = 1.8 \text{ mol NaBr}$$

Solution: $$1.4 \text{ mol Br}_2 \times \frac{2 \text{ mol NaBr}}{1 \text{ mol Br}_2} = 2.8 \text{ mol NaBr}$$

Na is Limiting Reactant

Check: Answer is reasonable since Na produced smallest amount of product.

c) **Given:** 2.5 mol Na; 1 mol Br_2 **Find:** Limiting Reactant

Conceptual Plan: mol Na → mol NaBr

$\frac{2 \text{ mol NaBr}}{2 \text{ mol Na}}$ **→ smallest mol amount determines limiting reactant mol Br_2 → mol NaBr**

$\frac{2 \text{ mol NaBr}}{1 \text{ mol } Br_2}$

$$2.5 \text{ mol Na} \times \frac{2 \text{ mol NaBr}}{2 \text{ mol Na}} = 2.5 \text{ mol NaBr}$$

Solution: $1 \text{ mol } Br_2 \times \frac{2 \text{ mol NaBr}}{1 \text{ mol } Br_2} = 2 \text{ mol NaBr}$

Br_2 is Limiting Reactant

Check: Answer is reasonable since Br_2 produced smallest amount of product.

d) **Given:** 12.6 mol Na; 6.9 mol Br_2 **Find:** Limiting Reactant

Conceptual Plan: mol Na → mol NaBr

$\frac{2 \text{ mol NaBr}}{2 \text{ mol Na}}$ **→ smallest mol amount determines limiting reactant mol Br_2 → mol NaBr**

$\frac{2 \text{ mol NaBr}}{1 \text{ mol } Br_2}$

$$12.6 \text{ mol Na} \times \frac{2 \text{ mol NaBr}}{2 \text{ mol Na}} = 12.6 \text{ mol NaBr}$$

Solution: $6.9 \text{ mol } Br_2 \times \frac{2 \text{ mol NaBr}}{1 \text{ mol } Br_2} = 13.8 \text{ mol NaBr}$

Na is Limiting Reactant

Check: Answer is reasonable since Na produced smallest amount of product.

39. The greatest number of Cl_2 molecules will be formed from reaction mixture b and would be 3 molecules Cl_2.

a) **Given:** 7 molecules HCl, 1 molecule O_2 **Find:** Theoretical yield Cl_2

Conceptual Plan: molecule HCl → mol Cl_2 $\frac{2 \text{ molecule } Cl_2}{4 \text{ molecule HCl}}$ **→ smallest molecules amount determines limiting reactant molecules O_2 → molecules Cl_2**

$\frac{2 \text{ molecules } Cl_2}{1 \text{ molecules } O_2}$

Solution:

$$7 \text{ molecules HCl} \times \frac{2 \text{ molecules } Cl_2}{4 \text{ molecules HCl}} = 3 \text{ molecules } Cl_2$$

$$1 \text{ molecules } O_2 \times \frac{2 \text{ molecules } Cl_2}{1 \text{ molecules } O_2} = 2 \text{ molecules } Cl_2$$

Theoretical Yield = 2 molecules Cl_2

b) **Given:** 6 molecules HCl, molecule O_2 **Find:** Theoretical yield Cl_2

Conceptual Plan: molecule HCl → mol Cl_2

$\frac{2 \text{ molecule } Cl_2}{4 \text{ molecule HCl}}$ **→ smallest molecules amount determines limiting reactant molecules O_2 → molecules Cl_2**

$\frac{2 \text{ molecules } Cl_2}{1 \text{ molecules } O_2}$

Solution:

$$6 \text{ molecules HCl} \times \frac{2 \text{ molecules } Cl_2}{4 \text{ molecules HCl}} = 3 \text{ molecules } Cl_2$$

$$3 \text{ molecules } O_2 \times \frac{2 \text{ molecules } Cl_2}{1 \text{ molecules } O_2} = 6 \text{ molecules } Cl_2$$

Theoretical Yield = 3 molecules Cl_2

c) **Given:** 4 molecules HCl, 5 molecule O_2 **Find:** Theoretical yield Cl_2

Conceptual Plan: molecule HCl → mol Cl_2

$$\frac{2 \text{ molecule } Cl_2}{4 \text{ molecule HCl}}$$

→ smallest molecules amount determines limiting reactant molecules O_2 → molecules Cl_2

$$\frac{2 \text{ molecules } Cl_2}{1 \text{ molecules } O_2}$$

Solution:

$$4 \text{ molecules HCl} \times \frac{2 \text{ molecules } Cl_2}{4 \text{ molecules HCl}} = 2 \text{ molecules } Cl_2$$

$$5 \text{ molecules } O_2 \times \frac{2 \text{ molecules } Cl_2}{1 \text{ molecules } O_2} = 10 \text{ molecules } Cl_2$$

Theoretical Yield = 2 molecules Cl_2

Check: The units of the answer, molecules Cl_2, is correct. The answer is reasonable based on the limiting reactant in each mixture.

41. a) **Given:** 4 mol Ti, 4 mol Cl_2 **Find:** Theoretical yield $TiCl_4$

Conceptual Plan: mol Ti → mol $TiCl_4$

$$\frac{1 \text{ mol } TiCl_4}{1 \text{ mol Ti}}$$

→ smallest mol amount determines limiting reactant mol Cl_2 → mol $TiCl_4$

$$\frac{1 \text{ mol } TiCl_4}{2 \text{ mol } Cl_2}$$

Solution:

$$4 \text{ mol Ti} \times \frac{1 \text{ mol } TiCl_4}{1 \text{ mol Ti}} = 4 \text{ mol } TiCl_4$$

$$4 \text{ mol } Cl_2 \times \frac{1 \text{ mol } TiCl_4}{2 \text{ mol } Cl_2} = 2 \text{ mol } TiCl_4$$

Theoretical Yield = 2 mol $TiCl_4$

Check: Units of the answer, mol $TiCl_4$, are correct. Answer is reasonable since Cl_2 produced smallest amount of product and is the limiting reactant.

b) **Given:** 7 mol Ti, 17 mol Cl_2 **Find:** Theoretical yield $TiCl_4$

Conceptual Plan: mol Ti → mol $TiCl_4$

$$\frac{1 \text{ mol } TiCl_4}{1 \text{ mol Ti}}$$

→ smallest mol amount determines limiting reactant mol Cl_2 → mol $TiCl_4$

$$\frac{1 \text{ mol } TiCl_4}{2 \text{ mol } Cl_2}$$

Solution:

$$7 \text{ mol Ti} \times \frac{1 \text{ mol } TiCl_4}{1 \text{ mol Ti}} = 7 \text{ mol } TiCl_4$$

$$17 \text{ mol } Cl_2 \times \frac{1 \text{ mol } TiCl_4}{2 \text{ mol } Cl_2} = 8.5 \text{ mol } TiCl_4$$

Theoretical Yield = 7 mol $TiCl_4$

Check: Units of the answer, mol $TiCl_4$, are correct. Answer is reasonable since Ti produced smallest amount of product and is the limiting reactant.

c) **Given:** 12.4 mol Ti, 18.8 mol Cl_2 **Find:** Theoretical yield $TiCl_4$

Conceptual Plan: mol Ti → mol $TiCl_4$

$\frac{1 \text{ mol } TiCl_4}{1 \text{ mol Ti}}$ **→ smallest mol amount determines limiting reactant mol Cl_2 → mol $TiCl_4$**

$\frac{1 \text{ mol } TiCl_4}{2 \text{ mol } Cl_2}$

Solution:

$$12.4 \text{ mol Ti} \times \frac{1 \text{ mol } TiCl_4}{1 \text{ mol Ti}} = 12.4 \text{ mol } TiCl_4$$

$$18.8 \text{ mol } Cl_2 \times \frac{1 \text{ mol } TiCl_4}{2 \text{ mol } Cl_2} = 9.4 \text{ mol } TiCl_4$$

Theoretical Yield = 9.4 mol $TiCl_4$

Check: Units of the answer, mol $TiCl_4$, are correct. Answer is reasonable since Cl_2 produced smallest amount of product and is the limiting reactant.

43. a) **Given:** 2.0 g Al, 2.0 g Cl_2 **Find:** Theoretical yield in g $AlCl_3$

Conceptual Plan: g Al → mol Al → mol $AlCl_3$

$\frac{1 \text{ mol Al}}{26.98 \text{ g Al}} \quad \frac{2 \text{ mol } AlCl_3}{2 \text{ mol Al}}$ **→ smallest mol amount determines limiting reactant g Cl_2 → mol Cl_2 → mol $AlCl_3$**

$\frac{1 \text{ mol } Cl_2}{70.91 \text{ g } Cl_2} \quad \frac{2 \text{ mol } AlCl_3}{3 \text{ mol } Cl_2}$

then: mol $AlCl_3$ → g $AlCl_3$

$\frac{133.34 \text{ g } AlCl_3}{\text{mol } AlCl_3}$

Solution:

$$2.0 \text{ g Al} \times \frac{1 \text{ mol Al}}{26.98 \text{ g Al}} \times \frac{2 \text{ mol } AlCl_3}{2 \text{ mol Al}} = 0.074 \text{ mol } AlCl_3$$

$$2.0 \text{ g } Cl_2 \times \frac{1 \text{ mol } Cl_2}{70.90 \text{ g } Cl_2} \times \frac{2 \text{ mol } AlCl_3}{3 \text{ mol } Cl_2} = 0.01\underline{8}8 \text{ mol } AlCl_3$$

Check: Units of the answer, g $AlCl_3$, are correct. Answer is reasonable since Cl_2 produced smallest amount of product and is the limiting reactant. $0.01\underline{8}8 \text{ mol } AlCl_3 \times \frac{133.34 \text{ g } AlCl_3}{\text{mol } AlCl_3} = 2.5 \text{ g } AlCl_3$

b) **Given:** 7.5 g Al, 24.8 g Cl_2 **Find:** Theoretical yield in g $AlCl_3$

Conceptual Plan: g Al → mol Al → mol $AlCl_3$

$\frac{1 \text{ mol Al}}{26.98 \text{ g Al}} \quad \frac{2 \text{ mol } AlCl_3}{2 \text{ mol Al}}$ **→ smallest mol amount determines limiting reactant g Cl_2 → mol Cl_2 → mol $AlCl_3$**

$\frac{1 \text{ mol } Cl_2}{70.91 \text{ g } Cl_2} \quad \frac{2 \text{ mol } AlCl_3}{3 \text{ mol } Cl_2}$

then: mol $AlCl_3$ → g $AlCl_3$

$\frac{133.34 \text{ g } AlCl_3}{\text{mol } AlCl_3}$

Solution:

$$7.5\ \text{g Al} \times \frac{1\ \text{mol Al}}{26.98\ \text{g Al}} \times \frac{2\ \text{mol AlCl}_3}{2\ \text{mol Al}} = 0.2780\ \text{mol AlCl}_3$$

$$24.8\ \text{g Cl}_2 \times \frac{1\ \text{mol Cl}_2}{70.90\ \text{g Cl}_2} \times \frac{2\ \text{mol AlCl}_3}{3\ \text{mol Cl}_2} = 0.23\underline{3}2\ \text{mol AlCl}_3$$

$$0.23\underline{3}2\ \text{mol AlCl}_3 \times \frac{133.34\ \text{g AlCl}_3}{\text{mol AlCl}_3} = 31.1\ \text{g AlCl}_3$$

Check: Units of the answer, g $AlCl_3$, are correct. Answer is reasonable since Cl_2 produced smallest amount of product and is the limiting reactant.

c) **Given:** 0.235 g Al, 1.15 g Cl_2 **Find:** Theoretical yield in g $AlCl_3$

Conceptual Plan: g Al → mol Al → mol $AlCl_3$

$$\frac{1\ \text{mol Al}}{26.98\ \text{g Al}} \quad \frac{2\ \text{mol AlCl}_3}{2\ \text{mol Al}}$$

→ smallest mol amount determines limiting reactant

g Cl_2 → mol Cl_2 → mol $AlCl_3$

$$\frac{1\ \text{mol Cl}_2}{70.91\ \text{g Cl}_2} \quad \frac{2\ \text{mol AlCl}_3}{3\ \text{mol Cl}_2}$$

then: mol $AlCl_3$ → g $AlCl_3$

$$\frac{133.34\ \text{g AlCl}_3}{\text{mol AlCl}_3}$$

Solution:

$$0.235\ \text{g Al} \times \frac{1\ \text{mol Al}}{26.98\ \text{g Al}} \times \frac{2\ \text{mol AlCl}_3}{2\ \text{mol Al}} = 0.0087\underline{1}0\ \text{mol AlCl}_3$$

$$1.15\ \text{g Cl}_2 \times \frac{1\ \text{mol Cl}_2}{70.90\ \text{g Cl}_2} \times \frac{2\ \text{mol AlCl}_3}{3\ \text{mol Cl}_2} = 0.01081\ \text{mol AlCl}_3$$

$$0.0087\underline{1}0\ \text{mol AlCl}_3 \times \frac{133.34\ \text{g AlCl}_3}{\text{mol AlCl}_3} = 1.16\ \text{g AlCl}_3$$

Check: Units of the answer, g $AlCl_3$, are correct. Answer is reasonable since Al produced smallest amount of product and is the limiting reactant.

45. **Given:** 28.5 g KCl; 25.7 g Pb^{2+}; 29.4 g $PbCl_2$ **Find:** limiting reactant, theoretical yield $PbCl_2$, % yield

Conceptual Plan: g KCl → mol KCl → mol $PbCl_2$

$$\frac{1\ \text{mol KCl}}{74.55\ \text{g KCl}} \quad \frac{1\ \text{mol PbCl}_2}{2\ \text{mol KCl}}$$

→ smallest mol amount determines limiting reactant

g Pb^{2+} → mol Pb^{2+} → mol $PbCl_2$

$$\frac{1\ \text{mol Pb}^{2+}}{207.2\ \text{g Pb}^{2+}} \quad \frac{1\ \text{mol PbCl}_2}{1\ \text{mol Pb}^{2+}}$$

then: mol $PbCl_2$ → g $PbCl_2$

$$\frac{278.1\ \text{g PbCl}_2}{\text{mol PbCl}_2}$$

then: determine % yield

$$\frac{\text{actual yield g PbCl}_2}{\text{theoretical yield g PbCl}_2} \times 100$$

Solution:

$$28.5 \text{ g KCl} \times \frac{1 \text{ mol KCl}}{74.55 \text{ g KCl}} \times \frac{1 \text{ mol PbCl}_2}{2 \text{ mol KCl}} = 0.1911 \text{ mol PbCl}_2$$

$$25.7 \text{ g Pb}^{2+} \times \frac{1 \text{ mol Pb}^{2+}}{207.2 \text{ g Pb}^{2+}} \times \frac{1 \text{ mol PbCl}_2}{1 \text{ mol Pb}^{2+}} = 0.1240 \text{ mol PbCl}_2$$

$$0.1240 \text{ mol PbCl}_2 \times \frac{278.1 \text{ g PbCl}_2}{1 \text{ mol PbCl}_2} = 34.5 \text{ g PbCl}_2$$

$$\frac{34.5 \text{ g PbCl}_2}{53.14 \text{ g PbCl}_2} \times 100 = 85.3\%$$

Check: The theoretical yield has the correct units, g $PbCl_2$, and has a reasonable magnitude compared to the mass of KCl, the limiting reactant. The % yield is reasonable, under 100%.

47. **Given:** 136.4 kg NH_3; 211.4 kg CO_2; 168.4 kg CH_4N_2O **Find:** limiting reactant, theoretical yield CH_4N_2O, % yield

Conceptual Plan: kg NH_3 → g NH_3 → mol NH_3 → mol CH_4N_2O

$$\frac{1000 \text{ g}}{1 \text{ kg}} \quad \frac{1 \text{ mol NH}_3}{17.03 \text{ g NH}_3} \quad \frac{1 \text{ mol CH}_4\text{N}_2\text{O}}{2 \text{ mol NH}_3}$$

→ smallest mol amount determines limiting reactant

kg CO_2 → g CO_2 → mol CO_2 → mol CH_4N_2O

$$\frac{1000 \text{ g}}{1 \text{ kg}} \quad \frac{1 \text{ mol CO}_2}{44.01 \text{ g CO}_2} \quad \frac{1 \text{ mol CH}_4\text{N}_2\text{O}}{1 \text{ mol CO}_2}$$

then: mol CH_4N_2O → g CH_4N_2O → kg CH_4N_2O **then: determine % yield**

$$\frac{60.06 \text{ g CH}_4\text{N}_2\text{O}}{1 \text{ mol CH}_4\text{N}_2\text{O}} \quad \frac{1 \text{ kg}}{1000 \text{ g}}$$

$$\frac{\text{actual yield kg CH}_4\text{N}_2\text{O}}{\text{theoretical yield kg CH}_4\text{N}_2\text{O}} \times 100$$

Solution:

$$136.4 \text{ kg NH}_3 \times \frac{1000 \text{ g}}{\text{kg}} \times \frac{1 \text{ mol NH}_3}{17.03 \text{ g NH}_3} \times \frac{1 \text{ mol CH}_4\text{N}_2\text{O}}{2 \text{ mol NH}_3} = 4004.7 \text{ mol CH}_4\text{N}_2\text{O}$$

$$211.4 \text{ g CO}_2 \times \frac{1000 \text{ g}}{\text{kg}} \times \frac{1 \text{ mol CO}_2}{44.01 \text{ g CO}_2} \times \frac{1 \text{ mol CH}_4\text{N}_2\text{O}}{1 \text{ mol CO}_2} = 4803.2 \text{ mol CH}_4\text{N}_2\text{O}$$

$$4004.7 \text{ mol CH}_4\text{N}_2\text{O} \times \frac{60.06 \text{ g CH}_4\text{N}_2\text{O}}{1 \text{ mol CH}_4\text{N}_2\text{O}} \times \frac{\text{kg}}{1000 \text{ g}} = 240.52 \text{ kg CH}_4\text{N}_2\text{O}$$

$$\frac{168.4 \text{ kg CH}_4\text{N}_2\text{O}}{240.52 \text{ kg CH}_4\text{N}_2\text{O}} \times 100 = 70.01\%$$

Check: The theoretical yield has the correct units, kg CH_4N_2O, and has a reasonable magnitude compared to the mass of NH_3, the limiting reactant. The % yield is reasonable, under 100%.

49. a) **Given:** 4.3 mol LiCl; 2.8 L solution **Find:** Molarity LiCl

Conceptual Plan: mol LiCl, L solution → Molarity

$$\text{molarity (M)} = \frac{\text{amount of solute (in moles)}}{\text{volume of solution (in L)}}$$

Solution: $\frac{4.3 \text{ mol LiCl}}{2.8 \text{ L solution}} = 1.5 \text{ M}$

Check: The units of the answer, M, are correct. The magnitude of the answer is reasonable. Concentrations are usually between 0 M and 18 M.

b) **Given:** 22.6 g $C_6H_{12}O_6$; 1.08 L solution **Find:** Molarity $C_6H_{12}O_6$

Conceptual Plan: g $C_6H_{12}O_6$ →mol $C_6H_{12}O_6$, L solution → Molarity

$$\frac{\text{mol } C_6H_{12}O_6}{180.16 \text{ g } C_6H_{12}O_6}$$

$$\text{molarity (M)} = \frac{\text{amount of solute (in moles)}}{\text{volume of solution (in L)}}$$

Solution: $22.6 \text{ g } C_6H_{12}O_6 \times \frac{1 \text{ mol } C_6H_{12}O_6}{180.16 \text{ g } C_6H_{12}O_6} = 0.12\underline{5}4 \text{ mol } C_6H_{12}O_6$

$$\frac{0.12\underline{5}4 \text{ mol } C_6H_{12}O_6}{1.08 \text{ L solution}} = 0.116 \text{ M}$$

Check: The units of the answer, M, are correct. The magnitude of the answer is reasonable. Concentrations are usually between 0 M and 18 M.

c) **Given:** 45.5 mg NaCl; 154.4 mL solution **Find:** Molarity NaCl

Conceptual Plan: mg NaCl → g NaCl → mol NaCl, and mL solution →L solution then Molarity

$$\frac{\text{g NaCl}}{1000 \text{ mg NaCl}} \qquad \frac{\text{mol NaCl}}{58.45 \text{ g NaCl}} \qquad \frac{\text{L solution}}{1000 \text{ mL solution}}$$

$$\text{molarity (M)} = \frac{\text{amount of solute (in moles)}}{\text{volume of solution (in L)}}$$

Solution: $45.5 \text{ mg NaCl} \times \frac{1 \text{ g}}{1000 \text{ mg}} \times \frac{1 \text{ mol NaCl}}{58.45 \text{ g NaCl}} = 7.7\underline{8}4 \times 10^{-4} \text{ mol NaCl}$

$$154 \text{ mL solution} \times \frac{1 \text{ L}}{1000 \text{ mL}} = 0.154 \text{ L}$$

$$\frac{7.784 \times 10^{-4} \text{ mol NaCl}}{0.154 \text{ L}} = 0.00505 \text{ M NaCl}$$

Check: The units of the answer, M, are correct. The magnitude of the answer is reasonable. Concentrations are usually between 0 M and 18 M.

51. a) **Given:** 0.556 L; 2.3 M KCl **Find:** mol KCl

Conceptual Plan: volume solution x M = mol

volume solution (L) x M = mol

Solution: $0.556 \text{ L solution} \times \frac{2.3 \text{ mol KCl}}{\text{L soluion}} = 1.3 \text{ mol KCl}$

Check: Units of answer, mol KCl, are correct. The magnitude is reasonable since it is less than 1 L solution.

b) **Given:** 1.8 L; 0.85 M KCl **Find:** mol KCl

Conceptual Plan: volume solution x M = mol

volume solution (L) x M = mol

Solution: $1.8 \text{ L solution} \times \frac{0.85 \text{ mol KCl}}{\text{L soluion}} = 1.5 \text{ mol KCl}$

Check: Units of answer, mol KCl, are correct. The magnitude is reasonable since it is less than 2 L solution.

c) **Given:** 114 mL; 1.85 M KCl **Find:** mol KCl

Conceptual Plan: mL solution → L solution, then volume solution x M = mol

$$\frac{1\text{ L}}{1000\text{ mL}} \qquad \text{volume solution (L) x M = mol}$$

Solution: $114 \cancel{\text{mL solution}} \times \frac{1\ \cancel{\text{L}}}{1000\ \cancel{\text{mL}}} \times \frac{1.85\text{ mol KCl}}{\cancel{\text{L solution}}} = 0.2109\text{ mol KCl}$

Check: Units of answer, mol KCl, are correct. The magnitude is reasonable since it is less than 1 L solution.

53. **Given:** 400.0 mL; 1.1 M $NaNO_3$ **Find:** g $NaNO_3$

Conceptual Plan: mL solution → L solution, then volume solution x M = mol $NaNO_3$

volume solution (L) x M = mol

then mol $NaNO_3$ → g $NaNO_3$

$$\frac{85.01\text{ g NaNO}_3}{\text{mol NaNO}_3}$$

Solution: $400.0 \cancel{\text{mL solution}} \times \frac{1\ \cancel{\text{L}}}{1000\ \cancel{\text{mL}}} \times \frac{1.1\ \cancel{\text{mol NaNO}_3}}{\cancel{\text{L soluion}}} \times \frac{85.01\text{ g}}{\cancel{\text{mol NaNO}_3}} = 37\text{ g NaNO}_3$

Check: Units of answer, g $NaNO_3$, are correct. The magnitude is reasonable for the concentration and volume of solution.

55. **Given:** $V_1 = 123$ mL;$M_1 = 1.1$ M;$V_2 = 500.0$ mL **Find:** M_2

Conceptual Plan: mL → L then V_1, M_1, V_2 → M_2

$$\frac{1\text{ L}}{1000\text{ mL}} \qquad V_1M_1 = V_2M_2$$

Solution: $123 \cancel{\text{mL}} \times \frac{1\text{ L}}{1000\ \cancel{\text{mL}}} = 0.123\text{ L} \qquad 500.0 \cancel{\text{mL}} \times \frac{1\text{ L}}{1000\ \cancel{\text{mL}}} = 0.5000\text{ L}$

$$M_2 = \frac{V_1M_1}{V_2} = \frac{(0.123\ \cancel{\text{L}})(1.1\text{ M})}{(0.5000\ \cancel{\text{L}})} = 0.27\text{ M}$$

Check: Units of the answer, M, are correct. The magnitude of the answer is reasonable since it is less than the original concentration.

57. **Given:** $V_1 = 50$ mL;$M_1 = 12$ M; $M_2 = 0.100$ M **Find:** V_2

Conceptual Plan: mL → L then V_1, M_1, M_2 → V_2

$$\frac{1\text{ L}}{1000\text{ mL}} \qquad V_1M_1 = V_2M_2$$

Solution: $50 \cancel{\text{mL}} \times \frac{1\text{ L}}{1000\ \cancel{\text{mL}}} = 0.050\text{ L}$

$$V_2 = \frac{V_1M_1}{M_2} = \frac{(0.050\ \cancel{\text{L}})(12\text{ M})}{(0.100\ \cancel{\text{M}})} = 6.0\text{ L}$$

Check: Units of the answer, L, are correct. The magnitude of the answer is reasonable since the new concentration is much less than the original, the volume must be larger.

59. **Given:** 95.4 mL, 0.102 M $CuCl_2$; 0.175 M Na_3PO_4 **Find:** volume Na_3PO_4

Conceptual Plan: mL $CuCl_2$→ L $CuCl_2$ → mol $CuCl_2$ → mol Na_3PO_4 → L Na_3PO_4→ mL Na_3PO_4

$$\frac{1\text{ L}}{1000\text{ mL}} \qquad \frac{0.102\text{ mol CuCl}_2}{\text{L}} \qquad \frac{2\text{ mol Na}_3\text{PO}_4}{3\text{ mol CuCl}_2} \qquad \frac{1\text{ L}}{0.175\text{ mol Na}_3\text{PO}_4} \qquad \frac{1000\text{ mL}}{\text{L}}$$

Solution:

$$95.4 \text{ mL CuCl}_2 \times \frac{1 \text{ L}}{1000 \text{ mL}} \times \frac{0.102 \text{ mol CuCl}_2}{1 \text{ L}} \times \frac{2 \text{ mol Na}_3\text{PO}_4}{3 \text{ mol CuCl}_2} \times \frac{1 \text{ L}}{0.175 \text{ mol Na}_3\text{PO}_4} \times \frac{1000 \text{ mL}}{1 \text{ L}} = 37.1 \text{ mL Na}_3\text{PO}_4$$

Check: Units of answer, mL Na_3PO_4, are correct. The magnitude of the answer is reasonable since the concentration of Na_3PO_4 is greater.

61. **Given:** 25.0 g H_2; 6.0 M H_2SO_4 **Find:** volume H_2SO_4

Conceptual Plan: g H_2 → mol H_2 → mol H_2SO_4 → L H_2SO_4

$$\frac{2.016 \text{ g H}_2}{1 \text{ mol H}_2} \quad \frac{3 \text{ mol H}_2\text{SO}_4}{3 \text{ mol H}_2} \quad \frac{1 \text{ L}}{6.0 \text{ mol H}_2\text{SO}_4}$$

Solution: $$25.0 \text{ g H}_2 \times \frac{1 \text{ mol H}_2}{2.016 \text{ g H}_2} \times \frac{3 \text{ mol H}_2\text{SO}_4}{3 \text{ mol H}_2} \times \frac{1 \text{ L}}{6.0 \text{ mol H}_2\text{SO}_4} = 2.0 \text{ L H}_2\text{SO}_4$$

Check: The units, L H_2SO_4, are correct. The magnitude is reasonable since there are approximately 12 mol H_2 and the mole ratio is 1:1.

63. a) CsCl is an ionic compound. An aqueous solution is an electrolyte solution so, it conducts electricity.

b) CH_3OH is a molecular compound. An aqueous solution is a non electrolyte solution, so it does not conduct electricity.

c) $Ca(NO_3)_2$ is an ionic compound. An aqueous solution is an electrolyte solution so, it conducts electricity.

d) $C_6H_{12}O_6$ is a molecular compound. An aqueous solution is a non electrolyte solution, so it does not conduct electricity.

65. a) $AgNO_3$ is soluble. Compounds containing NO_3^- are always soluble with no exceptions. The ions in solution are $Ag^+(aq)$ and $NO_3^-(aq)$.

b) $Pb(C_2H_3O_2)_2$ is soluble. Compounds containing $C_2H_3O_2^-$ are always soluble with no exceptions. The ions in solution are $Pb^{+2}(aq)$ and $C_2H_3O_2^-(aq)$.

c) KNO_3 is soluble. Compounds containing K^+ are always soluble with no exceptions. The ions in solution are $K^+(aq)$ and $NO_3^-(aq)$.

d) $(NH_4)_2S$ is soluble. Compounds containing NH_4^+ are always soluble with no exceptions. The ions in solution are $NH_4^+(aq)$ and $S^{-2}(aq)$.

67. a) LiI(aq) + BaS(aq) → Possible products: Li_2S and BaI_2. Li_2S is soluble. Compounds containing S^{2-} are normally insoluble but Li^+ is and exception. BaI_2 is soluble. Compounds containing I^- are normally soluble and Ba^{2+} is not an exception. LiI(aq) + BaS(aq) → No Reaction.

b) KCl(aq) + CaS(aq) → Possible products: K_2S and $CaCl_2$. K_2S is soluble. Compounds containing $S\ 2^-$ are normally insoluble but K^+ is an exception. $CaCl_2$ is soluble. Compounds containing Cl^- are normally soluble and Ca^{2+} is not an exception. KCl(aq) + CaS(aq) → No Reaction.

c) $CrBr_2(aq) + Na_2CO_3(aq)$ → Possible products: $CrCO_3$ and NaBr. $CrCO_3$ is insoluble. Compounds containing CO_3^{2-} are normally insoluble and Cr^{2+} is not an exception. NaBr is soluble. Compounds containing Br^- are normally soluble and Na^+ is not an exception. $CrBr_2(aq) + Na_2CO_3(aq) \rightarrow CrCO_3(s) + 2\ NaBr(aq)$

d) $NaOH(aq) + FeCl_3(aq)$ → Possible products NaCl and $Fe(OH)_3$. NaCl is soluble. Compounds containing Na^+ are normally soluble, no exceptions. $Fe(OH)_3$ is insoluble. Compounds containing OH^- are normally insoluble and Fe^{3+} is not an exception. $3\ NaOH(aq) + FeCl_3(aq) \rightarrow 3\ NaCl(aq) + Fe(OH)_3(s)$

69. a) $K_2CO_3(aq) + Pb(NO_3)_2(aq)$ → Possible products: KNO_3 and $PbCO_3$. KNO_3 is soluble. Compounds

containing K^+ are always soluble, no exceptions. $PbCO_3$ is insoluble. Compounds containing $CO_3^{\;2-}$ are normally insoluble and Pb^{2+} is not an exception. $K_2CO_3(aq) + Pb(NO_3)_2(aq) \rightarrow 2\ KNO_3(aq)$ and $PbCO_3(s)$

b) $Li_2SO_4(aq) + Pb(C_2H_3O_2)_2(aq) \rightarrow$ Possible products: $LiC_2H_3O_2$ and $PbSO_4$. $LiC_2H_3O_2$ is soluble. Compounds containing Li^+ are always soluble, no exceptions. $PbSO_4$ is insoluble. Compounds containing $SO_4^{\;2-}$ are normally soluble but, $Pb^{\;2+}$ is an exception. $Li_2SO_4(aq) + Pb(C_2H_3O_2)_2(aq) \rightarrow 2\ LiC_2H_3O_2(aq)$ and $PbSO_4(s)$

c) $Cu(NO_3)_2(aq) + MgS(s) \rightarrow$ Possible products: CuS and $Mg(NO_3)_2$. CuS is insoluble. Compounds containing S^{-2} are normally insoluble and Cu^{2+} is not an exception. $Mg(NO_3)_2$ is soluble. Compounds containing $NO_3^{\;-}$ are always soluble, no exceptions. $Cu(NO_3)_2(aq) + MgS(s) \rightarrow CuS(s)$ and $Mg(NO_3)_2(aq)$

d) $Sr(NO_3)_2(aq) + KI(aq) \rightarrow$ Possible products: SrI_2 and KNO_3. SrI_2 is soluble. Compounds containing I^- are normally soluble and Sr^{2+} is not an exception. KNO_3 is soluble. Compounds containing K^+ are always soluble, no exceptions. $Sr(NO_3)_2(aq) + KI(aq) \rightarrow$ No Reaction.

71. a) $H^+(aq)$ + ~~Cl^-~~(aq) + ~~Li^+~~(aq) + $OH^-(aq) \rightarrow H_2O(l)$ + ~~Li^+~~(aq) + ~~Cl^-~~(aq)
$H^+(aq) + OH^-(aq) \rightarrow H_2O(l)$

b) ~~Mg^{2+}~~$(aq) + S^{2-}(aq) + Cu^{2+}(aq) + 2$ ~~Cl^-~~$(aq) \rightarrow CuS(s)$ + ~~Mg^{2+}~~$(aq) + 2$ ~~Cl^-~~(aq)
$Cu^{2+}(aq) + S^{2-}(aq) \rightarrow CuS(s)$

c) ~~K^+~~$(aq) + OH^-(aq) + H^+(aq)$ + ~~NO_3^-~~$(aq) \rightarrow H_2O(l)$ + ~~K^+~~(aq) + ~~NO_3^-~~(aq)
$H^+(aq) + OH^-(aq) + \rightarrow H_2O(l)$

d) 6 ~~Na^+~~$(aq) + 2\ PO_4^{\;3-}(aq) + 3\ Ni^{2+}(aq) + 6$ ~~Cl^-~~$(aq) \rightarrow Ni_3(PO_4)_2(s) + 6$ ~~Na^+~~$(aq) + 6$ ~~Cl^-~~(aq)
$3\ Ni^{2+}(aq) + 2\ PO_4^{\;3-}(aq) + \rightarrow Ni_3(PO_4)_2(s)$

73. $Hg_2^{\;2+}(aq)$ + ~~NO_3^-~~(aq) + ~~Na^+~~$(aq) + 2\ Cl^-(aq) \rightarrow Hg_2Cl_2(s)$ + ~~Na^+~~(aq) + ~~NO_3^-~~(aq)
$Hg_2^{\;2+}(aq) + 2\ Cl^-(aq) \rightarrow Hg_2Cl_2(s)$

75. Skeletal reaction: $HBr(aq) + KOH(aq) \rightarrow H_2O(l) + KBr(aq)$
acid base water salt
Net Ionic Equation: $H^+(aq) + OH^-(aq) \rightarrow H_2O(l)$

77. a) Skeletal reaction: $H_2SO_4(aq) + Ca(OH)_2(aq) \rightarrow H_2O(l) + CaSO_4(s)$
acid base water salt
Balanced Reaction: $H_2SO_4(aq) + Ca(OH)_2(aq) \rightarrow 2\ H_2O(l) + CaSO_4(s)$

b) Skeletal reaction: $HClO_4(aq) + KOH(aq) \rightarrow H_2O(l) + KClO_4(aq)$
acid base water salt
Balanced Reaction: $HClO_4(aq) + KOH(aq) \rightarrow H_2O(l) + KClO_4(aq)$

c) Skeletal reaction: $H_2SO_4(aq) + NaOH(aq) \rightarrow H_2O(l) + Na_2SO_4(aq)$
acid base water salt
Balanced Reaction: $H_2SO_4(aq) + 2\ NaOH(aq) \rightarrow H_2O(l) + Na_2SO_4(aq)$

79. **Given:** 25.3 mL, 0.200 M NaOH solution; 15.0 mL $HClO_4$ solution **Find:** M $HClO_4$ solution

Conceptual Plan: **mL NaOH → L NaOH → mol NaOH → mol $HClO_4$**

$$\frac{1\text{ L}}{1000\text{ mL}} \quad \frac{0.200\text{ mol NaOH}}{\text{L NaOH}} \quad \frac{1\text{ mol HClO}_4}{1\text{ mol NaOH}}$$

mol $HClO_4$, volume $HClO_4$ solution → M

$$M = \frac{1\text{ mol HClO}_4}{\text{L HClO}_4\text{ solution}}$$

Solution:

$$25.3\text{ mL NaOH} \times \frac{1\text{ L}}{1000\text{ mL}} \times \frac{0.200\text{ mol NaOH}}{\text{L NaOH}} \times \frac{1\text{ mol HClO}_4}{1\text{ mol NaOH}} = 0.0050\underline{6}0\text{ mol HClO}_4$$

$$\frac{0.0050\underline{6}0\text{ mol HClO}_4}{15.0\text{ mL HClO}_4} \times \frac{1000\text{ mL}}{1\text{ L}} = 0.337\text{ M HClO}_4$$

Check: The units of the answer, M $HClO_4$, are correct. The magnitude of the answer is reasonable since it is greater than the M of NaOH.

81. a) Skeletal Reaction: $HBr(aq) + NiS(aq) \rightarrow NiBr_2(aq) + H_2S(g)$
gas
Balanced Reaction: $2\ HBr(aq) + NiS(aq) \rightarrow NiBr_2(aq) + H_2S(g)$

b) Skeletal Reaction: $NH_4I(aq) + NaOH(aq) \rightarrow NH_4OH(aq) + NaI(aq) \rightarrow H_2O(l) + NH_3(g) + NaI(aq)$
decomposes gas
Balanced Reaction: $NH_4I(aq) + NaOH(aq) \rightarrow H_2O(l) + NH_3(g) + NaI(aq)$

c) Skeletal Reaction: $HBr(aq) + Na_2S(aq) \rightarrow NaBr(aq) + H_2S(g)$
gas
Balanced Reaction: $2\ HBr(aq) + Na_2S(aq) \rightarrow 2\ NaBr(aq) + H_2S(g)$

d) Skeletal Reaction: $HClO_4(aq)+Li_2CO_3(aq) \rightarrow H_2CO_3(aq) + LiClO_4(aq) \rightarrow H_2O(l) + CO_2(g) + LiClO_4(aq)$
decomposes gas
Balanced Reaction: $2\ HClO_4(aq) + Li_2CO_3(aq) \rightarrow H_2O(l) + CO_2(g) + 2\ LiClO_4(aq)$

83. a) Ag. The oxidation state of Ag = 0. The oxidation state of an atom in a free element is 0.

b) Ag^+. The oxidation state of $Ag^+ = +1$. The oxidation state of a monatomic ion is equal to its charge.

c) CaF_2. The oxidation state of Ca = +2, the oxidation state of F = – 1. The oxidation state of a Group 2A metal always has an oxidation state of +2, the oxidation of F is – 1 since the sum of the oxidation states in a neutral formula unit = 0.

d) H_2S. The oxidation state of H = + 1, the oxidation state of S = – 2. The oxidation state of H when listed first is +1, the oxidation state of S is – 2 since S is in group 6A and the sum of the oxidation states in a neutral molecular unit = 0.

e) CO_3^{2-}. The oxidation state of C = +4, the oxidation state of O = – 2. The oxidation state of O is normally – 2, the oxidation state of C is deduced from the formula since the sum of the oxidation states must equal the charge on the ion. (C ox state) + 4(O ox state) = – 2: (C ox state) + 2(– 2) = – 2, so C ox state = + 4.

f) CrO_4^{2-}. The oxidation state of Cr = +6, the oxidation state of O = – 2. The oxidation state of O is normally – 2, the oxidation state of Cr is deduced from the formula since the sum of the oxidation states must equal the charge on the ion. (Cr ox state) + 4(O ox state) = – 2; (Cr ox state) + 2(– 2) = – 2, so Cr ox state = + 6.

85. a) CrO. The oxidation state of Cr = +2, the oxidation state of O = – 2. The oxidation state of O is normally – 2, the oxidation state of Cr is deduced from the formula since the sum of the oxidation states must = 0.
(Cr ox state) + (O ox state) = 0; (Cr ox state) + (– 2) = 0, so Cr = +2.

b) CrO_3. The oxidation state of Cr = +6, the oxidation state of O = – 2. The oxidation state of O is normally – 2, the oxidation state of Cr is deduced from the formula since the sum of the oxidation states must = 0.
(Cr ox state) + 3(O ox state) = 0; (Cr ox state) +3 (– 2) = 0, so Cr = +6.

c) Cr_2O_3. The oxidation state of Cr = +3, the oxidation state of O = – 2. The oxidation state of O is normally – 2, the oxidation state of Cr is deduced from the formula since the sum of the oxidation states must = 0.
2(Cr ox state) +3 (O ox state) = 0; 2(Cr ox state) + 3(– 2) = 0, so Cr = +3.

87. a) $4\,Li(s) + O_2(g) \rightarrow 2\,Li_2O(s)$
oxidation states; o o +1 – 2
This is a redox reaction since Li increases in oxidation number (oxidation) and O decreases in number (reduction). O_2 is the oxidizing agent, Li is the reducing agent.

b) $Mg(s) + Fe^{2+}(aq) \rightarrow Mg^{2+}(aq) + Fe(s)$
oxidation states; o +2 +2 0
This is a redox reaction since Mg increases in oxidation number (oxidation) and Fe decreases in number (reduction). Fe^{2+} is the oxidizing agent, Mg is the reducing agent.

c) $Pb(NO_3)_2(aq) + Na_2SO_4(aq) \rightarrow PbSO_4(s) + 2\,NaNO_3(aq)$
oxidation states; +2 +5 - 2 +1 +6 - 2 +2 +6 -2 +1 +5 -2
This is a not a redox reaction since none of the atoms undergoes a change in oxidation number.

d) $HBr(aq) + KOH(aq) \rightarrow H_2O(l) + KBr(aq)$
oxidation states; +1 - 1 +1 – 2 +1 +1 -2 +1 -2
This is a not a redox reaction since none of the atoms undergoes a change in oxidation number.

89. a) Skeletal reaction: $S(s) + O_2(g) \rightarrow SO_2(g)$
Balanced reaction: $S(s) + O_2(g) \rightarrow SO_2(g)$

b) Skeletal reaction: $C_3H_6(g) + O_2(g) \rightarrow CO_2(g) + H_2O(g)$
Balance C: $C_3H_6(g) + O_2(g) \rightarrow 3CO_2(g) + H_2O(g)$
Balance H: $C_3H_6(g) + O_2(g) \rightarrow 3CO_2(g) + 3H_2O(g)$
Balance O: $C_3H_6(g) + 9/2\,O_2(g) \rightarrow 3CO_2(g) + 3H_2O(g)$
Clear fraction: $2C_3H_6(g) + 9O_2(g) \rightarrow 6CO_2(g) + 6H_2O(g)$

c) Skeletal reaction: $Ca(s) + O_2(g) \rightarrow CaO$
Balance O: $Ca(s) + O_2(g) \rightarrow 2CaO$
Balance Ca: $2Ca(s) + O_2(g) \rightarrow 2CaO$

d) Skeletal reaction: $C_5H_{12}S(l) + O_2(g) \rightarrow CO_2(g) + H_2O(g) + SO_2(g)$
Balance C: $C_5H_{12}S(l) + O_2(g) \rightarrow 5CO_2(g) + H_2O(g) + SO_2(g)$
Balance H: $C_5H_{12}S(l) + O_2(g) \rightarrow 5CO_2(g) + 6H_2O(g) + SO_2(g)$
Balance S: $C_5H_{12}S(l) + O_2(g) \rightarrow 5CO_2(g) + 6H_2O(g) + SO_2(g)$
Balance O: $C_5H_{12}S(l) + 9O_2(g) \rightarrow 5CO_2(g) + 6H_2O(g) + SO_2(g)$

91. **Given:** In 100 g solution, 20.0 g $C_2H_6O_2$; density of solution = 1.03 g/mL **Find:** M of solution
Conceptual Plan: **g $C_2H_6O_2 \rightarrow$ mol $C_2H_6O_2$ and g solution $\rightarrow$ mL solution $\rightarrow$ L solution**

$$\frac{1\ \text{mol}\ C_2H_6O_2}{62.06\ \text{g}\ C_2H_6O_2} \qquad \frac{1.00\ \text{mL}}{1.03\ \text{g}} \qquad \frac{1\ \text{L}}{1000\ \text{mL}}$$

then M $C_2H_6O_2$

$$M = \frac{\text{mol}\ C_2H_6O_2}{\text{L solution}}$$

Solution:

$$20.0\ \cancel{g\ C_2H_6O_2} \times \frac{1\ mol\ C_2H_6O_2}{62.06\ \cancel{g\ C_2H_6O_2}} = 0.3222\ mol\ C_2H_6O_2$$

$$100.0\ \cancel{g\ solution} \times \frac{1.00\ \cancel{mL\ solution}}{1.03\ \cancel{g\ solution}} \times \frac{1\ L}{1000\ mL} = 0.09708\ L$$

$$M = \frac{0.3222\ mol\ C_2H_6O_2}{0.09708\ L} = 3.32\ M$$

Check: The units of the answer, M $C_2H_6O_2$, are correct. The magnitude of the answer is reasonable since the concentration of solutions is usually between 0 and 18 M.

93. **Given:** 2.5 g $NaHCO_3$ **Find:** g HCl
Conceptual Plan: g $NaHCO_3$ → mol $NaHCO_3$ → mol HCl → g HCl

$$\frac{1\ mol\ NaHCO_3}{84.02\ g\ NaHCO_3} \quad \frac{1\ mol\ HCl}{1\ mol\ NaHCO_3} \quad \frac{36.46\ g\ HCl}{1\ mol\ HCl}$$

Solution: $HCl(aq) + NaHCO_3(aq) \rightarrow H_2O(l) + CO_2(g) + NaCl(aq)$

$$2.5\ \cancel{g\ NaHCO_3} \times \frac{1\ \cancel{mol\ NaHCO_3}}{84.02\ \cancel{g\ NaHCO_3}} \times \frac{1\ \cancel{mol\ HCl}}{1\ \cancel{mol\ NaHCO_3}} \times \frac{36.46\ g\ HCl}{1\ \cancel{mol\ HCl}} = 1.1\ g\ HCl$$

Check: The units of the answer, g HCl, are correct. The magnitude of the answer is reasonable since the molar mass of HCl is less than the molar mass of $NaHCO_3$.

95. **Given:** 1.0 kg C_8H_{18} **Find:** kg CO_2
Conceptual Plan: kg C_8H_{18} → g C_8H_{18} → mol C_8H_{18} → mol CO_2 → g CO_2 → kg CO_2

Solution: $2\ C_8H_{18}(g) + 25\ O_2(g) \rightarrow 16\ CO_2(g) + 18\ H_2O(g)$

$$1.0\ \cancel{kg\ C_8H_{18}} \times \frac{1000\ \cancel{g}}{\cancel{kg}} \times \frac{1\ \cancel{mol\ C_8H_{18}}}{114.22\ \cancel{g\ C_8H_{18}}} \times \frac{16\ \cancel{mol\ CO_2}}{2\ \cancel{mol\ C_8H_{18}}} \times \frac{44.01\ \cancel{g\ CO_2}}{1\ \cancel{mol\ CO_2}} \times \frac{kg}{1000\ \cancel{g}} = 3.1\ kg\ CO_2$$

Check: The units of the answer, kg CO_2, are correct. The magnitude of the answer is reasonable since the ratio of CO_2 to C_8H_{18} is 8:1.

97. **Given:** 3.00 mL $C_4H_6O_3$, d = 1.08 g/mL; 1.25 g $C_7H_6O_3$; 1.22 g $C_9H_8O_4$ **Find:** limiting reactant, theoretical yield $C_9H_8O_4$ and % yield $C_9H_8O_4$
Conceptual Plan: **mL $C_4H_6O_3$ → g $C_4H_6O_3$ → mol $C_4H_6O_3$ → mol $C_9H_8O_4$**

$$\frac{1.08\ g\ C_4H_6O_3}{1.00\ mL\ C_4H_6O_3} \quad \frac{1\ mol\ C_4H_6O_3}{102.09\ g\ C_4H_6O_3} \quad \frac{1\ mol\ C_9H_8O_4}{1\ mol\ C_4H_6O_3}$$

→ smallest amount determines limiting reactant

g $C_7H_6O_3$ → mol $C_7H_6O_3$ → mol $C_9H_8O_4$

$$\frac{1\ mol\ C_7H_6O_3}{138.12\ g\ C_7H_6O_3} \quad \frac{1\ mol\ C_9H_8O_4}{1\ mol\ C_7H_6O_3}$$

then: mol $C_9H_8O_4$ → g $C_9H_8O_4$ **then: determine % yield**

$$\frac{180.1\ g\ C_9H_8O_4}{mol\ C_9H_8O_4}$$

$$\frac{\text{actual yield g } C_9H_8O_4}{\text{theoretical yield g } C_9H_8O_4} \times 100$$

Solution:

$$3.00\ \cancel{mL\ C_4H_6O_3} \times \frac{1.08\ \cancel{g\ C_4H_6O_3}}{\cancel{mL\ C_4H_6O_3}} \times \frac{1\ \cancel{mol\ C_4H_6O_3}}{102.09\ \cancel{g\ C_4H_6O_3}} \times \frac{1\ mol\ C_9H_8O_4}{1\ \cancel{mol\ C_4H_6O_3}} = 0.031\underline{7}4\ mol\ C_9H_8O_4$$

$$1.25\ \cancel{g\ C_7H_6O_3} \times \frac{1\ \cancel{mol\ C_7H_6O_3}}{138.12\ \cancel{g\ C_7H_6O_3}} \times \frac{1\ mol\ C_9H_8O_4}{1\ \cancel{mol\ C_7H_6O_3}} = 0.0090\underline{5}0\ mol\ C_9H_8O_4$$

$$0.0090\underline{5}0\ \cancel{mol\ C_9H_8O_4} \times \frac{180.1\ g\ C_9H_8O_4}{1\ \cancel{mol\ C_9H_8O_4}} = 1.6\underline{3}0\ g\ C_9H_8O_4$$

$$\frac{1.22\ \cancel{g\ C_9H_8O_4}}{1.6\underline{3}0\ \cancel{g\ C_9H_8O_4}} \times 100 = 74.8\%$$

Check: The theoretical yield has the correct units, g $C_9H_8O_4$, and has a reasonable magnitude compared to the mass of $C_7H_6O_3$, the limiting reactant. The % yield is reasonable, under 100%.

99. **Given:** (a) 11 molecules H_2, 2 molecules O_2; (b) 8 molecules H_2, 4 molecules O_2; (c) 4 molecules H_2, 5 molecules O_2; (d) 3 molecules H_2, 6 molecules O_2 **Find:** loudest explosion based on equation
Conceptual Plan: loudest explosion will occur in the balloon with the mol ratio closest to the balanced equation and that contains the most H_2
Solution: $2H_2(g) + O_2(g) \rightarrow H_2O(l)$
Balloon (a) has enough O_2 to react with 4 molecules H_2; balloon (b) has enough O_2 to react with 8 molecules H_2; balloon (c) has enough O_2 to react with 10 molecules H_2; and balloon (d) has enough O_2 for 3 molecules of H_2 to react. Therefore, balloon (b) will have the loudest explosion because it has the most H_2 that will react.
Check: Answer seems correct since it has the most H_2 with enough O_2 in the balloon to completely react.

101.a) Skeletal Reaction: $HCl(aq) + Hg_2(NO_3)_2(aq) \rightarrow Hg_2Cl_2(s) + HNO_3(aq)$
Balance Cl: $2HCl(aq) + Hg_2(NO_3)_2(aq) \rightarrow Hg_2Cl_2(s) + 2HNO_3(aq)$

b) Skeletal Reaction: $KHSO_3(aq) + HNO_3(aq) \rightarrow H_2O(l) + SO_2(g) + KNO_3(aq)$
Balanced Reaction: $KHSO_3(aq) + HNO_3(aq) \rightarrow H_2O(l) + SO_2(g) + KNO_3(aq)$

c) Skeletal Reaction: $NH_4Cl(aq) + Pb(NO_3)_2(aq) \rightarrow PbCl_2(s) + NH_4NO_3(aq)$
Balance Cl: $2NH_4Cl(aq) + Pb(NO_3)_2(aq) \rightarrow PbCl_2(s) + NH_4NO_3(aq)$
Balance N: $2NH_4Cl(aq) + Pb(NO_3)_2(aq) \rightarrow PbCl_2(s) + 2NH_4NO_3(aq)$

d) Skeletal Reaction: $NH_4Cl(aq) + Ca(OH)_2(aq) \rightarrow NH_3(g) + H_2O(l) + CaCl_2(aq)$
Balance Cl: $2NH_4Cl(aq) + Ca(OH)_2(aq) \rightarrow NH_3(g) + H_2O(l) + CaCl_2(aq)$
Balance N: $2NH_4Cl(aq) + Ca(OH)_2(aq) \rightarrow 2NH_3(g) + H_2O(l) + CaCl_2(aq)$
Balance H: $2NH_4Cl(aq) + Ca(OH)_2(aq) \rightarrow 2NH_3(g) + 2H_2O(l) + CaCl_2(aq)$

103. **Given:** 1.5 L solution, 0.050 M $CaCl_2$, 0.085 M $Mg(NO_3)_2$ **Find:** g Na_3PO_4
Conceptual Plan: V,M $CaCl_2$ → mol $CaCl_2$ and V,M $Mg(NO_3)_2$ → mol $Mg(NO_3)_2$
$V \times M = mol$ $\quad$ $V \times M = mol$
then (mol $CaCl_2$ + mol $Mg(NO_3)_2$ → Na_3PO_4 → g Na_3PO_4

$$\frac{2\ mol\ Na_3PO_4}{3\ mol\ (CaCl_2 + Mg(NO_3)_2)} \qquad \frac{163.97\ g\ Na_3PO_4}{1\ mol\ Na_3PO_4}$$

Solution: $3CaCl_2(aq) + 2Na_3PO_4(aq) \rightarrow Ca_3(PO_4)_2(s) + 6NaCl(aq)$

$3\ Mg(NO_3)_2(aq) + 2Na_3PO_4(aq) \rightarrow Mg_3(PO_4)_2(s) + 6\ NaCl(aq)$

$1.5\ L \times 0.050\ M\ CaCl_2 = 0.07\underline{5}\ mol\ CaCl_2$

$1.5\ L \times 0.085\ M\ Mg(NO_3)_2 = 0.1\underline{2}75\ mol\ Mg(NO_3)_2$

$$0.2\underline{0}25\ \cancel{mol\ CaCl_2\ and\ Mg(NO_3)_2} \times \frac{2\ \cancel{mol\ Na_2PO_4}}{3\ mol\ \cancel{CaCl_2\ and\ Mg(NO_3)_2}} \times \frac{163.97\ g\ mol\ Na_2PO_4}{\cancel{mol\ Na_2PO_4}} = 22\ g\ mol\ Na_2PO_4$$

Check: The units of the answer, g Na_3PO_4, are correct. The magnitude of the answer is reasonable since it is needed to remove both the Ca and Mg ions.

105. **Given:** 1.0 L, 0.10 M OH^- **Find:** g Ba

Conceptual Plan: VM → mol OH^- → mol $Ba(OH)_2$ → mol BaO → mol Ba → g Ba

$$V \times M = mol \quad \frac{1\ mol\ Ba(OH)_2}{2\ mol\ OH^-} \quad \frac{1\ mol\ BaO}{1\ mol\ Ba(OH)_2} \quad \frac{1\ mol\ Ba}{1\ mol\ BaO} \quad \frac{137.3\ g\ Ba}{1\ mol\ Ba}$$

Solution: $BaO(s) + H_2O(l) \rightarrow Ba(OH)_2(aq)$

$$1.0\ \cancel{L} \times \frac{0.10\ \cancel{mol\ OH^-}}{\cancel{L}} \times \frac{1\ \cancel{mol\ Ba(OH)_2}}{2\ \cancel{mol\ OH^-}} \times \frac{1\ \cancel{mol\ BaO}}{1\ \cancel{mol\ Ba(OH)_2}} \times \frac{1\ \cancel{mol\ Ba}}{1\ \cancel{mol\ BaO}} \times \frac{137.3\ g\ Ba}{1\ mol\ \cancel{Ba}} = 6.9\ g\ Ba$$

Check: The units of the answer, g Ba, are correct. The magnitude is reasonable since the molar mass of Ba is large and there are 2 moles hydroxide per mole Ba.

107. **Given:** 30.0% $NaNO_3$, \$9.00/ 100 lb; 20.0 % $(NH_4)_2SO_4$, \$8.10/ 100 lb **Find:** cost / lb N

Conceptual Plan: **mass fertilizer → mass $NaNO_3$ → mass N → cost/lb N**
and: mass fertilizer → mass $(NH_4)_2SO_4$ → mass N → cost/lb N

Solution:

$$100\ \cancel{lb\ fertilizer} \times \frac{30.0\ lb\ \cancel{NaNO_3}}{100\ lb\ \cancel{fertilizer}} \times \frac{16.48\ lb\ N}{100\ \cancel{lb\ NaNO_3}} = 4.9\underline{4}4\ lb\ N$$

$$\frac{\$9.00}{100\ \cancel{lb\ fertilizer}} \times \frac{100\ lb\ fertilizer}{4.9\underline{4}4\ lb\ N} = \$1.82/\ lb\ N$$

$$100\ \cancel{lb\ fertilizer} \times \frac{20.0\ lb\ \cancel{(NH_4)_2SO_4}}{100\ lb\ \cancel{fertilizer}} \times \frac{21.2\ lb\ N}{100\ \cancel{lb\ (NH_4)_2SO_4}} = 4.2\underline{4}0\ lb\ N$$

$$\frac{\$8.10}{100\ \cancel{lb\ fertilizer}} \times \frac{100\ lb\ fertilizer}{4.2\underline{4}\ lb\ N} = \$1.91/\ lb\ N$$

The more economical fertilizer is the $NaNO_3$ because it costs less/ lb N.

Check: The units of the cost, \$/lb N, are correct. The answer is reasonable because you compare the cost/lb N directly.

109. **Given:** 24.5 g Au, 24.5 g BrF_3, 24.5 g KF **Find:** g $KAuF_4$

Conceptual Plan: g Au → mol Au → mol $KAuF_4$

$$\frac{1 \text{ mol Au}}{196.97 \text{ g Au}} \quad \frac{2 \text{ mol } KAuF_4}{2 \text{ mol Au}}$$

g BrF_3 → mol BrF_3 → mol $KAuF_4$ **→ smallest amount determines limiting reactant**

$$\frac{1 \text{ mol } BrF_3}{136.9 \text{ g } BrF_3} \quad \frac{2 \text{ mol } KAuF_4}{2 \text{ mol } BrF_3}$$

g KF → mol KF → mol $KAuF_4$

$$\frac{1 \text{ mol KF}}{58.10 \text{ g KF}} \quad \frac{2 \text{ mol } KAuF_4}{2 \text{ mol KF}}$$

then: mol $KAuF_4$ → g $KAuF_4$

$$\frac{312.07 \text{ g } KAuF_4}{\text{mol } KAuF_4}$$

$$2\,Au(s) + 2BrF_3(l) + 2KF(s) \rightarrow Br_2(l) + 2KAuF_4(s)$$

oxidation states; o +3 – 1 +1 – 1 0 +1 +3 – 1

This is a redox reaction since Au increases in oxidation number (oxidation) and Br decreases in number (reduction). BrF_3 is the oxidizing agent, Au is the reducing agent.

Solution:

$$24.5 \text{ g Au} \times \frac{1 \text{ mol Au}}{196.97 \text{ g Au}} \times \frac{2 \text{ mol } KAuF_4}{2 \text{ mol Au}} = 0.12\underline{4}4 \text{ mol } KAuF_4$$

$$24.5 \text{ g } BrF_3 \times \frac{1 \text{ mol } BrF_3}{196.97 \text{ g } BrF_3} \times \frac{2 \text{ mol } KAuF_4}{2 \text{ mol } BrF_3} = 0.17\underline{9}0 \text{ mol KAuF}$$

$$24.5 \text{ g KF} \times \frac{1 \text{ mol KF}}{196.97 \text{ g KF}} \times \frac{2 \text{ mol } KAuF_4}{2 \text{ mol KF}} = 0.42\underline{1}7 \text{ mol } KAuF_4$$

$$0.12\underline{4}4 \text{ mol } KAuF_4 \times \frac{312.07 \text{ g } KAuF_4}{1 \text{ mol } KAuF_4} = 38.8 \text{ g } KAuF_4$$

Check: Units of the answer, g $KAuF_4$, are correct. The magnitude of the answer is reasonable compared to the mass of the limiting reactant Au.

111.**Given:** solution may contain Ag^+, Ca^{2+},and Cu^{2+} **Find:** determine which ions are present.

Conceptual Plan: test the solution sequentially with NaCl, Na_2SO_4 and Na_2CO_3 and see if precipitates form.

Solution: original solution + NaCl yields no reaction: Ag^+ not present: since Cl is normally soluble but Ag^+ is an exception.

Original solution with Na_2SO_4 yields a precipitate and solution 2. The precipitate is $CaSO_4$, so Ca^{2+} is present. SO^{2-} are normally soluble but Ca^{2+} is an exception.

Solution 2 with Na_2CO_3 yields a precipitate. The precipitate is $CuCO_3$, so Cu^{2+} is present. All CO_3^{2-} are insoluble.

NET IONIC EQUATIONS:

$$Ca^{2+}(aq) + SO_4^{2-}(aq) \rightarrow CaSO_4(s)$$
$$Cu^{2+}(aq) + CO_3^{2-}(aq) \rightarrow CuCO_3(s)$$

Check: the answer is reasonable since two different precipitates formed and all the Ca^{2+} was removed before the carbonate was added.

113 **Given:** 15.2 billion L lake water, 1.8×10^{-5} M H_2SO_4, 8.7×10^{-6} M HNO_3 **Find:** kg $CaCO_3$ needed to neutralize

Conceptual Plan: Vol lake → mol H_2SO_4 → mol H^+ and vol lake → mol HNO_3 → mol H^+

$$\text{vol x M = mol} \quad \frac{2 \text{ mol } H^+}{1 \text{ mol } H_2SO_4} \qquad \text{vol x M = mol} \quad \frac{1 \text{ mol } H^+}{1 \text{ mol } HNO_3}$$

Then: total mol H^+ → mol CO_3^{2-} → mol $CaCO_3$ → g $CaCO_3$ → kg $CaCO_3$

$$\frac{1 \text{ mol } CO_3^{2-}}{1 \text{ mol } H^+} \quad \frac{1 \text{ mol } CaCO_3}{1 \text{ mol } CO_3^{2-}} \quad \frac{100.09 \text{ g } CaCO_3}{1 \text{ mol } CaCO_3} \quad \frac{\text{kg}}{1000 \text{ g}}$$

Solution: $2H^+(aq) + CO_3^{2-}(aq) \rightarrow H_2O(l) + CO_2(g)$

$$15.2 \times 10^9 \text{ L} \times \frac{1.8 \times 10^{-5} \text{ mol } H_2SO_4}{\text{L soln}} \times \frac{2 \text{ mol } H^+}{\text{mol } H_2SO_4} = 5\underline{4}7200 \text{ mol } H^+$$

$$15.2 \times 10^9 \text{ L} \times \frac{8.7 \times 10^{-6} \text{ mol } HNO_3}{\text{L soln}} \times \frac{1 \text{ mol } H^+}{\text{mol } HNO_3} = 1\underline{3}2240 \text{ mol } H^+$$

$$6\underline{7}9440 \text{ mol } H^+ \times \frac{1 \text{ mol } CO_3^{2-}}{\text{mol } H^+} \times \frac{1 \text{ mol } CaCO_3}{\text{mol } CO_3^{2-}} \times \frac{100.09 \text{ g } CaCO_3}{1 \text{ mol } CaCO_3} \times \frac{\text{kg}}{1000 \text{ g}} = 3.4 \times 10^4 \text{ kg } CaCO_3$$

Check: Units of the answer, kg $CaCO_3$, are correct. The magnitude of the answer is reasonable based on the size of the lake.

115. **Given:** 45 μg Pb/dL blood, Vol = 5.0 L, 1 mol succimer($C_4H_6O_4S_2$) = 1 mol Pb **Find:** mass $C_4H_6O_4S_2$ in mg

Conceptual Plan: Volume blood L → Volume blood dL → μg Pb → g Pb → mol Pb →

$$\frac{10 \text{ dL}}{\text{L}} \quad \frac{45 \text{ μg}}{\text{dL}} \quad \frac{10^6 \text{ μg}}{\text{g}} \quad \frac{\text{mol Pb}}{207.2 \text{ g Pb}} \quad \frac{1 \text{ mol succimer}}{1 \text{ mol Pb}}$$

mol succimer → g succimer → mg succimer

$$\frac{182.23 \text{ g succimer}}{1 \text{ mol succimer}} \quad \frac{1000 \text{ mg succimer}}{1 \text{ g succimer}}$$

Solution:

$$5.0 \text{ L blood} \times \frac{10 \text{ dL}}{\text{L}} \times \frac{45 \text{ μg}}{\text{dL}} \times \frac{1 \text{ g}}{10^6 \text{ μg}} \times \frac{1 \text{ mol Pb}}{207.2 \text{ g}} \times \frac{1 \text{ mol succimer}}{1 \text{ mol Pb}} \times \frac{182.23 \text{ g succimer}}{1 \text{ mol succimer}} \times \frac{1000 \text{ mg}}{\text{g}} = 2.0 \text{ mg succimer}$$

Check: The units of the answer, mg succimer, are correct. The magnitude is reasonable for the volume of blood and the concentration.

117. **Given:** 250 g sample, 67.2 mol % Al **Find:** Theoretical yield in g of Mn

Conceptual Plan: mol % Al → g Al and mol % MnO_2 → g MnO_2, then mass % Al

$$\frac{26.98 \text{ g Al}}{\text{mol Al}} \quad \frac{86.94 \text{ g } MnO_2}{\text{mol } MnO_2} \quad \frac{\text{g Al}}{\text{total g}} \times 100$$

then: sample → g Al → mol Al → mol Mn

$$\frac{38.86 \text{ g Al}}{100 \text{ g sample}} \quad \frac{\text{mol Al}}{26.98 \text{ g Al}} \quad \frac{3 \text{ mol Mn}}{4 \text{ mol Al}}$$

→ smallest mol amount determines limiting reactant

sample → g MnO_2 → mol MnO_2 → mol Mn

$$\frac{61.14 \text{ g } MnO_2}{100 \text{ g sample}} \quad \frac{\text{mol } MnO_2}{86.94 \text{ g } MnO_2} \quad \frac{3 \text{ mol Mn}}{\text{mol } MnO_2}$$

then mol Mn → g Mn

$$\frac{54.94 \text{ g Mn}}{\text{mol Mn}}$$

Solution: $4Al(s) + 3\ MnO_2(s) \rightarrow 3Mn + 2Al_2O_3(s)$

Assume 1 mole: $0.672 \text{ mol Al} \times \frac{26.98 \text{ g Al}}{\text{mol Al}} = 18.13 \text{ g Al}$

$0.328 \text{ mol } MnO_2 \times \frac{26.98 \text{ g } MnO_2}{\text{mol } MnO_2} = 28.52 \text{ g } MnO_2$

$\frac{18.13 \text{ g Al}}{(18.13 \text{ g Al} + 28\ 52 \text{ g } MnO_2)} \times 100 = 38.86\ \%\ \text{Al}$ So: 61.14 % MnO_2

$250 \text{ g } \cancel{\text{sample}} \times \frac{38.86 \cancel{\text{g Al}}}{100 \text{ g } \cancel{\text{sample}}} \times \frac{\cancel{\text{mol Al}}}{26.98 \cancel{\text{g Al}}} \times \frac{3 \text{ mol Mn}}{4 \cancel{\text{mol Al}}} = 1.7\underline{5}8 \text{ mol Mn}$

$250 \text{ g } \cancel{\text{sample}} \times \frac{61.14 \cancel{\text{g } MnO_2}}{100 \text{ g } \cancel{\text{sample}}} \times \frac{\cancel{\text{mol } MnO_2}}{26.98 \cancel{\text{g } MnO_2}} \times \frac{3 \text{ mol Mn}}{4 \cancel{\text{mol } MnO_2}} = 3.6\underline{0} \text{ mol Mn}$

$1.7\underline{5}8 \cancel{\text{mol Mn}} \times \frac{54.94 \text{ g Mn}}{1 \cancel{\text{mol Mn}}} = 96.6 \text{ g Mn}$

Check: the units of the answer, g Mn, are correct. The magnitude of the answer is reasonable based on the amount of the limiting reactant, Al.

119. The correct answer is d. The molar mass of K and O_2 are comparable. Since the stoichiometry has a ratio of 4 mol K to 1 mol O_2, K will be the limiting reactant when mass of K is less than 4 times the mass of O_2.

121. **Given:** 1 M solution contains 8 particles **Find:** amount of solute or solvent needed to obtain new concentration

Conceptual Plan: determine amount of solute particles in each new solution, then determine if solute (if the number is greater), or solvent (if the number is less) needs to be added to obtain the new concentration.

Solution: Solution (a) contains 12 particles solute. Concentration is greater than the original, so solute needs to be added. $12 \cancel{\text{particles}} \times \frac{1 \text{ mol}}{8 \cancel{\text{particles}}} = 1.5 \text{ mol}$ (1.5 mol – 1.0 mol) = 0.5 mol solute added.

$0.5 \text{ mol solute} \times \frac{8 \text{ particles}}{1 \text{ mol solute}} = 4 \text{ solute particles added}$

Solution (a) is obtained by adding 4 particles solute to 1 L of original solution

Solution (b) contains 4 particles. Concentration is less than the original so solvent needs to be added.

$4 \text{ particles} \times \frac{1 \text{ mol}}{8 \text{ particles}} = 0.5 \text{ mol solute}$ So, 1 L solution contains 0.5 mol = 0.5M

$(1\ M)(1\ L) = (0.5\ M)(x)$ $x = 2\ L$

Solution (b) is obtained by diluting 1 L of the original solution to 2 L

Solution (c) contains 6 particles. Concentration is less than the original so solvent needs to be added.

$6 \text{ particles} \times \frac{1 \text{ mol}}{8 \text{ particles}} = 0.75 \text{ mol solute}$ So, 1 L solution contains 0.75 mol = 0.75M

$(1\ M)(1\ L) = (0.75\ M)(x)$ $x = 2\ L$

Solution (c) is obtained by diluting 1 L of the original solution to 1.3 L

Chapter 5
Gases

1. Pressure is the force exerted per unit area by gas molecules as they strike the surfaces around them. Pressure is caused by collisions of gas molecules with any surface or other gas molecule.

3. In the shallow well, atmospheric pressure does the work of pushing the water up the pipe. The pump acts to reduce the pressure inside the pipe, allowing atmospheric pressure to drive the water to the surface, against the force of gravity. The maximum depth of the well, therefore, depends on the total atmospheric pressure. Even if the pump could create a perfect vacuum (zero pressure) within the pipe, normal atmospheric pressure can only push the water to a total height of about 10.3 m because a 10.3-meter column of water exerts the same pressure—101,325 newtons per square meter or 14.7 pounds per square inch—as the gas molecules in our atmosphere.

5. A manometer is a U-shaped tube containing a dense liquid, usually mercury. In an open-ended manometer, one end of the tube is open to atmospheric pressure and the other is attached to a flask containing the gas sample. If the pressure of the gas sample is exactly equal to atmospheric pressure, then the mercury levels on both sides of the tube are the same. If the pressure of the sample is greater than atmospheric pressure, the mercury level on the sample side of the tube is lower than on the side open to the atmosphere. If the pressure of the sample is less than atmospheric pressure, the mercury level on the sample side is higher than on the side open to the atmosphere. This type of manometer always measures the pressure of the gas sample relative to atmospheric pressure. The difference in height between the two levels is equal to the pressure difference from atmospheric pressure.

7. This pain is caused by air-containing cavities within your ear. When you ascend a mountain, the external pressure (the pressure that surrounds you) drops, while the pressure within your ear cavities (the internal pressure) remains the same. This creates an imbalance—the greater internal pressure forces your eardrum to bulge outward, causing pain. With time, and the help of a yawn or two, the excess air within your ear's cavities escapes, equalizing the internal and external pressure and relieving the pain.

9. When we breathe, we expand the volume of our chest cavity, reducing the pressure on the outer surface of the lungs to less than 1 atm (Boyle's law). Because of this pressure differential, the lungs expand, the pressure in them falls, and air from outside our lungs then flows into them. Extra-long snorkels do not work because of the pressure exerted by water at an increased depth. A diver at 10 m experiences an external pressure of 2 atm. This is more than the muscles of the chest cavity can overcome—the chest cavity and lungs are compressed, resulting in an air pressure within them of more than 1 atm. If the diver had a snorkel that went to the surface—where the air pressure is 1 atm—air would flow out of his lungs, not into them. It would be impossible to breathe.

11. The ideal gas law (PV = nRT) combines all of the relationships between the four variables (pressure, volume, number of moles and temperature (in Kelvins)) discussed in one simple expression.

13. The molar volume of an ideal gas is the volume occupied by one mole of gas at T = 0 °C (273 K) and P = 1.00 atm. Substituting these values into the ideal gas law, one can calculate that this value as 22.414 L.

15. The pressure due to any individual component in a gas mixture is called the partial pressure (P_n) of that component and can be calculated from the ideal gas law by assuming that each gas component acts independently. The sum of the partial pressures of the components in a gas mixture must equal the total pressure: $P_{total} = P_a + P_b + P_c + \ldots..$ where P_{total} is the total pressure and P_a, P_b, P_c... are the partial pressures of the components.

17. No, when collecting a gas over water, it will contain some water molecules. The vapor pressure of water can be gotten from Table 5.4. Therefore, $P_{Gas} = P_{total} - P_{H2O}$.

19. The basic postulates of kinetic molecular theory are: (1) The size of a particle is negligibly small. (2) The average kinetic energy of a particle is proportional to the temperature in kelvins. and (3) The collision of one particle with another (or with the walls) is completely elastic. Pressure is defined as force divided by area. According to kinetic molecular theory, a gas is a collection of particles in constant motion. The motion results in collisions between the particles and the surfaces around them. As each particle collides

with a surface, it exerts a force upon that surface. The result of many particles in a gas sample exerting forces on the surfaces around them is a constant pressure.

21. Postulate 2 of kinetic molecular theory states that the average kinetic energy is proportional to the temperature in kelvins. The root mean square velocity of a collection of gas particles is inversely proportional to the square root of the molar mass of the particles in kilograms per mole.

23. The process by which gas molecules spread out in response to a concentration gradient is called diffusion. Effusion, the process by which a gas escapes from a container into a vacuum through a small hole. The rate of effusion is inversely proportional to the square root of the molar mass of the gas.

25. Sulfur Oxides (SO_x) – Sulfur oxides include SO_2 and SO_3, which are produced chiefly during coal-fired electricity generation and industrial metal refining. Carbon Monoxide (CO) – Carbon monoxide is formed by the incomplete combustion of fossil fuels (petroleum, natural gas, and coal). It is emitted mainly by motor vehicles. Nitrogen Oxides (NO_x) – Nitrogen oxides include NO and NO_2, which are emitted by motor vehicles, by fossil-fuel based electricity generation plants, and by any high temperature combustion process that occurs in air. Ozone (O_3) – Ozone is produced when some of the products of fossil-fuel combustion, especially nitrogen oxides and unburned volatile organic compounds (VOCs), react in the presence of sunlight. The levels of all of these pollutants are decreasing over U.S. cities.

27. Chlorofluorocarbons (CFCs) are blamed for destroying stratospheric ozone. When CFCs reach the stratosphere, UV light (which is less abundant below the ozone layer because the ozone absorbs it) breaks a carbon-chlorine bond in the CFC, generating a very reactive chlorine atom. This chlorine atom then reacts with ozone in a cyclic reaction that destroys two ozone molecules and regenerates itself to repeat the process. In this way, a single chlorine atom can destroy hundreds of ozone molecules. Legislation has been passed in many nations calling for a complete ban on CFC production beginning in 1996.

29. a) **Given:** 24.9 in Hg **Find:** atm

Conceptual plan: in Hg → atm

$$\frac{1\,atm}{29.92\ in\,Hg}$$

Solution: $$24.9\ \cancel{in\,Hg}\ x\ \frac{1\,atm}{29.92\ \cancel{in\,Hg}} = 0.832\,atm$$

Check: The units (atm) are correct. The magnitude of the answer (<1) makes physical sense because we started with less than 29.92 in Hg.

b) **Given:** 24.9 in Hg **Find:** mm Hg

Conceptual plan: Use answer from part a) then convert atm → mm Hg

$$\frac{760\,mm\,Hg}{1\,atm}$$

Solution: $$0.832\ \cancel{atm}\ x\ \frac{760\,mm\,Hg}{1\ \cancel{atm}} = 632\ mm\ Hg$$

Check: The units (mm Hg) are correct. The magnitude of the answer (< 760 mm Hg) makes physical sense because we started with less than 1 atm.

c) **Given:** 24.9 in Hg **Find:** psi

Conceptual plan: Use answer from part a) then convert atm → psi

$$\frac{14.7\,psi}{1\,atm}$$

Solution: $$0.832\ \cancel{atm}\ x\ \frac{14.7\,psi}{1\ \cancel{atm}} = 12.2\ psi$$

Check: The units (psi) are correct. The magnitude of the answer (< 14.7 psi) makes physical sense because we started with less than 1 atm.

d) **Given:** 24.9 in Hg **Find:** Pa

Conceptual plan: **Use answer from part a) then convert** **atm → Pa**

$$\frac{101{,}325\,Pa}{1\,atm}$$

Solution: $0.832\,\cancel{atm}\ x\ \dfrac{101{,}325\,Pa}{1\,\cancel{atm}} = 8.43x10^4\ Pa$

Check: The units (mm Hg) are correct. The magnitude of the answer (< 760 mm Hg) makes physical sense because we started with less than 1 atm.

31. a) **Given:** 31.85 in Hg **Find:** mm Hg
Conceptual plan: **in Hg → mm Hg**

$$\frac{25.4\ mm\,Hg}{1\ in\,Hg}$$

Solution: $31.85\,\cancel{in\,Hg}\ x\ \dfrac{25.4\ mm\,Hg}{1\ \cancel{in\,Hg}} = 809.0\ mm\ Hg$

Check: The units (mm Hg) are correct. The magnitude of the answer (809) makes physical sense because inches are larger than mm.

b) **Given:** 31.85 in Hg **Find:** atm
Conceptual plan: **Use answer from part a) then convert** **mm Hg → atm**

$$\frac{1\,atm}{760\,mm\,Hg}$$

Solution: $809.0\ \cancel{mm\,Hg}\ x\ \dfrac{1\,atm}{760\ \cancel{mm\,Hg}} = 1.064\ atm$

Check: The units (atm) are correct. The magnitude of the answer (>1) makes physical sense because we started with more than 760 mm Hg.

c) **Given:** 31.85 in Hg **Find:** torr
Conceptual plan: **Use answer from part a) then convert** **mm Hg → torr**

$$\frac{1\,torr}{760\,mm\,Hg}$$

Solution: $809.0\ \cancel{mm\,Hg}\ x\ \dfrac{1\,torr}{760\ \cancel{mm\,Hg}} = 809.0\ torr$

Check: The units (torr) are correct. The magnitude of the answer (809) makes physical sense because both units are of the same size.

d) **Given:** 31.85 in Hg **Find:** kPa
Conceptual plan: **Use answer from part b) then convert** **atm → Pa → kPa**

$$\frac{101{,}325\,Pa}{1\,atm}\qquad\frac{1\ kPa}{1000\,Pa}$$

Solution: $1.064\ \cancel{atm}\ x\ \dfrac{101{,}325\ \cancel{Pa}}{1\,\cancel{atm}}\ x\ \dfrac{1\ kPa}{1000\ \cancel{Pa}} = 107.9\ kPa$

Check: The units (kPa) are correct. The magnitude of the answer (108) makes physical sense because we started with more than 1 atm and there are ~101 kPa in an atm.

33. a) **Given:** P_{bar} = 755.3 mm Hg and figure **Find:** P_{gas}
Conceptual plan: **Measure height difference then convert** **cm Hg → mm Hg → mm Hg**

$$\frac{10\,mm\ Hg}{1\,cm\ Hg}\qquad P_{gas} = h + P_{bar}$$

Solution: h = $7.2\ \cancel{cm\ Hg} \times \frac{10\ mm\ Hg}{1\ \cancel{cm\ Hg}} = 72\ mm\ Hg$ P_{gas} = 72 mm Hg + 755.3 mm Hg = 827 mm Hg

Check: The units (mm Hg) are correct. The magnitude of the answer (827 mm Hg) makes physical sense because mercury column is higher on the right, indicating that the pressure is above barometric pressure. No significant figures to the right of the decimal point can be reported since the mercury height is known only to the 1's place.

b) **Given:** P_{bar} = 755.3 mm Hg and figure **Find:** P_{gas}
Conceptual plan: **Measure height difference then convert** **cm Hg → mm Hg → mm Hg**

$$\frac{10\ mm\ Hg}{1\ cm\ Hg} \qquad P_{gas} = h + P_{bar}$$

Solution: h = $-4.4\ \cancel{cm\ Hg} \times \frac{10\ mm\ Hg}{1\ \cancel{cm\ Hg}} = -44\ mm\ Hg$

P_{gas} = – 44 mm Hg + 755.3 mm Hg = 711 mm Hg
Check: The units (mm Hg) are correct. The magnitude of the answer (711 mm Hg) makes physical sense because mercury column is higher on the left, indicating that the pressure is below barometric pressure. No significant figures to the right of the decimal point can be reported since the mercury height is known only to the 1's place.

35. **Given:** V_1 = 2.8 L, P_1 = 755 mm Hg and V_2 = 3.7 L **Find:** P_2
Conceptual plan: $\mathbf{V_1, P_1, V_2 \rightarrow P_2}$

$$P_1 V_1 = P_2 V_2$$

Solution: $P_1 V_1 = P_2 V_2$ Rearrange to solve for P_2.

$$P_2 = P_1 \frac{V_1}{V_2} = 755\ \text{mm Hg} \times \frac{2.8\ \cancel{L}}{3.7\ \cancel{L}} = 570\ \text{mm Hg} = 5.7 \times 10^2\ \text{mm Hg}$$

Check: The units (mm Hg) are correct. The magnitude of the answer (570 mm Hg) makes physical sense because Boyle's Law indicates that as the volume increases, the pressure decreases.

37. **Given:** V_1 = 48.3 mL, T_1 = 22 °C and T_2 = 87 °C **Find:** V_2
Conceptual plan: **°C → K then $\mathbf{V_1, T_1, T_2 \rightarrow V_2}$**

$$K = {}^\circ C + 273.15 \qquad \frac{V_1}{T_1} = \frac{V_2}{T_2}$$

Solution: T_1 = 22 °C + 273.15 = 295 K and T_2 = 87 °C + 273.15 = 360. K

$\frac{V_1}{T_1} = \frac{V_2}{T_2}$ Rearrange to solve for V_2. $V_2 = V_1 \frac{T_2}{T_1} = 48.3\ \text{mL} \times \frac{360\ \cancel{K}}{295\ \cancel{K}} = 58.9\ \text{mL}$

Check: The units (mL) are correct. The magnitude of the answer (59 mL) makes physical sense because Charles' Law indicates that as the volume increases, the temperature increases.

39. **Given:** V_1 = 2.76 L, n_1 = 0.128 mol and Δn = 0.073 mol **Find:** V_2
Conceptual plan: $\mathbf{n_1 \rightarrow n_2}$ **then** $\mathbf{V_1, n_1, n_2 \rightarrow V_2}$

$$n_1 + \Delta n = n_2 \qquad \frac{V_1}{n_1} = \frac{V_2}{n_2}$$

Solution: n_2 = 0.128 mol + 0.073 mol = 0.201 mol

$\frac{V_1}{n_1} = \frac{V_2}{n_2}$ Rearrange to solve for V_2. $V_2 = V_1 \frac{n_2}{n_1} = 2.76\ \text{L} \times \frac{0.201\ \cancel{mol}}{0.128\ \cancel{mol}} = 4.33\ \text{L}$

Check: The units (L) are correct. The magnitude of the answer (4.33 L) makes physical sense because Avogadro's Law indicates that as the number of moles increases, the volume increases.

41. **Given:** n = 0.118 mol, P = 0.97 atm and T = 305 K **Find:** V
Conceptual plan: **n, P, T → V**

$$P\,V = nRT$$

Solution: $PV = nRT$ Rearrange to solve for V.

$$V = \frac{nRT}{P} = \frac{0.118 \text{ mol} \times 0.08206 \frac{\text{L} \cdot \text{atm}}{\text{mol} \cdot \text{K}} \times 305 \text{ K}}{0.97 \text{ atm}} = 3.0 \text{ L}$$

Check: The units (L) are correct. The magnitude of the answer (3 L) makes sense because, as you will see in the next section, one mole of an ideal gas under standard conditions (273 K and 1 atm) occupies 22.4 L. Although these are not standard conditions, they are close enough for a ballpark check of the answer. Since this gas sample contains 0.118 moles, a volume of 3 L is reasonable.

43. **Given:** V = 28.5 L, P = 1.8 atm, and T = 298 K **Find:** n

Conceptual plan: **V, P, T → n**

$PV = nRT$

Solution: $PV = nRT$ Rearrange to solve for n.

$$n = \frac{PV}{RT} = \frac{1.8 \text{ atm} \times 28.5 \text{ L}}{0.08206 \frac{\text{L} \cdot \text{atm}}{\text{mol} \cdot \text{K}} \times 298 \text{ K}} = 2.1 \text{ mol}$$

Check: The units (mol) are correct. The magnitude of the answer (2 mol) makes sense because, as you will see in the next section, one mole of an ideal gas under standard conditions (273 K and 1 atm) occupies 22.4 L. Although these are not standard conditions, they are close enough for a ballpark check of the answer. Since this gas sample has a volume of 28.5 L, and a pressure of 1.8 atm, ~ 2 mol is reasonable.

45. **Given:** P_1 = 36.0 psi (gauge P), V_1 = 11.8 L, T_1 = 12.0 °C, V_2 = 12.2 L and T_2 = 65.0 °C
Find: P_2 and compare to P_{max} = 38.0 psi (gauge P)
Conceptual plan: **°C → K and gauge P → psi → atm then P_1, V_1, T_1, V_2, T_2 → P_2**

$K = °C + 273.15$ psi = gauge P + 14.7 $\frac{1 \text{ atm}}{14.7 \text{ psi}}$ $\frac{P_1V_1}{T_1} = \frac{P_2V_2}{T_2}$

Solution: T_1 = 12.0 °C + 273.15 = 285.2 K and T_2 = 65.0 °C + 273.15 = 338.2 K

$$P_1 = 36.0 \text{ psi (gauge P)} + 14.7 = 50.7 \text{ psi} \times \frac{1 \text{ atm}}{14.7 \text{ psi}} = 3.4\underline{4}898 \text{ atm}$$

$$P_{max} = 38.0 \text{ psi (gauge P)} + 14.7 = 52.7 \text{ psi} \times \frac{1 \text{ atm}}{14.7 \text{ psi}} = 3.59 \text{ atm}$$

$\frac{P_1V_1}{T_1} = \frac{P_2V_2}{T_2}$ Rearrange to solve for P_2.

$$P_2 = P_1 \frac{V_1}{V_2}\frac{T_2}{T_1} = 3.4\underline{4}898 \text{ atm} \times \frac{11.8 \text{ L}}{12.2 \text{ L}} \times \frac{338.2 \text{ K}}{285.2 \text{ K}} = 3.95 \text{ atm}$$

This exceeds the maximum tire rating of 3.59 atm or 38.0 psi (gauge P).
Check: The units (atm) are correct. The magnitude of the answer (3.95 atm) makes physical sense because the relative increase in T is greater than the relative increase in V, so P should increase.

47. **Given:** m (CO_2) = 28.8 g, P = 742 mm Hg and T = 22 °C **Find:** V
Conceptual plan: **°C → K and mm Hg → atm and g → mol then n, P, T → V**

$K = °C + 273.15$ $\frac{1 \text{ atm}}{760 \text{ mm Hg}}$ $\frac{1 \text{ mol}}{44.01 \text{ g}}$ $PV = nRT$

Solution: T_1 = 22 °C + 273.15 = 295 K, $P = 742 \text{ mm Hg} \times \frac{1 \text{ atm}}{760 \text{ mm Hg}} = 0.97\underline{6}316 \text{ atm}$,

$$n = 28.8 \text{ g} \times \frac{1 \text{ mol}}{44.01 \text{ g}} = 0.65\underline{4}397 \text{ mol}$$

$PV = nRT$ Rearrange to solve for V.

$$V = \frac{nRT}{P} = \frac{0.65\underline{4}397 \text{ mol} \times 0.08206 \frac{\text{L} \cdot \text{atm}}{\text{mol} \cdot \text{K}} \times 295 \text{ K}}{0.97\underline{6}316 \text{ atm}} = 16.2 \text{ L}$$

Check: The units (L) are correct. The magnitude of the answer (16 L) makes sense because, one mole of an ideal gas under standard conditions (273 K and 1 atm) occupies 22.4 L. Although these are not standard conditions, they are close enough for a ballpark check of the answer. Since this gas sample contains 0.65 moles, a volume of 16 L is reasonable.

49. **Given:** sample a = 5 gas particles, sample b = 10 gas particles, and sample c = 8 gas particles, with all temperatures and volumes the same   **Find:** sample with largest P

    Conceptual plan: n, V, T → P
    $$P V = nRT$$

    Solution: $P V = nRT$ Since V and T are constant this means that P α n. The sample with the largest number of gas particles will have the highest P. $P_b > P_c > P_a$.

51. **Given:** $P_1 = 755$ mm Hg, $T_1 = 25\ °C$, and $T_2 = 1155\ °C$ **Find:** P_2

    Conceptual plan: °C → K   and   mm Hg → atm   then P_1, T_1, T_2 → P_2
    $$K = °C + 273.15 \qquad \frac{1\ atm}{760\ mm\ Hg} \qquad \frac{P_1}{T_1} = \frac{P_2}{T_2}$$

    Solution: $T_1 = 25\ °C + 273.15 = 298\ K$ and $T_2 = 1155\ °C + 273.15 = 1428\ K$

    $P = 755\ \cancel{mm\ Hg}\ \text{x}\ \dfrac{1\ atm}{760\ \cancel{mm\ Hg}} = 0.99\underline{3}421\ atm$ $\dfrac{P_1}{T_1} = \dfrac{P_2}{T_2}$ Rearrange to solve for P_2.

    $P_2 = P_1 \dfrac{T_2}{T_1} = 0.99\underline{3}421\ atm\ \text{x}\ \dfrac{1428\ \cancel{K}}{298\ \cancel{K}} = 4.76\ atm$

    Check: The units (atm) are correct.   The magnitude of the answer (5 atm) makes physical sense because there is a significant increase in T, which will increase P significantly.

53. **Given:** STP and m (Ne) = 10.0 g   **Find:** V

    Conceptual plan: g → mol   →   V
    $$\frac{1\ mol}{20.18\ g} \qquad \frac{22.414\ L}{1\ mol}$$

    Solution: $10.0\ \cancel{g}\ \text{x}\ \dfrac{1\ \cancel{mol}}{20.18\ \cancel{g}}\ \text{x}\ \dfrac{22.414\ L}{1\ \cancel{mol}} = 11.1\ L$

    Check: The units (L) are correct.   The magnitude of the answer (11 L) makes sense because, one mole of an ideal gas under standard conditions (273 K and 1 atm) occupies 22.4 L and we have about 0.5 mol.

55. **Given:** H_2, P = 1655 psi, and T = 20.0 °C   **Find:** d

    Conceptual plan: °C → K   and   psi → atm   then   P, T, $\mathfrak{M}$ → d
    $$K = °C + 273.15 \qquad \frac{1\ atm}{14.7\ psi} \qquad d = \frac{P\mathfrak{M}}{RT}$$

    Solution: $T = 20.0\ °C + 273.15 = 293.2\ K$ $P = 1655\ \cancel{psi}\ \text{x}\ \dfrac{1\ atm}{14.7\ \cancel{psi}} = 11\underline{2}.585\ atm$

    $d = \dfrac{P\mathfrak{M}}{RT} = \dfrac{11\underline{2}.585\ \cancel{atm}\ \text{x}\ 2.016\ \dfrac{g}{\cancel{mol}}}{0.08206\ \dfrac{L\ \cancel{atm}}{\cancel{K}\ \cancel{mol}}\ \text{x}\ 293.2\ \cancel{K}} = 9.43\ \dfrac{g}{L}$

    Check: The units (g/L) are correct.   The magnitude of the answer (9 g/L) makes physical sense because this is a high pressure, so the gas density will be on the high side.

57. **Given:** V = 248 mL, m = 0.433 g, P = 745 mm Hg, and T = 28 °C   **Find:** $\mathfrak{M}$

    Conceptual plan: °C → K   mm Hg → atm   mL → L   then   V, m → d then   d, P, T, → $\mathfrak{M}$
    $$K = °C + 273.15 \qquad \frac{1\ atm}{760\ mm\ Hg} \qquad \frac{1\ L}{1000\ mL} \qquad d = \frac{m}{V} \qquad d = \frac{P\mathfrak{M}}{RT}$$

Solution: $T = 28\ °C + 273.15 = 301\ K$ $P = 745\ \text{mm Hg} \times \frac{1\ \text{atm}}{760\ \text{mm Hg}} = 0.98\underline{0}263\ \text{atm}$

$V = 248\ \text{mL} \times \frac{1\ L}{1000\ \text{mL}} = 0.248\ L$ $d = \frac{m}{V} = \frac{0.433\ g}{0.248\ L} = 1.7\underline{4}597\ g/L$ $d = \frac{P\mathfrak{M}}{RT}$ Rearrange to solve for $\mathfrak{M}$.

$$\mathfrak{M} = \frac{dRT}{P} = \frac{1.7\underline{4}597\ \frac{g}{L} \times 0.08206 \frac{L\ atm}{K\ mol} \times 301\ K}{0.98\underline{0}263\ atm} = 44.0\ g/mol$$

Check: The units (g/mol) are correct. The magnitude of the answer (44 g/mol) makes physical sense because this is a reasonable number for a molecular weight of a gas.

59. **Given:** m = 38.8 mg, V = 224 mL, T = 55 °C, and P = 886 torr **Find:** $\mathfrak{M}$

Conceptual plan: mg → g mL → L °C → K torr → atm then V, m → d then d, P, T → $\mathfrak{M}$

$$\frac{1\ g}{1000\ mg} \quad \frac{1\ L}{1000\ mL} \quad K = °C + 273.15 \quad \frac{1\ atm}{760\ torr} \quad d = \frac{m}{V} \quad d = \frac{P\mathfrak{M}}{RT}$$

Solution: $m = 38.8\ \text{mg} \times \frac{1\ g}{1000\ \text{mg}} = 0.0388\ g$, $V = 224\ \text{mL} \times \frac{1\ L}{1000\ \text{mL}} = 0.224\ L$, $T = 55\ °C + 273.15 = 328\ K$

$P = 886\ \text{torr} \times \frac{1\ atm}{760\ \text{torr}} = 1.1\underline{6}5789\ atm$ $d = \frac{m}{V} = \frac{0.0388\ g}{0.224\ L} = 0.17\underline{3}214\ g/L$ $d = \frac{P\mathfrak{M}}{RT}$

Rearrange to solve for $\mathfrak{M}$. $$\mathfrak{M} = \frac{dRT}{P} = \frac{0.17\underline{3}214\ \frac{g}{L} \times 0.08206 \frac{L\ atm}{K\ mol} \times 328\ K}{1.1\underline{6}5789\ atm} = 4.00\ g/mol$$

Check: The units (g/mol) are correct. The magnitude of the answer (4 g/mol) makes physical sense because this is a reasonable number for a molecular weight of a gas, especially since the density is on the low side.

61. **Given:** P_{N2} = 325 torr, P_{O2} = 124 torr, P_{He} = 209 torr, V = 1.05 L, and T = 25.0 °C

Find: m_{N2}, m_{O2}, m_{He}

Conceptual plan: °C → K and torr → atm and P, V, T → n then mol → g

$$K = °C + 273.15 \quad \frac{1\ atm}{760\ torr} \quad PV = nRT \quad \mathfrak{M}$$

and P_{N2}, P_{O2}, P_{He} → P_{Total}

$$P_{Total} = P_{N2} + P_{O2} + P_{He}$$

Solution: $T_1 = 25.0\ °C + 273.15 = 298.2\ K$, $PV = nRT$ Rearrange to solve for n. $n = \frac{PV}{RT}$

$$P_{N2} = 325\ \text{torr} \times \frac{1\ atm}{760\ \text{torr}} = 0.42\underline{7}632\ atm$$

$$n_{N2} = \frac{0.42\underline{7}632\ atm \times 1.05\ L}{0.08206 \frac{L \cdot atm}{mol \cdot K} \times 298.2\ K} = 0.018\underline{3}493\ mol$$

$$0.018\underline{3}493\ mol \times \frac{28.02\ mol}{1\ mol} = 0.514\ g\ N_2$$

$$P_{O2} = 124\ \text{torr} \times \frac{1\ atm}{760\ \text{torr}} = 0.16\underline{3}158\ atm$$

$$n_{O2} = \frac{0.16\underline{3}158\ atm \times 1.05\ L}{0.08206 \frac{L \cdot atm}{mol \cdot K} \times 298.2\ K} = 0.0070\underline{0}097\ mol$$

$$0.0070\underline{0}097\ mol \times \frac{32.00\ mol}{1\ mol} = 0.224\ g\ O_2$$

$$P_{He} = 209 \text{ torr} \times \frac{1 \text{ atm}}{760 \text{ torr}} = 0.275 \text{ atm} \qquad n_{He} = \frac{0.275 \text{ atm} \times 1.05 \text{ L}}{0.08206 \frac{\text{L} \cdot \text{atm}}{\text{mol} \cdot \text{K}} \times 298.2 \text{ K}} = 0.0118000 \text{ mol}$$

$$0.0118000 \text{ mol} \times \frac{4.003 \text{ mol}}{1 \text{ mol}} = 0.0472 \text{ g He} \quad \text{and}$$

$P_{Total} = P_{N2} + P_{O2} + P_{He} = 0.428 \text{ atm} + 0.163 \text{ atm} + 0.275 \text{ atm} = 0.866 \text{ atm}$

Check: The units (g and atm) are correct. The magnitude of the answer (1 g) makes sense because, gases are not very dense and these pressures are < 1 atm. Since all of the pressures are small, the total is < 1 atm.

63. **Given:** m (CO_2) = 1.20 g, V = 755 mL, P_{N2} = 725 mm Hg, and T = 25.0 °C **Find:** P_{Total}

Conceptual plan: mL → L and °C → K and g → mol and n, P, T → V then atm → mm Hg

$$\frac{1 \text{ L}}{1000 \text{ mL}} \qquad K = °C + 273.15 \qquad \frac{1 \text{ mol}}{44.01 \text{ g}} \qquad PV = nRT \qquad \frac{760 \text{ mm Hg}}{1 \text{ atm}}$$

finally $P_{CO2}, P_{N2} \rightarrow P_{Total}$

$$P_{Total} = P_{CO2} + P_{N2}$$

Solution: $V = 755 \text{ mL} \times \frac{1 \text{ L}}{1000 \text{ mL}} = 0.755 \text{ L}$ T = 25.0 °C + 273.15 = 298.2 K,

$n = 1.20 \text{ g} \times \frac{1 \text{ mol}}{44.01 \text{ g}} = 0.0272665 \text{ mol}$, $PV = nRT$ Rearrange to solve for P.

$$P = \frac{nRT}{V} = \frac{0.0272665 \text{ mol} \times 0.08206 \frac{\text{L} \cdot \text{atm}}{\text{mol} \cdot \text{K}} \times 298.2 \text{ K}}{0.755 \text{ L}} = 0.883735 \text{ atm}$$

$$P_{CO2} = 0.883735 \text{ atm} \times \frac{760 \text{ mm Hg}}{1 \text{ atm}} = 672 \text{ mm Hg},$$

$$P_{Total} = P_{CO2} + P_{N2} = 672 \text{ mm Hg} + 725 \text{ mm Hg} = 1397 \text{ mm Hg}$$

Check: The units (mm Hg) are correct. The magnitude of the answer (1400 mm Hg) makes sense because, it must be greater than 725 mm Hg.

65. **Given:** m (N_2) = 1.25 g, m (O_2) = 0.85 g, V = 1.55 L, and T = 18 °C **Find:** χ_{N2}, χ_{O2}, P_{N2}, P_{O2}

Conceptual plan: g → mol then $n_{N2}, n_{O2} \rightarrow \chi_{N2}$ and $n_{N2}, n_{O2} \rightarrow \chi_{O2}$ °C → K then n, V, T → P

$$\mathfrak{M} \qquad \chi_{N2} = \frac{n_{N2}}{n_{N2} + n_{O2}} \qquad \chi_{O2} = \frac{n_{O2}}{n_{N2} + n_{O2}} \qquad K = °C + 273.15 \qquad PV = nRT$$

Solution: $n_{N2} = 1.25 \text{ g} \times \frac{1 \text{ mol}}{28.02 \text{ g}} = 0.0446110 \text{ mol}$, $n_{O2} = 0.85 \text{ g} \times \frac{1 \text{ mol}}{32.00 \text{ g}} = 0.026563 \text{ mol}$,

T = 18 °C + 273.15 = 291 K, $\chi_{N2} = \frac{n_{N2}}{n_{N2} + n_{O2}} = \frac{0.0446110 \text{ mol}}{0.0446110 \text{ mol} + 0.026563 \text{ mol}} = 0.626792 = 0.627$,

$\chi_{O2} = \frac{n_{O2}}{n_{N2} + n_{O2}} = \frac{0.026563 \text{ mol}}{0.0446110 \text{ mol} + 0.026563 \text{ mol}} = 0.38$ We can also calculate this as

$\chi_{O2} = 1 - \chi_{N2} = 1 - 0.626792 = 0.373208 = 0.373$ $PV = nRT$ Rearrange to solve for P.

$$P = \frac{nRT}{V}$$

$$P_{N2} = \frac{0.044611 \text{ mol} \times 0.08206 \frac{\text{L} \cdot \text{atm}}{\text{mol} \cdot \text{K}} \times 291 \text{ K}}{1.55 \text{ L}} = 0.687 \text{ atm}$$

$$P_{O2} = \frac{0.026563 \text{ mol} \times 0.08206 \frac{\text{L} \cdot \text{atm}}{\text{mol} \cdot \text{K}} \times 291 \text{ K}}{1.55 \text{ L}} = 0.409 \text{ atm}$$

Check: The units (none and atm) are correct. The magnitude of the answers makes sense because, the mole fractions should total 1 and since the weight of N_2 is greater than O_2, its mole fraction is larger. The number of moles is <<1, so we expect the pressures to be <1 atm, given the V (1.55 L).

67. **Given:** T = 30.0 °C, P_{Total} = 732 mm Hg, and V = 722 mL **Find:** P_{H2} and m_{H2}

Conceptual plan: $T \rightarrow P_{H2O}$ **then** $P_{Total}, P_{H2O} \rightarrow P_{H2}$ **then** **mm Hg → atm** **and** **mL → L**

Table 5.4 $\quad P_{Total} = P_{H2O} + P_{H2} \quad \frac{1\text{ atm}}{760\text{ torr}} \quad \frac{1\text{ L}}{1000\text{ mL}}$

and °C → K **P, V, T → n** **then mol → g**

$K = °C + 273.15 \quad PV = nRT \quad \frac{2.016\text{ g}}{1\text{ mol}}$

Solution: Table 5.4 states that P_{H2O} = 31.86 mm Hg $\quad P_{Total} = P_{H2O} + P_{H2}$ Rearrange to solve for P_{H2}.
$P_{H2} = P_{Total} - P_{H2O}$ = 732 mm Hg − 31.86 mm Hg = 700. mm Hg

$$P_{He} = 700.\ \text{torr} \times \frac{1\text{ atm}}{760\text{ torr}} = 0.92\underline{1}052\text{ atm},\ V = 722\text{ mL} \times \frac{1\text{ L}}{1000\text{ mL}} = 0.722\text{ L},\ T = 30.0\ °C + 273.15 = 303.2\text{ K},$$

$PV = nRT$ Rearrange to solve for n. $n = \frac{PV}{RT}$

$$n_{H2} = \frac{0.92\underline{1}052\text{ atm} \times 0.722\text{ L}}{0.08206\ \frac{\text{L}\cdot\text{atm}}{\text{mol}\cdot\text{K}} \times 303.2\text{ K}} = 0.026\underline{7}277\text{ mol}$$

$$0.026\underline{7}277\text{ mol} \times \frac{2.016\text{ g}}{1\text{ mol}} = 0.0539\text{ g } H_2$$

Check: The units (g) are correct. The magnitude of the answer (<< 1 g) makes sense because, gases are not very dense, hydrogen is light, the volume is small and the pressure is ~1 atm.

69. **Given:** T = 25 °C, P_{Total} = 748 mm Hg, and V = 0.951 L **Find:** P_{H2} and m_{H2}
Conceptual plan: $T \rightarrow P_{H2O}$ **then** $P_{Total}, P_{H2O} \rightarrow P_{H2}$ **then** **mm Hg → atm** **and** **mL → L**

Table 5.4 $\quad P_{Total} = P_{H2O} + P_{H2} \quad \frac{1\text{ atm}}{760\text{ torr}} \quad \frac{1\text{ L}}{1000\text{ mL}}$

and °C → K **P, V, T → n** **then mol → g**

$K = °C + 273.15 \quad PV = nRT \quad \frac{2.016\text{ g}}{1\text{ mol}}$

Solution: Table 5.4 states that P_{H2O} = 23.78 mm Hg $\quad P_{Total} = P_{H2O} + P_{H2}$ Rearrange to solve for P_{H2}.
$P_{H2} = P_{Total} - P_{H2O}$ = 748 mm Hg − 23.78 mm Hg = 724 mm Hg

$$P_{H2} = 724\text{ torr} \times \frac{1\text{ atm}}{760\text{ torr}} = 0.95\underline{2}632\text{ atm}$$

T = 25 °C + 273.15 = 298 K, $\quad PV = nRT \quad$ Rearrange to solve for n.

$$n_{H2} = \frac{PV}{RT} = \frac{0.95\underline{2}632\text{ atm} \times 0.951\text{ L}}{0.08206\ \frac{\text{L}\cdot\text{atm}}{\text{mol}\cdot\text{K}} \times 298\text{ K}} = 0.037\underline{0}474\text{ mol} \qquad 0.037\underline{0}474\text{ mol} \times \frac{2.016\text{ g}}{1\text{ mol}} = 0.0747\text{ g } H_2$$

Check: The units (g) are correct. The magnitude of the answer (<< 1 g) makes sense because, gases are not very dense, hydrogen is light, the volume is small and the pressure is ~1 atm.

71. **Given:** m (C) = 15.7 g, P = 1.0 atm, and T = 355 K **Find:** V
Conceptual plan: **g C → mol C → mol H_2** **then** **n (mol H_2), P, T → V**

$\frac{1\text{ mol}}{12.01\text{ g C}} \quad \frac{1\text{ mol } H_2}{1\text{ mol C}} \quad PV = nRT$

Solution: $15.7\text{ g C} \times \frac{1\text{ mol C}}{12.01\text{ g C}} \times \frac{1\text{ mol } H_2}{1\text{ mol C}} = 1.3\underline{0}724\text{ mol } H_2$, $PV = nRT$ Rearrange to solve for V.

$$V = \frac{nRT}{P} = \frac{1.3\underline{0}724\text{ mol} \times 0.08206\ \frac{\text{L}\cdot\text{atm}}{\text{mol}\cdot\text{K}} \times 355\text{ K}}{1.0\text{ atm}} = 38\text{ L}$$

Check: The units (L) are correct. The magnitude of the answer (38 L) makes sense because, we have more than one mole of gas and so we expect more than 22 L.

73. **Given:** P = 748 mm Hg, T = 86 °C, and m (CH_3OH) = 25.8 g, and **Find:** V_{H2} and V_{CO}

Conceptual plan: **g CH_3OH → mol CH_3OH → mol H_2 and mm Hg → atm and °C → K then**

$$\frac{1\text{ mol }CH_3OH}{32.04\text{ g }CH_3OH} \quad \frac{2\text{ mol }H_2}{1\text{ mol }CH_3OH} \quad \frac{1\text{ atm}}{760\text{ torr}} \quad K = °C + 273.15$$

n (mol H_2), P, T → V **and mol H_2 → mol CO** **then** **n (mol CO), P, T → V**

$$PV = nRT \quad \frac{1\text{ mol CO}}{2\text{ mol }H_2} \quad PV = nRT$$

Solution: $25.8\text{ g }CH_3OH \times \frac{1\text{ mol }CH_3OH}{32.04\text{ g }CH_3OH} \times \frac{2\text{ mol }H_2}{1\text{ mol }CH_3OH} = 1.6\underline{1}049\text{ mol }H_2$,

$P_{H2} = 748\text{ torr} \times \frac{1\text{ atm}}{760\text{ torr}} = 0.98\underline{4}211\text{ atm}$, T = 86 °C + 273.15 = 359 K, $PV = nRT$ Rearrange to solve for V.

$$V = \frac{nRT}{P} \quad V_{H2} = \frac{1.6\underline{1}049\text{ mol} \times 0.08206\frac{\text{L}\cdot\text{atm}}{\text{mol}\cdot\text{K}} \times 359\text{ K}}{0.98\underline{4}211\text{ atm}} = 48.2\text{ L }H_2$$

$$1.6\underline{1}049\text{ mol }H_2 \times \frac{1\text{ mol CO}}{2\text{ mol }H_2} = 0.80\underline{5}25\text{ mol CO},$$

$$V_{CO} = \frac{0.80\underline{5}25\text{ mol} \times 0.08206\frac{\text{L}\cdot\text{atm}}{\text{mol}\cdot\text{K}} \times 359\text{ K}}{0.98\underline{4}211\text{ atm}} = 24.1\text{ L CO}$$

Check: The units (L) are correct. The magnitude of the answer (48 L & 24 L) makes sense because, we have more than one mole of hydrogen gas and half that of CO and so we expect significantly more than 22L for hydrogen and half that for CO.

75. **Given:** V = 11.8 L, and STP **Find:** m (NaN_3)
Conceptual plan: **V_{N2} → mol N_2 → mol NaN_3 → g NaN_3**

$$\frac{1\text{ mol }N_2}{22.414\text{ L }N_2} \quad \frac{2\text{ mol }NaN_3}{3\text{ mol }N_2} \quad \frac{65.01\text{ g }NaN_3}{1\text{ mol }NaN_3}$$

Solution: $11.8\text{ L }N_2 \times \frac{1\text{ mol }N_2}{22.414\text{ L }N_2} \times \frac{2\text{ mol }NaN_3}{3\text{ mol }N_2} \times \frac{65.01\text{ g }NaN_3}{1\text{ mol }NaN_3} = 22.8\text{ g }NaN_3$,

Check: The units (g) are correct. The magnitude of the answer (23 g) makes sense because, we have about a half a mole of nitrogen gas, which translates to even fewer moles of NaN_3 and so we expect significantly less than 65 g.

77. **Given:** V_{CH4} = 25.5 L, P_{CH4} = 732 torr, and T = 25 °C; mixed with V_{H2O} = 22.8 L, P_{H2O} = 702 torr, and T = 125 °C; forms P_{H2} = 26.2 L at STP **Find:** % Yield
Conceptual plan: **CH_4: torr → atm and °C → K and P, V, T → n_{CH4} → n_{H2}**

$$\frac{1\text{ atm}}{760\text{ torr}} \quad K = °C + 273.15 \quad PV = nRT \quad \frac{3\text{ mol }H_2}{1\text{ mol }CH_4}$$

H_2O: torr → atm and °C → K and P, V, T → n_{H2O} → n_{H2} Select smaller n_{H2} as theoretical yield.

$$\frac{1\text{ atm}}{760\text{ torr}} \quad K = °C + 273.15 \quad PV = nRT \quad \frac{3\text{ mol }H_2}{1\text{ mol }H_2O}$$

then L_{H2} → mol H_2 (actual yield) finally **actual yield, theoretical yield → % Yield**

$$\frac{1\text{ mol }H_2}{22.414\text{ L }H_2}$$

$$\%\ Yield = \frac{actual\ yield}{theoretical\ yield} \times 100\ \%$$

Solution: CH_4: $P_{N2} = 732 \text{ torr} \times \frac{1 \text{ atm}}{760 \text{ torr}} = 0.96\underline{3}158 \text{ atm}$, $T = 25\ °C + 273.15 = 298\text{ K}$,

$PV = nRT$

Rearrange to solve for n. $n = \frac{PV}{RT}$ $\quad n_{CH4} = \dfrac{0.96\underline{3}158 \text{ atm} \times 25.5 \text{ L}}{0.08206 \frac{\text{L} \cdot \text{atm}}{\text{mol} \cdot \text{K}} \times 298 \text{ K}} = 1.0\underline{0}436 \text{ mol } CH_4$

$$1.0\underline{0}436 \text{ mol } CH_4 \times \frac{3 \text{ mol } H_2}{1 \text{ mol } CH_4} = 3.0\underline{1}308 \text{ mol } H_2$$

H_2O: $P_{N2} = 702 \text{ torr} \times \frac{1 \text{ atm}}{760 \text{ torr}} = 0.92\underline{3}684 \text{ atm}$, $T = 125\ °C + 273.15 = 398\text{ K}$, $\quad n = \frac{PV}{RT}$

$$n_{H2O} = \frac{0.92\underline{3}684 \text{ atm} \times 22.8 \text{ L}}{0.08206 \frac{\text{L} \cdot \text{atm}}{\text{mol} \cdot \text{K}} \times 398 \text{ K}} = 0.64\underline{4}828 \text{ mol } H_2O$$

$$0.64\underline{4}828 \text{ mol } H_2O \times \frac{3 \text{ mol } H_2}{1 \text{ mol } H_2O} = 1.9\underline{3}448 \text{ mol } H_2$$

Water is the limiting reagent since the moles of hydrogen generated is lower.

Theoretical yield = $1.9\underline{3}448$ mol H_2. $26.2 \text{ L } H_2 \times \frac{1 \text{ mol } H_2}{22.414 \text{ L } H_2} = 1.1\underline{6}891 \text{ mol } H_2$ = actual yield,

$$\% \text{ Yield} = \frac{\text{actual yield}}{\text{theoretical yield}} \times 100\ \% = \frac{1.1\underline{6}891 \text{ mol } H_2}{1.9\underline{3}448 \text{ mol } H_2} \times 100\ \% = 60.4\ \%$$

Check: The units (%) are correct. The magnitude of the answer (60 %) makes sense because; it is between 0 and 100 %.

79. a) Yes, since the average kinetic energy of a particle is proportional to the temperature in kelvins and the two gases are at the same temperature, they have the same average kinetic energy.

b) No, since the helium atoms are lighter, they must move faster to have the same kinetic energy as argon atoms.

c) No, since the Ar atoms are moving slower to compensate for their larger mass, they will exert the same pressure on the walls of the container.

d) Since He is lighter, it will have the faster rate of effusion.

81. **Given:** F_2, Cl_2, Br_2, and T = 298 K **Find:** u_{rms} KE_{avg} for each gas and relative rates of effusion

Conceptual plan: $\mathfrak{M}$, T → u_{rms} → KE_{avg}

$$u_{rms} = \sqrt{\frac{3RT}{\mathfrak{M}}} \qquad KE_{avg} = \frac{1}{2} N_A m u_{rms}^2 = \frac{3}{2} RT$$

Solution: F_2: $\mathfrak{M} = \frac{38.00 \text{ g}}{1 \text{ mol}} \times \frac{1 \text{ kg}}{1000 \text{ g}} = 0.03800 \text{ kg/mol}$,

$$u_{rms} = \sqrt{\frac{3RT}{\mathfrak{M}}} = \sqrt{\frac{3 \times 8.314 \frac{\text{J}}{\text{K} \times \text{mol}} \times 298 \text{ K}}{0.03800 \frac{\text{kg}}{\text{mol}}}} = 442 \text{ m/s}$$

Cl_2: $\mathfrak{M} = \frac{70.90 \text{ g}}{1 \text{ mol}} \times \frac{1 \text{ kg}}{1000 \text{ g}} = 0.07090 \text{ kg/mol}$,

$$u_{rms} = \sqrt{\frac{3RT}{\mathfrak{M}}} = \sqrt{\frac{3 \times 8.314 \frac{\text{J}}{\text{K} \times \text{mol}} \times 298 \text{ K}}{0.07090 \frac{\text{kg}}{\text{mol}}}} = 324 \text{ m/s}$$

Br_2: $\mathfrak{M} = \frac{159.80\ \text{g}}{1\ \text{mol}} \times \frac{1\ \text{kg}}{1000\ \text{g}} = 0.15980\ \text{kg/mol}$, $u_{rms} = \sqrt{\frac{3RT}{\mathfrak{M}}} = \sqrt{\frac{3 \times 8.314 \frac{\text{J}}{\text{K} \times \text{mol}} \times 298\ \text{K}}{0.15980 \frac{\text{kg}}{\text{mol}}}} = 216\ \text{m/s}$

All molecules have the same kinetic energy:

$$KE_{avg} = \frac{3}{2}RT = \frac{3}{2} \times 8.314 \frac{\text{J}}{\text{K} \times \text{mol}} \times 298\ \text{K} = 3.72 \times 10^3\ \text{J/mol}$$

Since rate of effusion is proportional to $\sqrt{\frac{1}{\mathfrak{M}}}$, F_2 will have the fastest rate and Br_2 will have the slowest rate.

Check: The units (m/s) are correct. The magnitude of the answer (200 – 450 m/s) makes sense because, they are consistent with what was seen in the text and the heavier the molecule, the slower the molecule.

83. **Given:** $^{238}UF_6$ and $^{235}UF_6$ U-235 = 235.054 amu, U-238 = 238.051 amu
Find: ratio of effusion rates $^{238}UF_6$ / $^{235}UF_6$
Conceptual plan: $\mathfrak{M}$ ($^{238}UF_6$) , $\mathfrak{M}$ ($^{235}UF_6$) → Rate ($^{238}UF_6$) / Rate ($^{235}UF_6$)

$$\frac{Rate(^{238}UF_6)}{Rate(^{235}UF_6)} = \sqrt{\frac{\mathfrak{M}(^{235}UF_6)}{\mathfrak{M}(^{238}UF_6)}}$$

Solution: $^{238}UF_6$: $\mathfrak{M} = \frac{352.05\ \text{g}}{1\ \text{mol}} \times \frac{1\ \text{kg}}{1000\ \text{g}} = 0.35205\ \text{kg/mol}$, $^{235}UF_6$:

$$\mathfrak{M} = \frac{349.05\ \text{g}}{1\ \text{mol}} \times \frac{1\ \text{kg}}{1000\ \text{g}} = 0.34905\ \text{kg/mol},$$

$$\frac{Rate(^{238}UF_6)}{Rate(^{235}UF_6)} = \sqrt{\frac{\mathfrak{M}(^{235}UF_6)}{\mathfrak{M}(^{238}UF_6)}} = \sqrt{\frac{0.34905\ \text{kg/mol}}{0.35205\ \text{kg/mol}}} = 0.99574$$

Check: The units (none) are correct. The magnitude of the answer (<1) makes sense because, they the heavier molecule has the lower effusion rate because it moves slower.

85. **Given:** Ne and unknown gas; and Ne effusion in 76 s and unknown in 155 s **Find:** identify unknown gas
Conceptual plan: $\mathfrak{M}$ (Ne), Rate (Ne), Rate (U) → $\mathfrak{M}$ (Kr)

$$\frac{Rate(Ne)}{Rate(U)} = \sqrt{\frac{\mathfrak{M}(U)}{\mathfrak{M}(Ne)}}$$

Solution: Ne: $\mathfrak{M} = \frac{20.18\ \text{g}}{1\ \text{mol}} \times \frac{1\ \text{kg}}{1000\ \text{g}} = 0.02018\ \text{kg/mol}$, $\frac{Rate(Ne)}{Rate(U)} = \sqrt{\frac{\mathfrak{M}(U)}{\mathfrak{M}(Ne)}}$ Rearrange to solve for $\mathfrak{M}(U)$.

$$\mathfrak{M}(U) = \mathfrak{M}(Ne)\left(\frac{Rate(Ne)}{Rate(U)}\right)^2$$ Since Rate α 1/(effusion time)

$$\mathfrak{M}(U) = \mathfrak{M}(Ne)\left(\frac{Time(U)}{Time(Ne)}\right)^2 = 0.02018 \frac{\text{kg}}{\text{mol}} \times \left(\frac{155\ \text{s}}{76\ \text{s}}\right)^2 = 0.084 \frac{\text{kg}}{\text{mol}} \times \frac{1000\ \text{g}}{1\ \text{kg}} = 84\ \text{g/mol}$$ or Kr.

Check: The units (g/mol) are correct. The magnitude of the answer (>Ne) makes sense because, Ne effused faster and so must be lighter.

87. Gas A has the higher molar mass, since it has the slower velocity. Gas B will have the higher effusion rate, since it has the higher velocity.

89. The postulate that the volume of the gas particles is small compared to the space between them breaks down at high pressure. At high pressures the number of molecules increases, so the volume of the gas particles becomes larger and since the spacing between the particles is smaller, the molecules themselves occupy a significant portion of the volume.

91. **Given:** Ne, n = 1.000 mol, P = 500.0 atm and T = 355.0 K **Find:** V(ideal) and V(van der Waals)

Conceptual plan: **n, P, T → V** **and** **n, P, T → V**

$$P\,V = nRT \qquad \left(P + \frac{an^2}{V^2}\right)(V - nb) = nRT$$

Solution: $P\,V = nRT$ Rearrange to solve for V.

$$V = \frac{nRT}{P} = \frac{1.000\ \text{mol} \times 0.08206\ \frac{\text{L}\cdot\text{atm}}{\text{mol}\cdot\text{K}} \times 355.0\ \text{K}}{500.0\ \text{atm}} = 0.05826\ \text{L} \qquad \left(P + \frac{an^2}{V^2}\right)(V - nb) = nRT$$

Rearrange to solve to: $V = \dfrac{nRT}{\left(P + \dfrac{an^2}{V^2}\right)} + nb$

Using a = 0.211 L^2 atm/mol^2 and b = 0.0171 L/mol from Table 5.5, and the V from the ideal gas law above, solve for V by successive approximations.

$$V = \frac{1.000\ \text{mol} \times 0.08206\ \frac{\text{L}\cdot\text{atm}}{\text{mol}\cdot\text{K}} \times 355.0\ \text{K}}{500.0\ \text{atm} + \dfrac{0.211\frac{\text{L}^2\cdot\text{atm}}{\text{mol}^2} \times (1.000\ \text{mol})^2}{(0.05826\ \text{L})^2}} + \left(1.000\ \text{mol} \times 0.0171\ \frac{\text{L}}{\text{mol}}\right) = 0.068\underline{9}15\ \text{L}$$

Plug in this new value.

$$V = \frac{1.000\ \text{mol} \times 0.08206\ \frac{\text{L}\cdot\text{atm}}{\text{mol}\cdot\text{K}} \times 355.0\ \text{K}}{500.0\ \text{atm} + \dfrac{0.211\frac{\text{L}^2\cdot\text{atm}}{\text{mol}^2} \times (1.000\ \text{mol})^2}{(0.068915\ \text{L})^2}} + \left(1.000\ \text{mol} \times 0.0171\ \frac{\text{L}}{\text{mol}}\right) = 0.070\underline{6}09\ \text{L}$$

Plug in this new value.

$$V = \frac{1.000\ \text{mol} \times 0.08206\ \frac{\text{L}\cdot\text{atm}}{\text{mol}\cdot\text{K}} \times 355.0\ \text{K}}{500.0\ \text{atm} + \dfrac{0.211\frac{\text{L}^2\cdot\text{atm}}{\text{mol}^2} \times (1.000\ \text{mol})^2}{(0.070609\ \text{L})^2}} + \left(1.000\ \text{mol} \times 0.0171\ \frac{\text{L}}{\text{mol}}\right) = 0.070\underline{8}17\ \text{L}$$

Plug in this new value.

$$V = \frac{1.000\ \text{mol} \times 0.08206\ \frac{\text{L}\cdot\text{atm}}{\text{mol}\cdot\text{K}} \times 355.0\ \text{K}}{500.0\ \text{atm} + \dfrac{0.211\frac{\text{L}^2\cdot\text{atm}}{\text{mol}^2} \times (1.000\ \text{mol})^2}{(0.070817\ \text{L})^2}} + \left(1.000\ \text{mol} \times 0.0171\ \frac{\text{L}}{\text{mol}}\right) = 0.070\underline{8}42\ \text{L} = 0.0708\ \text{L}$$

The two values are different because we are at very high pressures. The pressure is corrected from 500.0 atm to 542.1 atm and the final volume correction is 0.0171 L.

Check: The units (L) are correct. The magnitude of the answer (~0.06 L) makes sense because, we are at such a high pressure and have, one mole of gas.

93. **Given:** m (penny) = 2.482 g, T = 25 °C, V = 0.899 L, and P_{Total} = 791 mm Hg **Find:** % Zn in penny

Conceptual plan: **T → P_{H2O} then P_{Total}, P_{H2O} → P_{H2} then mm Hg → atm and °C → K**

Table 5.4 $\qquad P_{Total} = P_{H2O} + P_{H2} \qquad \dfrac{1\ \text{atm}}{760\ \text{torr}} \qquad K = °C + 273.15$

and P, V, T → n_{H2} → n_{Zn} → g_{Zn} → % Zn

$$P\,V = nRT \qquad \frac{1\ \text{mol Zn}}{1\ \text{mol H}_2} \qquad \frac{65.39\ \text{g Zn}}{1\ \text{mol Zn}} \qquad \%\ Zn = \frac{g_{Zn}}{g_{penny}} \times 100\ \%$$

Solution: Table 5.4 states that P_{H2O} = 23.78 mm Hg $P_{Total} = P_{H2O} + P_{H2}$ Rearrange to solve for P_{H2}.
$P_{H2} = P_{Total} - P_{H2O}$ = 791 mm Hg – 23.78 mm Hg = 767 mm Hg

$$P_{H2} = 767\ \text{torr} \times \frac{1\ \text{atm}}{760\ \text{torr}} = 1.0\underline{0}95\ \text{atm}$$

T = 25 °C + 273.15 = 298 K, $PV = nRT$ Rearrange to solve for n.

$$n_{H2} = \frac{PV}{RT} = \frac{1.0\underline{0}95\ \text{atm} \times 0.899\ \text{L}}{0.08206\ \frac{\text{L} \cdot \text{atm}}{\text{mol} \cdot \text{K}} \times 298\ \text{K}} = 0.037\underline{1}123\ \text{mol}$$

$$0.037\underline{1}123\ \text{mol H}_2 \times \frac{1\ \text{mol Zn}}{1\ \text{mol H}_2} \times \frac{65.39\ \text{g Zn}}{1\ \text{mol Zn}} = 2.4\underline{2}677\ \text{g Zn}$$

$$\%\ Zn = \frac{g_{Zn}}{g_{penny}} \times 100\ \% = \frac{2.4\underline{2}677\ \text{g}}{2.482\ \text{g}} \times 100\ \% = 97.8\ \%\ \text{Zn}$$

Check: The units (% Zn) are correct. The magnitude of the answer (98 %) makes sense because, it should be between 0 and 100 %. We expect about 1/22 a mole of gas, since our conditions are close to STP and we have ~ 1 L of gas.

95. **Given:** V = 255 mL, m (flask) = 143.187 g, m (flask + gas) = 143.289 g, P = 267 torr, and T = 25 °C
Find: $\mathfrak{M}$
Conceptual plan: **°C → K**   **torr → atm**   **mL → L**   **m (flask), m (flask + gas) → m (gas)**

$K = °C + 273.15$ $\frac{1\ \text{atm}}{760\ \text{torr}}$ $\frac{1\ \text{L}}{1000\ \text{mL}}$ $m\ (gas) = m\ (flask + gas) - m\ (flask)$

then V, m → d then d, P, T, → $\mathfrak{M}$

$d = \frac{m}{V}$ $d = \frac{P\mathfrak{M}}{RT}$

Solution: T = 25 °C + 273.15 = 298 K, $P = 267\ \text{torr} \times \frac{1\ \text{atm}}{760\ \text{torr}} = 0.35\underline{1}216\ \text{atm}$,

$$V = 255\ \text{mL} \times \frac{1\ \text{L}}{1000\ \text{mL}} = 0.255\ \text{L},$$

$$m\ (gas) = m\ (flask + gas) - m\ (flask) = 143.289\ \text{g} - 143.187\ \text{g} = 0.102\ \text{g},$$

$d = \frac{m}{V} = \frac{0.102\ \text{g}}{0.255\ \text{L}} = 0.400\ \text{g/L}$, $d = \frac{P\mathfrak{M}}{RT}$ Rearrange to solve for $\mathfrak{M}$.

$$\mathfrak{M} = \frac{dRT}{P} = \frac{0.400\ \frac{\text{g}}{\text{L}} \times 0.08206\ \frac{\text{L}\ \text{atm}}{\text{K}\ \text{mol}} \times 298\ \text{K}}{0.35\underline{1}216\ \text{atm}} = 27.9\ \text{g/mol}$$

Check: The units (g/mol) are correct. The magnitude of the answer (28 g/mol) makes physical sense because this is a reasonable number for a molecular weight of a gas.

97. **Given:** V = 158 mL, m (gas) = 0.275 g,, P = 556 mm Hg, T = 25 °C, gas = 82.666 % C and 17.34 % H.
Find: $\mathfrak{M}$

Conceptual plan: **°C → K mm Hg → atm mL → L then V, m → d then d, P, T, →** $\mathfrak{M}$

$K = °C + 273.15$ $\frac{1\ \text{atm}}{760\ \text{mm Hg}}$ $\frac{1\ \text{L}}{1000\ \text{mL}}$ $d = \frac{m}{V}$ $d = \frac{P\mathfrak{M}}{RT}$

then % C, % H, $\mathfrak{M}$ **→ formula**

$$\#C = \frac{\mathfrak{M}\ 0.8266\ g\ C}{12.01\ \frac{g\ C}{mol\ C}} \quad \#H = \frac{\mathfrak{M}\ 0.1734\ g\ H}{1.008\ \frac{g\ H}{mol\ H}}$$

Solution: T = 25 °C + 273.15 = 298 K, $P = 556\ \text{torr} \times \frac{1\ \text{atm}}{760\ \text{torr}} = 0.73\underline{1}579\ \text{atm}$,

$V = 158\ \text{mL} \times \frac{1\ \text{L}}{1000\ \text{mL}} = 0.158\ \text{L}$, $d = \frac{m}{V} = \frac{0.275\ \text{g}}{0.158\ \text{L}} = 1.7\underline{4}051\ \text{g/L}$, $d = \frac{P\mathfrak{M}}{RT}$ Rearrange to solve for $\mathfrak{M}$.

$$\mathfrak{M} = \frac{dRT}{P} = \frac{1.7\underline{4}051\ \frac{\text{g}}{\text{L}} \times 0.08206 \frac{\text{L atm}}{\text{K mol}} \times 298\ \text{K}}{0.73\underline{1}579\ \text{atm}} = 58.2\ \text{g/mol},$$

$$\#C = \frac{\mathfrak{M} \times 0.8266\ g\ C}{12.01 \frac{g\ C}{mol\ C}} = \frac{58.2 \frac{\text{g HC}}{\text{mol HC}} \times \frac{0.8266\ \text{g C}}{1\ \text{g HC}}}{12.01 \frac{\text{g C}}{\text{mol C}}} = 4.00 \frac{\text{mol C}}{\text{mol HC}}$$

$$\#H = \frac{\mathfrak{M} \times 0.1734\ g\ H}{1.008 \frac{g\ H}{mol\ H}} = \frac{58.2 \frac{\text{g HC}}{\text{mol HC}} \times \frac{0.1734\ \text{g H}}{1\ \text{g HC}}}{1.008 \frac{\text{g H}}{\text{mol H}}} = 10.0 \frac{\text{mol H}}{\text{mol HC}}$$

Formula is C_4H_{10} or butane.

Check: The answer came up with integer number of C and H atoms in the formula and a molecular weight (58 g/mol) that is reasonable for a gas.

99. **Given:** m (NiO) = 24.78 g, T = 40.0 °C, and P_{Total} = 745 mm Hg **Find:** V_{O2}

Conceptual plan: T → P_{H2O} then P_{Total}, P_{H2O} → P_{O2} then mm Hg → atm and °C → K

Table 5.4 ; $P_{Total} = P_{H2O} + P_{O2}$; $\frac{1\ \text{atm}}{760\ \text{torr}}$; $K = °C + 273.15$

and g_{NiO} → n_{NiO} → n_{O2} **then P, V, T →** n_{O2}

$\frac{1\ \text{mol NiO}}{74.69\ \text{g NiO}}$ $\frac{1\ \text{mol } O_2}{2\ \text{mol NiO}}$; $PV = nRT$

Solution: Table 5.4 states that P_{H2O} = 55.40 mm Hg $P_{Total} = P_{H2O} + P_{O2}$ Rearrange to solve for P_{O2}.
$P_{O2} = P_{Total} - P_{H2O}$ = 745 mm Hg − 55.40 mm Hg = 689.$\underline{6}$ mm Hg

$P_{O2} = 689.\underline{6}\ \text{torr} \times \frac{1\ \text{atm}}{760\ \text{torr}} = 0.90\underline{7}368\ \text{atm}$ T = 40.0 °C + 273.15 = 313.2 K,

$24.78\ \text{g NiO} \times \frac{1\ \text{mol NiO}}{74.69\ \text{mol NiO}} \times \frac{1\ \text{mol } O_2}{2\ \text{mol NiO}} = 0.165\underline{8}857\ \text{mol } O_2$ $PV = nRT$ Rearrange to solve for V.

$$V_{O2} = \frac{nRT}{P} = \frac{0.165\underline{8}857\ \text{mol} \times 0.08206 \frac{\text{L} \cdot \text{atm}}{\text{mol} \cdot \text{K}} \times 313.2\ \text{K}}{0.90\underline{7}368\ \text{atm}} = 4.70\ \text{L}$$

Check: The units (L) are correct. The magnitude of the answer (5 L) makes sense because, we have much less than 0.5 mole of NiO, so we get less than a mole of oxygen. Thus we expect a volume much less than 22 L.

101. **Given:** HCl, K_2S to H_2S, V_{H2S} = 42.9 mL, P_{H2S} = 752 mm Hg, and T = 25.8 °C **Find:** m(K_2S)

Conceptual plan: read description of reaction and convert words to equation then °C → K and ($K = °C + 273.15$)

mm Hg → atm and mL → L then P, V, T → n_{H2S} → n_{K2S} → g_{K2S}

$\frac{1\ \text{atm}}{760\ \text{torr}}$; $\frac{1\ \text{L}}{1000\ \text{mL}}$; $PV = nRT$; $\frac{1\ \text{mol } K_2S}{1\ \text{mol } H_2S}$; $\frac{1\ \text{mol } K_2S}{110.27\ \text{g } K_2S}$

Solution: $2\ HCl\ (aq) + K_2S\ (s) \rightarrow H_2S\ (g) + 2\ KCl\ (aq)$

T = 25.8 °C + 273.15 = 299.0 K, $P_{H2S} = 752\ \text{torr} \times \frac{1\ \text{atm}}{760\ \text{torr}} = 0.98\underline{9}474\ \text{atm}$,

$V_{H2S} = 42.9\ \text{mL} \times \frac{1\ \text{L}}{1000\ \text{mL}} = 0.0429\ \text{L}$ $PV = nRT$ Rearrange to solve for n_{H2S}.

$$n_{H2S} = \frac{PV}{RT} = \frac{0.98\underline{9}474\ \text{atm} \times 0.0429\ \text{L}}{0.08206 \frac{\text{L} \cdot \text{atm}}{\text{mol} \cdot \text{K}} \times 299.0\ \text{K}} = 0.0017\underline{3}003\ \text{mol}$$

$$0.00173003 \text{ mol } H_2S \times \frac{1 \text{ mol } K_2S}{1 \text{ mol } H_2S} \times \frac{110.27 \text{ g } K_2S}{1 \text{ mol } K_2S} = 0.191 \text{ g } K_2S$$

Check: The units (g) are correct. The magnitude of the answer (0.1 g) makes sense because, we have such a small volume of gas generated.

103. **Given:** T = 22 °C, P = 1.02 atm, and m = 11.82 g **Find:** V_{Total}

Conceptual plan: **°C → K and $g_{(NH4)2CO3}$ → $n_{(NH4)2CO3}$ → n_{Gas} then P, n, T → V**

$K = °C + 273.15$ $\quad \frac{1 \text{ mol } (NH_4)_2CO_3}{96.09 \text{ g } (NH_4)_2CO_3} \quad \frac{(2+1+1=4) \text{ mol gas}}{1 \text{ mol } NH_4CO_3} \quad PV = nRT$

Solution: T = 22 °C + 273.15 = 295 K,

$$11.83 \text{ g } NH_4CO_3 \times \frac{1 \text{ mol } (NH_4)_2CO_3}{96.09 \text{ g } (NH_4)_2CO_3} \times \frac{4 \text{ mol gas}}{1 \text{ mol } (NH_4)_2CO_3} = 0.492434 \text{ mol gas}$$

$PV = nRT$ Rearrange to solve for V_{gas}.

$$V_{gas} = \frac{nRT}{P} = \frac{0.492434 \text{ mol gas} \times 0.08206 \frac{L \cdot atm}{mol \cdot K} \times 295 \text{ K}}{1.02 \text{ atm}} = 11.7 \text{ L}$$

Check: The units (L) are correct. The magnitude of the answer (12 L) makes sense because, we have about a half a mole of gas generated.

105. **Given:** He and air; V = 855 mL, P = 125 psi, T = 25 °C, $\mathfrak{M}$ (air) = 28.8 g/mol **Find:** Δ = m(air) – m(He)

Conceptual Plan: **mL → L and psi → atm and °C → K then P, T, $\mathfrak{M}$ → d**

$\frac{1 \text{ L}}{1000 \text{ mL}} \quad \frac{1 \text{ atm}}{14.7 \text{ psi}} \quad K = °C + 273.15 \quad d = \frac{P\mathfrak{M}}{RT}$

then d, V → m m(air), m(He) → Δ

$d = \frac{m}{V} \quad \Delta = m(air) - m(He)$

Solution: $V = 855 \text{ mL} \times \frac{1 \text{ L}}{1000 \text{ mL}} = 0.855 \text{ L}$, $P = 125 \text{ psi} \times \frac{1 \text{ atm}}{14.7 \text{ psi}} = 8.50340 \text{ atm}$,

$T = 25 °C + 273.15 = 298K$, $d_{air} = \frac{P\mathfrak{M}}{RT} = \frac{8.50340 \text{ atm} \times 28.8 \frac{\text{g air}}{\text{mol air}}}{0.08206 \frac{L \times atm}{K \times mol} \times 298 \text{ K}} = 10.0147 \frac{\text{g air}}{L}$, $d = \frac{m}{V}$

Rearrange to solve for m. $m = dV$

$$m_{air} = 10.0147 \frac{\text{g air}}{L} \times 0.8554 \text{ L} = 8.56257 \text{ g air},$$

$$d_{He} = \frac{P\mathfrak{M}}{RT} = \frac{8.50340 \text{ atm} \times 4.03 \frac{\text{g He}}{\text{mol He}}}{0.08206 \frac{L \times atm}{K \times mol} \times 298 \text{ K}} = 1.40136 \frac{\text{g He}}{L},$$

$$m_{air} = 1.40136 \frac{\text{g He}}{L} \times 0.855 \text{ L} = 1.19816 \text{ g He},$$

$$\Delta = m(air) - m(He) = 8.56257 \text{ g air} - 1.19816 \text{ g He} = 7.36 \text{ g}$$

Check: The units (g) are correct. We expect the difference to be less than the difference in the molecular weights since we have less than a mole of gas.

107. **Given:** $Flow_{NO}$ = 335 L/s, P_{NO} = 22.4 torr, T_{NO} = 955 K, P_{NH3} = 755 torr, and T_{NO} = 298 K, and NH_3 purity = 65.2 % **Find:** $Flow_{NH3}$

Conceptual plan: torr → atm then P_{NO}, V_{NO}/s, T_{NO} → n_{NO}/s → n_{NH3}/s (pure) → n_{NH3}/s (impure)

$$\frac{1\text{ atm}}{760\text{ torr}} \qquad PV = nRT \qquad \frac{4\text{ mol NH}_3}{4\text{ mol NO}} \qquad \frac{100\text{ mol NH}_3\text{ impure}}{65.2\text{ mol NH}_3\text{ pure}}$$

then n_{NH3}/s (impure), P_{NH3}, T_{NH3} → V_{NH3}/s

$$PV = nRT$$

Solution: $P_{NO} = 22.4\ \cancel{\text{torr}} \times \frac{1\text{ atm}}{760\ \cancel{\text{torr}}} = 0.02947\underline{3}7\text{ atm}$, $P_{NH3} = 755\ \cancel{\text{torr}} \times \frac{1\text{ atm}}{760\ \cancel{\text{torr}}} = 0.99\underline{3}421\text{ atm}$

$PV = nRT$ Rearrange to solve for n_{NO}. Note that we can substitute V/s for V and get n/s as a result.

$$\frac{n_{NO}}{s} = \frac{PV}{RT} = \frac{0.0294737\ \cancel{\text{atm}} \times 335\ \cancel{\text{L}}/\text{s}}{0.08206\ \frac{\cancel{\text{L}}\cdot\cancel{\text{atm}}}{\text{mol}\cdot\cancel{\text{K}}} \times 955\ \cancel{\text{K}}} = 0.12\underline{5}992\ \frac{\text{mol NO}}{\text{s}}$$

$$0.12\underline{5}992\ \frac{\cancel{\text{mol NO}}}{\text{s}} \times \frac{4\ \cancel{\text{mol NH}_3}}{4\ \cancel{\text{mol NO}}} \times \frac{100\text{ mol NH}_3\text{ impure}}{65.2\ \cancel{\text{mol NH}_3\text{ pure}}} = 0.19\underline{3}240\ \frac{\text{mol NH}_3\text{ impure}}{\text{s}} \qquad PV = nRT$$

Rearrange to solve for V_{NH3}. Note that we can substitute n/s for n and get V/s as a result.

$$\frac{V_{NH3}}{s} = \frac{nRT}{P} = \frac{0.19\underline{3}240\ \frac{\cancel{\text{mol NH}_3\text{ impure}}}{\text{s}} \times 0.08206\ \frac{\text{L}\cdot\cancel{\text{atm}}}{\cancel{\text{mol}}\cdot\cancel{\text{K}}} \times 298\ \cancel{\text{K}}}{0.99\underline{3}421\ \cancel{\text{atm}}} = 4.76\ \frac{\text{L}}{\text{s}}\text{ impure NH}_3$$

Check: The units (L) are correct. The magnitude of the answer (5 L/s) makes sense because, we expect it to be less than for the NO. The NO is at a very low concentration and a high temperature, when this converts to a much higher pressure and lower temperature this will go down significantly, even though the ammonia is impure. From a practical standpoint, you would like a low flow rate to make it economical.

109. **Given:** l = 30.0 cm, w = 20.0 cm, h = 15.0 cm, 14.7 psi **Find:** Force (lbs)
Conceptual plan: l, w, h → Surface Area(cm^2) → Surface Area(in^2) → Force

$$SA = 2(lxh) + 2(wxh) + 2(lxw) \qquad \frac{(1\text{ in})^2}{(2.54\text{ cm})^2} \qquad \frac{14.7\text{ lbs}}{1\text{ in}^2}$$

Solution:

$$SA = 2(lxh) + 2(wxh) + 2(lxw) = 2(30.0\text{ cm x }15.0\text{ cm}) + 2(20.0\text{ cm x }15.0\text{ cm}) + 2(30.0\text{ cm x }20.0\text{ cm}) = 2700$$

$$27\underline{0}0\ \cancel{\text{cm}^2} \times \frac{(1\text{ in})^2}{(2.54\ \cancel{\text{cm}})^2} = 41\underline{8}.50\text{ in}^2,\quad 41\underline{8}.50\ \cancel{\text{in}^2} \times \frac{14.7\text{ lbs}}{1\ \cancel{\text{in}^2}} = 6150\text{ lbs}$$ The can would be crushed.

Check: The units (lbs) are correct. The magnitude of the answer (6150 lbs) is not unreasonable since there is a large surface area.

111. **Given:** V_1 = 160.0 L, P_1 =1855 psi, 3.5 L/balloon, P_2 = 1.0 atm = 14.7 psi and T = 298 K **Find:** # balloons
Conceptual plan: V_1, P_1, P_2 → V_2 then L → # balloons

$$P_1V_1 = P_2V_2 \qquad \frac{1\text{ balloon}}{3.5\text{ L}}$$

Solution: $P_1V_1 = P_2V_2$ Rearrange to solve for V_2. $V_2 = \frac{P_1}{P_2}V_1 = \frac{1855\ \cancel{\text{psi}}}{14.7\ \cancel{\text{psi}}} \times 160.0\text{ L} = 20\underline{1}90.5\text{ L}$,

$$20\underline{1}90.5\ \cancel{\text{L}} \times \frac{1\text{ balloon}}{3.5\ \cancel{\text{L}}} = 5800\text{ balloons}$$

Check: The units (balloons) are correct. The magnitude of the answer (5800) is reasonable since a store does not want to buy a new helium tank very often.

113. **Given:** r_1 = 2.5 cm, P_1 = 4.00 atm. T = 298 K, P_2 = 1.00 atm **Find:** r_2
Conceptual plan: r_1 → V_1 V_1, P_1, P_2 → V_2 then V_2 → r_2

$$V = \frac{4}{3}\pi r^3 \qquad P_1V_1 = P_2V_2 \qquad V = \frac{4}{3}\pi r^3$$

Solution: $V = \frac{4}{3}\pi r^3 = \frac{4}{3}$ x π x (2.5 cm)^3 = 65.450 cm^3 $P_1V_1 = P_2V_2$ Rearrange to solve for V_2.

$V_2 = \frac{P_1}{P_2}V_1 = \frac{4.00 \text{ atm}}{1.00 \text{ atm}}$ x 65.450 cm^3 = 261.80 cm^3, $V = \frac{4}{3}\pi r^3$ Rearrange to solve for r.

$r = \sqrt[3]{\frac{3V}{4\pi}} = \sqrt[3]{\frac{3\text{x } 261.80 \text{ cm}^3}{4 \text{ x } \pi}} = 4.0 \text{ cm}$

Check: The units (cm) are correct. The magnitude of the answer (4 cm) is reasonable since the bubble will expand as the pressure is decreased.

115. **Given:** 2.0 mol CO :1.0 mol O_2, V = 2.45 L, P_1 = 745 torr, P_2 = 552 torr, and T = 5552 °C **Find:** % reacted
Conceptual plan: from PV = nRT we know that P α n, looking at the chemical reaction we see that 2 + 1 = 3 moles of gas gets converted to 2 moles of gas. If all the gas reacts, P_2 = 2/3 P_1.
Calculate –ΔP for 100 % reacted and for actual case. Then calculate % reacted.

$-\Delta P \text{ 100 \% reacted} = P_1 - \frac{2}{3}P_1 \qquad -\Delta P \text{ actual} = P_1 - P_2 \qquad \% \text{ reacted} = \frac{\Delta P \text{ actual}}{\Delta P \text{ 100\% reacted}} \text{ x } 100 \%$

Solution: $-\Delta P \text{ 100 \% reacted} = P_1 - \frac{2}{3}P_1 = 745 \text{ torr} - \frac{2}{3} 745 \text{ torr} = 248.333 \text{ torr}$,

$-\Delta P \text{ actual} = P_1 - P_2 = 745 \text{ torr} - 552 \text{ torr} = 193 \text{ torr}$,

$$\% \text{ reacted} = \frac{\Delta P \text{ actual}}{\Delta P \text{ 100\% reacted}} \text{ x } 100 \% = \frac{193 \text{ torr}}{248.333 \text{ torr}} \text{ x } 100 \% = 77.7 \%$$

Check: The units (%) are correct. The magnitude of the answer (78 %) makes sense because; the pressure dropped most of the way to the pressure if all of the reactants had reacted. Note: There are many ways to solve this problem, including calculating the moles of reactants and products using PV = nRT.

117. **Given:** $P(\text{Total})_1$ = 2.2 atm = CO + O_2, $P(\text{Total})_2$ = 1.9 atm = CO + O_2 + CO_2, V = 1.0 L, T = 1.0 x 10^3 K
Find: CO_2 made
Conceptual plan: $P(\text{Total})_1$ = 2.2 atm = $P(CO)_1$ + $P(O_2)_1$, $P(\text{Total})_2$ = 1.9 atm = $P(CO)_2$ + $P(O_2)_2$ + $P(CO_2)_2$, Let x = amount of $P(O_2)$ reacted. From stoichiometry: $P(CO)_2$ = $P(CO)_1$ – 2x, $P(O_2)_2$ = $P(O_2)_1$ – x, $P(CO_2)_2$ = 2x. Thus $P(\text{Total})_2$ = 1.9 atm = $P(CO)_1$ – 2x + $P(O_2)_1$ – x + 2x = $P(\text{Total})_1$ x. Using the initial conditions: 1.9 atm = 2.2 atm – x. So x = 0.3 atm and since 2x = $P(CO_2)_2$ = 0.6 atm then P, V, T → n

$PV = nRT$

Solution: $PV = nRT$ Rearrange to solve for n. $n = \frac{PV}{RT} = \frac{0.6 \text{ atm x } 1.0 \text{ L}}{0.08206 \frac{\text{L} \cdot \text{atm}}{\text{mol} \cdot \text{K}} \text{ x } 1000 \text{ K}} = 0.007 \text{ mol}$

Check: The units (mol) are correct. The magnitude of the answer (0.007 mol) makes sense because, we have such a small volume, at a very high temperature and such a small pressure. All of these lead us to expect a very small number of moles.

119. **Given:** h_1 = 22.6 m, T_1 = 22 °C, and h_2 = 23.8 m **Find:** T_2
Conceptual Plan: °C →K since $V_{cylinder}$ α h we do not need to know r to use V_1, T_1, T_2 → V_2

$K = °C + 273.15 \qquad V = \pi r^2 h \qquad \frac{V_1}{T_1} = \frac{V_2}{T_2}$

Solution: T_1 = 22 °C + 273.15 = 295K, $\frac{V_1}{T_1} = \frac{V_2}{T_2}$ Rearrange to solve for T_2.

$$T_2 = T_1 \text{ x } \frac{V_2}{V_1} = T_1 \text{ x } \frac{\pi r^2 l_2}{\pi r^2 l_1} = 295 \text{ K x } \frac{23.8 \text{ m}}{22.6 \text{ m}} = 311 \text{ K}$$

Check: The units (K) are correct. We expect the temperature to increase since the volume increased.

121. **Given:** He, V = 0.35 L, P_{max} = 88 atm, and T = 298 K **Find:** m_{He}
Conceptual plan: P, V, T → n then mol → g

$PV = nRT \qquad \mathfrak{M}$

Solution: $PV = nRT$ Rearrange to solve for n.

$$n_{He} = \frac{PV}{RT} = \frac{88 \text{ atm} \times 0.35 \text{ L}}{0.08206 \frac{\text{L} \cdot \text{atm}}{\text{mol} \cdot \text{K}} \times 298 \text{ K}} = 1.2595 \text{ mol}$$

$$1.2595 \text{ mol} \times \frac{4.003 \text{ g}}{1 \text{ mol}} = 5.0 \text{ g He}$$

Check: The units (g) are correct. The magnitude of the answer (5 g) makes sense because, the high pressure and the low volume cancel out (remember 22 L/mol at STP) and so we expect ~ 1 mol and so ~ 4 g.

123. **Given:** 15.0 mL HBr in 1.0 min; and 20.3 mL unknown hydrocarbon gas in 1.0 min
Find: formula of unknown gas
Conceptual plan: Since these are gases under the same conditions V α n, V, time → Rate then

$$Rate = \frac{V}{time}$$

$\mathfrak{M}$ (HBr), Rate (HBr), Rate (U) → $\mathfrak{M}$ (U)

$$\frac{Rate(HBr)}{Rate(U)} = \sqrt{\frac{\mathfrak{M}(U)}{\mathfrak{M}(HBr)}}$$

Solution: $Rate(HBr) = \frac{V}{time} = \frac{15.0 \text{ mL}}{1.0 \text{ min}} = 15.0 \frac{\text{mL}}{\text{min}}$ $Rate(U) = \frac{V}{time} = \frac{20.3 \text{ mL}}{1.0 \text{ min}} = 20.3 \frac{\text{mL}}{\text{min}}$,

$\frac{Rate(HBr)}{Rate(U)} = \sqrt{\frac{\mathfrak{M}(U)}{\mathfrak{M}(HBr)}}$ Rearrange to solve for $\mathfrak{M}(U)$.

$$\mathfrak{M}(U) = \mathfrak{M}(HBr)\left(\frac{Rate(HBr)}{Rate(U)}\right)^2 = 80.91 \frac{\text{g}}{\text{mol}} \times \left(\frac{15.0 \frac{\text{mL}}{\text{min}}}{20.3 \frac{\text{mL}}{\text{min}}}\right)^2 = 44.2 \frac{\text{g}}{\text{mol}}$$

, formula is C_3H_8, propane

Check: The units (g/mol) are correct. The magnitude of the answer (< HBr) makes sense because, the unknown diffused faster and so must be lighter.

125. **Given:** CH_4: V = 155 mL at STP; O_2: V = 885 mL at STP; NO: V = 55.5 mL at STP; mixed is a flask: V = 2.0 L, T = 275 K, and 90.0 % of limiting reagent used. **Find:** P's of all components and P_{Total}.

Conceptual plan: CH_4: mL → L → mol_{CH4} → mol_{CO2} and O_2: mL → L → mol_{O2} → mol_{CO2}

$\frac{1 \text{ L}}{1000 \text{ mL}}$ $\frac{1 \text{ mol}}{22.414 \text{ L}}$ $\frac{1 \text{ mol } CO_2}{5 \text{ mol NO}}$ $\frac{1 \text{ L}}{1000 \text{ mL}}$ $\frac{1 \text{ mol}}{22.414 \text{ L}}$ $\frac{1 \text{ mol } CO_2}{5 \text{ mol } O_2}$

and NO: mL → L → mol_{NO} → mol_{CO2}

$\frac{1 \text{ L}}{1000 \text{ mL}}$ $\frac{1 \text{ mol}}{22.414 \text{ L}}$ $\frac{1 \text{ mol } CO_2}{5 \text{ mol NO}}$

the smallest yield to determines the limiting reagent then initial mol_{NO} → reacted mol_{NO} → final mol_{NO}

NO is the limiting reagent 90.0 % 0.100 x initial mol_{no}

reacted mol_{NO} → reacted mol_{CH4} then initial mol_{CH4}, reacted mol_{CH4} → final mol_{CH4} then

$\frac{1 \text{ mol } CH_4}{5 \text{ mol NO}}$ initial mol_{CH4} – reacted mol_{CH4} = final mol_{CH4}

final mol_{CH4}, V, T → final P_{CH4} and reacted mol_{NO} → reacted mol_{O2} then

$PV = nRT$ $\frac{5 \text{ mol } O_2}{5 \text{ mol NO}}$

initial mol_{O2}, reacted mol_{O2} → final mol_{O2} then final mol_{O2}, V, T → final P_{O2} and

initial mol_{O2} – reacted mol_{O2} = final mol_{O2} $PV = nRT$

final mol_{NO}, V, T → final P_{NO} and theoretical mol_{CO2} from NO → final mol_{CO2}

$PV = nRT$ 90.0 %

final mol_{CO2}, V, T → P_{CO2} then final mol_{CO2}→ mol_{H2O} then mol_{H2O},V, T → P_{H2O} and

$PV = nRT$ $\frac{1 \text{ mol } H_2O}{1 \text{ mol } CO_2}$ $PV = nRT$

final mol$_{CO2}$ → mol$_{NO2}$ then mol$_{NO2}$, V, T → P$_{NO2}$ and final mol$_{CO2}$ → mol$_{OH}$ then

$$\frac{1\text{ mol NO}_2}{1\text{ mol CO}_2} \qquad PV = nRT \qquad \frac{2\text{ mol OH}}{1\text{ mol CO}_2}$$

mol$_{OH}$, V, T→ P$_{OH}$ finally P$_{CH4}$, P$_{O2}$, P$_{NO}$, P$_{CO2}$, P$_{H2O}$, P$_{NO2}$, P$_{OH}$ →P$_{Ttoal}$

$$PV = nRT \qquad P_{Total} = \sum P$$

Solution: CH₄: $155\text{ mL} \times \frac{1\text{ L}}{1000\text{ mL}} \times \frac{1\text{ mol CH}_4}{22.414\text{ L}} \times \frac{1\text{ mol CO}_2}{1\text{ mol CH}_4} = 0.0069\underline{1}532\text{ mol CO}_2$,

O₂: $885\text{ mL} \times \frac{1\text{ L}}{1000\text{ mL}} \times \frac{1\text{ mol O}_2}{22.414\text{ L}} = 0.039\underline{4}842\text{ mol O}_2 \times \frac{1\text{ mol CO}_2}{5\text{ mol O}_2} = 0.0078\underline{9}685\text{ mol CO}_2$

NO: $55.5\text{ mL} \times \frac{1\text{ L}}{1000\text{ mL}} \times \frac{1\text{ mol NO}}{22.414\text{ L}} \times \frac{1\text{ mol CO}_2}{5\text{ mol NO}} = 0.00049\underline{5}226\text{ mol CO}_2$.

$0.00049\underline{5}226\text{ mol CO}_2$ is the smallest yield, so NO is the limiting reagent.

$$55.5\text{ mL} \times \frac{1\text{ L}}{1000\text{ mL}} \times \frac{1\text{ mol NO}}{22.414\text{ L}} = 0.0024\underline{7}613\text{ mol NO}$$

reacted mol NO = 0.900 *x mol NO* = 0.900 x 0.0024$\underline{7}$613 mol NO = 0.0022$\underline{2}$852 mol NO ,

unreacted mol NO = 0.100 *x mol NO* = 0.100 x 0.0024$\underline{7}$613 mol NO = 0.00024$\underline{7}$613 mol NO ,

$$0.0022\underline{2}852\text{ mol NO} \times \frac{1\text{ mol CH}_4}{5\text{ mol NO}} = 0.00044\underline{5}704\text{ mol CH}_4\text{ reacted},$$

$0.0069\underline{1}532\text{ mol CH}_4 - 0.00044\underline{5}704\text{ mol CH}_4\text{ reacted} = 0.0064\underline{6}962\text{ mol CH}_4$ then $PV = nRT$

Rearrange to solve for P. $P = \frac{nRT}{V} = \frac{0.0064\underline{6}962\text{ mol} \times 0.08206\,\frac{\text{L}\cdot\text{atm}}{\text{mol}\cdot\text{K}} \times 275\text{ K}}{2.0\text{ L}} = 0.0730\text{ atm CH}_4$

$$0.0022\underline{2}852\text{ mol NO} \times \frac{5\text{ mol O}_2}{5\text{ mol NO}} = 0.0022\underline{2}852\text{ mol O}_2\text{ reacted},$$

$$0.039\underline{4}842\text{ mol O}_2 - 0.0022\underline{2}852\text{ mol O}_2\text{ reacted} = 0.037\underline{2}557\text{ mol O}_2$$

$$P = \frac{nRT}{V} = \frac{0.037\underline{2}557\text{ mol} \times 0.08206\,\frac{\text{L}\cdot\text{atm}}{\text{mol}\cdot\text{K}} \times 275\text{ K}}{2.0\text{ L}} = 0.420\text{ atm O}_2$$

$$P = \frac{nRT}{V} = \frac{0.00024\underline{7}613\text{ mol} \times 0.08206\,\frac{\text{L}\cdot\text{atm}}{\text{mol}\cdot\text{K}} \times 275\text{ K}}{2.0\text{ L}} = 0.00279\text{ atm NO},$$

$$0.0022\underline{2}852\text{ mol NO} \times \frac{1\text{ mol CO}_2}{5\text{ mol NO}} = 0.00044\underline{5}704\text{ mol CO}_2$$

$$P = \frac{nRT}{V} = \frac{0.00044\underline{5}704\text{ mol} \times 0.08206\,\frac{\text{L}\cdot\text{atm}}{\text{mol}\cdot\text{K}} \times 275\text{ K}}{2.0\text{ L}} = 0.00503\text{ atm CO}_2$$

$$0.0022\underline{2}852\text{ mol NO} \times \frac{1\text{ mol H}_2\text{O}}{5\text{ mol NO}} = 0.00044\underline{5}704\text{ mol H}_2\text{O}$$

$$P = \frac{nRT}{V} = \frac{0.00044\underline{5}704\text{ mol} \times 0.08206\,\frac{\text{L}\cdot\text{atm}}{\text{mol}\cdot\text{K}} \times 275\text{ K}}{2.0\text{ L}} = 0.00503\text{ atm H}_2\text{O}$$

$$0.0022\underline{2}852\text{ mol NO} \times \frac{5\text{ mol NO}_2}{5\text{ mol NO}} = 0.0022\underline{2}852\text{ mol NO}_2$$

$$P = \frac{nRT}{V} = \frac{0.0022\underline{2}852\text{ mol} \times 0.08206\,\frac{\text{L}\cdot\text{atm}}{\text{mol}\cdot\text{K}} \times 275\text{ K}}{2.0\text{ L}} = 0.0251\text{ atm NO}_2$$

$$0.0022\underline{2}852\text{ mol NO} \times \frac{2\text{ mol OH}}{5\text{ mol NO}} = 0.00089\underline{1}408\text{ mol OH}$$

$$P = \frac{nRT}{V} = \frac{0.000891408\ \text{mol} \times 0.08206\ \frac{\text{L} \cdot \text{atm}}{\text{mol} \cdot \text{K}} \times 275\ \text{K}}{2.0\ \text{L}} = 0.0101\ \text{atm OH}$$

$$P_{Total} = \sum P = 0.0730\ \text{atm} + 0.420\ \text{atm} + 0.00279\ \text{atm} + 0.00503\ \text{atm} + 0.00503\ \text{atm} + 0.0251\ \text{atm} + 0.0101\ \text{atm} = 0.533\ \text{atm}$$

Check: The units (atm) are correct. The magnitude of the answers is reasonable. The limiting reagent has the lowest pressure. The product pressures are in line with the ratios of the stoichiometric coefficients.

127. **Given:** $P_{CH4} + P_{C2H6} = 0.53$ atm, $P_{CO2} + P_{H2O} = 2.2$ atm **Find:** χ_{CH4}

Conceptual plan: Write balanced reactions to determine change in moles of gas for CH_4 and C_2H_6.

$2\ CH_4\ (g) + 4\ O_2\ (g) \rightarrow 4\ H_2O\ (g) + 2\ CO_2\ (g)$ and $2\ C_2H_6\ (g) + 5\ O_2\ (g) \rightarrow 6\ H_2O\ (g) + 4\ CO_2\ (g)$ thus

$$\frac{6\ \text{mol gases}}{2\ \text{mol}\ CH_4} \qquad \frac{10\ \text{mol gases}}{2\ \text{mol}\ C_2H_6}$$

write expression for final pressure, substituting in data given → χ_{CH4}

$$\chi_{CH4} = \frac{n_{CH4}}{n_{CH4} + n_{C2H6}} \text{ and } \chi_{C2H6} = 1 - \chi_{CH4} \quad P_{CH4} = \chi_{CH4}P_{Total} \quad P_{C2H6} = \chi_{C2H6}P_{Total}$$

$$P_{Final} = \left(\chi_{CH4}P_{Total}\ x\ \frac{6\ \text{mol gases}}{1\ \text{mol}\ CH_4}\right) + \left((1 - \chi_{CH4})P_{Total}\ x\ \frac{10\ \text{mol gases}}{1\ \text{mol}\ C_2H_6}\right)$$

Solution:

$$P_{Final} = \left(\chi_{CH4}\ x\ 0.53\ \text{atm}\ x\ \frac{6\ \text{mol gases}}{2\ \text{mol}\ CH_4}\right) + \left((1 - \chi_{CH4})\ x\ 0.53\ \text{atm}\ x\ \frac{10\ \text{mol gases}}{2\ \text{mol}\ C_2H_6}\right) = 2.2\ \text{atm}$$

Rearrange to solve for $\chi_{CH4} = 0.42$.
Check: The units (none) are correct. The magnitude of the answer (0.42) makes sense because, if it were all methane the final pressure would have been 1.59 atm and if it were all ethane the final pressure would have been 2.65 atm. Since we are closer to the latter pressure we expect the mole fraction of methane to be less than 0.5.

129. Since the passengers have more mass than the balloon they have more momentum than the balloon. The passengers will continue to travel in their original direction longer. The car is slowing so the relative position of the passengers is to move forward and the balloons to move backwards. The opposite happens upon acceleration.

131. B is the limiting reactant (2.0 L of B requires 1.0 L A to completely react). The final container will have 0.5 L A and 2.0 L C, so the final volume will be 2.5 L. The change will be 100 % - ((2.5 L/3.5 L) x 100 %) = - 29 %.

133. a) False – all gases have the same average kinetic energy at the same temperature.

b) False – the gases will have the same partial pressures since we have the same number of moles of each.

c) False – the average velocity of the B molecules will be less than that of the A molecules since the B's are heavier.

d) True – since B molecules are heavier they will contribute more to the density (d = m/V)

Chapter 6

Thermochemistry

1. Thermochemistry is the study of the relationships between chemistry and energy. It is so important because energy and its uses are so critical to our society. It is important to understand how much energy is required or released in a process.

3. Kinetic energy is energy associated with the motion of an object. Potential energy is energy associated with the position or composition of an object. Examples of kinetic energy are a moving billiard ball, gas molecules and a raging river. Examples of potential energy are a billiard ball raised above the surface of a billiard table, a compressed spring and molecules.

5. The SI unit of energy is $kg\frac{m^2}{s^2}$, defined as the joule (J), named after the English scientist James Joule. Other units of energy are the kilojoule (kJ), the calorie (cal) the Calorie (Cal) and the kilowatt-hour (kWh).

7. According to the first law, a device that would continually produce energy with no energy input, sometimes known as a perpetual motion machine, cannot exist because the best we can do with energy is break even.

9. The internal energy (E) of a system is the sum of the kinetic and potential energies of all of the particles that compose the system.

11. If the reactants have a lower internal energy than the products, ΔE_{sys} is positive and energy flows into the system from the surroundings.

13. The internal energy (E) of a system is the sum of the kinetic and potential energies of all of the particles that compose the system. The change in the internal energy of the system (ΔE) must be the sum of the heat transferred (q) and the work done (w): $\Delta E = q + w$.

15. The heat capacity of a system is usually defined as the quantity of heat required to change its temperature by 1 °C. Heat capacity (C) is a measure of the system's ability to hold thermal energy without undergoing a large change in temperature. Since water has such a high heat capacity, it can moderate temperature changes, keeping inland temperatures more constant.

17. In calorimetry, the thermal energy exchanged between the reaction (defined as the system) and the surroundings is measured by observing the change in temperature of the surroundings. A bomb calorimeter is used to measure the ΔE_{rxn} for combustion reactions. The calorimeter includes a tight fitting, sealed container forces the reaction to occur at constant volume. A coffee-cup calorimeter is used to measure ΔH_{rxn} for many aqueous reactions. The calorimeter consists of two Styrofoam coffee cups, one inserted into the other, to provide insulation from the laboratory environment. Since the reaction happens under conditions of constant pressure (open to the atmosphere), $q_{rxn} = q_p = \Delta H_{rxn}$.

19. An endothermic reaction has a positive ΔH and absorbs heat from the surroundings. An endothermic reaction feels cold to the touch. An exothermic reaction has a negative ΔH and gives off heat to the surroundings. An exothermic reaction feels warm to the touch.

21. The internal energy of a chemical system is the sum of its kinetic energy and its potential energy. It is this potential energy that absorbs the energy in an endothermic chemical reaction. In an endothermic reaction, as some bonds break and others form, the protons and electrons go from an arrangement of lower potential energy to one of higher potential energy, absorbing thermal energy in the process.

23. a) If a reaction is multiplied by a factor, the ΔH is multiplied by the same factor.

    b) If a reaction is reversed, the sign of ΔH is reversed.

    The relationships hold because H is a state function. Twice as much energy is contained in twice as much reactants or products. If the reaction is reversed, the final and initial states have been switched and the direction of heat flow is reversed.

25. The standard state is defined as : for a gas –the pure gas at a pressure of exactly 1 atmosphere; for a liquid or solid –the pure substance in its most stable form at a pressure of 1 atm and the temperature of interest (often taken to be 25 °C); and for a substance in solution – a concentration of exactly 1 M. The standard enthalpy change (ΔH°) is the change in enthalpy for a process when all reactants and products are in their standard states. The superscript degree sign indicates standard states.

27. To calculate $\Delta H_{rxn}°$, subtract the heats of formations of the reactants multiplied by their stoichiometric coefficients from the heats of formation of the products multiplied by their stoichiometric coefficients. In the form of an equation:

$$\Delta H^o_{rxn} = \sum n_p \Delta H_f^\circ(\text{products}) - \sum n_r \Delta H_f^\circ(\text{reactants})$$

29. One of the main problems associated with the burning of fossil fuels is that, even though they are abundant in the Earth's crust, they are a finite and non-renewable energy source. The other major problems associated with fossil fuel use are related to the products of combustion. Three major environmental problems associated with the emissions of fossil fuel combustion are air pollution, acid rain, and global warming. One of the main products of fossil fuel combustion is carbon dioxide (CO_2), which is a greenhouse gas.

31. a) **Given:** 3.55×10^4 J **Find:** cal

Conceptual plan: **J → cal**

$$\frac{1\,cal}{4.184\,J}$$

Solution: $3.55 \times 10^4 \cancel{J} \times \frac{1\,cal}{4.814 \cancel{J}} = 8.48 \times 10^3\ cal$

Check: The units (cal) are correct. The magnitude of the answer (8000) makes physical sense because a calorie is larger than a Joule, so the answer decreases.

b) **Given:** 1025 Cal **Find:** J

Conceptual plan: **Cal → J**

$$\frac{4184\,J}{1\,Cal}$$

Solution: $1025 \cancel{Cal} \times \frac{4184\ J}{1 \cancel{Cal}} = 4.289 \times 10^6\ J$

Check: The units (J) are correct. The magnitude of the answer (10^6) makes physical sense because a Calorie is much larger than a Joule, so the answer increases.

c) **Given:** 355 kJ **Find:** cal

Conceptual plan: **kJ → J → cal**

$$\frac{1000\,J}{1\ kJ} \quad \frac{1\,cal}{4.184\ J}$$

Solution: $355 \cancel{kJ} \times \frac{1000 \cancel{J}}{1 \cancel{kJ}} \times \frac{1\,cal}{4.814 \cancel{J}} = 8.48 \times 10^4\ cal$

Check: The units (cal) are correct. The magnitude of the answer (10^4) makes physical sense because a calorie is much smaller than a kJ, so the answer increases.

d) **Given:** 125 kWh **Find:** J

Conceptual plan: **kWh → J**

$$\frac{3.60 \times 10^6\ J}{1\,kWh}$$

Solution: $125 \cancel{kWh} \times \frac{3.60 \times 10^6\ J}{1 \cancel{kWh}} = 4.50 \times 10^8\ J$

Check: The units (J) are correct. The magnitude of the answer (10^8) makes physical sense because a kWh is much larger than a Joule, so the answer increases.

33. a) **Given:** 2155 Cal **Find:** J

Conceptual plan: Cal → J

$$\frac{4184\,J}{1\,Cal}$$

Solution: $2155\ \cancel{Cal}\ x\ \dfrac{4184\,J}{1\ \cancel{Cal}} = 9.017\ x\ 10^6\ J$

Check: The units (J) are correct. The magnitude of the answer (10^6) makes physical sense because a Calorie is much larger than a Joule, so the answer increases.

b) **Given:** 2155 Cal   **Find:** kJ
Conceptual plan: Cal → J → kWh

$$\frac{4184\,J}{1\,Cal} \qquad \frac{1\,kJ}{1000\,J}$$

Solution: $2155\ \cancel{Cal}\ x\ \dfrac{4184\ \cancel{J}}{1\ \cancel{Cal}}\ x\ \dfrac{1\,kJ}{1000\ \cancel{J}} = 3.62\ kJ = 9.017\ x\ 10^3\ kJ$

Check: The units (kJ) are correct. The magnitude of the answer (10^3) makes physical sense because a Calorie is larger than a kJ, so the answer increases.

c) **Given:** 2155 Cal   **Find:** kWh
Conceptual plan: Cal → J → kWh

$$\frac{4184\,J}{1\,Cal} \qquad \frac{1\,kWh}{3.60\,x\,10^6\,J}$$

Solution: $2155\ \cancel{Cal}\ x\ \dfrac{4184\ \cancel{J}}{1\ \cancel{Cal}}\ x\ \dfrac{1\ kWh}{3.60\ x10^6\ \cancel{J}} = 2.50\ kWh$

Check: The units (kWh) are correct. The magnitude of the answer (3) makes physical sense because a Calorie is much smaller than a kWh, so the answer decreases.

35. d) $\Delta E_{sys} = -\ \Delta E_{surr}$

37. a)   The energy exchange is primarily heat since the skin (part of the surroundings) is cooled. There is a small expansion (work) since water is being converted from a liquid to a gas. The sign of ΔE_{sys} is positive since the surroundings cool.

b)   The energy exchange is primarily work. The sign of ΔE_{sys} is negative since the system is expanding (doing work on the surroundings).

c)   The energy exchange is primarily heat. The sign of ΔE_{sys} is positive since the system is being heated by the flame.

39. **Given:** 415 kJ heat released; 125 kJ work done on surroundings   **Find:** ΔE_{sys}

Conceptual plan: interpret language to determine the sign of the two terms then q, w → ΔE_{sys}

$$\Delta E = q + w$$

Solution: since heat is released from the system to the surroundings, q = − 415 kJ; since the system is doing work on the surroundings, w = − 125 kJ.

$\Delta E = q + w = -\ 415\ kJ\ -\ 125\,kJ = -\ 540.\ kJ = -\ 5.40\ x\ 10^2\ kJ$

Check: The units (kJ) are correct. The magnitude of the answer (-540) makes physical sense because both terms are negative.

41. **Given:** 655 J heat absorbed; 344 J work done on surroundings   **Find:** ΔE_{sys}
Conceptual plan: interpret language to determine the sign of the two terms then q, w → ΔE_{sys}

$$\Delta E = q + w$$

Solution: since heat is absorbed by the system, q = + 655 J; since the system is doing work on the surroundings, w = − 344 J. $\Delta E = q + w = 655\ J\ -344\ J = 311\ J$

Check: The units (J) are correct. The magnitude of the answer (+300) makes physical sense because heat term dominates over the work term.

43. Cooler A had more ice after 3 hours because most of the ice in cooler B was melted in order to cool the soft drinks that started at room temperature. In cooler A the drinks were already cold and so the ice only needed to maintain this cool temperature.

45. **Given:** 1.50 L water, T_i = 25.0 °C, T_f = 100.0 °C, d = 1.0 g/mL **Find:** q
Conceptual plan: L → mL → g and pull C_s from table and T_i, T_f → ΔT then m, C_s, ΔT → q

$$\frac{1000\,mL}{1\,L} \quad \frac{1.0\,g}{1.0\,mL} \quad 4.18\frac{J}{g\cdot°C} \quad \Delta T = T_f - T_i \quad q = m\,C_s\,\Delta T$$

Solution: $1.50\,L \times \frac{1000\,mL}{1\,L} \times \frac{1.0\,g}{1.0\,mL} = 1\underline{5}00\,g$ and $\Delta T = T_f - T_i$ = 100.0 °C - 25.0 °C = 75.0 °C

then

$$q = m\,C_s\,\Delta T = 1\underline{5}00\,g \times 4.18\frac{J}{g\cdot°C} \times 75.0\,°C = 4.7 \times 10^5\,J$$

Check: The units (J) are correct. The magnitude of the answer (10^6) makes physical sense because there is such a large mass, a significant temperature change and a high specific heat capacity material.

47. a) **Given:** 25 g gold, T_i = 27.0 °C, q = 2.35 kJ **Find:** T_f
Conceptual plan: kJ → J and pull C_s from table then m, C_s, q → ΔT then T_i, ΔT → T_f

$$\frac{1000\,g}{1\,g} \quad 0.128\frac{J}{g\cdot°C} \quad q = m\,C_s\,\Delta T \quad \Delta T = T_f - T_i$$

Solution: $2.35\,kJ \times \frac{1000\,J}{1\,kJ} = 23\underline{5}0\,J$ then $q = m\,C_s\,\Delta T$ Rearrange to solve for ΔT.

$$\Delta T = \frac{q}{m\,C_s} = \frac{23\underline{5}0\,J}{25\,g \times 0.128\frac{J}{g\times°C}} = 7\underline{3}4.375\,°C$$ finally $\Delta T = T_f - T_i$ Rearrange to solve for T_f.

$T_f = \Delta T + T_i$ = 7$\underline{3}$4.375 °C + 27.0 °C = 760 °C

Check: The units (°C) are correct. The magnitude of the answer (760) makes physical sense because there is such a large heat absorbed, such a small mass, and specific heat capacity. The temperature change should be very large.

b) **Given:** 25 g silver, T_i = 27.0 °C, q = 2.35 kJ **Find:** T_f
Conceptual plan: kJ → J and pull C_s from table then m, C_s, q → ΔT then T_i, ΔT → T_f

$$\frac{1000\,g}{1\,g} \quad 0.235\frac{J}{g\cdot°C} \quad q = m\,C_s\,\Delta T \quad \Delta T = T_f - T_i$$

Solution: $2.35\,kJ \times \frac{1000\,J}{1\,kJ} = 23\underline{5}0\,J$ then $q = m\,C_s\,\Delta T$ Rearrange to solve for ΔT.

$$\Delta T = \frac{q}{m\,C_s} = \frac{23\underline{5}0\,J}{25\,g \times 0.235\frac{J}{g\times°C}} = 4\underline{0}0\,°C$$ finally $\Delta T = T_f - T_i$ Rearrange to solve for T_f.

$T_f = \Delta T + T_i$ = 4$\underline{0}$0 °C + 27.0 °C = 430 °C

Check: The units (°C) are correct. The magnitude of the answer (430) makes physical sense because there is such a large heat absorbed, such a small mass, and specific heat capacity. The temperature change should be very large. The temperature change should be less than that of the gold because the specific heat capacity is greater.

c) **Given:** 25 g aluminum, T_i = 27.0 °C, q = 2.35 kJ **Find:** T_f
Conceptual plan: kJ → J and pull C_s from table then m, C_s, q → ΔT then T_i, ΔT → T_f

$$\frac{1000\,g}{1\,g} \quad 0.903\frac{J}{g\cdot°C} \quad q = m\,C_s\,\Delta T \quad \Delta T = T_f - T_i$$

Solution: $2.35\,\cancel{kJ} \times \frac{1000\text{ J}}{1\,\cancel{kJ}} = 23\underline{5}0\text{ J}$ then $q = m\,C_s\,\Delta T$ Rearrange to solve for ΔT.

$$\Delta T = \frac{q}{m\,C_s} = \frac{23\underline{5}0\,\cancel{J}}{25\,\cancel{g} \times 0.903\,\frac{\cancel{J}}{\cancel{g}\times°C}} = 1\underline{0}4.10\text{ °C}$$ finally $\Delta T = T_f - T_i$ Rearrange to solve for T_f.

$T_f = \Delta T + T_i = 1\underline{0}4.10\text{ °C} + 27.0\text{ °C} = 130\text{ °C}$

Check: The units (°C) are correct. The magnitude of the answer (130) makes physical sense because there is such a large heat absorbed, and such a small mass. The temperature change should be less than that of the silver because the specific heat capacity is greater.

d) **Given:** 25 g water, T_i = 27.0 °C, q = 2.35 kJ **Find:** T_f

Conceptual plan: kJ → J and pull C_s from table then m, C_s, q → ΔT then T_i, ΔT → T_f

$\frac{1000\,g}{1\,g}$ $\quad 4.18\,\frac{\text{J}}{\text{g}\cdot°\text{C}}$ $\quad q = m\,C_s\,\Delta T$ $\quad \Delta T = T_f - T_i$

Solution: $2.35\,\cancel{kJ} \times \frac{1000\text{ J}}{1\,\cancel{kJ}} = 23\underline{5}0\text{ J}$ then $q = m\,C_s\,\Delta T$ Rearrange to solve for ΔT.

$$\Delta T = \frac{q}{m\,C_s} = \frac{23\underline{5}0\,\cancel{J}}{25\,\cancel{g} \times 4.18\,\frac{\cancel{J}}{\cancel{g}\times°C}} = 2\underline{2}.491\text{ °C}$$ finally $\Delta T = T_f - T_i$ Rearrange to solve for T_f.

$T_f = \Delta T + T_i = 2\underline{2}.491\text{ °C} + 27.0\text{ °C} = 49\text{ °C}$

Check: The units (°C) are correct. The magnitude of the answer (130) makes physical sense because there is such a large heat absorbed, and such a small mass. The temperature change should be less than that of the aluminum because the specific heat capacity is greater.

49. **Given:** V_i = 0.0 L, V_f = 2.5 L, P = 1.1 atm **Find:** w (J)

Conceptual plan: V_i, V_f → ΔV then then P, ΔV → w (L atm) → w (J)

$\Delta V = V_f - V_i$ $\quad w = -P\,\Delta V$ $\quad \frac{101.3\text{ J}}{1\text{ L atm}}$

Solution: $\Delta V = V_f - V_i$ = 2.5 L - 0.0 L = 2.5 L then

$$w = -P\,\Delta V = -1.1\,\cancel{atm} \times 2.5\,\cancel{L} \times \frac{101.3\text{ J}}{1\,\cancel{L}\,\cancel{atm}} = -280\text{ J}$$

Check: The units (J) are correct. The magnitude of the answer (-280) makes physical sense because this is an expansion (negative work) and we have ~ atmospheric pressure and a small volume of expansion.

51. **Given:** q = 565 J absorbed, V_i = 0.10 L, V_f = 0.85 L, P = 1.0 atm **Find:** ΔE_{sys}

Conceptual plan: V_i, V_f → ΔV and interpret language to determine the sign of the heat then

$\Delta V = V_f - V_i$ $\quad$ q = + 565 J

then P, ΔV → w (L atm) → w (J) finally q, w → ΔE_{sys}

$w = -P\,\Delta V$ $\quad \frac{101.3\text{ J}}{1\text{ L atm}}$ $\quad \Delta E = q + w$

Solution: $\Delta V = V_f - V_i$ = 0.85 L - 0.10 L =0.75 L then

$$w = -P\,\Delta V = -1.0\,\cancel{atm} \times 0.75\,\cancel{L} \times \frac{101.3\text{ J}}{1\,\cancel{L}\,\cancel{atm}} = -7\underline{5}.975\text{ J} \quad \Delta E = q + w = +565\text{ J} - 7\underline{5}.975\text{ J} = 489\text{ J}$$

Check: The units (J) are correct. The magnitude of the answer (500) makes physical sense because the heat absorbed dominated the small expansion work (negative work).

53. **Given:** 1 mol fuel, 3452 kJ heat produced; 11 kJ work done on surroundings **Find:** ΔE_{sys}, ΔH

Conceptual plan: interpret language to determine the sign of the two terms then q → ΔH and

$\Delta H = q_P$

q, w → ΔE_{sys}

$\Delta E = q + w$

Solution: since heat is produced by the system to the surroundings, q = – 3452 kJ; since the system is doing work on the surroundings, w = – 11 kJ. $\Delta H = q_P$ = - 3452 kJ and

$\Delta E = q + w$ = - 3452 kJ - 11kJ = - 3463 kJ

Check: The units (kJ) are correct. The magnitude of the answer (-3500) makes physical sense because both terms are negative. We expect significant amounts of energy from fuels.

55. a) Combustion is an exothermic process, ΔH is negative

b) Evaporation requires an input of energy and so it is endothermic, ΔH is positive

c) Condensation is the reverse of evaporation, so it is exothermic, ΔH is negative

57. **Given:** 177 mL acetone (C_3H_6O), ΔH°_{rxn} = – 1790 kJ; d = 0.788 g/mL **Find:** q
Conceptual plan: mL acetone → g acetone → mol acetone → q

$$\frac{0.788\text{ g}}{1\text{ mL}} \qquad \frac{1\text{ mol}}{58.08\text{ g}} \qquad \frac{-1790\text{ kJ}}{1\text{ mol}}$$

Solution: $177\ \cancel{mL} \times \frac{0.788\ \cancel{g}}{1\ \cancel{mL}} \times \frac{1\ \cancel{mol}}{58.08\ \cancel{g}} \times \frac{-1790\ kJ}{1\ \cancel{mol}} = -4.30 \times 10^3\ kJ$ *or* $4.30 \times 10^3\ kJ$ *released*

Check: The units (kJ) are correct. The magnitude of the answer (-10^3) makes physical sense because the enthalpy change is negative and we have more than a mole of acetone. We expect more than 1790 kJ to be released.

59. **Given:** pork roast, ΔH°_{rxn} = – 2217 kJ; q used = 1.6×10^3 kJ, 10 % efficiency **Find:** m(CO_2)
Conceptual plan: q used → q generated → mol CO_2 → g CO_2

$$\frac{100\text{ kJ generated}}{10\text{ kJ used}} \qquad \frac{3\text{ mol}}{2217\text{ kJ}} \qquad \frac{44.01\text{ g}}{1\text{ mol}}$$

Solution: $1.6 \times 10^3\ \cancel{kJ} \times \frac{100\ \cancel{kJ\ generated}}{10\ \cancel{kJ\ used}} \times \frac{3\ \cancel{mol\ CO_2}}{2217\ \cancel{kJ}} \times \frac{44.01\ g\ CO_2}{1\ \cancel{mol\ CO_2}} = 950\ g\ CO_2$

Check: The units (g) are correct. The magnitude of the answer (~1000) makes physical sense because the process is not very efficient and a lot of energy is needed.

61. $\Delta H_{rxn} = q_P$ and $\Delta E_{rxn} = q_V = \Delta H - P\,\Delta V$. Since combustions always involve expansions; expansions do work and so have a negative value. Combustions are always exothermic and so have a negative value. This means that ΔE_{rxn} is more negative than ΔH°_{rxn} and so A (-25.9 kJ) is the constant volume process and B (-23.3 kJ) is the constant pressure process.

63. **Given:** 0.514 g biphenyl ($C_{12}H_{10}$), bomb calorimeter, T_i = 25.8 °C, T_f = 29.4 °C, C_{cal} = 5.86 kJ/°C
Find: ΔE_{rxn}
Conceptual plan: T_i, T_f → ΔT then ΔT, C_{cal} → q_{cal} → q_{rxn} then g $C_{12}H_{10}$ → mol $C_{12}H_{10}$

$$\Delta T = T_f - T_i \qquad q_{cal} = C_{cal}\,\Delta T \qquad q_{cal} = -\,q_{rxn} \qquad \frac{1\text{ mol}}{154.20\text{ g}}$$

then q_{rxn}, mol $C_{12}H_{10}$ → ΔE_{rxn}

$$\Delta E_{rxn} = \frac{q_V}{mol\ C_{12}H_{10}}$$

Solution: $\Delta T = T_f - T_i$ = 29.4 °C – 25.8 °C = 3.6 °C then

$$q_{cal} = C_{cal}\,\Delta T = 5.86\ \frac{\text{kJ}}{\cancel{°C}} \times 3.6\ \cancel{°C} = 21.096\text{ kJ then}$$

$$q_{cal} = -\,q_{rxn} = -\,21.096\text{ kJ and } 0.514\ \cancel{g\ C_{12}H_{10}} \times \frac{1\text{ mol } C_{12}H_{10}}{154.20\ \cancel{g\ C_{12}H_{10}}} = 0.00333333\text{ mol } C_{12}H_{10}\text{ then}$$

$$\Delta E_{rxn} = \frac{q_V}{mol\ C_{12}H_{10}} = \frac{-\,21.096\text{ kJ}}{0.00333333\text{ mol } C_{12}H_{10}} = -6.3 \times 10^3\text{ kJ/mol}$$

Check: The units (kJ/mol) are correct. The magnitude of the answer (-6000) makes physical sense because there is such a large heat generated from a very small amount of biphenyl.

65. **Given:** 0.103 g zinc, coffee-cup calorimeter, T_i = 22.5°C, T_f = 23.7 °C, 50.0 mL solution, d (solution) = 1.0 g/mL, $C_{sol'n}$ = 4.18 kJ/g °C **Find:** ΔH_{rxn}

Conceptual plan: T_i, T_f → ΔT and mL sol'n → g sol'n then ΔT, C_{cal} → q_{cal} → q_{rxn} then

$\Delta T = T_f - T_i$ $\qquad \dfrac{1.0\text{ g}}{1.0\text{ mL}}$ $\qquad q_{cal} = m\,C_{sol'n}\,\Delta T$ $\qquad q_{sol'n} = -q_{rxn}$

g Zn → mol Zn then q_{rxn}, mol Zn → ΔH_{rxn}

$\dfrac{1\text{ mol}}{65.37\text{ g}}$ $\qquad \Delta H_{rxn} = \dfrac{q_P}{mol\ Zn}$

Solution: $\Delta T = T_f - T_i =$ 23.7 °C – 22.5 °C = 1.2 °C and $50.0\text{ mL} \times \dfrac{1.0\text{ g}}{1.0\text{ mL}} = 50.0\text{ g}$ then

$q_{sol'n} = m\,C_{sol'n}\,\Delta T = 50.0\text{ g} \times 4.18\dfrac{\text{J}}{\text{g}\,°\text{C}} \times 1.2\,°\text{C} = 25\underline{0}.8\text{ J}$ then $q_{sol'n} = -q_{rxn} = -25\underline{0}.8\text{ J}$ and

$0.103\text{ g Zn} \times \dfrac{1\text{ mol Zn}}{65.37\text{ g Zn}} = 0.0015\underline{7}565\text{ mol Zn}$ then

$\Delta H_{rxn} = \dfrac{q_P}{mol\ Zn} = \dfrac{-25\underline{0}.8\text{ J}}{0.0015\underline{7}565\text{ mol Zn}} = -1.6 \times 10^5\text{ J/mol} = -1.6 \times 10^2\text{ kJ/mol}$

Check: The units (kJ/mol) are correct. The magnitude of the answer (-160) makes physical sense because there is such a large heat generated from a very small amount of zinc.

67. a) Since A + B → 2 C has ΔH_1 then 2 C → A + B will have a $\Delta H_2 = -\Delta H_1$. When the reaction direction is reversed it changes from exothermic to endothermic (or vice versa), so the sign of ΔH changes.

b) Since A + ½ B → C has ΔH_1 then 2 A + B → 2 C will have a $\Delta H_2 = 2\,\Delta H_1$. When the reaction amount doubles, the amount of heat (or ΔH) doubles.

c) Since A → B + 2 C has ΔH_1 then ½ A → ½ B + C will have a $\Delta H_{1'} = ½\,\Delta H_1$. When the reaction amount is cut in half the amount of heat (or ΔH) is cut in half. Then ½ B + C → ½ A will have a $\Delta H_2 = -\Delta H_{1'} = -½\,\Delta H_1$. When the reaction direction is reversed it changes from exothermic to endothermic (or vice versa), so the sign of ΔH changes.

69. Since the first reaction has Fe_2O_3 as a product and the reaction of interest has it as a product, we need to reverse the first reaction. When the reaction direction is reversed ΔH changes.

$Fe_2O_3\ (s) \rightarrow 2\ Fe\ (s) + 3/2\ O_2\ (g)$ $\qquad$ ΔH = + 824.2 kJ

Since the second reaction has 1 mole CO as a reactant and the reaction of interest has 3 moles of CO as a reactant, we need to multiply the second reaction and the ΔH by 3.

$3\ [CO\ (g) + 1/2\ O_2\ (g) \rightarrow CO_2\ (g)]$ $\qquad$ ΔH = 3(- 282.7 kJ) = - 848.1 kJ

Hess' Law states the ΔH of the net reaction is the sum of the ΔH of the steps.
The rewritten reactions are:

$Fe_2O_3\ (s) \rightarrow 2\ Fe\ (s) + \cancel{3/2\ O_2\ (g)}$	ΔH = + 824.2 kJ
$3\ CO\ (g) + \cancel{3/2\ O_2\ (g)} \rightarrow 3\ CO_2\ (g)$	ΔH = - 848.1 kJ
$Fe_2O_3\ (s) + 3\ CO\ (g) \rightarrow 2\ Fe\ (s) + 3\ CO_2\ (g)$	ΔH_{rxn} = - 23.9 kJ

71. Since the first reaction has C_5H_{12} as a reactant and the reaction of interest has it as a product, we need to reverse the first reaction. When the reaction direction is reversed ΔH changes.

$5\ CO_2\ (g) + 6\ H_2O\ (g) \rightarrow C_5H_{12}\ (l) + 8\ O_2\ (g)$ $\qquad$ ΔH = + 3505.8 kJ

Since the second reaction has 1 mole C as a reactant and the reaction of interest has 5 moles of C as a reactant, we need to multiply the second reaction and the ΔH by 5.

$5\ [C\ (s) + O_2\ (g) \rightarrow CO_2\ (g)]$ $\qquad$ ΔH = 5(–393.5 kJ) = –1967.5 kJ

Since the third reaction has 2 moles H_2 as a reactant and the reaction of interest has 6 moles of H_2 as a reactant, we need to multiply the third reaction and the ΔH by 3.

$3[2\ H_2\ (g) + O_2\ (g) \rightarrow 2\ H_2O\ (g)]$ $\qquad$ ΔH = 3(- 483.5 kJ) = - 1450.5 kJ

Hess' Law states the ΔH of the net reaction is the sum of the ΔH of the steps.
The rewritten reactions are:

$\cancel{5CO_2(g)} + \cancel{6H_2O(g)} \rightarrow C_5H_{12}(l) + \cancel{8O_2(g)}$	$\Delta H = +3505.8$ kJ
$5C(s) + \cancel{5O_2(g)} \rightarrow \cancel{5CO_2(g)}$	$\Delta H = -1967.5$ kJ
$6H_2(g) + \cancel{3O_2(g)} \rightarrow \cancel{6H_2O(g)}$	$\Delta H = -1450.5$ kJ
$5C(s) + 6H_2(g) \rightarrow C_5H_{12}(l)$	$\Delta H_{rxn} = +87.8$ kJ

73. a) $\frac{1}{2}N_2(g) + 3/2\,H_2(g) \rightarrow NH_3(g)$ $\quad \Delta H^\circ_f = -45.9$ kJ/mol

b) $C(s) + O_2(g) \rightarrow CO_2(g)$ $\quad \Delta H^\circ_f = -393.5$ kJ/mol

c) $Fe(s) + 3/2\,O_2(g) \rightarrow Fe_2O_3(g)$ $\quad \Delta H^\circ_f = -824.2$ kJ/mol

d) $C(s) + 2H_2(g) \rightarrow CH_4(g)$ $\quad \Delta H^\circ_f = -74.6$ kJ/mol

75. **Given:** $N_2H_4(l) + N_2O_4(g) \rightarrow 2N_2O(g) + 2H_2O(g)$ **Find:** ΔH°_{rxn}

Conceptual plan: $\Delta H^\circ_{rxn} = \sum n_P \Delta H^\circ_f(products) - \sum n_R \Delta H^\circ_f(reactants)$

Solution:

Reactant/Product	ΔH^0_f (kJ/mol from Appendix IIB)
$N_2H_4(l)$	50.6
$N_2O_4(g)$	11.1
$N_2O(g)$	81.6
$H_2O(g)$	– 241.8

Be sure to pull data for the correct formula and phase.

$$\Delta H^0_{rxn} = \sum n_P \Delta H^0_f(products) - \sum n_R \Delta H^0_f(reactants)$$
$$= [2(\Delta H^0_f(N_2O(g)) + 2(\Delta H^0_f(H_2O(g))] - [1(\Delta H^0_f(N_2H_4(l)) + 1(\Delta H^0_f(N_2O_4(g))]$$
$$= [2(81.6\text{ kJ}) + 2(-241.8\text{ kJ})] - [1(50.6\text{ kJ}) + 1(11.1\text{ kJ})]$$
$$= [-320.4\text{ kJ}] - [61.7\text{ kJ}]$$
$$= -382.1\text{ kJ}$$

Check: The units (kJ) are correct. The answer is negative, which means that the reaction is exothermic. The answer is dominated by the negative heat of formation of water.

77. a) **Given:** $C_2H_4(g) + H_2(g) \rightarrow C_2H_6(g)$ **Find:** ΔH°_{rxn}

Conceptual plan: $\Delta H^\circ_{rxn} = \sum n_P \Delta H^\circ_f(products) - \sum n_R \Delta H^\circ_f(reactants)$

Solution:

Reactant/Product	ΔH^0_f (kJ/mol from Appendix IIB)
$C_2H_4(g)$	52.4
$H_2(g)$	0.0
$C_2H_6(g)$	– 84.68

Be sure to pull data for the correct formula and phase.

$$\Delta H^0_{rxn} = \sum n_P \Delta H^0_f(products) - \sum n_R \Delta H^0_f(reactants)$$
$$= [1(\Delta H^0_f(C_2H_6(g))] - [1(\Delta H^0_f(C_2H_4(g)) + 1(\Delta H^0_f(H_2(g))]$$
$$= [1(-84.68\text{ kJ})] - [1(52.4\text{ kJ}) + 1(0.0\text{ kJ})]$$
$$= [-84.68\text{ kJ}] - [52.4\text{ kJ}]$$
$$= -137.1\text{ kJ}$$

Check: The units (kJ) are correct. The answer is negative, which means that the reaction is exothermic. Both hydrocarbon terms are negative, so the final answer is negative.

b) **Given:** $CO\ (g)\ +\ H_2O\ (g)\ \rightarrow\ H_2\ (g) + CO_2\ (g)$ **Find:** ΔH°_{rxn}

Conceptual plan: $\Delta H^\circ_{rxn} = \sum n_P \Delta H^\circ_f(products) - \sum n_R \Delta H^\circ_f(reactants)$

Solution:

Reactant/Product	ΔH^0_f(kJ/mol from Appendix IIB)
$CO\ (g)$	−110.5
$H_2O\ (g)$	− 241.8
$H_2\ (g)$	0.0
$CO_2\ (g)$	− 393.5

Be sure to pull data for the correct formula and phase.

$$\Delta H^0_{rxn} = \sum n_P \Delta H^0_f(products) - \sum n_R \Delta H^0_f(reactants)$$
$$= [1(\Delta H^0_f(H_2\ (g)) + 1(\Delta H^0_f(CO_2\ (g))] - [1(\Delta H^0_f(CO\ (g)) + 1(\Delta H^0_f(H_2O\ (g))]$$
$$= [1(0.0\text{ kJ}) + 1(-393.5\text{ kJ})] - [1(-110.5\text{ kJ}) + 1(-241.8\text{ kJ})]$$
$$= [-393.5\text{ kJ}] - [-352.3\text{ kJ}]$$
$$= -41.2\text{ kJ}$$

Check: The units (kJ) are correct. The answer is negative, which means that the reaction is exothermic.

c) **Given:** $3\ NO_2\ (g)\ +\ H_2O\ (l)\ \rightarrow\ 2\ HNO_3\ (aq) + NO\ (g)$ **Find:** ΔH°_{rxn}

Conceptual plan: $\Delta H^\circ_{rxn} = \sum n_P \Delta H^\circ_f(products) - \sum n_R \Delta H^\circ_f(reactants)$

Solution:

Reactant/Product	ΔH^0_f(kJ/mol from Appendix IIB)
$NO_2\ (g)$	33.2
$H_2O\ (l)$	− 285.8
$HNO_3\ (aq)$	−207
$NO\ (g)$	91.3

Be sure to pull data for the correct formula and phase.

$$\Delta H^0_{rxn} = \sum n_P \Delta H^0_f(products) - \sum n_R \Delta H^0_f(reactants)$$
$$= [2(\Delta H^0_f(HNO_3\ (aq)) + 1(\Delta H^0_f(NO\ (g))] - [3(\Delta H^0_f(NO_2\ (g)) + 1(\Delta H^0_f(H_2O\ (l))]$$
$$= [2(-207\text{ kJ}) + 1(91.3\text{ kJ})] - [3(33.2\text{ kJ}) + 1(-285.8\text{ kJ})]$$
$$= [-32\underline{2}.7\text{ kJ}] - [-186.2\text{ kJ}]$$
$$= -137\text{ kJ}$$

Check: The units (kJ) are correct. The answer is negative, which means that the reaction is exothermic.

d) **Given:** $Cr_2O_3\ (s)\ +\ 3\ CO\ (g)\ \rightarrow\ 2\ Cr\ (s) + 3\ CO_2\ (g)$ **Find:** ΔH°_{rxn}

Conceptual plan: $\Delta H^\circ_{rxn} = \sum n_P \Delta H^\circ_f(products) - \sum n_R \Delta H^\circ_f(reactants)$

Solution:

Reactant/Product	ΔH^0_f(kJ/mol from Appendix IIB)
$Cr_2O_3\ (s)$	− 1139.7
$CO\ (g)$	−110.5
$Cr\ (s)$	0.0
$CO_2\ (g)$	− 393.5

Be sure to pull data for the correct formula and phase.

$$\Delta H^0_{rxn} = \sum n_P \Delta H^0_f(products) - \sum n_R \Delta H^0_f(reactants)$$
$$= [2(\Delta H^0_f(Cr\ (s)) + 3(\Delta H^0_f(CO_2\ (g))] - [1(\Delta H^0_f(Cr_2O_3\ (s)) + 3(\Delta H^0_f(CO\ (g))]$$
$$= [2(0.0\text{ kJ}) + 3(-393.5\text{ kJ})] - [1(-1139.7\text{ kJ}) + 3(-110.5\text{ kJ})]$$
$$= [-1180.5\text{ kJ}] - [-1471.2\text{ kJ}]$$
$$= -290.7\text{ kJ}$$

Check: The units (kJ) are correct. The answer is negative, which means that the reaction is exothermic.

79. **Given:** form glucose ($C_6H_{12}O_6$) and oxygen from sunlight, carbon dioxide and water **Find:** ΔH°_{rxn}

Conceptual plan: write balanced reaction then $\Delta H^0_{rxn} = \sum n_p \Delta H^0_f(products) - \sum n_R \Delta H^0_f(reactants)$

Solution: $6\ CO_2\ (g) + 6\ H_2O\ (l) \rightarrow C_6H_{12}O_6\ (s) + 6\ O_2\ (g)$

Reactant/Product	ΔH^0_f (kJ/mol from Appendix IIB)
$CO_2\ (g)$	– 393.5
$H_2O\ (l)$	– 285.8
$C_6H_{12}O\ (s)$	– 1273.3
$O_2\ (g)$	0.0

Be sure to pull data for the correct formula and phase.

$$\Delta H^0_{rxn} = \sum n_p \Delta H^0_f(products) - \sum n_R \Delta H^0_f(reactants)$$
$$= [1(\Delta H^0_f(C_6H_{12}O\ (s)) + 6(\Delta H^0_f(O_2\ (g))\ ] - [6(\Delta H^0_f(CO_2\ (g)) + 6(\Delta H^0_f(H_2O\ (l))]$$
$$= [1(-1273.3\text{ kJ}) + 6(0.0\text{ kJ})] - [6(-393.5\text{ kJ}) + 6(-285.8\text{ kJ})]$$
$$= [-1273.3\text{ kJ}] - [-4076\text{ kJ}]$$
$$= +2803\text{ kJ}$$

Check: The units (kJ) are correct. The answer is positive, which means that the reaction is endothermic. The reaction requires the input of light energy, so we expect that this will be an endothermic reaction.

81. **Given:** $2\ CH_3NO_2\ (l) + 3/2\ O_2\ (g) \rightarrow 2\ CO_2\ (g) + 3\ H_2O\ (g) + N_2\ (g)$ and $\Delta H^\circ_{rxn} = -709.2$ kJ/mol
Find: $\Delta H^\circ_f\ (CH_3NO_2\ (l))$
Conceptual plan: fill known values into $\Delta H^0_{rxn} = \sum n_p \Delta H^0_f(products) - \sum n_R \Delta H^0_f(reactants)$ **and rearrange to solve for** $\Delta H^\circ_f\ (CH_3NO_2\ (l))$
Solution:

Reactant/Product	ΔH^0_f (kJ/mol from Appendix IIB)
$O_2\ (g)$	0.0
$CO_2\ (g)$	– 393.5
$H_2O\ (g)$	– 241.8
$N_2\ (g)$	0.0

Be sure to pull data for the correct formula and phase.

$$H^0_{rxn} = \sum n_p \Delta H^0_f(products) - \sum n_R \Delta H^0_f(reactants)$$
$$= [2(\Delta H^0_f(CO_2\ (g)) + 3(\Delta H^0_f(H_2O\ (g)) + 1(\Delta H^0_f(N_2\ (g))] - [2(\Delta H^0_f(CH_3NO_2\ (l)) + 3/2(\Delta H^0_f(O_2\ (g))$$
$$-709.2\text{ kJ} = [2(-393.5\text{ kJ}) + 3(-241.8\text{ kJ}) + 1(0.0\text{ kJ})] - [2(\Delta H^0_f(CH_3NO_2\ (l)) + 3/2(0.0\text{ kJ})]$$
$$-709.2\text{ kJ} = [-1512.4\text{ kJ}] - [2(\Delta H^0_f(CH_3NO_2\ (l))]$$
$$\Delta H^0_f(CH_3NO_2\ (l)) = -401.6\text{ kJ/mol}$$

Check: The units (kJ/mol) are correct. The answer is negative, which means that the reaction is exothermic, which is typical for combustion reactions.

83. a) **Given:** methane, $\Delta H^\circ_{rxn} = -802.3$ kJ; $q = 1.00 \times 10^2$ kJ **Find:** $m(CO_2)$
Conceptual plan: q $\rightarrow$ mol CO_2 $\rightarrow$ g CO_2

$$\frac{1\text{ mol}}{-802.3\text{ kJ}} \qquad \frac{44.01\text{ g}}{1\text{ mol}}$$

Solution: $-1.00 \times 10^2\ \text{kJ} \times \frac{1\text{ mol } CO_2}{-802.3\text{ kJ}} \times \frac{44.01\text{ g } CO_2}{1\text{ mol } CO_2} = 5.49\text{ g } CO_2$

Check: The units (g) are correct. The magnitude of the answer (~5) makes physical sense because less than a mole of fuel is used.

b) **Given:** propane, $\Delta H^\circ_{rxn} = -2217$ kJ; $q = 1.00 \times 10^2$ kJ **Find:** $m(CO_2)$
Conceptual plan: **q → mol CO_2 → g CO_2**

$$\frac{3\text{ mol}}{-2217\text{ kJ}} \qquad \frac{44.01\text{ g}}{1\text{ mol}}$$

Solution: $$-1.00 \times 10^2\ \text{kJ} \times \frac{3\text{ mol } CO_2}{-2217\text{ kJ}} \times \frac{44.01\text{ g } CO_2}{1\text{ mol } CO_2} = 5.96\text{ g } CO_2$$

Check: The units (g) are correct. The magnitude of the answer (~6) makes physical sense because less than a mole of fuel is used.

c) **Given:** octane, $\Delta H^\circ_{rxn} = -5074.1$ kJ; $q = 1.00 \times 10^2$ kJ **Find:** $m(CO_2)$
Conceptual plan: **q → mol CO_2 → g CO_2**

$$\frac{8\text{ mol}}{-5074.1\text{ kJ}} \qquad \frac{44.01\text{ g}}{1\text{ mol}}$$

Solution: $$-1.00 \times 10^2\ \text{kJ} \times \frac{8\text{ mol } CO_2}{-5074.1\text{ kJ}} \times \frac{44.01\text{ g } CO_2}{1\text{ mol } CO_2} = 6.94\text{ g } CO_2$$

Check: The units (g) are correct. The magnitude of the answer (~7) makes physical sense because less than a mole of fuel is used. The methane generated the least carbon dioxide for the given amount of heat, while the octane generated the most carbon dioxide for the given amount of heat.

85. **Given:** 7×10^{12} kg/yr octane (C_8H_{18}), $\Delta H^\circ_{rxn} = -5074.1$ kJ (using #83), 3×10^{15} kg CO_2 in atmosphere,
Find: $m(CO_2)$ in kg and time to double atmospheric CO_2
Conceptual plan: use reaction from #83 $C_8H_{18}\,(l) + 25/2\ O_2\,(g) \rightarrow 8\ CO_2\,(g) + 9\ H_2O\,(g)$
kg (C_8H_{18}) → g (C_8H_{18}) → mol (C_8H_{18}) → mol CO_2 → g CO_2 → kg CO_2 produced

$$\frac{1000\text{ g}}{1\text{ kg}} \qquad \frac{1\text{ mol } C_8H_{18}}{114.22\text{ g}} \qquad \frac{8\text{ mol } CO_2}{1\text{ mol } C_8H_{18}} \qquad \frac{44.01\text{ g}}{1\text{ mol}} \qquad \frac{1\text{ kg}}{1000\text{ g}}$$

then kg CO_2 → yr

$$\frac{3 \times 10^{15}\text{ kg } CO_2}{\left(\frac{\text{kg } CO_2\text{ produced}}{\text{yr}}\right)}$$

Solution:

$$7 \times 10^{12}\text{ kg} \times \frac{1000\text{ g}}{1\text{ kg}} \times \frac{1\text{ mol } C_8H_{18}}{114.22\text{ g}} \times \frac{8\text{ mol } CO_2}{1\text{ mol } C_8H_{18}} \times \frac{44.01\text{ g } CO_2}{1\text{ mol } CO_2} \times \frac{1\text{ kg}}{1000\text{ g}} = \underline{2}.158 \times 10^{13}\text{ kg } CO_2$$

$$= 2 \times 10^{13}\text{ kg } CO_2 \text{ per year}$$

$$\frac{3 \times 10^{15}\text{ kg } CO_2}{\left(\frac{\underline{2}.158 \times 10^{13}\text{ kg } CO_2\text{ produced}}{\text{yr}}\right)} = \underline{1}39\text{ yr} = 100\text{ yr}$$

Check: The units (kg and yr) are correct. The magnitude of the answer (10^{13} kg) makes physical sense because the mass of CO_2 is larger than the hydrocarbon mass in a combustion (since O is much heavier than H). The time is consistent with Chapter 5 problems where ~1 % of the atmospheric carbon monoxide is generated each year.

87. **Given:** billiard ball$_A$ = system: $m_A = 0.17$ kg, $v_{A1} = 4.5$ m/s slows to $v_{A2} = 3.8$ m/s and $v_{A3} = 0$; ball$_B$: $m_B = 0.17$ kg, $v_{B1} = 0$ and $v_{B2} = 3.8$ m/s and $KE = \frac{1}{2} mv^2$ **Find:** w, q, ΔE_{sys}
Conceptual plan: m, v → KE then KE_{A3}, KE_{A1} → ΔE_{sys} and KE_{A2}, KE_{A1} → q and KE_{B2}, KE_{B1} → w_B

$$KE = \tfrac{1}{2} mv^2 \qquad \Delta E_{sys} = KE_{A3} - KE_{A1} \qquad q = KE_{A2} - KE_{A1} \qquad w_B = KE_{B2} - KE_{B1}$$

ΔE_{sys}, q → w_A verify that $w_A = -w_B$ so that no heat is transferred to ball$_B$

$$\Delta E = q + w$$

Solution: $KE = \frac{1}{2} mv^2$ since m is in kg and v is in m/s, KE will be in kgm^2/s^2 which is joule.

$$KE_{A1} = ½\,(0.17\text{ kg})\left(4.5\frac{\text{m}}{\text{s}}\right)^2 = 1.\underline{7}213\,\frac{\text{kg m}^2}{\text{s}^2} = 1.\underline{7}213\text{ J},$$

$$KE_{A2} = ½\,(0.17\text{ kg})\left(3.8\frac{\text{m}}{\text{s}}\right)^2 = 1.\underline{2}274\,\frac{\text{kg m}^2}{\text{s}^2} = 1.\underline{2}274\text{ J},$$

$$KE_{A3} = ½\,(0.17\text{ kg})\left(0\frac{\text{m}}{\text{s}}\right)^2 = 0\,\frac{\text{kg m}^2}{\text{s}^2} = 0\text{ J},\quad KE_{B1} = ½\,(0.17\text{ kg})\left(0\frac{\text{m}}{\text{s}}\right)^2 = 0\,\frac{\text{kg m}^2}{\text{s}^2} = 0\text{ J and}$$

$$KE_{B2} = ½\,(0.17\text{ kg})\left(3.8\frac{\text{m}}{\text{s}}\right)^2 = 1.\underline{2}274\,\frac{\text{kg m}^2}{\text{s}^2} = 1.\underline{2}274\text{ J}.$$

$\Delta E_{sys} = KE_{A3} - KE_{A1} = 0\text{ J} - 1.\underline{7}213\text{ J} = -1.\underline{7}213\text{ J} = -1.7\text{ J}$,

$q = KE_{A2} - KE_{A1} = 1.\underline{2}274\text{ J} - 1.\underline{7}213\text{ J} = -0.\underline{4}939\text{ J} = -0.5\text{ J}$,

$w_B = KE_{B2} - KE_{B1} = 1.\underline{2}274\text{ J} - 0\text{ J} = 1.\underline{2}274\text{ J}$ and

$w = \Delta E - q = -1.\underline{7}213\text{ J} - -0.\underline{4}939\text{ J} = -1.\underline{2}274\text{ J} = -1.2\text{ J}$.

Since $w_A = -w_B$ so that no heat is transferred to ball_B.

Check: The units (kJ) are correct. Since the ball is initially moving and is stopped at the end, it has lost energy (negative ΔE_{sys}). As the ball slows due to friction it is releasing heat (negative q). The kinetic energy is transferred to a second ball, so it does work (w negative).

89. **Given:** $H_2O\,(l) \rightarrow H_2O\,(g)$ $\Delta H°_{rxn} = +44.01$ kJ/mol; $\Delta T_{body} = -0.50$ °C, $m_{body} = 95$ kg, $C_{body} = 4.0$ J/g °C
Find: m_{H2O}
Conceptual plan: kg → g then m_{body}, ΔT, C_{body} → q_{body} → q_{rxn} (J) → q_{rxn} (kJ) → mol H_2O → g H_2O

$$\frac{1000\text{ g}}{1\text{ kg}} \quad q_{body} = m_{body}\,C_{body}\,\Delta T_{body} \quad q_{rxn} = -\,q_{body} \quad \frac{1\text{ kJ}}{1000\text{ J}} \quad \frac{1\text{ mol}}{44.01\text{ kJ}} \quad \frac{18.01\text{ g}}{1\text{ mol}}$$

Solution: $95\text{ kg} \times \frac{1000\text{ g}}{1\text{ kg}} = 95000\text{ g}$ then

$$q_{body} = m_{body}\,C_{body}\,\Delta T_{body} = 95000\text{ g} \times 4.0\,\frac{\text{J}}{\text{g °C}} \times (-0.50\text{ °C}) = -1\underline{9}0000\text{ J then}$$

$$q_{rxn} = -q_{body} = 1\underline{9}0000\text{ J} \times \frac{1\text{ kJ}}{1000\text{ J}} \times \frac{1\text{ mol}}{44.01\text{ kJ}} \times \frac{18.01\text{ g}}{1\text{ mol}} = 78\text{ g}\,H_2O$$

Check: The units (g) are correct. The magnitude of the answer (78) makes physical sense because a person can sweat this much on a hot day.

91. **Given:** $H_2O\,(s) \rightarrow H_2O\,(l)$ $\Delta H°_f\,(H_2O\,(s)) = -291.8$ kJ/mol; 355 mL beverage $T_{Bevi} = 25.0$ °C, $T_{Bevf} = 0.0$ °C, $C_{Bev} = 4.184$ J/g °C, $d_{Bev} = 1.0$ g/mL **Find:** $\Delta H°_{rxn}$ (ice melting) and m_{ice}
Conceptual plan: $\Delta H^0_{rxn} = \sum n_P \Delta H^0_f(products) - \sum n_R \Delta H^0_f(reactants)$ **mL → g and T_i, T_f → ΔT then**

$$\frac{1.0\text{ g}}{1.0\text{ mL}} \qquad \Delta T = T_f - T_i$$

m_{H2O}, ΔT_{H2O}, C_{H2O} → q_{H2O} → q_{rxn} (J) → q_{rxn} (kJ) → mol ice → g ice

$$q_{Bev} = m_{Bev}\,C_{Bev}\,\Delta T_{Bev} \quad q_{rxn} = -\,q_{Bev} \quad \frac{1\text{ kJ}}{1000\text{ J}} \quad \frac{1\text{ mol}}{\Delta H^0_{rxn}} \quad \frac{18.01\text{ g}}{1\text{ mol}}$$

Solution:

Reactant/Product	ΔH^0_f (kJ/mol from Appendix IIB)
$H_2O\,(s)$	– 291.8
$H_2O\,(l)$	– 285.8

Be sure to pull data for the correct formula and phase.

$$\begin{aligned}\Delta H^0_{rxn} &= \sum n_P \Delta H^0_f(products) - \sum n_R \Delta H^0_f(reactants)\\ &= [1(\Delta H^0_f(H_2O\,(l))] - [1(\Delta H^0_f(H_2O\,(s))]\\ &= [1(-285.8\text{ kJ})] - [1(-291.8\text{ kJ})]\\ &= +6.0\text{ kJ}\end{aligned}$$

$$355\ \text{mL} \times \frac{1.0\ \text{g}}{1.0\ \text{mL}} = 355\ \text{g} \text{ and } \Delta T = T_f - T_i = 0.0\ °\text{C} - 25.0\ °\text{C} = -25.0\ °\text{C} \text{ then}$$

$$q_{Bev} = m_{Bev}\, C_{Bev}\, \Delta T_{Bev} = 355\ \text{g} \times 4.184 \frac{\text{J}}{\text{g}\ °\text{C}} \times (-25.0\ °\text{C}) = 37\underline{1}33\ \text{J then}$$

$$q_{rxn} = -q_{Bev} = -37\underline{1}33\ \text{J} \times \frac{1\ \text{kJ}}{1000\ \text{J}} \times \frac{1\ \text{mol}}{-6.0\ \text{kJ}} \times \frac{18.01\ \text{g}}{1\ \text{mol}} = 110\ \text{g ice}$$

Check: The units (kJ and g) are correct. The answer is positive, which means that the reaction is endothermic. We expect an endothermic reaction because we know that heat must be added to melt ice. The magnitude of the answer (110 g) makes physical sense because it is much smaller than the weight of the beverage and it would fit in a glass with the beverage.

93. **Given:** 25.5 g aluminum, $T_{Ali} = 65.4$ °C, 55.2 g water, $T_{H2Oi} = 22.2$ °C, **Find:** T_f
Conceptual plan: pull C_s values from table then m, C_s, T_i → T_f

$$\text{Al: } 0.903 \frac{\text{J}}{\text{g}\cdot°\text{C}} \quad \text{H}_2\text{O: } 4.18 \frac{\text{J}}{\text{g}\cdot°\text{C}} \qquad q = m\, C_s \left(T_f - T_i\right) \text{ then set } q_{Al} = -q_{H2O}$$

Solution: $q = m\, C_s \left(T_f - T_i\right)$ substitute in values and set $q_{Al} = -q_{H2O}$.

$$q_{Al} = m_{Al}\, C_{Al}\left(T_f - T_{Ali}\right) = 25.5\ \text{g} \times 0.903 \frac{\text{J}}{\text{g} \times °\text{C}} \times \left(T_f - 65.4\ °\text{C}\right) =$$

Rearrange to

$$-q_{H2O} = -m_{H2O}\, C_{H2O}\left(T_f - T_{H2Oi}\right) = -55.2\ \text{g} \times 4.18 \frac{\text{J}}{\text{g} \times °\text{C}} \times \left(T_f - 22.2\ °\text{C}\right)$$

solve for T_f.

$$23.\underline{0}265 \frac{\text{J}}{°\text{C}} \times \left(T_f - 65.4\ °\text{C}\right) = -23\underline{0}.736 \frac{\text{J}}{°\text{C}} \times \left(T_f - 22.2\ °\text{C}\right) \rightarrow$$

$$23.\underline{0}265 \frac{\text{J}}{°\text{C}} T_f - 15\underline{0}5.93\ \text{J} = -23\underline{0}.736 \frac{\text{J}}{°\text{C}} T_f + 51\underline{6}8.49\ \text{J} \rightarrow$$

$$-51\underline{6}8.49\ \text{J} - 15\underline{0}5.93\ \text{J} = -23\underline{0}.736 \frac{\text{J}}{°\text{C}} T_f - 23.\underline{0}265 \frac{\text{J}}{°\text{C}} T_f \rightarrow 66\underline{7}4.42\ \text{J} = 25\underline{3}.7625 \frac{\text{J}}{°\text{C}} T_f \rightarrow$$

$$T_f = \frac{66\underline{7}4.42\ \text{J}}{25\underline{3}.7625 \frac{\text{J}}{°\text{C}}} = 26.3\ °\text{C}$$

Check: The units (°C) are correct. The magnitude of the answer (26) makes physical sense because the heat transfer is dominated by the water (larger mass and larger specific heat capacity). The final temperature should be closer to the initial temperature of water than of aluminum.

95. **Given:** palmitic acid ($C_{16}H_{32}O_2$) combustion $\Delta H°_f$ ($C_{16}H_{32}O_2$ *(s)*) = – 208 kJ/mol; sucrose ($C_{12}H_{22}O_{11}$) combustion $\Delta H°_f$ ($C_{12}H_{22}O_{11}$ *(s)*) = – 2222 kJ/mol **Find:** $\Delta H°_{rxn}$ in kJ/mol and Cal/g
Conceptual plan: write balanced reaction then $\Delta H^0_{rxn} = \sum n_p \Delta H^0_f(products) - \sum n_R \Delta H^0_f(reactants)$ **then kJ/mol → J/mol → Cal/mol → Cal/g**

$$\frac{1000\ \text{J}}{1\ \text{kJ}} \quad \frac{1\ \text{Cal}}{4184\ \text{J}} \quad \text{PA: } \frac{1\ \text{mol}}{256.42\ \text{g}} \quad \text{S: } \frac{1\ \text{mol}}{342.30\ \text{g}}$$

Solution: combustion is combination with oxygen to form carbon dioxide and water *(l)*
$C_{16}H_{32}O_2$ *(s)* + 23 O_2 *(g)* → 16 CO_2 *(g)* + 16 H_2O *(l)*

Reactant/Product	ΔH^0_f (kJ/mol from Appendix IIB)
$C_{16}H_{32}O_2$ *(s)*	– 208
O_2 *(g)*	0.0
CO_2 *(g)*	– 393.5
H_2O *(g)*	– 285.8

Be sure to pull data for the correct formula and phase.

$\Delta H^0_{rxn} = \sum n_P \Delta H^0_f(products) - \sum n_R \Delta H^0_f(reactants)$

$= [16(\Delta H^0_f(CO_2\ (g)) + 16(\Delta H^0_f(H_2O\ (l))] - [1(\Delta H^0_f(C_{16}H_{32}O_2\ (s)) + 23(\Delta H^0_f(O_2\ (g))]$

$= [16(-393.5\ kJ) + 16(-285.8\ kJ)] - [1(-208\ kJ) + 23(0.0\ kJ)]$

$= [-10868.8\ kJ] - [-208\ kJ]$

$= -10,66\underline{0}.8$ kJ/mol = -10,661 kJ/mol

$$-10,66\underline{0}.8\ \frac{kJ}{mol} \times \frac{1000\ J}{1\ kJ} \times \frac{1\ Cal}{4184\ J} \times \frac{1\ mol}{256.42\ g} = 9.9378\ Cal/g$$

$C_{12}H_{22}O_{11}\ (s) + 12\ O_2\ (g) \rightarrow 12\ CO_2\ (g) + 11\ H_2O\ (l)$

Reactant/Product	ΔH^0_f (kJ/mol from Appendix IIB)
$C_{12}H_{22}O_{11}\ (s)$	– 2226.1
$O_2\ (g)$	0.0
$CO_2\ (g)$	– 393.5
$H_2O\ (g)$	– 285.8

Be sure to pull data for the correct formula and phase.

$\Delta H^0_{rxn} = \sum n_P \Delta H^0_f(products) - \sum n_R \Delta H^0_f(reactants)$

$= [12(\Delta H^0_f(CO_2\ (g)) + 11(\Delta H^0_f(H_2O\ (l))] - [1(\Delta H^0_f(C_{12}H_{22}O_{11}\ (s)) + 12(\Delta H^0_f(O_2\ (g))]$

$= [12(-393.5\ kJ) + 11(-285.8\ kJ)] - [1(-2226.1\ kJ) + 12(0.0\ kJ)]$

$= [-7865.8\ kJ] - [-2226.1\ kJ]$

$= -5639.7$ kJ/mol

$$-5639.7\ \frac{kJ}{mol} \times \frac{1000\ J}{1\ kJ} \times \frac{1\ Cal}{4184\ J} \times \frac{1\ mol}{342.30\ g} = -3.938\ Cal/g$$

Check: The units (kJ/mol and Cal/g) are correct. The magnitude of the answers are consistent with the food labels we see everyday.

97. At constant P $\Delta H_{rxn} = q_P$ and at constant V $\Delta E_{rxn} = q_V = \Delta H_{rxn} - P\,\Delta V$. $PV = nRT$ at constant P, and a constant number of moles of gas as we change the T the only variable that can change is V, so $P\Delta V = nR\Delta T$. Substituting into the equation for ΔE_{rxn} we get $\Delta E_{rxn} = \Delta H_{rxn} - nR\Delta T$ or $\Delta H_{rxn} = \Delta E_{rxn} + nR\Delta T$.

99. **Given:** 16 g peanut butter, bomb calorimeter, T_i = 22.2 °C, T_f = 25.4 °C, C_{cal} = 120.0 kJ/°C
Find: calories in peanut butter
Conceptual plan: T_i, T_f → ΔT then ΔT, C_{cal} → q_{cal} → q_{rxn} (kJ) → q_{rxn} (kJ) → q_{rxn} (Cal)

$\Delta T = T_f - T_i$ $\quad q_{cal} = C_{cal}\,\Delta T \quad q_{rxn} = -q_{cal} \quad \frac{1000\,J}{1\,kJ} \quad \frac{1\,Cal}{4184\,J}$

then q_{rxn} (Cal) → Cal/g
÷16 g peanut butter

Solution: $\Delta T = T_f - T_i =$ 25.4 °C – 22.2 °C = 3.2 °C then

$$q_{cal} = C_{cal}\,\Delta T = 120.0\ \frac{kJ}{°C} \times 3.2\ °C = 3\underline{8}4\ kJ \text{ then}$$

$$q_{rxn} = -q_{cal} = -3\underline{8}4\ kJ \times \frac{1000\ J}{1\ kJ} \times \frac{1\,Cal}{4184\ J} = 9\underline{1}.778\ Cal \text{ then } \frac{9\underline{1}.778\ Cal}{16\ g} = 5.7\ Cal/g$$

Check: The units (Cal/g) are correct. The magnitude of the answer (6) makes physical sense because there is a significant percentage of fat and sugar in peanut butter. The answer is in line with the answers in #93.

101. **Given:** V_1 = 20.0 L at P_1 = 3.0 atm; P_2 = 1.5 atm let expand at constant T. **Find:** w, q, ΔE_{sys}
Conceptual plan: V_1, P_1, P_2 → V_2 then V_1, V_2 → ΔV then $P, \Delta V$ → w (L atm) → w (J)

$P_1 V_1 = P_2 V_2 \quad \Delta V = V_2 - V_1 \quad w = -P\,\Delta V \quad \frac{101.3\ J}{1\ L\ atm}$

for an ideal gas ΔE_{sys} α T, so since this is a constant temperature process ΔE_{sys} = 0 finally ΔE_{sys}, w → q
$\Delta E = q + w$

Solution: $P_1 V_1 = P_2 V_2$ Rearrange to solve for V_2. $V_2 = V_1 \frac{P_1}{P_2} = (20.0 \text{ L}) \times \frac{3.0 \text{ atm}}{1.5 \text{ atm}} = 40. \text{ L}$ and

$\Delta V = V_2 - V_1 = 40. \text{ L} - 20.0 \text{ L} = 20. \text{ L}$ then

$w = -P\,\Delta V = -1.5 \text{ atm} \times 20. \text{ L} \times \frac{101.3 \text{ J}}{1 \text{ L atm}} = -3\underline{0}39 \text{ J} = -3.0 \times 10^3 \text{ J}$ $\quad \Delta E = q + w \quad$ Rearrange to solve for q.

$q = \Delta E_{sys} - w = +0 \text{ J} - -3\underline{0}39 \text{ J} = 3.0 \times 10^3 \text{ J}$

Check: The units (J) are correct. Since there is no temperature change, we expect no energy change ($\Delta E_{sys} = 0$). The piston expands and so does work (negative work) and so heat is absorbed (positive q).

103. **Given:** 655 kWh/yr, coal is 3.2 % S, remainder is C, S emitted as SO_2 *(g)* and gets converted to H_2SO_4 when reacting with water **Find:** m (H_2SO_4)/yr

Conceptual plan: write balanced reaction then $\Delta H^\circ_{rxn} = \sum n_P \Delta H^\circ_f(products) - \sum n_R \Delta H^\circ_f(reactants)$

(since the form of sulfur is not given, assume all heat is from combustion of only carbon) then kWh → J → kJ → mol (C) → g (C) → g (S) → mol (H_2SO_4) → mol (H_2SO_4) → g (H_2SO_4)

$$\frac{3.60 \times 10^6 \ J}{1 kWh} \quad \frac{1 \text{ kJ}}{1000 \text{ J}} \quad \frac{\text{mol C}}{\Delta H^\circ_f (CO_2\ (g))} \quad \frac{12.01 \text{ g}}{1 \text{ mol}} \quad \frac{3.2 \text{ g S}}{(100.0 - 3.2) \text{ g C}} \quad \frac{1 \text{ mol}}{32.06 \text{ g}} \quad \frac{1 \text{ mol } H_2SO_4}{1 \text{ mol S}} \quad \frac{98.09 \text{ g}}{1 \text{ mol}}$$

Solution: C *(s)* + O_2 *(g)* → CO_2 *(g)* This reaction is the heat of formation of CO_2 *(g)* so $\Delta H^0_{rxn} = \Delta H^0_f (CO_2\ (g)) = -393.5$ kJ/mol then

$$655 \text{ kWh} \times \frac{3.60 \times 10^6 \text{ J}}{1 \text{ kWh}} \times \frac{1 \text{ kJ}}{1000 \text{ J}} \times \frac{\text{mol C}}{393.5 \text{ kJ}} \times \frac{12.01 \text{ g C}}{1 \text{ mol C}} \times \frac{3.2 \text{ g S}}{(100.0 - 3.2) \text{ g C}} \times \frac{1 \text{ mol S}}{32.06 \text{ g S}} \times \frac{1 \text{ mol } H_2SO_4}{1 \text{ mol S}} \times \frac{98.09 \text{ g } H_2SO_4}{1 \text{ mol } H_2SO_4} = 7.3 \times 10^3 \text{ g } H_2SO_4$$

Check: The units (g) are correct. The magnitude (7300) is reasonable, considering this is just 1 home.

105. **Given:** methane combustion, 100 % efficiency, $\Delta T = 10.0$ °C, house = 30.0 m x 30.0 m x 3.0 m, Cs (air) = 30 J/K·mol, 1.00 mol air = 22.4 L **Find:** m (CH_4)

Conceptual plan: l, w, h → V(m^3) → V(cm^3) → V(L) → mol (air) then m, C_s, ΔT → q_{air} (J)

$$V = l \times w \times h \quad \frac{(100 \text{ cm})^3}{(1 \text{ m})^3} \quad \frac{1 \text{ L}}{1000 \text{ cm}^3} \quad \frac{1 \text{ mol air}}{22.4 \text{ L}} \qquad q = m\ C_s\ \Delta T$$

then q_{air} (J) → q_{rxn} (J) → q(kJ) → mol (CH_4) → g (CH_4)

$$q_{rxn} = -q_{air} \quad \frac{1 \text{ kJ}}{1000 \text{ J}} \quad \frac{1 \text{ mol } CH_4}{-802.3 \text{ kJ}} \text{ from \#83} \quad \frac{16.04 \text{ g}}{1 \text{ mol}}$$

Solution: $V = l \times w \times h = 30.0 \text{ m} \times 30.0 \text{ m} \times 3.0 \text{ m} = 2\underline{7}00 \text{ m}^3$ then

$$2\underline{7}00 \text{ m}^3 \times \frac{(100 \text{ cm})^3}{(1 \text{ m})^3} \times \frac{1 \text{ L}}{1000 \text{ cm}^3} \times \frac{1 \text{mol air}}{22.4 \text{ L}} = 1.\underline{2}0536 \times 10^5 \text{ mol air then}$$

$$q = m\ C_s\ \Delta T = 1.\underline{2}0536 \times 10^5 \text{ mol} \times 30 \frac{\text{J}}{\text{mol} \times {}^\circ\text{C}} \times 10.0\ {}^\circ\text{C} = \underline{3}.6161 \times 10^7 \text{ J}$$

$$q_{rxn} = -q_{air} = -\underline{3}.6161 \times 10^7 \text{ J} \times \frac{1 \text{ kJ}}{1000 \text{ J}} \times \frac{1 \text{ mol } CH_4}{-802.3 \text{ kJ}} \times \frac{16.04 \text{ g } CH_4}{1 \text{ mol } CH_4} = \underline{7}22.9 \text{ g } CH_4 = 700 \text{ g } CH_4$$

Check: The units (g) are correct. The magnitude (700) is not surprising since the volume of a house is large.

107. **Given:** m (ice) = 9.0 g; Coffee: T_1 = 90.0 °C, m = 120.0 g, $C_s = C_{H2O}$, $\Delta H°_{fus}$ = 6.0 kJ/mol
Find: T_f of coffee
Conceptual plan: $q_{ice} = - q_{coffee}$ **so g (ice) → mol (ice) → q_{fus}(kJ) → q_{fus} (J) → q_{coffee} (J) then**

$$\frac{1\text{ mol}}{18.01\text{ g}} \qquad \frac{6.0\text{ kJ}}{1\text{ mol}} \qquad \frac{1000\text{ J}}{1\text{ kJ}} \qquad q_{coffee} = - q_{ice}$$

q, m, C_s → ΔT then T_i, ΔT → T_2 now we have slightly cooled coffee in contact with 0.0 °C water

$$q = m C_S \Delta T \qquad \Delta T = T_2 - T_i$$

so $q_{ice} = - q_{coffee}$ **with m, C_s, T_i → T_f**

$$q = m\, C_s\left(T_f - T_i\right) \quad \text{then set } q_{Al} = - q_{H2O}$$

Solution: $9.0\ \text{g} \times \dfrac{1\text{ mol}}{18.01\text{ g}} \times \dfrac{6.0\text{ kJ}}{1\text{ mol}} \times \dfrac{1000\text{ J}}{1\text{ kJ}} = 2.9982 \times 10^3\text{ J}$, $q_{coffee} = - q_{ice} = - 2.9982 \times 10^3$ J

$q = m C_S \Delta T$ Rearrange to solve for ΔT. $\Delta T = \dfrac{q}{m C_S} = \dfrac{-2.9982 \times 10^3\text{ J}}{120.0\text{ g} \times 4.18 \dfrac{\text{J}}{\text{g} \times °\text{C}}} = -5.977\ °\text{C}$ then

$\Delta T = T_2 - T_i$.

Rearrange to solve for T_2. $T_2 = \Delta T + T_i = -5.977\ °\text{C} + 90.0\ °\text{C} = 84.0225\ °\text{C}$ $q = m\, C_s\left(T_f - T_i\right)$

substitute in values and set $q_{H2O} = - q_{coffee}$.

$$q_{H2O} = m_{H2O}\, C_{H2O}\left(T_f - T_{H2Oi}\right) = 9.0\text{ g} \times 4.18 \frac{\text{J}}{\text{g} \times °\text{C}} \times \left(T_f - 0.0\ °\text{C}\right) =$$

$$- q_{coffee} = - m_{coffee}\, C_{coffee}\left(T_f - T_{coffee2}\right) = -120.0\text{ g} \times 4.18 \frac{\text{J}}{\text{g} \times °\text{C}} \times \left(T_f - 84.0225\ °\text{C}\right)$$

Rearrange to solve for T_f.

$$9.0\text{ g } T_f = -120.0\text{ g}\left(T_f - 84.0225\ °\text{C}\right) \rightarrow 9.0\text{ g } T_f = -120.0\text{ g } T_f + 10082.7\text{ g} \rightarrow$$

$$-10082.7\text{ g} = -129.0 \frac{\text{g}}{°\text{C}} T_f \rightarrow T_f = \frac{-10082.7\text{ g}}{-129.0 \dfrac{\text{g}}{°\text{C}}} = 78.2\ °\text{C}$$

Check: The units (°C) are correct. The temperature is closer to the original coffee temperature since the mass of coffee is so much larger than the ice mass.

109. $KE = ½\, mv^2$, for an ideal gas $v = u_{rms} = \sqrt{\dfrac{3RT}{\mathfrak{M}}}$ and so $KE_{avg} = \dfrac{1}{2} N_A m u_{rms}^2 = \dfrac{3}{2} RT$ then

$\Delta E_{sys} = KE_2 - KE_1 = \dfrac{3}{2} RT_2 - \dfrac{3}{2} RT_1 = \dfrac{3}{2} R\Delta T$. At constant V $\Delta E_{sys} = C_V \Delta T$ so $C_V = \dfrac{3}{2} R$. At constant P, $\Delta E_{sys} = q + w = q_P - P\Delta V = \Delta H - P\Delta V$ but since $PV = nRT$, for one mole of an ideal gas at constant P $P\,\Delta V = R\Delta T$ so $\Delta E_{sys} = q + w = q_P - P\Delta V = \Delta H - P\Delta V = \Delta H - R\Delta T$ then

$\dfrac{3}{2} R\Delta T = \Delta H - R\Delta T$ or

$\Delta H = \dfrac{5}{2} R\Delta T = C_P \Delta T$ so $C_P = \dfrac{5}{2} R$.

111. d) Only one answer is possible. $\Delta E_{sys} = - \Delta E_{surr}$

113. a) At constant P, $\Delta E_{sys} = q + w = q_P + w = \Delta H + w$ so $\Delta E_{sys} - w = \Delta H$

115. The aluminum cylinder will be cooler after one hour because it has a lower heat capacity than water (less heat needs to be pulled out for every °C temperature change).

117. The internal energy of a chemical system is the sum of its kinetic energy and its potential energy. It is this potential energy that is the energy source in an exothermic chemical reaction. Under normal circumstances, chemical potential energy (or simply chemical energy) arises primarily from the electrostatic forces between

the protons and electrons that compose the atoms and molecules within the system. In an exothermic reaction, some bonds break and new ones form, and the protons and electrons go from an arrangement of higher potential energy to one of lower potential energy. As they rearrange, their potential energy is converted into kinetic energy, the heat emitted in the reaction and so it feels hot to the touch.

Chapter 7
The Quantum - Mechanical Model of the Atom

1. When a particle is absolutely small it means that you cannot observe it without disturbing it. When you observe the particle, it behaves differently than when you do not observe it. Electrons fit this description.

3. The quantum – mechanical model of the atom is important because it explains how electrons exist in atoms and how those electrons determine the chemical and physical properties of elements.

5. The wavelength (λ) of the wave is the distance in space between adjacent crests and is measured in units of distance. The amplitude of the wave is the vertical height of a crest. The more closely spaced the waves, that is, the shorter the wavelength, the more energy there is. The amplitude of the electric and magnetic field wave in light determines the intensity or brightness of the light. The higher the amplitude, the more energy the wave has.

7. For visible light, wavelength determines the color. Red light has a wavelength of 750 nm, the longest wavelength of visible light, and blue has a wavelength of 500 nm.

9. a) Gamma rays(γ) – the wavelength range is 10^{-11} to 10^{-15} m. Gamma rays are produced by the sun, other stars, and certain unstable atomic nuclei of Earth. Human exposure to gamma rays is dangerous because the high energy of gamma rays can damage biological molecules.

   b) X – rays – the wavelength range is 10^{-8} to 10^{-11} m. X – rays are used in medicine. X – rays pass through many substances that block visible light and are therefore used to image bones and internal organs. X – rays are sufficiently energetic to damage biological molecules so, while several yearly exposures to X – rays are harmless, excessive exposure increases cancer risk.

   c) Ultraviolet radiation (UV) – the wavelength range is 0.4 x 10^{-6} to 10^{-8} m. Ultraviolet radiation is most familiar as the component of sunlight that produces a sunburn or suntan. While not as energetic as gamma rays or X – rays, ultraviolet light still carries enough energy to damage biological molecules. Excessive exposure to ultraviolet light increases the risk of skin cancer and cataracts and causes premature wrinkling of the skin.

   d) Visible light – the wavelength range is 0.75 x 10^{-6} to 0.4 x 10^{-6} m (750 nm to 400 nm). Visible light – as long as the intensity is not too high – does not carry enough energy to damage biological molecules. It does, however, cause certain molecules in our eyes to change their shape, sending a signal to brains that results in vision.

   e) Infrared radiation (IR) – the wavelength range is 0.75 x 10^{-6} to 10^{-3} m. The heat you feel when you place your hand near a hot object is infrared radiation. All warm objects, including human bodies, emit infrared light. Although infrared light is invisible to our eyes, infrared sensors can detect it and are often used in night vision technology to "see" in the dark.

   f) Microwave radiation – the wavelength range is 10^{-3} to 10^{-1} m. Microwave radiation is used in radar and in microwave ovens. Microwave radiation is efficiently absorbed by water and can therefore heat substance that contain water.

   g) Radio waves – the wavelength range is 10^{-1} to 10^{5} m. Radio waves are used to transmit the signals responsible for AM and FM radio, cellular telephones, television, and other forms of communication.

11. Diffraction occurs when a wave encountes an obstacle or a slit that is comparable in size to its wavelength. The wave bends around the slit. The diffraction of light through two slits separated by a distance comparable to the wavelength of the light results in an interference pattern. Each slit acts as a new wave source, and the two new waves interfere with each other. Which result in a pattern of bright and dark lines.

Interference From Two Slits

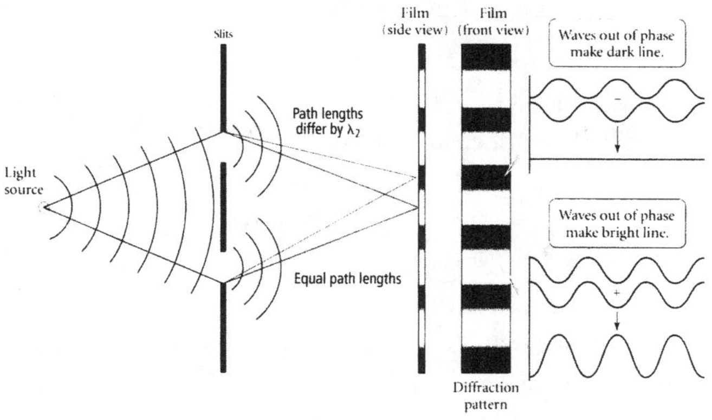

13. Because of the results of the experiments with the photoelectric effect, Einstein proposed that light energy must come in packets. The amount of energy in a light packet depends on its frequency (wavelength). The emission of electrons depends on whether or not a single photon has sufficient energy to dislodge a single electron.

15. An emission spectrum occurs when an atom absorbs energy and reemits that energy as light. The light emitted contains distinct wavelengths for each element. The emission spectrum of a particular element is always the same and can be used to identify the element. A white light spectrum is continuous, meaning that there are no sudden interruptions in the intensity of the light as a function of wavelengths. It consists of all wavelengths. Emission spectra are not continuous. They consist of bright lines at specific wavelengths, with complete darkness in between.

17. Electron diffraction occurs when an electron beam is aimed at two closely spaced slits, and a series of detectors is arranged to detect the electrons after they pass through the slits. An interference pattern similar to that observed for light is recorded behind the slits. Electron diffraction was evidence of the wave nature of electrons.

19. Complementary properties are those that exclude one another. The more you know about one, the less you know about the other. Which of two complementary properties you observe depends on the experiment you perform. In electron diffraction, when you try to observe which hole the electron goes through (particle nature) you lose the interference pattern (wave nature). When you try to observe the interference pattern, you cannot determine which hole the electron goes through.

21. A trajectory is a path that is determined by the particle's velocity (the speed and direction of travel), its position, and the forces acting on it. Both position and velocity are required to predict a trajectory.

23. Deterministic means that the present determines the future. That means that under identical condition, identical results will occur.

25. A probability distribution map is a statistical map that shows where an electron is likely to be found under a given set of conditions.

27. An orbital is a probability distribution map showing where the electron is likely to be found.

29. The principal quantum number (n) is an integer and has possible values of 1,2,3,... The principal quantum number determines the overall size and energy of an orbital.

31. The magnetic quantum number (m_l) is an integer ranging from $-l$ to $+l$. For example, if $l = 1$, $m_l = -1, 0, +1$. The magnetic quantum number specifies the orientation of the orbital.

33. The probability density is the probability per unit volume of finding the electron at a point in space. The radial distribution function represents the total probability of finding the electron within a thin spherical shell at a distance r from the nucleus. In contrast to probability density, which has a maximum at the nucleus for an s orbital, the radial distribution function has a value of zero at the nucleus. It increases to a maximum and then decreases again with increasing r.

35. The sublevels are s (l=0) which can hold a maximum of 2 electrons; p (l = 1) which can hold a maximum of 6 electrons, d (l = 2), which can hold a maximum of 10 electrons; and f (l = 3) which can hold a maximum of 14 electrons.

37. **Given:** distance to sun = 1.496×10^8km **Find:** time for light to travel from sun to Earth
Conceptual Plan: distance km → distance m → time

$$time = \frac{\text{distance}}{3.00 \times 10^8 \text{ m/s}}$$

Solution: $1.496 x 10^8 \cancel{km}\, x \frac{1000 \cancel{m}}{\cancel{km}} x \frac{s}{3.00 x 10^8 \cancel{m}} = 499\, s$

Check: The units of the answer, seconds, is correct. The magnitude of the answer is reasonable, since it corresponds to about 8 min.

39. i) By increasing wavelength the order is: d) ultraviolet < c) infrared < b) microwave < a) radio waves

ii) By increasing energy the order is: a) radio waves < b) microwaves < c) infrared < d) ultraviolet

41. a) **Given:** λ = 632.8 nm **Find:** frequency (ν)
Conceptual Plan: nm → m → ν

$$\frac{m}{10^9\, nm} \quad \mathbf{c}\, \nu = \frac{c}{\lambda}$$

Solution:

$632.8 \cancel{nm}\, x \frac{m}{10^9 \cancel{nm}} = 6.328 x 10^{-7}\, m$ $\nu = \frac{3.00 x 10^8 \cancel{m}}{s} x \frac{1}{6.328 x 10^{-7} \cancel{m}} = 4.74 x 10^{14}\, s^{-1}$

Check: The units of the answer, s^{-1}, are correct. The magnitude of the answer seems reasonable since wavelength and frequency are inversely proportional.

b) **Given:** λ = 503 nm **Find:** frequency (ν)
Conceptual Plan: nm → m → ν
o

Solution: $503 \cancel{nm}\, x \frac{m}{10^9 \cancel{nm}} = 5.03 x 10^{-7}\, m$ $\nu = \frac{3.00 x 10^8 \cancel{m}}{s} x \frac{1}{5.03 x 10^{-7} \cancel{m}} = 5.96 x 10^{14}\, s^{-1}$

Check: The units of the answer, s^{-1}, are correct. The magnitude of the ans $E = h\nu$ wer seems reasonable since wavelength and frequency are inversely proportional.

c) **Given:** λ = 0.052 nm **Find:** frequency (ν)
Conceptual Plan: nm → m → ν

$$\frac{m}{10^9\, nm} \quad \nu = \frac{c}{\lambda}$$

Solution: $0.052 \cancel{nm}\, x \frac{m}{10^9 \cancel{nm}} = 5.2 x 10^{-9}\, m$ $\nu = \frac{3.00 x 10^8 \cancel{m}}{s} x \frac{1}{5.2 x 10^{-9} \cancel{m}} = 5.8 x 10^{18}\, s^{-1}$

Check: The units of the answer, s^{-1}, are correct. The magnitude of the answer seems reasonable since wavelength and frequency are inversely proportional.

43. a) **Given:** frequency (ν) from 41 a. = $4.74 \times 10^{14}\ s^{-1}$ **Find:** Energy
Conceptual Plan: ν → E

$h = 6.626 \times 10^{-34}$ J s

Solution: $6.626 x 10^{-34}\, J \cancel{s}\, x \frac{4.74 x 10^{14}}{\cancel{s}} = 3.14 x 10^{-19}\, J$

Check: The units of the answer, J, are correct. The magnitude of the answer is reasonable since we are talking about the energy of one photon.

b) **Given:** frequency (ν) from 41 b. = $5.96 \times 10^{14}\ s^{-1}$ **Find:** Energy
Conceptual Plan: ν → E

$E = h\nu$ h = 6.626 x 10^{-34} J s

Solution: $6.626 x 10^{-34}\ J \cancel{s}\ x \frac{5.96 x 10^{14}}{\cancel{s}} = 3.95 x 10^{-19}\ J$

Check: The units of the answer, J, are correct. The magnitude of the answer is reasonable since we are talking about the energy of one photon.

c) **Given:** frequency (ν) from 41 c. = 5.8 x 10^{18} s^{-1} **Find:** Energy
Conceptual Plan: $\nu \rightarrow E$

$E = h\nu$ h = 6.626 x 10^{-34} J s

Solution: $6.626 x 10^{-34}\ J \cancel{s}\ x \frac{5.8 x 10^{18}}{\cancel{s}} = 3.8 x 10^{-15}\ J$

Check: The units of the answer, J, are correct. The magnitude of the answer is reasonable since we are talking about the energy of one photon.

45. **Given:** λ = 532 nm and E_{pulse} = 4.67 mJ **Find:** number of photons
Conceptual Plan: nm → m → E_{photon} → number of photons

$$\frac{m}{10^9\ nm} \quad E = \frac{hc}{\lambda}; h = 6.626 x 10^{-34}\ J\ s \quad \frac{E_{pulse}}{E_{photon}}$$

Solution:

$$532\ \cancel{nm}\ x \frac{m}{10^9\ \cancel{nm}} = 5.32 x 10^{-7}\ m \qquad E = \frac{6.626 \text{ x } 10^{-34}\ J\ \cancel{s}\ x\ \frac{3.00 \text{ x } 10^8\ \cancel{m}}{\cancel{s}}}{5.32 x 10^{-7}\ \cancel{m}} = 3.7\underline{3}64 x 10^{-19}\ J / photon$$

$$4.67\ \cancel{mJ}\ x \frac{\cancel{J}}{1000\ \cancel{mJ}}\ x \frac{1\ photon}{3.7\underline{3}64 x 10^{-19}\ \cancel{J}} = 1.25 x 10^{16}\ photons$$

Check: The units of the answer, number of photons, is correct. The magnitude of the answer is reasonable for the amount of energy involved.

47. a) **Given:** λ = 1500 nm **Find:** E for 1 mol photons
Conceptual Plan: nm → m → E_{photon} → $E(J)_{mol}$ → $E(kJ)_{mol}$

$$\frac{m}{10^9\ nm} \quad E = \frac{hc}{\lambda} \quad \frac{mol}{6.022 x 10^{23}\ photons} \quad \frac{kJ}{1000\ J}$$

Solution:

$$1500\ \cancel{nm}\ x \frac{m}{10^9\ \cancel{nm}} = 1.500 x 10^{-6}\ m \qquad E = \frac{6.626 \text{ x } 10^{-34}\ J\ \cancel{s}\ x\ \frac{3.00 \text{ x } 10^8\ \cancel{m}}{\cancel{s}}}{1.500 x 10^{-6}\ \cancel{m}} = 1.3\underline{2}52 x 10^{-19}\ J / photon$$

$$\frac{1.3\underline{2}52 x 10^{-19}\ \cancel{J}}{\cancel{photon}}\ x \frac{6.022 x 10^{23}\ \cancel{photons}}{mol}\ x \frac{kJ}{1000\ \cancel{J}} = 79.8\ kJ / mol$$

Check: The units of the answer, kJ/mol, are correct. The magnitude of the answer is reasonable for a wavelength in the infrared region.

b) **Given:** λ = 500 nm **Find:** E for 1 mol photons
Conceptual Plan: nm → m → E_{photon} → E_{mol} → $E(kJ)_{mol}$

$$\frac{m}{10^9\ nm} \quad E = \frac{hc}{\lambda} \quad \frac{mol}{6.022 x 10^{23}\ photons} \quad \frac{kJ}{1000\ J}$$

Solution:

$$500\ nm\ x\frac{m}{10^9\ nm}=5.00x10^{-7}\ m \qquad E=\frac{6.626\ x\ 10^{-34}\ J\ s\ x\ \dfrac{3.00\ x\ 10^8\ m}{s}}{5.00x10^{-7}\ m}=3.9\underline{7}56x10^{-19}\ J/photon$$

$$\frac{3.9\underline{7}56x10^{-19}\ J}{photon}x\frac{6.022x10^{23}\ photons}{mol}x\frac{kJ}{1000\ J}=239\ kJ/mol$$

Check: The units of the answer, kJ/mol, are correct. The magnitude of the answer is reasonable for a wavelength in the visible region.

c) **Given:** λ = 150 nm **Find:** E for 1 mol photons
Conceptual Plan: nm → m → E_{photon} → E_{mol} → $E(kJ)_{mol}$

$$\frac{m}{10^9 nm} \quad E=\frac{hc}{\lambda} \quad \frac{mol}{6.022x10^{23}\ photons} \quad \frac{kJ}{1000\,J}$$

Solution:

$$1.50\ nm\ x\frac{m}{10^9\ nm}=1.50x10^{-7}\ m \qquad E=\frac{6.626\ x\ 10^{-34}\ J\ s\ x\ \dfrac{3.00\ x\ 10^8\ m}{s}}{1.50x10^{-7}\ m}=1.3\underline{2}52x10^{-18}\ J/photon$$

$$\frac{1.3\underline{2}52x10^{-18}\ J}{photon}x\frac{6.022x10^{23}\ photons}{mol}x\frac{kJ}{1000\ J}=798\ kJ/mol$$

Check: The units of the answer, kJ/mol, are correct. The magnitude of the answer is reasonable for a wavelength in the ultraviolet region. Note: the energy increases from the IR to the Vis to the UV as expected.

49. The interference pattern would be a series of light and dark lines.

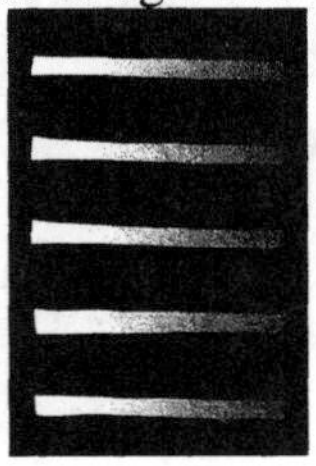

51. **Given:** v = 1.15 x 10^5 m/s **Find:** λ
Conceptual Plan: v → λ

$$\lambda=\frac{h}{mv}$$

Solution: $$\frac{6.626x10^{-34}\ \dfrac{kg\bullet m^2}{s^2}\bullet s}{(9.11x10^{-31}\ kg)(1.15x10^5\ \dfrac{m}{s})}=6.32x10^{-9}\ m$$

Check: The units of the answer, m, are correct. The magnitude of the answer is very small, as expected for the wavelength of an electron.

53. **Given:** m = 143 g; v = 95 mph **Find:** λ
Conceptual Plan: m,v → λ

$$\lambda=\frac{h}{mv}$$

Solution: $$\frac{6.626x10^{-34}\ \dfrac{kg\bullet m^2}{s^2}\bullet s}{(143\ g)\left(\dfrac{kg}{1000\ g}\right)\left(\dfrac{95\ mi}{hr}\right)\left(\dfrac{1.609\ km}{mi}\right)\left(\dfrac{1000\ m}{km}\right)\left(\dfrac{hr}{3600\ s}\right)}=1.1x10^{-34}\ m$$

The value of the wavelength, 1.1×10^{-34} m, is so small it will not have an effect on the trajectory of the baseball.

Check: The units of the answer, m, are correct. The magnitude of the answer is very small as would be expected for the de Broglie wavelength of a baseball.

55. Since the size of the orbital is determined by the n quantum, with the size increasing with increasing n, an electron in a 2s orbital is closer, on average, to the nucleus than an electron in a 3s orbital.

57. The value of l is an integer that lies between 0 and $n - 1$.
    a) When $n = 1$, l can only be: $l = 0$.

    b) When $n = 2$, l can be: $l = 0$ or $l = 1$.

    c) When $n = 3$, l can be: $l = 0$, $l = 1$, or $l = 2$.

    d) When $n = 4$, l can be: $l = 0$, $l = 1$, $l = 2$, or $l = 3$.

59. The quantum number is set c cannot occur together to specify an orbital. l must lie between 0 and $n - 1$, so for $n = 3$, l can only be as high as 2.

61. The 2s orbital would be the same shape as the 1s orbital but would be larger in size and the 3p orbitals would have the same shape as the 2p orbitals but would be larger in size. Also, the 2s and 3p orbitals would have more nodes.

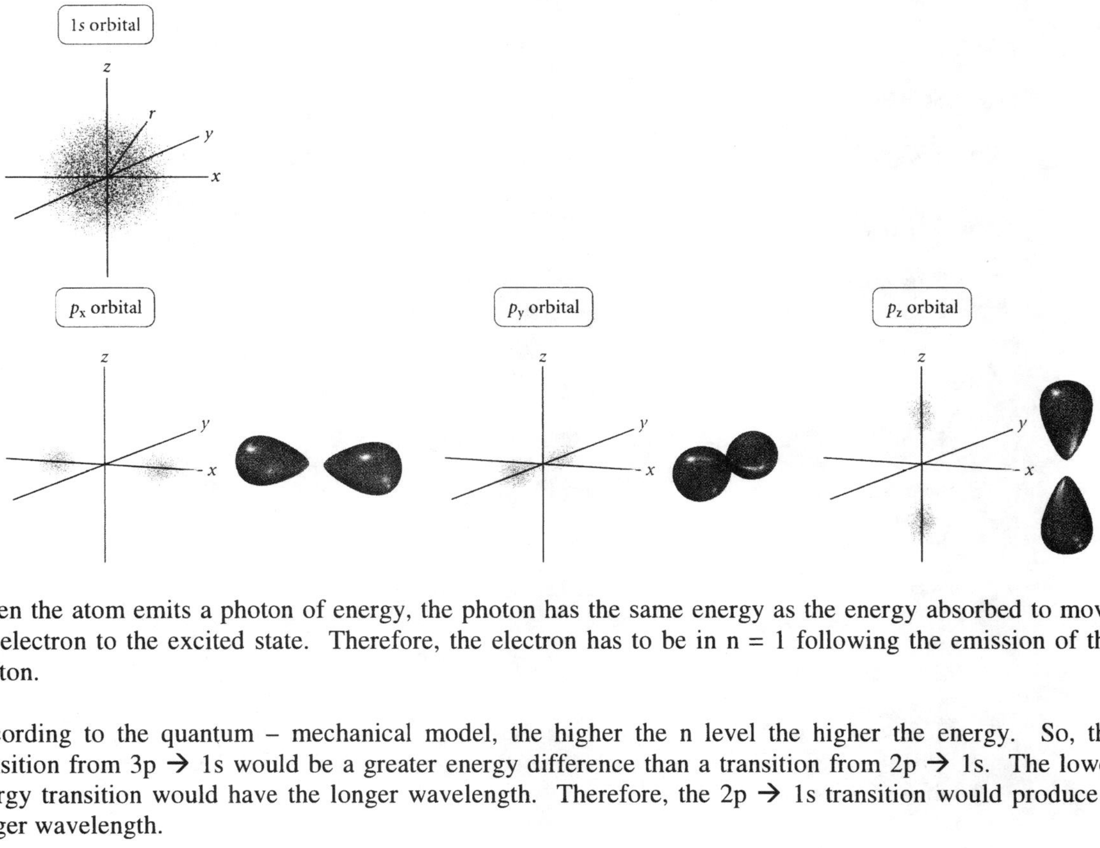

63. When the atom emits a photon of energy, the photon has the same energy as the energy absorbed to move the electron to the excited state. Therefore, the electron has to be in n = 1 following the emission of the photon.

65. According to the quantum – mechanical model, the higher the n level the higher the energy. So, the transition from 3p → 1s would be a greater energy difference than a transition from 2p → 1s. The lower energy transition would have the longer wavelength. Therefore, the 2p → 1s transition would produce a longer wavelength.

67. a) **Given:** n = 2 → n = 1 **Find:** λ

Conceptual Plan: n =1, n = 2 → ΔE_{atom} → ΔE_{photon} → λ

$$\Delta E_{atom} = E_1 - E_2 \quad \Delta E_{atom} \rightarrow -\Delta E_{photon} \qquad E = \frac{hc}{\lambda}$$

Solution:

$$\Delta E = E_1 - E_2 \quad = \quad -2.18 x10^{-18}\, J\left(\frac{1}{1^2}\right) - \left[-2.18 x10^{-18}\left(\frac{1}{2^2}\right)\right] = -2.18 x10^{-18}\, J\left[\left(\frac{1}{1^2}\right) - \left(\frac{1}{2^2}\right)\right] = -1.635 x10^{-18}\, J$$

$\Delta E_{photon} = -\Delta E_{atom} = 1.6\underline{3}5 x 10^{-18} J$ $\lambda = \frac{hc}{E} = \frac{(6.626 x 10^{-34} J \cdot s)(3.00 x 10^8 m/s)}{1.6\underline{3}5 x 10^{-18} J} = 1.22 x 10^{-7} m$

This transition would produce a wavelength in the UV region.
Check: The units of the answer, m, are correct. The magnitude of the answer is reasonable since it is in the region of UV radiation.

b) **Given:** n = 3 → n = 1 **Find:** λ
Conceptual Plan: n =1, n = 3 → ΔE_{atom} → ΔE_{photon} → λ

$\Delta E_{atom} = E_1 - E_3$ $\Delta E_{atom} \rightarrow -\Delta E_{photon}$ $E = \frac{hc}{\lambda}$

Solution:

$$\Delta E = E_1 - E_3 = -2.18 x 10^{-18} J\left(\frac{1}{1^2}\right) - \left[-2.18 x 10^{-18}\left(\frac{1}{3^2}\right)\right] = -2.18 x 10^{-18} J\left[\left(\frac{1}{1^2}\right) - \left(\frac{1}{3^2}\right)\right] = -1.9\underline{3}8 x 10^{-18} J$$

$\Delta E_{photon} = -\Delta E_{atom} = 1.9\underline{3}8 x 10^{-18} J$ $\lambda = \frac{hc}{E} = \frac{(6.626 x 10^{-34} J \cdot s)(3.00 x 10^8 m/s)}{1.9\underline{3}8 x 10^{-18} J} = 1.03 x 10^{-7} m$

This transition would produce a wavelength in the UV region.
Check: The units of the answer, m, are correct. The magnitude of the answer is reasonable since it is in the region of UV radiation.

c) **Given:** n = 4 → n = 2 **Find:** λ
Conceptual Plan: n =2, n = 4 → ΔE_{atom} → ΔE_{photon} → λ

$\Delta E_{atom} = E_2 - E_4$ $\Delta E_{atom} \rightarrow -\Delta E_{photon}$ $E = \frac{hc}{\lambda}$

Solution:

$$\Delta E = E_2 - E_4 = -2.18 x 10^{-18} J\left(\frac{1}{2^2}\right) - \left[-2.18 x 10^{-18}\left(\frac{1}{4^2}\right)\right] = -2.18 x 10^{-18} J\left[\left(\frac{1}{2^2}\right) - \left(\frac{1}{4^2}\right)\right] = -4.0\underline{8}7 x 10^{-18} J$$

$\Delta E_{photon} = -\Delta E_{atom} = 4.0\underline{8}7 x 10^{-19} J$ $\lambda = \frac{hc}{E} = \frac{(6.626 x 10^{-34} J \cdot s)(3.00 x 10^8 m/s)}{4.0\underline{8}7 x 10^{-18} J} = 4.86 x 10^{-7} m$

This transition would produce a wavelength in the Visible region.
Check: The units of the answer, m, are correct. The magnitude of the answer is reasonable since it is in the region of Visible light.

d) **Given:** n = 5 → n = 2 **Find:** λ
Conceptual Plan: n =2, n = 5 → ΔE_{atom} → ΔE_{photon} → λ

$\Delta E_{atom} = E_2 - E_5$ $\Delta E_{atom} \rightarrow -\Delta E_{photon}$ $E = \frac{hc}{\lambda}$

Solution:

$$\Delta E = E_2 - E_5 = -2.18 x 10^{-18} J\left(\frac{1}{2^2}\right) - \left[-2.18 x 10^{-18}\left(\frac{1}{5^2}\right)\right] = -2.18 x 10^{-18} J\left[\left(\frac{1}{2^2}\right) - \left(\frac{1}{5^2}\right)\right] = -4.5\underline{7}8 x 10^{-19} J$$

$\Delta E_{photon} = -\Delta E_{atom} = 4.5\underline{7}8 x 10^{-18} J$ $\lambda = \frac{hc}{E} = \frac{(6.626 x 10^{-34} J \cdot s)(3.00 x 10^8 m/s)}{4.5\underline{7}8 x 10^{-18} J} = 4.34 x 10^{-7} m$

This transition would produce a wavelength in the Visible region.
Check: The units of the answer, m, are correct. The magnitude of the answer is reasonable since it is in the region of Visible light.

69. **Given:** n(initial) = 7 λ = 397 nm **Find:** n(final)
Conceptual Plan: λ → ΔE_{photon} → ΔE_{atom} → n = x, n = 7

$E = \frac{hc}{\lambda}$ $\Delta E_{photon} \rightarrow -\Delta E_{atom}$ $\Delta E_{atom} = E_x - E_7$

Solution: $E = \frac{hc}{\lambda} = \frac{(6.626 x 10^{-34}\ J \cdot \cancel{s})(3.00 x 10^{8}\ \cancel{m}/\cancel{s})}{(397\ \cancel{nm})\left(\frac{\cancel{m}}{10^9\ \cancel{nm}}\right)} = 5.0\underline{0}7 x 10^{-19}\ J$

$\Delta E_{atom} = -\ \Delta E_{photon} = -5.0\underline{0}7 x 10^{-19}\ J$

$\Delta E = E_x - E_7 \quad = -5.0\underline{0}7 x 10^{-19} = -2.18 x 10^{-18}\ J\left(\frac{1}{x^2}\right) - \left[-2.18 x 10^{-18}\left(\frac{1}{7^2}\right)\right] = -2.18 x 10^{-18}\ J\left[\left(\frac{1}{x^2}\right) - \left(\frac{1}{7^2}\right)\right]$

$0.2297 = \left(\frac{1}{x^2}\right) - \left(\frac{1}{7^2}\right) \qquad 0.25229 = \left(\frac{1}{x^2}\right) \qquad x^2 = 3.998 \qquad x = 2$

Check: The answer is reasonable since it is an integer less than the initial value of 7.

71. **Given:** 348 kJ/mol **Find:** λ
Conceptual Plan: kJ/mol → kJ/molec → J/molec → λ

$$\frac{6.022 x 10^{23}\ molec}{mol} \quad \frac{1000\ J}{kJ} \quad E = \frac{hc}{\lambda}$$

Solution: $\frac{348\ \cancel{kJ}}{\cancel{mol}} x \frac{\cancel{mol}}{6.022 x 10^{23}\ \cancel{molecules}} x \frac{1000\ J}{\cancel{kJ}} = 5.7\underline{7}9 x 10^{-19}\ J$

$\lambda = \frac{(6.626 x 10^{-34}\ \cancel{J} \cdot \cancel{s})(3.00 x 10^{8}\ m/\cancel{s})}{5.7\underline{7}9 x 10^{-19}\ \cancel{J}} = 3.44 x 10^{-7}\ m = 344\ nm$

Check: The units of the answer, m or nm, is correct. The magnitude of the answer is reasonable since this wavelength is in the UV region.

73. **Given:** E_{pulse} = 5.0 watts; d = 5.5 mm; hole = 1.2 mm; λ = 532 nm **Find:** photons/s
Conceptual Plan: fraction of beam through hole → fraction of power and then E_{photon} → number photons/s

$$\frac{area\ hole}{area\ beam} \quad fraction\ x\ power \quad E = \frac{hc}{\lambda} \quad \frac{power/s}{E/photon}$$

Solution: $A = \pi r^2 \qquad \frac{\pi(0.60 mm)^2}{\pi(2.75 mm)^2} = 0.0476 \qquad 0.0476 x 5.0\ watts\ x \frac{J/s}{watt} = 0.2\underline{3}8\ J/s$

$E_{photon} = \frac{(6.626 x 10^{-34}\ J \cdot \cancel{s})(3.00 x 10^{8}\ \cancel{m}/\cancel{s})}{(532\ \cancel{nm})\left(\frac{\cancel{m}}{10^9\ \cancel{nm}}\right)} = 3.7\underline{3}6 x 10^{-19}\ J/photon$

$\frac{0.2\underline{3}8\ J/s}{3.7\underline{3}6 x 10^{-19}\ J/photon} = 6.4 x 10^{17}\ photons/s$

Check: The units of the answer, number of photons /s, is correct. The magnitude of the answer is reasonable.

75. **Given:** KE = 506 eV **Find:** λ
Conceptual Plan: KE_{ev} → KE_J → v → λ

$$\frac{1.602 x 10^{-19}\ J}{eV} \quad KE = 1/2\, mv^2 \quad \lambda = \frac{h}{mv}$$

Solution:

$$506\,eV\left(\frac{1.602x10^{-19}\,J}{eV}\right)\left(\frac{\frac{kg\bullet m^2}{s^2}}{J}\right)=\frac{1}{2}\left(9.11x10^{-31}\,kg\right)v^2 \qquad v^2=\frac{506\,\cancel{eV}\left(\frac{1.602x10^{-19}\,\cancel{J}}{\cancel{eV}}\right)\left(\frac{\frac{\cancel{kg}\bullet m^2}{s^2}}{\cancel{J}}\right)}{\frac{1}{2}\left(9.11x10^{-31}\,\cancel{kg}\right)}=1.7796x10^{14}\,\frac{m^2}{s^2}$$

$$v=1.33x10^7\,m/s \qquad \lambda=\frac{h}{mv}=\frac{6.626x10^{-34}\,\frac{\cancel{kg}\bullet m^{\cancel{2}}}{\cancel{s}^{\cancel{2}}}\bullet\cancel{s}}{(9.11x10^{-31}\,\cancel{kg})(1.33x10^7\,\cancel{m}/\cancel{s})}=5.47\,x10^{-11}\,m=0.0547\,nm$$

Check: The units of the answer, m or nm, are correct. The magnitude of the answer is reasonable because a deBroglie wavelength is usually a very small number.

77. **Given:** n = 1 → n = ∞ **Find:** E; λ

Conceptual Plan: n = ∞, n = 1 → ΔE_{atom} → ΔE_{photon} → λ

$\Delta E_{atom} = E_\infty - E_1 \Delta E_{atom}$ → ΔE_{photon} $\quad E=\frac{hc}{\lambda}$

Solution: $\Delta E = E_\infty - E_1 \;=\; 0-\left[-2.18x10^{-18}\left(\frac{1}{1^2}\right)\right]=+2.18x10^{-18}\,J$

$\Delta E_{photon} = -\Delta E_{atom} = +2.18x10^{-18}\,J$

$$\lambda=\frac{hc}{E}=\frac{(6.626x10^{-34}\,\cancel{J}\bullet\cancel{s})(3.00x10^8\,m/\cancel{s})}{2.18x10^{-18}\,\cancel{J}}=9.12x10^{-8}\,m=91.2\,nm$$

Check: The units of the answers, J for E and m or nm for part 1, are correct. The magnitude of the answer is reasonable because it would require more energy to completely remove the electron than just moving it to a higher n level. This results in a shorter wavelength.

79. a. **Given:** n = 1 **Find:** number of orbitals if l = 0 → n

Conceptual Plan: value n → values l → values m_l → number of orbitals

l = 0 → n $\quad m_l$ = - 1 → + 1 total m_l

Solution:	n =	1	
	l =	0	1
	m_l =	0	- 1, 0, +1
	total 4 orbitals		

Check: The total orbitals will be equal to the number of l sublevels^2

b. **Given:** n = 2 **Find:** number of orbitals if l = 0 → n

Conceptual Plan: value n → values l → values m_l → number of orbitals

1 = 0 → n $\quad m_l$ = - 1 → + 1 total m_l

Solution:	n =	2		
	l =	0	1	2
	m_l =	0	- 1, 0, +1	- 2,- 1,0,1,2
	total 9 orbitals			

Check: The total orbitals will be equal to the number of l sublevels^2

c. **Given:** n = 3 **Find:** number of orbitals if l = 0 → n

Conceptual Plan: value n → values l → values m_l → number of orbitals

1 = 0 → n $\quad m_l$ = - 1 → + 1 total m_l

Solution:	n =	3			
	l =	0	1	2	3
	m_l =	0	- 1, 0, +1	- 2,- 1,0,1,2	- 3,- 2,- 1,0,1,2,3
	total 16 orbitals				

Check: The total orbitals will be equal to the number of l sublevels^2

81. **Given:** λ = 1875 nm;1282 nm; 1093 nm **Find:** equivalent transitions

Conceptual Plan: $\lambda \rightarrow E_{photon} \rightarrow E_{atom} \rightarrow n$

$$E = \frac{hc}{\lambda} \qquad E_{photon} = -E_{atom} \qquad E = -2.18 x 10^{-18} J\left(\frac{1}{n_f^2} - \frac{1}{n_i^2}\right)$$

Solution: Since the wavelength of the transitions are longer wavelengths than those obtained in the visual region, the electron must relax to a higher n level. Therefore, we can assume that the electron returns to the n = 3 level.

For λ = 1875 nm:

$$E = \frac{(6.626 x 10^{-34} J \bullet s)(3.00 x 10^8 m/s)}{1875 nm\left(\frac{m}{10^9 nm}\right)} = 1.060 x 10^{-19} J \qquad 1.060 x 10^{-19} J = -1.060 x 10^{-19} J$$

$$-1.060 x 10^{-19} J = -2.18 x 10^{-18}\left(\frac{1}{3^2} - \frac{1}{n^2}\right); \ n = 4$$

For λ = 1282 nm:

$$E = \frac{(6.626 x 10^{-34} J \bullet s)(3.00 x 10^8 m/s)}{1282 nm\left(\frac{m}{10^9 nm}\right)} = 1.551 x 10^{-19} J \qquad 1.551 x 10^{-19} J = -1.551 x 10^{-19} J$$

$$-1.551 x 10^{-19} = -2.18 x 10^{-18}\left(\frac{1}{3^2} - \frac{1}{n^2}\right); \ n = 5$$

For λ = 1093 nm:

$$E = \frac{(6.626 x 10^{-34} J \bullet s)(3.00 x 10^8 m/s)}{1093 nm\left(\frac{m}{10^9 nm}\right)} = 1.819 x 10^{-19} J \qquad 1.819 x 10^{-19} J = -1.819 x 10^{-19} J$$

$$-1.819 x 10^{-19} J = -2.18 x 10^{-18}\left(\frac{1}{3^2} - \frac{1}{n^2}\right); \ n = 6$$

Check: The values obtained are all integers, which is correct. The values of n; 4,5,6, are reasonable. The values of n increase as the wavelength decreases because the two n levels involved are further apart and more energy is released as the electron relaxes to the n = 3 level.

83. **Given:** Φ = 193 kJ/mol **Find:** threshold frequency(ν)

Conceptual Plan: Φ **kJ/ mol** $\rightarrow$ Φ **kJ/ atom** $\rightarrow$ Φ **J/ atom** $\rightarrow$ ν

$$\frac{6.022 x 10^{23} molecules}{mol} \qquad \frac{1000 J}{kJ} \qquad \Phi = h\nu$$

Solution: $$\nu = \frac{\Phi}{h} = \frac{\left(\frac{193 kJ}{mol}\right)\left(\frac{mol}{6.022 x 10^{23} atoms}\right)\left(\frac{1000 J}{kJ}\right)}{6.626 x 10^{-34} J \bullet s} = 4.84 x 10^{14} s^{-1}$$

Check: The units of the answer, s^{-1}, are correct. The magnitude of the answer puts the frequency in the infrared range and is a reasonable answer.

85. **Given:** $\nu_{low} = 30 s^{-1}$ $\nu_{hi} = 1.5 \text{ x } 10^4 s^{-1}$; speed = 344 m/s **Find:** $\lambda_{low} - \lambda_{hi}$

Conceptual Plan: $\nu_{low} \rightarrow \lambda_{low}$ and $\nu_{hi} = \lambda_{hi}$ then $\lambda_{low} - \lambda_{hi}$

$\lambda\nu$ = speed

Solution: $$\lambda = \frac{speed}{\nu} \qquad \lambda_{low} = \frac{344 m/s}{30 s^{-1}} = 11 m \qquad \lambda_{hi} = \frac{344 m/s}{1.5 x 10^4 s^{-1}} = 0.023 m \qquad 11 m - 0.023 m = 11 m$$

Check: The units of the answer, m, are correct. The magnitude is reasonable since the value is only determined by the low frequency value because of significant figures.

87. a) **Given:** n = 1, n = 2, n = 3, L = 155 pm **Find:** E_1, E_2, E_3

Conceptual Plan: n → E

$$E_n = \frac{n^2 h^2}{8mL^2}$$

Solution:

$$E_1 = \frac{1^2(6.626x10^{-34}\ J \cdot s)^2}{8(9.11x10^{-31}\ kg)(155\ \cancel{pm})^2\left(\frac{m}{10^{12}\ \cancel{pm}}\right)^2} = \frac{1(6.626x10^{-34})^2 J^2 s^2}{8(9.11x10^{-31} kg)(155x10^{-12})^2 m^2} = \frac{1(6.626x10^{-34})^2\left(\frac{\cancel{kg} \cdot \cancel{m^2}}{\cancel{s^2}}\right) J\ \cancel{s^2}}{8(9.11x10^{-31}\ \cancel{kg})(155x10^{-12})^2\ \cancel{m^2}} = 2.51x10^{-18}\ J$$

$$E_2 = \frac{2^2(6.626x10^{-34}\ J \cdot s)^2}{8(9.11x10^{-31}\ kg)(155\ \cancel{pm})^2\left(\frac{m}{10^{12}\ \cancel{pm}}\right)^2} = \frac{4(6.626x10^{-34})^2 J^2 s^2}{8(9.11x10^{-31} kg)(155x10^{-12})^2 m^2} = \frac{4(6.626x10^{-34})^2\left(\frac{\cancel{kg} \cdot \cancel{m^2}}{\cancel{s^2}}\right) J\ \cancel{s^2}}{8(9.11x10^{-31}\ \cancel{kg})(155x10^{-12})^2\ \cancel{m^2}} = 1.00x10^{-17}\ J$$

$$E_3 = \frac{3^2(6.626x10^{-34}\ J \cdot s)^2}{8(9.11x10^{-31}\ kg)(155\ \cancel{pm})^2\left(\frac{m}{10^{12}\ \cancel{pm}}\right)^2} = \frac{9(6.626x10^{-34})^2 J^2 s^2}{8(9.11x10^{-31} kg)(155x10^{-12})^2 m^2} = \frac{9(6.626x10^{-34})^2\left(\frac{\cancel{kg} \cdot \cancel{m^2}}{\cancel{s^2}}\right) J\ \cancel{s^2}}{8(9.11x10^{-31}\ \cancel{kg})(155x10^{-12})^2\ \cancel{m^2}} = 2.26x10^{-17}\ J$$

Check: The units of the answers, J, are correct. The answers seem reasonable since the energy is increasing with increasing n level.

b) **Given:** n = 1 → n = 2 and n = 2 → n = 3 **Find:** λ
Conceptual Plan: n =1, n = 2 → ΔE_{atom} → ΔE_{photon} → λ

$\Delta E_{atom} = E_2 - E_1$ $\quad$ $\Delta E_{atom} \rightarrow -\Delta E_{photon}$ $\quad$ $E = \frac{hc}{\lambda}$

Solution: Using the energies calculated in part a:

$E_2 - E_1 = (1.00 \times 10^{-17}\ J - 2.51 \times 10^{-18}\ J) = 7.49 \times 10^{-18}\ J$

$$\lambda = \frac{(6.626 \times 10^{-34}\ \cancel{J} \cdot \cancel{s})(3.00x10^{8}\ m/\cancel{s})}{7.49 \times 10^{-18}\ \cancel{J}} = 2.65x10^{-8}\ m = 26.5\ nm$$

$E_3 - E_2 = (2.26 \times 10^{-17}\ J - 1.00 \times 10^{-17}\ J) = 1.26 \times 10^{-17}\ J$

$$\lambda = \frac{(6.626 \times 10^{-34}\ \cancel{J} \cdot \cancel{s})(3.00x10^{8}\ m/\cancel{s})}{1.26 \times 10^{-17}\ \cancel{J}} = 1.58x10^{-8}\ m = 15.8\ nm$$

These wavelengths would lie in the UV region.

Check: The units of the answers, m, are correct. The magnitude of the answers is reasonable based on the energies obtained for the levels.

89. For the 1s orbital: In the excel spread sheet, call column A: r; and column B: ψ(1s). Make the values for r column A: 0 – 200. In the column B put the equation for the wavefunction written as: =(POWER(1/3.1415,1/2))*(1/POWER(53,3/2))*(EXP((-A2/53))). Go to make chart, chose xy scatter.

e.g. sample values

r	ψ(1s)
0	0.00146224
1	0.00143491
2	0.00140809
3	0.00138177
4	0.00135594
5	0.0013306
6	0.00130573

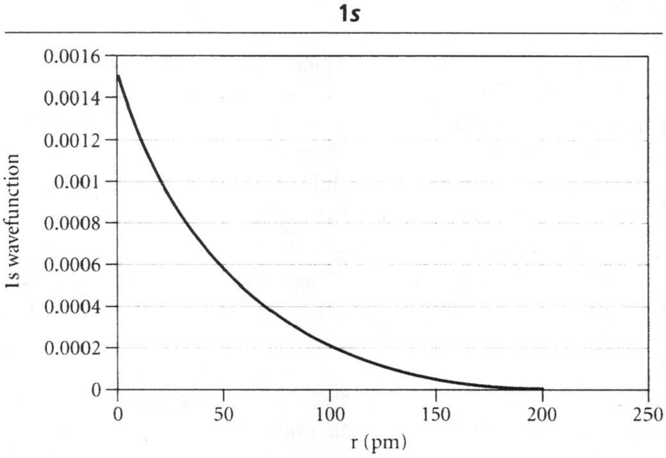

For the 2s orbital: In the same excel spread sheet, call column A: r; and column C: ψ(2s). Use the same values for r in column A: 0 – 200. In the column C put the equation for the wavefunction written as: =(POWER(1/((32)*(3.1415)),1/2))*(1/POWER(53,3/2))*(2-(A2/53))*(EXP((-A2/53))). Go to make chart, chose xy scatter.

e.g.sample values

r	ψ(2s)
0	0.000516979
1	0.00050253
2	0.000488441
3	0.000474702
4	0.000461307
5	0.000448247
6	0.000435513

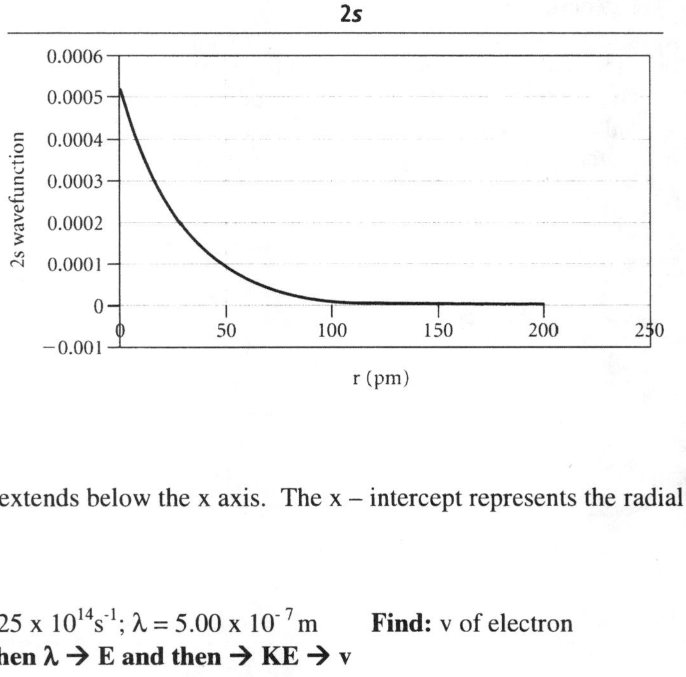

Note: The plot for the 2s orbital extends below the x axis. The x – intercept represents the radial node of the orbital.

91. **Given:** threshold frequency = $2.25 \times 10^{14} s^{-1}$; $\lambda = 5.00 \times 10^{-7}$ m **Find:** v of electron

Conceptual Plan: ν → Φ and then λ → E and then → KE → v

$$\Phi = h\nu \qquad E = \frac{hc}{\lambda} \qquad KE = E - \Phi \qquad KE = 1/2\, mv^2$$

Solution:

$$\Phi = (6.626 x10^{-34} J\bullet \cancel{s})(2.25 x10^{14} \cancel{s^{-1}}) = 1.4\underline{9}1 x10^{-19} J \qquad E = \frac{(6.626 x10^{-34} J\bullet \cancel{s})(3.00 x10^{8} \cancel{m}/\cancel{s})}{5.00 x10^{-7} \cancel{m}} = 3.9\underline{7}6 x10^{-19} J$$

$$KE = 3.976 x10^{-19}\ J - 1.491 x10^{-19}\ J = 2.485 x10^{-19}\ J \qquad v^2 = \frac{2.485 x10^{-19} \frac{kg \cdot m^2}{s^2}}{\frac{1}{2}(9.11x10^{-31}\ kg)} = 5.455 x10^{11} \frac{m^2}{s^2}$$

v = 7.39 x 10^5 m/s

Check: The units of the answer, m/s, are correct. The magnitude of the answer is reasonable for the speed of an electron.

93. In the Bohr model of the atom, the electron travels in a circular orbit around the nucleus. It is a 2 – dimensional model. The electron is constrained to move only from one orbit to another orbit. But, the electron is treated as a particle that behaves according to the laws of classical physics. The quantum – mechanical model of the atom is 3 – dimensional. In this model, we treat the electron, an absolutely small particle, differently than we treat particles with classical physics. The electron is in an orbital, which gives us the probability of finding the electron within a volume of space.
Because the electron in the Bohr model is constrained to a circular orbit, it would theoretically be possible to know both the position and the velocity of the electron simultaneously. This contradicts the Heisenberg uncertainty principle which states that position and velocity are complementary terms which can not both be known with precision.

95. a) Since the interference pattern is caused by single electrons interfering with themselves, the pattern remains the same even when the rate of the electrons passing through the slits is one electron per minute. It will simply take longer for the full pattern to develop.

b) When a light is placed behind the slits to determine which hole the electron passes through, the light flashes to indicate which hole the electron passed through, but the interference pattern is now absent. With the laser on the electrons hit positions directly behind each slit, as if they were ordinary particles.

c) Diffraction occurs when a wave encounter an obstacle of a slit that is comparable in size to its wavelength. The wave bends around the slit. The diffraction of light through two slits separated by a distance comparable to the wavelength of the light results in an interference pattern. Each slit acts as a new wave source, and the two new waves interfere with each other. Which result in a pattern of bright and dark lines.

d) Since the mass of the bullets and their particle size, are not absolutely small, the bullets will not produce an interference pattern when they pass through the slits. The de Broglie wavelength produced by the bullets will not be sufficiently large enough to interfere with the bullet trajectory and no interference pattern will be observed.

Chapter 8
Periodic Properties of the Elements

1. A periodic property is one that is predictable based on the element's position within the periodic table.

3. The first attempt to organize the elements according to similarities in their properties was bade by the German chemist Johann Dobereiner. He grouped elements into triads; three elements with similar properties. A more complex approach was attempted by the English chemist John Newlands. He organized elements into octaves, in analogy to musical notes. When arranged this way, the properties of every eighth element were similar.

5. Meyer proposed an organization of the known elements based on some periodic properties. Moseley listed elements according to the atomic number rather than atomic mass. This resolved the problems in Mendeleev's table where an increase in atomic mass did not correlate with similar properties.

7. Electron spin is a fundamental property of electrons. It is more correctly expressed as saying the electron has inherent angular momentum. The value m_s is the spin quantum number. An electron with $m_s = +1/2$ has a spin opposite of an electron with $m_s = - 1/2$.

9. An electron configuration shows the particular orbitals that are occupied by electrons in an atom. Some examples are: $H = 1s^1$; $He = 1s^2$; $Li = 1s^2 2s^1$

11. The sublevels within a principle level split in multielectron atoms because of penetration of the outer electrons into the region of the core electrons. The sublevels in hydrogen are not split because they are empty in the ground state.

13. The Pauli Exclusion Principle states: No two electrons in an atom can have the same four quantum numbers. Since two electrons occupying the same orbital have three identical quantum numbers (n, l, m_l), they must have different spin quantum numbers. The Pauli exclusion principle implies that each orbital can have a maximum of only two electrons, with opposing spins.

15. In order of increasing energy the orbitals are: 1s < 2s < 2p < 3s < 3p < 4s < 3d < 4p < 5s. The 4s orbital fills before the 3d and the 5s fills before the 4d because they are lower in energy because of greater penetration of the 4s and 5s orbitals.

17.

Orbital Blocks of the Periodic Table

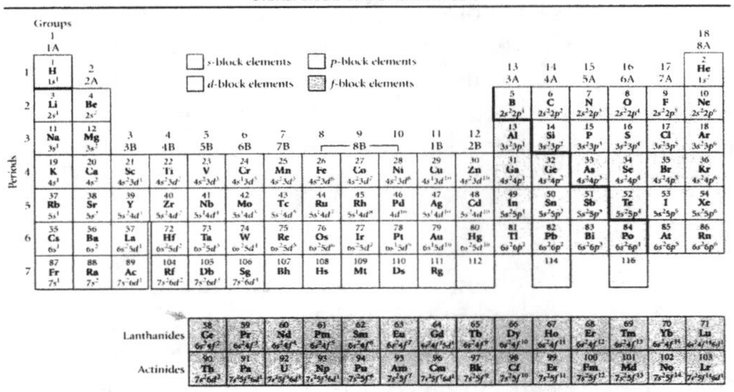

19. The rows in the periodic table grow progressively longer because you are adding sublevels as the n level increases.

21. The row number of a main – group element is equal to the highest principal quantum number of that element. However, the principal quantum number of the d orbital being filled across each row in the transition series is equal to the row number minus one. For the inner transition elements, the principal quantum number of the f orbital being filled across each row is the row number minus two.

23. To use the periodic table to write the electron configuration: find the noble gas that precedes the element. The element has the inner electron configuration of that noble gas. Place the symbol for the noble gas in [ ].

Obtain the outer electron configuration by tracing the element across the period and assigning electrons in the appropriate orbitals.

25. a) The alkali metals (group 1A) have 1 valence electron and are among the most reactive metals because their outer electron configuration (ns^1) is one electron beyond a noble gas configuration. They react to lose the ns^1 electron, obtaining a noble gas configuration. This is why the group 1A metals tend to form 1+ cations.

    b) The alkaline earth metals (group 2A) have 2 valence electrons and have an outer electron configuration of ns^2 and also tend to be reactive metals. They lose their ns^2 electrons to form 2+ cations.

    c) The halogens (group 7A) have 7 valence electrons and have an outer electron configuration of ns^2np^5. They are among the most reactive non – metals. They are only one electron short of a noble gas configuration and tend to react to gain that one electron, forming 1– anions.

    d) The oxygen family (group 6A) has 6 valence electrons and have an outer electron configuration of ns^2np^4. They are 2 electrons short of a noble gas configuration and tend to react to gain those two electrons, forming 2– anions.

27. The effective nuclear charge (Z_{eff}) is the average or net charge from the nucleus experienced by the electrons in the outermost levels. Shielding is the blocking of nuclear charge from the outermost electrons. The shielding is primarily due to the inner (core) electrons although there is some interaction and shielding from the electron repulsions of the outer electrons with each other.

29. a) The radii of transition elements stay roughly constant across each row instead of decreasing in size as in the main group elements. The difference is that, across a row of transition elements, the number of electrons in the outermost principal energy level is nearly constant. As another proton is added to the nucleus with each successive element, another electron is added as well, but the electron goes into an $n_{highest} - 1$ orbital. The number of outermost electrons stays constant and they experience a roughly constant effective nuclear charge, keeping the radius approximately constant.

    b) As you go down the first two rows of a column within the transition metals, the elements follow the same general trend in atomic radii and the main – group elements, i.e., the radii get larger because you are adding outermost electrons into higher n levels.

31. An important exception to simply subtracting the number of electrons occurs for transition metal cations. When writing the electron configuration of a transition metal cation, remove the electrons in the highest n – value orbitals first, even if this does not correspond to the reverse order of filling. Normally, even though the d orbital electrons add after the s orbital electrons, the s orbital electrons are lost first. This is because; 1) the ns and (n – 1)d orbitals are extremely close in energy and depending on the exact configuration can vary in relative energy ordering and 2) as the (n – 1)d orbitals begin to fill in the first transition series, the increasing nuclear charge stabilizes the (n – 1)d orbitals relative to the ns orbitals. This happens because the (n – 1)d orbitals are not outermost orbitals and are therefore not effectively shielded from the increasing nuclear charge by the ns orbitals.

33. The ionization energy (IE) of an atom or ion is the energy required to remove an electron from the atom or ion in the gaseous state. The ionization energy is always positive because removing an electron always takes energy. The energy required to remove the first electron is called the first ionization energy (IE_1). The energy required to remove the second electron is called the second ionization energy (IE_2).

35. Exceptions occur with elements Be, Mg and Ca in group 2A having a higher first ionization energy than elements B, Al, and Ga group 3A. This exception is caused by the change in going form the s block to the p block. The result is that the electrons in the s orbital shield the electron in the p orbital from nuclear charge making it easier to remove.
    Another exception occurs with N, P, and As in group 5A having a higher first ionization energy than O, S, and Se in group 6A. This exception is caused by the repulsion between electrons when they must occupy the same orbital. Group 5A has 3 p electrons while group 6A has 4 p electrons. In the group 5A elements the p orbitals are half – filled which makes the configuration particularly stable. The 4th group 6A electron must pair with another electron making it easier to remove.

37. The electron affinity (EA) of an atom or ion is the energy change associated with the gaining of an electron by the atom in the gaseous state. The electron affinity is usually – though not always – negative because an atom or ion usually releases energy when it gains an electron. The trends in electron affinity are not as regular as trends in other properties. For main – group elements, electron affinity generally becomes more negative as you move to the right across a row in the periodic table. There is not a corresponding trend in electron affinity going down a column.

39. a) The reactions of the alkali metals with halogens results in the formation of metal halides
$2\ M(s) + X_2 \rightarrow 2\ MX(s)$

b) Alkali metals react with water to form the dissolved alkali metal ion, the hydroxide ion and hydrogen gas.
$2\ M(s) + 2\ H_2O(l) \rightarrow 2\ M^+(aq) + 2\ OH^-(aq) + H_2(g)$

41. a) P Phosphorus has 15 electrons. Distribute two of these into the 1s orbital, two into the 2s orbital, six into the 2p orbital, two into the 3s orbital, and three into the 3p orbital. $1s^22s^22p^63s^23p^3$

b) C Carbon has 6 electrons. Distribute two of these into the 1s orbital, two into the 2s orbital, and two into the 2p orbital. $1s^22s^22p^2$

c) Na Sodium has 11 electrons. Distribute two of these into the 1s orbital, two into the 2s orbital, six into the 2p orbital and one into the 3s orbital. $1s^22s^22p^63s^1$

d) Ar Argon has 18 electrons. Distribute two of these into the 1s orbital, two into the 2s orbital, six into the 2p orbital, two into the 3s orbital, and six into the 3p orbital. $1s^22s^22p^63s^23p^6$

43. a) N Nitrogen has 7 electrons and has the electron configuration: $1s^22s^22p^3$. Draw a box for each orbital putting the lowest energy orbital (1s) on the far left and proceeding to orbitals of higher energy to the right. Distribute the 7 electrons into the boxes representing the orbitals, allowing a maximum of two electrons per orbital and remembering Hund's rule. You can see from the diagram that nitrogen has 3 unpaired electrons.

1s	2s	2p		
↓↑	↓↑	↑	↑	↑

b) F Fluorine has 9 electrons and has the electron configuration: $1s^22s^22p^5$. Draw a box for each orbital putting the lowest energy orbital (1s) on the far left and proceeding to orbitals of higher energy to the right. Distribute the 9 electrons into the boxes representing the orbitals, allowing a maximum of two electrons per orbital and remembering Hund's rule. You can see from the diagram that fluorine has 1 unpaired electrons.

1s	2s	2p		
↓↑	↓↑	↓↑	↓↑	↑

c) Mg Magnesium has 12 electrons and has the electron configuration: $1s^22s^22p^63s^2$. Draw a box for each orbital putting the lowest energy orbital (1s) on the far left and proceeding to orbitals of higher energy to the right. Distribute the 12 electrons into the boxes representing the orbitals, allowing a maximum of two electrons per orbital and remembering Hund's rule. You can see from the diagram that magnesium has no unpaired electrons.

1s	2s	2p			3s
↓↑	↓↑	↓↑	↓↑	↓↑	↓↑

d) Al Aluminum has 13 electrons and has the electron configuration: $1s^22s^22p^63s^23p^1$. Draw a box for each orbital putting the lowest energy orbital (1s) on the far left and proceeding to orbitals of higher energy to the right. Distribute the 13 electrons into the boxes representing the orbitals, allowing a maximum of two electrons per orbital and remembering Hund's rule. You can see from the diagram that aluminum has 1 unpaired electron.

1s	2s	2p			3s	3p		
↓↑	↓↑	↓↑	↓↑	↓↑	↓↑	↑		

45. a) P The atomic number of P is 15. The noble gas that precedes P in the periodic table is neon, so the inner electron configuration is [Ne]. Obtain the outer electron configuration by tracing the elements between Ne and P and assigning electrons to the appropriate orbitals. Begin with [Ne]. Because P is in

row 3, add two 3s electrons. Next add three 3p electrons as you trace across the p block to P which is in the third column of the p block.

P $[Ne]3s^23p^3$

b) Ge The atomic number of Ge is 32. The noble gas that precedes Ge in the periodic table is argon, so the inner electron configuration is [Ar]. Obtain the outer electron configuration by tracing the elements between Ar and Ge and assigning electrons to the appropriate orbitals. Begin with [Ar]. Because Ge is in row 4, add two 4s electrons. Next, add ten 3d electrons as you trace across the d block. Finally add two 4p electrons as you trace across the p block to Ge which is in the second column of the p block.

Ge $[Ar]4s^23d^{10}4p^2$

c) Zr The atomic number of Zr is 40. The noble gas that precedes Zr in the periodic table is krypton, so the inner electron configuration is [Kr]. Obtain the outer electron configuration by tracing the elements between Kr and Zr and assigning electrons to the appropriate orbitals. Begin with [Kr]. Because Zr is in row 5, add two 5s electrons. Next, add two 4d electrons as you trace across the d block to Zr which is in the second column.

Zr $[Kr]5s^24d^2$

d) I The atomic number of I is 53. The noble gas that precedes I in the periodic table is Krypton, so the inner electron configuration is [Kr]. Obtain the outer electron configuration by tracing the elements between Kr and I and assigning electrons to the appropriate orbitals. Begin with [Kr]. Because I is in row 5, add two 5s electrons. Next, add ten 4d electrons as you trace across the d block. Finally add five 5p electrons as you trace across the p block to I which is in the fifth column of the p block.

I $[Kr]5s^24d^{10}5p^5$

47. a) Li is in period 2, and the first column in the s block so Li has one 2s electon.

b) Cu is in period 4, and the ninth column in the d block (n – 1) so Cu should have nine 3d electrons, however, it is one of our exceptions so it has ten 3d electrons.

c) Br is in period 4, and the fifth column of the p block, so Br has five 4p electrons.

d) Zr is in period 5, and the second column of the d block (n – 1), so Zr has two 4d electrons.

49. a) In period 4, an element with five valence electrons could be V or As.

b) In period 4, an element with four 4p electrons would be in the fourth column of the p block and is Se.

c) In period 4, an element with three 3d electrons would be in the third column of the d block (n – 1) and is V.

d) In period 4, an element with a complete outer shell would be in the sixth column of the p block and is Kr.

51. a) Ba is in column 2A, so it has two valence electrons.

b) Cs is in column 1A, so it has one valence electron.

c) Ni is in column 8 of the d block, so it has 10 valence electrons (8 from the d block and 2 from the s block)

d) S is in column 6A, so it has six valence electrons.

53. a) The outer electron configuration ns^2 would belong to a reactive metal in the alkaline earth family.

b) The outer electron configuration ns^2np^6 would belong to an unreactive nonmetal in the noble gas family.

c) The outer electron configuration ns^2np^5 would belong to a reactive nonmetal in the halogen family.

d) The outer electron configuration ns^2np^2 would belong to an element in the carbon family. If n = 2, the element is a nonmetal, if n = 3 or 4, the element is a metalloid, and if n = 5 or 6, the element is a metal.

55. The valence electrons in nitrogen would experience a greater effective nuclear charge. Be has 4 protons and N has 7 protons. Both atoms have 2 core electrons that predominately contribute to the shielding while the valence electrons will contribute a slight shielding effect. So, Be has an effective nuclear charge of slightly more than 2+ and N has an effective nuclear charge of slightly more than 5+.

57. a) K(19) $[Ar]4s^1$ Z_{eff} = Z – core electrons = 19 – 18 = 1+

b) Ca(20) $[Ar]4s^2$ Z_{eff} = Z – core electrons = 20 – 18 = 2+

c) O(8) $[He]2s^22p^4$ Z_{eff} = Z – core electrons = 8 – 2 = 6+

d) C(6) $[He]2s^22p^2$ Z_{eff} = Z – core electrons = 6 – 2 = 4+

59. a) Al or In — In atoms are larger than Al atoms because, as you trace the path between Al and In on the periodic table, you move down a column. Atomic size increases as you move down a column because the outermost electrons occupy orbitals with a higher principal quantum number that are therefore larger, resulting in a larger atom.

b) Si or N — Si atoms are larger than N atoms because, as you trace the path between N and Si on the periodic table, you move down a column (atomic size increases) and then to the left across a period (atomic size increases). These effects add together for an overall increase.

c) P or Pb — Pb atoms are larger than P atoms because, as you trace the path between P and Pb on the periodic table, you move down a column (atomic size increases) and then to the left across a period (atomic size increases). These effects add together for an overall increase.

d) C or F — C atoms are larger than F atoms because, as you trace the path between C and F on the periodic table, you move to the right within the same period. As you move to the right across a period, the effective nuclear charge experienced by the outermost electrons increase, resulting in a smaller size.

61. Ca, Rb, S, Si, Ge, F F is above and to the right of the other elements so we start with F as the smallest atom. As you trace a path from F to S you move to the left (size increases) and down (size increases), then you move to Si to the left (size increases) then down to Ge (size increases) then to the left to Ca (size increases) then to the left and down to Rb (size increases). So, in order of increasing atomic radii; F < S < Si < Ge < Ca < Rb

63. a) O^{2-} Begin by writing the electron configuration of the neutral atom.

O $1s^22s^22p^4$

Since this ion has a 2 – charge, add two electrons to write the electron configuration of the ion.

O^{2-} $1s^22s^22p^6$ or [Ar]

b) Br^- Begin by writing the electron configuration of the neutral atom.

Br $[Ar]4s^23d^{10}4p^5$

Since this ion has a 1 – charge, add one electrons to write the electron configuration of the ion.

Br^- $[Ar]4s^23d^{10}4p^6$ or [Kr]

c) Sr^{2+} Begin by writing the electron configuration of the neutral atom.

Sr $[Kr]5s^2$

Since this ion has a 2+ charge, remove two electrons to write the electron configuration of the ion.

Sr^{2+} [Kr]

d) Co^{3+} Begin by writing the electron configuration of the neutral atom.

Co $[Ar]4s^23d^7$

Since this ion has a 3+ charge, remove three electrons to write the electron configuration of the ion. Since it is a transition metal, remove the electrons from the 4s orbital before removing electrons from the 3d orbitals.

Co^{3+} [Ar]$4s^03d^6$

e) Cu^{2+} Begin by writing the electron configuration of the neutral atom. Remember, Cu is one of our exceptions.

Cu [Ar]$4s^13d^{10}$

Since this ion has a 2+ charge, remove two electrons to write the electron configuration of the ion. Since it is a transition metal, remove the electrons from the 4s orbital before removing electrons from the 3d orbitals.

Cu^{2+} [Ar]$4s^03d^9$

65. a) V^{5+} Begin by writing the electron configuration of the neutral atom.

V [Ar]$4s^23d^3$

Since this ion has a 5+ charge, remove five electrons to write the electron configuration of the ion. Since it is a transition metal, remove the electrons from the 4s orbital before removing electrons from the 3d orbitals.

V^{5+} [Ar]$4s^03d^0$ = [Ne]$3s^23p^6$

[Ne] [↓↑] [↓↑ | ↓↑ | ↓↑]

3s 3p

V^{5+} is diamagnetic

b) Cr^{3+} Begin by writing the electron configuration of the neutral atom. Remember, Cr is one of our exceptions.

Cr [Ar]$4s^13d^5$

Since this ion has a 3+ charge, remove three electrons to write the electron configuration of the ion. Since it is a transition metal, remove the electrons from the 4s orbital before removing electrons from the 3d orbitals.

Cr^{3+} [Ar]$4s^03d^3$

[Ar] [ ] [↑ | ↑ | ↑ | |]

4s 3d

Cr^{3+} is paramagnetic

c) Ni^{2+} Begin by writing the electron configuration of the neutral atom.

Ni [Ar]$4s^23d^8$

Since this ion has a 2+ charge, remove two electrons to write the electron configuration of the ion. Since it is a transition metal, remove the electrons from the 4s orbital before removing electrons from the 3d orbitals.

Ni^{2+} [Ar]$4s^03d^8$

[Ar] [ ] [↓↑ | ↓↑ | ↓↑ | ↑ | ↑]

4s 3d

Ni^{2+} is paramagnetic

d) Fe^{3+} Begin by writing the electron configuration of the neutral atom.

Fe [Ar]$4s^23d^6$

Since this ion has a 3+ charge, remove three electrons to write the electron configuration of the ion. Since it is a transition metal, remove the electrons from the 4s orbital before removing electrons from the 3d orbitals.

Fe^{3+} [Ar]$4s^03d^5$

[Ar] [ ] [↑ | ↑ | ↑ | ↑ | ↑]

4s 3d Fe^{3+} is paramagnetic

67. a) Li or Li^+ A Li atom is larger than Li^+ because cations are smaller than the atoms from which they are formed.

b) I^- or Cs^+ An I^- ion is larger than a Cs^+ ion because, although they are isoelectronic, I^- has two fewer protons than Cs^+ resulting in a lesser pull on the electrons and therefore larger radius.

c) Cr or Cr^{3+} A Cr atom is larger than Cr^{3+} because cations are smaller than the atoms from which they are formed.

d) O or O^{2-} An O^{2-} ion is larger than an O atom because anions are larger than the atoms from which they are formed.

69. Since all the species are isoelectronic, the radius will depend on the number of protons in each species. The fewer the protons the larger the radius.
F: Z = 9; Ne: Z = 10; O: Z = 8; Mg: Z = 12; Na: Z = 11
So: $O^{2-} > F^- > Ne > Na^+ > Mg^{2+}$

71. a) Br or Bi Br has a higher ionization energy than Bi because, as you trace the path between Br and Bi on the periodic table, you move down a column (ionization energy decreases) and then to the left across a period (ionization energy decreases). Theses effects sum together for an overall decrease.

b) Na or Rb Na has a higher ionization energy than Rb because, as you trace a path between Na and Rb on the periodic table, you move down a column. Ionization energy decreases as you go down a column because of the increasing size of orbitals with increasing n.

c) As or At Based on periodic trends alone, it is impossible to tell which has a higher ionization energy because, as trace the path between As and At you go to the right across a period (ionization energy increases) and then down a column (ionization energy decreases). These effects tend to oppose each other, and it is not obvious which will dominate.

d) P or Sn P has a higher ionization energy than Sn because, as you trace the path between P and Sn on the periodic table, you move down a column (ionization energy decreases) and then to the left across a period (ionization energy decreases). Theses effects sum together for an overall decrease.

73. Since ionization energy increases as you move to the right across a period and increases as you move up a column, the element with the smallest first ionization energy would be the element farthest to the left and lowest down on the periodic table. So, In has the smallest ionization energy, as you trace a path to the right and up on the periodic table, the next element reached is Si, continuing up and to the right you reach N and then continuing to the right you reach F. So, in the order of increasing first ionization energy the elements are: In < Si < N < F.

75. The jump in ionization energy occurs when you change from removing a valence electron to removing a core electron. To determine where this jump occurs you need to look at the electron configuration of the atom.

a) Be $1s^22s^2$ The first and second ionization energies involve removing 2s electrons, the third ionization energy removes a core electron, so the jump will occur between the second and third ionization energies.

b) N $1s^22s^22p^3$ The first five ionization energies involve removing the 2p and 2s electrons, the sixth ionization energy removes a core electron, so the jump will occur between the fifth and sixth ionization energies.

c) O $1s^22s^22p^4$ The first six ionization energies involve removing the 2p and 2s electrons, the seventh ionization energy removes a core electron, so the jump will occur between the sixth and seventh ionization energies.

d) Li $1s^22s^1$ The first ionization energy involve removing a 2s electrons, the second ionization energy removes a core electron, so the jump will occur between the first and second ionization energies.

77. a) Na or Rb Na has a more negative electron affinity than Rb. In column 1A electron affinity decreases as you go down the column.

b) B or S   S has a more negative electron affinity than B. As you trace from B to S in the periodic table, you move to the right which shows the value of the electron affinity becoming more negative. Also, as you move from period 2 to period 3, the value of the electron affinity becomes more negative. Both of these trends sum together for the value of the electron affinity te become more negative.

c) C or N   C has the more negative electron affinity. As you trace from C to N across the periodic table, you would normally expect N to have the more negative electron affinity. However, N has a half – filled p sublevel which lends it extra stability therefore, it is harder to add an electron.

d) Li or F   F has the more negative electron affinity. As you trace from Li to F on the periodic table you move to the right in the period. As you go to the right across a period, the value of the electron affinity generally becomes more negative.

79. a) Sr or Sb   Sr is more metallic than Sb because, as we trace the path between Sr and Sb on the periodic table, we move to the right within the same period. Metallic character decreases as you go to the right.

b) As or Bi   Bi is more metallic because as we trace a path between As and Bi on the periodic table, we move down a column in the same family (metallic character increases).

c) Cl or O   Based on periodic trends alone, we cannot tell which is more metallic because as we trace the path between O and Cl, we go to the right across a period (metallic character decreases) and then down a column (metallic character increases). These effects tend to oppose each other, and it is not easy to tell which will predominate.

d) S or As   As is more metallic than S because, as we trace the path between S and As on the periodic table, we move down a column (metallic character increases) and then to the left across a period (metallic character increases). These effects add together for an overall increase.

81. The order of increasing metallic character is: S < Se < Sb < In < Ba < Fr.
Metallic character decreases as you move left to right across a period and decreases as you move up a column, therefore, the element with the least metallic character will be to the top – right of the periodic table. So, of these elements, S has the least metallic character. As you move down the column, the next element is Se, as you continue down and then to the right, you reach Sb, continuing to the right goes to In, going down the column and then to the right comes to Ba and then down the column and to the right is Fr.

83. Alkaline earth metals react with halogens to form metal halides. Write the formulas for the reactants and the metal halide product. $Sr(s) + I_2(g) \rightarrow SrI_2(s)$

85. Alkali metals react with water to form the dissolved metal ion, the hydroxide ion, and hydrogen gas. Write the skeletal equation including each of these and then balance it:

$$Li(s) + H_2O(l) \rightarrow Li^+(aq) + OH^-(aq) + H_2(g)$$
$$2\ Li(s) + 2\ H_2O(l) \rightarrow 2\ Li^+(aq) + 2\ OH^-(aq) + H_2(g)$$

87. The halogens react with hydrogen to form hydrogen halides. Write the skeletal reaction with each of the halogen and hydrogen as the reactants and the hydrogen halide compound as the product and balance the equation.

$$H_2(g) + Br_2(g) \rightarrow HBr(g)$$
$$H_2(g) + Br_2(g) \rightarrow 2\ HBr(g)$$

89. Br: $1s^22s^22p^63s^23p^64s^23d^{10}4p^5$
Kr: $1s^22s^22p^63s^23p^64s^23d^{10}4p^6$
Krypton has a completely filled p sublevel giving it chemical stability. Bromine needs one electron to achieve a completely filled p sublevel and therefore has a high electron affinity. It therefore, easily takes on an electron and is reduced to the bromide ion giving it the added stability of the filled p sublevel.

91. Write the electron configuration of vanadium V: $[Ar]\ 4s^23d^3$
Since this ion has a 3+ charge, remove three electrons to write the electron configuration of the ion. Since it is a transition metal, remove the electrons from the 4s orbital before removing electrons from the 3d orbitals.
V^{3+}: $[Ar]\ 4s^03d^2$

Both vanadium and the V^{3+} ion have unpaired electrons and are paramagnetic.

93. Since K^+ has a 1+ charge you would need a cation with a similar size and a 1+ charge. Looking at the ions in the same family, Na^+ would be too small and Rb^+ would be too large. If we then consider Ar^+ and Ca^+ we would have ions of similar size and charge. Between these two Ca^+ would be the easier to achieve because the first ionization energy of Ca is similar to that of K while the first ionization energy of Ar is much larger. However, the second ionization energy of Ca is relatively low making it easy to lose the second electron.

95. C has an outer shell electron configuration of ns^2np^2 based on this you would expect Si and Ge which are in the same family to be most like carbon. Ionization energies for both Si and Ge are similar and tend to be slightly lower than C but all are intermediate in the range of first ionization energies. The electron affinities of Si and Ge are close to that of C.

97. a) N: $[He]2s^22p^3$ Mg: $[Ne]3s^2$ O: $[He]2s^22p^4$
F: $[He]2s^22p^5$ Al: $[Ne]3s^23p^1$

b) Mg > Al > N > O > F Mg and Al would have the largest radius because they are in period n = 3, Al is smaller than Mg because radius decreases as you move to the right across the period. F is smaller than O is smaller than N because as you move to the right across the period, radius decreases.

c) Al < Mg < O < N < F (from the table)

d) The first ionization energy of Al is smaller than the first ionization energy of Mg because Al loses the electron from the 3p orbital which is shielded by the electrons in the 3s orbital, while Mg loses the electron from the filled 3s orbital which has added stability because it is a filled orbital. The first ionization energy of O is lower than the first ionization energy of N because N has a half – filled 2p orbital which adds extra stability thus, making it harder to remove the electron and the fourth electron in the O 2p orbital experiences added electron – electron repulsion from the other electron in its same orbital.

99. As you move to the right across a row in the periodic table for the main – group elements, the effective nuclear charge (Z_{eff}) experienced by the electrons in the outermost principal energy level increases, resulting in a stronger attraction between the outermost electrons and the nucleus and therefore, smaller atomic radii. Across the row of transition elements, the number of electron in the outermost principal energy level (highest n value) is nearly constant. As another proton is added to the nucleus with each successive element, another electron is added but, that electron goes into an $n_{highest} - 1$ orbital (a core level). The number of outermost electrons stays constant and they experience a roughly constant effective nuclear charge, keeping the radius approximately constant after the first couple of elements in the series.

101. The noble gases all have a filled outer quantum level, very high first ionization energies, and positive values for the electron affinity and are thus particularly unreactive. The lighter noble gases will not form any compounds because the ionization energies of He and Ne are both over 2000 kJ/mol. Since ionization energy decreases as you move down a column we find that the heavier noble gases, Ar, Kr, and Xe do form some compounds, they have ionization energies that are close to the ionization energy of H and can thus be forced to lose an electron.

103. Group 6A: ns^2np^4 Group 7A: ns^2np^5
The electron affinity of the group 7A elements are more negative than the group 6A elements in the same period because group 7A requires only one electron to achieve the noble gas configuration ns^2np^6 while the group 6A elements require 2 electrons. Adding one electron to the group 6A element will not give them any added stability and leads to extra electron – electron repulsions so the value of the electron affinity is less negative than that for group 7A.

105. 35 = Br = $[Ar]4s^23d^{10}4p^5$ 53 = I = $[Kr]5s^24d^{10}5p^6$

Br and I are both halogens with an outermost electron configuration of ns^2np^5, the next element with the same outermost electron configuration is 85, At.

107.a) Using Excel, make a table of radius, atomic number, and density. Using xy scatter, make a chart of radius vs. density. With an exponential trendline, estimate the density of argon and xenon. Also, make a chart or atomic number vs. density. With a linear trendline, estimate the density of argon and xenon.

element	radius(pm)	atomic number	density
He	32	2	0.18
Ne	70	10	0.90
Ar	98	18	
Kr	112	36	3.75
Xe	130	54	
Rn		86	9.73
		118	

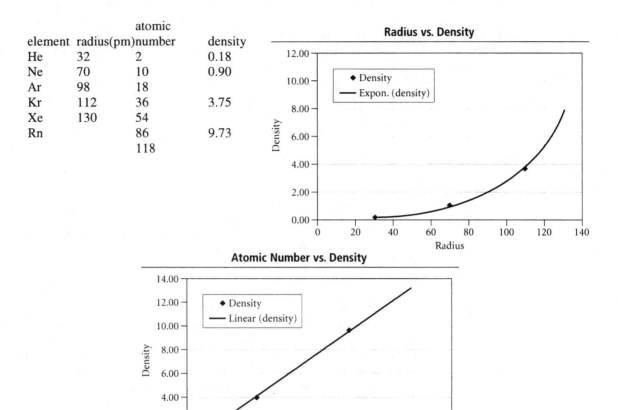

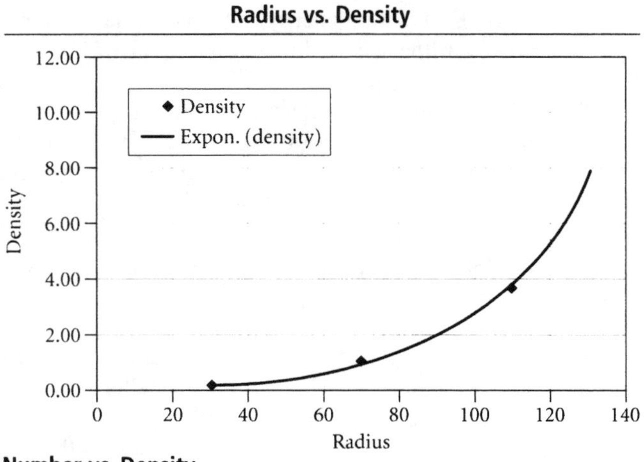

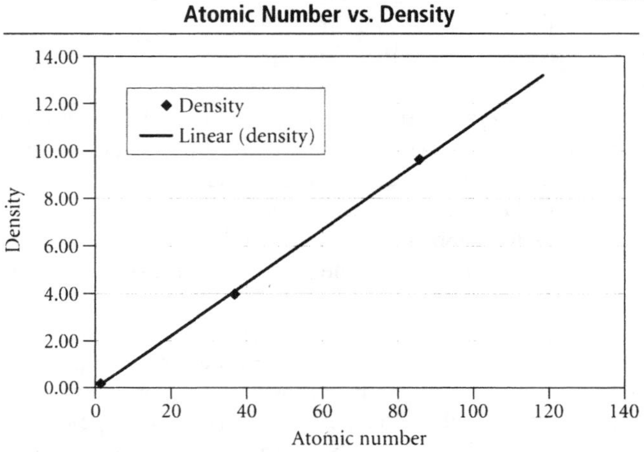

From the radius vs. density chart: Ar has a density of ~ 2 g/L and Xe has a density of ~ 6.5 g/L. From the atomic number vs. density chart: Ar has a density of ~1.8 g/L and Xe has a density of ~6 g/L.

b) Using the chart of atomic number vs. density, element 118 would be predicted to have a density of ~ 13 g/L.

c) **Given:** Ne: M = 20.18 g/mol; r = 70 pm **Find:** mass of neon; d neon
Conceptual Plan: M → $\mathbf{m_{atom}}$ and then r → $\mathbf{vol_{atom}}$ and then → d

$$\frac{6.022 \times 10^{23}\ \text{atoms}}{\text{mol}} \qquad V = \frac{4}{3}\pi r^3 \qquad d = \frac{\text{mass}}{\text{vol}}$$

Solution: $\frac{20.18\ \text{g}}{\text{mol}} \times \frac{\text{mol}}{6.022 \times 10^{23}\ \text{atoms}} = 3.35 \times 10^{-23}\ \text{g/atom}$

$$V = \frac{4}{3} \times 3.14 \times (70\ \text{pm})^3 \times \left(\frac{\text{m}}{10^{12}\ \text{pm}}\right)^3 \times \frac{\text{L}}{0.0010\ \text{m}^3} = 1.\underline{4}4 \times 10^{-27}\ \text{L}$$

$$d = \frac{3.35 \times 10^{-23}\ \text{g}}{1.\underline{4}4 \times 10^{-27}\ \text{L}} = 2.\underline{3}3 \times 10^4 = 2.3 \times 10^4\ \text{g/L}$$

Check: The units of the answer, g/L, are correct. This density is significantly larger than the actual density of neon gas. This suggests that a L of neon is composed of primarily empty space.

d) **Given:** Ne: M = 20.18 g/ mol, d = 0.90 g/L; Kr: M = 83.30 g/ mol, d = 3.75 g/L; Ar: M = 39.95 g/ mol
Find: d of argon in g/L

Conceptual Plan: d → mol/L → atoms/L for Kr and Ne and then atoms/L → mol/L → d for Ar

$$\text{mol} = \frac{\text{mass}}{\text{molar mass}} \qquad \frac{6.022 \times 10^{23}\ \text{atoms}}{\text{mol}} \qquad\qquad \frac{\text{mol}}{6.022 \times 10^{23}\ \text{atoms}} \qquad \frac{39.95\ \text{g}}{\text{mol}}$$

Solution: for Ne: $\frac{0.90\ \cancel{\text{g}}}{\text{L}} \times \frac{\cancel{\text{mol}}}{20.18\ \cancel{\text{g}}} \times \frac{6.022 \times 10^{23}\ \text{atoms}}{\cancel{\text{mol}}} = 2.69 \times 10^{22}\ \text{atoms/ L}$

for Kr: $\frac{3.75\ \cancel{\text{g}}}{\text{L}} \times \frac{\cancel{\text{mol}}}{83.80\ \cancel{\text{g}}} \times \frac{6.022 \times 10^{23}\ \text{atoms}}{\cancel{\text{mol}}} = 2.69 \times 10^{22}\ \text{atoms/ L}$

for Ar: $\frac{2.69 \times 10^{22}\ \cancel{\text{atoms}}}{\text{L}} \times \frac{\cancel{\text{mol}}}{6.022 \times 10^{23}\ \cancel{\text{atoms}}} \times \frac{39.95\ \text{g}}{\cancel{\text{mol}}} = 1.78\ \text{g/ L}$

Check: The units of the answer, g/L , are correct. The value of the answer agrees with published values.

109. The density increases as you move to the right across the first transition series. For the first transition series, the mass increases as you move to the right across the periodic table. However, the radius of the transition series elements stays nearly constant as you move to the right across the periodic table thus, the volume will remain nearly constant. Since density is mass/ volume the density of the elements increases.

111. The longest wavelength would be associated with the lowest energy state next to the ground state of carbon which has 2 unpaired electrons:

Ground state of carbon:

1s [↓↑]   2s [↓↑]   2p [↑][↑][ ]

longest wavelength: one of the p electrons flipped in its orbital which requires the least amount of energy.

1s [↓↑]   2s [↓↑]   2p [↑][↓][ ]

the next wavelength would be associated with the pairing of the two p electrons in the same orbital because this requires energy and raises the energy.

1s [↓↑]   2s [↓↑]   2p [↓↑][ ][ ]

the next wavelength would be associated with the energy needed to promote one of the s electrons to a p orbital.

1s [↓↑]   2s [↑]   2p [↑][↑][↑]

113. The element that would fill the 8s and 8p orbitals would have atomic number 168. The element is in the noble gas family and would have the properties of noble gases. It would have the electron configuration of $[118]8s^2 5g^{18} 6f^{14} 7d^{10} 8p^6$. The outer shell electron (highest n level) configuration would be $8s^2 8p^6$. The element would be relatively inert, have a first ionization energy less than 1037 kJ/mol (the first ionization energy of Rn), and have a positive electron affinity. It would be difficult to form compounds with other elements but would be able to form compounds with fluorine.

115. When you move down the column from Al to Ga, the size of the atom actually decreases because not much shielding is contributed by the 3d electrons in the Ga atom while there is a large increase in the nuclear charge, therefore, the effective nuclear charge is greater for Ga than for Al, so the ionization energy does not decrease. As you go from In to Tl, the ionization energy actually increases because the 4f electrons do not contribute to the shielding of the outermost electrons and there is a large increase in the effective nuclear charge.

117. The second electron is added to an ion with a 1 – charge, so there is a large repulsive force that has to be overcome to add the second electron. Thus, it will require energy to add the second electron and the second electron affinity will have a positive value.

119. If six electrons rather than eight electrons led to a stable configuration, the electron configuration of the stable configuration would be: ns^2np^4

a) A noble gas: ns^2np^4

b) A reactive nonmetal: would have one less electron than the stable configuration: ns^2np^3
c) A reactive metal: would have one have one more electron than the stable configuration: ns^1

121. An electron in a 5p orbital could have any one of the following combinations of quantum numbers.
5,1, -1,+1/2 5,1,-1,-1/2 5,1,0,+1/2 5,1,0,-1/2 5,1,1,+1/2 5,1,1,-1/2
An electron in a 6d orbital could have any one of the following combinations of quantum numbers.
6,2,-2,+1/2 6,2,-2,-1/2 6,2,-1,+1/2 6,2,-1,-1/2 6,2,0,+1/2 6,2,0,-1/2
6,2,1,+1/2 6,2,1,-1/2 6,2,2,+1/2 6,2,2,-1/2

Chapter 9
Chemical Bonding I: Lewis Theory

1. Bonding theories are central to chemistry because they explain how atoms bond together to form molecules. Bonding theories explain why some combinations of atoms are stable and others are not.

3. Chemical bonds form because they lower the potential energy between the charged particles that compose the atom. Bonds involve the attraction and repulsion of charged particles.

5. The three types of bonds are: Ionic bonds – which occurs between a metal and nonmetal and is characterized by the transfer of electrons; Covalent bonds – which occurs between a nonmetal and nonmetal and is characterized by the sharing of electrons; and Metallic bonds – which occurs between a metal and a metal and is characterized by the electron being pooled.

7. Bonds are formed when atoms obtain a stable electron configuration since the stable configuration is usually eight electrons in the outermost shell, this is known as the octet rule.

9. In Lewis theory, we represent ionic bonding by moving electron dots from the metal to the nonmetal and then allowing the resultant ions to form a crystalline lattice composed of alternating cations and anions. The cation loses its valence electron(s) and is left with an octet in the previous principal energy level, the anion gains electron(s) to form an octet. The Lewis structure of the anion is usually written within brackets with the charge in the upper right – hand corner, outside the brackets. The positive and negative charges attract one another, resulting in the compound.

11. Lattice energy is the energy associated with forming a crystalline lattice of alternating cations and anions from the gaseous ions. Since the cations are positively charged and the anions are negatively charged there is a lowering of potential – as prescribed by Coulomb's law – when the ions come together to form a lattice. That energy is emitted as heat when the lattice forms.

13. The Born – Haber cycle is a hypothetical series of steps that represents the formation of an ionic compound from its constituent elements. The steps are chosen so that the change in enthalpy of each step is known except for the last one, which is the lattice energy. In terms of the formation of NaCl, the steps are:
    Step 1: The formation of gaseous sodium from solid sodium (heat of sublimation of sodium).
    Step 2; The formation of a chlorine atom from a chlorine molecule (bond energy of chlorine).
    Step 3: The ionization of gaseous sodium (ionization energy of sodium).
    Step 4: The addition of an electron to gaseous chlorine (the electron affinity of chlorine).
    Step 5: The formation of the crystalline solid from the gaseous ions (the lattice energy)
    The overall reaction is the formation of NaCl(s), so we can use Hess' law to determine the lattice energy.
    $$\Delta H^\circ_f = \Delta H_{step\ 1} + \Delta H_{step\ 2} + \Delta H_{step\ 3} + \Delta H_{step\ 4} + \Delta H_{step\ 5}$$
    $$\Delta H^\circ_f = \text{heat of sublimation} + \tfrac{1}{2}\ \text{bond energy} + \text{ionization energy} + \text{electron affinity} + \text{lattice energy}$$
    Since all the terms are know except the lattice energy, we can calculate the lattice energy.

15. We modeled ionic solids as a lattice of individual ions held together by coulombic forces, which are equal in all directions. To melt the solid, these forces must be overcome, which requires a significant amount of heat. Therefore, the model accounts for the high melting points of ionic solids.

17. A pair of electrons that is shared between two atoms is called a bonding pair, while a pair of electrons that is associated with only one atom – and therefore, not involved in bonding – is called a lone pair.

19. Generally, combinations of atoms that can satisfy the octet rule on each atom are stable, while those combinations that do not satisfy the octet rule are not stable.

21. Electronegativity is the ability of an atom to attract electrons to itself in a chemical bond. This results in a polar bond. Electronegativity generally increases across a period in the periodic table. And, electronegativity generally decreases down a column (group) in the periodic table. The most electronegative element is fluorine.

23. Percent ionic character is defined as the ratio of a bond's actual dipole moment to the dipole moment it would have if the electron were completely transferred from one atom to the other, multiplied by 100.

A bond in which an electron is completely transferred form one atom to another would have 100% ionic character. However, no bond is 100% ionic. Percent ionic character generally increases as the electronegativity difference increases. In general, bonds with greater than 50% ionic character are referred to as ionic bonds.

25. To calculate the dipole moment we use $\mu = qr$:

For 100 pm: $\mu = 1.6 \times 10^{-19}\ \cancel{C} \times 100\ \cancel{pm} \times \dfrac{\cancel{m}}{10^{12}\ \cancel{pm}} \times \dfrac{D}{3.34 \times 10^{-30}\ \cancel{C}\cancel{m}} = 4.8\ D$

For 200 pm: $\mu = 1.6 \times 10^{-19}\ \cancel{C} \times 200\ \cancel{pm} \times \dfrac{\cancel{m}}{10^{12}\ \cancel{pm}} \times \dfrac{D}{3.34 \times 10^{-30}\ \cancel{C}\cancel{m}} = 9.6\ D$

27. The total number of electrons for a Lewis structure of a molecule is the sum of the valence electrons of each atom in the molecule.
The total number of electrons for the Lewis structure of an ion is found by summing the number of valence electrons for each atom and then subtracting 1 electron for each positive charge or adding 1 electron for each negative charge.

29. In some cases we can write resonance structures that are not equivalent.. One possible resonance structure may be somewhat better than another. In such cases the true structure may still be represented as an average of the resonance structures, but with the better resonance structure contributing more to the true structure. Multiple nonequivalent resonance structures may be weighted differently in their contributions to the true overall structure of a molecule.

31. The octet rule has some exceptions because not all atoms always have eight electrons surrounding them. The three major categories are: 1) odd – electron species, molecules or ions with an odd number of electron. For example: NO ; 2) incomplete octets, molecules or ions with fewer than eight electrons around an atom. For example: BF_3; and 3) expanded octets, molecules or ions with more than eight electrons around an atom. For example: AsF_5.

33. The bond energy of a chemical bond is the energy required to break 1 mole of the bond in the gas phase. Since breaking bonds is endothermic and forming bonds is exothermic we can calculate the overall enthalpy change as a sum of the enthalpy changes associated with breaking the required bonds in the reactants and forming the required bonds in the products.

35. When metal atoms bond together to form a solid, each metal atom donates one or more electrons to an electron sea.

37. N : $1s^2 2s^2 2p^3$  The electrons included in the Lewis structure are: $2s^2 2p^3$.

39. a) Al: $1s^2 2s^2 2p^6 3s^2 3p^1$

b) Na^+: $1s^2 2s^2 2p^6$

c) Cl: $1s^2 2s^2 2p^6 3s^2 3p^5$

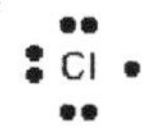

d) Cl^-: $1s^2 2s^2 2p^6 3s^2 3p^6$

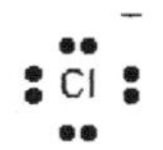

41. a) NaF: Draw the Lewis structures for Na and F based on their valence electrons. Na: $3s^1$ F: $2s^2 2p^5$

Na $\cdot$ $:\overset{\bullet\bullet}{\underset{\bullet\bullet}{F}}\cdot$

Sodium must lose one electron and be left with the octet from the previous shell, while fluorine needs to gain one electron to get an octet.

$\text{Na}^+ \left[:\overset{\bullet\bullet}{\underset{\bullet\bullet}{F}}: \right]^-$

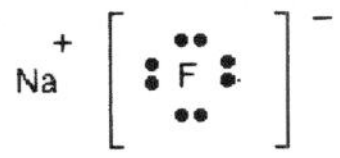

b) CaO: Draw the Lewis structures for Ca and O based on their valence electrons. Ca: $4s^2$ O: $2s^2 2p^4$

Ca$\vdots$ $:\overset{\bullet\bullet}{\underset{\bullet}{O}}\cdot$

Calcium must lose two electrons and be left with the octet from the previous shell, while oxygen needs to gain two electrons to get an octet.

$\text{Ca}^{2+} \left[:\overset{\bullet\bullet}{\underset{\bullet\bullet}{O}}: \right]^{2-}$

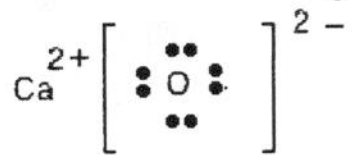

c) SrBr$_2$: Draw the Lewis structures for Sr and Br based on their valence electrons. Sr: $5s^2$ Br: $4s^2 4p^5$

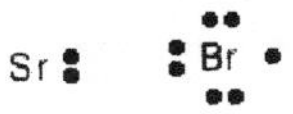

Sr$\vdots$ $:\overset{\bullet\bullet}{\underset{\bullet\bullet}{Br}}\cdot$

Strontium must lose two electrons and be left with the octet from the previous shell, while bromine needs to gain one electron to get an octet.

$\text{Sr}^{2+} 2\left[:\overset{\bullet\bullet}{\underset{\bullet\bullet}{Br}}\cdot \right]^-$

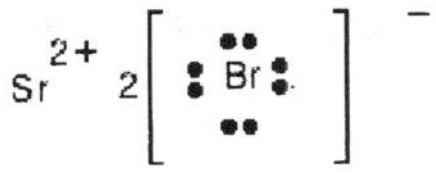

d) K$_2$O: Draw the Lewis structures for K and O based on their valence electrons. K: $4s^1$ O: $2s^2 2p^4$

K $\cdot$ $:\overset{\bullet\bullet}{\underset{\bullet}{O}}\cdot$

Potassium must lose one electron and be left with the octet from the previous shell, while oxygen needs to gain two electrons to get an octet.

$2\text{K}^+ \left[:\overset{\bullet\bullet}{\underset{\bullet\bullet}{O}}: \right]^{2-}$

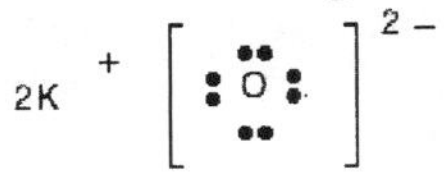

43. a) Sr and Se: Draw the Lewis structures for Sr and Se based on their valence electrons. Sr: $5s^2$ Se: $4s^2 4p^4$

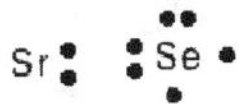

Sr$\vdots$ $:\overset{\bullet\bullet}{\underset{\bullet}{Se}}\cdot$

Strontium must lose two electrons and be left with the octet from the previous shell, while selenium needs to gain two electrons to get an octet.

$\text{Sr}^{2+} \left[:\overset{\bullet\bullet}{\underset{\bullet\bullet}{Se}}: \right]^{2-}$

Thus, we need one Sr^{2+} and one Se^{2-}. Write the formula with subscripts (if necessary) to indicate the number of atoms.

SrSe

b) Ba and Cl: Draw the Lewis structures for Ba and Cl based on their valence electrons. Ba: $6s^2$ Cl: $3s^2 3p^5$

Ba$\vdots$ $:\overset{\bullet\bullet}{\underset{\bullet\bullet}{Cl}}\cdot$

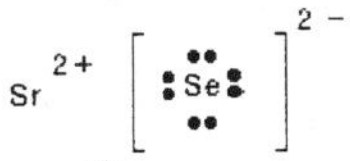

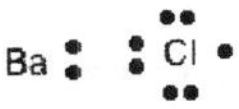

Barium must lose two electrons and be left with the octet from the previous shell, while chlorine needs to gain one electron to get an octet.

$\text{Ba}^{2+} 2\left[:\overset{\bullet\bullet}{\underset{\bullet\bullet}{Cl}}: \right]^-$

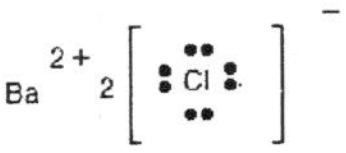

Thus, we need one Ba^{2+} and two Cl^-. Write the formula with subscripts (if necessary) to indicate the number of atoms.

$BaCl_2$

c) Na and S: Draw the Lewis structures for Na and S based on their valence electrons. Na: $3s^1$ S: $3s^2 3p^4$

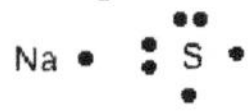

Sodium must lose one electron and be left with the octet from the previous shell, while sulfur needs to gain two electrons to get an octet.

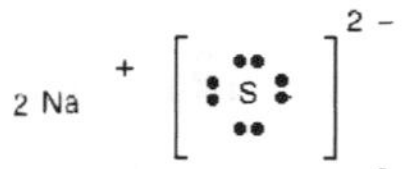

Thus, we need two Na^+ and one S^{2-}. Write the formula with subscripts (if necessary) to indicate the number of atoms.

Na_2S

d) Al and O: Draw the Lewis structures for Al and O based on their valence electrons. Al: $3s^2 3p^1$ O: $2s^2 2p^4$

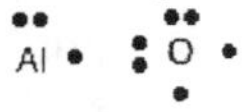

Aluminum must lose three electrons and be left with the octet from the previous shell, while oxygen needs to gain two electrons to get an octet.

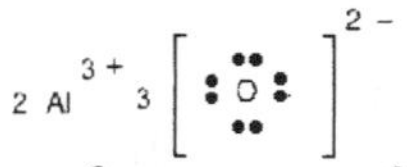

Thus, we need two Al^{3+} and three O^{2-} in order to lose and gain the same number of electrons. Write the formula with subscripts (if necessary) to indicate the number of atoms.

Al_2O_3

45. As the size of the alkaline metal ions increases down the column, so does the distance between the metal cations and the oxide anion. Therefore, the magnitude of the lattice energy of the oxides decreases making the formation of the oxides less exothermic and the compounds less stable. Since the ions cannot get as close to each other they therefore, do not release as much energy.

47. Cesium is slightly larger than barium, but oxygen is slightly larger than fluorine so, we cannot use size to explain the difference in the lattice energy. However, the charge on cesium ion is 1+ and the charge on fluoride ion is 1 –, while the charge on barium ion is 2+ and the charge on oxide ion is 2 –. The coulombic equation states that the magnitude of the potential also depends on the product of the charges. Since the product of the charges for CsF = 1 –, and the product of the charges for BaO = 4 –, the stabilization for BaO relative to CsF should be about four times greater which is what we see in its much more exothermic lattice energy.

49. **Given:** $\Delta H_f^\circ KCl = -436.5$ kJ/mol; $IE_1(K) = 419$ kJ/mol; $\Delta H_{sub}(K) = 89.0$ kJ/mol; $Cl_2(g)$ bond energy = 243 kJ/mol; EA(Cl) = – 349 kJ/mol.
Find: lattice energy
Conceptual Plan:

$$K(s) + 1/2Cl_2(g) \rightarrow K(g) + 1/2Cl_2(g) \rightarrow K^+(g) + 1/2Cl_2(g) \rightarrow K^+(g) + Cl(g) \rightarrow K^+(g) + Cl^-(g) \rightarrow KCl(s)$$

ΔH_{sub} **IE_1** **Bond energy** **EA** **lattice energy**

ΔH_f°

Solution: $\Delta H_f^\circ = \Delta H_{sub} + IE_1 + 1/2$ **Bond energy + EA + lattice energy**

$$-436.5\frac{kJ}{mol} = +89.0\frac{kJ}{mol} + 419\frac{kJ}{mol} + \frac{1}{2}(243)\frac{kJ}{mol} + (-349)\frac{kJ}{mol} + \text{lattice energy}$$

lattice energy = – 717 kJ/mol

51. a) Hydrogen: Write the Lewis structure of each atom based on the number of valence electrons.

When the two hydrogen atoms share their electrons, they each get a duet, which is a stable configuration for hydrogen

b) The halogens: Write the Lewis structure of each atom based on the number of valence electrons.

If the two halogens pair together they can each achieve an octet, which is a stable configuration. So, the halogens are predicted to exist as diatomic molecules.

c) Oxygen: Write the Lewis structure of each atom based on the number of valence electrons.

In order to achieve a stable octet on each oxygen, the oxygen atoms will need to share two electron pairs. So, oxygen is predicted to exist as a diatomic molecule with a double bond.

d) Nitrogen: Write the Lewis structure of each atom based on the number of valence electrons.

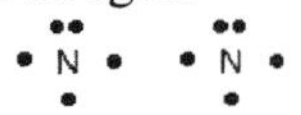

In order to achieve a stable octet on each nitrogen, the nitrogen atoms will need to share three electron pairs. So, nitrogen is predicted to exist as a diatomic molecule with a triple bond.

53. a) PH_3: Write the Lewis structure for each atom based on the valence electron.

The phosphorus will share an electron pair with each hydrogen in order to achieve a stable octet.

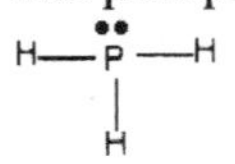

b) SCl_2: Write the Lewis structure for each atom based on the valence electron.

The sulfur will share an electron pair with each chlorine in order to achieve a stable octet.

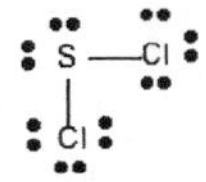

c) HI: Write the Lewis structure for each atom based on the valence electron.

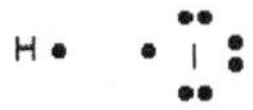

The iodine will share an electron pair with hydrogen in order to achieve a stable octet.

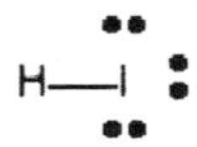

d) CH_4: Write the Lewis structure for each atom based on the valence electron.

The carbon will share an electron pair with each hydrogen in order to achieve a stable octet.

$$
\begin{array}{c}
\ \ \ \ \overset{\displaystyle H}{|} \\
H \!-\! \overset{|}{\underset{|}{C}} \!-\! H \\
\ \ \ \ \overset{|}{\displaystyle H}
\end{array}
$$

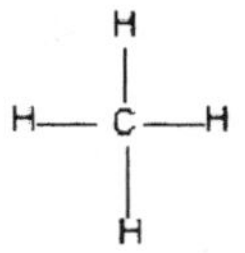

55. a) Br and Br: pure covalent   From figure 9.10 we find the electronegativity of Br is 2.5. Since both atoms are the same, the electronegativity difference (ΔEN) = 0, and using table 9.1 we classify this bond as pure covalent.

     b) C and Cl: polar covalent   From figure 9.10 we find the electronegativity of C is 2.5 and Cl is 3.0. The electronegativity difference (ΔEN) is ΔEN = 3.0 − 2.5 = 0.5. Using table 9.1 we classify this bond as polar covalent.

     c) C and S: pure covalent   From figure 9.10 we find the electronegativity of C is 2.5 and S is 2.5. The electronegativity difference (ΔEN) is ΔEN = 2.5 − 2.5 = 0. Using table 9.1 we classify this bond as pure covalent.

     d) Sr and O: ionic   From figure 9.10 we find the electronegativity of Sr is 1.0 and O is 3.5. The electronegativity difference (ΔEN) is ΔEN = 3.5 − 1.0 = 2.5. Using table 9.1 we classify this bond as ionic.

57. CO: Write the Lewis structure for each atom based on the valence electron.

The carbon will share three electron pair with oxygen in order to achieve a stable octet.
The oxygen atom is more electronegative than the carbon atom, so the oxygen will have a partial negative charge and the carbon will have a partial positive charge.

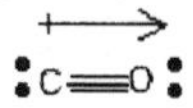

To estimate the percent ionic character, determine the difference in electronegativity between carbon and oxygen.
From figure 9.10 we find the electronegativity of C is 2.5 and O is 3.5. The electronegativity difference (ΔEN) is ΔEN = 3.5 − 2.5 = 1.0.
From figure 9.12, we can estimate a percent ionic character of 25%

59. a) CI$_4$:   Write the correct skeletal structure for the molecule

$$
\begin{array}{c}
\ \ \ \ | \\
| \!-\! \overset{|}{\underset{|}{C}} \!-\! | \\
\ \ \ \ |
\end{array}
$$

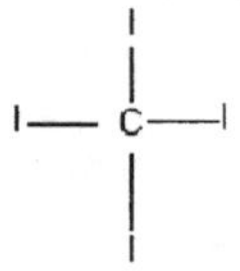

Calculate the total number of electrons for the Lewis structure by summing the valence electron of each atom in the molecule
(number of valence e$^-$ for C) + 4(number of valence e$^-$ for I) = 4 + 4(7) = 32
Distribute the electrons among the atoms, giving octets (or duets for H) to as many atoms as possible. Begin with the bonding electrons, and then proceed to lone pairs on terminal atoms and finally to lone pairs of the central atom.

$$
\begin{array}{c}
:\overset{..}{I}: \\
\ \ | \\
:\overset{..}{I} \!-\! \overset{|}{\underset{|}{C}} \!-\! \overset{..}{I}: \\
\ \ | \\
:\overset{..}{I}:
\end{array}
$$

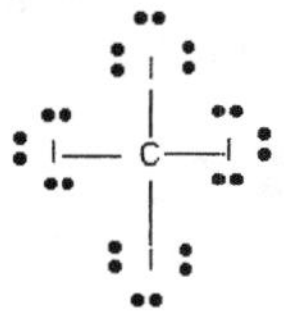

All 32 valence electrons are used
If any atom lacks an octet, form double or triple bonds as necessary to give them octets. All atoms have octets, duets for H, structure is complete.

     b) N$_2$O: Write the correct skeletal structure for the molecule
N is the less electronegative, so it is central
N$-$N$-$O

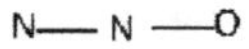

Calculate the total number of electrons for the Lewis structure by summing the valence electron of each atom in the molecule
2(number of valence e^- for N) + (number of valence e^- for O) = 2(5) + 6 = 16
Distribute the electrons among the atoms, giving octets (or duets for H) to as many atoms as possible. Begin with the bonding electrons, and then proceed to lone pairs on terminal atoms and finally to lone pairs of the central atom.

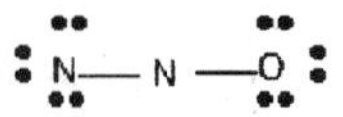

All 16 valence electrons are used
If any atom lacks an octet, form double or triple bonds as necessary to give them octets.

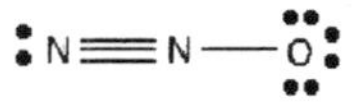

All atoms have octets, duets for H, structure is complete

c) SiH_4: Write the correct skeletal structure for the molecule
H is always terminal so Si is the central atom

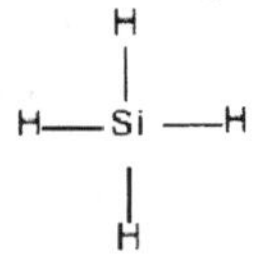

Calculate the total number of electrons for the Lewis structure by summing the valence electrons of each atom in the molecule
(number of valence e^- for Si) + 4(number of valence e^- for H) = 4 + 4(1) = 8
Distribute the electrons among the atoms, giving octets (or duets for H) to as many atoms as possible. Begin with the bonding electrons, and then proceed to lone pairs on terminal atoms and finally to lone pairs of the central atom.

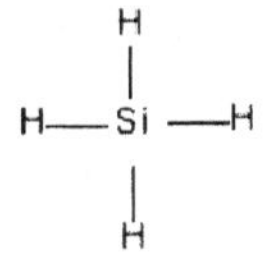

All 8 valence electrons are used.
If any atom lacks an octet, form double or triple bonds as necessary to give them octets.All atoms have octets, duets for H, structure is complete.

d) Cl_2CO: Write the correct skeletal structure for the molecule
C is the least electronegative, so it is the central atom

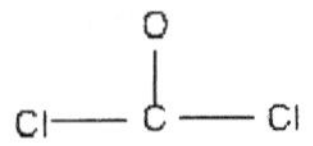

Calculate the total number of electrons for the Lewis structure by summing the valence electrons of each atom in the molecule
(number of valence e^- for C) + 2(number of valence e^- for Cl) + (number of valence e^- for O) = 4 + 2(7) + 6 = 24
Distribute the electrons among the atoms, giving octets (or duets for H) to as many atoms as possible. Begin with the bonding electrons, and then proceed to lone pairs on terminal atoms and finally to lone pairs of the central atom.

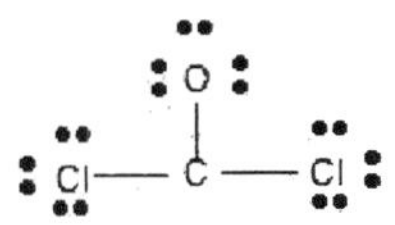

All 24 valence electrons are used.
If any atom lacks an octet, form double or triple bonds as necessary to give them octets.

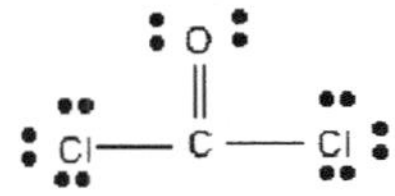

All atoms have octets, duets for H, structure is complete.

e) H_3COH: Write the correct skeletal structure for the molecule
C is less electronegative and H is terminal

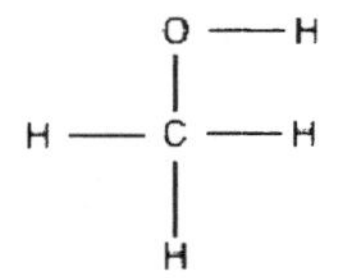

Calculate the total number of electrons for the Lewis structure by summing the valence electrons of each atom in the molecule

(number of valence e^- for C) + 4(number of valence e^- for H) + (number of valence e^- for O)
$= 4 + 4(1) + 6 = 14$

Distribute the electrons among the atoms, giving octets (or duets for H) to as many atoms as possible. Begin with the bonding electrons, and then proceed to lone pairs on terminal atoms and finally to lone pairs of the central atom.

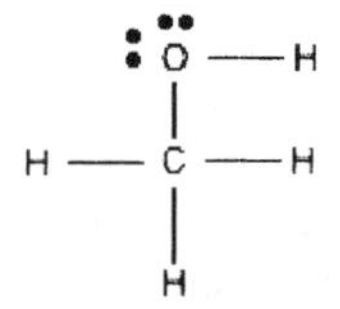

All 14 valence electrons are used.

If any atom lacks an octet, form double or triple bonds as necessary to give them octets.All atoms have octets, duets for H, structure is complete.

f) OH^-: Write the correct skeletal structure for the ion

O — H

Calculate the total number of electrons for the Lewis structure by summing the valence electrons of each atom in the ion and adding 1 for the 1 – charge.

(number of valence e^- for O) + (number of valence e^- for H) + 1 = 6 + 1 + 1 = 8

Distribute the electrons among the atoms, giving octets (or duets for H) to as many atoms as possible. Begin with the bonding electrons, and then proceed to lone pairs on terminal atoms and finally to lone pairs of the central atom.

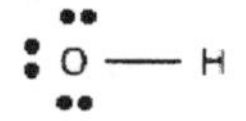

All 8 valence electrons are used.

If any atom lacks an octet, form double or triple bonds as necessary to give them octets.

Lastly, write the Lewis structure in brackets with the charge of the ion in the upper right – hand corner.

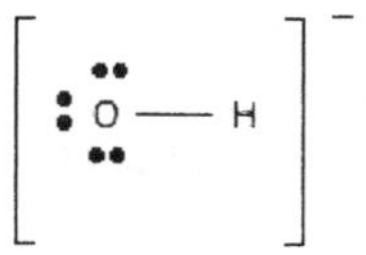

g) BrO^-: Write the correct skeletal structure for the ion

Br — O

Calculate the total number of electrons for the Lewis structure by summing the valence electrons of each atom in the ion and adding 1 for the 1 – charge.

(number of valence e^- for O) + (number of valence e^- for Br) + 1 = 6 + 7 + 1 = 14

Distribute the electrons among the atoms, giving octets (or duets for H) to as many atoms as possible. Begin with the bonding electrons, and then proceed to lone pairs on terminal atoms and finally to lone pairs of the central atom.

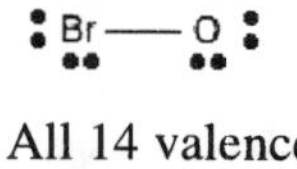

All 14 valence electrons are used.

If any atom lacks an octet, form double or triple bonds as necessary to give them octets.

Lastly, write the Lewis structure in brackets with the charge of the ion in the upper right – hand corner.

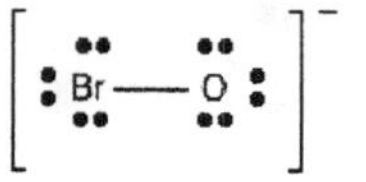

61. a) SeO_2: Write the correct skeletal structure for the molecule

Se is the less electronegative, so it is central

O—Se—O

Calculate the total number of electrons for the Lewis structure by summing the valence electrons of each atom in the molecule

(number of valence e$^-$ for Se) + 2(number of valence e$^-$ for O) = 6 +2(6) = 18
Distribute the electrons among the atoms, giving octets (or duets for H) to as many atoms as possible. Begin with the bonding electrons, and then proceed to lone pairs on terminal atoms and finally to lone pairs of the central atom.

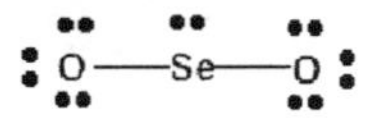

All 18 valence electrons are used.

If any atom lacks an octet, form double or triple bonds as necessary to give them octets.

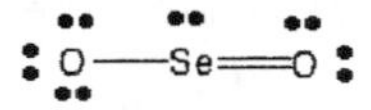

All atoms have octets, duets for H, structure is complete. However, the double bond can form from either oxygen atom, so there are two resonance forms.

Calculate the formal charge on each atom by finding the number of valence electrons and subtracting the number of lone pair electrons and one – half the number of bonding electrons.

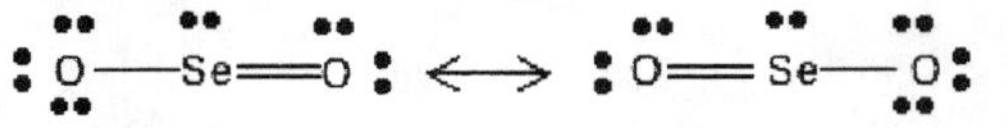

	O	Se	O	O	Se	O
number of valence electrons	6	6	6	6	6	6
- number of lone pair electrons	6	2	4	4	2	6
- 1/2(number of bonding electrons)	1	3	2	2	3	1
Formal charge	–1	+1	0	0	+1	–1

b) CO_3^{2-}: Write the correct skeletal structure for the ion

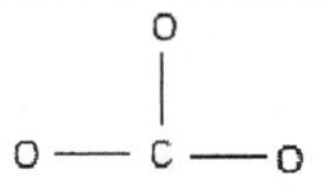

Calculate the total number of electrons for the Lewis structure by summing the valence electrons of each atom in the ion and adding 2 for the 2 – charge.
3(number of valence e$^-$ for O) + (number of valence e$^-$ for C) + 2 = 3(6) + 4 + 2 = 24
Distribute the electrons among the atoms, giving octets (or duets for H) to as many atoms as possible. Begin with the bonding electrons, and then proceed to lone pairs on terminal atoms and finally to lone pairs of the central atom.

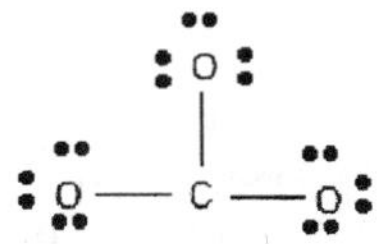

All 24 valence electrons are used.

If any atom lacks an octet, form double or triple bonds as necessary to give them octets.

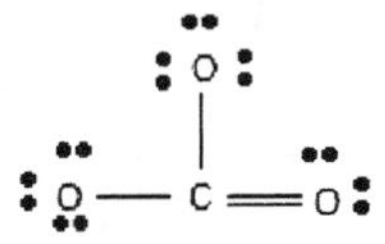

Lastly, write the Lewis structure in brackets with the charge of the ion in the upper right – hand corner.

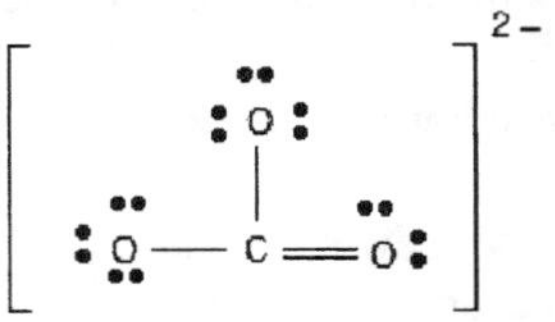

All atoms have octets, duets for H, structure is complete. However, the double bond can form from any oxygen atom, so there are three resonance forms.

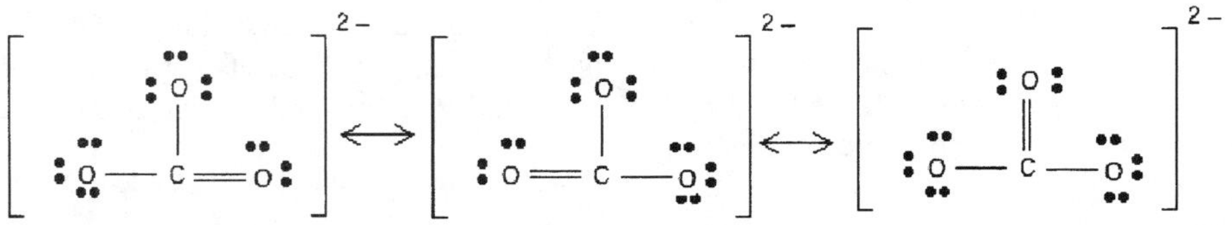

Calculate the formal charge on each atom by finding the number of valence electrons and subtracting the number of lone pair electrons and one – half the number of bonding electrons.

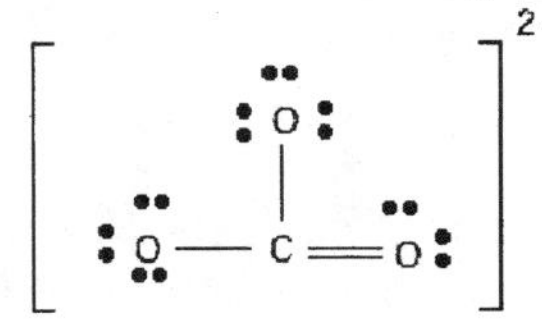

	O_{left}	O_{top}	O_{right}	C
number of valence electrons	6	6	6	4
- number of lone pair electrons	6	6	4	0
- 1/2(number of bonding electrons)	1	1	2	4
Formal charge	–1	–1	0	0

The sum of the formal charges is – 2 which is the overall charge of the ion. The other resonance forms would be the same.

c) ClO^-: Write the correct skeletal structure for the ion

Cl — O

Calculate the total number of electrons for the Lewis structure by summing the valence electrons of each atom in the ion and adding 1 for the 1 – charge.

(number of valence e^- for O) + (number of valence e^- for Cl) + 1 = 6 + 7 + 1 = 14

Distribute the electrons among the atoms, giving octets (or duets for H) to as many atoms as possible. Begin with the bonding electrons, and then proceed to lone pairs on terminal atoms and finally to lone pairs of the central atom.

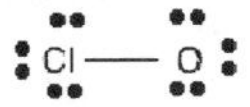

All 14 valence electrons are used.

If any atom lacks an octet, form double or triple bonds as necessary to give them octets.

Lastly, write the Lewis structure in brackets with the charge of the ion in the upper right – hand corner.

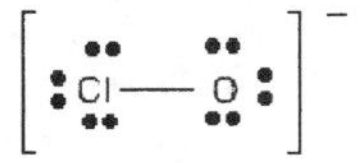

All atoms have octets, duets for H, structure is complete.

Calculate the formal charge on each atom by finding the number of valence electrons and subtracting the number of lone pair electrons and one – half the number of bonding electrons.

	Cl	O
number of valence electrons	7	6
- number of lone pair electrons	6	6
- 1/2(number of bonding electrons)	1	1
Formal charge	0	–1

The sum of the formal charges is – 1 which is the overall charge of the ion.

d) NO_2^-: Write the correct skeletal structure for the ion

O — N — O

Calculate the total number of electrons for the Lewis structure by summing the valence electrons of each atom in the ion and adding 1 for the 1 – charge.

2(number of valence e^- for O) + (number of valence e^- for N) + 1 = 2(6) + 5 + 1 = 18

Distribute the electrons among the atoms, giving octets (or duets for H) to as many atoms as possible. Begin with the bonding electrons, and then proceed to lone pairs on terminal atoms and finally to lone pairs of the central atom.

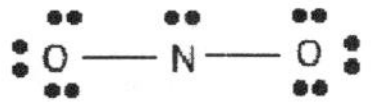

All 14 valence electrons are used.

If any atom lacks an octet, form double or triple bonds as necessary to give them octets.

$$:\underset{\bullet\bullet}{O} = \overset{\bullet\bullet}{N} - \overset{\bullet\bullet}{\underset{\bullet\bullet}{O}}:$$

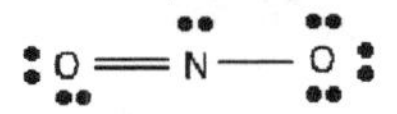

Lastly, write the Lewis structure in brackets with the charge of the ion in the upper right – hand corner.

$$\left[:\underset{\bullet\bullet}{O} = \overset{\bullet\bullet}{N} - \overset{\bullet\bullet}{\underset{\bullet\bullet}{O}}: \right]^{-}$$

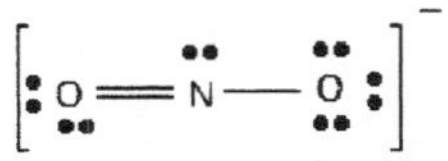

All atoms have octets, duets for H, structure is complete. However, the double bond can form from either oxygen atom, so there are two resonance forms.

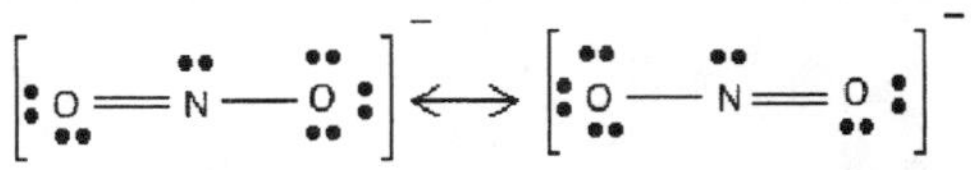

$$\left[:\underset{\bullet\bullet}{O} = \overset{\bullet\bullet}{N} - \overset{\bullet\bullet}{\underset{\bullet\bullet}{O}}: \right]^{-} \longleftrightarrow \left[:\overset{\bullet\bullet}{\underset{\bullet\bullet}{O}} - \overset{\bullet\bullet}{N} = \underset{\bullet\bullet}{O}: \right]^{-}$$

Calculate the formal charge on each atom by finding the number of valence electrons and subtracting the number of lone pair electrons and one – half the number of bonding electrons. Using the left side structure:

	O	N	O
number of valence electrons	6	5	6
- number of lone pair electrons	4	2	6
- 1/2(number of bonding electrons)	2	3	1
Formal charge	0	0	– 1

The sum of the formal charges is – 1 which is the overall charge of the ion.

63.

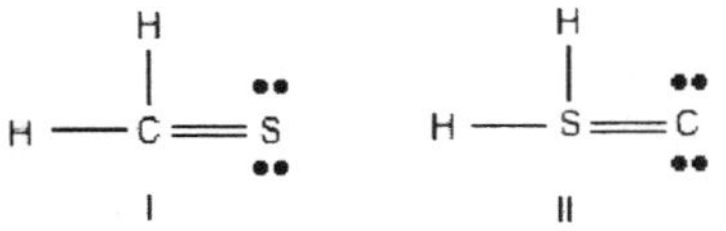

Calculate the formal charge on each atom in structure I by finding the number of valence electrons and subtracting the number of lone pair electrons and one – half the number of bonding electrons.

	H_{left}	H_{top}	C	S
number of valence electrons	1	1	4	6
- number of lone pair electrons	0	0	0	4
- 1/2(number of bonding electrons)	1	1	4	2
Formal charge	0	0	0	0

The sum of the formal charges is 0 which is the overall charge of the molecule.
Calculate the formal charge on each atom in structure II by finding the number of valence electrons and subtracting the number of lone pair electrons and one – half the number of bonding electrons

	H_{left}	H_{top}	S	C
number of valence electrons	1	1	6	4
- number of lone pair electrons	0	0	0	4
- 1/2(number of bonding electrons)	1	1	4	2
Formal charge	0	0+	2	– 2

The sum of the formal charges is 0 which is the overall charge of the molecule.
Structure I is the better Lewis structure because it has the least amount of formal charge on each atom.

65. $:O \equiv C - \overset{\bullet\bullet}{\underset{\bullet\bullet}{O}}:$ does not provide a significant contribution to the resonance hybrid as it has a +1 formal charge on a very electronegative oxygen.

	O_{left}	O_{right}	C
number of valence electrons	6	6	4
- number of lone pair electrons	2	6	0
- 1/2(number of bonding electrons)	3	1	4
Formal charge	+1	–1	0

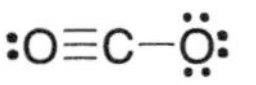

67. a) BCl_3: Write the correct skeletal structure for the molecule.
B is the less electronegative, so it is central

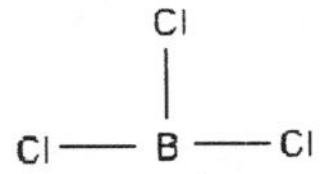

Calculate the total number of electrons for the Lewis structure by summing the valence electrons of each atom in the molecule
(number of valence e^- for B) + 3(number of valence e^- for Cl) = 3 +3(7) = 24
Distribute the electrons among the atoms, giving octets (or duets for H) to as many atoms as possible. Begin with the bonding electrons, and then proceed to lone pairs on terminal atoms and finally to lone pairs of the central atom.

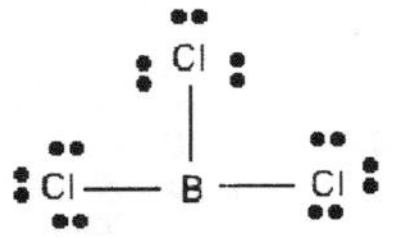

All 24 valence electrons are used.
B has an incomplete octet. If we complete the octet, there is a formal charge of – 1 on the B which is less electronegative than Cl.

b) NO_2: Write the correct skeletal structure for the molecule.
N is the less electronegative, so it is central

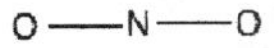

Calculate the total number of electrons for the Lewis structure by summing the valence electrons of each atom in the molecule
(number of valence e^- for N) + 2(number of valence e^- for O) = 5 +2(6) = 17
Distribute the electrons among the atoms, giving octets (or duets for H) to as many atoms as possible. Begin with the bonding electrons, and then proceed to lone pairs on terminal atoms and finally to lone pairs of the central atom.

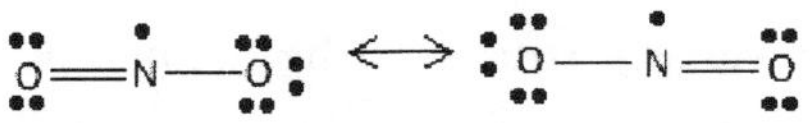

All 17 valence electrons are used.
N has an incomplete octet. It has 7 electrons because we have an odd number of valence electrons.

c) BH_3: Write the correct skeletal structure for the molecule.
B is the less electronegative, so it is central

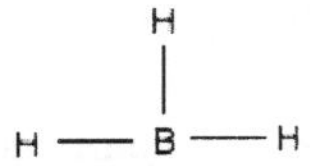

Calculate the total number of electrons for the Lewis structure by summing the valence electrons of each atom in the molecule
(number of valence e^- for B) + 3(number of valence e^- for H) = 3 +3(1) = 6
Distribute the electrons among the atoms, giving octets (or duets for H) to as many atoms as possible. Begin with the bonding electrons, and then proceed to lone pairs on terminal atoms and finally to lone pairs of the central atom.

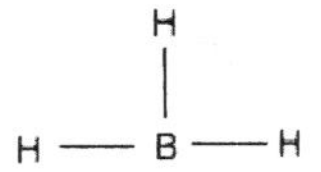

All 6 valence electrons are used.
B has an incomplete octet. H cannot double bond, so it is not possible to complete the octet on B with a double bond.

octet. It has 7 electrons because we have an odd number of valence electrons.

69. a) PO_4^{3-}: Write the correct skeletal structure for the ion.

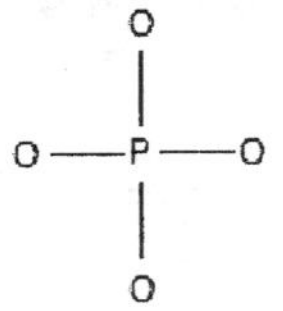

Calculate the total number of electrons for the Lewis structure by summing the valence electrons of each atom in the ion and adding 3 for the 3 – charge.
4(number of valence e^- for O) + (number of valence e^- for P) + 3 = 4(6) + 5 + 3 = 32
Distribute the electrons among the atoms, giving octets (or duets for H) to as many atoms as possible. Begin with the bonding electrons, and then proceed to lone pairs on terminal atoms and finally to lone pairs of the central atom.

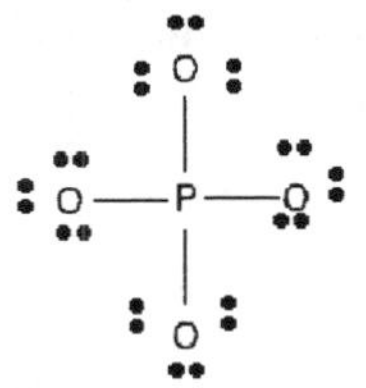

All 32 valence electrons are used.
Lastly, write the Lewis structure in brackets with the charge of the ion in the upper right – hand corner.

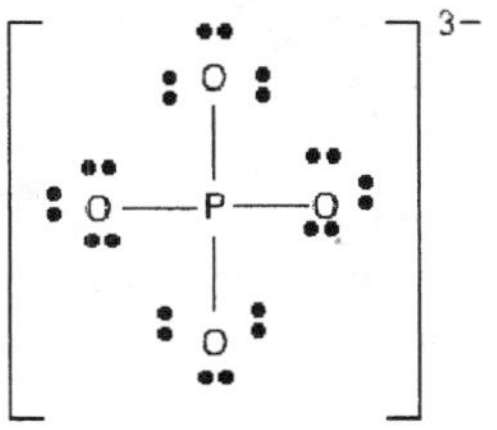

All atoms have octets, duets for H, structure is complete.
Calculate the formal charge on each atom by finding the number of valence electrons and subtracting the number of lone pair electrons and one – half the number of bonding electrons.

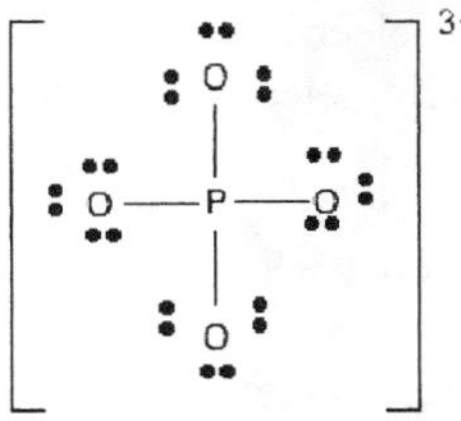

	O_{left}	O_{top}	O_{right}	O_{bottom}	P
number of valence electrons	6	6	6	6	5
- number of lone pair electrons	6	6	6	6	0
- 1/2(number of bonding electrons)	1	1	1	1	4
Formal charge	−1	−1	−1	−1	+1

The sum of the formal charges is 3–which is the overall charge of the ion. However, we can write a resonance structure with a double bond to an oxygen because P can expand its octet. This leads to lower formal charges on P and O

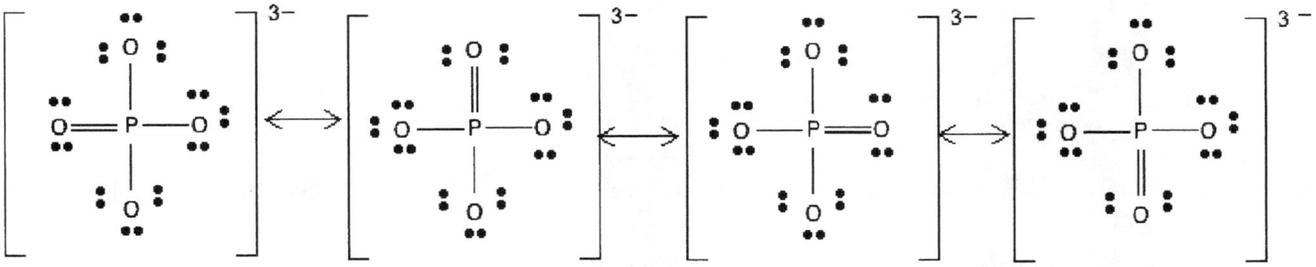

Using the leftmost structure calculate the formal charge on each atom by finding the number of valence electrons and subtracting the number of lone pair electrons and one – half the number of bonding electrons.

	O_{left}	O_{top}	O_{right}	O_{bottom}	P
number of valence electrons	6	6	6	6	5
- number of lone pair electrons	4	6	6	6	0
- 1/2(number of bonding electrons)	2	1	1	1	5
Formal charge	0	−1	−1	−1	0

The sum of the formal charges is 3–which is the overall charge of the ion. These resonance forms would all have the lower formal charges associated with the double bonded O and P.

b) CN^-: Write the correct skeletal structure for the ion.

Calculate the total number of electrons for the Lewis structure by summing the valence electrons of each atom in the ion and adding 1 for the 1 – charge.
(number of valence e^- for C) + (number of valence e^- for N) + 1 = 4 + 5 + 1 = 10
Distribute the electrons among the atoms, giving octets (or duets for H) to as many atoms as possible. Begin with the bonding electrons, and then proceed to lone pairs on terminal atoms and finally to lone pairs of the central atom.

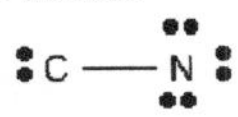

All 10 valence electrons are used.
If any atom lacks an octet, form double or triple bonds as necessary to give them octets.

Lastly, write the Lewis structure in brackets with the charge of the ion in the upper right – hand corner.

All atoms have octets, duets for H, structure is complete.
Calculate the formal charge on each atom be finding the number of valence electrons and subtracting the number of lone pair electrons and one – half the number of bonding electrons.

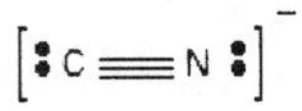

	C	N
number of valence electrona	4	5
- number of lone pair electrons	2	2
- 1/2(number of bonding electrons)	3	3
Formal charge	–1	0

The sum of the formal charges is – 1 which is the overall charge of the ion

c) SO_3^{2-}: Write the correct skeletal structure for the ion.

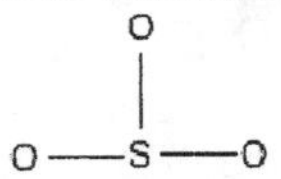

Calculate the total number of electrons for the Lewis structure by summing the valence electrons of each atom in the ion and adding 2 for the 2 – charge.
3(number of valence e^- for O) + (number of valence e^- for S) + 2 = 3(6) + 6 + 2 = 26
Distribute the electrons among the atoms, giving octets (or duets for H) to as many atoms as possible. Begin with the bonding electrons, and then proceed to lone pairs on terminal atoms and finally to lone pairs of the central atom.

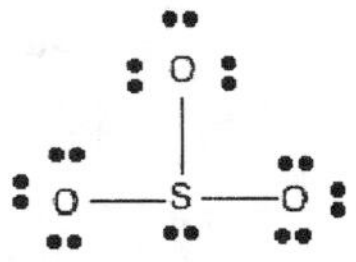

All 26 valence electrons are used.
Lastly, write the Lewis structure in brackets with the charge of the ion in the upper right – hand corner.

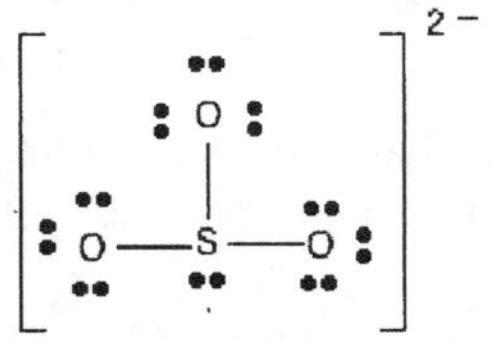

Calculate the formal charge on each atom be finding the number of valence electrons and subtracting the number of lone pair electrons and one – half the number of bonding electrons.

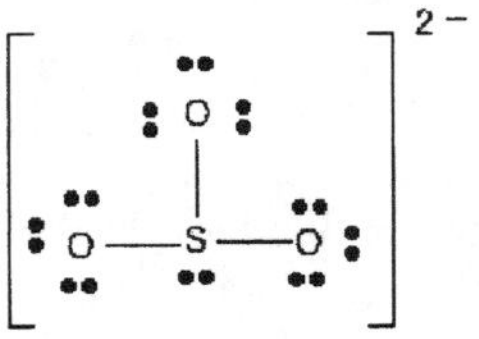

	O_{left}	O_{top}	O_{right}	S
number of valence electrons	6	6	6	6
- number of lone pair electrons	6	6	6	2
- 1/2(number of bonding electrons)	1	1	1	3
Formal charge	–1	–1	–1	+1

The sum of the formal charges is – 1 which is the overall charge of the ion. However, we can write a resonance structure with a double bond to an oxygen because S can expand its octet. This leads to a lower formal charge

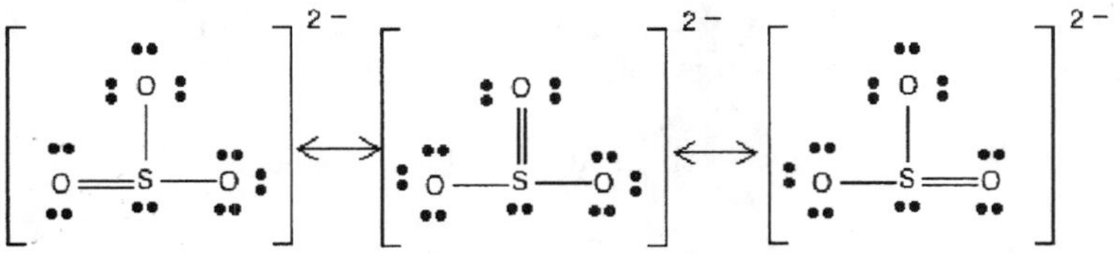

Using the leftmost resonance form: Calculate the formal charge on each atom be finding the number of valence electrons and subtracting the number of lone pair electrons and one – half the number of bonding electrons.

	O_{left}	O_{top}	O_{right}	S
number of valence electrons	6	6	6	6
- number of lone pair electrons	4	6	6	2
- 1/2(number of bonding electrons)	2	1	1	4
Formal charge	0	–1	–1	0

The sum of the formal charges is 2–which is the overall charge of the ion. These resonance forms would all have the lower formal charge on the double bonded O and S.

d) ClO_2^-: Write the correct skeletal structure for the ion.

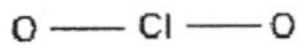

Calculate the total number of electrons for the Lewis structure by summing the valence electrons of each atom in the ion and adding 1 for the 1 – charge.

2(number of valence e^- for O) + (number of valence e^- for Cl) + 1 = 2(6) + 7 + 1 = 20

Distribute the electrons among the atoms, giving octets (or duets for H) to as many atoms as possible. Begin with the bonding electrons, and then proceed to lone pairs on terminal atoms and finally to lone pairs of the central atom.

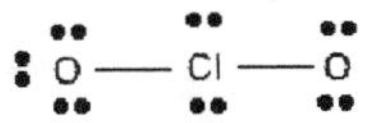

All 20 valence electrons are used.

Lastly, write the Lewis structure in brackets with the charge of the ion in the upper right – hand corner.

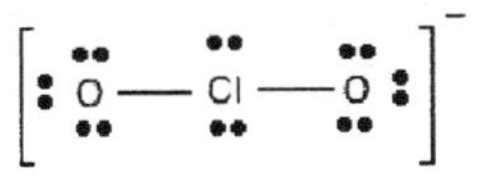

All atoms have octets, duets for H, structure is complete.

Calculate the formal charge on each atom be finding the number of valence electrons and subtracting the number of lone pair electrons and one – half the number of bonding electrons.

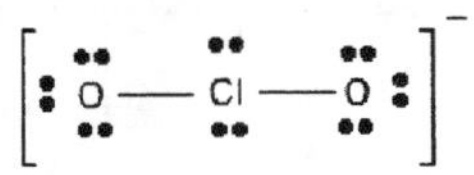

	O_{left}	O_{righ}	$_{t}$Cl
number of valence electrons	6	6	7
- number of lone pair electrons	6	6	4
- 1/2(number of bonding electrons)	1	1	2
Formal charge	–1	–1	+1

The sum of the formal charges is – 1 which is the overall charge of the ion. However, we can write a resonance structure with a double bond to an oxygen because Cl can expand its octet. This leads to a lower formal charge

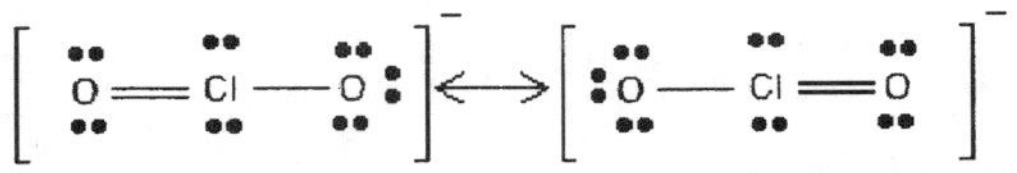

Using the leftmost resonance form: Calculate the formal charge on each atom be finding the number of valence electrons and subtracting the number of lone pair electrons and one – half the number of bonding electrons.

	O_{left}	O_{right}	Cl
number of valence electrons	6	6	7
- number of lone pair electrons	4	6	4
- 1/2(number of bonding electrons)	2	1	3
Formal charge	0	–1	0

The sum of the formal charges is 1–which is the overall charge of the ion. These resonance forms would all have the lower formal charge on the double bonded O and Cl.

71. a) PF_5: Write the correct skeletal structure for the molecule

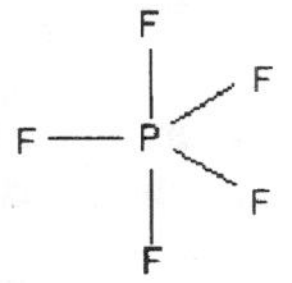

Calculate the total number of electrons for the Lewis structure by summing the valence electrons of each atom in the molecule

(number of valence e^- for P) + 5(number of valence e^- for F) = 5 +5(7) = 40

Distribute the electrons among the atoms, giving octets (or duets for H) to as many atoms as possible. Begin with the bonding electrons, and then proceed to lone pairs on terminal atoms and finally to lone pairs of the central atom. Arrange additional electrons around the central atom, giving it an expanded octet of up to 12 electrons.

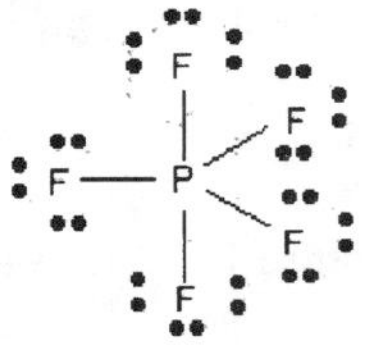

b) I_3^-: Write the correct skeletal structure for the ion

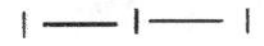

Calculate the total number of electrons for the Lewis structure by summing the valence electrons of each atom in the ion and adding 1 for the 1 – charge.

3(number of valence e^- for I) + 1 = 3(7) + 1 = 22

Distribute the electrons among the atoms, giving octets (or duets for H) to as many atoms as possible. Begin with the bonding electrons, and then proceed to lone pairs on terminal atoms and finally to lone pairs of the central atom. Arrange additional electrons around the central atom, giving it an expanded octet of up to 12 electrons.

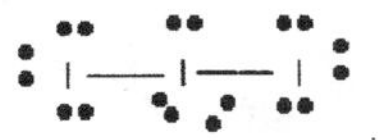

Lastly, write the Lewis structure in brackets with the charge of the ion in the upper right – hand corner.

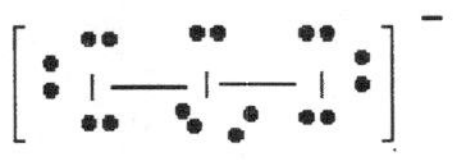

c) SF_4: Write the correct skeletal structure for the molecule

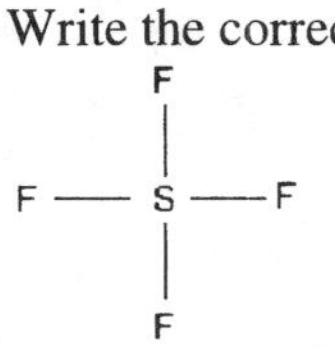

Calculate the total number of electrons for the Lewis structure by summing the valence electrons of each atom in the molecule

(number of valence e^- for S) + 4(number of valence e^- for F) = 6 + 4(7) = 34

Distribute the electrons among the atoms, giving octets (or duets for H) to as many atoms as possible. Begin with the bonding electrons, and then proceed to lone pairs on terminal atoms and finally to lone pairs of the central atom. Arrange additional electrons around the central atom, giving it an expanded octet of up to 12 electrons.

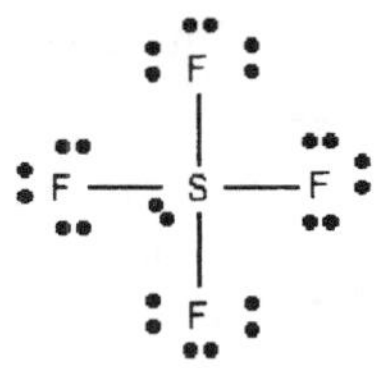

d) GeF_4: Write the correct skeletal structure for the molecule

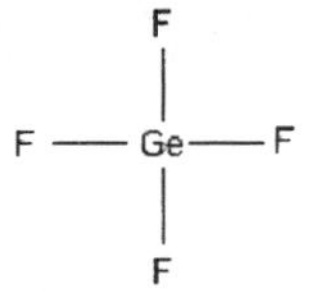

Calculate the total number of electrons for the Lewis structure by summing the valence electrons of each atom in the molecule
(number of valence e^- for Ge) + 4(number of valence e^- for F) = 4 + 4(7) = 32
Distribute the electrons among the atoms, giving octets (or duets for H) to as many atoms as possible. Begin with the bonding electrons, and then proceed to lone pairs on terminal atoms and finally to lone pairs of the central atom. Arrange additional electrons around the central atom, giving it an expanded octet of up to 12 electrons.

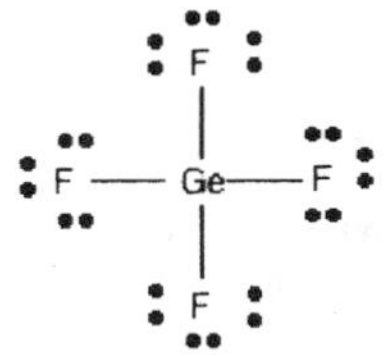

73. Bond strength: $H_3CCH_3 < H_2CCH_2 < HCCH$
Bond length: $H_3CCH_3 > H_2CCH_2 > HCCH$
Write the Lewis structures for the three compounds. Compare the C – C bonds. Triple bonds are stronger than double bonds which are stronger than single bonds. Also, single bonds are longer than double bonds are longer than triple bonds.

HCCH (10 e^-) H_2CCH_2 (12 e^-) H_3CCH_3(14 e^-)

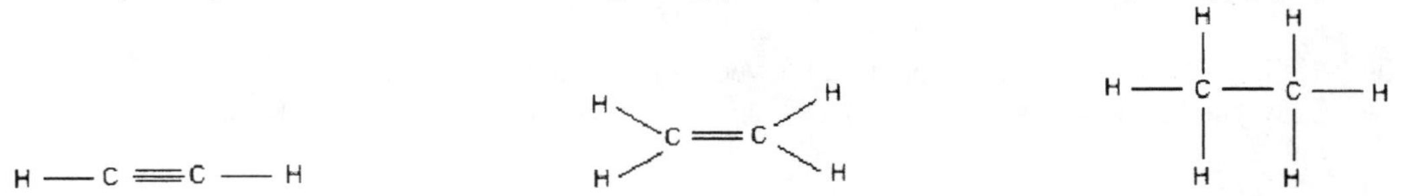

75. Rewrite the reaction using the Lewis structures of the molecules involved.

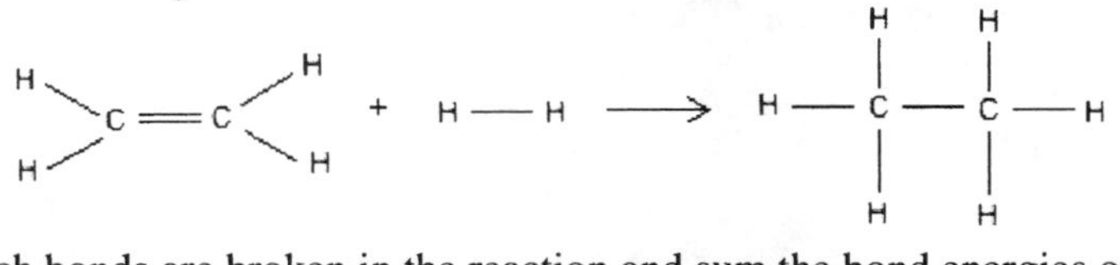

Determine which bonds are broken in the reaction and sum the bond energies of these

$\Sigma(\Delta H$'s bonds broken)
= 4(C – H) + (C = C) + (H – H)
= 4(414 kJ/mol) + (611 kJ/mol) + (436 kJ/mol)
= 2703 kJ/mol

Determine which bonds are formed in the reaction and sum the negatives of the bond energies of these

$\Sigma(-\Delta H$'s of bonds formed)
= – 6(C – H) – (C – C)
= – 6(414 kJ/mol) – (347 kJ/mol)
= –2831 kJ/mol

Find ΔH_{rxn} by summing the results of the two steps.

ΔH_{rxn} = Σ(ΔH's bonds broken) + Σ(-ΔH's of bonds formed)
= 2703 kJ/mol – 2831 kJ/mol
= – 128 kJ/mol

77. Rewrite the reaction using the Lewis structures of the molecules involved.

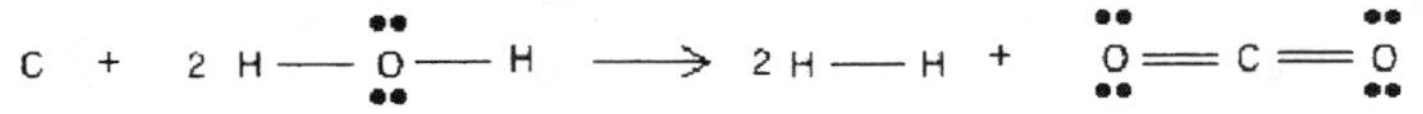

Determine which bonds are broken in the reaction and sum the bond energies of these

Σ(ΔH's bonds broken)
= 4(O – H)
= 4(464 kJ/mol)
= 1856 kJ/mol

Determine which bonds are formed in the reaction and sum the negatives of the bond energies of these

Σ(-ΔH's of bonds formed)
= – 2(C = O) – 2(H – H)
= – 2(799 kJ/mol) – 2(436 kJ/mol)
= –2470 kJ/mol

Find ΔH_{rxn} by summing the results of the two steps.

ΔH_{rxn} = Σ(ΔH's bonds broken) + Σ(-ΔH's of bonds formed)
= 1856 kJ/mol – 2470 kJ/mol
= – 614 kJ/mol

79. a) BI_3: This is a covalent compound between two nonmetals.
Write the correct skeletal structure for the molecule

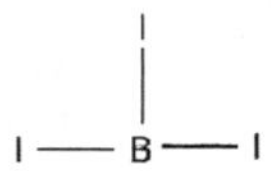

Calculate the total number of electrons for the Lewis structure by summing the valence electrons of each atom in the molecule

(number of valence e^- for B) + (number of valence e^- for I) = 3 +3(7) = 24

Distribute the electrons among the atoms, giving octets (or duets for H) to as many atoms as possible. Begin with the bonding electrons, and then proceed to lone pairs on terminal atoms and finally to lone pairs of the central atom.

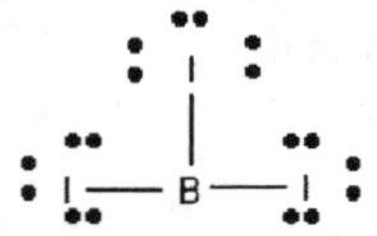

b) K_2S: This is an ionic compound between a metal and nonmetal
Draw the Lewis structures for K and S based on their valence electrons. K: $4s^1$ S: $3s^23p^4$

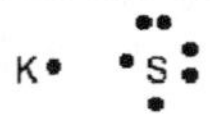

Potassium must lose one electron and be left with the octet from the previous shell, while sulfur needs to gain two electrons to get an octet.

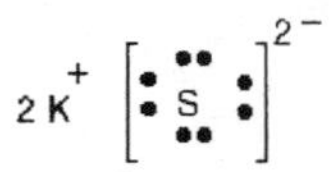

c) HCFO: This is a covalent compound between nonmetals.
Write the correct skeletal structure for the molecule

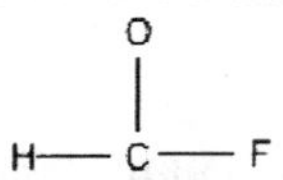

Calculate the total number of electrons for the Lewis structure by summing the valence electrons of each atom in the molecule

(number of valence e^- for H)+(number of valence e^- for C)+(number of valence e^- for F)+(number of valence e^- for O) = 1 + 4 + 7 +6 = 18

Distribute the electrons among the atoms, giving octets (or duets for H) to as many atoms as possible. Begin with the bonding electrons, and then proceed to lone pairs on terminal atoms and finally to lone pairs of the central atom.

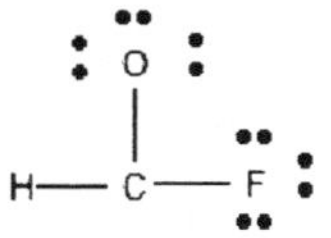

If any atom lacks an octet, form double or triple bonds as necessary to give them octets.

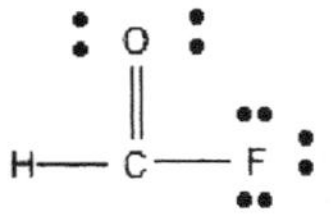

d) PBr_3: This is a covalent compound between two nonmetals.
Write the correct skeletal structure for the molecule

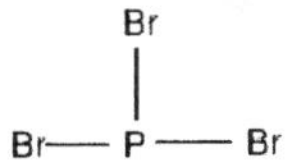

Calculate the total number of electrons for the Lewis structure by summing the valence electrons of each atom in the molecule
(number of valence e^- for P) + 3(number of valence e^- for Br) = 5 + 3(7) = 26
Distribute the electrons among the atoms, giving octets (or duets for H) to as many atoms as possible. Begin with the bonding electrons, and then proceed to lone pairs on terminal atoms and finally to lone pairs of the central atom.

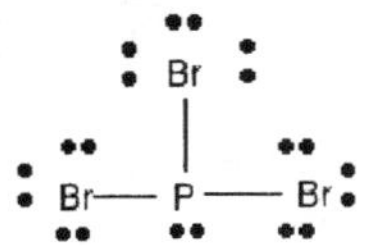

81. a) $BaCO_3$: Ba^{2+}

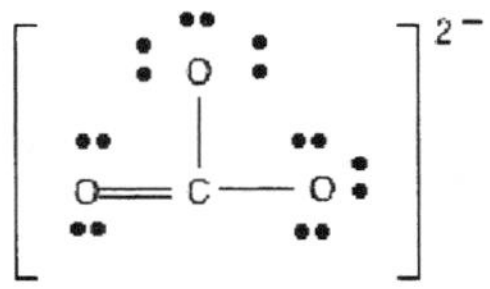

Determine the cation and anion
Ba^{2+} CO_3^{2-}
Write the Lewis structure for the barium cation based on the valence electrons
Ba $5s^2$ Ba^{2+} $5s^0$
Ba : Ba^{2+}
Ba must lose two electrons and be left with the octet from the previous shell
Write the Lewis structure for the covalent anion.
Write the correct skeletal structure for the ion

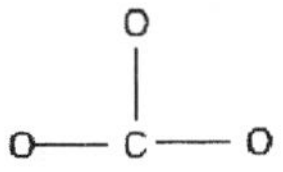

Calculate the total number of electrons for the Lewis structure by summing the valence electrons of each atom in the ion and adding two for the 2 – charge.
(number of valence e^- for C) + 3(number of valence e^- for O) = 4 + 3(6) + 2 = 24
Distribute the electrons among the atoms, giving octets (or duets for H) to as many atoms as possible. Begin with the bonding electrons, and then proceed to lone pairs on terminal atoms and finally to lone pairs of the central atom.

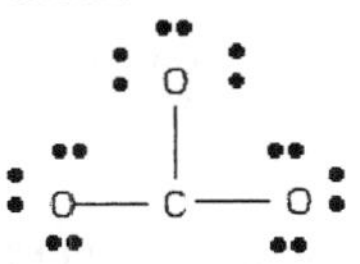

If any atom lacks an octet, form double or triple bonds as necessary to give them octets.

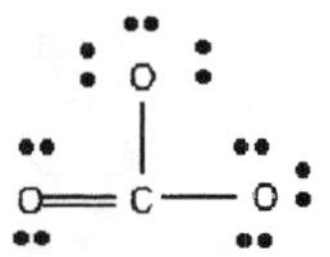

Lastly, write the Lewis structure in brackets with the charge of the ion in the upper right – hand corner.

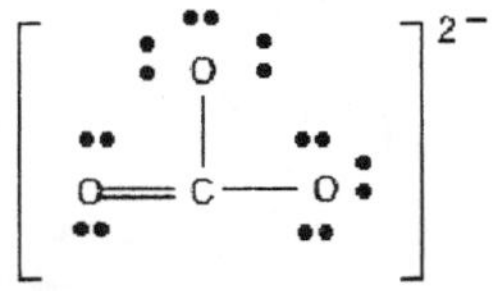

The double bond can be between the C and any of the oxygen atoms, so there are resonance structures

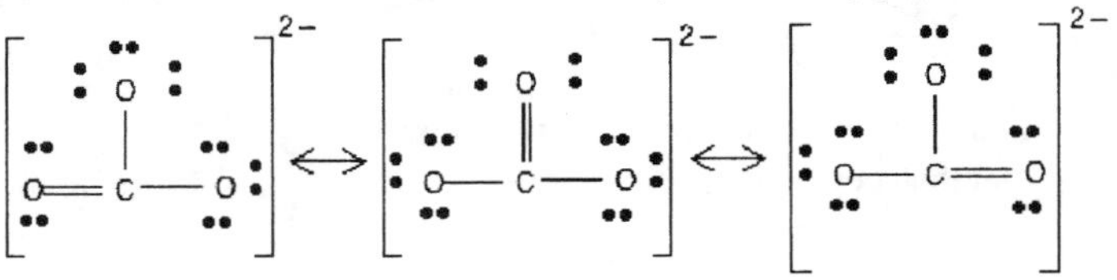

b) $Ca(OH)_2$: Ca^{2+} 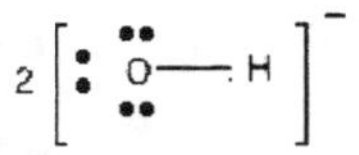

Determine the cation and anion

Ca^{2+} OH^-

Write the Lewis structure for the calcium cation based on the valence electrons

Ca $4s^2$ Ca^{2+} $4s^0$

Ca : Ca^{2+}

Ca must lose two electrons and be left with the octet from the previous shell

Write the Lewis structure for the covalent anion.

Write the correct skeletal structure for the ion

O—H

Calculate the total number of electrons for the Lewis structure by summing the valence electrons of each atom in the ion and adding one for the 1 – charge.
(number of valence e^- for H) + (number of valence e^- for O) +1 = 1 + 6 +1 = 8
Distribute the electrons among the atoms, giving octets (or duets for H) to as many atoms as possible. Begin with the bonding electrons, and then proceed to lone pairs on terminal atoms and finally to lone pairs of the central atom.

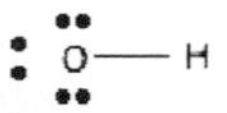

Lastly, write the Lewis structure in brackets with the charge of the ion in the upper right – hand corner.

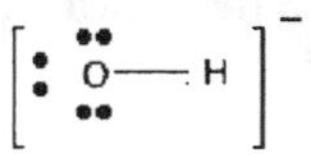

c) KNO_3: K^+ 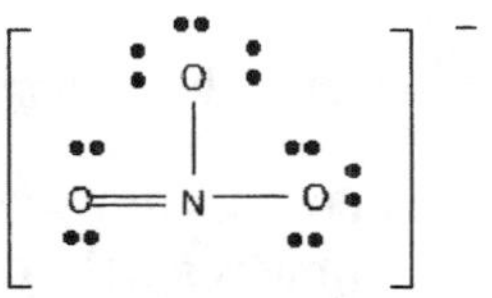

Determine the cation and anion

K^+ NO_3^-

Write the Lewis structure for the potassium cation based on the valence electrons

K $4s^1$ K^+ $4s^0$

K• $K^=$

K must lose one electron and be left with the octet from the previous shell

Write the Lewis structure for the covalent anion.

Write the correct skeletal structure for the ion

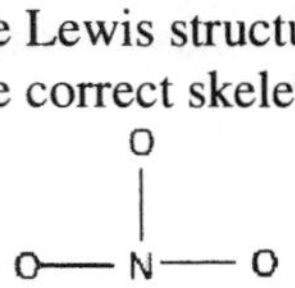

Calculate the total number of electrons for the Lewis structure by summing the valence electrons of each atom in the ion and adding one for the 1 – charge.
(number of valence e^- for N) + (number of valence e^- for O) = 5 + 3(6) +1 = 24
Distribute the electrons among the atoms, giving octets (or duets for H) to as many atoms as possible. Begin with the bonding electrons, and then proceed to lone pairs on terminal atoms and finally to lone pairs of the central atom.

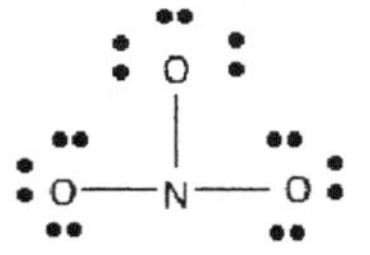

If any atom lacks an octet, form double or triple bonds as necessary to give them octets.

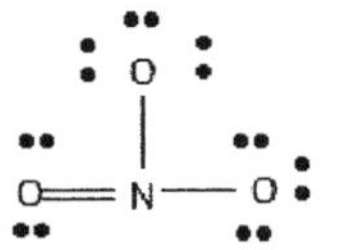

Lastly, write the Lewis structure in brackets with the charge of the ion in the upper right – hand corner.

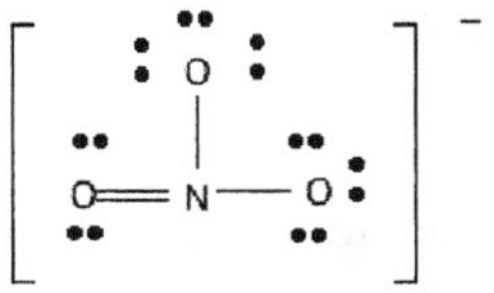

The double bond can be between the N and any of the oxygen atoms, so there are resonance structures

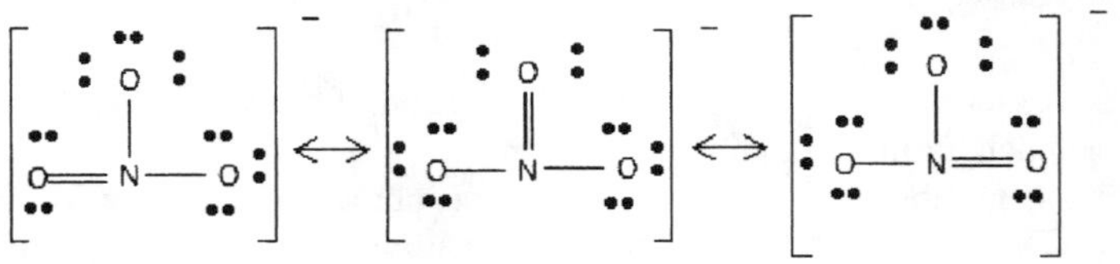

d) LiIO: Li^+ 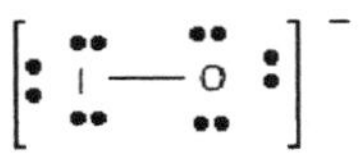

Determine the cation and anion

Li^+ IO^-

Write the Lewis structure for the lithium cation based on the valence electrons

Li $2s^1$ Li^+ $2s^0$

Li•

Li^+

Li must lose one electrons and be left with the octet from the previous shell

Write the Lewis structure for the covalent anion.

Write the correct skeletal structure for the ion

I — O

Calculate the total number of electrons for the Lewis structure by summing the valence electrons of each atom in the ion and adding one for the 1 – charge.
(number of valence e^- for I) + (number of valence e^- for O) = 7 +6 + 1 = 14
Distribute the electrons among the atoms, giving octets (or duets for H) to as many atoms as possible. Begin with the bonding electrons, and then proceed to lone pairs on terminal atoms and finally to lone pairs of the central atom.

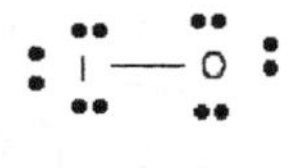

Lastly, write the Lewis structure in brackets with the charge of the ion in the upper right – hand corner.

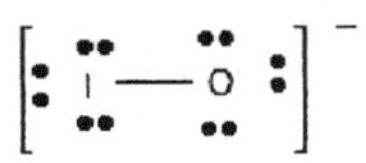

83. a) C₄H₈: Write the correct skeletal structure for the molecule

$$\begin{array}{ccc}
 & H & H \\
 & | & | \\
H- & C-C & -H \\
 & | & | \\
H- & C-C & -H \\
 & | & | \\
 & H & H
\end{array}$$

Calculate the total number of electrons for the Lewis structure by summing the valence electrons of each atom in the molecule.

4 (number of valence e⁻ for C) + 8(number of valence e⁻ for H) = 4(4) + 8(1) = 24

Distribute the electrons among the atoms, giving octets (or duets for H) to as many atoms as possible.

$$\begin{array}{ccc}
 & H & H \\
 & | & | \\
H- & C-C & -H \\
 & | & | \\
H- & C-C & -H \\
 & | & | \\
 & H & H
\end{array}$$

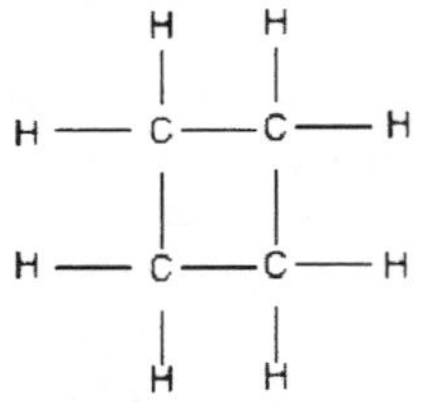

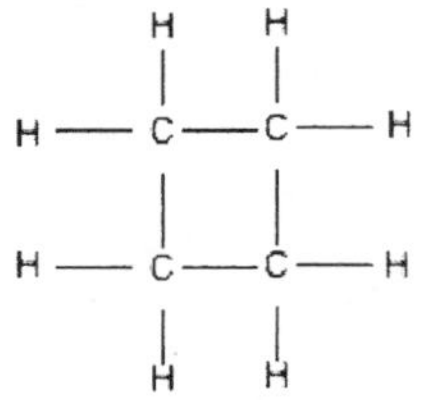

All atoms have octets or duets for H,

b) C₄H₄: Write the correct skeletal structure for the molecule

$$\begin{array}{ccc}
H- & C-C & -H \\
 & | & | \\
H- & C-C & -H
\end{array}$$

Calculate the total number of electrons for the Lewis structure by summing the valence electrons of each atom in the molecule.

4 (number of valence e⁻ for C) + 4(number of valence e⁻ for H) = 4(4) + 4(1) = 20

Distribute the electrons among the atoms, giving octets (or duets for H) to as many atoms as possible.

$$\begin{array}{ccc}
 & \cdot\cdot & \cdot\cdot \\
H- & C-C & -H \\
 & | & | \\
H- & C-C & -H
\end{array}$$

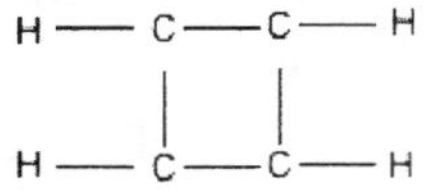

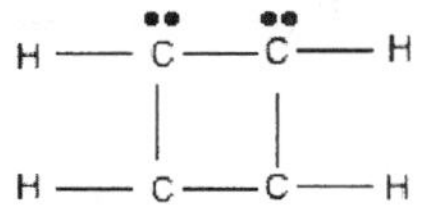

Complete octets by forming double bonds on alternating carbons, draw resonance structures.

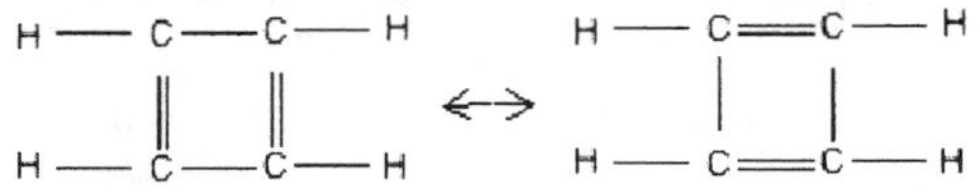

$$\begin{array}{ccc}
H- & C-C & -H \\
 & \| & \| \\
H- & C-C & -H
\end{array} \longleftrightarrow \begin{array}{ccc}
H- & C=C & -H \\
 & | & | \\
H- & C=C & -H
\end{array}$$

c) C₆H₁₂: Write the correct skeletal structure for the molecule.

Calculate the total number of electrons for the Lewis structure by summing the valence electrons of each atom in the molecule.

6 (number of valence e⁻ for C) + 12(number of valence e⁻ for H) = 6(4) + 12(1) = 36

Distribute the electrons among the atoms, giving octets (or duets for H) to as many atoms as possible. Begin with the bonding electrons, and then proceed to lone pairs on terminal atoms and finally to lone pairs of the central atom.

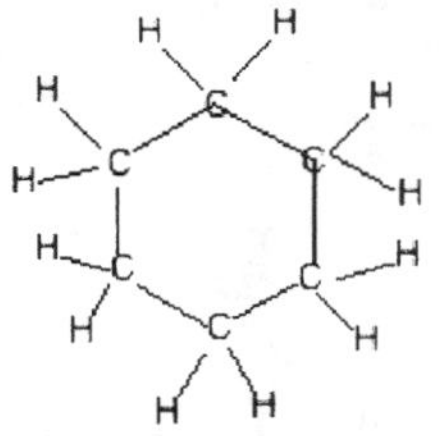

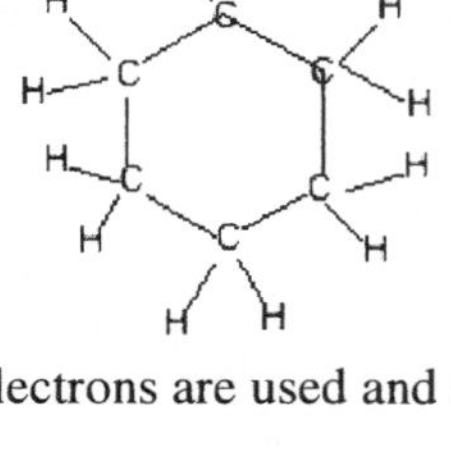

All 36 electrons are used and all atoms have octets or duets for H.

d) C_6H_6: Write the correct skeletal structure for the molecule.

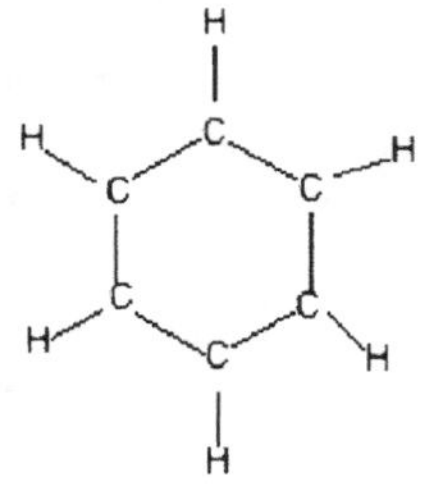

Calculate the total number of electrons for the Lewis structure by summing the valence electrons of each atom in the molecule.

6 (number of valence e^- for C) + 6(number of valence e^- for H) = 6(4) + 6(1) = 30

Distribute the electrons among the atoms, giving octets (or duets for H) to as many atoms as possible.

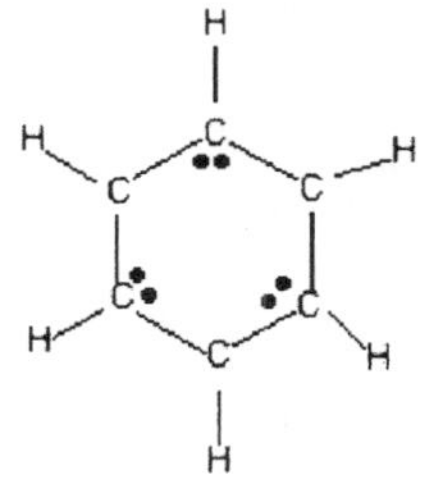

Complete octets by forming double bonds on alternating carbons, draw resonance structures.

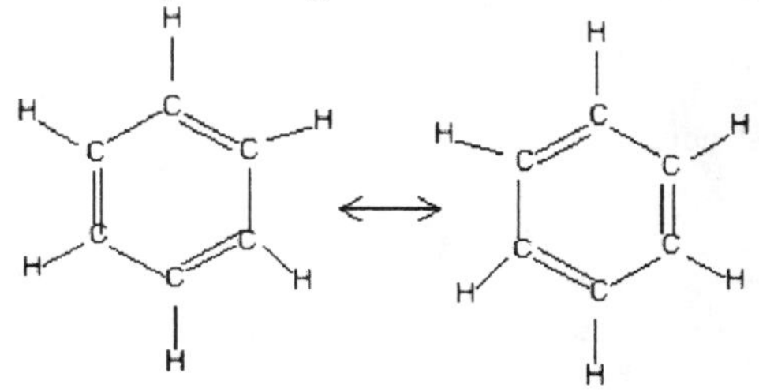

85. **Given:** 26.01% C; 4.38 % H; 69.52 % O; molar mass = 46.02 g/mol **Find:** molecular formula and Lewis structure

Conceptual Plan: convert mass to mol of each element→pseudoformula→empirical formula→molecular formula→Lewis structure

$\frac{1 \text{ mol C}}{12.01 \text{ g C}}$ $\frac{1 \text{ mol H}}{1.008 \text{ g H}}$ $\frac{1 \text{ mol O}}{16.00 \text{ g O}}$ divide by smallest number empirical formula x n

Solution: $26.01 \text{ g C} \times \frac{1 \text{ mol C}}{12.01 \text{ g C}} = 2.166 \text{ mol C}$

$4.38 \text{ g H} \times \frac{1 \text{ mol H}}{1.008 \text{ g H}} = 4.3\underline{4}5 \text{ mol H}$

$69.52 \text{ g O} \times \frac{1 \text{ mol O}}{16.00 \text{ g O}} = 4.345 \text{ mol O}$

$C_{2.166}H_{4.345}O_{4.345}$

$C_{\frac{2.166}{2.166}}H_{\frac{4.345}{2.166}}O_{\frac{4.345}{2.166}} \rightarrow CH_2O_2$

The correct empirical formula is CH_2O_2

empirical formula mass = (12.01 g/mol) + 2(1.008 g/mol) + 2(16.00 g/mol) = 46.03 g/mol

$$n = \frac{\text{molar mass}}{\text{formula molar mass}} = \frac{46.02 \text{ g/mol}}{46.03 \text{ g/mol}} = 1$$

molecular formula $= CH_2O_2 \times 1$

$= CH_2O_2$

Write the correct skeletal structure for the molecule

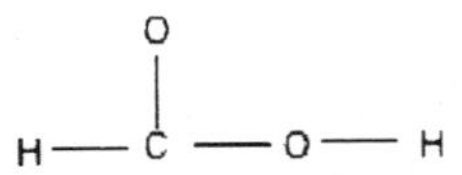

$$\begin{array}{c} O \\ \mid \\ H-C-O-H \end{array}$$

Calculate the total number of electrons for the Lewis structure by summing the valence electrons of each atom in the molecule.

(number of valence e⁻ for C) + 2(number of valence e⁻ for O) + 2(number of valence e⁻ for H) = 4 + 2(6) + 2(1) = 18

Distribute the electrons among the atoms, giving octets (or duets for H) to as many atoms as possible. Begin with the bonding electrons, and then proceed to lone pairs on terminal atoms and finally to lone pairs of the central atom.

$$\begin{array}{c} :\!\ddot{O}\!: \\ \mid \\ H-C-\ddot{O}-H \end{array}$$

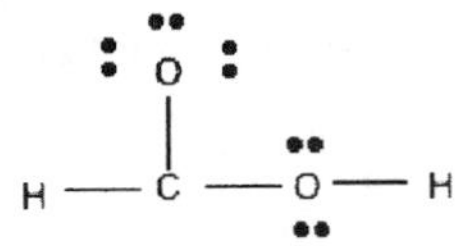

Complete the octet on C by forming a double bond

$$\begin{array}{c} :\!\ddot{O}\!: \\ \parallel \\ H-C-\ddot{O}-H \end{array}$$

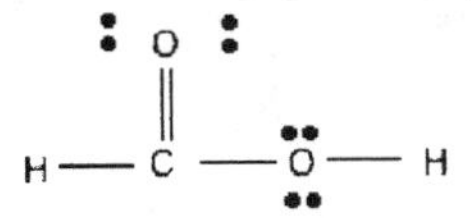

87. The lattice energy of Al_2O_3 is −15,916 kJ/mol. The thermite reaction is exothermic due to the energy released when the Al_2O_3 lattice forms. The lattice energy of Al_2O_3 is much more negative than the lattice energy of Fe_2O_3.

89. HNO_3

Write the correct skeletal structure for the molecule

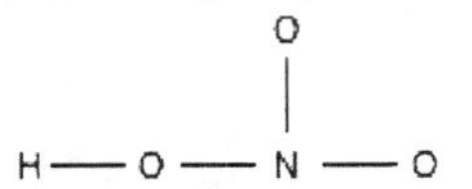

$$\begin{array}{c} O \\ \mid \\ H-O-N-O \end{array}$$

Calculate the total number of electrons for the Lewis structure by summing the valence electrons of each atom in the molecule.

3(number of valence e⁻ for O) + (number of valence e⁻ for N) + (number of valence e⁻ for H) = 3(6) + 5 +1 = 24

Distribute the electrons among the atoms, giving octets (or duets for H) to as many atoms as possible. Begin with the bonding electrons, and then proceed to lone pairs on terminal atoms and finally to lone pairs of the central atom.

$$\begin{array}{c} :\!\ddot{O}\!: \\ \mid \\ H-\ddot{O}-N-\ddot{O}: \end{array}$$

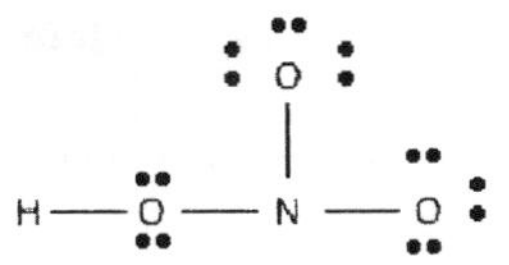

All 24 valence electrons are used

If any atom lacks an octet, form double or triple bonds as necessary to give them octets. The double bond can be formed to any of the three oxygen atoms, so there are three resonance forms.

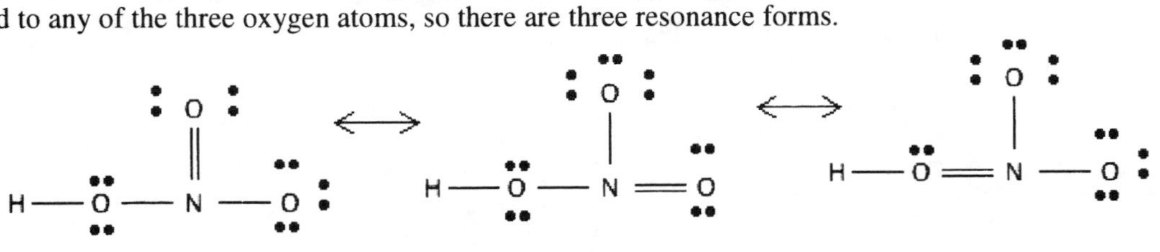

$$\begin{array}{ccccc} :\!\ddot{O}\!: & & :\!\ddot{O}\!: & & :\!\ddot{O}\!: \\ \parallel & & \mid & & \mid \\ H-\ddot{O}-N-\ddot{O}: & \longleftrightarrow & H-\ddot{O}-N=\ddot{O} & \longleftrightarrow & H-\ddot{O}=N-\ddot{O}: \\ \text{I} & & \text{II} & & \text{III} \end{array}$$

All atoms have octets, duets for H, structures is complete.

To determine which resonance hybrid(s) is most important, calculate the formal charge on each atom in each structure by finding the number of valence electrons and subtracting the number of lone pair electrons and one – half the number of bonding electrons.

	Structure I					Structure II				
	O_{left}	O_{top}	O_{right}	N	H	O_{left}	O_{top}	O_{right}	N	H
number of valence electrons	6	6	6	5	1	6	6	6	5	1
- number of lone pair electrons	4	4	6	0	0	4	6	4	0	0
- 1/2(number of bonding electrons)	2	2	1	4	1	2	1	2	4	1
Formal charge	0	0	−1	+1	0	0	−1	0	+1	0

	Structure III				
	O_{left}	O_{top}	O_{right}	N	H
number of valence electrons	6	6	6	5	1
- number of lone pair electrons	2	6	6	0	0
- 1/2(number of bonding electrons)	3	1	1	4	1
Formal charge	+1	−1	−1	+1	0

The sum of the formal charges is 0 for each structure which is the overall charge of the molecule. However, in structures I and II, the individual formal charges are lower and these two forms would contribute equally to the structure of HNO_3 and structure III would be less important since the individual formal charges are higher.

91. CNO^-
Write the skeletal structure: C – N – O
Determine the number of valence electrons.
(valence e^- from C) + (valence e^- from N) + (valence e^- from O) +1(from the negative charge)
4 + 5 + 6 + 1 = 16

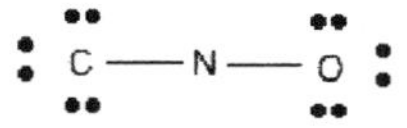

Distribute the electrons to complete octets if possible.

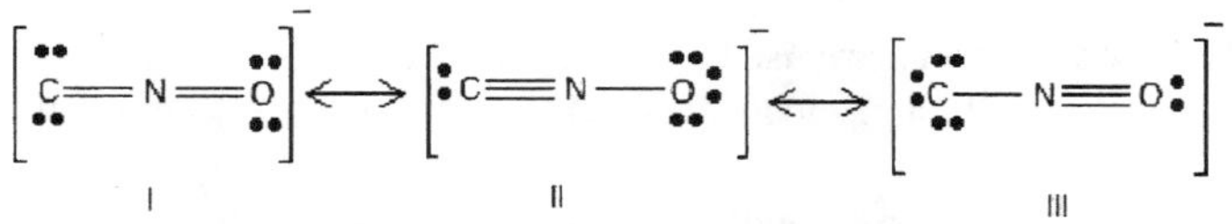

Determine the formal charge on each atom for each structure

	Structure I			Structure II		
	C	N	O	C	N	O
number of valence electrons	4	5	6	4	5	6
- number of lone pair electrons	4	0	4	2	0	6
- 1/2(number of bonding electrons)	2	4	2	3	4	1
Formal charge	−2	+1	0	−1	+1	−1

	Structure III		
	C	N	O
number of valence electrons	4	5	6
- number of lone pair electrons	6	0	2
- 1/2(number of bonding electrons)	1	4	3
Formal charge	−3	+1	+1

Structures I,II,and III all follow the octet rule but have varying degrees of negative formal charge on carbon, which is the least electronegative atom. Also the amount of formal charge is very high in all three resonance forms. Therefore, none of these resonance forms contribute to the stability of the fulminate ion and the ion is not very stable.

93. $HCSNH_2$: Write the correct skeletal structure for the molecule.

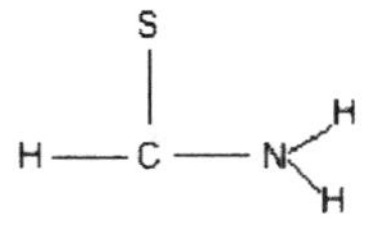

Calculate the total number of electrons for the Lewis structure by summing the valence electrons of each atom in the molecule.
(number of valence e^- for N) + (number of valence e^- for S) + (number of valence e^- for C) + 3(number of valence e^- for H) = 5 + 6 + 4 + 3(1) = 18
Distribute the electrons among the atoms, giving octets (or duets for H) to as many atoms as possible. Begin with the bonding electrons, and then proceed to lone pairs on terminal atoms and finally to lone pairs of the central atom.

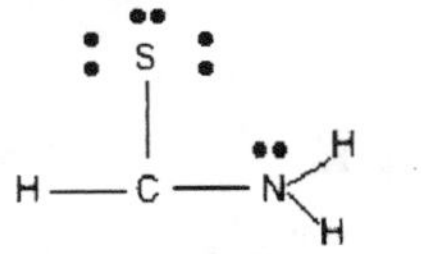

Complete the octet on C by forming a double bond.

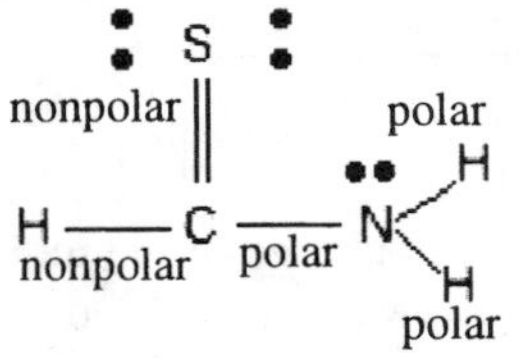

95. a) O_2^-: Write the correct skeletal structure for the radical

O — O

Calculate the total number of electrons for the Lewis structure by summing the valence electrons of each atom in the radical and adding 1 for the 1 – charge.
2(number of valence e^- for O) + 1 = 2(6) + 1 = 13
Distribute the electrons among the atoms, giving octets to as many atoms as possible. Begin with the bonding electrons, and then proceed to lone pairs on terminal atoms and finally to lone pairs of the central atom.

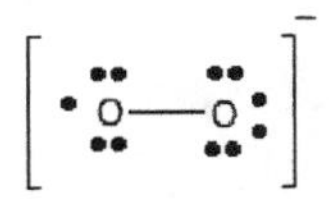

All 13 valence electrons are used.
O has an incomplete octet. It has 7 electrons because we have an odd number of valence electrons.

b) O^-: Write the Lewis structure based on the valence electrons: $2s^22p^5$.

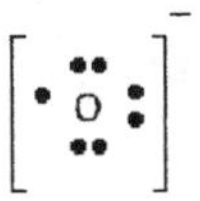

c) OH: Write the correct skeletal structure for the molecule

H — O

Calculate the total number of electrons for the Lewis structure by summing the valence electrons of each atom in the molecule
(number of valence e^- for O) + (number of valence e^- for H) = 6 + 1 = 7
Distribute the electrons among the atoms, giving octets (or duets for H) to as many atoms as possible. Begin with the bonding electrons, and then proceed to lone pairs on terminal atoms and finally to lone pairs of the central atom.

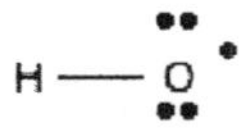

All 7 valence electrons are used.
O has an incomplete octet. It has 7 electrons because we have an odd number of valence electrons.

d) CH_3OO: Write the correct skeletal structure for the radical
C is the less electronegative, so it is central

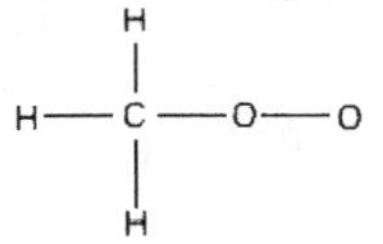

Calculate the total number of electrons for the Lewis structure by summing the valence electrons of each atom in the molecule

3(number of valence e^- for H) + (number of valence e^- for C) + 2(number of valence e^- for O)
= 3(1) + 4 +2(6) = 19
Distribute the electrons among the atoms, giving octets (or duets for H) to as many atoms as possible. Begin with the bonding electrons, and then proceed to lone pairs on terminal atoms and finally to lone pairs of the central atom.

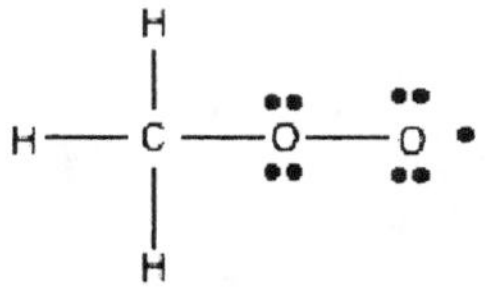

All 19 valence electrons are used.
O has an incomplete octet. It has 7 electrons because we have an odd number of valence electrons.

97. Rewrite the reaction using the Lewis structures of the molecules involved.

$$H - H\,(g) + 1/2\,O = O\,(g) \rightarrow H - O - H$$

(with two lone pairs shown above and below each O)

Determine which bonds are broken in the reaction and sum the bond energies of these
Σ(ΔH's bonds broken)
= (H – H) + 1/2(O = O)
= (436 kJ/mol) + 1/2(498)
= 685 kJ/mol
Determine which bonds are formed in the reaction and sum the negatives of the bond energies of these
Σ(-ΔH's of bonds formed)
= – 2(O – H)
= – 2(464 kJ/mol)
= –928 kJ/mol
Find ΔH_{rxn} by summing the results of the two steps.
ΔH_{rxn} = Σ(ΔH's bonds broken) + Σ(-ΔH's of bonds formed)
= 685 kJ/mol – 928 kJ/mol
= – 243 kJ/mol

$CH_4(g) + 2O_2(g) \rightarrow CO_2(g) + 2H_2O(g)$
Rewrite the reaction using the Lewis structures of the molecules involved.

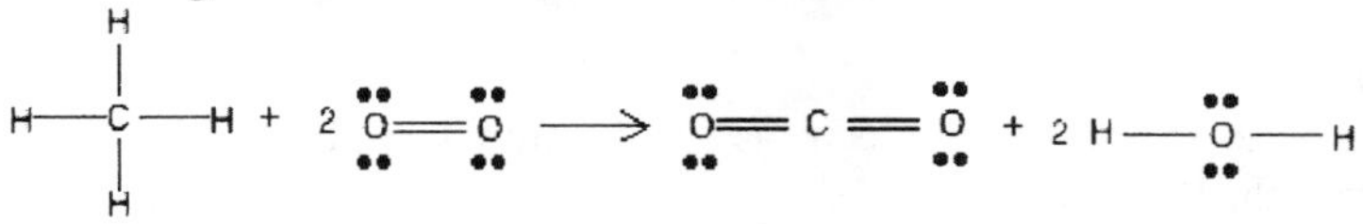

Determine which bonds are broken in the reaction and sum the bond energies of these
Σ(ΔH's bonds broken)
= 4(C – H) + 2(O = O)
= (414 kJ/mol) + 2(498)
= 2652 kJ/mol
Determine which bonds are formed in the reaction and sum the negatives of the bond energies of these
Σ(-ΔH's of bonds formed)
= – 2(C = O) – 4(O – H)
= – 2(799 kJ/mol) – 4(464 kJ/mol)
= –3454 kJ/mol
Find ΔH_{rxn} by summing the results of the two steps.
ΔH_{rxn} = Σ(ΔH's bonds broken) + Σ(-ΔH's of bonds formed)
= 2653 kJ/mol – 3454 kJ/mol
= – 802 kJ/mol

Compare

	kJ/mol	kJ/g
H_2	–243	–120
CH_4	–802	–50.1

So, methane yields more energy per mole but hydrogen yields more energy per gram.

99. a) Cl_2O_7: Write the correct skeletal structure for the molecule

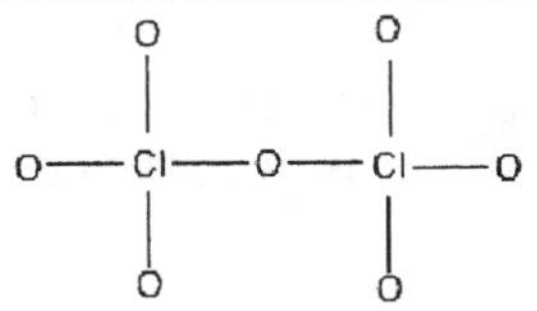

Calculate the total number of electrons for the Lewis structure by summing the valence electrons of each atom in the molecule
2(number of valence e^- for Cl) + 7(number of valence e^- for O) = 2(7)5 + 7(6) = 56
Distribute the electrons among the atoms, giving octets (or duets for H) to as many atoms as possible. Begin with the bonding electrons, and then proceed to lone pairs on terminal atoms and finally to lone pairs of the central atom.

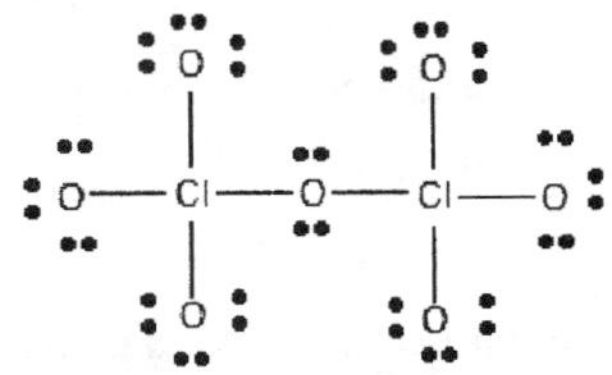

Form double bonds to minimize formal charge.

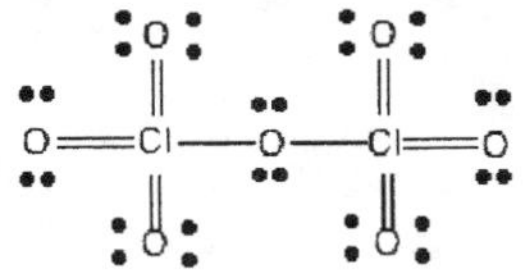

b) H_3PO_3: Write the correct skeletal structure for the molecule

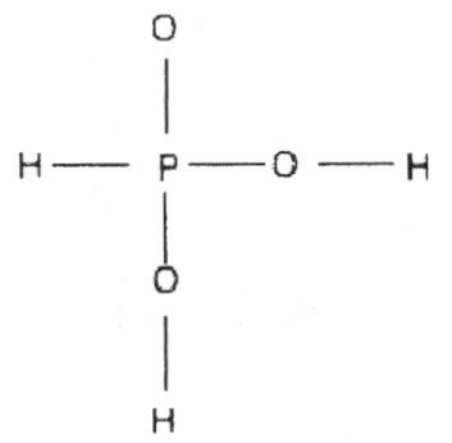

Calculate the total number of electrons for the Lewis structure by summing the valence electrons of each atom in the molecule
(number of valence e^- for P) + 3(number of valence e^- for O) + 3(number of valence e^- for H)
= 5 + 3(6) + 3(1) = 26
Distribute the electrons among the atoms, giving octets (or duets for H) to as many atoms as possible. Begin with the bonding electrons, and then proceed to lone pairs on terminal atoms and finally to lone pairs of the central atom.

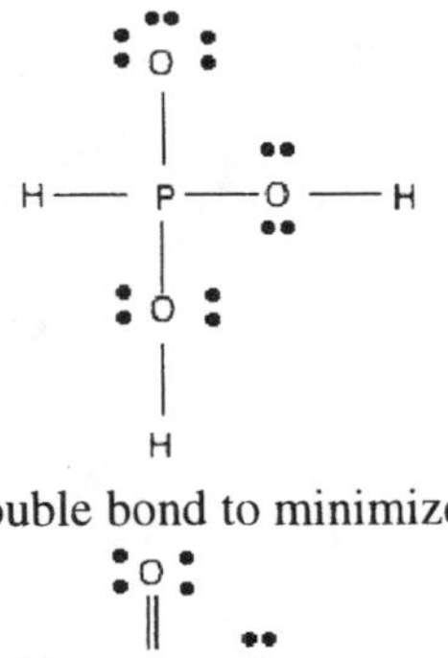

Form double bond to minimize formal charge.

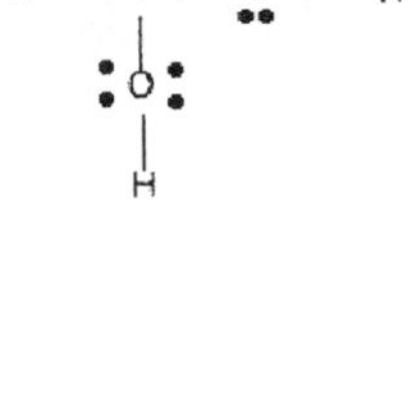

c) H_3AsO_4: Write the correct skeletal structure for the radical

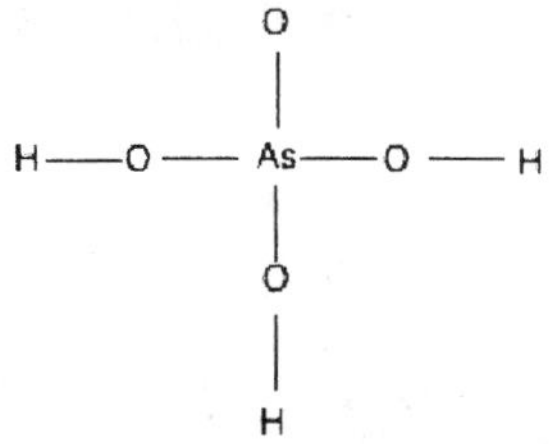

Calculate the total number of electrons for the Lewis structure by summing the valence electrons of each atom in the molecule
(number of valence e^- for As) + 4(number of valence e^- for O) + 3(number of valence e^- for H)
$= 5 + 4(6) + 3(1) = 32$
Distribute the electrons among the atoms, giving octets (or duets for H) to as many atoms as possible. Begin with the bonding electrons, and then proceed to lone pairs on terminal atoms and finally to lone pairs of the central atom.

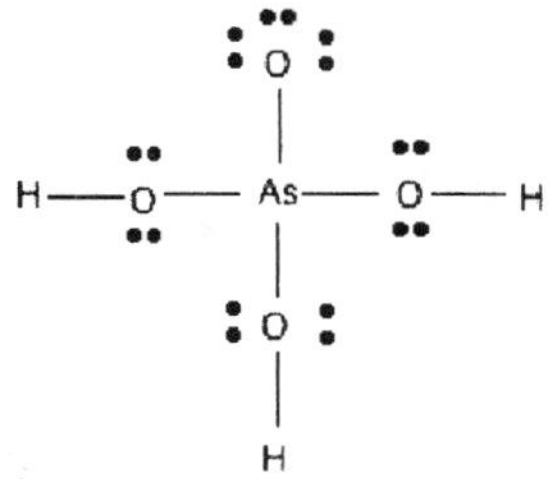

Form double bond to minimize formal charge.

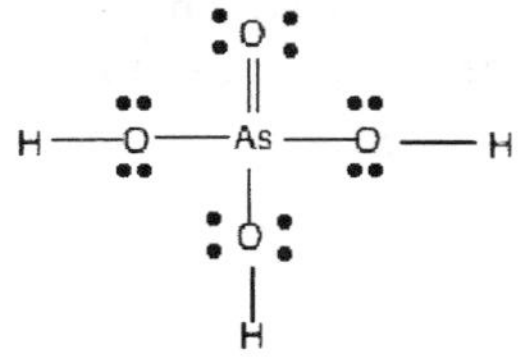

101. $Na^+F^- < Na^+O^{2-} < Mg^{2+}F^- < Mg^{2+}O^{2-} < Al^{3+}O^{2-}$

The lattice energy is proportional to the magnitude of the charge and inversely proportional to the distance between the atoms. Na^+F^- would have the smallest lattice energy because the magnitude of the charges on Na and F are the smallest. $Mg^{2+}F^-$ and Na^+O^{2-} both have the same magnitude formal charge, the O^{2-} is larger than F^- in size, Na^+ is larger than Mg^{2+}, so Na^+O^{2-} should be less than $Mg^{2+}F^-$. The magnitude of the charge makes $Mg^{2+}O^{2-} < Al^{3+}O^{2-}$.

103.

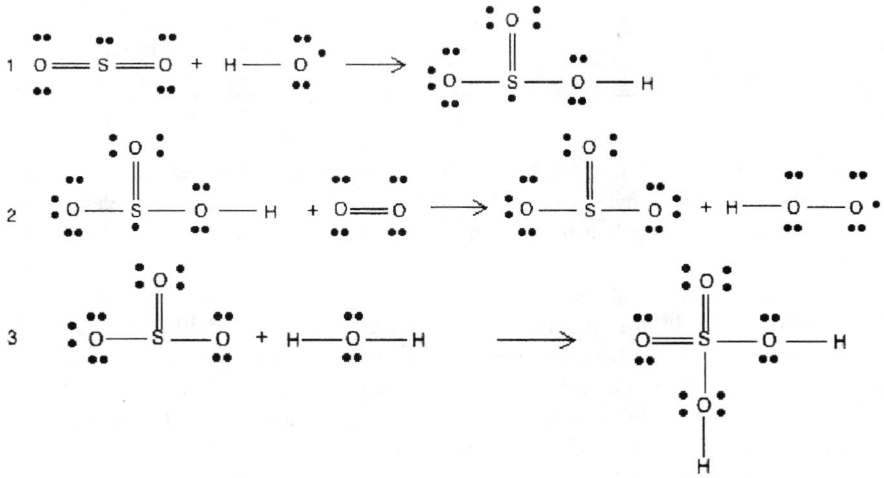

Step 1:

Bonds broken: 2(S = O) + (H – O) = 2(523) + (464) = 1510 kJ/mol

Bonds formed: –2(S – O) –(S = O) –(O – H) = –2(265) –(523) –(464) = –1517 kJ/mol

$\Delta H_{step} = -7$ kJ/mol

Step 2:

Bonds broken: $2(S-O) + (S=O) + (O-H) + (O=O) = 2(265)+(523)+(464)+(498) = 2015$ kJ/mol
Bonds formed: : $-2(S-O) -(S=O) -(O-H) -(O-O) = -2(265) -(523) -(464) -(142) = -1659$ kJ/mol

$\Delta H_{step} = +356$ kJ/mol

Step 3:

Bonds broken: $2(S-O) + (S=O) + 2(O-H) = 2(265)+(523)+2(464) = 1981$ kJ/mol
Bonds formed: $-2(S-O) - 2(S=O) - 2(O-H) = -2(265) + -2(523) + -2(464) = -2504$ kJ/mol

$\Delta H_{step} = -523$ kJ/mol

Hess's law states that ΔH for the reaction is the sum of ΔH of the steps:
$\Delta H_{rxn} = (-7 \text{ kJ/mol}) + (+356 \text{ kJ/mol}) + (-523 \text{ kJ/mol}) = -174$ kJ/mol

105. **Given:** $\mu = 1.08$ D HCl, 20% ionic and $\mu = 1.82$ D HF, 45% ionic **Find:** r
Conceptual Plan: $\mu \rightarrow \mu_{calc} \rightarrow r$

$$\% \text{ ionic character} = \frac{\mu}{\mu_{calc}} \qquad \mu_{calc} = qr$$

Solution: For HCl $\quad \mu_{calc} = \frac{1.08}{0.20} = 5.4\,D \qquad \frac{5.4\,D \times \frac{3.34 \times 10^{-30}\ C\,m}{D} \times \frac{10^{12}\ pm}{m}}{1.6 \times 10^{-19}\ C} = 113\ pm$

For HF $\quad \mu_{calc} = \frac{1.82}{0.45} = 4.04\,D \qquad \frac{4.04\,D \times \frac{3.34 \times 10^{-30}\ C\,m}{D} \times \frac{10^{12}\ pm}{m}}{1.6 \times 10^{-19}\ C} = 84\ pm$

107. In order for the four P atoms to be equivalent, they must all be in the same electronic environment. That is, they must all see the same number of bonds and lone pair electrons. The only way to achieve this is with a tetrahedral configuration where the P atoms are at the four points of the tetrahedron.

109. **Given:** $\Delta H_f^\circ PI_3(s) = -24.7$kJ/mol; $P-I = 184$ kJ/mol; $I-I = 151$kJ/mol; $\Delta H_f^\circ P(g) = 334$ kJ/mol; $\Delta H_f^\circ I_2(g) = 62$ kJ/mol
Find: $\Delta H_{sub} PI_3(s)$
Conceptual Plan: $PI_3(s) \rightarrow PI_3(g)$; **use Hess's law**
Solution:

Reaction	ΔH(kJ/mol)	
$PI_3(s) \rightarrow P(s) + 3/2\ I_2(s)$	+ 24.7	(this is the reverse of the formation reaction)
$P(s) \rightarrow P(g)$	+334	(formation of P(g))
$3/2\ I_2(s) \rightarrow 3/2\ I_2(g)$	3/2(62)	(formation of $I_2(g)$)
$3/2\ I_2(g) \rightarrow 3\ I(g)$	3/2(151)	(breaking I – I bond)
$P(g) + 3I(g) \rightarrow PI_3(g)$	– 3(184)	(forming P – I bond)
$PI_3(s) \rightarrow PI_3(g)$	+126	(sublimation of $PI_3(s)$)

111. When we say that a compound is "energy rich" we mean that it gives off a great amount of energy when it reacts. It means that there is a lot of energy stored in the compound. This energy is released when the weak bonds in the compound break and much stronger bonds are formed in the product thereby releasing energy.

113. Lewis theory is successful because it allows us to understand and predict many chemical observations. We can use it to determine the formula of ionic compounds, to account for low melting points and boiling points of molecular compounds compared to ionic compounds. Lewis theory allows us to predict what molecules or ions will be stable, which will be more reactive, and which will not exist. Lewis theory, however, does not really tell us anything about how the bonds in the molecules and ions form. It does not give us a way to account for the paramagnetism of oxygen. And, by itself, Lewis theory does not really tell us anything about the shape of the molecule or ion.

Chapter 10
Chemical Bonding II: Molecular Shapes, Valence Bond Theory, and Molecular Orbital Theory

1. The properties of molecules are directly related to their shape. The sensation of taste, immune response, the sense of smell, and many types of drug action all depend on the shape – specific interactions between molecules and proteins.

3. The five basic electron geometries are:
   1) Linear, which has two electron groups
   2) Trigonal planar, which has three electron groups
   3) Tetrahedral, which has four electron groups
   4) Trigonal bypyramid, which has five electron groups
   5) Octahedral, which has six electron groups.
   An electron group is at least one pair of electrons.

(a) Linear geometry

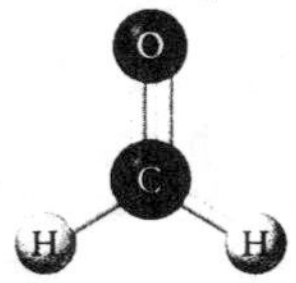

(b) Trigonal planar geometry

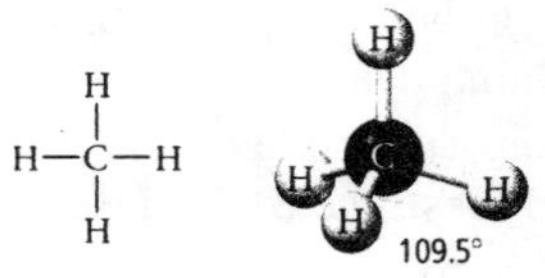

Tetrahedral geometry

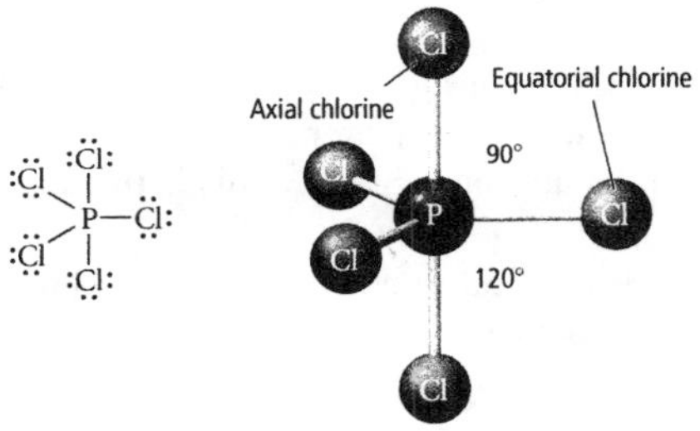

Trigonal bipyramidal geometry

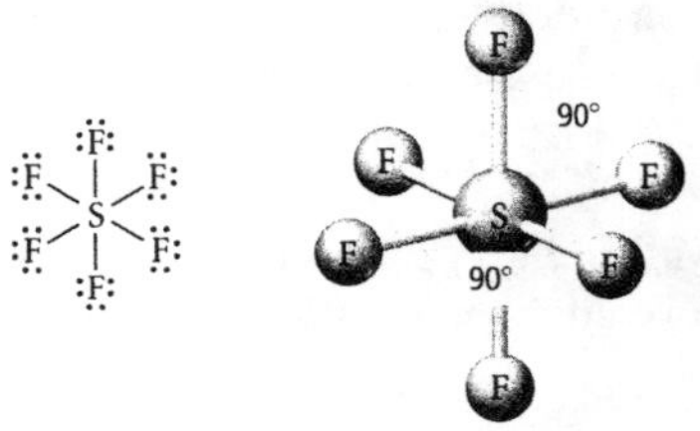

Octahedral geometry

5. a) Four electron groups gives tetrahedral electron geometry, while three bonding groups and one lone pair give a trigonal pyramidal molecular geometry.

   b) Four electron groups gives a tetrahedral electron geometry, while two bonding groups and two lone pair give a bent molecular geometry.

   c) Five electron groups gives a trigonal bipyramidal electron geometry, while four bonding groups and one lone pair give a seesaw molecular geometry.

   d) Five electron groups gives a trigonal bipyramidal electron geometry, while three bonding groups and two lone pair give a T – shaped molecular geometry.

   e) Five electron groups gives a trigonal bipyramidal electron geometry, while two bonding groups and three lone pair give a linear geometry.

   f) Six electron groups gives an octahedral electron geometry, while five bonding groups and one lone pair give a square pyramidal molecular geometry.

   g) Six electron groups gives an octahedral electron geometry, while four bonding groups and two lone pair gives a square planar molecular geometry.

7. To determine if a molecule is polar:
   1. Draw the Lewis structure for the molecule and determine the molecular geometry

2. Determine whether the molecule contains polar bonds.
3. Determine whether the polar bonds add together to form a net dipole moment

Polarity is important because polar and nonpolar molecules have different properties. Polar molecules interact strongly with other polar molecules while they do not interact with nonpolar molecules.

9. According to valence bond theory, the shape of the molecule is determined by the geometry of the overlapping orbitals.

11. Hybridization is a mathematical procedure in which the standard atomic orbitals are combined to form new atomic orbitals called hybrid orbitals. Hybrid orbitals are still localized on individual atoms, but they have different shapes and energies from those of standard atomic orbitals. They are necessary in valence bond theory because they correspond more closely to the actual distribution of electron in chemically bonded atoms.

13. The number of standard atomic orbitals added together always equals the number of hybrid orbitals formed. The total number of orbitals is conserved.

15. The double bond in Lewis theory is simply two pair of electrons between the same two atoms. However, in valence bond theory we see that it is made up of two different kinds of bonds. The double bond in valence bond theory consists of one σ bond and one π bond. Valence bond theory shows us that rotation about a double bond is severely restricted. Because of the side – by – side overlap of the p orbitals, the π bond must essentially break for rotation to occur. The single bond consists of overlap that results in a σ bond. Since the overlap is linear, rotation is not restricted.

17. In molecular orbital theory, atoms will bond when the electrons in the atoms can lower their energy by occupying the molecular orbitals of the resultant molecule.

19. Constructive interference between two atomic orbitals gives rise to a molecular orbital that is lower in energy than the atomic orbitals. This is the bonding orbital. Destructive interference between two atomic orbitals gives rise to a molecular orbital that is higher in energy than the atomic orbitals. This is the antibonding orbital.

21. Molecular orbitals can be approximated by a linear combination of atomic orbitals. The total number of MO's formed from a particular set of AO's will always equal the number of AO's used.

23.

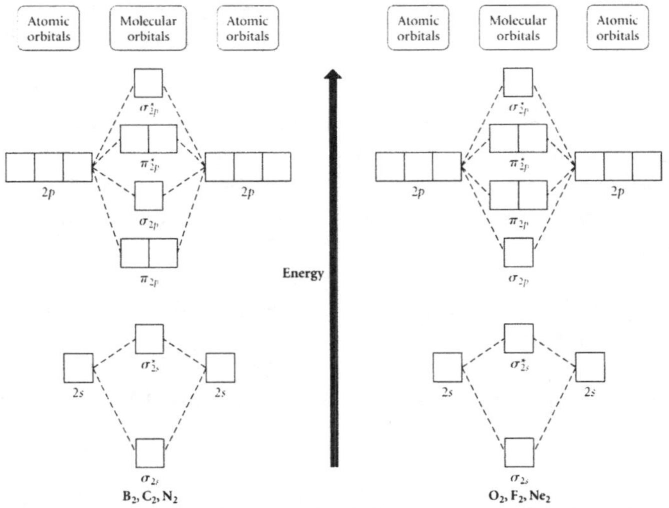

25. A paramagnetic species has unpaired electrons in molecular orbitals of equal energy. A paramagnetic species is attracted to a magnetic field. The magnetic property is a direct result of the unpaired electrons. The spin and angular momentum of the electrons generate tiny magnetic fields. A diamagnetic species has all of the electrons paired. The magnetic fields caused by the electron spin and orbital angular momentum tend to cancel each other. A diamagnetic species is not attracted to a magnetic field, and is, in fact, slightly repelled.

27. Nonbonding orbitals are atomic orbitals not involved in a bond and will remain localized on the atom.

29. 4 electron pairs A trigonal pyramidal molecular geometry has three bonding groups and one lone pair of electrons, so there are four electron pairs on atom A.

31. a) 4 total electron groups, 4 bonding groups, 0 lone pair
A tetrahedral molecular geometry has four bonding groups and no lone pair. So, there are four total electron groups, four bonding groups and no lone pair.

b) 5 total electron groups, 3 bonding groups, 2 lone pair.
A T – shaped molecular geometry has three bonding groups and two lone pair. So, there are five total electron groups, three bonding groups and two lone pair.

c) 6 total electron groups, 5 bonding groups, 1 lone pair.
A square pyramidal molecular geometry has five bonding groups and one lone pair. So, there are six total electron groups, five bonding groups and one lone pair.

33. a) PF_3: Electron geometry – tetrahedral; molecular geometry – trigonal pyramidal; bond angle = 109.5°
Because of the lone pair, the bond angle will be less than 109.5°
Draw a Lewis structure for the molecule
PF_3 has 26 valence electrons

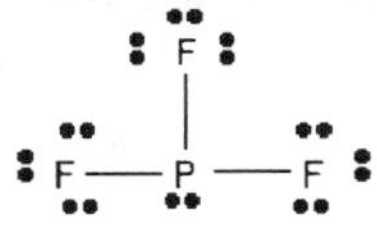

Determine the total number of electron groups around the central atom
There are 4 electron groups on P
Determine the number of bonding groups and the number of lone pairs around the central atom
There are three bonding groups and one lone pair
Use Table 10.1 to determine the electron geometry and molecular geometry and bond angles.
Four electron groups is tetrahedral electron geometry, three bonding groups and one lone pair is trigonal pyramidal molecular geometry, the idealized bond angles for tetrahedral are 109.5° however, the lone pair will make the bond angle less than idealized.

b) SBr_2: Electron geometry – tetrahedral; molecular geometry – bent; bond angle = 109.5°
Because of the lone pairs, the bond angle will be less than 109.5°
Draw a Lewis structure for the molecule
SBr_3 has 20 valence electrons

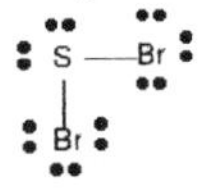

Determine the total number of electron groups around the central atom
There are 4 electron groups on S
Determine the number of bonding groups and the number of lone pairs around the central atom
There are 2 bonding groups and two lone pair
Use Table 10.1 to determine the electron geometry and molecular geometry and bond angles.
Four electron groups is tetrahedral electron geometry, two bonding groups and two lone pair is a bent molecular geometry, the idealized bond angles for tetrahedral are 109.5° however, the lone pairs will make the bond angle less than idealized.

c) $CHCl_3$: Electron geometry – tetrahedral; molecular geometry – tetrahedral; bond angle = 109.5°
Because there are no lone pair, the bond angle will be 109.5°
Draw a Lewis structure for the molecule
$CHCl_3$ has 26 valence electrons

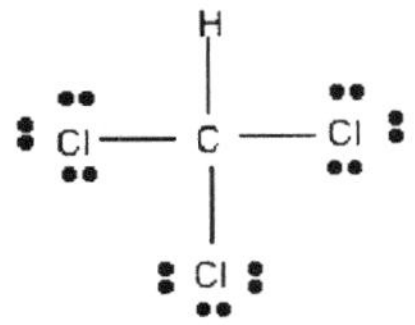

Determine the total number of electron groups around the central atom
There are 4 electron groups on C
Determine the number of bonding groups and the number of lone pairs around the central atom
There are four bonding groups and no lone pair

Use Table 10.1 to determine the electron geometry and molecular geometry and bond angles.
Four electron groups is tetrahedral electron geometry, four bonding groups and no lone pair is a tetrahedral molecular geometry, the idealized bond angles for tetrahedral are 109.5° however, because the attached atoms have different electronegativities the bond angles are less than idealized.

d) CS_2: Electron geometry – linear; molecular geometry – linear; bond angle = 180°
Because there are no lone pair, the bond angle will 180°
Draw a Lewis structure for the molecule
CS_2 has 16 valence electrons

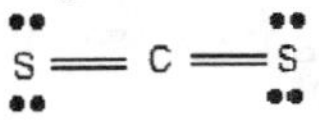

Determine the total number of electron groups around the central atom
There are 2 electron groups on C
Determine the number of bonding groups and the number of lone pairs around the central atom
There are 2 bonding groups and no lone pair
Use Table 10.1 to determine the electron geometry and molecular geometry and bond angles.
Two electron groups is linear geometry, two bonding groups and no lone pair is linear molecular geometry, the idealized bond angle is 180°

35. H_2O will have the smaller bond angle because lone pair – lone pair repulsions are greater than lone – pair bonding pair repulsions.
Draw the Lewis structures for both structures
H_3O^+ has 8 valence electrons H_2O has 8 valence electrons

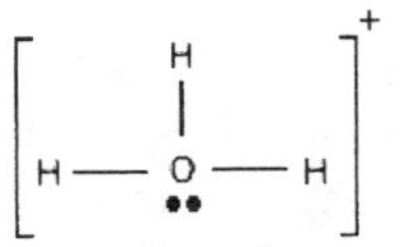

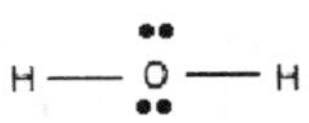

3 bonding groups and 1 lone pair 2 bonding groups and 2 lone pair
Both have 4 electron groups, but the 2 lone pair in H_2O will cause the bond angle to be smaller because of the lone pair – lone pair repulsions.

37. a) SF_4 Draw a Lewis structure for the molecule
SF_4 has 34 valence electrons

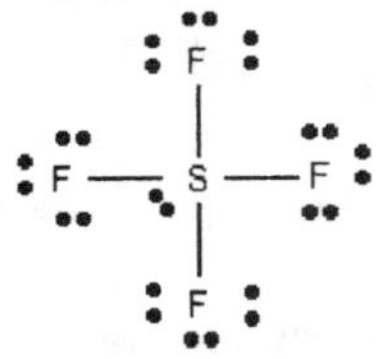

Determine the total number of electron groups around the central atom
There are 5 electron groups on S
Determine the number of bonding groups and the number of lone pairs around the central atom
There are four bonding groups and one lone pair
Use Table 10.1 to determine the electron geometry and molecular geometry
The electron geometry is trigonal bipyramidal so the molecular geometry is seesaw
Sketch the molecule

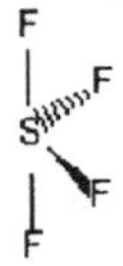

b) ClF_3 Draw a Lewis structure for the molecule
ClF_3 has 28 valence electrons

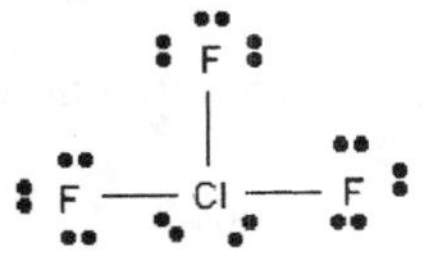

Determine the total number of electron groups around the central atom
There are 5 electron groups on Cl
Determine the number of bonding groups and the number of lone pairs around the central atom
There are three bonding groups and two lone pair
Use Table 10.1 to determine the electron geometry and molecular geometry
The electron geometry is trigonal bipyramidal so the molecular geometry is T – shape
Sketch the molecule

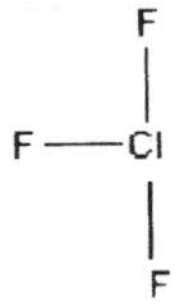

c) IF_2^- Draw a Lewis structure for the ion
IF_2^- has 22 valence electrons

$$\left[:\!\overset{\bullet\bullet}{\underset{\bullet\bullet}{F}}\! - \overset{\bullet\bullet}{\underset{\bullet\bullet}{I}} - \!\overset{\bullet\bullet}{\underset{\bullet\bullet}{F}}\!: \right]^-$$

Determine the total number of electron groups around the central atom
There are 5 electron groups on I
Determine the number of bonding groups and the number of lone pairs around the central atom
There are two bonding groups and three lone pair
Use Table 10.1 to determine the electron geometry and molecular geometry
The electron geometry is trigonal bipyramidal so the molecular geometry is linear
Sketch the ion

$$[F \!-\! I \!-\! F\,]^-$$

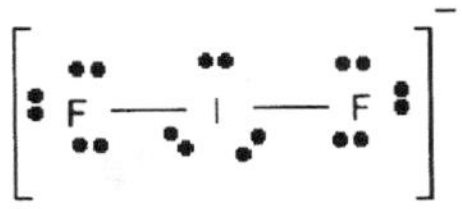

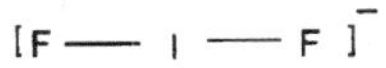

d) IBr_4^- Draw a Lewis structure for the ion
IBr_4^- has 36 valence electrons

Determine the total number of electron groups around the central atom
There are 6 electron groups on I
Determine the number of bonding groups and the number of lone pairs around the central atom
There are four bonding groups and two lone pair
Use Table 10.1 to determine the electron geometry and molecular geometry
The electron geometry is octahedral so the molecular geometry is square planar
Sketch the ion

$$\left[\begin{matrix} Br & & Br \\ & \diagdown\;\diagup & \\ & I & \\ & \diagup\;\diagdown & \\ Br & & Br \end{matrix} \right]^-$$

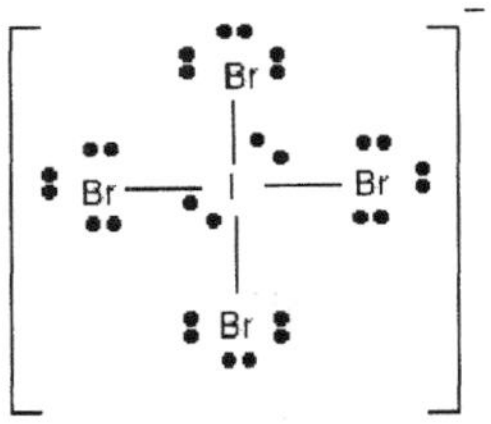

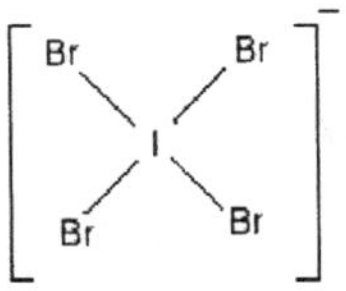

39. a) C_2H_2 Draw the Lewis structure

$$H \!-\! C \!\equiv\! C \!-\! H$$

Atom	Number of Electron Groups	Number of Lone Pairs	Molecular Geometry
Left C	2	0	Linear
Right C	2	0	Linear

Sketch the molecule

b) C_2H_4 Draw the Lewis structure

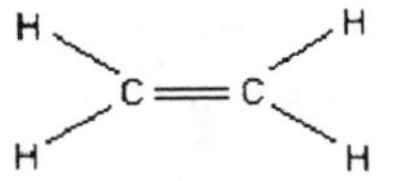

Atom	Number of Electron Groups	Number of Lone Pairs	Molecular Geometry
Left C	3	0	Trigonal planar
Right C	3	0	Trigonal planar

Sketch the molecule

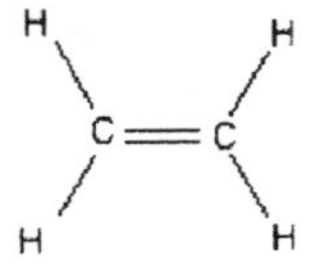

c) C_2H_6 Draw the Lewis structure

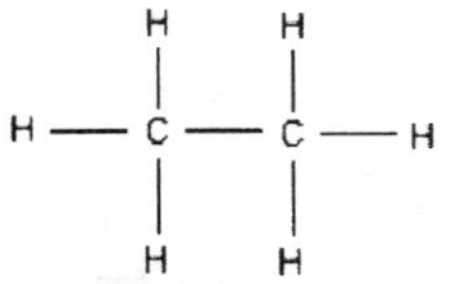

Atom	Number of Electron Groups	Number of Lone Pairs	Molecular Geometry
Left C	4	0	Tetrahedral
Right C	4	0	Tetrahedral

Sketch the molecule

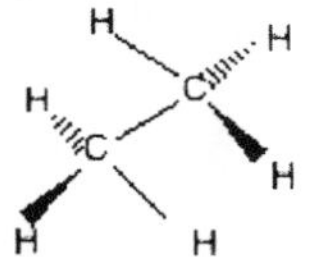

41. a) Four pair of electrons gives a tetrahedral electron geometry, the lone pair would cause lone pair – bonded pair repulsions and would have a trigonal pyramidal molecular geometry.

b) Five pair of electrons gives a trigonal bipyramidal electron geometry, the lone pair occupies an equatorial position in order to minimize lone pair – bonded pair repulsions and the molecule would have a seesaw molecular geometry.

c) Six pair of electrons gives an octahedral electron geometry, the two lone pair would occupy opposite position in order to minimize lone pair – lone pair repulsions. The molecular geometry would be square planar

43. a) CH_3OH Draw the Lewis structure, determine the geometry about each interior atom

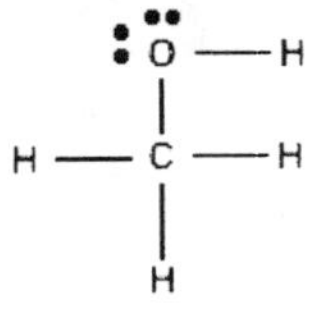

Atom	Number of Electron Groups	Number of Lone Pairs	Molecular Geometry
C	4	0	Tetrahedral
O	4	2	Bent

Sketch the molecule

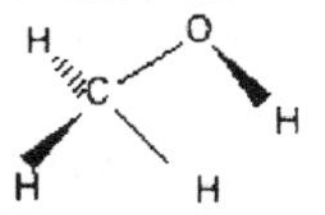

b) CH_3OCH_3 Draw the Lewis structure, determine the geometry about each interior atom

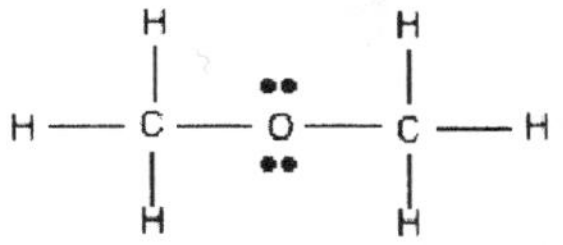

Atom	Number of Electron Groups	Number of Lone Pairs	Molecular Geometry
C	4	0	Tetrahedral
O	4	2	Bent
C	4	0	Tetrahedral

Sketch the molecule

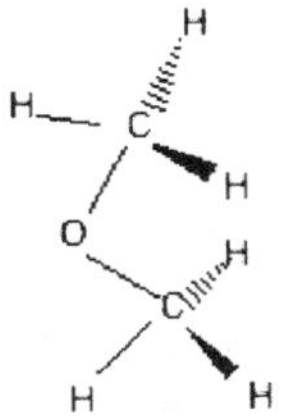

c) H_2O_2 Draw the Lewis structure, determine the geometry about each interior atom

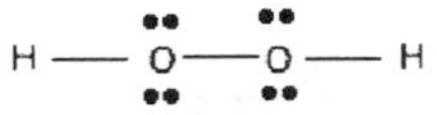

Atom	Number of Electron Groups	Number of Lone Pairs	Molecular Geometry
O	4	2	Bent
O	4	2	Bent

Sketch the molecule

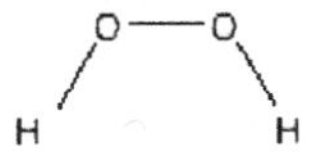

45. Draw the Lewis structure for CO_2 and CCl_4 and determine the molecular geometry.

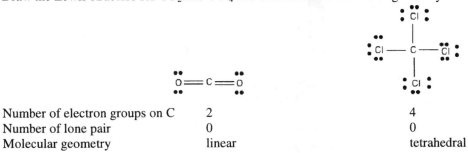

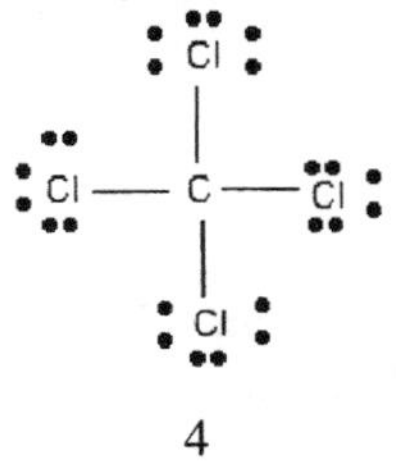

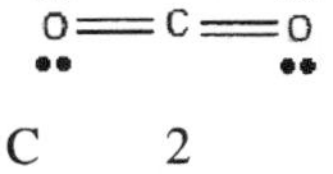

	CO_2	CCl_4
Number of electron groups on C	2	4
Number of lone pair	0	0
Molecular geometry	linear	tetrahedral

Even though each molecule contains polar bonds, the sum of the bond dipoles gives a net dipole of zero for each molecule.

47. a) PF_3 – polar

Draw the Lewis structure and determine the molecular geometry

The molecular geometry from problem 33 is trigonal pyramidal

Determine if the molecule contains polar bonds.
The electronegativities of P = 2.1 and F = 4. Therefore the bonds are polar

Determine whether the polar bond add together to form a net dipole.
Because the molecule is trigonal pyramidal the three dipole moments sum to a net dipole moment. The molecule is polar

b) SBr_2 – polar

Draw the Lewis structure and determine the molecular geometry
The molecular geometry from problem 33 is bent

Determine if the molecule contains polar bonds.
The electronegativities of S = 2.5 and Br = 2.0. Therefore the bonds are polar

Determine whether the polar bond add together to form a net dipole.
Because the molecule is bent, the two dipole moments sum to a net dipole moment. The molecule is polar

c) $CHCl_3$ – polar

Draw the Lewis structure and determine the molecular geometry
The molecular geometry from problem 33 is tetrahedral

Determine if the molecule contains polar bonds.
The electronegativities of C = 2.5, H = 2.1 and Cl = 3.0. Therefore the bonds are polar

Determine whether the polar bond add together to form a net dipole.
Because the bonds have different dipole moments because of the different atoms involved, the four dipole moments sum to a net dipole moment. The molecule is polar

d) CS_2 – nonpolar

Draw the Lewis structure and determine the molecular geometry
The molecular geometry from problem 33 is linear

Determine if the molecule contains polar bonds.
The electronegativities of C = 2.5 and S = 2.5. Therefore the bonds are nonpolar. Also, the molecule is linear which would result in a zero net dipole even if the bonds were polar.

The molecule is nonpolar

49. a) ClO_3^- – polar

Draw the Lewis structure and determine the molecular geometry

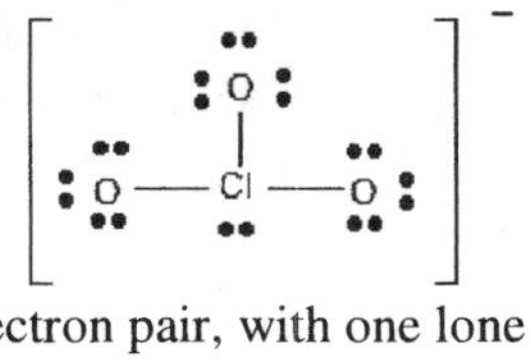

Four electron pair, with one lone pair gives a trigonal pyramidal molecular geometry

Determine if the molecule contains polar bonds.
The electronegativities of Cl = 3.0 and O = 3.5 Therefore the bonds are polar

Determine whether the polar bond add together to form a net dipole.
Because the molecular geometry is trigonal pyramidal, the three dipole moments sum to a net dipole moment. The molecule is polar.

b) SCl_2 – polar

Draw the Lewis structure and determine the molecular geometry

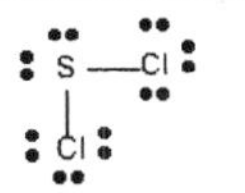

Four electron pair with two lone pair gives a bent molecular geometry

Determine if the molecule contains polar bonds.
The electronegativities of S = 2.5 and Cl = 3.0 Therefore the bonds are polar

Determine whether the polar bond add together to form a net dipole.
Because the molecular geometry is bent, the two dipole moments sum to a net dipole moment. The molecule is polar.

c) SCl_4 – polar

Draw the Lewis structure and determine the molecular geometry

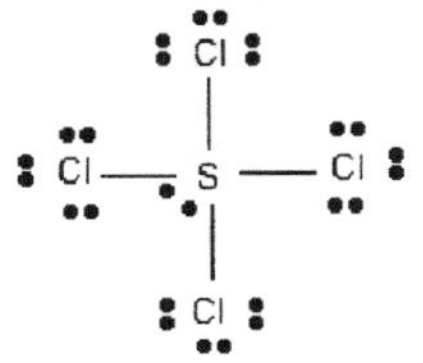

Five electron pair with one lone pair give a seesaw molecular geometry

Determine if the molecule contains polar bonds.
The electronegativities of S = 2.5 and Cl = 3.0 Therefore the bonds are polar

Determine whether the polar bond add together to form a net dipole.
Because the molecular geometry is seesaw, the four dipole moments sum to a net dipole moment. The molecule is polar.

d) $BrCl_5$ – nonpolar

Draw the Lewis structure and determine the molecular geometry

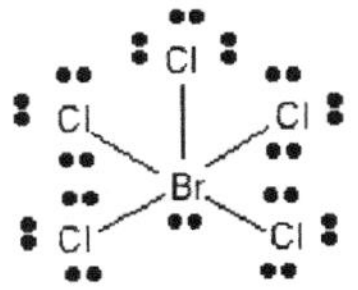

Six electron pair with one lone pair give square pyramidal molecular geometry

Determine if the molecule contains polar bonds.
The electronegativity of Br = 2.8 and Cl = 3.0 The difference is only 0.2, therefore the bonds are nonpolar
Even though the molecular geometry is square pyramidal, the five bonds are nonpolar so there is no net dipole. The molecule is nonpolar.

51. a) Be $2s^2$ — 0 bonds can form. Beryllium contains no unpaired electrons, so no bonds can form without hybridization.

b) P $3s^23p^3$ — 3 bonds can form. Phosphorus contains 3 unpaired electrons, so 3 bonds can form without hybridization.

c) F $2s^22p^5$ — 1 bond can form. Fluorine contains 1 unpaired electron, so 1 bond can form without hybridization.

53. PH_3

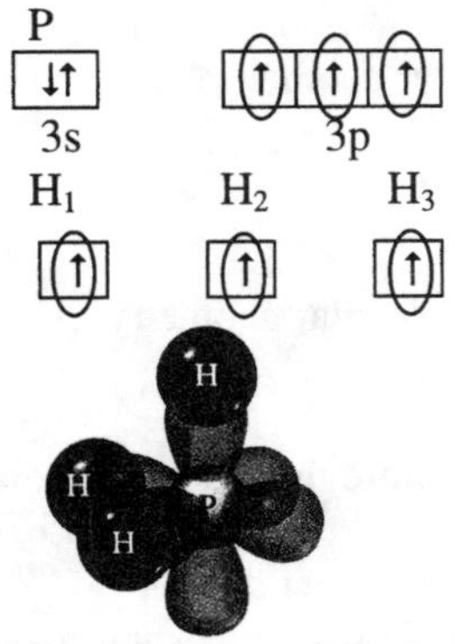

The unhybridized bond angles should be 90°. So, without hybridization, there is not very good agreement between valence bond theory and the actual bond angle of 93.3°

55. C $2s^22p^2$

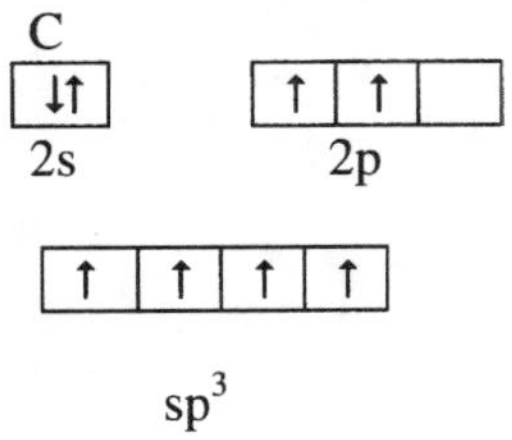

57. sp^2 Only sp^2 hybridization of this set of orbitals has a remaining p orbital to form a π bond.
sp^3 hybridization utilizes all 3 p orbitals
sp^3d^2 hybridization utilizes all 3 p orbitals and 2 d orbitals

59. a) CCl_4 Write the Lewis structure for the molecule

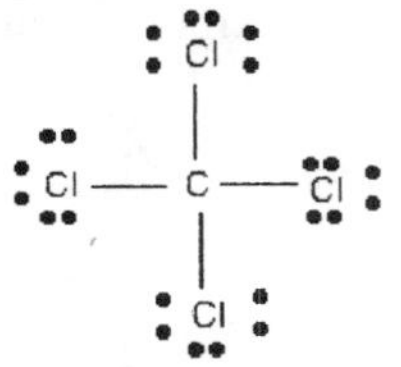

Use VSEPR to predict the electron geometry
Four electron groups around the central atom give tetrahedral electron geometry

Select the correct hybridization for the central atom based on the electron geometry
Tetrahedral electron geometry has sp^3 hybridization

Sketch the molecule and label the bonds

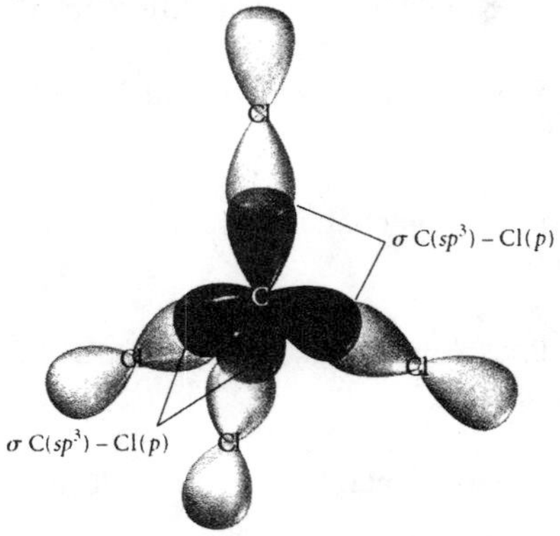

b) NH_3 Write the Lewis structure for the molecule

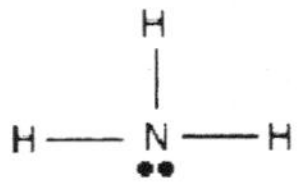

Use VSEPR to predict the electron geometry
Four electron groups around the central atom give tetrahedral electron geometry

Select the correct hybridization for the central atom based on the electron geometry
Tetrahedral electron geometry has sp^3 hybridization

Sketch the molecule and label the bonds

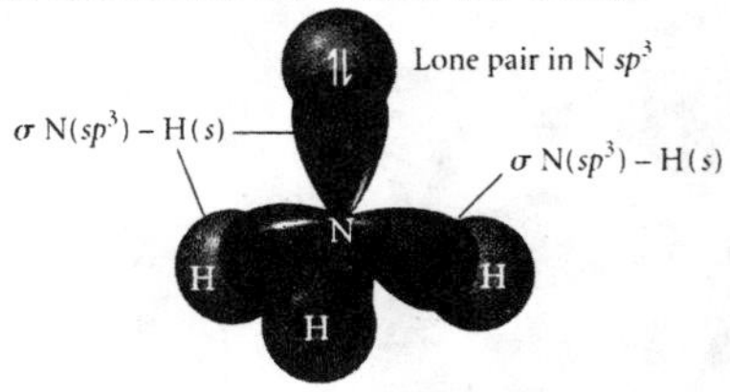

c) OF_2 Write the Lewis structure for the molecule

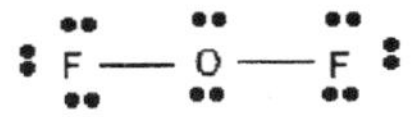

Use VSEPR to predict the electron geometry
Four electron groups around the central atom give tetrahedral electron geometry

Select the correct hybridization for the central atom based on the electron geometry
Tetrahedral electron geometry has sp^3 hybridization

Sketch the molecule and label the bonds

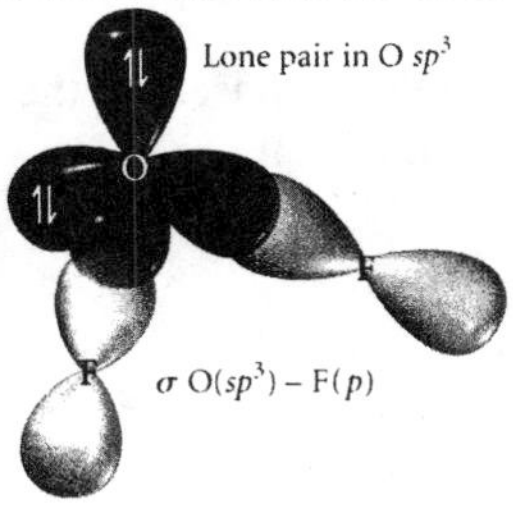

d) CO_2 Write the Lewis structure for the molecule

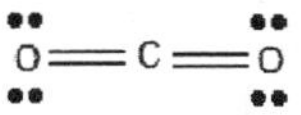

Use VSEPR to predict the electron geometry
Two electron groups around the central atom gives linear electron geometry

Select the correct hybridization for the central atom based on the electron geometry
Linear electron geometry has sp hybridization

Sketch the molecule and label the bonds

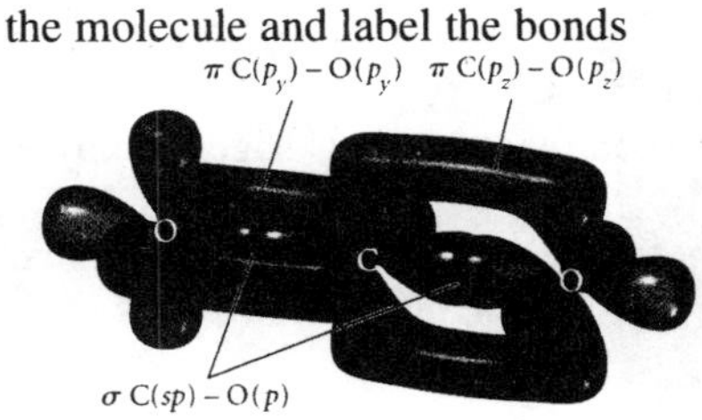

61. a) $COCl_2$ Write the Lewis structure for the molecule

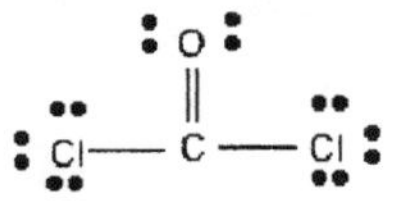

Use VSEPR to predict the electron geometry
Three electron groups around the central atom gives trigonal planar electron geometry

Select the correct hybridization for the central atom based on the electron geometry
Trigonal planar electron geometry has sp^2 hybridization

Sketch the molecule and label the bonds

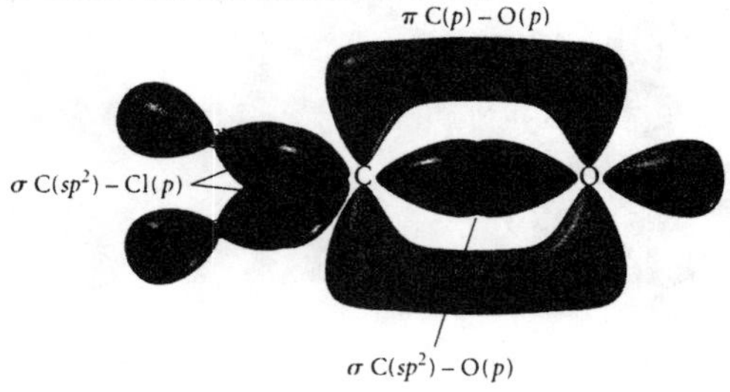

b) BrF_5 Write the Lewis structure for the molecule

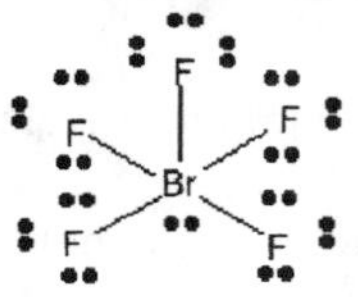

Use VSEPR to predict the electron geometry
Six electron pairs around the central atoms gives octahedral electron geometry

Select the correct hybridization for the central atom based on the electron geometry
Octahedral electron geometry has sp^3d^2 hybridization

Sketch the molecule and label the bonds

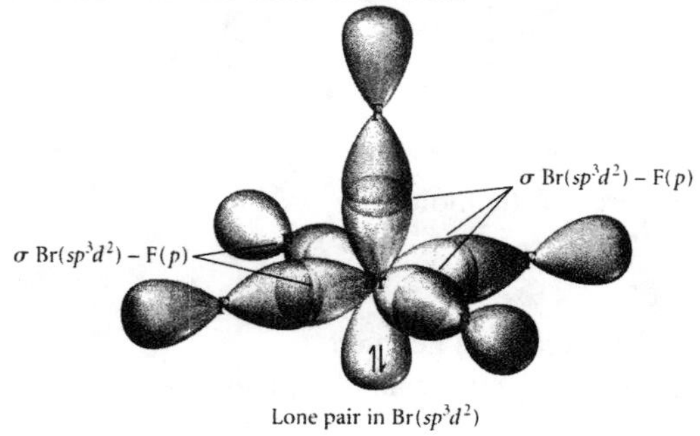

c) XeF_2 Write the Lewis structure for the molecule

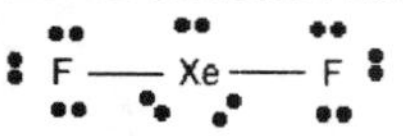

Use VSEPR to predict the electron geometry
Five electron groups around the central atom gives trigonal bipyramidal geometry

Select the correct hybridization for the central atom based on the electron geometry
Trigonal bipyramidal geometry has sp^3d hybridization

Sketch the molecule and label the bonds

3 lone pair in Xe
sp^3d equatorial

σ Xe(sp^3d) – F(p)

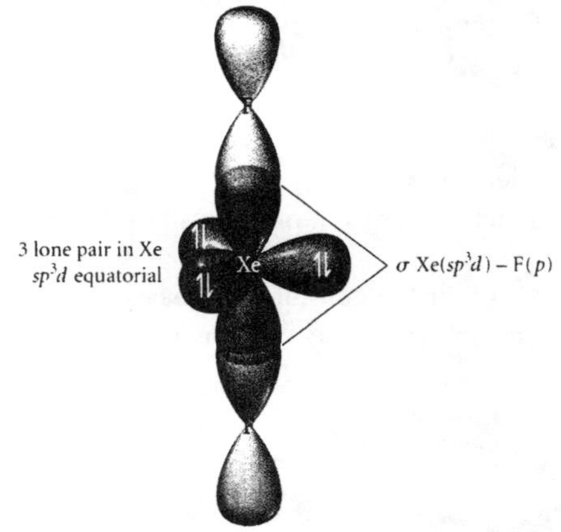

d) I_3^- Write the Lewis structure for the molecule

$$\left[\ddot{\underset{\cdot\cdot}{I}} - \ddot{\underset{\cdot\cdot}{I}} - \ddot{\underset{\cdot\cdot}{I}} \right]^-$$

Use VSEPR to predict the electron geometry
Five electron groups around the central atom gives trigonal bipyramidal geometry

Select the correct hybridization for the central atom based on the electron geometry
Trigonal bipyramidal geometry has sp^3d hybridization

Sketch the molecule and label the bonds

3 lone pair in I
sp^3d equatorial

σ I(sp^3d) – I(p)

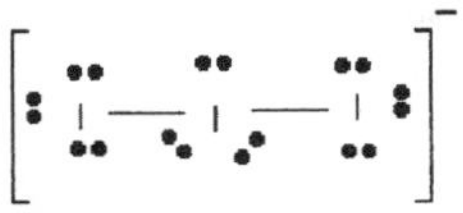

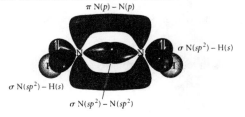

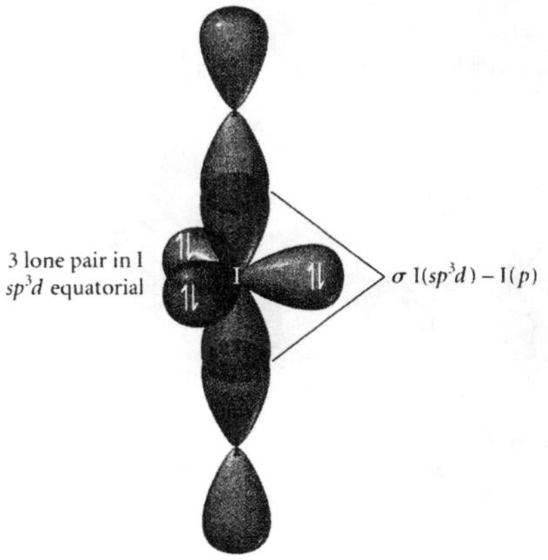

63. a) N_2H_2 Write the Lewis structure for the molecule

$$H - \ddot{N} = \ddot{N} - H$$

Use VSEPR to predict the electron geometry
Three electron groups around each interior atom gives trigonal planar electron geometry

Select the correct hybridization for the central atom based on the electron geometry
Trigonal planar electron geometry has sp^2 hybridization

Sketch the molecule and label the bonds

π N(p) – N(p)

σ N(sp^2) – H(s)

σ N(sp^2) – H(s)

σ N(sp^2) – N(sp^2)

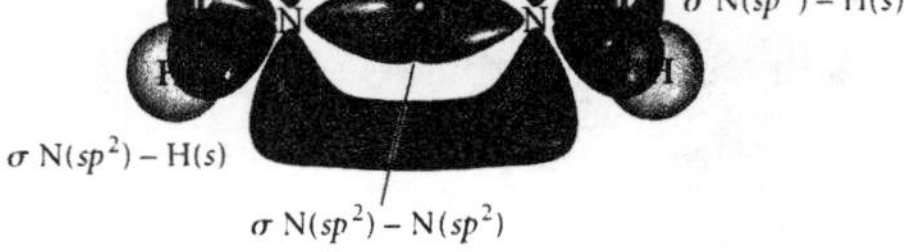

b) N_2H_4 Write the Lewis structure for the molecule

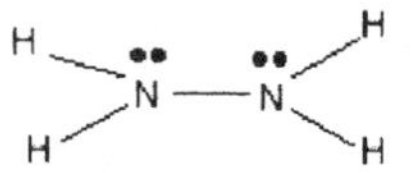

Use VSEPR to predict the electron geometry
Four electron groups around each interior atom gives tetrahedral electron geometry

Select the correct hybridization for the central atom based on the electron geometry
Tetrahedral electron geometry has sp^3 hybridization

Sketch the molecule and label the bonds

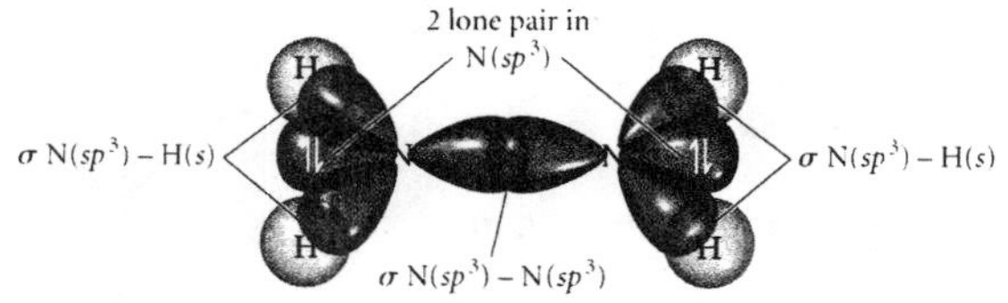

c) CH_3NH_2 Write the Lewis structure for the molecule

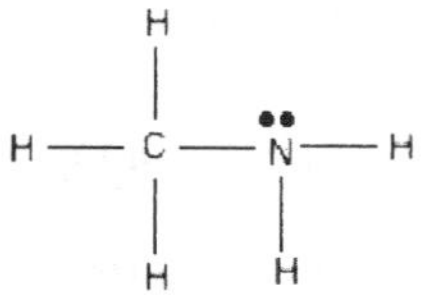

Use VSEPR to predict the electron geometry
Four electron groups around the C gives tetrahedral electron geometry, four electron groups around the N gives tetrahedral geometry.

Select the correct hybridization for the central atom based on the electron geometry
Tetrahedral electron geometry has sp^3 hybridization of both C and N

Sketch the molecule and label the bonds

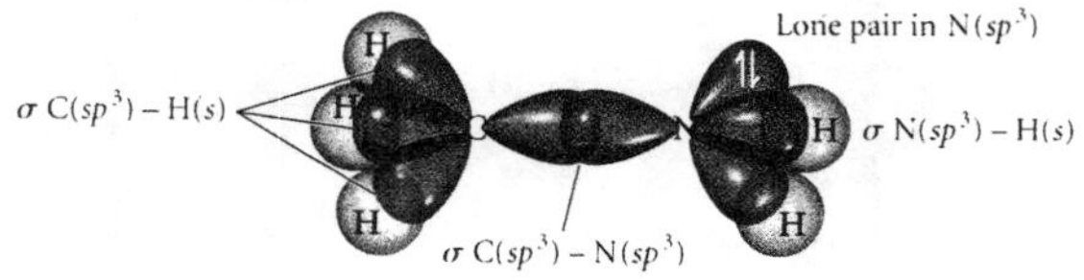

65.

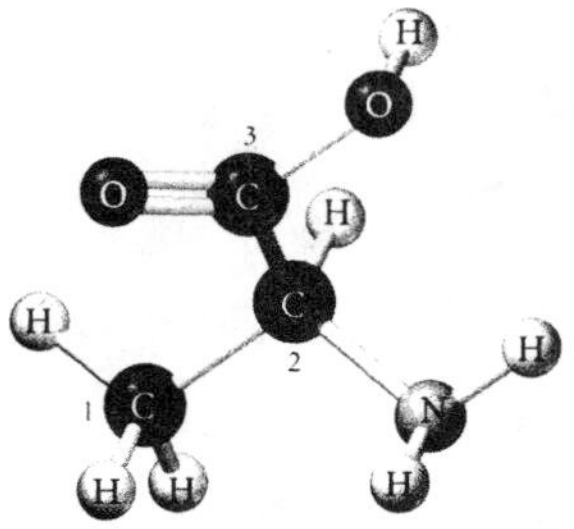

C – 1 and C – 2 each have four electron pairs around the atom, which is tetrahedral electron pair geometry. Tetrahedral electron pair geometry is sp^3 hybridization.
C – 3 has three electron pairs around the atom, which is trigonal planar electron pair geometry. Trigonal planar electron pair geometry is sp^2 hybridization.
O has four electron pairs around the atom, which is tetrahedral electron pair geometry. Tetrahedral electron pair geometry is sp^3 hybridization.
N has four electron pairs around the atom, which is tetrahedral electron pair geometry. Tetrahedral electron pair geometry is sp^3 hybridization.

67. 1s + 1s constructive interference results in a bonding orbital

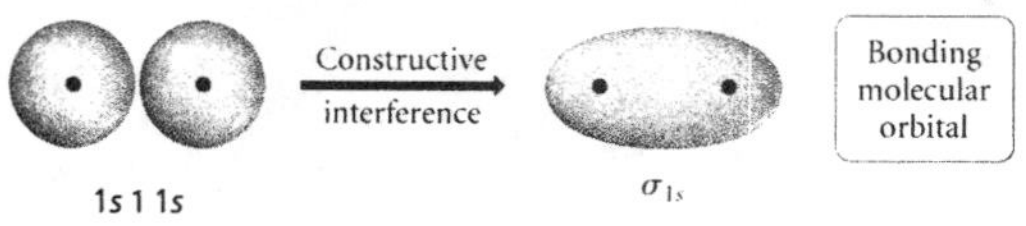

69. Be_2^+ has 7 electrons

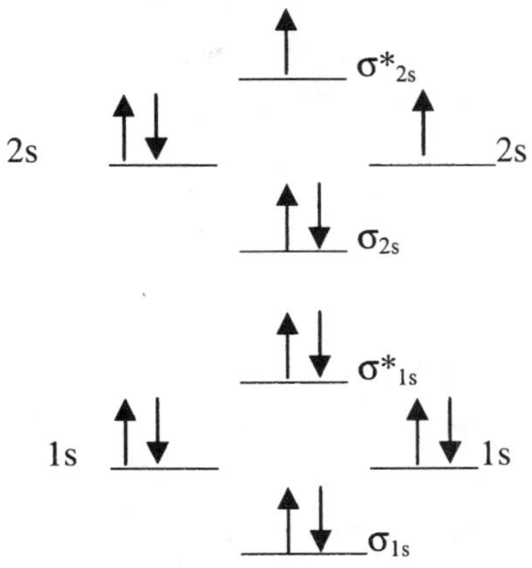

Bond order = $\frac{4-3}{2}=\frac{1}{2}$ stable

Be_2^- has 9 electrons

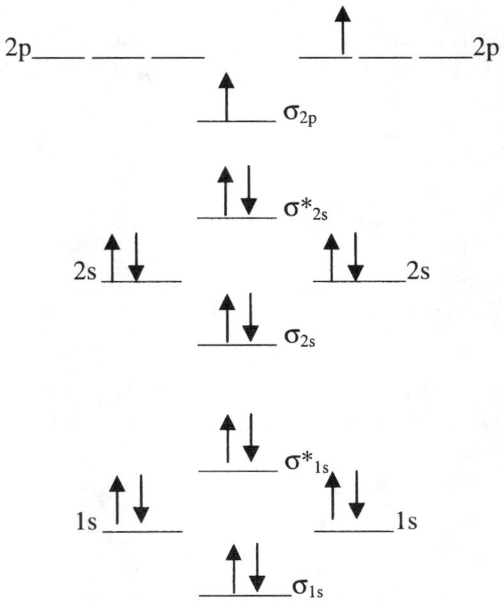

Bond order = $\frac{5-4}{2}=\frac{1}{2}$ stable

71. The bonding and antibonding molecular orbitals from the combination of p_x and p_x atomic orbitals lie along the internuclear axis.

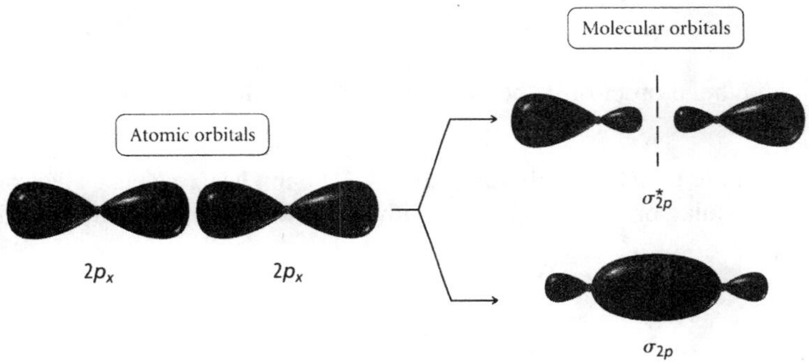

73. a) 4 valence electrons b) 6 valence electrons c) 8 valence electrons d) 9 valence electrons

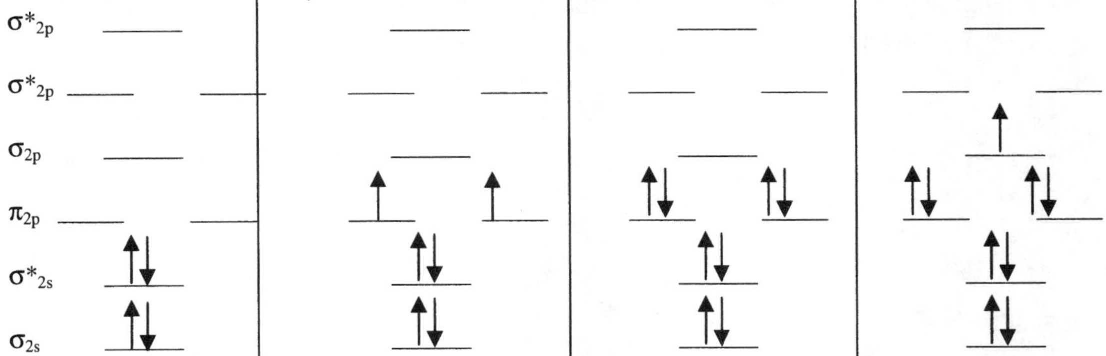

a) Bond order = $\frac{2-2}{2}=0$ diamagnetic

b) Bond order = $\frac{4-2}{2}=1$ paramagnetic

c) Bond order = $\frac{6-2}{2}=2$ diamagnetic

d) Bond order = $\frac{7-2}{2}=2.5$ paramagnetic

75. a) Write an energy level diagram for the molecular orbitals in H_2^{2-}. The ion has 4 valence electrons. Assign the electrons to the molecular orbitals beginning with the lowest energy orbitals and following Hund's rule.

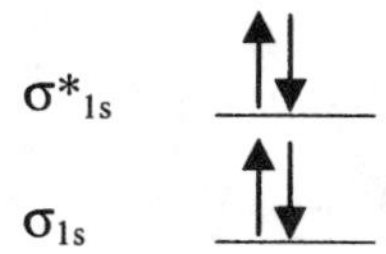

$$\text{Bond order} = \frac{2 - 2}{2} = 0$$

With a bond order of 0, the ion will not exist.

b) Write an energy level diagram for the molecular orbitals in Ne_2. The molecule has 16 valence electrons. Assign the electrons to the molecular orbitals beginning with the lowest energy orbitals and following Hund's rule.

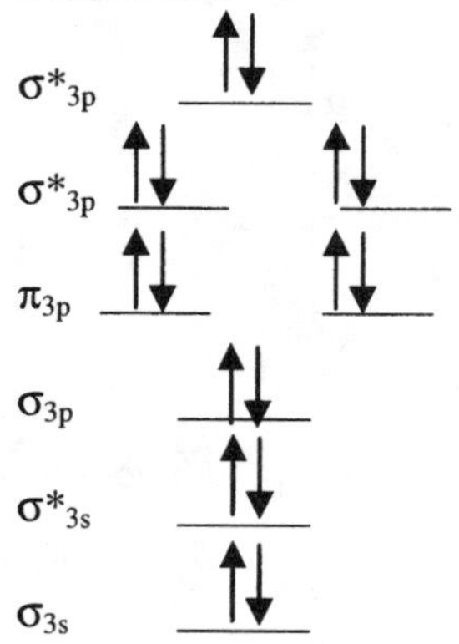

$$\text{Bond order} = \frac{8 - 8}{2} = 0$$

With a bond order of 0, the molecule will not exist.

c) Write an energy level diagram for the molecular orbitals in He_2^{2+}. The ion has 2 valence electrons. Assign the electrons to the molecular orbitals beginning with the lowest energy orbitals and following Hund's rule.

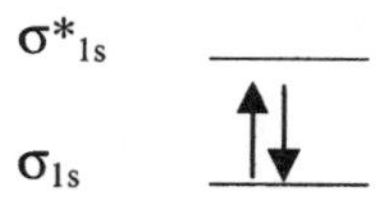

$$\text{Bond order} = \frac{2 - 0}{2} = 1$$

With a bond order of 1, the ion will exist.

d) Write an energy level diagram for the molecular orbitals in F_2^{2-}. The molecule has 16 valence electrons. Assign the electrons to the molecular orbitals beginning with the lowest energy orbitals and following Hund's rule.

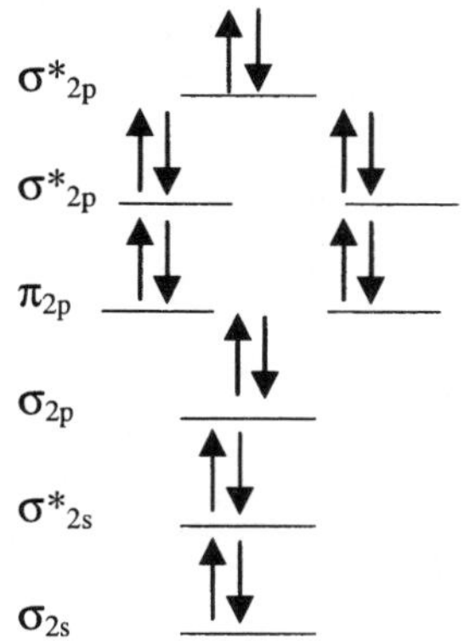

Bond order = $\frac{8-8}{2}=0$ With a bond order of 0, the ion will not exist.

77. C_2^- has the highest bond order, the highest bond energy and the shortest bond.
Write an energy level diagram for the molecular orbitals in C_2.
Assign the electrons to the molecular orbitals beginning with the lowest energy orbitals and following Hund's rule.
C_2 (8 valence electrons); C_2^+ (7 valence electrons): C_2^- (9 valence electrons)

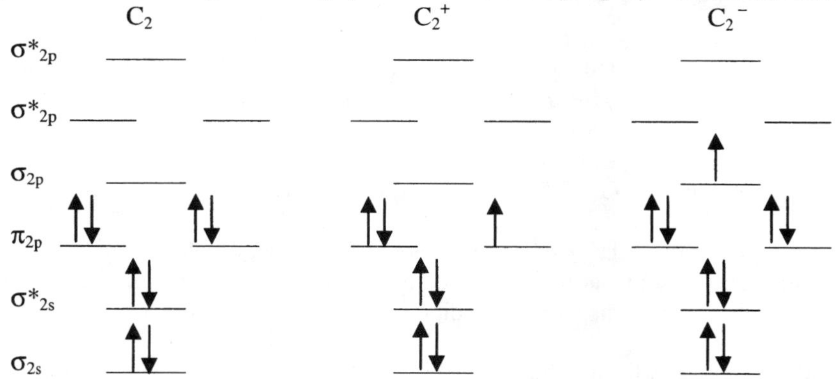

Bond order = $\frac{6-2}{2}=2$ Bond order = $\frac{5-2}{2}=1.5$ Bond order = $\frac{7-2}{2}=2.5$

C_2^- has the highest bond order at 2.5. Bond order is directly related to bond energy, so C_2^- has the largest bond energy and bond order is indirectly related to bond length, so C_2^- has the shortest bond length.

79. Write an energy level diagram for the molecular orbitals in CO using O_2 energy ordering.
Assign the electrons to the molecular orbitals beginning with the lowest energy orbitals and following Hund's rule.

CO has 10 valence electrons.

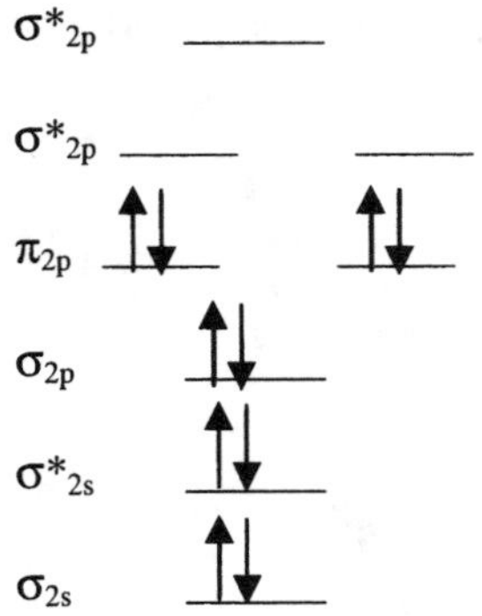

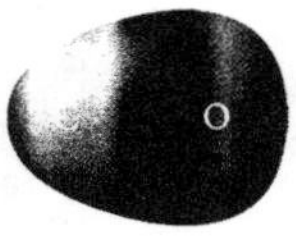

Bond order = $\frac{8-2}{2}=3$

The electron density toward the O atom since it is more electronegative.

81. a) COF_2 Write the Lewis structure for the molecule

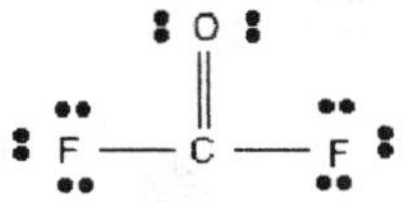

Use VSEPR to predict the electron geometry

Three electron groups around the central atom gives trigonal planar electron geometry. Three bonding pairs of electrons give trigonal planar molecular geometry.

Determine if the molecule contains polar bonds.

The electronegativities of C = 2.5, O = 3.5 and F = 4.0 Therefore the bonds are polar

Determine whether the polar bond add together to form a net dipole.

Even though a trigonal planar molecular geometry normally is nonpolar, because the bonds have different dipole moments the sum of the dipole moments is not zero. The molecule is polar

Select the correct hybridization for the central atom based on the electron geometry

Trigonal planar geometry has sp^2 hybridization

Sketch the molecule and label the bonds

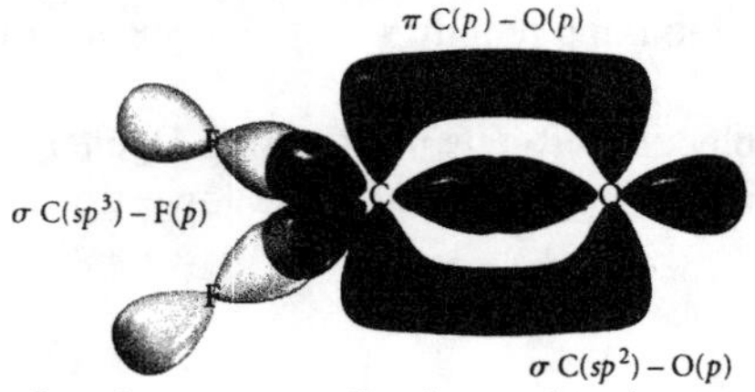

b) S_2Cl_2 Write the Lewis structure for the molecule

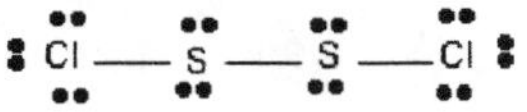

Use VSEPR to predict the electron geometry

Four electron groups around the central atom gives tetrahedral electron geometry. Two bonding pairs and two lone pair of electrons gives bent molecular geometry.

Determine if the molecule contains polar bonds.

The electronegativities of S = 2.5 and Cl = 3.0 Therefore the bonds are polar

Determine whether the polar bond add together to form a net dipole.
In a bent molecular geometry the sum of the dipole moments is not zero. The molecule is polar

Select the correct hybridization for the central atom based on the electron geometry
Tetrahedral geometry has sp^3 hybridization

Sketch the molecule and label the bonds

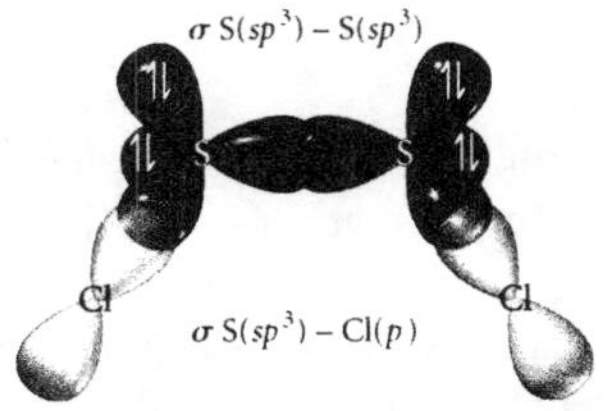

c) SF_4 Write the Lewis structure for the molecule

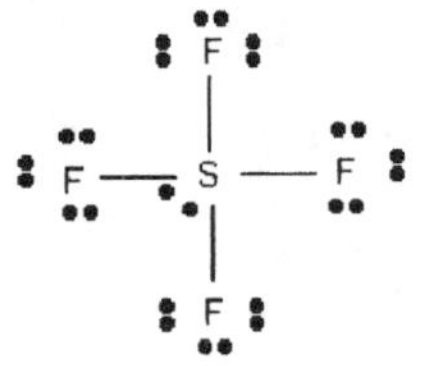

Use VSEPR to predict the electron geometry
Five electron groups around the central atom gives trigonal bipyramidal electron geometry. Four bonding pair and one lone pair of electrons gives seesaw molecular geometry.

Determine if the molecule contains polar bonds.
The electronegativities of S = 2.5 and F = 4.0 Therefore the bonds are polar

Determine whether the polar bond add together to form a net dipole.
In a seesaw molecular geometry the sum of the dipole moments is not zero. The molecule is polar

Select the correct hybridization for the central atom based on the electron geometry
Trigonal bipyramidal electron geometry has sp^3d hybridization

Sketch the molecule and label the bonds

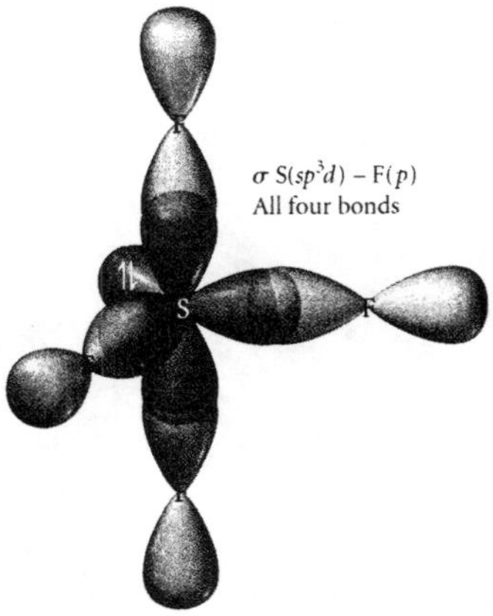

83. a) serine

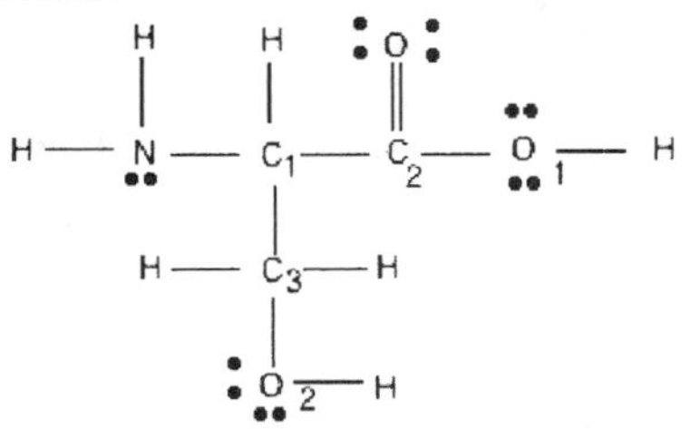

C – 1 and C – 3 each have four electron pair around the atom. Four electron pair gives tetrahedral electron geometry, tetrahedral electron geometry has sp^3 hybridization. Four bonding pair and zero lone pair give stetrahedral molecular geometry.
C – 2 has three electron pair around the atom. Three electron pair gives trigonal planar geometry, trigonal planar geometry has sp^2 hybridization. Three bonding pair and zero lone pair gives trigonal planar molecular geometry.
N has four electron pair around the atom. Four electron pair gives tetrahedral electron geometry, tetrahedral electron geometry has sp^3 hybridization. Three bonding pair and one lone pair gives trigonal pyramidal molecular geometry.
O – 1 and O – 2 each has four electron pair around the atom. Four electron pair gives tetrahedral electron geometry, tetrahedral electron geometry has sp^3 hybridization. Two bonding pair and two lone pair gives bent molecular geometry.

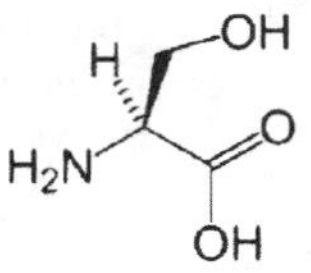

b) asparagine

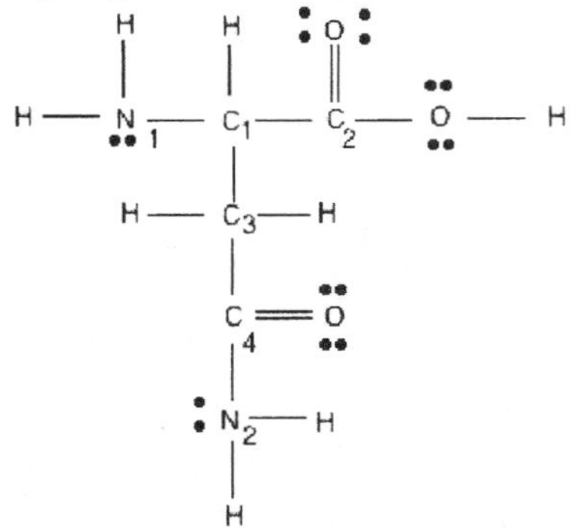

C – 1 and C – 3 each have four electron pair around the atom. Four electron pair gives tetrahedral electron geometry, tetrahedral electron geometry has sp^3 hybridization. Four bonding pair and zero lone pair gives tetrahedral molecular geometry.
C – 2 and C – 4 each have three electron pair around the atom. Three electron pair gives trigonal planar geometry, trigonal planar geometry has sp^2 hybridization. Three bonding pair and zero lone pair gives trigonal planar molecular geometry.
N – 1 and N – 2 each have four electron pair around the atom. Four electron pair gives tetrahedral electron geometry, tetrahedral electron geometry has sp^3 hybridization. Three bonding pair and one lone pair gives trigonal pyramidal molecular geometry.

O has four electron pair around the atom. Four electron pair gives tetrahedral electron geometry, tetrahedral electron geometry has sp^3 hybridization. Two bonding pair and two lone pair gives bent molecular geometry.

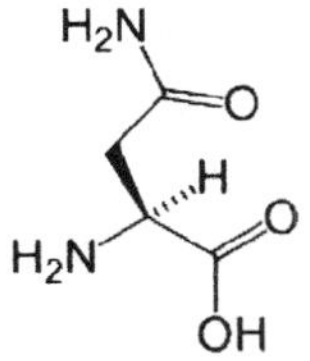

c) cysteine

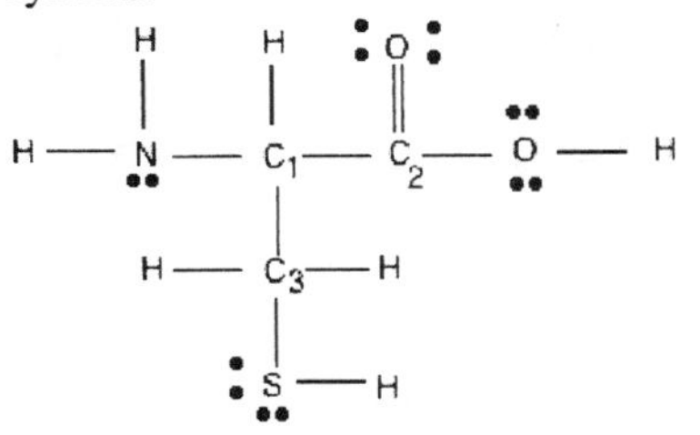

C – 1 and C – 3 each have four electron pair around the atom. Four electron pair gives tetrahedral electron geometry, tetrahedral electron geometry has sp^3 hybridization. Four bonding pair and zero lone pair gives tetrahedral molecular geometry.
C – 2 has three electron pair around the atom. Three electron pair gives trigonal planar geometry, trigonal planar geometry has sp^2 hybridization. Three bonding pair and zero lone pair gives trigonal planar molecular geometry.
N has four electron pair around the atom. Four electron pair gives tetrahedral electron geometry, tetrahedral electron geometry has sp^3 hybridization. Three bonding pair and one lone pair gives trigonal pyramidal molecular geometry.
O and S have four electron pair around the atom. Four electron pair gives tetrahedral electron geometry, tetrahedral electron geometry has sp^3 hybridization. Two bonding pair and two lone pair gives bent molecular geometry.

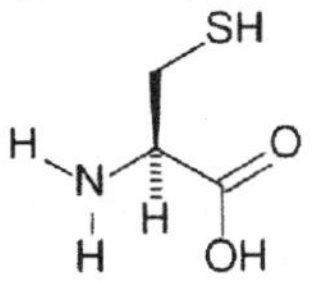

85. 4 π bonds; 25 σ bonds; the lone pair on the O and N – 2 occupy sp^2 orbitals, the lone pairs on N – 1, N – 3 and N – 4 occupy sp^3 orbitals.

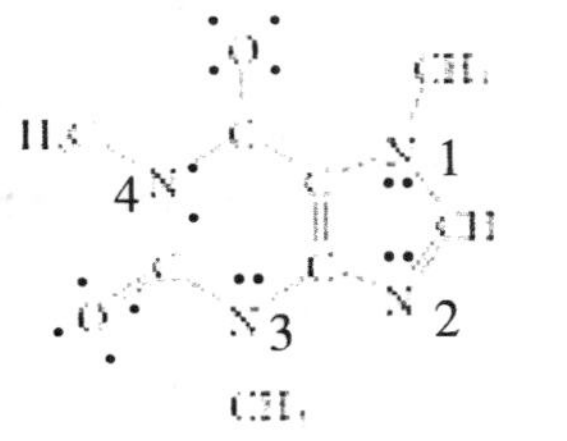

87. a) water soluble – the 4 C – OH bonds, the C = O bond and the C – O bonds in the ring, make the molecule polar. Because of the large electronegativity difference between the C and O, each of the bonds will have a dipole moment. The sum of the dipole moments does NOT give a net zero dipole moment, so the molecule is polar. Since it is polar, it will be water soluble.

b) fat soluble – There is only one C – O bond in the molecule. The dipole moment from this bond is not enough to make the molecule polar because of all of the non polar components of the molecule. The C – H bonds in the structure lead to a net dipole of zero for most of the sites in the molecule. Since the molecule is nonpolar, it is fat soluble.

c) water soluble – the carboxylic acid function (COOH group) along with the N atom in the ring make the molecule polar. Because of the electronegativity difference between the C and O and the C and N atoms, the bonds will have a dipole moment and the net dipole moment of the molecule is NOT zero, so the molecule is polar. Since the molecule is polar, it is water soluble.

d) fat soluble – The two O atoms in the structure contribute a very small amount to the net dipole moment of this molecule. The majority of the molecule is nonpolar because there is no net dipole moment at the interior C atoms. Because the molecule is nonpolar it is fat soluble.

89. ClF has 14 valence electron. Assign the electrons to the lowest energy MO's first and then follow Hund's rule.

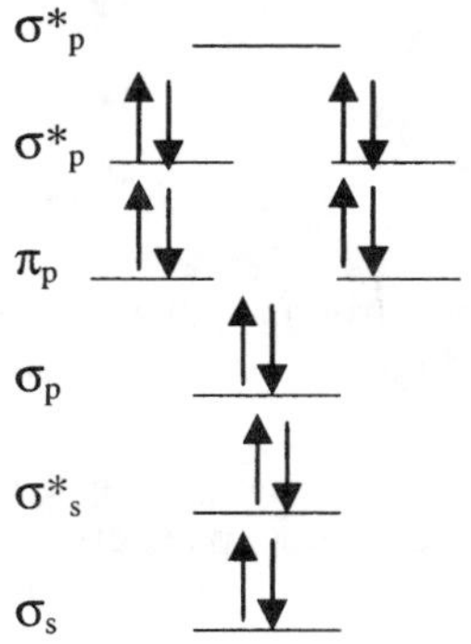

91. BrF (14 valence electrons)

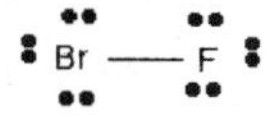

no central atom, no hybridization, no electron structure

BrF_2^- (22 valence electrons)

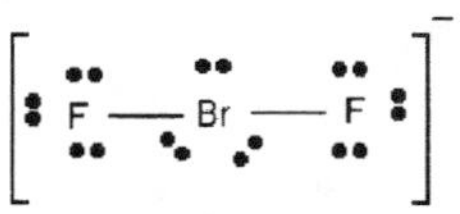

Five electron pairs on the central atom, electron geometry is trigonal bipyramidal, two bonding pairs and three lone pairs gives a linear molecular geometry. An electron geometry of trigonal bipyramidal has sp^3d hybridization.

BrF_3 (28 valence electrons)

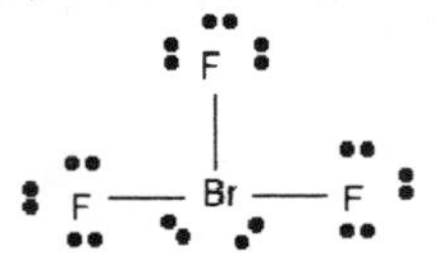

Five electron pairs on the central atom, electron geometry is trigonal bipyramidal, three bonding pairs and two lone pairs gives a T – shaped molecular geometry. An electron geometry of trigonal bipyramidal has sp^3d hybridization.

BrF_4^- (36 valence electrons)

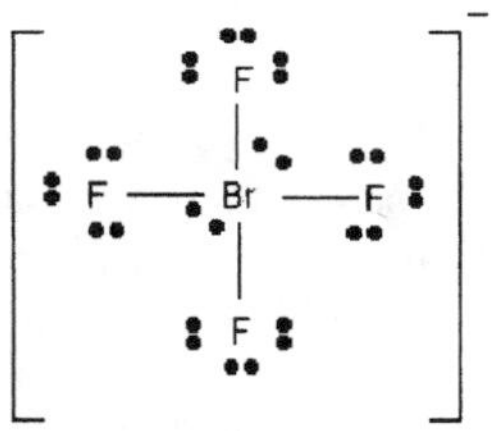

Six electron pairs on the central atom, electron geometry is octahedral, four bonding pairs and two lone pairs gives a square planar molecular geometry. An electron geometry of octahedral has sp^3d^2 hybridization.

BrF_5(42 valence electrons)

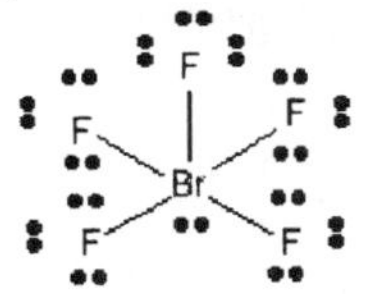

Six electron pairs on the central atom, electron geometry is octahedral, five bonding pairs and one lone pair gives a square pyramidal molecular geometry. An electron geometry of octahedral has sp^3d^2 hybridization.

93. According to valence bond theory, CH_4, NH_3, and H_2O are all sp^3 hybridized. This hybridization results in a tetrahedral electron group configuration with a 109.5° bond angle. NH_3 and H_2O deviate from this idealized bond angle because their lone electron pairs exist in their own sp^3 orbitals. The presence of lone pairs lowers the tendency for the central atom's orbitals to hybridize. As a result, as lone pairs are added, the bond angle moves further from the 109.5° hybrid angle to the 90° unhybridized angle.

95. Using the MO diagram for NH_3 assign the 8 valence electrons to the molecular orbitals. Start with the lowest energy orbital first and follow Hund's rule.

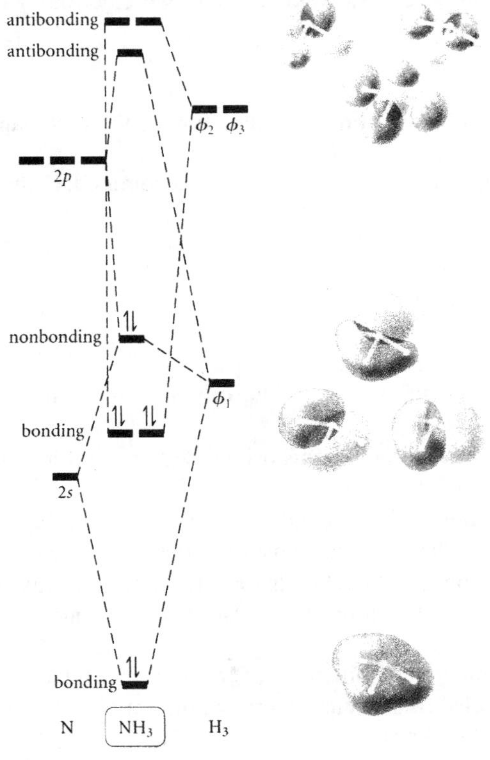

$$\text{Bond order} = \frac{6 - 0}{2} = 3$$

With a bond order of 3 the molecule is stable.
of the ultraviolet – visible region of the electromagnetic spectrum.

97. Write the Lewis structure for
Determine electron pair geometry around each central atom
Determine the molecular geometry, determine idealized bond angles and predict actual bond angles
NO_2

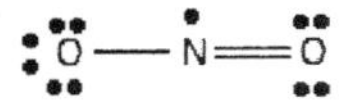

2 bonding pair and a lone electron gives trigonal planar electron geometry, the molecular geometry will be bent.

Trigonal planar electron geometry has idealized bond angles of 120°. The bond angle is expected to be slightly less than 120° because of the lone electron occupying the third sp^2 orbital.

NO_2^+

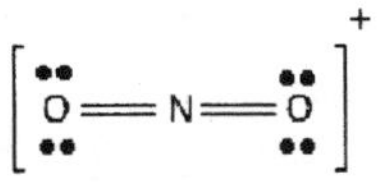

Two bonding pair of electrons and no lone pair gives linear electron geometry and molecular geometry. Linear electron geometry has a bond angle of 180°.

NO_2^-

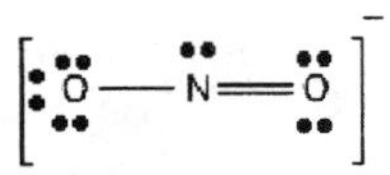

Two bonding pair of electrons and one lone pair gives trigonal planar electron geometry, the molecular geometry will be bent.

Trigonal planar electron geometry has idealized bond angles of 120°. The bond angle is expected to be less than 120° because of the lone pair electrons occupying the third sp^2 orbital. Further, the bond angle should be less than the bond angle in NO_2 because the presence of lone pairs lowers the tendency for the central atom's orbitals to hybridize. As a result, as lone pairs are added, the bond angle moves further from the 120° hybrid angle to the 90° unhybridized angle and the two electrons will increase this tendency.

99. Statement a is the best statement.

Statement b neglects the lowering of potential energy that arises from the interaction of the lone pair electrons with the bonding electrons.

Statement c neglects the interaction of the electrons altogether. The bonds form to accommodate the electrons, not the other way around.

101. In Lewis theory, a covalent bond comes from the sharing of electrons.

A single bond shares 2 electrons (one pair)

A double bond shares 4 electrons (two pair)

A triple bond shares 6 electrons (three pair)

In valence bond theory, a covalent bond forms when orbitals overlap. The orbitals can be unhybridzed or hybridized orbitals.

A single bond forms when a σ bond is formed from the overlap of an s orbital with an s orbital, an s orbital with a p orbital or a p orbital and a p orbital overlapping end to end.

A double bond is a combination of a σ bond and a π bond. The π bond forms from the sideways overlap of a p orbital on each of the atoms involved in the bond. The p orbitals must have the same orientation

A triple bond is a combination of a σ bond and a 2π bonds. The π bonds form from the sideways overlap of a p orbital on each of the atoms involved in the bond. The p orbitals must have the same orientation so each π bond is formed from a different set of p orbitals.

In molecular orbital theory, molecular orbitals form which are combinations of the atomic orbitals of the atoms involved in the bond. The bonds form when the valence electrons occupy more bonding molecular orbitals than antibonding molecular orbitals. This is calculated by the bond order.

A single bond, has a bond order of 1

A double bond, has a bond order of 2

A triple bond, has a bond order of 3

Chapter 11
Liquids, Solids, and Intermolecular Forces

1. The key to the gecko's sticky feet lies in the millions of microhairs, called setae, that line its toes. Each seta is between 30 and 130 μm long and branches out to end in several hundred flattened tips called spatula. This unique structure allows the gecko's toes to have unusually close contact with the surfaces it climbs. The close contact allows intermolecular forces—which are significant only at short distances—to hold the gecko to the wall.

3. The main properties of liquids are that: liquids have much higher densities in comparison to gases and generally have lower densities in comparison to solids; liquids have an indefinite shape and assume the shape of their container; liquids have a definite volume; and liquids are not easily compressed.

5. Solids may be crystalline, in which case the atoms or molecules that compose them are arranged in a well-ordered three-dimensional array, or they may be amorphous, in which case the atoms or molecules that compose them have no long-range order.

7. Since there is the most molecular motion in the gas phase and the least molecular motion in the solid phase (since atoms are pushed closer together), a substance will be converted from a solid then to a liquid and finally to a gas as the temperature increases. The strength of the intermolecular interactions is least in the gas phase, since there are large distances between particles and they are moving very fast. Intermolecular forces are the strongest in liquids and solids, where molecules are "touching" one another. The strength of the interactions in the condensed phases will determine at what temperature the substance will melt and boil.

9. Intermolecular forces, even the strongest ones, are generally much weaker than bonding forces. The reason for the relative weakness of intermolecular forces compared to bonding forces is also related to Coulomb's law $\left(E = \frac{1}{4\pi\varepsilon_o}\frac{q_1 q_2}{r}\right)$. Bonding forces are the result of large charges (the charges on protons and electrons) interacting at very close distances. Intermolecular forces are the result of smaller charges (as we shall see in the following discussion) interacting at greater distances.

11. The dipole-dipole force exists in all molecules that are polar. Polar molecules have permanent dipoles that interact with the permanent dipoles of neighboring molecules. The positive end of one permanent dipole is attracted to the negative end of another; this attraction is the dipole-dipole force.

13. The hydrogen bond is a sort of super dipole-dipole force. Polar molecules containing hydrogen atoms bonded directly to fluorine, oxygen, or nitrogen exhibit an intermolecular force called hydrogen bonding. The large electronegativity difference between hydrogen and these electronegative elements means that the H atoms will have fairly large partial positive charges (δ+), while the F, O, or N atoms will have fairly large partial negative charges (δ-). In addition, since these atoms are all are quite small, they can approach one another very closely. The result is a strong attraction between the hydrogen in each of these molecules and the F, O, or N on its neighbors, an attraction called a hydrogen bond.

15. Surface tension is the tendency of liquids to minimize their surface area. Molecules at the surface have relatively fewer neighbors with which to interact, because there are no molecules above it. Consequently, molecules at the surface are inherently less stable—they have higher potential energy—than those in the interior. In order to increase the surface area of the liquid, some molecules from the interior have to be moved to the surface, a process requiring energy. The surface tension of a liquid is the energy required to increase the surface area by a unit amount. Surface tension decreases with decreasing intermolecular forces.

17. Capillary action is the ability of a liquid to flow against gravity up a narrow tube. Capillary action results from a combination of two forces: the attraction between molecules in a liquid, called cohesive forces, and the attraction between these molecules and the surface of the tube, called adhesive forces. The adhesive forces cause the liquid to spread out over the surface of the tube, while the cohesive forces cause the liquid to stay together. If the adhesive forces are greater than the cohesive forces (as is the case for water in a glass tube), the attraction to the surface draws the liquid up the tube while the cohesive forces pull along those molecules not in direct contact with the tube walls. The water rises up the tube until the force of

gravity balances the capillary action—the thinner the tube, the higher the rise. If the adhesive forces are smaller than the cohesive forces (as is the case for liquid mercury), the liquid does not rise up the tube at all (and in fact will drop to a level below the level of the surrounding liquid).

19. The molecules that leave the liquid are the ones at the high end of the energy curve—the most energetic. If no additional heat enters the liquid, the average energy of the entire collection of molecules goes down—much as the class average on an exam goes down if you eliminate the highest-scoring students. So vaporization is an endothermic process; it takes energy to vaporize the molecules in a liquid. Also, vaporization requires overcoming the intermolecular forces that hold liquids together. Since energy must be absorbed to pull the molecules apart, the process is endothermic. Condensation is the opposite process, so it must be exothermic. Also, gas particles have more energy than those in the liquid. It is the least energetic of these that condense, adding energy to the liquid.

21. The heat of vaporization (ΔH_{vap}) is the amount of heat required to vaporize one mole of a liquid to a gas. The heat of vaporization of a liquid can be used to calculate the amount of heat energy required to vaporize a given mass of the liquid (or the amount of heat given off by the condensation of a given mass of liquid), and to compare the volatility of two substances.

23. When a system in dynamic equilibrium is disturbed, the system responds so as to minimize the disturbance and return to a state of equilibrium.

25. The boiling point of a liquid is the temperature at which its vapor pressure equals the external pressure. The normal boiling point of a liquid is the temperature at which its vapor pressure equals 1 atm.

27. As the temperature rises, more liquid vaporizes and the pressure within the container increases. As more and more gas is forced into the same amount of space, the density of the gas becomes higher and higher. At the same time, the increasing temperature causes the density of the liquid to become lower and lower. At the critical temperature, the meniscus between the liquid and gas disappears and the gas and liquid phases commingle to form a supercritical fluid.

29. Fusion, or melting, is the phase transition from solid to liquid. The term fusion is used for melting because, if you heat several crystals of a solid, they will fuse into a continuous liquid upon melting. Fusion is endothermic because solids have less kinetic energy than liquids, so energy must be added to the solid.

31. There are two horizontal lines (i.e. heat is added, but the temperature stays constant) in the heating curve because there are two endothermic phase changes. The heat that is added is used to change the phase from solid to liquid or liquid to gas.

33. A phase diagram is simply a map of the phase of a substance as a function of pressure (on the y-axis) and temperature (on the x-axis).

Phase Diagram for Water

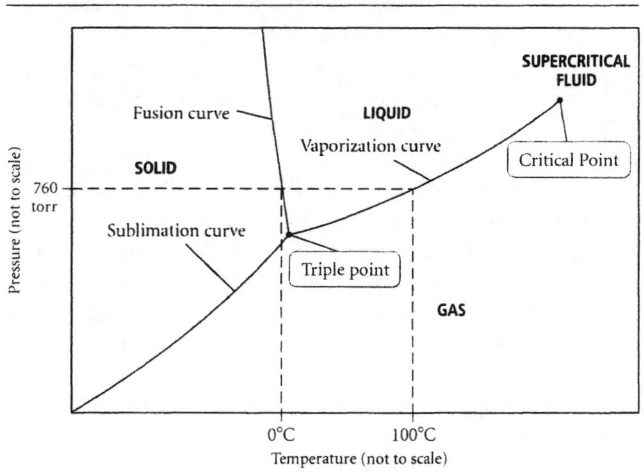

35. Water has a low molar mass (18.01 g/mol), yet it is a liquid at room temperature. Water's high boiling point for its molar mass can be understood by examining the structure of the water molecule. The bent geometry of the water molecule and the highly polar nature of the O–H bonds result in a molecule with a significant dipole moment. Water's two O–H bonds (hydrogen directly bonded to oxygen) allow a water molecule to

form strong hydrogen bonds with four other water molecules, resulting in a relatively high boiling point. Water's high polarity also allows it to dissolve many other polar and ionic compounds, and even a number of nonpolar gases such as oxygen and carbon dioxide (by inducing a dipole moment in their molecules). Water has an exceptionally high specific heat capacity. One significant difference between the phase diagram of water and that of other substances is that the fusion curve for water has a negative slope. The fusion curve within the phase diagrams for most substances has a positive slope because increasing pressure favors the denser phase, which for most substances is the solid phase. This negative slope means that ice is less dense than liquid water and so ice floats. The solids sink in the liquids of most other substances.

37. A crystalline lattice is the regular arrangements of atoms within a crystalline solid. The crystalline lattice can be represented by a small collection of atoms, ions, or molecules a fundamental building block called the unit cell. When the unit cell is repeated over and over—like the tiles of a floor or the pattern in a wallpaper design, but in three dimensions—the entire lattice can be reproduced.

39. Atoms in a simple cubic cell structure have a coordination number of 6, an edge length of 2r, and 1 atom in the unit cell. Atoms in a body-centered cubic cell structure have a coordination number of 8, an edge length of $4r/\sqrt{3}$, and 2 atoms in the unit cell. Atoms in a face-centered cubic cell structure have a coordination number of 12, an edge length of $2\sqrt{2}r$, and 4 atoms in the unit cell.

41. The three types of solids are molecular solids, ionic solids and atomic solids. Molecular solids are those solids whose composite units are molecules. The lattice sites in a crystalline molecular solid are therefore occupied by molecules. Ice (solid H_2O) and dry ice (solid CO_2) are examples of molecular solids. Molecular solids are held together by the kinds of intermolecular forces—dispersion forces, dipole-dipole forces, and hydrogen bonding. Ionic solids are those solids whose composite units are ions. Table salt (NaCl) and calcium fluoride (CaF_2) are good examples of ionic solids. Ionic solids are held together by the coulombic interactions that occur between the cations and anions occupying the lattice sites in the crystal. Atomic Solids are those solids whose composite units are individual atoms are called atomic solids. Atomic solids can themselves be divided into three categories—nonbonding atomic solids, metallic atomic solids, and network covalent atomic solids—each held together by a different kind of force. Nonbonding atomic solids, which include only the noble gases in their solid form, are held together by relatively weak dispersion forces. Metallic atomic solids, such as iron or gold, are held together by metallic bonds, which in the simplest model are represented by the interaction of metal cations with a sea of electrons that surround them. Network covalent atomic solids, such as diamond, graphite, and silicon dioxide, are held together by covalent bonds.

43. Cesium chloride (CsCl) is a good example of an ionic compound containing cations and anions of similar size (Cs^+ radius = 167 pm; Cl^- radius = 181 pm). In the cesium chloride structure, the chloride ions occupy the lattice sites of a simple cubic cell and one cesium ion lies in the very center of the cell, as shown in Figure 11.51. Notice that the cesium chloride unit cell contains one chloride anion (8 × 1/8 = 1) and one cesium cation (the cesium ion in the middle belongs entirely to the unit cell) for a ratio of Cs to Cl of 1:1, just as in the formula for the compound.

    The crystal structure of sodium chloride must accommodate the more disproportionate sizes of Na^+ (radius = 97 pm) and Cl^- (radius = 181 pm). The larger chloride anion could theoretically fit many of the smaller sodium cations around it, but charge neutrality requires that each sodium cation be surrounded by an equal number of chloride anions. The structure that minimizes the energy is shown in Figure 11.52 and has a coordination number of 6 (each chloride anion is surrounded by six sodium cations and vice versa). You can visualize this structure, called the rock salt structure, as chloride anions occupying the lattice sites of a face-centered cubic structure with the smaller sodium cations occupying the holes between the anions. (Alternatively, you can visualize this structure as the sodium cations occupying the lattice sites of a face-centered cubic structure with the larger chloride anions occupying the spaces between the cations.) Each unit cell contains four chloride anions ([8 × 1/8] + [6 × ½] = 4) and four sodium cations (12 × ¼) resulting in a ratio of 1:1, just as in the formula of the compound.

    You can visualize this structure, called the zinc blende structure, as sulfide anions occupying the lattice sites of a face-centered cubic structure with the smaller zinc cations occupying four of the eight tetrahedral holes located directly beneath each corner atom. A tetrahedral hole is the empty space that lies in the center of a tetrahedral arrangement of four atoms, as shown below. Each unit cell contains four sulfide anions ([8 × 1/8] + [6 × ½] = 4) and four zinc cations (each of the four zinc cations is completely contained within the unit cell), resulting in a ratio of 1:1, just as in the formula of the compound.

45. Atomic Solids can themselves be divided into three categories—nonbonding atomic solids, metallic atomic solids, and network covalent atomic solids. Nonbonding atomic solids, which include only the noble gases in their solid form, are held together by relatively weak dispersion forces. Metallic atomic solids, such as iron or gold, are held together by metallic bonds, which in the simplest model are represented by the interaction of metal cations with a sea of electrons that surround them. Network covalent atomic solids, such as diamond, graphite, and silicon dioxide, are held together by covalent bonds.

47. The band gap is an energy gap that exists between the valence band and conduction band. In metals, the valence band and conduction band are always energetically continuous—the energy difference between the top of the valence band and the bottom of the conduction band is infinitesimally small. In semiconductors, the band gap is small, allowing some electrons to be promoted at ordinary temperatures resulting in limited conductivity. In insulators, the band gap is large, and electrons are not promoted into the conduction band at ordinary temperatures, resulting in no electrical conductivity.

49. a) Dispersion forces

b) Dispersion forces and dipole-dipole forces

c) Dispersion forces

d) Dispersion forces, dipole-dipole forces and hydrogen bonding

e) Dispersion forces

f) Dispersion forces, dipole-dipole forces and hydrogen bonding

g) Dispersion forces and dipole-dipole forces

h) Dispersion forces

51. a) CH_4 < b) CH_3CH_3 < c) CH_3CH_2Cl < d) CH_3CH_2OH. The first two molecules only exhibit dispersion forces, so the boiling point increases with increasing molar mass. The third molecule also exhibits dipole-dipole forces, which are stronger than dispersion forces. The last molecule exhibits hydrogen bonding. Since these are the strongest intermolecular forces in this group, the last molecule has the highest boiling point.

53. a) CH_3OH has the higher boiling point since it exhibits hydrogen bonding.

b) CH_3CH_2OH has the higher boiling point since it exhibits hydrogen bonding.

c) CH_3CH_3 has the higher boiling point since it has the larger molar mass.

55. a) Br_2 has the higher vapor pressure since it has the smaller molar mass.

b) H_2S has the higher vapor pressure since it does not exhibit hydrogen bonding.

c) PH_3 has the higher vapor pressure since it does not exhibit hydrogen bonding.

57. a) This will not form a homogeneous solution, since one is polar and one is nonpolar.

b) This will form a homogeneous solution. There will be ion-dipole interactions between the K^+ and Cl^- ions and the water molecules. There will also be dispersion forces, dipole-dipole forces and hydrogen bonding between the water molecules.

c) This will form a homogeneous solution. There will be dispersion forces present.

d) This will form a homogeneous solution. There will be dispersion forces, dipole-dipole forces and hydrogen bonding.

59. Water will have the higher surface tension since it exhibits hydrogen bonding, a strong intermolecular force. Acetone can not form hydrogen bonds.

61. Compound A will have the higher viscosity since it can interact with other molecules along the entire molecule, not just at a single point. Also the molecule is very flexible and the molecules can get tangled with each other.

63. In a clean glass tube the water can generate strong adhesive interactions with the glass (due to the dipoles at the surface of the glass). Water experiences adhesive forces with glass that are stronger than its cohesive forces, causing it to climb the surface of a glass tube. When grease or oil coats the glass this interferes with the formation of these adhesive interactions with the glass, since oils are nonpolar and cannot interact strongly with the dipoles in the water. Without these experiencing these strong intermolecular forces with oil, the water's cohesive forces will be greater and it will be drawn away from the surface of the tube.

65. The water in the 12 cm diameter beaker will evaporate more quickly because there is more surface area for the molecules to evaporate from. The vapor pressure will be the same in the two containers because the vapor pressure is the pressure of the gas when it is in dynamic equilibrium with the liquid (evaporation rate = condensation rate). The vapor pressure is dependent only on the substance and the temperature. The 12 cm diameter container will reach this dynamic equilibrium faster.

67. The boiling point and higher heat of vaporizaion of oil is much higher than that of water, so it will not vaporize as quickly as the water. The evaporation of water cools your skin because evaporation is an endothermic process.

69. **Given:** 955 kJ from candy bar, water d = 1.00 g/ml **Find:** L(H_2O) vaporized at 100.0 °C
Other: $\Delta H°_{vap}$ = 40.7 kJ/mol
Conceptual plan: **q → mol H_2O → g H_2O → mL H_2O → L H_2O**

$$\frac{1\text{ mol}}{40.7\text{ kJ}} \qquad \frac{18.01\text{ g}}{1\text{ mol}} \qquad \frac{1.00\text{ mL}}{1.00\text{ g}} \qquad \frac{1\text{ L}}{1000\text{ mL}}$$

Solution: $$955\text{ kJ} \times \frac{1\text{ mol}}{40.7\text{ kJ}} \times \frac{18.01\text{ g}}{1\text{ mol}} \times \frac{1.00\text{ mL}}{1\text{ g}} \times \frac{1\text{ L}}{1000\text{ mL}} = 0.423\ L\ H_2O$$

Check: The units (L) are correct. The magnitude of the answer (< 1 L) makes physical sense because we are vaporizing about 20 moles of water.

71. **Given:** 0.88 g water condenses on iron block 75.0 g at T_i = 22 °C **Find:** T_f (iron block)
Other: $\Delta H°_{vap}$ = 44.0 kJ/mol; C_{Fe} = 0.449 J/g · °C from text
Conceptual plan: **g H_2O → mol H_2O → q_{H2O} (kJ) → q_{H2O} (J) → q_{Fe} then q_{Fe}, m_{Fe}, T_i → T_f**

$$\frac{1\text{ mol}}{18.01\text{ g}} \qquad \frac{-44.0\text{ kJ}}{1\text{ mol}} \qquad \frac{1000\text{ J}}{1\text{ kJ}} \qquad -q_{H2O} = q_{Fe} \qquad q = m\,C_s\left(T_f - T_i\right)$$

Solution: $$0.88\text{ g} \times \frac{1\text{ mol}}{18.01\text{ g}} \times \frac{-44.0\text{ kJ}}{1\text{ mol}} \times \frac{1000\text{ J}}{1\text{ kJ}} = -2\underline{1}49.92\ J \text{ then } -q_{H2O} = q_{Fe} = 2\underline{1}49.92\text{ J then}$$

$q = m\,C_s\left(T_f - T_i\right)$ Rearrange to solve for T_f.

$$T_f = \frac{m\,C_s\,T_i + q}{m\,C_s} = \frac{\left(75.0\text{ g} \times 0.449 \frac{\text{J}}{\text{g} \times °\text{C}} \times 22\,°\text{C}\right) + 2\underline{1}49.92\text{ J}}{75.0\text{ g} \times 0.449 \frac{\text{J}}{\text{g} \times °\text{C}}} = 86°\text{C}$$

Check: The units (°C) are correct. The temperature rose, which is consistent with heat being added to the block. The magnitude of the answer (86 °C) makes physical sense because even though we have ~ 1/20 th of a mole, the energy involved in condensation is very large.

73. **Given:**

Temperature (K)	Vapor Pressure (torr)
200	65.3
210	134.3
220	255.7
230	456.0
235	597.0

Find: ΔH°_{vap} (NH_3) and normal boiling point

Conceptual Plan: To find the heat of vaporization, use Excel or similar software to make a plot of the natural log of vapor pressure (ln *P*) as a function of the inverse of the temperature in K (1/*T*). Then fit the points to a line and determine the slope of the line. Since the slope = - $\Delta H_{vap}/R$, we find the heat of vaporization as follows:

slope = - $\Delta H_{vap}/R$ → ΔH_{vap} = - slope × *R* then J → kJ. For the normal boiling point,

$$\frac{1\ \text{kJ}}{1000\ \text{J}}$$

use the equation of the best fit line, substitute 760 torr for the pressure and calculate the temperature.

Solution: Data was plotted in Excel.

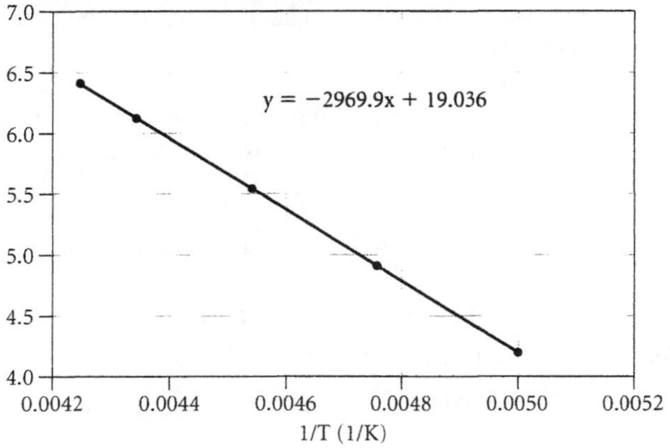

The slope of the best fitting line is – 2969.9 K.

$$\Delta H_{vap} = -\ slope\ x\ R = -(-\ 2969.9\ \cancel{K})\ x\ \frac{8.314\ J}{\cancel{K}\ mol} = \frac{2.46917\ x\ 10^4\ \cancel{J}}{mol}\ x\ \frac{1\ kJ}{1000\ \cancel{J}} = 24.7\ \frac{kJ}{mol}$$

$$\ln P = -\ 2969.9\,\text{K}\left(\frac{1}{T}\right) + 19.036 \rightarrow$$

$$\ln 760 = -\ 2969.9\,\text{K}\left(\frac{1}{T}\right) + 19.036 \rightarrow$$

$$2969.9\,\text{K}\left(\frac{1}{T}\right) = 19.036 - 6.63332 \rightarrow T = \frac{2969.9\,\text{K}}{12.40268} = 239\ \text{K}$$

Check: The units (kJ/mol) are correct. The magnitude of the answer (25) is consistent with other values in the text.

75. **Given:** ethanol, ΔH°_{vap} = 38.56 kJ/mol; normal boiling point = 78.4 °C **Find:** $P_{Ethanol}$ at 15 °C

Conceptual plan: **°C → K and kJ → J then ΔH°_{vap}, T_1, P_1, T_2 → P_2**

$$K = °C + 273.15 \qquad \frac{1000\ J}{1\ kJ} \qquad \ln\frac{P_2}{P_1} = \frac{-\Delta H_{vap}}{R}\left(\frac{1}{T_2} - \frac{1}{T_1}\right)$$

Solution: T_1 = 78.4 °C + 273.15 = 351.6 K; T_2 = 15 °C + 273.15 = 288 K;

$$\frac{38.56\ \cancel{kJ}}{mol}\ x\ \frac{1000\ J}{1\ \cancel{kJ}} = 3.856\ x\ 10^4\ \frac{J}{mol} \qquad P_1 = 760\ torr \qquad \ln\frac{P_2}{P_1} = \frac{-\Delta H_{vap}}{R}\left(\frac{1}{T_2} - \frac{1}{T_1}\right)$$

Substitute values in equation.

$$\ln\frac{P_2}{760\ torr} = \frac{-\ 3.856\ x\ 10^4\ \frac{\cancel{J}}{\cancel{mol}}}{8.314\ \frac{\cancel{J}}{\cancel{K}\ \cancel{mol}}}\left(\frac{1}{288\ \cancel{K}} - \frac{1}{351.6\ \cancel{K}}\right) = -2.91302 \rightarrow \frac{P_2}{760\ torr} = e^{-2.91302} = 0.054311 \rightarrow$$

P_2 = 0.054311 x 760 torr = 41 torr

Check: The units (torr) are correct. Since 15 °C is significantly below the boiling point, we expect the answer to be much less than 760 torr.

77. **Given:** 47.5 g water freezes **Find:** energy released **Other:** ΔH°_{fus} = 6.02 kJ/mol from text

Conceptual plan: **g H_2O → mol H_2O → q_{H_2O} (kJ) → q_{H_2O} (J)**

$$\frac{1\ mol}{18.01\ g} \qquad \frac{-\ 6.02\ kJ}{1\ mol} \qquad \frac{1000\ J}{1\ kJ}$$

Solution: $47.5\,\text{g} \times \frac{1\ \text{mol}}{18.01\ \text{g}} \times \frac{-6.02\ \text{kJ}}{1\ \text{mol}} \times \frac{1000\ J}{1\ \text{kJ}} = -15900\ J$ *or* 15900 *J released or* 15.9 *kJ released*

Check: The units (J) are correct. The magnitude (15900 J) makes sense since we are freezing about 3 moles of water. Freezing is exothermic, so heat is released.

79. **Given:** 8.5 g ice; 255 g water **Find:** ΔT of water
Other: $\Delta H^\circ_{fus} = 6.0$ kJ/mol; $C_{H_2O} = 4.18$ J/g · °C from text
Conceptual plan: $q_{ice} = -q_{water}$ **so g (ice) → mol (ice) →** q_{fus}**(kJ) →** q_{fus} **(J) →** q_{water} **(J) then**

$$\frac{1\ \text{mol}}{18.01\ \text{g}} \quad \frac{6.0\ \text{kJ}}{1\ \text{mol}} \quad \frac{1000\ \text{J}}{1\ \text{kJ}} \quad q_{water} = -\ q_{ice}$$

q, m, C_s → ΔT_1 now we have slightly cooled water in contact with 0.0 °C water

$$q = m C_S \Delta T_1$$

so $q_{ice} = -\ q_{water}$ with m, C_s → ΔT_2 with $\Delta T_1\ \Delta T_2$ → ΔT_{Total}

$$q = m\ C_s\ \Delta T_2 \quad \textit{then set } q_{ice} = -\ q_{H2O} \qquad \Delta T_{Total} = \Delta T_1 + \Delta T_2$$

Solution: $8.5\ \text{g} \times \frac{1\ \text{mol}}{18.01\ \text{g}} \times \frac{6.0\ \text{kJ}}{1\ \text{mol}} \times \frac{1000\ \text{J}}{1\ \text{kJ}} = 2.\underline{8}3176 \times 10^3\ \text{J}$, $q_{water} = -\ q_{ice} = -\ 2.\underline{8}3176 \times 10^3$ J

$q = m C_S \Delta T$ Rearrange to solve for ΔT. $\Delta T_1 = \frac{q}{m C_S} = \frac{-\ 2.\underline{8}3176 \times 10^3\ \text{J}}{255\ \text{g} \times 4.18\ \frac{\text{J}}{\text{g} \times {}^\circ\text{C}}} = -\ 2.\underline{6}567\ {}^\circ\text{C}$.

$q = m C_S \Delta T$ substitute in values and set $q_{ice} = -\ q_{water}$.

$$q_{ice} = m_{ice}\ C_{ice}\left(T_f - T_{icei}\right) = 8.5\ \text{g} \times 4.18 \frac{\text{J}}{\text{g} \times {}^\circ\text{C}} \times \left(\text{T}_\text{f} - 0.0\ {}^\circ\text{C}\right) = \rightarrow$$

$$-\ q_{water} = -\ m_{water}\ C_{water}\ \Delta T_{water2} = -\ 255\ \text{g} \times 4.18 \frac{\text{J}}{\text{g} \times {}^\circ\text{C}} \times \Delta T_{water2}$$

$8.5\ \text{T}_\text{f} = -\ 255\ \Delta T_{water2} = -\ 255\ (T_f - T_{f1})$ Rearrange to solve for T_f. $8.5\ \text{T}_\text{f} + 255\ T_f = 255\ T_{f1}$ → $26\underline{3}.5\ T_f = 255\ T_{f1}$ → $T_f = 0.96\underline{7}74\ T_{f1}$ but $\Delta T_1 = (T_{f1} - T_{i1}) = -\ 2.\underline{6}567$ °C which says that $T_{f1} = T_{i1} - 2.\underline{6}567$ °C and $\Delta T_{Total} = (T_f - T_{i1})$ so

$\Delta T_{Total} = 0.96\underline{7}74\ T_{f1} - T_{i1} = 0.96\underline{7}74(T_{i1} - 2.\underline{6}567\ {}^\circ\text{C}) - T_{i1} = -\ 2.\underline{6}567\ {}^\circ\text{C} - 0.03\underline{2}26 T_{i1}$.

This implies that the larger the initial temperature of the water, the larger the temperature drop. If the initial temperature was 90 °C, the temperature drop would be 5.6 °C. If the initial temperature was 25 °C, the temperature drop would be 3.5 °C. If the initial temperature was 5 °C, the temperature drop would be 2.8°C. This makes physical sense because the lower the initial temperature of the water, the less kinetic energy it initially has and the smaller the heat transfer from the water to the melted ice will be.
Check: The units (°C) are correct. The temperature drop form the melting of the ice is only 2.7 °C because the mass of the water is so much larger than the ice.

81. **Given:** 10.0 g ice $T_i = -10.0$ °C to steam at $T_f = 110.0$ °C **Find:** heat required (kJ)
Other: $\Delta H^\circ_{fus} = 6.02$ kJ/mol; $\Delta H^\circ_{vap} = 40.7$ kJ/mol; $C_{ice} = 2.09$ J/g · °C; $C_{H_2O} = 4.18$ J/g · °C; $C_{steam} = 2.01$ J/g · °C
Conceptual plan: Follow the heating curve in Figure 11.36. $q_{Total} = q_1 + q_2 + q_3 + q_4 + q_5$ where q_1, q_3, and q_5 are heating of a single phase then J → kJ and q_2 and q_4 are phase transitions.

$$q = m C_S (T_f - T_i) \qquad \frac{1\ \text{kJ}}{1000\ \text{J}} \qquad q = m \times \frac{1\ \text{mol}}{18.01\text{g}} \times \frac{\Delta \text{H}}{1\ \text{mol}}$$

Solution:

$$q_1 = m_{ice}\ C_{ice}\ (T_{icef} - T_{icei}) = 10.0\ \text{g} \times 2.09\ \frac{\text{J}}{\text{g} \times {}^\circ\text{C}} \times (0.0\ {}^\circ\text{C} - (-10.0\ {}^\circ\text{C})) = 209\ \text{J} \times \frac{1\ \text{kJ}}{1000\ \text{J}} = 0.209\ \text{kJ},$$

$$q_2 = m \times \frac{1\ \text{mol}}{18.01\text{g}} \times \frac{\Delta H_{fus}}{1\ \text{mol}} = 10.0\ \text{g} \times \frac{1\ \text{mol}}{18.01\ \text{g}} \times \frac{6.02\ \text{kJ}}{1\ \text{mol}} = 3.3\underline{4}3\ \text{kJ},$$

$$q_3 = m_{water}\ C_{water}\ (T_{waterf} - T_{wateri}) = 10.0\ \text{g} \times 4.18\ \frac{\text{J}}{\text{g} \times {}^\circ\text{C}} \times (100.0\ {}^\circ\text{C} - 0.0\ {}^\circ\text{C}) = 4180\ \text{J} \times \frac{1\ \text{kJ}}{1000\ \text{J}} = 4.18\ \text{kJ},$$

$$q_4 = m \times \frac{1\ \text{mol}}{18.01\text{g}} \times \frac{\Delta H_{vap}}{1\ \text{mol}} = 10.0\ \text{g} \times \frac{1\ \text{mol}}{18.01\ \text{g}} \times \frac{40.7\ \text{kJ}}{1\ \text{mol}} = 22.\underline{5}99\ \text{kJ},$$

$$q_5 = m_{steam}\ C_{steam}\ (T_{steamf} - T_{steami}) = 10.0\ \text{g} \times 2.01\ \frac{\text{J}}{\text{g} \times {}^\circ\text{C}} \times (110.0\ {}^\circ\text{C} - 100.0\ {}^\circ\text{C}) = 201\ \text{J} \times \frac{1\ \text{kJ}}{1000\ \text{J}} = 0.201\ \text{kJ}.$$

$$q_{Total} = q_1 + q_2 + q_3 + q_4 + q_5 = 0.209\ \text{kJ} + 3.3\underline{4}3\ \text{kJ} + 4.18\ \text{kJ} + 22.\underline{5}99\ \text{kJ} + 0.201\ \text{kJ} = 30.5\ \text{kJ}$$

Check: The units (kJ) are correct. The amount of heat is dominated the vaporization step. Since we have less than 1 mole we expect less than 41 kJ.

83. a) Solid

b) Liquid

c) Gas

d) Supercritical fluid

e) Solid/liquid equilibrium

f) Liquid/gas equilibrium

g) Solid/liquid/gas equilibrium

85. **Given:** nitrogen, normal boiling point = 77.3 K, normal melting point = 63.1 K, critical temperature = 126.2 K, critical pressure = 2.55 x 10^4 torr, triple point at 63.1 K and 94.0 torr

Find: Sketch phase diagram. Does nitrogen have a stable liquid phase at 1 atm?

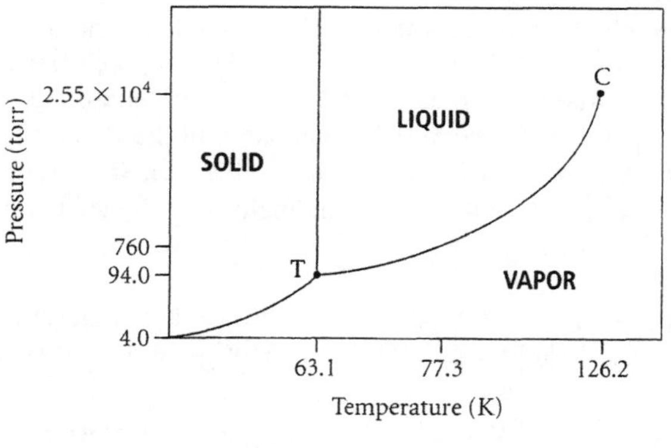

Nitrogen has a stable liquid phase at 1 atm.

87. a) 0.027 mmHg, the higher of the two triple points

b) The Rhombic phase is denser because if we start in the Monoclinic phase at 100 °C and increase the pressure, we will cross into the Rhombic phase.

89. Water has a low molar mass (18.01 g/mol), yet it is a liquid at room temperature. Water's high boiling point for its molar mass can be understood by examining the structure of the water molecule. The bent geometry of the water molecule and the highly polar nature of the O–H bonds result in a molecule with a significant dipole moment. Water's two O–H bonds (hydrogen directly bonded to oxygen) allow a water molecule to form strong hydrogen bonds with four other water molecules, resulting in a relatively high boiling point.

91. Water has an exceptionally high specific heat capacity, which has a moderating effect on the climate of coastal cities. Also, its high ΔH_{vap} causes water evaporation and condensation to have a strong effect on temperature. A tremendous amount of heat can be stored in large bodies of water. The heat will be absorbed or released from the large bodies of water preferentially over the land around it. In some cities, such as San Francisco, for example, the daily fluctuation in temperature can be less than 10 °C. This same moderating effect occurs over the entire planet, two-thirds of which is covered by water. In other words, without water, the daily temperature fluctuations on our planet might be more like those on Mars, where temperature fluctuations of 63 °C (113 °F) have been measured between midday and early morning.

93. **Given:** x-ray with λ = 154 pm, maximum reflection angle of θ = 28.3°; assume n = 1
Find: distance between layers
Conceptual plan: $\lambda, \theta, n \rightarrow d$

$$n\lambda = 2d \sin\theta$$

Solution: $n\lambda = 2d \sin\theta$ Rearrange to solve for *d*. $d = \dfrac{n\lambda}{2\sin\theta} = \dfrac{1 \text{ x } 154 \text{ pm}}{2 \sin 28.3°} = 162 \text{ pm}$

Check: The units (pm) are correct. The magnitude (164 pm) makes sense since n = 1 and the *sin* is always < 1. The number is consistent with interatomic distances.

95. a) 8 corner atoms x (1/8 atom / unit cell) = 1 atom / unit cell

b) 8 corner atoms x (1/8 atom / unit cell) + 1 atom in center = (1 + 1) atoms / unit cell = 2 atoms / unit cell

c) 8 corner atoms x (1/8 atom / unit cell) + 6 face-centered atoms x (1/2 atom / unit cell) = (1 + 3) atoms / unit cell = 4 atoms / unit cell

97. **Given:** platinum, face-centered cubic structure, r = 139 pm **Find:** edge length of unit cell and density (g/cm³) **Conceptual plan:** $r \rightarrow l$ **and** $l \rightarrow V(\text{pm}^3) \rightarrow V(\text{cm}^3)$ **and** $\mathfrak{M}$**, FCC structure** $\rightarrow m$ **then** $m, V \rightarrow d$

$$l = 2\sqrt{2}\, r \qquad V = l^3 \qquad \frac{(1 \text{ cm})^3}{(10^{10} \text{ pm})^3} \qquad m = \frac{4 \text{ atoms}}{\text{unit cell}} \text{ x } \frac{\mathfrak{M}}{N_A} \qquad d = m/V$$

Solution: $l = 2\sqrt{2}\, r = 2\sqrt{2} \text{ x } 139 \text{ pm} = 39\underline{3}.151 \text{ pm} = 393 \text{ pm}$ and

$$V = l^3 = (39\underline{3}.151 \text{ pm})^3 \text{ x } \frac{(1 \text{ cm})^3}{(10^{10} \text{ pm})^3} = 6.0\underline{7}682 \text{ x } 10^{-23} \text{ cm}^3 \text{ and}$$

$$m = \frac{4 \text{ atoms}}{\text{unit cell}} \text{ x } \frac{\mathfrak{M}}{N_A} = \frac{4 \text{ atoms}}{\text{unit cell}} \text{ x } \frac{195.09 \text{ g}}{1 \text{ mol}} \text{ x } \frac{1 \text{ mol}}{6.022 \text{ x } 10^{23} \text{ atoms}} = 1.29\underline{5}848 \text{ x } 10^{-21} \frac{\text{g}}{\text{unit cell}} \text{ then}$$

$$d = \frac{m}{V} = \frac{1.29\underline{5}848 \text{ x } 10^{-21} \dfrac{\text{g}}{\text{unit cell}}}{6.0\underline{7}682 \text{ x } 10^{-23} \dfrac{\text{cm}^3}{\text{unit cell}}} = 21.3 \frac{\text{g}}{\text{cm}^3}$$

Check: The units (pm and g/cm³) are correct. The magnitude (393 pm) makes sense because it must be larger than the radius of an atom. The magnitude (21 g/ cm³) is consistent for Pt from chapter 1.

99. **Given:** rhodium, face-centered cubic structure, d = 12.41 g/cm³ **Find:** r (Rh)
Conceptual plan: $\mathfrak{M}$ **, FCC structure** $\rightarrow$**m then m, V** $\rightarrow$ **d then V(cm³)** $\rightarrow l$ **(cm)** $\rightarrow l$ **(pm) then** $l \rightarrow r$

$$m = \frac{4 \text{ atoms}}{\text{unit cell}} \text{ x } \frac{\mathfrak{M}}{N_A} \qquad d = m/V \qquad V = l^3 \qquad \frac{10^{10} \text{ pm}}{1 \text{ cm}} \qquad l = 2\sqrt{2}\, r$$

Solution:

$m = \frac{4\text{ atoms}}{\text{unit cell}} \, x \, \frac{\mathfrak{M}}{N_A} = \frac{4\text{ atoms}}{\text{unit cell}} \text{ x } \frac{102.905\text{ g}}{1\text{ mol}} \text{ x } \frac{1\text{ mol}}{6.022\text{ x }10^{23}\text{ atoms}} = 6.835271\text{ x }10^{-22}\ \frac{\text{g}}{\text{unit cell}}$ then

$d = \frac{m}{V}$

Rearrange to solve for V. $V = \frac{m}{d} = \frac{6.835271\text{ x }10^{-22}\ \frac{\text{g}}{\text{unit cell}}}{12.41\ \frac{\text{g}}{\text{cm}^3}} = 5.507873\text{ x }10^{-23}\ \frac{\text{cm}^3}{\text{unit cell}}$ then $V = l^3$

Rearrange to solve for l.

$l = \sqrt[3]{V} = \sqrt[3]{5.507873\text{ x }10^{-23}\text{ cm}^3} = 3.804831\text{x }10^{-8}\text{ cm x }\frac{10^{10}\text{ pm}}{1\text{ cm}} = 380.4831\text{ pm}$ then $l = 2\sqrt{2}\, r$

Rearrange to solve for r. $r = \frac{l}{2\sqrt{2}} = \frac{380.4831\text{ pm}}{2\sqrt{2}} = 134.5\text{ pm}$

Check: The units (pm) are correct. The magnitude (135 pm) is consistent with atom diameters.

101. **Given:** polonium, simple cubic structure, d = 9.3 g/cm^3; r = 167 pm; $\mathfrak{M}$ = 209 g/mol **Find:** estimate N_A

Conceptual plan: $r \rightarrow l$ **and** $l \rightarrow V(\text{pm}^3) \rightarrow V(\text{cm}^3)$ **then** $d, V \rightarrow m$ **then** $\mathfrak{M}$**, SC structure** $\rightarrow m$

$l = 2r$ $\quad V = l^3 \quad \frac{(1\text{ cm})^3}{(10^{10}\text{ pm})^3} \quad d = m/V \quad m = \frac{1\text{ atom}}{\text{unit cell}} \, x \, \frac{\mathfrak{M}}{N_A}$

Solution: $l = 2\,r = 2\text{ x }167\text{ pm} = 334\text{ pm}$ and

$V = l^3 = (334\text{ pm})^3 \text{ x } \frac{(1\text{ cm})^3}{(10^{10}\text{ pm})^3} = 3.72597\text{ x }10^{-23}\text{ cm}^3$ then $d = \frac{m}{V}$ Rearrange to solve for m.

$m = d\,V = 9.3\ \frac{\text{g}}{\text{cm}^3} \text{ x } \frac{3.72597\text{ x }10^{-23}\text{ cm}^3}{\text{unit cell}} = 3.46515\text{ x }10^{-22}\ \frac{\text{g}}{\text{unit cell}}$ then

$m = \frac{1\text{ atom}}{\text{unit cell}} \, x \, \frac{\mathfrak{M}}{N_A}$ Rearrange to solve for N_A.

$N_A = \frac{1\text{ atom}}{\text{unit cell}} \, x \, \frac{\mathfrak{M}}{m} = \frac{1\text{ atom}}{\text{unit cell}} \text{ x } \frac{209\text{ g}}{1\text{ mol}} \text{ x } \frac{1\text{ unit cell}}{3.46515\text{ x }10^{-22}\text{ g}} = 6.03\text{ x }10^{23}\ \frac{\text{atom}}{\text{mol}}$

Check: The units (atoms/mol) are correct. The magnitude (6 x 10^{23}) is consistent with Avogadro's number.

103. a) Atomic, since Ar is an atom

b) Molecular, since water is a molecule

c) Ionic, since K_2O is an ionic solid

d) Atomic, since iron is an atom

105. LiCl has the highest melting point since it is the only ionic solid in the group. The other three solids are held together by intermolecular forces while LiCl is held together by stronger coulombic interactions between the cations and anions of the crystal lattice.

107. a) TiO_2 because it is an ionic solid

b) $SiCl_4$ because it is larger, and therefore, has stronger dispersion forces

c) Xe because it is larger, and therefore, has stronger dispersion forces

d) CaO because the ions have greater charge, stronger dipole-dipole interactions

109. The Ti atoms occupy the corner positions and the center of the unit cell - 8 corner atoms x (1/8 atom / unit cell) + 1 atom in center = (1 + 1) Ti atoms / unit cell = 2 Ti atoms / unit cell. The O atoms occupy four positions on the top and bottom faces and two positions inside the unit cell - 4 face-centered atoms x (1/2 atom / unit cell) + 2 atoms in the interior = (2 + 2) O atoms / unit cell = 4 O atoms / unit cell. Therefore there are 2 Ti atoms / unit cell and 4 O atoms / unit cell, so the ratio Ti:O is 2:4 or 1:2. The formula for the compound is TiO_2.

111. In CsCl: The Cs atoms occupy the center of the unit cell - 1 atom in center = 1 Cs atom / unit cell. The Cl atoms occupy corner positions of the unit cell - 8 corner atoms x (1/8 atom / unit cell) = 1 Cl atom / unit cell. Therefore there are 1 Cl atom / unit cell and 1 Cl atom / unit cell, so the ratio Cs:Cl is 1:1. The formula for the compound is CsCl, as expected.

In $BaCl_2$: The Ba atoms occupy the corner positions and the face-centered positions of the unit cell - 8 corner atoms x (1/8 atom / unit cell) + 6 face-centered atoms x (1/2 atom / unit cell) = (1 + 3) Ba atoms / unit cell = 4 Ba atoms / unit cell. The Cl atoms occupy eight positions inside the unit cell - 8 Cl atoms / unit cell. Therefore there are 4 Ba atoms / unit cell and 8 Cl atoms / unit cell, so the ratio Ba:Cl is 4:8 or 1:2. The formula for the compound is $BaCl_2$, as expected.

113. a) Zn should have little or no band gap because it is the only metal in the group.

115. a) p-type semiconductor – Ge is Group 4A and Ga is Group 3A, so the Ga will generate electron "holes".

b) n-type semiconductor – Si is Group 4A and As is Group 5A, so the As will add electrons to the conduction band.

117. The general trend is that melting point increases with increasing mass. This is due to the fact that the electrons of the larger molecules are held more loosely and a stronger dipole moment can be induced more easily. HF is the exception to the rule. It has a relatively high melting point due to hydrogen bonding.

119. **Given:** P_{H2O} = 23.76 torr at 25 °C; 1.25 g water in 1.5 L container **Find:** m (H_2O) as liquid
Conceptual plan: **°C → K and torr → atm then P, V, T → mol *(g)*→ g *(g)* then g *(g)*, g *(l)*$_i$ → g *(l)*$_f$**

$K = °C + 273.15$ $\frac{1 \text{ atm}}{760 \text{ torr}}$ $PV = nRT$ $\frac{18.01 \text{ g}}{1 \text{ mol}}$ $g\ (l)_f = g\ (l)_i - g\ (g)$

Solution: T = 25 °C + 273.15 = 298 K, $23.76 \text{ torr} \times \frac{1 \text{ atm}}{760 \text{ torr}} = 0.0312\underline{6}32 \text{ atm}$ then $PV = nRT$

Rearrange to solve for n. $n = \frac{PV}{RT} = \frac{0.0312\underline{6}32 \text{ atm} \times 1.5 \text{ L}}{0.08206 \frac{\text{L atm}}{\text{K mol}} \times 298 \text{ K}} = 0.001\underline{9}1768 \text{ mol}$ then

$0.001\underline{9}1768 \text{ mol} \times \frac{18.01 \text{ g}}{1 \text{ mol}} = 0.03\underline{4}5375 \text{ g in gas phase}$ then

$g\ (l)_f = g\ (l)_i - g\ (g) = 1.25 \text{ g} - 0.03\underline{4}5375 \text{ g} = 1.22$ g remaining as liquid Yes, there is 1.22 g of liquid.

Check: The units (g) are correct. The magnitude (1.2 g) is expected since very little material is expected to be in the gas phase.

121. gas → liquid → solid

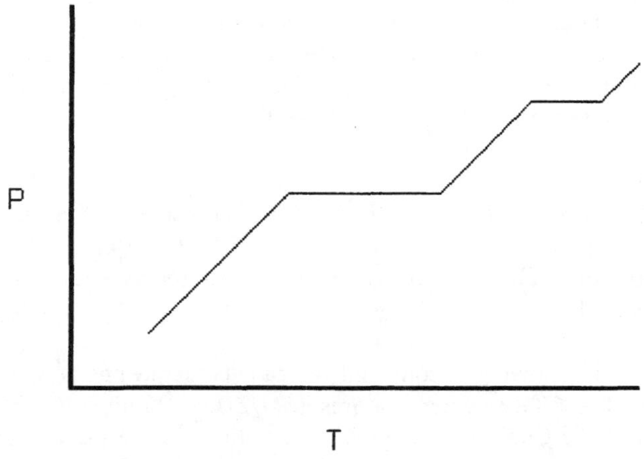

123. **Given:** Ice: T_1 = 0 °C exactly, m = 53.5 g; Water: T_1 = 75 °C, m = 115 g **Find:** T_f
Other: ΔH°_{fus} = 6.0 kJ/mol; C_{H2O} = 4.18 J/g ·°C
Conceptual plan: $q_{ice} = - q_{water}$ **so g (ice) → mol (ice) →** q_{fus}**(kJ) →** q_{fus} **(J) →** q_{water} **(J) then**

$$\frac{1 \text{ mol}}{18.01 \text{ g}} \qquad \frac{6.02 \text{ kJ}}{1 \text{ mol}} \qquad \frac{1000 \text{ J}}{1 \text{ kJ}} \qquad q_{water} = - q_{ice}$$

q, m, C_s → ΔT then T_i, ΔT → T_2 now we have slightly cooled water in contact with 0.0 °C water

$$q = m C_S \Delta T \qquad \Delta T = T_2 - T_i$$

so $q_{ice} = - q_{water}$ with m, C_s, T_i → T_f

$$q = m\, C_s \left(T_f - T_i\right) \quad \text{then set } q_{ice} = - q_{water}$$

Solution: $53.5 \text{ g} \times \frac{1 \text{ mol}}{18.01 \text{ g}} \times \frac{6.02 \text{ kJ}}{1 \text{ mol}} \times \frac{1000 \text{ J}}{1 \text{ kJ}} = 1.7\underline{8}828 \times 10^4 \text{ J}$, $q_{water} = - q_{ice} = - 1.7\underline{8}828 \times 10^4$ J

$q = m C_S \Delta T$ Rearrange to solve for ΔT. $\Delta T = \frac{q}{m C_S} = \frac{-1.7\underline{8}828 \times 10^4 \text{ J}}{115 \text{ g} \times 4.18 \frac{\text{J}}{\text{g} \times {}^\circ\text{C}}} = -37.\underline{2}017\ ^\circ\text{C}$ then $\Delta T = T_2 - T_i$.

Rearrange to solve for T_2. $T_2 = \Delta T + T_i = -37.\underline{2}017\ ^\circ\text{C} + 75\ ^\circ\text{C} = 3\underline{7}.798\ ^\circ\text{C}$ $q = m\, C_s\left(T_f - T_i\right)$

substitute in values and set $q_{ice} = - q_{water}$.

$$q_{ice} = m_{ice}\, C_{ice}\left(T_f - T_{icei}\right) = 53.5 \text{ g} \times 4.18 \frac{\text{J}}{\text{g} \times {}^\circ\text{C}} \times \left(T_f - 0.0\ ^\circ\text{C}\right) =$$

Rearrange to solve for T_f.

$$- q_{water} = - m_{water}\, C_{water}\left(T_f - T_{water2}\right) = -115 \text{ g} \times 4.18 \frac{\text{J}}{\text{g} \times {}^\circ\text{C}} \times \left(T_f - 3\underline{7}.798\ ^\circ\text{C}\right)$$

$$53.5\, T_f = -115\left(T_f - 3\underline{7}.798\ ^\circ\text{C}\right) \rightarrow 53.5\, T_f = -115\, T_f + 4\underline{3}46.8\ ^\circ\text{C} \rightarrow -4\underline{3}46.8\ ^\circ\text{C} = -16\underline{8}.5\, T_f$$

$$\rightarrow T_f = \frac{-4\underline{3}46.8\ ^\circ\text{C}}{-16\underline{8}.5} = 2\underline{5}.8\ ^\circ\text{C} = 26\ ^\circ\text{C}$$

Check: The units (°C) are correct. The temperature is between the two initial temperatures. Since the ice mass is about half the water mass, we are not surprised that the temperature is closer to the original ice temperature.

125. **Given:** Home: 6.0 m x 10.0 m x 2.2 m; T = 30 °C, P_{H2O} = 85 % of P°_{H2O} **Find:** m (H_2O) removed
Other: P°_{H2O} = 31.86 mm Hg from text

Conceptual plan: l, w, h → V (m^3) → V (cm^3) → V (L) and

$$V = l \times w \times h \qquad \frac{(100\ cm)^3}{(1\ m)^3} \qquad \frac{1\ L}{1000\ cm^3}$$

$P°_{H2O}$ → P_{H2O} (mm Hg) → P_{H2O} (atm) and °C → K then P, V, T → mol (H_2O) → g (H_2O)

$$P_{H2O} = 0.85\ P^0_{H2O} \qquad \frac{1\ atm}{760\ mm\ Hg} \qquad K = °C + 273.15 \qquad PV = nRT \qquad \frac{18.01\ g}{1\ mol}$$

Solution: $V = l \times w \times h = 6.0\ m \times 10.0\ m \times 2.2\ m = 132\ m^3 \times \frac{(100\ cm)^3}{(1\ m)^3} \times \frac{1\ L}{1000\ cm^3} = 1.32 \times 10^5\ L$,

$P_{H2O} = 0.85\ P^0_{H2O} = 0.85 \times 31.86\ mm\ Hg \times \frac{1\ atm}{760\ mm\ Hg} = 0.035633\ atm$, T = 30 °C + 273.15 = 303 K, then

$PV = nRT$ Rearrange to solve for n. $n = \frac{PV}{RT} = \frac{0.035633\ atm \times 1.32 \times 10^5\ L}{0.08206\ \frac{L\ atm}{K\ mol} \times 303\ K} = 189.17\ mol$ then

$189.17\ mol \times \frac{18.01\ g}{1\ mol} = 3400$ g to remove

Check: The units (g) are correct. The magnitude of the answer (3400 g) makes sense since the volume of the house is so large. We are removing almost 200 moles of water.

127. CsCl has a higher melting point than AgI because of its higher coordination number. In CsCl, one anion bonds to 8 cations (and vice versa) while in AgI, one anion bonds only to 4 cations.

129. a) Atoms are connected across the face diagonal (c), so $c = 4r$

b) From the Pythagorean Theorem $c^2 = a^2 + b^2$, from part a) $c = 4r$ and for a cubic structure $a = l, b = l$ so $(4r)^2 = l^2 + l^2$ → $16r^2 = 2l^2$ → $8r^2 = l^2$ → $l = \sqrt{8r^2}$ → $l = 2\sqrt{2}\,r$

131. **Given:** diamond, V (unit cell) = 0.0454 nm^3; d = 3.52 g/cm^3 **Find:** number of carbon atoms / unit cell
Conceptual plan: V(nm^3) → V(cm^3) then d, V → m → mol → atoms

$$\frac{(1\ cm)^3}{(10^7\ nm)^3} \qquad d = m/V \qquad \frac{1\ mol}{12.01\ g} \qquad \frac{6.022 \times 10^{23}\ atoms}{1\ mol}$$

Solution: $0.0454\ nm^3 \times \frac{(1\ cm)^3}{(10^7\ nm)^3} = 4.54 \times 10^{-23}\ cm^3$ then $d = \frac{m}{V}$ Rearrange to solve for *m*.

$m = d\,V = 3.52\ \frac{g}{cm^3} \times 4.54 \times 10^{-23}\ cm^3 = 1.59808 \times 10^{-22}\ g$ then

$$\frac{1.59808 \times 10^{-22}\ g}{unit\ cell} \times \frac{1\ mol}{12.01\ g} \times \frac{6.022 \times 10^{23}\ atoms}{1\ mol} = 8.01\ \frac{C\ atoms}{unit\ cell} = 8\ \frac{C\ atoms}{unit\ cell}$$

Check: The units (atoms) are correct. The magnitude (8) makes sense because it is a fairly small number and our answer within our calculation error of an integer.

133. a) $CO_2\,(s) \rightarrow CO_2\,(g)$ at 194.7 K

b) $CO_2\,(s) \rightarrow$ triple point at 216.5 K $\rightarrow CO_2\,(g)$ just above 216.5 K

c) $CO_2\,(s) \rightarrow CO_2\,(l)$ at somewhat above 216 K $\rightarrow CO_2\,(g)$ at around 250 K

d) CO_2 (s) → CO_2 above the critical point where there is no distinction between liquid and gas. This change occurs at about 300 K.

135. **Given:** KCl, rock salt structure **Find:** density (g/cm^3)
Other: r (K^+) = 133 pm; r (Cl^-) = 181 pm from Chapter 8
Conceptual plan: Rock salt structure is a face-centered cubic structure with anions at the lattice points and cations in the holes between lattice sites → assume r = r(Cl^-), but $\mathfrak{M} = \mathfrak{M}$(KCl)
r(K^+), r(Cl^-) → l and l → V(pm^3) → V(cm^3) and FCC structure → m then m, V → d

from Figure 11.52 $l = 2r(Cl^-) + 2r(K^+)$ $V = l^3$ $\frac{(1\text{ cm})^3}{(10^{10}\text{ pm})^3}$ $m = \frac{4\text{ formula units}}{\text{unit cell}} x \frac{\mathfrak{M}}{N_A}$ $d = m/V$

Solution: $l = 2r(Cl^-) + 2r(K^+) = 2(181\text{ pm}) + 2(133\text{ pm}) = 628\text{ pm}$ and

$$V = l^3 = (628\text{ pm})^3 \text{ x } \frac{(1\text{ cm})^3}{(10^{10}\text{ pm})^3} = 2.47673 \text{ x } 10^{-22}\text{ cm}^3$$ and

$$m = \frac{4\text{ formula units}}{\text{unit cell}} x \frac{\mathfrak{M}}{N_A} = \frac{4\text{ formula units}}{\text{unit cell}} \text{ x } \frac{74.55\text{ g}}{1\text{ mol}} \text{ x } \frac{1\text{ mol}}{6.022 \text{ x } 10^{23}\text{ formula units}} = 4.951976 \text{ x } 10^{-22}\,\frac{\text{g}}{\text{unit cell}}$$ then

$$d = \frac{m}{V} = \frac{4.951976 \text{ x } 10^{-22}\,\frac{\text{g}}{\text{unit cell}}}{2.47673 \text{ x } 10^{-22}\,\frac{\text{cm}^3}{\text{unit cell}}} = 1.99940\,\frac{\text{g}}{\text{cm}^3} = 2.00\,\frac{\text{g}}{\text{cm}^3}$$

Check: The units (g/cm^3) are correct. The magnitude (2 g/ cm^3) is reasonable for a salt density. The published value is 1.98 g/cm^3. This method of estimating the density gives a value that is close to the experimentally measured density.

137. Decreasing the pressure will decrease the temperature of liquid nitrogen. Because the nitrogen is boiling, its temperature must be constant at a given pressure. As the pressure decreases, the boiling point decreases, and therefore so does the temperature. Remember that vaporization is an endothermic process, so as the nitrogen vaporizes it will remove heat from the liquid, dropping its temperature. If the pressure drops below the pressure of the triple point, the phase change will shift from vaporization to sublimation and the liquid nitrogen will become solid.

139. **Given:** cubic closest packing structure = cube with touching spheres of radius = r on alternating corners of a cube **Find:** body diagonal of cube and radius of tetrahedral hole
Solution: The cell edge length = l and $l^2 + l^2 = (2r)^2$ → $2\,l^2 = 4\,r^2$ → $l^2 = 2\,r^2$. Since body diagonal = BD is the hypotenuse of the right triangle formed by the face diagonal and the cell edge we have $(BD)^2 = l^2 + (2r)^2 = 2\,r^2 + 4r^2 = 6\,r^2$ → $BD = \sqrt{6}\,r$. The radius of the tetrahedral hole = r_T is the half the body diagonal minus the radius of the sphere or

$$r_T = \frac{BD}{2} - r = \frac{\sqrt{6}\,r}{2} - r = \left(\frac{\sqrt{6}}{2} - 1\right) r = \left(\frac{\sqrt{6}-2}{2}\right) r = \left(\frac{\sqrt{3}\sqrt{2} - \sqrt{2}\sqrt{2}}{\sqrt{2}\sqrt{2}}\right) r = \left(\frac{\sqrt{3}-\sqrt{2}}{\sqrt{2}}\right) r \approx 0.22474\,r$$

141. Melting of an ice cube in a glass of water will not raise or lower the level of the liquid in the glass as long as the ice is always floating in the liquid. This is because the ice will displace a volume of water based on its mass. By the same logic, melting floating icebergs will not raise the ocean levels (assuming that the dissolved solids content, and thus the density, will not change when the icebergs melt). Dissolving ice formations that are supported by land will raise the ocean levels, just as pouring more water into the glass will raise the liquid level in the glass.

143. Substance A will have the larger change in vapor pressure with the same temperature change. To understand this consider the Clausius-Clapeyron Equation: $\ln \frac{P_2}{P_1} = \frac{-\Delta H_{vap}}{R}\left(\frac{1}{T_2} - \frac{1}{T_1}\right)$, if we use the same temperatures we see that $\frac{P_2}{P_1}\ \alpha\ e^{-\Delta H_{vap}}$. So the smaller the heat of vaporization the larger the final vapor pressure. We can also consider that the lower the heat of vaporization the easier it is to convert the substance from a liquid to a gas. This again leads to Substance A having the larger change in vapor pressure.

145. $\Delta H_{sub} = \Delta H_{fus} + \Delta H_{vap}$ as long as the heats of fusion and vaporization are measured at the same temperatures.

147. Water has an exceptionally high specific heat capacity, which has a moderating effect on the temperature of the root cellar. A large amount of heat can be stored in a large vat of water. The heat will be absorbed or released from the large bodies of water preferentially over the area around it. As the temperature of the air drops, the water will release heat keeping the temperature more constant. If the temperature of the cellar falls enough to begin to freeze the water, the heat given off during the freezing will further protect the food in the cellar.

Chapter 12
Solutions

1. As seawater moves through the intestine, it flows past cells that line the digestive tract, which consist of largely fluid interiors surrounded by membranes. Although cellular fluids themselves contain dissolved ions, including sodium and chloride, the fluids are more dilute than seawater. Nature's tendency towards mixing (which tends to produce solutions of uniform concentration), together with the selective permeability of the cell membranes (which allow water to flow in and out, but restrict the flow of dissolved solids) causes a flow of solvent out of the body's cells and into the seawater.

3. A substance is soluble in another substance if they can form a homogeneous mixture. The solubility of a substance is the amount of the substance that will dissolve in a given amount of solvent. Several different units can be used to express the solubility, such as, grams of solute per 100 grams of solvent, grams of solute per liter of solvent, or moles of solute per liter of solution.

5. Entropy is a measure of energy randomization or energy dispersal in a system. When two substances mix to form a solution there is an increase in randomness, due to the fact that the components are no longer segregated to separate regions.

7. A solution always forms if the solvent-solute interactions are comparable to, or stronger than, the solvent-solvent interactions and the solute-solute interactions.

9. Step 1: Separating the solute into its constituent particles. This step is always endothermic (positive ΔH) because energy is required to overcome the forces that hold the solute together.

   Step 2: Separating the solvent particles from each other to make room for the solute particles. This step is also endothermic because energy is required to overcome the intermolecular forces among the solvent particles.

   Step 3: Mixing the solute particles with the solvent particles. This step is exothermic, because energy is released as the solute particles interact with the solvent particles through the various types of intermolecular forces.

11. In any solution formation, the initial rate of dissolution far exceeds the rate of deposition. But as the concentration of dissolved solute increases, the rate of deposition also increases. Eventually the rate of dissolution and deposition become equal—dynamic equilibrium has been reached.

    A saturated solution is a solution in which the dissolved solute is in dynamic equilibrium with the solid (or undissolved) solute is called. If you add additional solute to a saturated solution, it will not dissolve.

    An unsaturated solution is a solution containing less than the equilibrium amount of solute is called. If you add additional solute to an unsaturated solution, it will dissolve.

    A supersaturated solution is a solution containing more than the equilibrium amount of solute. Such solutions are unstable and the excess solute normally precipitates out of the solution. However, in some cases, if left undisturbed, a supersaturated solution can exist for an extended period of time.

13. The solubility of gases in liquids decreases with increasing temperature. The decreasing solubility of gases with increasing temperature results in a lower oxygen concentration available for fish and other aquatic life in warm waters.

15. Henry's law quantifies the solubility of gases with increasing pressure as follows: $S_{gas} = k_H P_{gas}$, where S_{gas} is the solubility of the gas, k_H is a constant of proportionality (called the Henry's law constant) that depends on the specific solute and solvent and also on temperature, and P_{gas} is the partial pressure of the gas. The equation simply shows that the solubility of a gas in a liquid is directly proportional to the pressure of the gas above the liquid, so the solubility of a gas is known at a certain temperature; the solubility at another pressure at this temperature can be calculated.

17. Parts by mass and parts by volume are ratios of masses and volume, respectively. A parts by mass concentration is the ratio of the mass of the solute to the mass of the solution, all multiplied by a multiplication factor: where percent by mass (%) factor = 100, parts per million by mass (ppm) factor = 10^6, and parts per billion by mass (ppb) factor = 10^9. The size of the multiplication factor depends on the concentration of the solution. For example, in percent by mass, the multiplication factor is 100 %, so . If a solution with a concentration of 28 % by mass contains 28 g of solute per 100 g of solution.

19. Raoult's law quantifies the relationship between the vapor pressure of a solution and its concentration as: , where $P_{solution}$ is the vapor pressure of the solution, $\chi_{solvent}$ is the mole fraction of the solvent, and is the vapor pressure of the pure solvent. This equation allows you to calculate the vapor pressure of a solution or to calculate the concentration of a solution, given the vapor pressure of the solution.

21. An ideal solution is a solution that follows Raoult's law at all concentrations for both the solute and the solvent. A nonideal solution will exhibit deviations from Raoult's law in the vapor pressure of a component as mole fraction of this component decreases from 1 (i.e. the pure component). If the solute-solvent interactions are particularly strong (stronger than solvent-solvent interactions), then the solute tends to prevent the solvent from vaporizing as easily as it would otherwise and the vapor pressure of the solution will be less than that predicted by Raoult's law. If the solute-solvent interactions are particularly weak (weaker than solvent-solvent interactions), then the solute tends to allow more vaporization than would occur with just the solvent and the vapor pressure of the solution will be greater than predicted by Raoult's law.

23. Colligative properties are properties that depend on the amount of solute and not the type of solute are called. Examples of collogative properties are vapor pressure lowering, freezing point depression, boiling point elevation and osmotic pressure.

25. The van't Hoff factor (i) is the ratio of moles of particles in solution to moles of formula units dissolved: . The van't Hoff factor corrects for the fact that the ionic solute is not completely dissolved into the expected number of ions, leaving ion pairs in solution. The result is that the number of particles in the solution is not as high as theoretically expected.

27. The Tyndall effect is the scattering of light by a colloidal dispersion. The Tyndall effect is often used as a test to determine whether a mixture is a solution or a colloid, since solutions contain completely dissolved solute molecules that are too small to scatter light.

29. a) Hexane, toluene, or CCl_4; dispersion forces

    b) Water, methanol; dispersion, dipole-dipole, hydrogen bonding

    c) Hexane, toluene, or CCl_4; dispersion forces

    d) Water, acetone, methanol, ethanol; dispersion, ion-dipole

31. $HOCH_2CH_2CH_2OH$ would be more soluble in water because it has –OH groups on both ends of the molecule, so it can hydrogen bond on both ends, not just one end.

33. a) Water; dispersion, dipole-dipole, hydrogen bonding

    b) Hexane; dispersion forces

    c) Water; dispersion, dipole-dipole

    d) Water; dispersion, dipole-dipole, hydrogen bonding

35. a) Endothermic

    b) The lattice energy is greater in magnitude than the heat of hydration

c)

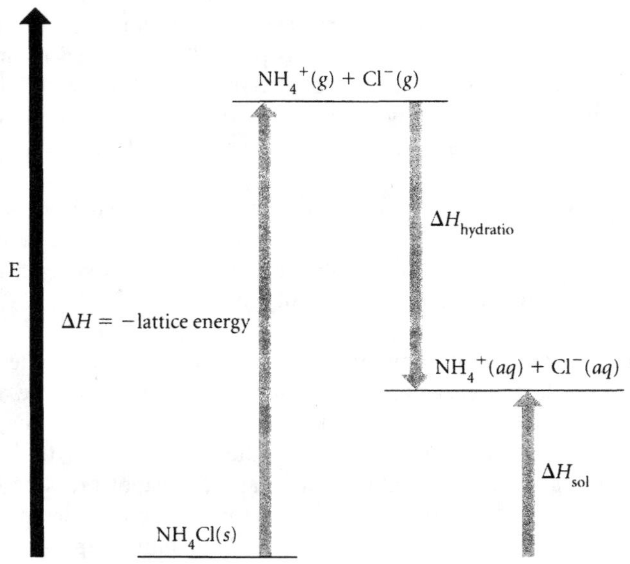

d) The solution forms because chemical systems tend towards greater entropy.

37. **Given:** $AgNO_3$: Lattice Energy 820. kJ/mol, ΔH_{soln} = +22.6 kJ/mol **Find:** $\Delta H_{hydration}$
Conceptual plan: Lattice Energy, ΔH_{soln} → $\Delta H_{hydration}$

$\Delta H_{soln} = \Delta H_{solute} + \Delta H_{hydration}$ where $\Delta H_{solute} = -\Delta H_{lattice}$

Solution: $\Delta H_{soln} = \Delta H_{solute} + \Delta H_{hydration}$ where $\Delta H_{solute} = -\Delta H_{lattice}$ so $\Delta H_{hydration} = \Delta H_{soln} + \Delta H_{lattice}$

$\Delta H_{hydration} = 22.6\,\text{kJ/mol} - 820.\,\text{kJ/mol} = -797\,\text{kJ/mol}$

Check: The units (kJ/mol) are correct. The magnitude of the answer (– 800) makes physical sense because the lattice energy is so negative and so it dominates the calculation.

39. **Given:** LiI: Lattice Energy = – 7.3 x 10^2 kJ/mol, $\Delta H_{hydration}$ = – 793 kJ/mol; 15.0 g LiI
Find: ΔH_{soln} and heat evolved
Conceptual plan: Lattice Energy, $\Delta H_{hydration}$ → ΔH_{soln} and g → mol then mol, ΔH_{soln} → q

$\Delta H_{soln} = \Delta H_{solute} + \Delta H_{hydration}$ where $\Delta H_{solute} = -\Delta H_{lattice}$ $\qquad \frac{1\text{ mol}}{133.843\text{ g}}$ $\qquad q = n\,\Delta H_{soln}$

Solution: $\Delta H_{soln} = \Delta H_{solute} + \Delta H_{hydration}$ where $\Delta H_{solute} = -\Delta H_{lattice}$ so $\Delta H_{soln} = \Delta H_{hydration} - \Delta H_{lattice}$
and

$$15.0\,\cancel{\text{g}} \times \frac{1\text{ mol}}{133.843\,\cancel{\text{g}}} = 0.11\underline{2}072\text{ mol}$$ then

$$q = n\,\Delta H_{soln} = 0.11\underline{2}072\,\cancel{\text{mol}} \times -6 \times 10^1\,\frac{\text{kJ}}{\cancel{\text{mol}}} = -7\text{ kJ or 7 kJ released}$$

Check: The units (kJ/mol and kJ) are correct. The magnitude of the answer (– 60) makes physical sense because the lattice energy and the heat of hydration are about the same. The magnitude of the heat (7) makes physical sense since 15 g is much less than a mole, and so the amount of heat released is going to be small.

41. The solution is unsaturated since we are dissolving 25g of NaCl per 100 g of water and the solubility from the figure is ~ 35 g NaCl per 100 g of water.

43. At 40 °C the solution has 45 g of KNO_3 per 100 g of water and it can contain up to 63 g of KNO_3 per 100 g of water. At 0 °C the solubility from the figure is ~ 14 g KNO_3 per 100 g of water, so ~ 31 g KNO_3 per 100 g of water will precipitate out of solution.

45. Since the solubility of gases decrease as the temperature increased, dissolved oxygen will be removed from the solution.

47. Henry's Law says that as pressure increases, nitrogen will more easily dissolve in blood. To reverse this process, divers should ascend to lower pressures.

49. **Given:** room temperature, 80.0 L aquarium, $P_{Total} = 1.0$ atm; $\chi_{N2} = 0.78$ **Find:** m (N_2)
Other: $k_H(N_2) = 6.1 \times 10^{-4}$ M/L at 25 °C
Conceptual plan: $P_{Total}, \chi_{N2} \rightarrow P_{N2}$ **then** $P_{N2}, k_H(N_2) \rightarrow S_{N2}$ **then** **L → mol → g**

$$P_{N2} = \chi_{N2}\, P_{Total} \qquad S_{N2} = k_H(N_2) P_{N2} \qquad S_{N2} \qquad \frac{28.01\ \text{g}}{1\ \text{mol}}$$

Solution: $P_{N2} = \chi_{N2}\, P_{Total} = 0.78 \times 1.0\ \text{atm} = 0.78\ \text{atm}$ then

$$S_{N2} = k_H(N_2) P_{N2} = 6.1 \times 10^{-4}\ \frac{\text{M}}{\text{atm}} \times 0.78\ \text{atm} = 4.\underline{7}58 \times 10^{-4}\ \text{M}$$ then

$$80.0\ \text{L} \times 4.\underline{7}58 \times 10^{-4}\ \frac{\text{mol}}{\text{L}} \times \frac{28.01\ \text{g}}{1\ \text{mol}} = 1.1\ \text{g}$$

Check: The units (g) are correct. The magnitude of the answer (1) seems reasonable since we have 80 L of water and expect much less than a mole of nitrogen.

51. **Given:** NaCl and water; 133 g NaCl in 1.00 L solution **Find:** M, *m*, and mass percent
Other: d = 1.08 g/mL
Conceptual plan: $g_{NaCl} \rightarrow$ **mol** **and** **L → mL →** g_{soln} **and** $g_{soln}\ g_{NaCl} \rightarrow g_{H2O} \rightarrow kg_{H2O}$ **then**

$$\frac{1\ \text{mol NaCl}}{58.44\ \text{g NaCl}} \qquad \frac{1000\ \text{mL}}{1\ \text{L}} \quad \frac{1.08\ \text{g}}{1\ \text{mL}} \qquad \frac{1\ \text{kg}}{1000\ \text{g}}$$

mol, V → M **and** **mol,** $kg_{H2O} \rightarrow m$ **and** $g_{soln}\ g_{NaCl} \rightarrow$ **mass percent**

$$\text{M} = \frac{\text{amount solute (moles)}}{\text{volume solution (L)}} \qquad m = \frac{\text{amount solute (moles)}}{\text{mass solvent (kg)}} \qquad mass\ \ percent = \frac{mass\ solute}{mass\ solution} \times 100\ \%$$

Solution: and $1.00\ \text{L} \times \dfrac{1000\ \text{mL}}{1\ \text{L}} \times \dfrac{1.08\ \text{g}}{1\ \text{mL}} = 10\underline{8}0\ \text{g soln}$ and

$$g_{H2O} = g_{soln} - g_{NaCl} = 10\underline{8}0\ \text{g} - 133\ \text{g} = 9\underline{4}7\ \text{g H}_2\text{O} \times \frac{1\ \text{kg}}{1000\ \text{g}} = 0.9\underline{4}7\ \text{kg H}_2\text{O}$$ then

$$\text{M} = \frac{\text{amount solute (moles)}}{\text{volume solution (L)}} = \frac{2.2\underline{7}584\ \text{mol NaCl}}{1.00\ \text{L soln}} = 2.28\ \text{M}$$ and

$$m = \frac{\text{amount solute (moles)}}{\text{mass solvent (kg)}} = \frac{2.2\underline{7}584\ \text{mol NaCl}}{0.9\underline{4}7\ \text{kg H}_2\text{O}} = 2.4\ m$$ and

$$mass\ \ percent = \frac{mass\ solute}{mass\ solution} \times 100\ \% = \frac{133\ \text{g NaCl}}{10\underline{8}0\ \text{g soln}} \times 100\ \% = 12.3\ \%\ \text{by mass}$$

Check: The units (M, m and percent by mass) are correct. The magnitude of the answer (2.28 M) seems reasonable since we have 133 g NaCl, which is a couple of moles and we have 1 L. The magnitude of the answer (2.4 *m*) seems reasonable since it is a little higher than the molarity, which we expect since we only use the solvent weight in the denominator. The magnitude of the answer (12 %) seems reasonable since we have 133 g NaCl and just over 1000 g of solution.

53. **Given:** initial solution: 50.0 mL of 5.00 M KI; final solution contains: 3.25 g KI in 25.0 mL
Find: final volume to dilute initial solution to
Conceptual plan: final solution: g_{KI} → mol and mL → L then mol, V → M_2 then M_1, V_1, M_2 → V_2

$$\frac{1\ \text{L}}{1000\ \text{mL}} \qquad M = \frac{\text{amount solute (moles)}}{\text{volume solution (L)}} \qquad M_1 V_1 = M_2 V_2$$

Solution: and $3.25\ \text{g KI} \times \frac{1\ \text{mol KI}}{166.006\ \text{g KI}} = 0.0195776\ \text{mol KI}$ and $25.0\ \text{mL} \times \frac{1\ \text{L}}{1000\ \text{mL}} = 0.0250\ \text{mL}$

then $M = \frac{\text{amount solute (moles)}}{\text{volume solution (L)}} = \frac{0.0195776\ \text{mol KI}}{0.0250\ \text{L soln}} = 0.783104\ \text{M}$ then $M_1 V_1 = M_2 V_2$. Rearrange to solve for V_2. $V_2 = \frac{M_1}{M_2} x\ V_1 = \frac{5.00\ \text{M}}{0.783104\ \text{M}} \text{x } 50.0\ \text{mL} = 319\ \text{mL diluted volume}$

Check: The units (mL) are correct. The magnitude of the answer (319 mL) seems reasonable since we are starting with a concentration of 5 M and ending with a concentration of less than 1 M.

55. **Given:** $AgNO_3$ and water; 3.4 % Ag by mass, 4.8 L solution **Find:** m (Ag) **Other:** d = 1.01 g/mL
Conceptual plan: L → mL → g_{soln} → g_{Ag}

$$\frac{1000\ \text{mL}}{1\ \text{L}} \quad \frac{1.01\ \text{g}}{1\ \text{mL}} \quad \frac{3.4\ \text{g Ag}}{100\ \text{g soln}}$$

Solution: $4.8\ \text{L} \times \frac{1000\ \text{mL}}{1\ \text{L}} \times \frac{1.01\ \text{g}}{1\ \text{mL}} = 4848\ \text{g soln}$ then

$4848\ \text{g soln} \times \frac{3.4\ \text{g Ag}}{100\ \text{g soln}} = 160\ \text{g Ag} = 1.6 \times 10^2\ \text{g Ag}$.

Check: The units (g) are correct. The magnitude of the answer (160 g) seems reasonable since we have almost 5000 g solution.

57. **Given:** Ca^{2+} and water; 0.0085 % Ca^{2+} by mass, 1.2 g Ca **Find:** m (water)
Conceptual plan: g_{Ca} → g_{soln} → g_{H2O}

$$\frac{100\ \text{g soln}}{0.0085\ \text{g Ca}} \qquad g_{H2O} = g_{soln} - g_{Ca}$$

Solution: $1.2\ \text{g Ca} \times \frac{100\ \text{g soln}}{0.0085\ \text{g Ca}} = 14118\ \text{g soln}$ then

$g_{H2O} = g_{soln} - g_{Ca} = 14118\ \text{g} - 1.2\ \text{g} = 1.4 \times 10^4\ \text{g water}$.

Check: The units (g) are correct. The magnitude of the answer (10^4 g) seems reasonable since we have such a low concentration of Ca.

59. **Given:** concentrated HNO_3: 70.3 % HNO_3 by mass, d = 1.41 g/mL; final solution: 1.15 L of 0.100 M HNO_3
Find: describe final solution preparation
Conceptual plan: M_2, V_2 → mol_{HNO3} → g_{HNO3} → $g_{conc\ acid}$ → $mL_{conc\ acid}$ then describe method

$$mol = M\ V \qquad \frac{63.02\ \text{g}\ HNO_3}{1\ \text{mol}\ HNO_3} \quad \frac{100\ \text{g conc acid}}{70.3\ \text{g}\ HNO_3} \quad \frac{1\ \text{mL}}{1.41\ \text{g}}$$

Solution: then

$0.115\ \text{mol}\ HNO_3 \times \frac{63.02\ \text{g}\ HNO_3}{1\ \text{mol}\ HNO_3} \times \frac{100\ \text{g conc acid}}{70.3\ \text{g}\ HNO_3} \times \frac{1\ \text{mL conc acid}}{1.41\ \text{g conc acid}} = 7.31\ \text{mL conc acid}$.

Prepare the solution by putting 1.00 L of distilled water in a container. Carefully pour in the 7.31 mL of the concentrated acid, mix the solution and allow it to cool. Finally add enough water to generate a total volume of solution (1.15 L). It is important to add acid to water, and not the reverse, since there is such a large amount of heat released upon mixing.
Check: The units (mL) are correct. The magnitude of the answer (7 g) seems reasonable since we are starting with such a very concentrated solution and diluting it to a low concentration.

61. a) **Given:** 1.00×10^2 mL of 0.500 M KCl **Find:** describe final solution preparation

Conceptual plan: mL → L then M, V → mol$_{KCl}$ → g$_{KCl}$ then describe method

$$\frac{1\,L}{1000\ mL} \qquad mol = M\ V \qquad \frac{74.56\ g\ KCl}{1\ mol\ KCl}$$

Solution: $1.00 \times 10^2\ \cancel{mL} \times \frac{1\,L}{1000\ \cancel{mL}} = 0.100\ L$

$mol = M\ V = 0.500\ \frac{mol\ KCl}{1\ \cancel{L\ soln}} \times 0.100\ \cancel{L\ soln} = 0.0500\ mol\ KCl$ then

$0.0500\ \cancel{mol\ KCl} \times \frac{74.56\,g\ KCl}{1\ \cancel{mol\ KCl}} = 3.73\ g\ KCl$. Prepare the solution carefully adding 3.73 g KCl in a 100-mL volumetric flask. Add ~ 75 mL of distilled water and agitate the solution until the salt dissolves completely. Finally add enough water to generate a total volume of solution (add water to the mark on the flask).

Check: The units (g) are correct. The magnitude of the answer (7 g) seems reasonable since we are making a small volume of solution and the formula weight of KCl is ~ 75 g/mol.

b) **Given:** 1.00×10^2 g of 0.500 *m* KCl **Find:** describe final solution preparation

Conceptual plan: *m*→mol$_{KCl}$ /1 kg solvent→g$_{KCl}$/1 kg solvent then

$$m = \frac{\text{amount solute (moles)}}{\text{mass solvent (kg)}} \qquad \frac{74.56\ g\ KCl}{1\ mol\ KCl}$$

g$_{KCl}$/1 kg solvent, g$_{soln}$→g$_{KCl}$, g$_{H2O}$ then describe method

$$g_{soln} = g_{KCl} + g_{H2O}$$

Solution: $m = \frac{\text{amount solute (moles)}}{\text{mass solvent (kg)}}$ so $0.500\ m = \frac{0.500\ mol\ KCl}{1\ kg\ H_2O}$ so

$\frac{0.500\ \cancel{mol\ KCl}}{1\ kg\,H_2O} \times \frac{74.56\,g\ KCl}{1\ \cancel{mol\ KCl}} = \frac{37.\underline{2}8\ g\ KCl}{1000\ g\,H_2O}$ $\quad g_{soln} = g_{KCl} + g_{H2O}$ $\quad$ so $\quad g_{soln} - g_{KCl} = g_{H2O}$

substitute into ratio $\frac{0.037\underline{2}8\ g\ KCl}{1\ g\,H_2O} = \frac{x\ g\ KCl}{100\ g\ soln - x\ g\ KCl}$ Rearrange and solve for x g KCl

$0.037\underline{2}8\,(100\ g\ soln - x\ g\ KCl) = x\ g\ KCl \rightarrow 3.7\underline{2}8 - 0.037\underline{2}8\,(x\ g\ KCl) = x\ g\ KCl \rightarrow$

$3.7\underline{2}8 = 1.037\underline{2}8\,(x\ g\ KCl) \rightarrow \frac{3.7\underline{2}8}{1.037\underline{2}8} = x\ g\ KCl = 3.59\ g\ KCl$ then

$g_{H2O} = g_{soln} - g_{KCl} = 100.\ g - 3.59\ g = 96.41\ g\ H_2O$. Prepare the solution carefully adding 3.59 g KCl to a container with 96.41 g of distilled water and agitate the solution until the salt dissolves completely.

Check: The units (g) are correct. The magnitude of the answer (3.6 g) seems reasonable since we are making a small volume of solution and the formula weight of KCl is ~ 75 g/mol.

c) **Given:** 1.00×10^2 g of 5.0 % KCl by mass **Find:** describe final solution preparation

Conceptual plan: g$_{soln}$ → g$_{KCl}$ then g$_{KCl}$, g$_{soln}$ → g$_{H2O}$ then describe method

$$\frac{5.0\ g\ KCl}{100\ g\ soln} \qquad g_{soln} = g_{KCl} + g_{H2O}$$

Solution: $1.00 \times 10^2\ g\ \cancel{soln} \times \frac{5.0\,g\ KCl}{100\ g\ \cancel{soln}} = 5.0\ g\ KCl$ then $g_{soln} = g_{KCl} + g_{H2O}$. So

$g_{H2O} = g_{soln} - g_{KCl} = 100.\ g - 5.0\ g = 95\ g\ H_2O$ Prepare the solution carefully adding 5.0 g KCl to a container with 95 g of distilled water and agitate the solution until the salt dissolves completely.

Check: The units (g) are correct. The magnitude of the answer (5 g) seems reasonable since we are making a small volume of solution and the solution is 5 % by mass KCl.

63. a) **Given:** 28.4 g of glucose ($C_6H_{12}O_6$) in 355 g water; final volume = 378 mL **Find:** molarity
Conceptual plan: **mL → L and $g_{C6H12O6}$ → $mol_{C6H12O6}$ then $mol_{C6H12O6}$, V → M**

$$\frac{1\ \text{L}}{1000\ \text{mL}} \qquad \frac{1\ \text{mol}\ C_6H_{12}O_6}{180.16\ \text{g}\ C_6H_{12}O_6} \qquad M = \frac{\text{amount solute (moles)}}{\text{volume solution (L)}}$$

Solution: $378\ \text{mL} \times \frac{1\ \text{L}}{1000\ \text{mL}} = 0.378\ \text{L}$ and

$$28.4\ \text{g}\ C_6H_{12}O_6 \times \frac{1\ \text{mol}\ C_6H_{12}O_6}{180.16\ \text{g}\ C_6H_{12}O_6} = 0.15\underline{7}638\ \text{mol}\ C_6H_{12}O_6$$

$$M = \frac{\text{amount solute (moles)}}{\text{volume solution (L)}} = \frac{0.15\underline{7}638\ \text{mol}\ C_6H_{12}O_6}{0.378\ \text{L}} = 0.417\ \text{M}$$

Check: The units (M) are correct. The magnitude of the answer (0.4 M) seems reasonable since we have 1/8 mole in about 1/3 L.

b) **Given:** 28.4 g of glucose ($C_6H_{12}O_6$) in 355 g water; final volume = 378 mL **Find:** molality
Conceptual plan: **g_{H2O} → kg_{H2O} and $g_{C6H12O6}$ → $mol_{C6H12O6}$ then $mol_{C6H12O6}$, kg_{H2O} → *m***

$$\frac{1\ \text{kg}}{1000\ \text{g}} \qquad \frac{1\ \text{mol}\ C_6H_{12}O_6}{180.16\ \text{g}\ C_6H_{12}O_6} \qquad m = \frac{\text{amount solute (moles)}}{\text{mass solvent (kg)}}$$

Solution: $355\ \text{g} \times \frac{1\ \text{kg}}{1000\ \text{g}} = 0.355\ \text{kg}$ and

$$28.4\ \text{g}\ C_6H_{12}O_6 \times \frac{1\ \text{mol}\ C_6H_{12}O_6}{180.16\ \text{g}\ C_6H_{12}O_6} = 0.15\underline{7}638\ \text{mol}\ C_6H_{12}O_6$$

$$m = \frac{\text{amount solute (moles)}}{\text{mass solvent (kg)}} = \frac{0.15\underline{7}638\ \text{mol}\ C_6H_{12}O_6}{0.355\ \text{kg}} = 0.444\ m$$

Check: The units (*m*) are correct. The magnitude of the answer (0.4 *m*) seems reasonable since we have 1/8 mole in about 1/3 kg.

c) **Given:** 28.4 g of glucose ($C_6H_{12}O_6$) in 355 g water; final volume = 378 mL **Find:** percent by mass
Conceptual plan: **$g_{C6H12O6}$, g_{H2O} → g_{soln} then $g_{C6H12O6}$, g_{soln} → percent by mass**

$$g_{soln} = g_{C6H12O6} + g_{H2O} \qquad \text{mass percent} = \frac{\text{mass solute}}{\text{mass solution}} \times 100\ \%$$

Solution: $g_{soln} = g_{C6H12O6} + g_{H2O} = 28.4\ \text{g} + 355\ \text{g} = 38\underline{3}.4\ \text{g soln}$ then

$$\text{mass percent} = \frac{\text{mass solute}}{\text{mass solution}} \times 100\ \% = \frac{28.4\ \text{g}\ C_6H_{12}O_6}{38\underline{3}.4\ \text{g soln}} \times 100\ \% = 7.41\ \text{percent by mass}$$

Check: The units (percent by mass) are correct. The magnitude of the answer (7 %) seems reasonable since we are dissolving 28 g in 355 g.

d) **Given:** 28.4 g of glucose ($C_6H_{12}O_6$) in 355 g water; final volume = 378 mL **Find:** mole fraction
Conceptual plan: **$g_{C6H12O6}$ → $mol_{C6H12O6}$ and g_{H2O} → mol_{H2O} then**

$$\frac{1\ \text{mol}\ C_6H_{12}O_6}{180.16\ \text{g}\ C_6H_{12}O_6} \qquad \frac{1\ \text{mol}\ H_2O}{18.02\ \text{g}\ H_2O}$$

$mol_{C6H12O6}$, mol_{H2O} → $\chi_{C6H12O6}$

$$\chi = \frac{\text{amount solute (in moles)}}{\text{total amount of solute and solvent (in moles)}}$$

Solution: $28.4\ \cancel{g\,C_6H_{12}O_6} \times \dfrac{1\ mol\,C_6H_{12}O_6}{180.16\ \cancel{g\,C_6H_{12}O_6}} = 0.15\underline{7}638\ mol\,C_6H_{12}O_6$ and

$355\ \cancel{g\,H_2O} \times \dfrac{1\ mol\,H_2O}{18.02\ \cancel{g\,H_2O}} = 19.\underline{7}003\ mol\,H_2O$ then

$$\chi = \frac{\text{amount solute (in moles)}}{\text{total amount of solute and solvent (in moles)}} = \frac{0.15\underline{7}638\ \cancel{mol}}{0.15\underline{7}638\ \cancel{mol} + 19.\underline{7}003\ \cancel{mol}} = 0.00794$$

Check: The units (none) are correct. The magnitude of the answer (0.008) seems reasonable since we have many more grams of water and water has a much lower molecular weight.

e) **Given:** 28.4 g of glucose ($C_6H_{12}O_6$) in 355 g water; final volume = 378 mL **Find:** mole percent
Conceptual plan: use answer from part d) then $\chi_{C6H12O6}$ **→ mole percent**
χ x 100 %

Solution: *mole percent* = χ x 100 % = 0.00794 x 100 % = 0.794 mole percent

Check: The units (%) are correct. The magnitude of the answer (0.8) seems reasonable since we have many more grams of water and water has a much lower molecular weight and we are just increasing the answer from part d) by a factor of 100.

65. The level has decreased more in the beaker filled with pure water. The dissolved salt in the seawater decreases the vapor pressure and subsequently lowers the rate of vaporization.

67. **Given:** 28.5 g of glycerin ($C_3H_8O_3$) in 125 mL water at 30 °C; $P°_{H2O}$ = 31.8 torr **Find:** P_{H2O}
Other: d (H_2O) = 1.00 g/mL; glycerin is a nonionic solid
Conceptual plan: g_{C3H8O3} **→** mol_{C3H8O3} **and** mL_{H2O} **→** g_{H2O} **→** mol_{H2O} **then** mol_{C3H8O3}, mol_{H2O} **→** χ_{H2O}

$\dfrac{1\ mol\ C_3H_8O_3}{92.09\ g\ C_3H_8O_3}$ $\dfrac{1.00\ g}{1\ mL}$ $\dfrac{1\ mol\ H_2O}{18.01\ g\ H_2O}$ $\chi = \dfrac{\text{amount solvent (in moles)}}{\text{total amount of solute and solvent (in moles)}}$

then $\chi_{H2O}, P°_{H2O}$ **→** P_{H2O}
$P_{solution} = \chi_{solvent}\, P^o_{solvent}$

Solution: $28.5\ \cancel{g\,C_3H_8O_3} \times \dfrac{1\ mol\,C_3H_8O_3}{92.09\ \cancel{g\,C_3H_8O_3}} = 0.30\underline{9}466\ mol\,C_3H_8O_3$ and

$125\ \cancel{mL} \times \dfrac{1.00\ \cancel{g}}{1\ \cancel{mL}} \times \dfrac{1\ mol\,H_2O}{18.01\ \cancel{g\,H_2O}} = 6.9\underline{4}059\ mol\,H_2O$ then

$$\chi = \frac{\text{amount solvent (in moles)}}{\text{total amount of solute and solvent (in moles)}} = \frac{6.9\underline{4}059\ \cancel{mol}}{0.30\underline{9}466\ \cancel{mol} + 6.9\underline{4}059\ \cancel{mol}} = 0.95\underline{7}316$$ then

$P_{solution} = \chi_{solvent}\, P^o_{solvent} = 0.95\underline{7}316 \times 31.8\ torr = 30.4\ torr$

Check: The units (torr) are correct. The magnitude of the answer (30 torr) seems reasonable since it is a drop from the pure vapor pressure. Very few moles of glycerin are added, so the pressure will not drop much.

69. **Given:** 5.50 % NaCl by mass in water at 25 °C; **Find:** P_{H2O} **Other:** $P°_{H2O}$ = 23.78 torr, i_{NaCl} = 1.9
Conceptual plan: % NaCl by mass → g_{NaCl}, g_{H2O} **then** g_{NaCl} **→** mol_{NaCl} **and**

$\dfrac{5.50\ g\ NaCl}{100\ g\ (NaCl + H_2O)}$ $\dfrac{1\ mol\ NaCl}{58.44\ g\ NaCl}$

g_{H2O} **→** mol_{H2O} **then** mol_{NaCl}, mol_{H2O} **→** χ_{H2O} **then** $\chi_{H2O}, P°_{H2O}$ **→** P_{H2O}

$\dfrac{1\ mol\,H_2O}{18.01\ g\ H_2O}$ $\chi = \dfrac{\text{amount sovent (in moles)}}{\text{total amount of solute and solvent (in moles)}}$ $P_{solution} = \chi_{solvent}\, P^o_{solvent}$

Solution: $\dfrac{5.50 \text{ g NaCl}}{100 \text{ g } (\text{NaCl} + H_2O)}$ means 5.50 g NaCl and (100 g - 5.50 g) = 94.5 g H_2O then

$$5.50 \text{ g NaCl} \times \frac{1 \text{ mol NaCl}}{58.44 \text{ g NaCl}} = 0.094\underline{1}136 \text{ mol NaCl} \text{ and}$$

$$94.5 \text{ g } H_2O \times \frac{1 \text{ mol } H_2O}{18.01 \text{ g } H_2O} = 5.2\underline{4}708 \text{ mol } H_2O \quad \text{then}$$

$$\chi_{solv} = \frac{\text{amount solvent (in moles)}}{\text{total amount solute and solvent (in moles)}} = \frac{5.2\underline{4}708 \text{ mol}}{5.2\underline{4}708 \text{ mol} + 1.9(0.094\underline{1}136 \text{ mol})} = 0.96\underline{7}04 \text{ then}$$

$$P_{soln} = \chi_{solv} P^{\circ}_{solv} = 0.96\underline{7}04 \times 23.78 \text{ torr} = 23.0 \text{ torr}$$

Check: The units (torr) are correct. The magnitude of the answer (23 torr) seems reasonable since it is a drop from the pure vapor pressure. Only a fraction of a mole of NaCl is added, so the pressure will not drop much.

71. **Given:** 50.0 g of heptane (C_7H_{16}) + 50.0 g of octane (C_8H_{18}) at 25°C; P°_{C7H16} = 45.8 torr; P°_{C8H18} = 10.9 torr

a) **Find:** P_{C7H16}, P_{C8H18}

Conceptual plan: g_{C7H16} → mol_{C7H16} and g_{C8H18} → mol_{C8H18} then

$$\frac{1 \text{ mol } C_7H_{16}}{100.20 \text{ g } C_7H_{16}} \qquad \frac{1 \text{ mol } C_8H_{18}}{114.22 \text{ g } C_8H_{18}}$$

mol_{C7H16}, mol_{C8H18} → χ_{C7H16}, χ_{C8H18} then

$$\chi_{C7H16} = \frac{\text{amount } C_7H_{16} \text{ (in moles)}}{\text{total amount (in moles)}} \qquad \chi_{C8H18} = 1 - \chi_{C7H16}$$

χ_{C7H16}, P°_{C7H16} → P_{C7H16} and χ_{C8H18}, P°_{C8H18} → P_{C8H18}

$$P_{C7H16} = \chi_{C7H16} P^{o}_{C7H16} \qquad P_{C8H18} = \chi_{C8H18} P^{o}_{C8H18}$$

Solution: $50.0 \text{ g } C_7H_{16} \times \dfrac{1 \text{ mol } C_7H_{16}}{100.20 \text{ g } C_7H_{16}} = 0.49\underline{9}002 \text{ mol } C_7H_{16}$ and

$$50.0 \text{ g } C_8H_{18} \times \frac{1 \text{ mol } C_8H_{18}}{114.22 \text{ g } C_8H_{18}} = 0.43\underline{7}752 \text{ mol } C_8H_{18} \text{ then}$$

$$\chi_{C7H16} = \frac{\text{amount } C_7H_{16} \text{ (in moles)}}{\text{total amount (in moles)}} = \frac{0.49\underline{9}002 \text{ mol}}{0.49\underline{9}002 \text{ mol} + 0.43\underline{7}752 \text{ mol}} = 0.53\underline{2}693 \text{ and}$$

$$\chi_{C8H18} = 1 - \chi_{C7H16} = 1 - 0.53\underline{2}693 = 0.46\underline{7}307 \text{ then}$$

$$P_{C7H16} = \chi_{C7H16} P^{o}_{C7H16} = 0.53\underline{2}693 \text{ x } 45.8 \text{ torr} = 24.4 \text{ torr} \quad \text{and}$$

$$P_{C8H18} = \chi_{C8H18} P^{o}_{C8H18} = 0.46\underline{7}307 \text{ x } 10.9 \text{ torr} = 5.09 \text{ torr}$$

Check: The units (torr) are correct. The magnitude of the answer (24 and 5 torr) seems reasonable since it we expect a drop in half from the pure vapor pressures since we have roughly a 50:50 mole ratio of the two components.

b) **Find:** P_{Total}

Conceptual plan: P_{C7H16}, P_{C8H18} → P_{Total}

$$P_{Total} = P_{C7H16} + P_{C8H18}$$

Solution: $P_{Total} = P_{C7H16} + P_{C8H18}$ = 24.4 torr + 5.09 torr = 29.5 torr

Check: The units (torr) are correct. The magnitude of the answer (30 torr) seems reasonable considering the two pressures.

c) **Find:** mass percent composition of the gas phase

Conceptual plan: since n α P and we are calculating a mass percent, which is a ratio of masses, we can simply convert 1 torr to 1 mole so P_{C7H16}, P_{C8H18} → n_{C7H16}, n_{C8H18} then mol_{C7H16} → g_{C7H16} and mol_{C8H18} → g_{C8H18}

$$\frac{100.20 \text{ g } C_7H_{16}}{1 \text{ mol } C_7H_{16}} \qquad \frac{114.22 \text{ g } C_8H_{18}}{1 \text{ mol } C_8H_{18}}$$

then g_{C7H16}, g_{C8H18} → **mass percents**

$$mass\ percent = \frac{mass\ solute}{mass\ solution} \times 100\ \%$$

Solution: so $n_{C7H16} = 24.4$ mol and $n_{C8H18} = 5.09$ mol then

$$24.4\ \cancel{mol\ C_7H_{16}} \times \frac{100.20\ g\ C_7H_{16}}{1\ \cancel{mol\ C_7H_{16}}} = 2444.88\ g\ C_7H_{16}\ \text{and}$$

$$5.09\ \cancel{mol\ C_8H_{18}} \times \frac{114.22\ g\ C_8H_{18}}{1\ \cancel{mol\ C_8H_{18}}} = 581.380\ g\ C_8H_{18}\ \text{then}$$

$$mass\ percent = \frac{mass\ solute}{mass\ solution} \times 100\ \% = \frac{2444.88\ \cancel{g\ C_7H_{16}}}{2444.88\ \cancel{g\ C_7H_{16}} + 581.380\ \cancel{g\ C_8H_{18}}} \times 100\ \% = 80.8\ \text{percent by mass}\ C_7H_{16}$$

then 100 % – 80.8 % = 19.2 percent by mass C_8H_{18}

Check: The units (%) are correct. The magnitudes of the answers (81 % and 19 %) seem reasonable considering the two pressures.

d) The two mass percents are different because the vapor is richer in the more volatile component (the lighter molecule).

73. **Given:** 4.08 g of chloroform ($CHCl_3$) and 9.29 g of acetone (CH_3COCH_3); at 35 °C $P°_{CHCl3} = 295$ torr; $P°_{CH3COCH3} = 332$ torr; assume ideal behavior; $P_{Total\ measured} = 312$ torr
Find: P_{CHCl3}, $P_{CH3COCH3}$ and P_{Total} then is soln ideal?
Conceptual plan: g_{CHCl3} → **mol** $_{CHCl3}$ **and** $g_{CH3COCH3}$ → **mol** $_{CH3COCH3}$ **then**

$$\frac{1\ mol\ CHCl_3}{119.38\ g\ CHCl_3} \qquad \frac{1\ mol\ CH_3COCH_3}{58.08\ g\ CH_3COCH_3}$$

mol_{CHCl3}, $mol_{CH3COCH3}$ → χ_{CHCl3}, $\chi_{CH3COCH3}$ **then** χ_{CHCl3}, $P°_{CHCl3}$ → P_{CHCl3} **and**

$$\chi_{CHCl3} = \frac{\text{amount } CHCl_3 \text{ (in moles)}}{\text{total amount (in moles)}} \qquad \chi_{CH3COCH3} = 1 - \chi_{CHCl3} \qquad P_{CHCl3} = \chi_{CHCl3}\ P^o_{CHCl3}$$

$\chi_{CH3COCH3}$, $P°_{CH3COCH3}$ → $P_{CH3COCH3}$ **then** P_{CHCl3}, $P_{CH3COCH3}$ → P_{Total} **then compare values**

$$P_{CH3COCH3} = \chi_{CH3COCH3}\ P^o_{CH3COCH3} \qquad P_{Total} = P_{CHCl3} + P_{CH3COCH3}$$

Solution: $$4.08\ \cancel{g\ CHCl_3} \times \frac{1\ mol\ CHCl_3}{119.38\ \cancel{g\ CHCl_3}} = 0.0341766\ mol\ CHCl_3\ \text{and}$$

$$9.29\ \cancel{g\ CH_3COCH_3} \times \frac{1\ mol\ CH_3COCH_3}{58.08\ \cancel{g\ CH_3COCH_3}} = 0.159952\ mol\ CH_3COCH_3\ \text{then}$$

$$\chi_{CHCl3} = \frac{\text{amount } CHCl_3 \text{ (in moles)}}{\text{total amount (in moles)}} = \frac{0.0341766\ \cancel{mol}}{0.0341766\ \cancel{mol} + 0.159952\ \cancel{mol}} = 0.176052\ \text{and}$$

$\chi_{CH3COCH3} = 1 - \chi_{CHCl3} = 1 - 0.176052 = 0.823948$ then

$P_{CHCl3} = \chi_{CHCl3}\ P^o_{CHCl3} = 0.176052 \times 295\ torr = 51.9\ torr$ and

$P_{CH3COCH3} = \chi_{CH3COCH3}\ P^o_{CH3COCH3} = 0.823948 \times 332\ torr = 274\ torr$ then

$P_{Total} = P_{CHCl3} + P_{CH3COCH3} = 51.9\ torr + 274\ torr = 326\ torr$. Since 326 torr ≠ 312 torr, the solution is not behaving ideally. The chloroform-acetone interactions are stronger than the chloroform-chloroform and acetone-acetone interactions.

Check: The units (torr) are correct. The magnitude of the answer seems reasonable since each is a fraction of the pure vapor pressures. We are not surprised that the solution is not ideal, since the typed of bonds in the two molecules are very different.

75. **Given:** 55.8 g of glucose ($C_6H_{12}O_6$) in 455g water **Find:** T_f and T_b **Other:** $K_f = 1.86 \ °C/m$; $K_b = 0.512 \ °C/m$;

Conceptual plan: $g_{H2O} \rightarrow kg_{H2O}$ and $g_{C6H12O6} \rightarrow mol_{C6H12O6}$ then $mol_{C6H12O6}, kg_{H2O} \rightarrow m$

$$\frac{1 \ kg}{1000 \ g} \qquad \frac{1 \ mol \ C_6H_{12}O_6}{180.16 \ g \ C_6H_{12}O_6} \qquad m = \frac{\text{amount solute (moles)}}{\text{mass solvent (kg)}}$$

$m, K_f \rightarrow \Delta T_f \rightarrow T_f$ and $m, K_b \rightarrow \Delta T_b \rightarrow T_b$

$$\Delta T_f = K_f \ m \quad T_f = T_f^o - \Delta T_f \qquad \Delta T_b = K_b \ m \quad \Delta T_b = T_b - T_b^o$$

Solution: $455 \ \cancel{g} \times \dfrac{1 \ kg}{1000 \ \cancel{g}} = 0.455 \ kg$ and

$55.8 \ \cancel{g \ C_6H_{12}O_6} \times \dfrac{1 \ mol \ C_6H_{12}O_6}{180.16 \ \cancel{g \ C_6H_{12}O_6}} = 0.30\underline{9}725 \ mol \ C_6H_{12}O_6$ then

$m = \dfrac{\text{amount solute (moles)}}{\text{mass solvent (kg)}} = \dfrac{0.30\underline{9}725 \ mol \ C_6H_{12}O_6}{0.455 \ kg} = 0.68\underline{0}714 \ m$ then $\Delta T_f = K_f \ m = 1.86 \ \dfrac{°C}{\cancel{m}} \times 0.68\underline{0}714 \ \cancel{m} = 1.27 \ °C$

then $T_f = T_f^o - \Delta T_f = 0.00 \ °C - 1.27 \ °C = -1.27 \ °C$ and

$\Delta T_b = K_b \ m = 0.512 \ \dfrac{°C}{\cancel{m}} \times 0.68\underline{0}714 \ \cancel{m} = 0.349 \ °C$ and $\Delta T_b = T_b - T_b^o$ so

$T_b = T_b^o + \Delta T_b = 100.000 \ °C + 0.349 \ °C = 100.349 \ °C$

Check: The units (°C) are correct. The magnitudes of the answers seem reasonable since the molality is ~ 2/3. The shift in boiling point less than the shift in freezing point because the constant is smaller for freezing.

77. **Given:** 17.5 g of unknown nonelectrolyte in 100.0 g water, $T_f = -1.8 \ °C$ **Find:** $\mathfrak{M}$
Other: $K_f = 1.86 \ °C/m$
Conceptual plan: $g_{H2O} \rightarrow kg_{H2O}$ and $T_f \rightarrow \Delta T_f$ then $\Delta T_f, K_f \rightarrow m$ then $m, kg_{H2O} \rightarrow mol_{Unk}$

$$\frac{1 \ kg}{1000 \ g} \qquad T_f = T_f^o - \Delta T_f \qquad \Delta T_f = K_f \ m \qquad m = \frac{\text{amount solute (moles)}}{\text{mass solvent (kg)}}$$

then $g_{Unk}, mol_{Unk} \rightarrow \mathfrak{M}$

$$\mathfrak{M} = \frac{g_{Unk}}{mol_{Unk}}$$

Solution: $100.0 \ \cancel{g} \times \dfrac{1 \ kg}{1000 \ \cancel{g}} = 0.1000 \ kg$ and $T_f = T_f^o - \Delta T_f$ so

$\Delta T_f = T_f^o - T_f = 0.00 \ °C - -1.8 \ °C = +1.8 \ °C$

$\Delta T_f = K_f \ m$ Rearrange to solve for m. $m = \dfrac{\Delta T_f}{K_f} = \dfrac{1.8 \ \cancel{°C}}{1.86 \ \dfrac{\cancel{°C}}{m}} = 0.9\underline{6}774 \ m$ then $m = \dfrac{\text{amount solute (moles)}}{\text{mass solvent (kg)}}$

so $mol_{Unk} = m_{Unk} \times kg_{H2O} = 0.9\underline{6}774 \ \dfrac{mol \ Unk}{\cancel{kg}} \times 0.1000 \ \cancel{kg} = 0.09\underline{6}774 \ mol \ Unk$ then

$\mathfrak{M} = \dfrac{g_{Unk}}{mol_{Unk}} = \dfrac{17.5 \ g}{0.09\underline{6}774 \ mol} = 180 \ \dfrac{g}{mol} = 1.8 \times 10^2 \ \dfrac{g}{mol}$

Check: The units (g/mol) are correct. The magnitude of the answer (180 g/mol) seems reasonable since the molality is ~ 0.1 and we have ~18 g. It is a reasonable molecular weight for a solid or liquid.

79. **Given:** 24.6 g of glycerin ($C_3H_8O_3$) in 250.0 mL of solution at 298 K **Find:** Π

Conceptual plan: $\mathbf{mL_{soln} \rightarrow L_{soln}}$ **and** $\mathbf{g_{C3H8O3} \rightarrow mol_{C3H8O3}}$ **then** $\mathbf{mol_{C3H8O3}, L_{soln} \rightarrow M}$ **then**

$$\frac{1\text{ L}}{1000\text{ mL}} \qquad \frac{1\text{ mol } C_3H_8O_3}{92.09\text{ g } C_3H_8O_3} \qquad M = \frac{\text{amount solute (moles)}}{\text{volume solution (L)}}$$

$\mathbf{M, T \rightarrow \Pi}$

$$\Pi = M\,RT$$

Solution: $250.0\text{ mL} \times \dfrac{1\text{L}}{1000\text{ mL}} = 0.2500\text{ L}$ and

$$24.6\text{ g } C_3H_8O_3 \times \frac{1\text{ mol } C_3H_8O_3}{92.09\text{ g } C_3H_8O_3} = 0.26\underline{7}130\text{ mol } C_3H_8O_3 \text{ then}$$

$$M = \frac{\text{amount solute (moles)}}{\text{volume solution (L)}} = \frac{0.26\underline{7}130\text{ mol } C_3H_8O_3}{0.2500\text{ L}} = 1.0\underline{6}852\text{ M then}$$

$$\Pi = M\,RT = 1.0\underline{6}852\,\frac{\text{mol}}{\text{L}} \times 0.08206\,\frac{\text{L atm}}{\text{K mol}} \times 298\text{ K} = 26.1\text{ atm}$$

Check: The units (atm) are correct. The magnitude of the answer (26 atm) seems reasonable since the molarity is ~ 1.

81. **Given:** 27.55 mg unknown protein in 25.0 mL solution; Π = 3.22 torr at 25 °C **Find:** $\mathfrak{M}_{\text{unknown protein}}$

Conceptual plan: **°C → K and torr → atm then** $\mathbf{\Pi, T \rightarrow M}$ **then** $\mathbf{mL_{soln} \rightarrow L_{soln}}$ **then**

$$K = °C + 273.15 \qquad \frac{1\text{ atm}}{760\text{ torr}} \qquad \Pi = M\,RT \qquad \frac{1\text{L}}{1000\text{ mL}}$$

$\mathbf{L_{soln}, M \rightarrow mol_{unknown\ protein}}$ **and mg → g then** $\mathbf{g_{unknown\ protein}, mol_{unknown\ protein} \rightarrow \mathfrak{M}_{unknown\ protein}}$

$$M = \frac{\text{amount solute (moles)}}{\text{volume solution (L)}} \qquad \frac{1\text{ g}}{1000\text{ mg}} \qquad \mathfrak{M} = \frac{g_{\text{unknown protein}}}{mol_{\text{unknown protein}}}$$

Solution: 25 °C + 273.15 = 298 K and $3.22\text{ torr} \times \dfrac{1\text{ atm}}{760\text{ torr}} = 0.0042\underline{3}684\text{ atm}$ $\Pi = M\,RT$ for M.

$$M = \frac{\Pi}{RT} = \frac{0.0042\underline{3}684\text{ atm}}{0.08206\,\frac{\text{L atm}}{\text{K mol}} \times 298\text{ K}} = 1.7\underline{3}258 \times 10^{-4}\,\frac{\text{mol}}{\text{L}} \text{ then } 25.0\text{ mL} \times \frac{1\text{L}}{1000\text{ mL}} = 0.0250\text{ L}$$

then $M = \dfrac{\text{amount solute (moles)}}{\text{volume solution (L)}}$ Rearrange to solve for $mol_{\text{unknown protein}}$.

$$mol_{\text{unknown protein}} = M \times L = 1.7\underline{3}258 \times 10^{-4}\,\frac{\text{mol}}{\text{L}} \times 0.0250\text{ L} = 4.3\underline{3}146 \times 10^{-4}\text{ mol and}$$

$$27.55\text{ mg} \times \frac{1\text{g}}{1000\text{ mg}} = 0.02755\text{ g then}$$

$$\mathfrak{M} = \frac{g_{\text{unknown protein}}}{mol_{\text{unknown protein}}} = \frac{0.02755\text{ g}}{4.3\underline{3}146 \times 10^{-4}\text{ mol}} = 6.36 \times 10^{3}\,\frac{\text{g}}{\text{mol}}$$

Check: The units (g/mol) are correct. The magnitude of the answer (6400 g/mol) seems reasonable for a large biological molecule. A small amount of material is put into 0.025 L, so the concentration is very small and the molecular weight is large.

83. a) **Given:** 0.100 m of K_2S, completely dissociated **Find:** T_f, T_b
Other: $K_f = 1.86\ °C/m$; $K_b = 0.512\ °C/m$;
Conceptual plan: $m, i, K_f \rightarrow \Delta T_f$ then $\Delta T_f \rightarrow T_f$ and $m, i, K_b \rightarrow \Delta T_b$ then $\Delta T_b \rightarrow T_b$

$$\Delta T_f = K_f\, i\, m \quad i = 3 \qquad T_f = T_f^o - \Delta T_f \qquad \Delta T_b = K_b\, i\, m \quad i = 3 \qquad T_b = T_b^o - \Delta T_b$$

Solution: $\Delta T_f = K_f\, i\, m = 1.86 \dfrac{°C}{\cancel{m}} \times 3 \times 0.100\ \cancel{m} = 0.558\ °C$ then

$T_f = T_f^o - \Delta T_f = 0.000\,°C - 0.558\ °C = -\,0.558\ °C$ and

$\Delta T_b = K_b\, i\, m = 0.512 \dfrac{°C}{\cancel{m}} \times 3 \times 0.100\ \cancel{m} = 0.154\ °C$ then

$T_b = T_b^o - \Delta T_b = 100.000\,°C + 0.154\ °C = 100.154\ °C$.

Check: The units (°C) are correct. The magnitude of the answer (- 0.6 °C and 100.2 °C) seems reasonable since the molality of the particles is 0.3. The shift in boiling point less than the shift in freezing point because the constant is smaller for freezing.

b) **Given:** 21.5 g $CuCl_2$ in 4.50 x 10^2 g water, completely dissociated **Find:** T_f, T_b
Other: $K_f = 1.86\ °C/m$; $K_b = 0.512\ °C/m$;
Conceptual plan: $g_{H2O} \rightarrow kg_{H2O}$ and $g_{CuCl2} \rightarrow mol_{CuCl2}$ then $mol_{CuCl2}, kg_{H2O} \rightarrow m$

$$\frac{1\,kg}{1000\ g} \qquad \frac{1\ mol\,CuCl_2}{134.46\ g\ CuCl_2} \qquad m = \frac{\text{amount solute (moles)}}{\text{mass solvent (kg)}}$$

$m, i, K_f \rightarrow \Delta T_f \rightarrow T_f$ and $m, i, K_b \rightarrow \Delta T_b \rightarrow T_b$

$$\Delta T_f = K_f\, i\, m \quad i = 3 \quad T_f = T_f^o - \Delta T_f \qquad \Delta T_b = K_b\, i\, m \quad i = 3 \quad T_b = T_b^o - \Delta T_b$$

Solution: $4.5 \times 10^2\ \cancel{g} \times \dfrac{1\,kg}{1000\ \cancel{g}} = 0.450\ kg$ and

$21.5\ \cancel{g\,CuCl_2} \times \dfrac{1\ mol\,CuCl_2}{134.46\ \cancel{g\ CuCl_2}} = 0.15\underline{9}904\ mol\,CuCl_2$ then

$m = \dfrac{\text{amount solute (moles)}}{\text{mass solvent (kg)}} = \dfrac{0.15\underline{9}904\ \ mol\,CuCl_2}{0.450\ \ kg} = 0.35\underline{5}341\ m$ then

$\Delta T_f = K_f\, i\, m = 1.86 \dfrac{°C}{\cancel{m}} \times 3 \times 0.35\underline{5}341\ \cancel{m} = 1.98\ °C$ then

$T_f = T_f^o - \Delta T_f = 0.000\,°C - 1.98\ °C = -\,1.98\ °C$ and

$\Delta T_b = K_b\, i\, m = 0.512 \dfrac{°C}{\cancel{m}} \times 3 \times 0.35\underline{5}341\ \cancel{m} = 0.546\ °C$ then

$T_b = T_b^o - \Delta T_b = 100.000\,°C + 0.546\ °C = 100.546\ °C$.

Check: The units (°C) are correct. The magnitude of the answer (- 2 °C and 100.5 °C) seems reasonable since the molality of the particles is ~ 1. The shift in boiling point less than the shift in freezing point because the constant is smaller for freezing.

c) **Given:** 5.5 % by mass $NaNO_3$, completely dissociated **Find:** T_f, T_b
Other: $K_f = 1.86$ °C/*m*; $K_b = 0.512$ °C/*m*;
Conceptual plan: percent by mass → g_{NaNO3}, g_{H2O} then g_{H2O} → kg_{H2O} and g_{NaNO3} → mol $_{NaNO3}$

$$mass\ percent = \frac{mass\ solute}{mass\ solution} \times 100\ \% \qquad \frac{1\,kg}{1000\ g} \qquad \frac{1\ mol\ NaNO_3}{84.99\ g\ NaNO_3}$$

then mol $_{NaNO3}$, kg_{H2O} → *m* then *m*, *i*, K_f → ΔT_f → T_f and *m*, *i*, K_b → ΔT_b → T_b

$$m = \frac{\text{amount solute (moles)}}{\text{mass solvent (kg)}} \qquad \Delta T_f = K_f\, i\, m \quad i = 2 \quad T_f = T_f^o - \Delta T_f \qquad \Delta T_b = K_b\, i\, m \quad i = 2 \quad T_b = T_b^o - \Delta T_b$$

Solution: $mass\ percent = \frac{mass\ solute}{mass\ solution} \times 100\ \%$ so 5.5 % by mass $NaNO_3$ means 5.5 g $NaNO_3$ and

100.0 g – 5.5 g = 94.5 g water. Then $94.5\ \cancel{g} \times \frac{1\,kg}{1000\ \cancel{g}} = 0.0945$ kg and

$$5.5\ \cancel{g\ NaNO_3} \times \frac{1\ mol\ NaNO_3}{84.99\ \cancel{g\ NaNO_3}} = 0.06\underline{4}713\ mol\ NaNO_3 \text{ then}$$

$$m = \frac{\text{amount solute (moles)}}{\text{mass solvent (kg)}} = \frac{0.06\underline{4}713\ mol\ NaNO_3}{0.0945\ kg} = 0.6\underline{8}480\ m \text{ then}$$

$$\Delta T_f = K_f\, i\, m = 1.86 \frac{°C}{\cancel{m}} \times 2 \times 0.6\underline{8}480\ \cancel{m} = 2.5\ °C \text{ then}$$

$$T_f = T_f^o - \Delta T_f = 0.000°C - 2.3\ °C = -2.5\ °C \text{ and}$$

$$\Delta T_b = K_b\, i\, m = 0.512 \frac{°C}{\cancel{m}} \times 2 \times 0.6\underline{8}480\ \cancel{m} = 0.70\ °C \text{ then}$$

$$T_b = T_b^o - \Delta T_b = 100.000°C + 0.64\ °C = 100.70\ °C.$$

Check: The units (°C) are correct. The magnitude of the answer (- 2 °C and 100.6 °C) seems reasonable since the molality of the particles is ~ 1. The shift in boiling point less than the shift in freezing point because the constant is smaller for freezing.

85. a) **Given:** 0.100 *m* of $FeCl_3$ **Find:** T_f **Other:** $K_f = 1.86$ °C/*m*; $i_{measured} = 3.4$
Conceptual plan: *m*, *i*, K_f → ΔT_f then ΔT_f → T_f

$$\Delta T_f = K_f\, i\, m \qquad T_f = T_f^o - \Delta T_f$$

Solution: $\Delta T_f = K_f\, i\, m = 1.86 \frac{°C}{\cancel{m}} \times 3.4 \times 0.100\ \cancel{m} = 0.632\ °C$ then

$$T_f = T_f^o - \Delta T_f = 0.000°C - 0.632\ °C = -0.632\ °C.$$

Check: The units (°C) are correct. The magnitude of the answer (- 0.6 °C) seems reasonable since the molality of the particles is 0.3.

b) **Given:** 0.085 M of K_2SO_4 at 298 K **Find:** Π **Other:** $i_{measured} = 2.6$
Conceptual plan: M, *i*, T → Π

$$\Pi = i\,M\,RT$$

Solution: $\Pi = i\,M\,RT = 2.6 \times 0.085 \frac{\cancel{mol}}{\cancel{L}} \times 0.08206 \frac{\cancel{L}\ atm}{\cancel{K}\ \cancel{mol}} \times 298\ \cancel{K} = 5.4\ atm$

Check: The units (atm) are correct. The magnitude of the answer (5 atm) seems reasonable since the molarity is ~ 0.2.

c) **Given:** 1.22 % by mass $MgCl_2$ **Find:** T_b **Other:** $K_b = 0.512$ °C/*m*; $i_{measured} = 2.7$
Conceptual plan: percent by mass → g_{MgCl2}, g_{H2O} then g_{H2O} → kg_{H2O} and g_{MgCl2} → mol $_{MgCl2}$

$$mass\ percent = \frac{mass\ solute}{mass\ solution} \times 100\ \% \qquad \frac{1\,kg}{1000\ g} \qquad \frac{1\ mol\ MgCl_2}{95.22\ g\ MgCl_2}$$

then mol $_{MgCl2}$, kg$_{H2O}$ → m then m, i, K_b → ΔT_b → T_b

$$m = \frac{\text{amount solute (moles)}}{\text{mass solvent (kg)}} \qquad \Delta T_b = K_b\, i\, m \quad T_b = T_b^{\circ} + \Delta T_b$$

Solution: $\text{mass percent} = \frac{\text{mass solute}}{\text{mass solution}} \times 100\,\%$ so 1.22 % by mass $MgCl_2$ means 1.22 g $MgCl_2$ and 100.00 g – 1.22 g = 98.78 g water. Then $98.78\ \text{g} \times \frac{1\,\text{kg}}{1000\ \text{g}} = 0.09878\ \text{kg}$ and

$$1.22\ \text{g}\ MgCl_2 \times \frac{1\ \text{mol}\ MgCl_2}{95.22\ \text{g}\ MgCl_2} = 0.0128124\ \text{mol}\ MgCl_2 \text{ then}$$

$$m = \frac{\text{amount solute (moles)}}{\text{mass solvent (kg)}} = \frac{0.0128124\ \text{mol}\ MgCl_2}{0.09878\ \text{kg}} = 0.129706\ m \text{ then}$$

$$\Delta T_b = K_b\, i\, m = 0.512\ \frac{^{\circ}\text{C}}{m} \times 2.7 \times 0.129706\ m = 0.18\ ^{\circ}\text{C then}$$

$T_b = T_b^{o} - \Delta T_b = 100.000\,^{\circ}\text{C} + 0.18\ ^{\circ}\text{C} = 100.18\ ^{\circ}\text{C}$.

Check: The units (°C) are correct. The magnitude of the answer (100.2 °C) seems reasonable since the molality of the particles is ~ 1/3.

87. **Given:** 0.100 M of ionic solution, Π = 8.3 atm at 25 °C **Find:** i_{measured}
Conceptual plan: °C → K then Π, M, T → i
$K = {}^{\circ}C + 273.15 \qquad \Pi = i\,M\,RT$

Solution: 25 °C + 273.15 = 298 K then $\Pi = i\,M\,RT$ Rearrange to solve for i.

$$i = \frac{\Pi}{M\,RT} = \frac{8.3\ \text{atm}}{0.100\ \frac{\text{mol}}{\text{L}} \times 0.08206\ \frac{\text{L atm}}{\text{K mol}} \times 298\ \text{K}} = 3.4$$

Check: The units (none) are correct. The magnitude of the answer (3) seems reasonable for an ionic solution with a high osmotic pressure.

89. Chloroform is polar and has stronger solute-solvent interactions than nonpolar carbon tetrachloride.

91. **Given:** $KClO_4$: Lattice Energy = – 599 kJ/mol, $\Delta H_{\text{hydration}}$ = – 548 kJ/mol; 10.0 g $KClO_4$ in 100.00 mL solution
Find: ΔH_{soln} and ΔT **Other:** C_s = 4.05 J/g °C; d = 1.05 g/mL
Conceptual plan: Lattice Energy, $\Delta H_{\text{hydration}}$ → ΔH_{soln} and g → mol then mol, ΔH_{soln} → q(kJ) → q(J)

$$\Delta H_{soln} = \Delta H_{solute} + \Delta H_{hydration} \text{ where } \Delta H_{solute} = -\Delta H_{lattice} \qquad \frac{1\ \text{mol}}{138.56\ \text{g}} \qquad q = n\,\Delta H_{soln} \qquad \frac{1000\ \text{J}}{1\ \text{kJ}}$$

then mL$_{\text{soln}}$ → g$_{\text{soln}}$ then q, g$_{\text{soln}}$, C_s → ΔT

$$\frac{1.05\ \text{g}}{1\ \text{mL}} \qquad q = m\, C_s\, \Delta T$$

Solution: $\Delta H_{soln} = \Delta H_{solute} + \Delta H_{hydration}$ where $\Delta H_{solute} = -\Delta H_{lattice}$ so $\Delta H_{soln} = \Delta H_{hydration} - \Delta H_{lattice}$

$\Delta H_{soln} = -548$ kJ/mol - (- 599 kJ/mol) = +51 kJ/mol and $10.0 \text{ g} \times \frac{1 \text{ mol}}{138.56 \text{ g}} = 0.072\underline{1}709$ mol then

$q = n\,\Delta H_{soln} = 0.072\underline{1}709 \text{ mol} \times 51 \frac{\text{kJ}}{\text{mol}} = +3.\underline{6}807 \text{ kJ} \times \frac{1000 \text{ J}}{1 \text{ kJ}} = +3\underline{6}80.7$ J absorbed then

$100.0 \text{ mL} \times \frac{1.05 \text{ g}}{1 \text{ mL}} = 105$ g since heat is absorbed in the when $KClO_4$ dissolves then the temperature will drop or q = $-3\underline{6}80.7$ J and $q = m\,C_s\,\Delta T$ Rearrange to solve for ΔT.

$$\Delta T = \frac{q}{m\,C_s} = \frac{-3\underline{6}80.7 \text{ J}}{105 \text{ g} \times 4.05 \frac{\text{J}}{\text{g}\ ^\circ\text{C}}} = -8.7\ ^\circ\text{C}$$

Check: The units (kJ/mol and °**C**) are correct. The magnitude of the answer (51 kJ/mol) makes physical sense because the lattice energy is larger than the heat of hydration. The magnitude of the temperature change (- 9 °**C**) makes physical sense since heat is absorbed and the heat of solution is fairly small.

93. **Given:** Argon, 0.0537 L; 25 °C, P_{Ar} = 1.0 atm to make 1.0 L saturated solution **Find:** k_H(Ar)
Conceptual plan: **°C → K and P_{Ar}, V, T → mol_{He} then mol_{He}, V_{soln}, P_{Ar} → k_H(He)**

$K = °C + 273.15$ $\quad PV = nRT \quad S_{Ar} = k_H(Ar)\,P_{Ar}$ with $S_{Ar} = \frac{mol_{Ar}}{L_{soln}}$

Solution: 25 °C + 273.15 = 298 K and $PV = nRT$ Rearrange to solve for n.

$$n = \frac{PV}{RT} = \frac{1.0 \text{ atm} \times 0.0537 \text{ L}}{0.08206 \frac{\text{L atm}}{\text{K mol}} \times 298 \text{ K}} = 0.00\underline{2}19597 \text{ mol}$$ then $S_{Ar} = k_H(Ar)\,P_{Ar}$ with $S_{Ar} = \frac{mol_{Ar}}{L_{soln}}$

Substitute in values and rearrange to solve for k_H. $k_H(Ar) = \frac{mol_{Ar}}{L_{soln} P_{He}} = \frac{0.00\underline{2}19597 \text{ mol}}{1.0 L_{soln} \times 1.0 \text{ atm}} = 2.2 \times 10^{-3} \frac{\text{M}}{\text{atm}}$

Check: The units (M/atm) are correct. The magnitude of the answer (10^{-3}) seems reasonable since it is consistent with other values in the text.

95. **Given:** 0.0020 ppm by mass Hg = legal limit; 0.0040 ppm by mass Hg = contaminated water; 50.0 mg Hg ingested
Find: volume of contaminated water
Conceptual plan: **mg_{Hg} → g_{Hg} → g_{H2O} → mL_{H2O} → L_{H2O}**

$\frac{1 \text{ g}}{1000 \text{ mg}} \quad \frac{10^6 \text{ g water}}{0.0040 \text{ g Hg}} \quad \frac{1 \text{ mL}}{1.00 \text{ g}} \quad \frac{1 \text{ L}}{1000 \text{ mL}}$

Solution:

$$50.0 \text{ mg Hg} \times \frac{1 \text{ g Hg}}{1000 \text{ mg Hg}} \times \frac{10^6 \text{ g water}}{0.0040 \text{ g Hg}} \times \frac{1 \text{ mL water}}{1.00 \text{ g water}} \times \frac{1 \text{ L water}}{1000 \text{ mL water}} = 1.3 \times 10^4 \text{ L water}.$$

Check: The units (L) are correct. The magnitude of the answer (10^4 L) seems reasonable since the concentration is so low.

97. **Given:** 12.5 % NaCl by mass in water at 55 °C; 2.5 L vapor **Find:** g_{H2O} in vapor
Other: $P°_{H2O}$ = 118 torr, i_{NaCl} = 1.9

Conceptual plan: **% NaCl by mass → g_{NaCl}, g_{H2O} then g_{NaCl} → mol_{NaCl} and**

$\frac{12.5 \text{ g NaCl}}{100 \text{ g (NaCl + } H_2O)} \quad \frac{1 \text{ mol NaCl}}{58.44 \text{ g NaCl}}$

g_{H2O} → mol_{H2O} then mol_{NaCl}, mol_{H2O} → χ_{NaCl} → χ_{H2O} then χ_{H2O}, $P°_{H2O}$ → P_{H2O}

$\frac{1 \text{ mol } H_2O}{18.02 \text{ g } H_2O} \quad \chi = \frac{\text{amount solute (in moles)}}{\text{total amount of solute and solvent (in moles)}} \quad \chi_{H2O} = 1 - i_{NaCl}\chi_{NaCl} \quad P_{solution} = \chi_{solvent}\,P^o_{solvent}$

then torr → atm and °C → K P, V, T → mol_{H2O} → g_{H2O}

$\frac{1\ \text{atm}}{760\ \text{torr}}$ $K = °C + 273.15$ $PV = nRT$ $\frac{18.02\ g\ H_2O}{1\ mol\ H_2O}$

Solution: $\frac{12.5\ g\ NaCl}{100\ g\ (NaCl + H_2O)}$ means 12.5 g NaCl and (100 g - 12.5 g) = 87.5 g H_2O then

$12.5\ g\ NaCl \times \frac{1\ mol\ NaCl}{58.44\ g\ NaCl} = 0.213895\ mol\ NaCl$ and

$87.5\ g\ H_2O \times \frac{1\ mol\ H_2O}{18.02\ g\ H_2O} = 4.85572\ mol\ H_2O$ then

$\chi = \frac{\text{amount solute (in moles)}}{\text{total amount of solute and solvent (in moles)}} = \frac{0.213895\ mol}{0.213895\ mol + 4.85572\ mol} = 0.0421916$ then

$\chi_{H2O} = 1 - i_{NaCl}\chi_{NaCl} = 1 - 1.9 \times 0.0421916 = 0.919836$ then

$P_{solution} = \chi_{solvent}\ P^o_{solvent} = 0.919836 \times 118\ torr = 108.541\ torr\ H_2O$ then

$108.541\ torr\ H_2O \times \frac{1\ atm}{760\ torr} = 0.142817\ atm$

and 55 °C + 273.15 = 328 K then $PV = nRT$ Rearrange to solve for n.

$n = \frac{PV}{RT} = \frac{0.142817\ atm \times 2.5\ L}{0.08206\ \frac{L\ atm}{K\ mol} \times 328\ K} = 0.013265\ mol$ then

$0.013265\ mol\ H_2O \times \frac{18.02\ g\ H_2O}{1\ mol\ H_2O} = 0.24\ g\ H_2O$

Check: The units (g) are correct. The magnitude of the answer (0.2 g) seems reasonable since there is very little mass in a vapor.

99. **Given:** T_b = 106.5 °C aqueous solution **Find:** T_f **Other:** K_f = 1.86 °C/*m*; K_b = 0.512 °C/*m*
Conceptual plan: $T_b \rightarrow \Delta T_b$ **then** $\Delta T_b, K_b \rightarrow m$ **then** $m, K_f \rightarrow \Delta T_f \rightarrow T_f$

$T_b = T_b^o + \Delta T_b$ $\Delta T_b = K_b m$ $\Delta T_f = K_f m$ $T_f = T_f^o - \Delta T_f$

Solution: $T_b = T_b^o + \Delta T_b$ so $\Delta T_b = T_b - T_b^o = 106.5°C - 100.0\ °C = 6.5\ °C$ then $\Delta T_b = K_b m$

Rearrange to solve for *m*. $m = \frac{\Delta T_b}{K_b} = \frac{6.5\ °C}{0.512\ \frac{°C}{m}} = 12.695\ m$ then

$\Delta T_f = K_f m = 1.86\ \frac{°C}{m} \times 12.695\ m = 23.6\ °C$ then $T_f = T_f^o - \Delta T_f = 0.000°C - 23.6\ °C\ °C = -24\ °C$.

Check: The units (°C) are correct. The magnitude of the answer (- 24 °C) seems reasonable since the shift in boiling point less than the shift in freezing point because the constant is smaller for freezing.

101. a) **Given:** 0.90 % NaCl by mass per volume; isotonic aqueous solution at 25 °C; KCl; *i* = 1.9
Find: % KCl by mass per volume
Conceptual plan: Isotonic solutions will have the same number of particles. Since *i* is the same, $\frac{1\ mol\ KCl}{1\ mol\ NaCl}$

the new % mass per volume will be the mass ratio of the two salts.

$\text{percent by mass per volume} = \frac{\text{mass solute}}{V} \times 100\ \%$ $\frac{1\ mol\ NaCl}{58.44\ g\ NaCl}$ and $\frac{74.56\ g\ KCl}{1\ mol\ KCl}$

Solution:

$$\text{percent by mass per volume} = \frac{\text{mass solute}}{V} \times 100\,\% =$$

$$= \frac{0.0090\ \text{g NaCl}}{V} \times \frac{1\ \text{mol NaCl}}{58.44\ \text{g NaCl}} \times \frac{1\ \text{mol KCl}}{1\ \text{mol NaCl}} \times \frac{74.56\ \text{g KCl}}{1\ \text{mol KCl}} \times 100\,\% = 1.1\,\%\ \text{KCl by mass per volume}$$

Check: The units (% KCl by mass per volume) are correct. The magnitude of the answer (1.1) seems reasonable since the molar mass of KCl is larger than the molar mass of NaCl.

b) **Given:** 0.90 % NaCl by mass per volume; isotonic aqueous solution at 25 °C; NaBr; $i = 1.9$
Find: % NaBr by mass per volume
Conceptual plan: Isotonic solutions will have the same number of particles. Since i is the same,

$$\frac{1\ \text{mol NaBr}}{1\ \text{mol NaCl}}$$

the new % mass per volume will be the mass ratio of the two salts.

$$\text{percent by mass per volume} = \frac{\text{mass solute}}{V} \times 100\,\% \qquad \frac{1\ \text{mol NaCl}}{58.44\ \text{g NaCl}} \text{ and } \frac{102.90\ \text{g NaBr}}{1\ \text{mol NaBr}}$$

Solution:

$$\text{percent by mass per volume} = \frac{\text{mass solute}}{V} \times 100\,\% =$$

$$= \frac{0.0090\ \text{g NaCl}}{V} \times \frac{1\ \text{mol NaCl}}{58.44\ \text{g NaCl}} \times \frac{1\ \text{mol NaBr}}{1\ \text{mol NaCl}} \times \frac{102.90\ \text{g NaBr}}{1\ \text{mol NaBr}} \times 100\,\% = 1.6\,\%\ \text{NaBr by mass per volume}$$

Check: The units (% NaBr by mass per volume) are correct. The magnitude of the answer (1.6) seems reasonable since the molar mass of NaBr is larger than the molar mass of NaCl.

c) **Given:** 0.90 % NaCl by mass per volume; isotonic aqueous solution at 25 °C; glucose ($C_6H_{12}O_6$); $i = 1.9$ **Find:** % glucose by mass per volume
Conceptual plan: Isotonic solutions will have the same number of particles. Since glucose is a nonelectrolyte, the i is not the same, then use the mass ratio of the two compounds.

$$\frac{1.9\ \text{mol}\ C_6H_{12}O_6}{1\ \text{mol NaCl}} \quad \text{percent by mass per volume} = \frac{\text{mass solute}}{V} \times 100\,\% \quad \frac{1\ \text{mol NaCl}}{58.44\ \text{g NaCl}} \text{ and } \frac{180.16\ \text{g}\ C_6H_{12}O_6}{1\ \text{mol}\ C_6H_{12}O_6}$$

Solution:

$$\text{percent by mass per volume} = \frac{\text{mass solute}}{V} \times 100\,\% =$$

$$= \frac{0.0090\ \text{g NaCl}}{V} \times \frac{1\ \text{mol NaCl}}{58.44\ \text{g NaCl}} \times \frac{1.9\ \text{mol}\ C_6H_{12}O_6}{1\ \text{mol NaCl}} \times \frac{180.16\ \text{g}\ C_6H_{12}O_6}{1\ \text{mol}\ C_6H_{12}O_6} \times 100\,\% =$$

$= 5.3\,\%\ C_6H_{12}O_6$ by mass per volume

Check: The units (% $C_6H_{12}O_6$ by mass per volume) are correct. The magnitude of the answer (1.6) seems reasonable since the molar mass of $C_6H_{12}O_6$ is larger than the molar mass of NaCl and we need more moles of $C_6H_{12}O_6$ since it is a nonelectrolyte.

103. **Given:** 4.5701 g of $MgCl_2$ and 43.238 g water, $P_{soln} = 0.3624$ atm, $P°_{soln} = 0.3804$ atm at 348.0 K
Find: $i_{measured}$
Conceptual plan: g_{MgCl2} → mol_{MgCl2} and g_{H2O} → mol_{H2O} then $P_{soln}, P°_{soln}$ → χ_{MgCl2}

$$\frac{1\ \text{mol}\ MgCl_2}{95.218\ \text{g}\ MgCl_2} \qquad \frac{1\ \text{mol}\ H_2O}{18.015\ \text{g}\ H_2O} \qquad P_{Soln} = (1 - \chi_{MgCl2})\, P^o_{H2O}$$

then $mol_{MgCl2}, mol_{H2O}, \chi_{MgCl2}$ → i

$$\chi_{MgCl2} = \frac{i\,(\text{moles } MgCl_2)}{\text{moles } H_2O + i\,(\text{moles } MgCl_2)}$$

Solution: $4.5701 \text{ g MgCl}_2 \times \dfrac{1 \text{ mol MgCl}_2}{95.218 \text{ g MgCl}_2} = 0.047996177 \text{ mol MgCl}_2$ and

$43.238 \text{ g H}_2\text{O} \times \dfrac{1 \text{ mol H}_2\text{O}}{18.015 \text{ g H}_2\text{O}} = 2.4001110 \text{ mol H}_2\text{O}$ then $P_{Soln} = (1 - \chi_{MgCl2})\, P^o_{H2O}$ so

$\chi_{MgCl2} = 1 - \dfrac{P_{Soln}}{P^o_{H2O}} = 1 - \dfrac{0.3624 \text{ atm}}{0.3804 \text{ atm}} = 0.04731861$

Solve for i. $i\,(0.047996177) = 0.04731861\,(2.4001110 + i\,(0.047996177)) \rightarrow i\,(0.047996177 - 0.002271112) = 0.1135699 \rightarrow$

$i = \dfrac{0.1135699}{0.04572506} = 2.484$.

Check: The units (none) are correct. The magnitude of the answer (2.5) seems reasonable for $MgCl_2$ since we expect i to be 3 if it completely dissociates. Since Mg is small and doubly charges, we expect a significant drop from 3.

105. **Given:** $T_b = 375.5$ K aqueous solution **Find:** P_{H2O} **Other:** $P^\circ_{H2O} = 0.2467$ atm; $K_b = 0.512$ °C/m
Conceptual plan: $\mathbf{T_b \rightarrow \Delta T_b}$ **then** $\mathbf{\Delta T_b, K_b \rightarrow m}$ **assume 1kg water** $\mathbf{kg_{H2O} \rightarrow mol_{H2O}}$ **then**

$T_b = T_b^\circ + \Delta T_b$ $\quad \Delta T_b = K_b\, m$ $\quad \dfrac{1 \text{ mol H}_2\text{O}}{18.02 \text{ g H}_2\text{O}}$

$\mathbf{m \rightarrow mol_{Solute}}$ **then** $\mathbf{mol_{H2O}, mol_{Solute} \rightarrow \chi_{H2O}}$ **then** $\mathbf{\chi_{H2O}, P^\circ_{H2O} \rightarrow P_{H2O}}$

$m = \dfrac{\text{amount solute (moles)}}{\text{mass solvent (kg)}}$ $\quad \chi_{H2O} = \dfrac{\text{moles H}_2\text{O}}{\text{moles H}_2\text{O} + \text{moles solute}}$ $\quad P_{H2O} = \chi_{H2O}\, P^o_{H2O}$

Solution: $T_b = T_b^o + \Delta T_b$ so $\Delta T_b = T_b - T_b^o = 375.5 \text{ K} - 373.15 \text{ K} = 2.4 \text{ K} = 2.4\ °\text{C}$ then

$\Delta T_b = K_b\, m$ Rearrange to solve for m. $m = \dfrac{\Delta T_b}{K_b} = \dfrac{2.4\ °\text{C}}{0.512 \dfrac{°\text{C}}{m}} = 4.6875\ m$ then

$1000 \text{ g H}_2\text{O} \times \dfrac{1 \text{ mol H}_2\text{O}}{18.02 \text{ g H}_2\text{O}} = 55.49390 \text{ mol H}_2\text{O}$ then

$m = \dfrac{\text{amount solute (moles)}}{\text{mass solvent (kg)}} = \dfrac{x \text{ mol}}{1 \text{ kg}} = 4.6875\ m$ so

$x = 4.6875$ mol then $\chi_{H2O} = \dfrac{\text{moles H}_2\text{O}}{\text{moles H}_2\text{O} + \text{moles solute}} = \dfrac{55.49390 \text{ mol}}{55.49390 \text{ mol} + 4.6875 \text{ mol}} = 0.9221106$

then $P_{H2O} = \chi_{H2O}\, P^o_{H2O} = 0.9221106 \times 0.2467 \text{ atm} = 0.227$ atm

Check: The units (atm) are correct. The magnitude of the answer (0.227 atm) seems reasonable since the mole fraction is lowered by ~ 8%.

107. **Given:** equal masses of carbon tetrachloride (CCl_4) and chloroform ($CHCl_3$) at 316 K; $P°_{CCl4} = 0.354$ atm; $P°_{CHCl3} = 0.526$ atm **Find:** χ_{CCl4}, χ_{CHCl3} in vapor; and P_{CHCl3} in flask of condensed vapor

Conceptual plan: assume 100 grams of each $g_{CCl4} \rightarrow mol_{CCl4}$ and $g_{CHCl3} \rightarrow mol_{CHCl3}$ then

$$\frac{1 \text{ mol } CCl_4}{153.82 \text{ g } CCl_4} \qquad \frac{1 \text{ mol } CHCl_3}{119.38 \text{ g } CHCl_3}$$

mol_{CCl4}, $mol_{CHCl3} \rightarrow \chi_{CCl4}, \chi_{CHCl3}$ then $\chi_{CCl4}, P°_{CCl4} \rightarrow P_{CCl4}$ and $\chi_{CHCl3}, P°_{CHCl3} \rightarrow P_{CHCl3}$ then

$$\chi_{CCl4} = \frac{\text{amount } CCl_4 \text{ (in moles)}}{\text{total amount (in moles)}} \qquad \chi_{CHCl3} = 1 - \chi_{CCl4} \qquad P_{CCl4} = \chi_{CCl4}\, P^o_{CCl4} \qquad P_{CHCl3} = \chi_{CHCl3}\, P^o_{CHCl3}$$

P_{CCl4}, $P_{CHCl3} \rightarrow P_{Total}$ then since n α P and we are calculating a mass percent,

$$P_{Total} = P_{CCl4} + P_{CHCl3}$$

which is a ratio of masses, we can simply convert 1 atm to 1 mole so P_{CCl4}, $P_{CHCl3} \rightarrow n_{CCl4}$, n_{CHCl3} then mol_{CCl4}, $mol_{CHCl3} \rightarrow \chi_{CCl4}, \chi_{CHCl3}$

$$\chi_{CCl4} = \frac{\text{amount } CCl_4 \text{ (in moles)}}{\text{total amount (in moles)}} \qquad \chi_{CHCl3} = 1 - \chi_{CCl4}$$

then for the second vapor $\chi_{CHCl3}, P°_{CHCl3} \rightarrow P_{CHCl3}$

$$P_{CHCl3} = \chi_{CHCl3}\, P^o_{CHCl3}$$

Solution: $100.00 \text{ g } CCl_4 \times \frac{1 \text{ mol } CCl_4}{153.82 \text{ g } CCl_4} = 0.65011051 \text{ mol } CCl_4$ and

$100.00 \text{ g } CHCl_3 \times \frac{1 \text{ mol } CHCl_3}{119.38 \text{ g } CHCl_3} = 0.83766125 \text{ mol } CHCl_3$ then

$$\chi_{CCl4} = \frac{\text{amount } CCl_4 \text{ (in moles)}}{\text{total amount (in moles)}} = \frac{0.65011051 \text{ mol}}{0.65011051 \text{ mol} + 0.83766125 \text{ mol}} = 0.43696925$$ and

$\chi_{CHCl3} = 1 - \chi_{CCl4} = 1 - 0.43696925 = 0.56303075$ then

$P_{CCl4} = \chi_{CCl4}\, P^o_{CCl4} = 0.43696925 \times 0.354 \text{ atm} = 0.154687$ atm and

$P_{CHCl3} = \chi_{CHCl3}\, P^o_{CHCl3} = 0.56303075 \times 0.526 \text{ atm} = 0.296154$ atm then

$P_{Total} = P_{CCl4} + P_{CHCl3} = 0.154687 \text{ atm} + 0.296154 \text{ atm} = 0.450841$ atm then

$mol_{CCl4} = 0.154687$ mol and $mol_{CHCl3} = 0.296154$ mol then

$$\chi_{CCl4} = \frac{\text{amount } CCl_4 \text{ (in moles)}}{\text{total amount (in moles)}} = \frac{0.154687 \text{ mol}}{0.154687 \text{ mol} + 0.296154 \text{ mol}} = 0.343108 = 0.343 \text{ in the first vapor}$$

and $\chi_{CHCl3} = 1 - \chi_{CCl4} = 1 - 0.343108 = 0.656892 = 0.657$ in the first vapor then in the second vapor

$P_{CHCl3} = \chi_{CHCl3}\, P^o_{CHCl3} = 0.656892 \times 0.526 \text{ atm} = 0.345525 \text{ atm} = 0.346$ atm

Check: The units (none and atm) are correct. The magnitude of the answers seems reasonable since it we expect the lighter component to be preferentially in the vapor phase. This effect is magnified in the second vapor.

109. **Given:** N_2: $k_H(N_2) = 6.1 \times 10^{-4}$ M/L at 25 °C; 14.6 mg/L at 50 °C and 1.00 atm; $P_{N2} = 0.78$ atm;
O_2: $k_H(O_2) = 1.3 \times 10^{-3}$ M/L at 25 °C; 27.8 mg/L at 50 °C and 1.00 atm; $P_{O2} = 0.21$ atm; and 1.5 L water
Find: V (N_2) and V (O_2)

Conceptual plan: at 25 °C: P_{Total}, χ_{N2} → P_{N2} then P_{N2}, $k_H(N_2)$ → S_{N2} then L → mol

$P_{N2} = \chi_{N2}\, P_{Total}$ $\quad$ $S_{N2} = k_H(N_2)P_{N2}$ $\quad$ S_{N2}

at 50 °C: L → mL → mg → g → mol then $mol_{25\,°C}$, $mol_{25\,°C}$ → $mol_{removed}$ then °C → K

$\frac{1000 \text{ mL}}{1 \text{ L}}$ $\quad \frac{14.6 \text{ mg}}{1 \text{ L}}$ $\quad \frac{1 \text{ g}}{1000 \text{ mg}}$ $\quad \frac{1 \text{ mol}}{28.01 \text{ g}}$ $\quad mol_{removed} = mol_{25°C} - mol_{50°C}$ $\quad K = °C + 273.15$

then P, n, T → V

$PV = nRT$

at 25 °C: P_{Total}, χ_{O2} → P_{O2} then P_{O2}, $k_H(O_2)$ → S_{O2} then L → mol

$P_{O2} = \chi_{O2}\, P_{Total}$ $\quad$ $S_{O2} = k_H(O_2)P_{O2}$ $\quad$ S_{O2}

at 50 °C: L → mL → mg → g → mol then $mol_{25\,°C}$, $mol_{25\,°C}$ → $mol_{removed}$ then °C → K

$\frac{1000 \text{ mL}}{1 \text{ L}}$ $\quad \frac{27.8 \text{ mg}}{1 \text{ L}}$ $\quad \frac{1 \text{ g}}{1000 \text{ mg}}$ $\quad \frac{1 \text{ mol}}{32.00 \text{ g}}$ $\quad mol_{removed} = mol_{25°C} - mol_{50°C}$ $\quad K = °C + 273.15$

then P, n, T → V

$PV = nRT$

Solution: at 25 °C: $P_{N2} = \chi_{N2}\, P_{Total} = 0.78 \times 1.0 \text{ atm} = 0.78 \text{ atm}$ then

$S_{N2} = k_H(N_2)P_{N2} = 6.1 \times 10^{-4} \frac{\text{M}}{\text{atm}} \times 0.78 \text{ atm} = 4.758 \times 10^{-4} \text{ M}$ then

$1.5 \text{ L} \times 4.758 \times 10^{-4} \frac{\text{mol}}{\text{L}} = 0.00071371 \text{ mol}$

at 50 °C: $1.5 \text{ L} \times \frac{14.6 \text{ mg}}{1 \text{ L atm}} \times 0.78 \text{ atm} \times \frac{1 \text{ g}}{1000 \text{ mg}} \times \frac{1 \text{ mol}}{28.01 \text{ g}} = 0.00060985 \text{ mol}$ then

$mol_{removed} = mol_{25°C} - mol_{50°C} = 0.00071371 \text{ mol} - 0.00060985 \text{ mol} = 1.039 \times 10^{-4} \text{ mol}$.
then 50 °C + 273.15 = 323 K then $PV = nRT$ Rearrange to solve for V.

$$V = \frac{nRT}{P} = \frac{1.039 \times 10^{-4} \text{ mol} \times 0.08206 \frac{\text{L atm}}{\text{K mol}} \times 323 \text{ K}}{1.00 \text{ atm}} = 0.0027526 \text{ L}$$

at 25 °C: $P_{O2} = \chi_{O2}\, P_{Total} = 0.21 \times 1.0 \text{ atm} = 0.21 \text{ atm}$ then

$S_{O2} = k_H(O_2)P_{O2} = 1.3 \times 10^{-3} \frac{\text{M}}{\text{atm}} \times 0.21 \text{ atm} = 2.73 \times 10^{-4} \text{ M}$ then

$1.5 \text{ L} \times 2.73 \times 10^{-4} \frac{\text{mol}}{\text{L}} = 0.0004095 \text{ mol}$

at 50 °C: $1.5 \text{ L} \times \frac{27.8 \text{ mg}}{1 \text{ L atm}} \times 0.21 \text{ atm} \times \frac{1 \text{ g}}{1000 \text{ mg}} \times \frac{1 \text{ mol}}{32.00 \text{ g}} = 0.00027366 \text{ g}$ then

$mol_{removed} = mol_{25°C} - mol_{50°C} = 0.0004095 \text{ mol} - 0.00027366 \text{ mol} = 1.358 \times 10^{-4} \text{ mol}$
then 50 °C + 273.15 = 323 K then $PV = nRT$ Rearrange to solve for V.

$$V = \frac{nRT}{P} = \frac{1.358 \times 10^{-4} \text{ mol} \times 0.08206 \frac{\text{L atm}}{\text{K mol}} \times 323 \text{ K}}{1.00 \text{ atm}} = 0.0035994 \text{ L} \quad \text{finally}$$

$V_{Total} = V_{N2} + V_{O2} = 0.0027526 \text{ L} + 0.0035994 \text{ L} = 0.0064 \text{ L}$

Check: The units (L) are correct. The magnitude of the answer (0.006 L) seems reasonable since we have so little dissolves gas at room temperature and most is still soluble at 50 °C.

111. **Given:** 1.10 g glucose ($C_6H_{12}O_6$) and sucrose ($C_{12}H_{22}O_{11}$) mixture in 25.0 mL solution and Π = 3.78 atm at 298 K **Find:** percent composition of mixture

Conceptual plan: Π, **T → M then mL$_{soln}$ → L$_{soln}$ then L$_{soln}$, M → mol$_{mixture}$ then**

$$\Pi = MRT \qquad \frac{1\text{ L}}{1000\text{ mL}} \qquad M = \frac{\text{amount solute (moles)}}{\text{volume solution (L)}}$$

mol$_{mixture}$, g$_{mixture}$ → mol$_{C6H12O6}$, mol$_{C12H22O11}$ then

$$g_{mixture} = mol\ C_6H_{12}O_6 \times \frac{180.16\ g\ C_6H_{12}O_6}{1\ mol\ C_6H_{12}O_6} + mol\ C_{12}H_{22}O_{11} \times \frac{342.30\ g\ C_{12}H_{22}O_{11}}{1\ mol\ C_{12}H_{22}O_{11}} \text{ with}$$

$$mol_{mixture} = mol\ C_6H_{12}O_6 + mol\ C_{12}H_{22}O_{11}$$

mol$_{C6H12O6}$ → g$_{C6H12O6}$ and mol$_{C12H22O11}$ → g$_{C12H22O11}$ and g$_{C6H12O6}$, g$_{C12H22O11}$ → mass percents

$$\frac{180.16\text{ g } C_6H_{12}O_6}{1\text{ mol } C_6H_{12}O_6} \qquad \frac{342.30\text{ g } C_{12}H_{22}O_{11}}{1\text{ mol } C_{12}H_{22}O_{11}} \qquad mass\ percent = \frac{mass\ solute}{mass\ solution} \times 100\ \%$$

Solution: $\Pi = MRT$ Rearrange to solve for M.

$$M = \frac{\Pi}{RT} = \frac{3.78\ \text{atm}}{0.08206\ \frac{\text{L atm}}{\text{K mol}} \times 298\ \text{K}} = 0.15\underline{4}577\ \frac{\text{mol mixture}}{\text{L}} \quad \text{then}$$

$$25.0\ \text{mL} \times \frac{1\text{L}}{1000\ \text{mL}} = 0.0250\text{ L} \quad \text{then} \quad M = \frac{\text{amount solute (moles)}}{\text{volume solution (L)}} \quad \text{so}$$

$$mol_{mixture} = M \times L_{soln} = 0.15\underline{4}577\ \frac{\text{mol mixture}}{\text{L}} \times 0.0250\ \text{L} = 0.0038\underline{6}442\ \text{mol mixture} \quad \text{then}$$

$$g_{mixture} = mol\ C_6H_{12}O_6 \times \frac{180.16\ g\ C_6H_{12}O_6}{1\ mol\ C_6H_{12}O_6} + mol\ C_{12}H_{22}O_{11} \times \frac{342.30\ g\ C_{12}H_{22}O_{11}}{1\ mol\ C_{12}H_{22}O_{11}} \text{ with}$$

$$mol_{mixture} = mol\ C_6H_{12}O_6 + mol\ C_{12}H_{22}O_{11} \quad \text{so}$$

$$1.10\text{ g} = \text{mol } C_6H_{12}O_6 \times \frac{180.16\text{ g } C_6H_{12}O_6}{1\text{ mol } C_6H_{12}O_6} + (0.0038\underline{6}442\text{ mol} - \text{mol } C_6H_{12}O_6) \times \frac{342.30\text{ g } C_{12}H_{22}O_{11}}{1\text{ mol } C_{12}H_{22}O_{11}}$$

$$\rightarrow 1.10 = 180.16 \times \text{mol } C_6H_{12}O_6 + 1.3\underline{2}228 - 342.30 \times \text{mol } C_6H_{12}O_6 \rightarrow$$

$$162.14\ x\ \text{mol } C_6H_{12}O_6 = 0.2\underline{2}228 \rightarrow x\ \text{mol } C_6H_{12}O_6 = \frac{0.2\underline{2}228}{162.14} = 0.001\underline{3}7091\text{ mol } C_6H_{12}O_6 \quad \text{then}$$

$$mol\ C_{12}H_{22}O_{11} = mol_{mixture} - mol\ C_6H_{12}O_6 = 0.0038\underline{6}442\text{ mol} - 0.001\underline{3}7091\text{ mol} = 0.002\underline{4}935\text{ mol } C_{12}H_{22}O_{11}$$

then

$$0.001\underline{3}7091\ \text{mol } C_6H_{12}O_6 \times \frac{180.16\text{ g } C_6H_{12}O_6}{1\ \text{mol } C_6H_{12}O_6} = 0.2\underline{4}698\text{ g } C_6H_{12}O_6 \quad \text{and}$$

$$0.002\underline{4}935\ \text{mol } C_{12}H_{22}O_{11} \times \frac{342.30\text{ g } C_{12}H_{22}O_{11}}{1\ \text{mol } C_{12}H_{22}O_{11}} = 0.8\underline{5}353\text{ g } C_{12}H_{22}O_{11} \quad \text{finally}$$

$$mass\ percent = \frac{mass\ solute}{mass\ solution} \times 100\ \% = \frac{0.2\underline{4}698\ g\ C_6H_{12}O_6}{0.2\underline{4}698\ g\ C_6H_{12}O_6 + 0.8\underline{5}353\ g\ C_{12}H_{22}O_{11}} \times 100\ \% = 2\underline{2}.44\ \%\ C_6H_{12}O_6 \text{ by mass}$$

and $100.00\ \% - 2\underline{2}.44\ \% = 7\underline{7}.56\ \%\ C_{12}H_{22}O_{11}$ by mass

Check: The units (% by mass) are correct. We expect the percent by $C_6H_{12}O_6$ to be larger than that for $C_{12}H_{22}O_{11}$ since the g $_{mixture}$ /mol $_{mixture}$ = 285 g/mol, which is closer to $C_{12}H_{22}O_{11}$ than $C_6H_{12}O_6$ and the molar mass of $C_{12}H_{22}O_{11}$ is larger than the molar mass of $C_6H_{12}O_6$.

113. **Given:** isopropyl alcohol ($(CH_3)_2CHOH$) and propyl alcohol ($CH_3CH_2CH_2OH$) at 313 K; solution 2/3 by mass isopropyl alcohol $P_{2/3}$ = 0.110 atm; solution 1/3 by mass isopropyl alcohol $P_{1/3}$ = 0.089 atm; **Find:** P°_{iso} and P°_{pro} and explain why they are different
Conceptual plan: since these are isomers, they have the same molar mass and so the fraction by mass is the same as the mole fraction so mole fractions, P_{soln}'s → P°'s

$$\chi_{iso} = \frac{\text{amount iso (in moles)}}{\text{total amount (in moles)}} \quad \chi_{pro} = 1 - \chi_{iso} \quad P_{iso} = \chi_{iso} P^\circ_{iso} \quad P_{pro} = \chi_{pro} P^\circ_{pro} \text{ and } P_{soln} = P_{iso} + P_{pro}$$

Solution: Solution 1: χ_{iso} = 2/3 and χ_{iso} = 1/3 $P_{soln} = P_{iso} + P_{pro}$ so 0.110 atm = $2/3P^\circ_{iso} + 1/3P^\circ_{pro}$ and Solution 2: χ_{iso} = 1/3 and χ_{iso} = 2/3 $P_{soln} = P_{iso} + P_{pro}$ so 0.089 atm = $1/3P^\circ_{iso} + 2/3P^\circ_{pro}$. We now have 2 equations and 2 unknowns and a number of ways to solve this. One way is to rearrange the first equation for P°_{iso} and then substitute into the other equation. Thus, P°_{iso} = 3/2 (0.110 atm – $1/3P^\circ_{pro}$) and

$$0.089 \text{ atm} = \frac{1}{3}\frac{3}{2}(0.110 \text{ atm} - 1/3P^\circ_{pro}) + \frac{2}{3}P^\circ_{pro} \rightarrow 0.089 \text{ atm} = 0.0550 \text{ atm} - \frac{1}{6}P^\circ_{pro} + \frac{2}{3}P^\circ_{pro} \rightarrow$$

$$\frac{1}{2}P^\circ_{pro} = 0.03\underline{4}0 \text{ atm} \rightarrow P^\circ_{pro} = 0.06\underline{8}0 \text{ atm} = 0.068 \text{ atm and then}$$

$$P^\circ_{iso} = 3/2\,(0.110 \text{ atm} - 1/3P^\circ_{pro}) = 3/2\,(0.110 \text{ atm} - 1/3(0.06\underline{8}0 \text{ atm})) = 0.131 \text{ atm}$$

The major intermolecular attractions are between the OH groups. The OH group at the end of the chain in propyl alcohol is more accessible than the one in the middle of the chain in isopropyl alcohol. In addition, the molecular shape of propyl alcohol is a straight chain of carbon atoms, while that of isopropyl alcohol has a branched chain and is more like a ball. The contact area between two ball-like objects is smaller than that of two chain-like objects. The smaller contact area in isopropyl alcohol means the molecules don't attract each other as strongly as do those of propyl alcohol. As a result of both of these factors, the vapor pressure of isopropyl alcohol is higher.
Check: The units (atm) are correct. The magnitude of the answers seems reasonable since the solution partial pressures are both ~0.1 atm.

115. a) The two substances mix because their intermolecular forces between themselves are roughly equal to the forces between each other and there is a pervasive tendency to increase randomness, which happens when the two substances mix.

b) $\Delta H_{soln} \approx 0$, since the intermolecular forces between themselves are roughly equal to the forces between each other.

c) ΔH_{solute} and $\Delta H_{solvent}$ are positive, ΔH_{mix} is negative and equals the sum of ΔH_{solute} and $\Delta H_{solvent}$.

117. d) More solute particles are found in an ionic solution, since the solute breaks apart into its ions. The vapor pressure is lowered the more solute particle are present.

119. The balloon not only loses He, it also takes in N_2 and O_2 from the air surrounding the balloon (due to the tendency for mixing), increasing the density of the balloon.

Chapter 13
Chemical Kinetics

1. Unlike mammals, which actively regulate their body temperature through metabolic activity, lizards are ectotherms—their body temperature depends on their surroundings. When splashed with cold water, a lizard's body simply gets colder. The drop in body temperature immobilizes the lizard because its movement depends on chemical reactions that occur within its muscles, and the rates of those reactions—how fast they occur—are highly sensitive to temperature. In other words, when the temperature drops, the reactions that produce movement occur more slowly; therefore the movement itself slows down. Cold reptiles are lethargic, unable to move very quickly. For this reason, reptiles try to maintain their body temperature in a narrow range by moving between sun and shade.

3. The rate of a chemical reaction is measured as a change in the amounts of reactants or products (usually in terms of concentration) divided by the change in time. Typical units are molarity per second (M/s), molarity per minute (M/min), and molarity per year (M/yr), depending on how fast the reaction proceeds.

5. The average rate of the reaction can be calculated for any time interval as: $\text{Rate} = -\frac{1}{a}\frac{[A]_{t_2} - [A]_{t_1}}{t_2 - t_1} = -\frac{1}{b}\frac{[B]_{t_2} - [B]_{t_1}}{t_2 - t_1} = -\frac{1}{c}\frac{[C]_{t_2} - [C]_{t_1}}{t_2 - t_1} = -\frac{1}{d}\frac{[D]_{t_2} - [D]_{t_1}}{t_2 - t_1}$ for the chemical reaction: $aA + bB \rightarrow cC + dD$. The instantaneous rate of the reaction is the rate at any one point in time, represented by the instantaneous slope of the curve at that point. We can obtain the instantaneous rate from the slope of the tangent to the curve at the point of interest.

7. The reaction order can not be determined by the stoichiometry of the reaction. It can only be determined by running controlled experiments where the concentrations of the reactants are varied and the reaction rates are measured and analyzed.

9. The rate law shows the relationship between the rate of a reaction and the concentrations of the reactants. The integrated rate law for a chemical reaction is a relationship between the concentration of a reactant and time.

11. The half-life ($t_{1/2}$) of a reaction is the time required for the concentration of a reactant to fall to one-half of its initial value. For a zero order reaction, $t_{1/2} = \frac{[A]_0}{2k}$. For a first order reaction, $t_{1/2} = \frac{0.693}{k}$. For a second order reaction, $t_{1/2} = \frac{1}{k[A]_0}$.

13. The modern form of the Arrhenius equation, which relates the rate constant (k) and the temperature in Kelvin (T), is as follows: $k = A\,e^{-E_a/RT}$, where R is the gas constant (8.314 J/mol K), A is a constant called the frequency factor (or the pre-exponential factor), and E_a is called the activation energy (or activation barrier). The frequency factor is the number of times that the reactants approach the activation barrier per unit time. The exponential factor is the fraction of approaches that are successful in surmounting the activation barrier and forming products. The exponential factor increases with increasing temperature, but decreases with an increasing value for the activation energy. As the temperature increases, the number of collisions increases and the number of molecules having enough thermal energy to surmount the activation barrier increases. At any given temperature, a sample of molecules will have a distribution of energies, as shown in Figure 13.14. Under common circumstances, only a small fraction of the molecules have enough energy to make it over the activation barrier. Because of the shape of the energy distribution curve, however, a small change in temperature results in a large difference in the number of molecules having enough energy to surmount the activation barrier.

15. In the collision model, a chemical reaction occurs after a sufficiently energetic collision between the two reactant molecules. In collision theory, therefore, each approach to the activation barrier is a collision between the reactant molecules. The value of the frequency factor should simply be the number of collisions that occur per second. In the collision model $k = p\,z\,e^{-E_a/RT}$, where the frequency factor, A, has been separated into two separate parts, where p is called the orientation factor and z is the collision

frequency. The collision frequency is simply the number of collisions that occur per unit time. The orientation factor says that if two molecules are to react with each other, they must collide in such a way that allows the necessary bonds to break and form. The small orientation factor indicates that the orientational requirements for this reaction are fairly stringent—the molecules must be aligned in a very specific way for the reaction to occur. When two molecules with sufficient energy and the correct orientation collide, something unique happens. The electrons on one of the atoms or molecules are attracted to the nuclei of the other; some bonds begin to weaken while other bonds begin to form and, if all goes well, the reactants go through the transition state and are transformed into the products and a chemical reaction occurs.

17. An elementary step is a step in a reaction mechanism. Elementary steps cannot be broken down into simpler steps—they occur as they are written. Elementary steps are characterized by their molecularity, the number of reactant particles involved in the step. The molecularity of the three most common types of elementary steps are as follows: unimolecular - A → products and $\text{Rate} = k\,[A]$; bimolecular - A + A → products and $\text{Rate} = k\,[A]^2$; and bimolecular - A + B → products and $\text{Rate} = k\,[A][B]$. Elementary steps in which three reactant particles collide, called termolecular steps, are very rare because the probability of three particles simultaneously colliding is small.

19. Reaction intermediates are species that are formed in one step of a mechanism and consumed in another step. An intermediate is not found in the balanced equation for the overall reaction, but plays a key role in the mechanism.

21. In homogeneous catalysis, the catalyst exists in the same phase as the reactants. In heterogeneous catalysis, the catalyst exists in a phase different from the reactants.

23. Enzymes are biological catalysts that increase the rates of biochemical reactions. Enzymes are large protein molecules with complex three-dimensional structures. Within that structure is a specific area called the active site. The properties and shape of the active site are just right to bind the reactant molecule, usually called the substrate. The substrate fits into the active site in a manner that is analogous to a key fitting into a lock.

25. a) $\text{Rate} = -\frac{1}{2}\frac{\Delta[HBr]}{\Delta t} = \frac{\Delta[H_2]}{\Delta t} = \frac{\Delta[Br_2]}{\Delta t}$

b) **Given:** first 15.0 s; 0.500 M to 0.455 M **Find:** average rate
Conceptual plan: **t_1, t_2, $[HBr]_1$, $[HBr]_2$ → average rate**

$$\text{Rate} = -\frac{1}{2}\frac{\Delta[HBr]}{\Delta t}$$

Solution: $\text{Rate} = -\frac{1}{2}\frac{[HBr]_{t_2} - [HBr]_{t_1}}{t_2 - t_1} = -\frac{1}{2}\frac{0.455\text{ M} - 0.500\text{ M}}{15.0\text{s} - 0.0\text{ s}} = 1.5 \times 10^{-3}\text{ M s}^{-1}$

Check: The units (M s^{-1}) are correct. The magnitude of the answer (10^{-3}M s^{-1}) makes physical sense because rates are always positive and we are not changing the concentration much in 15 s.

c) **Given:** 0.500 L vessel and part b) data **Find:** mol_{Br2} formed
Conceptual plan: **average rate, t_1, t_2, → Δ $[Br_2]$ then Δ $[Br_2]$, L → mol_{Br2} formed**

$$\text{Rate} = \frac{\Delta[Br_2]}{\Delta t} \qquad M = \frac{mol_{Br2}}{L}$$

Solution: $\text{Rate} = 1.5 \times 10^{-3}\text{ M s}^{-1} = \frac{\Delta[Br_2]}{\Delta t} = \frac{\Delta[Br_2]}{15.0\text{s} - 0.0\text{ s}}$ Rearrange to solve for Δ $[Br_2]$.

$\Delta[Br_2] = 1.5 \times 10^{-3}\frac{\text{M}}{\text{s}} \times 15.0\text{ s} = 0.02\underline{2}5\text{ M}$ then $M = \frac{mol_{Br2}}{L}$. Rearrange to solve for mol_{Br2}.

$0.2\underline{2}5\frac{\text{mol Br}_2}{\text{L}} \times 0.500\text{ L} = 0.011\text{ mol Br}_2$.

Check: The units (mol) are correct. The magnitude of the answer (0.01 mol) makes physical sense because the times are the same as in part b) and we need to rates divide by 2 twice because of the stoichiometric coefficient difference and the volume of the vessel.

27\. a) Rate = $-\frac{1}{2}\frac{\Delta[A]}{\Delta t} = -\frac{\Delta[B]}{\Delta t} = \frac{1}{3}\frac{\Delta[C]}{\Delta t}$

b) **Given:** $\frac{\Delta[A]}{\Delta t} = -0.100$ M/s **Find:** $\frac{\Delta[B]}{\Delta t}$, and $\frac{\Delta[C]}{\Delta t}$

Conceptual plan: $\frac{\Delta[A]}{\Delta t} \rightarrow \frac{\Delta[B]}{\Delta t}$**, and** $\frac{\Delta[C]}{\Delta t}$

Rate = $-\frac{1}{2}\frac{\Delta[A]}{\Delta t} = -\frac{\Delta[B]}{\Delta t} = \frac{1}{3}\frac{\Delta[C]}{\Delta t}$

Solution: Rate = $-\frac{1}{2}\frac{\Delta[A]}{\Delta t} = -\frac{\Delta[B]}{\Delta t} = \frac{1}{3}\frac{\Delta[C]}{\Delta t}$ Sustitiute in value and solve for the two desired values.

$-\frac{1}{2}\frac{-0.100\ M}{s} = -\frac{\Delta[B]}{\Delta t}$ so $\frac{\Delta[B]}{\Delta t} = -0.0500\ M\,s^{-1}$ and $-\frac{1}{2}\frac{-0.100\ M}{s} = \frac{1}{3}\frac{\Delta[C]}{\Delta t}$ so $\frac{\Delta[C]}{\Delta t} = 0.150\ M\,s^{-1}$

Check: The units (M s^{-1}) are correct. The magnitude of the answer (- 0.05 M s^{-1}) makes physical sense because fewer moles of B are reacting for every mole of A and the change in concentration with time is negative since this is a reactant. The magnitude of the answer (0.15 M s^{-1}) makes physical sense because more moles of C are being formed for every mole of A reacting and the change in concentration with time is positive since this is a product.

29\. a) **Given:** [C_4H_8] versus time data **Find:** average rate between 0 and 10 s, and between 40 and 50 s
Conceptual plan: **t_1, t_2, $[C_4H_8]_1$, $[C_4H_8]_2$ → average rate**

$$\text{Rate} = -\frac{\Delta[C_4H_8]}{\Delta t}$$

Solution: for 0 to 10 s: Rate = $-\frac{[C_4H_8]_{t_2} - [C_4H_8]_{t_1}}{t_2 - t_1} = -\frac{0.913\ M - 1.000\ M}{10.\ s - 0.\ s} = 8.7 \times 10^{-3}\ M\ s^{-1}$ and

for 40 to 50 s: Rate = $-\frac{[C_4H_8]_{t_2} - [C_4H_8]_{t_1}}{t_2 - t_1} = -\frac{0.637\ M - 0.697\ M}{50.\ s - 40.\ s} = 6.0 \times 10^{-3}\ M\ s^{-1}$

Check: The units (M s^{-1}) are correct. The magnitude of the answer (10^{-3}M s^{-1}) makes physical sense because rates are always positive and we are not changing the concentration much in 10 s. Also reactions slow as they proceed because the concentration of the reactants is decreasing.

b) **Given:** [C_4H_8] versus time data **Find:** $\frac{\Delta[C_2H_4]}{\Delta t}$ between 20 and 30 s

Conceptual plan: **t_1, t_2, $[C_4H_8]_1$, $[C_4H_8]_2$ →** $\frac{\Delta[C_2H_4]}{\Delta t}$

$$\text{Rate} = -\frac{\Delta[C_4H_8]}{\Delta t} = \frac{1}{2}\frac{\Delta[C_2H_4]}{\Delta t}$$

Solution: Rate = $-\frac{[C_4H_8]_{t_2} - [C_4H_8]_{t_1}}{t_2 - t_1} = -\frac{0.763\ M - 0.835\ M}{30.\ s - 20.\ s} = 7.2 \times 10^{-3}\ M\ s^{-1} = \frac{1}{2}\frac{\Delta[C_2H_4]}{\Delta t}$ Rearrange to solve for $\frac{\Delta[C_2H_4]}{\Delta t}$. So $\frac{\Delta[C_2H_4]}{\Delta t} = 2(7.2 \times 10^{-3}\ M\ s^{-1}) = 1.4 \times 10^{-2}\ M\ s^{-1}$.

Check: The units (M s^{-1}) are correct. The magnitude of the answer (10^{-2}M s^{-1}) makes physical sense because rate of product formation is always positive and we are not changing the concentration much in 10 s. The rate of change of the product is faster than the decline of the reactant because of the stoichiometric coefficients.

31\. a) **Given:** [Br_2] versus time plot
Find: (i) average rate between 0 and 25 s; (ii) instantaneous rate at 25 s; and (iii) instantaneous rate of HBr formation at 50 s

Conceptual plan: **(i)** **t_1, t_2, $[Br_2]_1$, $[Br_2]_2$ → average rate** **then**

$$\text{Rate} = -\frac{\Delta[Br_2]}{\Delta t}$$

(ii) draw tangent at 25 s and determine slope → instantaneous rate **then**

$$\text{Rate} = -\frac{\Delta[Br_2]}{\Delta t}$$

(iii) draw tangent at 50 s and determine slope → instantaneous rate → $\frac{\Delta[HBr]}{\Delta t}$

$$\text{Rate} = -\frac{\Delta[Br_2]}{\Delta t} \qquad \text{Rate} = \frac{1}{2}\frac{\Delta[HBr]}{\Delta t}$$

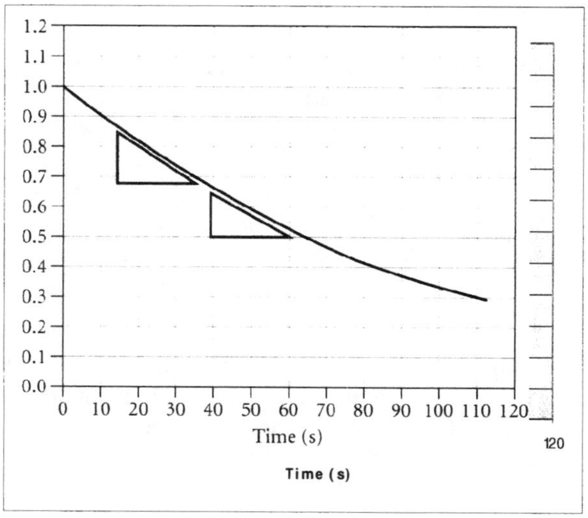

Solution: (i)

$$\text{Rate} = -\frac{[Br_2]_{t_2} - [Br_2]_{t_1}}{t_2 - t_1} = -\frac{0.75\text{ M} - 1.00\text{ M}}{25\text{ s} - 0.\text{ s}} = 1.0 \times 10^{-2}\text{ M s}^{-1}$$

and

(ii) at 25 s:

$$\text{Slope} = \frac{\Delta y}{\Delta x} = \frac{0.68\text{ M} - 0.85\text{ M}}{35\text{ s} - 15\text{ s}} = -8.5 \times 10^{-3}\text{ M s}^{-1}$$

since the slope $= \frac{\Delta[Br_2]}{\Delta t}$ and $\text{Rate} = -\frac{\Delta[Br_2]}{\Delta t}$,

then Rate $= -\,-8.5 \times 10^{-3}\text{ M s}^{-1} = 8.5 \times 10^{-3}\text{ M s}^{-1}$

(iii)) at 50 s:

$$\text{Slope} = \frac{\Delta y}{\Delta x} = \frac{0.53\text{ M} - 0.66\text{ M}}{60.\text{ s} - 40.\text{ s}} = -6.5 \times 10^{-3}\text{ M s}^{-1}$$

since the slope $= \frac{\Delta[Br_2]}{\Delta t}$ and

$\text{Rate} = -\frac{\Delta[Br_2]}{\Delta t} = \frac{1}{2}\frac{\Delta[HBr]}{\Delta t}$ then

$$\frac{\Delta[HBr]}{\Delta t} = -2\frac{\Delta[Br_2]}{\Delta t} = -2(-6.5 \times 10^{-3}\text{ M s}^{-1}) =$$

$= 1.3 \times 10^{-2}\text{ M s}^{-1}$

Check: The units (M s^{-1}) are correct. The magnitude of the first answer is larger than the second answer because the rate is slowing down and the first answer includes the initial portion of the data. The magnitudes of the answers (10^{-3}M s^{-1}) makes physical sense because rates are always positive and we are not changing the concentration much.

b) **Given:** $[Br_2]$ versus time data; and $[HBr]_0 = 0$ M **Find:** plot [HBr] with time

Conceptual plan: **Since** $\text{Rate} = -\frac{\Delta[Br_2]}{\Delta t} = \frac{1}{2}\frac{\Delta[HBr]}{\Delta t}$**. The rate of change of [HBr] will be twice that of $[Br_2]$. The plot will start at the origin.**

Check: The units (M versus s) are correct. The plot makes sense because the has the same general shape of the original plot, only we are increasing instead of decreasing and out concentration axis has changed by a factor of two (to account for the difference in stoichiometric coefficients.

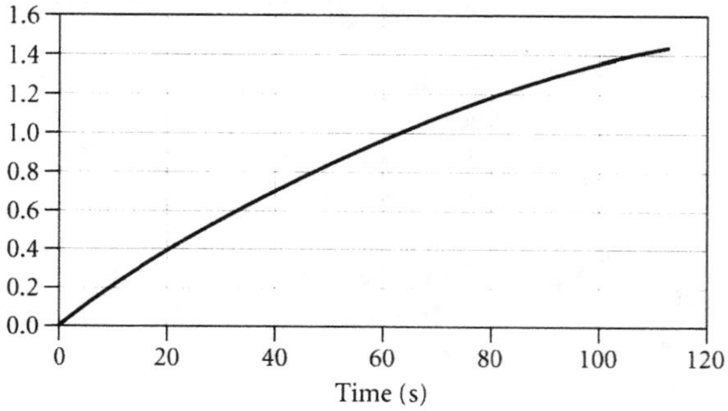

33. a) **Given:** Rate versus [A] plot **Find:** reaction order

Conceptual Plan: Look at shape of plot and match to possibilities.
Solution: The plot is a linear plot, so Rate α [A] or the reaction is first order.
Check: The order of the reaction is a common reaction order.

b) **Given:** part a) **Find:** sketch plot of [A] versus time
Conceptual plan: Using result from part a), shape plot of [A] versus time should be curved with [A] decreasing. Use 1.0 M as initial concentration.
Solution:

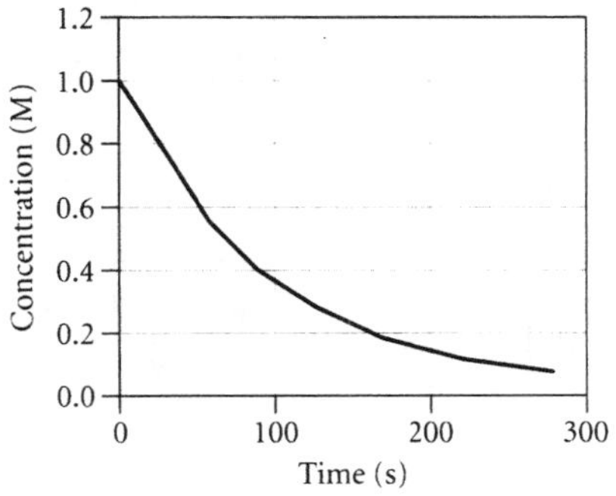

Check: The plot has a shape that matches the one in the text for first order plots.

c) **Given: part a)** **Find:** write a rate law and estimate k
Conceptual plan: Using result from part a), the slope of the plot is the rate constant.

Solution: $\text{Slope} = \frac{\Delta y}{\Delta x} = \frac{0.010\ \frac{\cancel{M}}{s} - 0.00\ \frac{\cancel{M}}{s}}{1.0\ \cancel{M} - 0.0\ \cancel{M}} = 0.010\ s^{-1}$ so Rate = k $[A]^1$ or Rate = k [A] or Rate = $0.010\ s^{-1}$ [A]
Check: The units (s^{-1}) are correct. The magnitude of the answer ($10^{-2}\ s^{-1}$) makes physical sense because rate and concentration data. Remember that concentration is in units of M, so plugging the rate constant into the equation has the units of the rate as $M\ s^{-1}$, which is correct.

35. **Given:** reaction order: a) first-order; b) second-order; and c) zero-order **Find:** units of k
Conceptual plan: Using rate law, rearrange to solve for k.

$\text{Rate} = k\ [A]^n$, where n = reaction order

Solution: for all cases Rate has units of $M\ s^{-1}$ and [A] has units of M

a) Rate = $k\ [A]^1 = k$ [A] so $k = \frac{\text{Rate}}{[A]} = \frac{\frac{\cancel{M}}{s}}{\cancel{M}} = s^{-1}$; b) Rate = $k\ [A]^2$ so $k = \frac{\text{Rate}}{[A]^2} = \frac{\frac{\cancel{M}}{s}}{\cancel{M}M} = M^{-1}\ s^{-1}$;

c) Rate = $k\ [A]^0 = k = M\ s^{-1}$.

Check: The units (s^{-1}, $M^{-1}s^{-1}$ and Ms^{-1}) are correct. The units for k change with the reaction order so that the units on the rate remain as $M\ s^{-1}$.

37. **Given:** A, B and C react to form products. reaction is first order in A, second order in B and zero order in C
Find: a) rate law; b) overall order of reaction; c) factor change in rate if [A] doubled; d) factor change in rate if [B] doubled; e) factor change in rate if [C] doubled; and f) factor change in rate if [A], [B] and [C] doubled.
Conceptual plan:
a) Using general rate law form, substitute in values for orders.

$\text{Rate} = k\ [A]^m\ [B]^n\ [C]^p$, where m, n and p = reaction orders

b) Using rate law in part a) add up all reaction orders.

overall reaction order $= m + n + p$

c) Through f) Using rate law from part a) substitute in concentration changes.

$$\frac{\text{Rate 2}}{\text{Rate 1}} = \frac{k\ [A]_2^1\ [B]_2^2}{k\ [A]_1^1\ [B]_1^2}$$

Solution:

a) m = 1, n = 2, and p = 0 so $\text{Rate} = k\,[A]^1[B]^2[C]^0$ or $\text{Rate} = k\,[A][B]^2$.

b) *overall reaction order* $= m + n + p = 1 + 2 + 0 = 3$ so it is a third order reaction overall.

c) $\dfrac{\text{Rate 2}}{\text{Rate 1}} = \dfrac{k\,[A]_2^1\,[B]_2^2}{k\,[A]_1^1\,[B]_1^2}$ and $[A]_2 = 2\,[A]_1$, $[B]_2 = [B]_1$, $[C]_2 = [C]_1$, so $\dfrac{\text{Rate 2}}{\text{Rate 1}} = \dfrac{\cancel{k}(2\,\cancel{[A]_1})^1\,\cancel{[B]_1^2}}{\cancel{k}\,\cancel{[A]_1^1}\,\cancel{[B]_1^2}} = 2$ so the reaction rate doubles (factor of 2).

d) $\dfrac{\text{Rate 2}}{\text{Rate 1}} = \dfrac{k\,[A]_2^1\,[B]_2^2}{k\,[A]_1^1\,[B]_1^2}$ and $[A]_2 = [A]_1$, $[B]_2 = 2\,[B]_1$, $[C]_2 = [C]_1$, so

$\dfrac{\text{Rate 2}}{\text{Rate 1}} = \dfrac{\cancel{k}\,\cancel{[A]_1}\,(2\,\cancel{[B]_1})^2}{\cancel{k}\,\cancel{[A]_1^1}\,\cancel{[B]_1^2}} = 2^2 = 4$ so the reaction rate quadruples (factor of 4).

e) $\dfrac{\text{Rate 2}}{\text{Rate 1}} = \dfrac{k\,[A]_2^1\,[B]_2^2}{k\,[A]_1^1\,[B]_1^2}$ and $[A]_2 = [A]_1$, $[B]_2 = [B]_1$, $[C]_2 = 2\,[C]_1$, so $\dfrac{\text{Rate 2}}{\text{Rate 1}} = \dfrac{\cancel{k}\,\cancel{[A]_1^1}\,\cancel{[B]_1^2}}{\cancel{k}\,\cancel{[A]_1^1}\,\cancel{[B]_1^2}} = 1$ so the reaction rate is unchanged (factor of 1).

f) $\dfrac{\text{Rate 2}}{\text{Rate 1}} = \dfrac{k\,[A]_2^1\,[B]_2^2}{k\,[A]_1^1\,[B]_1^2}$ and $[A]_2 = 2\,[A]_1$, $[B]_2 = 2\,[B]_1$, $[C]_2 = 2\,[C]_1$, so

$\dfrac{\text{Rate 2}}{\text{Rate 1}} = \dfrac{\cancel{k}\,(2\,\cancel{[A]_1})^1(2\,\cancel{[B]_1})^2}{\cancel{k}\,\cancel{[A]_1^1}\,\cancel{[B]_1^2}} = 2 \times 2^2 = 8$ so the reaction rate goes up by a factor of 8.

Check: The units (none) are correct. The rate law is consistent with the orders given and the overall order is larger that any of the individual orders. The factors are consistent with the reaction orders. The larger the order the larger the factor. When all concentrations are changes the rate changes the most. If a reactant is not in the rate law, then changing its concentration has no effect on the reaction rate.

39. **Given:** table of [A] versus initial rate **Find:** rate law and k

Conceptual plan: Using general rate law form, compare rate ratios to determine reaction order.

$$\frac{\text{Rate 2}}{\text{Rate 1}} = \frac{k\,[A]_2^n}{k\,[A]_1^n}$$

Then use one of the concentration/ initial rate pairs to determine *k*.

$$\text{Rate} = k[A]^n$$

Solution: $\dfrac{\text{Rate 2}}{\text{Rate 1}} = \dfrac{k\,[A]_2^n}{k\,[A]_1^n}$ Comparing the first two sets of data: $\dfrac{0.210\,\cancel{M/s}}{0.053\,\cancel{M/s}} = \dfrac{\cancel{k}\,(0.200\,\cancel{M})^n}{\cancel{k}\,(0.100\,\cancel{M})]^n}$ and $3.9623 = 2^n$ so n = 2. If we compare the first and the last data sets: $\dfrac{0.473\,\cancel{M/s}}{0.053\,\cancel{M/s}} = \dfrac{\cancel{k}\,(0.300\,\cancel{M})^n}{\cancel{k}\,(0.100\,\cancel{M})]^n}$ and $8.9245 = 3^n$ so n = 2. This second comparison is not necessary, but it increases our confidence in the reaction order. So $\text{Rate} = k[A]^2$. Selecting the second data set and rearranging the rate equation

$$k = \frac{\text{Rate}}{[A]^2} = \frac{0.210\dfrac{M}{s}}{(0.200\,M)^2} = 5.25\ M^{-1}\ s^{-1}$$ so $\text{Rate} = 5.25\ M^{-1}\ s^{-1}[A]^2$

Check: The units (none and $M^{-1}\ s^{-1}$) are correct. The rate law is a common form. The rate is changing more rapidly that the concentration, so second order is consistent. The rate constant is consistent with the units necessary to get rate as M/s and the magnitude is reasonable since we have a second order reaction.

41. **Given:** table of $[NO_2]$ and $[F_2]$ versus initial rate **Find:** rate law, k, and overall order

Conceptual plan: Using general rate law form, compare rate ratios to determine reaction order of each reactant. Be sure to choose data that changes only one concentration at a time.

$$\frac{\text{Rate 2}}{\text{Rate 1}} = \frac{k\,[NO_2]_2^m\,[F_2]_2^n}{k\,[NO_2]_1^m\,[F_2]_1^n}$$

Then use one of the concentration/ initial rate pairs to determine *k*.

$$\text{Rate} = k\,[NO_2]^m\,[F_2]^n$$

Solution: $\dfrac{\text{Rate 2}}{\text{Rate 1}} = \dfrac{k\,[NO_2]_2^m\,[F_2]_2^n}{k\,[NO_2]_1^m\,[F_2]_1^n}$ Comparing the first two sets of data:

$\dfrac{0.051\ \cancel{M/s}}{0.026\ \cancel{M/s}} = \dfrac{\cancel{k}\,(0.200\ \cancel{M})^m\,(\cancel{0.100\ M})^n}{\cancel{k}\,(0.100\ \cancel{M})^m\,(\cancel{0.100\ M})^n}$ and $1.\underline{9}15 = 2^m$ so $m = 1$. If we compare the second and the third

data sets: $\dfrac{0.103\ \cancel{M/s}}{0.051\ \cancel{M/s}} = \dfrac{\cancel{k}\,(\cancel{0.200\ M})^m\,(0.200\ \cancel{M})^n}{\cancel{k}\,(\cancel{0.200\ M})^m\,(0.100\ \cancel{M})^n}$ and $2 = 2^n$ so $n = 1$. Other comparisons can be made, but

are not necessary. They should reinforce these values of the reaction orders. So Rate $= k\,[NO_2]\,[F_2]$.

Selecting the last data set and rearranging the rate equation

$k = \dfrac{\text{Rate}}{[NO_2]\,[F_2]} = \dfrac{0.411\,\dfrac{\cancel{M}}{s}}{(0.400\ \cancel{M})\,(0.400\ M)} = 2.57\ M^{-1}s^{-1}$ so Rate $= 2.57\ M^{-1}s^{-1}[NO_2]\,[F_2]$ and the reactions is second

order overall.

Check: The units (none and $M^{-1}s^{-1}$) are correct. The rate law is a common form. The rate is changing as rapidly as the each concentration is consistent with first order in each reactant. The rate constant is consistent with the units necessary to get rate as M/s and the magnitude is reasonable since we have a second order reaction.

43. a) The reaction is zero order. Since the slope of the plot is independent of the concentration, there is no dependence of the concentration of the reactant in the rate law.

b) The reaction is first order. The expression for the half life of a first order reaction is $t_{1/2} = \dfrac{0.693}{k}$, which is independent of the reactant concentration.

c) The reaction is second order. The integrated rate expression for the a second order reaction is $\dfrac{1}{[A]_t} = kt + \dfrac{1}{[A]_0}$, which is linear when the inverse of the concentration is plotted versus time.

45. **Given:** table of [AB] versus time **Find:** reaction order, k, and [AB] at 25 s

Conceptual plan: **Look at the data and see if any common reaction orders can be eliminated. If the data does not show an equal concentration drop with time, then zero order can be eliminated. Look for changes in the half life (compare time for concentration to drop to one half of any value). If the half life is not constant, then the first order can be eliminated. If the half life is getting longer as the concentration drops, this might suggest second order. Plot the data as indicated by the appropriate rate law. Determine k from the slope of the plot. Finally calculate the [AB] at 25 s by using the appropriate integrated rate expression.**

Solution: By the above logic, we can eliminate both the zero order and the first order reactions. (alternatively, you could make all three plots and only one should be linear.) This suggests that we should have a second order reaction. Plot 1/[AB] versus time. Since $\dfrac{1}{[AB]_t} = kt + \dfrac{1}{[AB]_0}$, the slope will be the rate constant. The slope can be determined by measuring $\Delta y / \Delta x$ on the plot or by using functions, such as "add trendline" in Excel. Thus the rate constant is 0.0225 M^{-1} s^{-1} and the rate law is Rate $= 0.0225\ M^{-1}s^{-1}[AB]^2$.

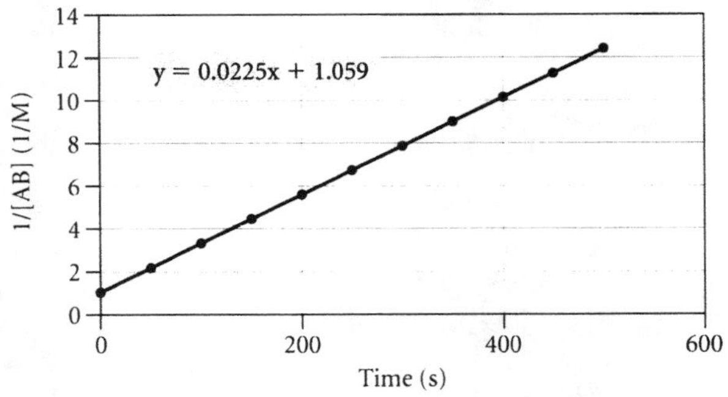

Finally, use $\dfrac{1}{[AB]_t} = kt + \dfrac{1}{[AB]_0}$, substitute in the values of $[AB]_0$, 25 s and k and rearrange to solve for [AB]

at 25 s. $[AB]_t = \dfrac{1}{kt + \dfrac{1}{[AB]_0}} = \dfrac{1}{(0.0225\ M^{-1}\cancel{s^{-1}})\,(25\ \cancel{s}) + \left(\dfrac{1}{0.950\ M}\right)} = 0.619\ M$

Check: The units (none, $M^{-1}s^{-1}$, and M) are correct. The rate law is a common form. The plot was extremely linear, confirming second order kinetics. The rate constant is consistent with the units necessary to get rate as M/s and the magnitude is reasonable since we have a second order reaction. The [AB] at 25 s is in between the values at 0 s and 50 s.

47. **Given:** table of [C_4H_8] versus time **Find:** reaction order, k, and reaction rate when [C_4H_8] = 0.25 M
Conceptual plan: Look at the data and see if any common reaction orders can be eliminated. If the data does not show an equal concentration drop with time, then zero order can be eliminated. Look for changes in half life (compare time for concentration to drop to one half of any value). If the half life is not constant, then the first order can be eliminated. If the half life is getting longer as the concentration drops, this might suggest second order. Plot the data as indicated by the appropriate rate law. Determine k from the slope of the plot. Finally calculate the reaction rate when [C_4H_8] = 0.25 M by using the rate law.
Solution: By the above logic, we can see that the reaction is most likely first order. It takes about 60 s for the concentration to be cut in half for any concentration. Plot ln [C_4H_8] versus time. Since $ln[A]_t = -kt + ln[A]_0$, the negative of the slope will be the rate constant. The slope can be determined by measuring Δy/Δx on the plot or by using functions, such as "add trendline" in Excel. Thus the rate constant is 0.0112 s^{-1} and the rate law is Rate = $0.0112\ s^{-1}[C_4H_8]$.

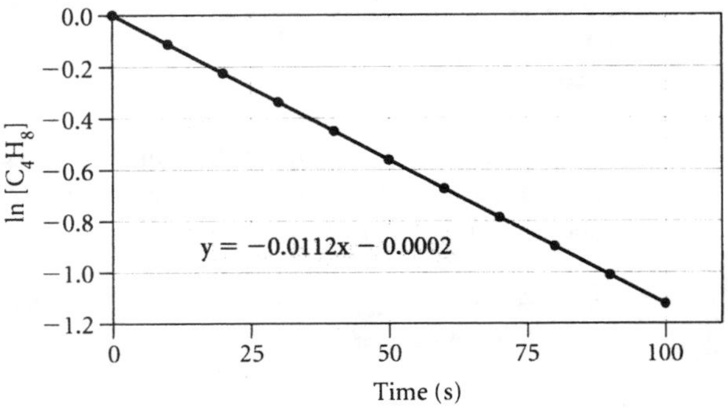

Finally, use Rate = $0.0112\ s^{-1}[C_4H_8]$, substitute in the values of [C_4H_8]. Rate = $0.0112\ s^{-1}[0.25\ M] = 2.8 \times 10^{-3}\ M\ s^{-1}$.
Check: The units (none, s^{-1}, and M s^{-1}) are correct. The rate law is a common form. The plot was extremely linear, confirming first order kinetics. The rate constant is consistent with the units necessary to get rate as M/s and the magnitude is reasonable since we have a first order reaction. The rate when [C_4H_8] = 0. 25 M s is consistent with the average rate using 90 s and 100 s.

49. **Given:** plot of ln [A] versus time has slope = - 0.0045/s; $[A]_0$ = 0.250 M
Find: a) k; b) rate law; c) $t_{1/2}$; and d) [A] after 225 s
Conceptual plan:
a) **A plot of ln [A] versus time is linear for a first order reaction. Using $ln[A]_t = -kt + ln[A]_0$, the rate constant is the negative of the slope.**
b) **Rate law is first order. Add rate constant from part a).**
c) **For a first order reaction, $t_{1/2} = \dfrac{0.693}{k}$. Substitute in k from part a).**
d) **Use the integrated rate law, $ln[A]_t = -kt + ln[A]_0$, and substitute in k and the initial concentration.**
Solution:
a) Since the rate constant is the negative of the slope, $k = 4.5 \times 10^{-3}\ s^{-1}$.
b) Since the reaction is first order, Rate = $4.5 \times 10^{-3}\ s^{-1}$ [A].
c) $t_{1/2} = \dfrac{0.693}{k} = \dfrac{0.693}{0.0045/s} = 1.5 \times 10^2\ s$.
d) $ln[A]_t = -kt + ln[A]_0$, and substitute in k and the initial concentration. So $ln[A]_t = -(0.0045/\cancel{s})(225\ \cancel{s}) + \ln 0.250\ M = -\ 2.3\underline{9}879$ and $[A]_{250s} = e^{-2.39879} = 0.0908\ M$
Check: The units (s^{-1}, none, s, and M) are correct. The rate law is a common form. The rate constant is consistent with value of the slope. The half life is consistent with a small value of k. The concentration at 225 s is consistent with being between one and two half lives.

51. **Given:** decomposition of SO_2Cl_2, first order; $k = 1.4 \times 10^{-4}\ s^{-1}$ **Find:** a) $t_{1/2}$; b) t to decrease to 25 % of $[SO_2Cl_2]_0$; c) t to 0.78 M when $[SO_2Cl_2]_0$ = 1.00 M; and d) [SO_2Cl_2] after 2.00×10^2 s and 5.00×10^2 s when $[SO_2Cl_2]_0$ = 0.150 M

Conceptual plan:

a) $k \rightarrow t_{1/2}$

$$t_{1/2} = \frac{0.693}{k}$$

b) $[SO_2Cl_2]_0$, **25 % of** $[SO_2Cl_2]_0$, $k \rightarrow t$

$$ln[A]_t = -kt + ln[A]_0$$

c) $[SO_2Cl_2]_0$, $[SO_2Cl_2]_t$, $k \rightarrow t$

$$ln[A]_t = -kt + ln[A]_0$$

d) $[SO_2Cl_2]_0$, **t**, $k \rightarrow [SO_2Cl_2]_t$

$$ln[A]_t = -kt + ln[A]_0$$

Solution:

a) $t_{1/2} = \frac{0.693}{k} = \frac{0.693}{1.42 \text{ x } 10^{-4} \text{ s}^{-1}} = 4.88 \text{ x } 10^3 \text{ s}$

b) $[SO_2Cl_2]_t = 0.25\ [SO_2Cl_2]_0$. Since $ln[SO_2Cl_2]_t = -kt + ln[SO_2Cl_2]_0$ rearrange to solve for t

$$t = -\frac{1}{k} \ln\frac{[SO_2Cl_2]_t}{[SO_2Cl_2]_0} = -\frac{1}{1.42 \text{ x } 10^{-4} \text{ s}^{-1}} \ln\frac{0.25\ \cancel{[SO_2Cl_2]_0}}{\cancel{[SO_2Cl_2]_0}} = 9.8 \text{ x } 10^3 \text{ s}$$

c) $[SO_2Cl_2]_t = 0.78$ M; $[SO_2Cl_2]_0 = 1.00$ M. Since $ln[SO_2Cl_2]_t = -kt + ln[SO_2Cl_2]_0$ rearrange to solve for t

$$t = -\frac{1}{k} \ln\frac{[SO_2Cl_2]_t}{[SO_2Cl_2]_0} = -\frac{1}{1.42 \text{ x } 10^{-4} \text{ s}^{-1}} \ln\frac{0.78\ \cancel{M}}{1.00\ \cancel{M}} = 1.8 \text{ x } 10^3 \text{ s}$$

d) $[SO_2Cl_2]_0 = 0.150$ M and $2.00 \text{ x } 10^2$ s in

$$ln[SO_2Cl_2]_t = -\left(1.42 \text{ x } 10^{-4}\ \cancel{\text{s}^{-1}}\right)\left(2.00 \text{ x } 10^2\ \cancel{\text{s}}\right) + \ln 0.150 \text{ M} = -1.9\underline{2}552 \rightarrow [SO_2Cl_2]_t = e^{-1.9\underline{2}552} = 0.146 \text{ M}$$

and $[SO_2Cl_2]_0 = 0.150$ M and $5.00 \text{ x } 10^2$ s in

$$ln[SO_2Cl_2]_t = -\left(1.42 \text{ x } 10^{-4}\ \cancel{\text{s}^{-1}}\right)\left(5.00 \text{ x } 10^2\ \cancel{\text{s}}\right) + \ln 0.150 \text{ M} = -1.9\underline{6}812 \rightarrow [SO_2Cl_2]_t = e^{-1.9\underline{6}812} = 0.140 \text{ M}$$

Check: The units (s, s, s, and M) are correct. The rate law is a common form. The half life is consistent with a small value of *k*. The time to 25% is consistent with two half lives. The time to 0.78 M is consistent with being less than one half life. The final concentrations are consistent with the time being less than one half life.

53. **Given:** $t_{1/2}$ for radioactive decay of U-238 = 4.5 billion years and independent of $[\text{U-238}]_0$
Find: t to decrease by 10 %; number U-238 atoms today, when $1.5 \text{ x } 10^{18}$ atoms formed 13.8 billion years ago
Conceptual plan: $t_{1/2}$ **independent of concentration implies first order kinetics,** $t_{1/2} \rightarrow k$ **then**

$$t_{1/2} = \frac{0.693}{k}$$

90 % of $[\text{U-238}]_0$, $k \rightarrow t$ **and** $[\text{U-238}]_0$, **t**, $k \rightarrow [\text{U-238}]_t$

$$ln[A]_t = -kt + ln[A]_0 \qquad ln[A]_t = -kt + ln[A]_0$$

Solution: $t_{1/2} = \frac{0.693}{k}$ rearrange to solve for *k*. $k = \frac{0.693}{t_{1/2}} = \frac{0.693}{4.5 \text{ x } 10^9 \text{ yr}} = 1.\underline{5}4 \text{ x } 10^{-10} \text{ yr}^{-1}$ then

$[\text{U-238}]_t = 0.10\ [\text{U-238}]_0$. Since $ln[\text{U-238}]_t = -kt + ln[\text{U-238}]_0$ rearrange to solve for t

$$t = -\frac{1}{k} \ln\frac{[\text{U-238}]_t}{[\text{U-238}]_0} = -\frac{1}{1.\underline{5}4 \text{ x } 10^{-10} \text{ yr}^{-1}} \ln\frac{0.90\ \cancel{[\text{U-238}]_0}}{\cancel{[\text{U-238}]_0}} = 6.8 \text{ x } 10^8 \text{ yr} \quad \text{and}$$

$[\text{U-238}]_0 = 1.5 \text{ x } 10^{18}$ atoms; $t = 13.8 \text{ x } 10^9$ yr in

$$ln[\text{U-238}]_t = -kt + \ln[\text{U-238}]_0 = -(1.\underline{5}4 \text{ x } 10^{-10}\ \cancel{\text{yr}^{-1}})(13.8 \text{ x } 10^9\ \cancel{\text{yr}}) + \ln\left(1.5 \text{ x } 10^{18} \text{ atoms}\right) = 39.7\underline{2}6797 \rightarrow$$

$$[\text{U-238}]_t = e^{39.7\underline{2}6797} = 1.8 \text{ x } 10^{17} \text{ atoms}$$

Check: The units (yr and atoms) are correct. The time to 10 % decay is consistent with less than one half life. The final concentration is consistent with the time being about three half lives.

55.

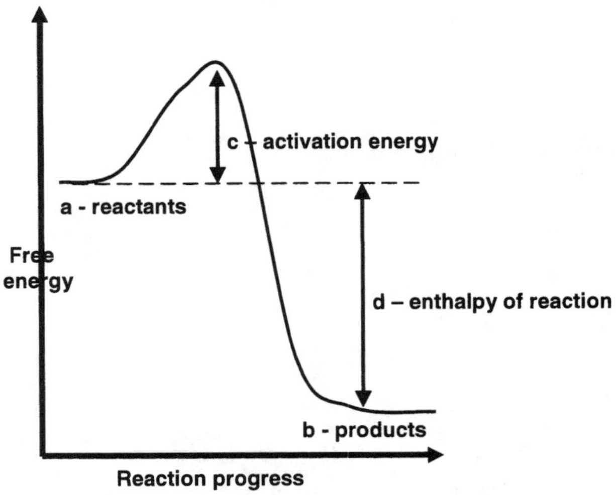

57. **Given:** activation energy = 56.8 kJ/mol, frequency factor = 1.5×10^{11} /s, 25 °C **Find:** rate constant
Conceptual plan: **°C → K and kJ/mol → J/mol then E_a, T, A → k**

$$K = °C + 273.15 \qquad \frac{1000\ J}{1\ kJ} \qquad k = A\,e^{-E_a/RT}$$

Solution: T = 25°C + 273.15 = 298 K and $\frac{56.8\ kJ}{mol} \times \frac{1000\ J}{1\ kJ} = 5.68 \times 10^4\ \frac{J}{mol}$ then

$$k = A\,e^{-E_a/RT} = (1.5 \times 10^{11}\ s^{-1})\ e^{\frac{-5.68 \times 10^4\ \frac{J}{mol}}{\left(8.314\ \frac{J}{K\ mol}\right) 298\ K}} = 17\ s^{-1}$$

Check: The units (s^{-1}) are correct. The rate constant is consistent with a large activation energy and a large frequency factor.

59. **Given:** table of rate constant versus T **Find:** E_a, and A
Conceptual plan: **Since** $\ln k = \frac{-E_a}{R}\left(\frac{1}{T}\right) + \ln A$ **a plot of ln k versus 1/T will have a slope = $-E_a/R$ and an intercept = ln A.**

Solution: The slope can be determined by measuring Δy/Δx on the plot or by using functions, such as "add trendline" in Excel. Since the slope = - 30189 K = $-E_a/R$ then

$$E_a = -(slope)R = -(-30189\ K)\left(8.314\ \frac{J}{K\ mol}\right)\left(\frac{1\ kJ}{1000\ J}\right) = 251\ \frac{kJ}{mol}$$

and intercept = 27.399 = ln A then
$A = e^{intercept} = e^{27.399} = 7.93 \times 10^{11}\ s^{-1}$.

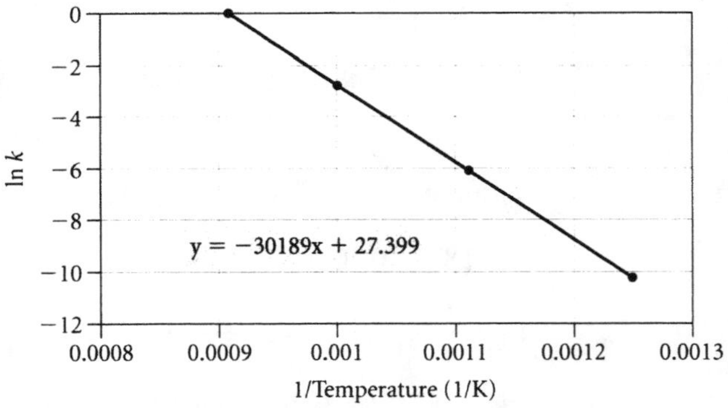

Check: The units (kJ/mol and s^{-1}) are correct. The plot was extremely linear, confirming Arrhenius behavior. The activation and frequency factor are typical form many reactions.

61. **Given:** table of rate constant versus T **Find:** E_a, and A
Conceptual plan: **Since** $\ln k = \frac{-E_a}{R}\left(\frac{1}{T}\right) + \ln A$ **a plot of ln k versus 1/T will have a slope = $-E_a/R$ and an intercept = ln A.**

Solution: The slope can be determined by measuring Δy/Δx on the plot or by using functions, such as "add trendline" in Excel. Since the slope = - 2767.2 K = $-E_a/R$ then $E_a = -(slope)R =$

$$-(-2767.2\ \text{K})\left(8.314\frac{\text{J}}{\text{K mol}}\right)\left(\frac{1\ \text{kJ}}{1000\ \text{J}}\right) = \text{ and } 23.0\ \frac{\text{kJ}}{\text{mol}}$$

intercept = 25.112 = ln A then

$A = e^{\text{intercept}} = e^{25.112} = 8.05 \times 10^{10}\ \text{s}^{-1}$.

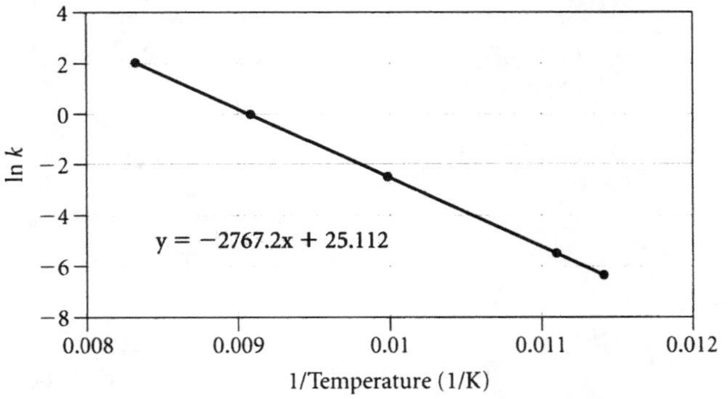

Check: The units (kJ/mol and s^{-1}) are correct. The plot was extremely linear, confirming Arrhenius behavior. The activation and frequency factor are typical form many reactions.

63. **Given:** rate constant = 0.0117/s at 400. K, and 0.689/s at 450. K
Find: a) E_a and b) rate constant at 425 K
Conceptual plan: a) k_1, T_1, k_2, T_2 → E_a then J/mol → kJ/mol b) E_a, k_1, T_1, T_2 → k_2

$$\ln\left(\frac{k_2}{k_1}\right) = \frac{E_a}{R}\left(\frac{1}{T_1} - \frac{1}{T_2}\right) \qquad \frac{1\ \text{kJ}}{1000\ \text{J}} \qquad \ln\left(\frac{k_2}{k_1}\right) = \frac{E_a}{R}\left(\frac{1}{T_1} - \frac{1}{T_2}\right)$$

Solution: $\ln\left(\frac{k_2}{k_1}\right) = \frac{E_a}{R}\left(\frac{1}{T_1} - \frac{1}{T_2}\right)$ Rearrange to solve for E_a.

$$E_a = \frac{R\ln\left(\frac{k_2}{k_1}\right)}{\left(\frac{1}{T_1} - \frac{1}{T_2}\right)} = \frac{8.314\frac{\text{J}}{\text{K mol}}\ln\left(\frac{0.689\ \text{s}^{-1}}{0.0117\ \text{s}^{-1}}\right)}{\left(\frac{1}{400.\ \text{K}} - \frac{1}{450.\ \text{K}}\right)} = 1.22 \times 10^5\ \frac{\text{J}}{\text{mol}} \times \frac{1\ \text{kJ}}{1000\ \text{J}} = 122\ \frac{\text{kJ}}{\text{mol}} \text{ and}$$

$\ln\left(\frac{k_2}{k_1}\right) = \frac{E_a}{R}\left(\frac{1}{T_1} - \frac{1}{T_2}\right)$ with k_1, = 0.0117/s, T_1 = 400. K, T_2 = 425 K Rearrange to solve for k_2.

$$\ln k_2 = \frac{E_a}{R}\left(\frac{1}{T_1} - \frac{1}{T_2}\right) + \ln k_1 = \frac{1.22 \times 10^5\ \frac{\text{J}}{\text{mol}}}{8.314\frac{\text{J}}{\text{K mol}}}\left(\frac{1}{400.\ \text{K}} - \frac{1}{425\ \text{K}}\right) + \ln 0.0117\ s^{-1} = -2.2902 \rightarrow$$

$k_2 = e^{-2.2902} = 0.101\ \text{s}^{-1}$

Check: The units (kJ/mol and s^{-1}) are correct. The activation energy is typical for a reaction. The rate constant at 425 K is in between the values given at 400 K and 450 K.

65. **Given:** rate constant doubles from at 10.0 °C to 20.0 °C **Find:** E_a
Conceptual plan: °C → K then k_1, T_1, k_2, T_2 → E_a then J/mol → kJ/mol;

$$K = °C + 273.15 \qquad \ln\left(\frac{k_2}{k_1}\right) = \frac{E_a}{R}\left(\frac{1}{T_1} - \frac{1}{T_2}\right) \qquad \frac{1\ \text{kJ}}{1000\ \text{J}}$$

Solution: T_1= 10.0 °C + 273.15 = 283.2 K and T_2 = 20.0 °C + 273.15 = 293.2 K and $k_2 = 2\ k_1$ then

$\ln\left(\frac{k_2}{k_1}\right) = \frac{E_a}{R}\left(\frac{1}{T_1} - \frac{1}{T_2}\right)$ Rearrange to solve for E_a.

$$E_a = \frac{R\ln\left(\frac{k_2}{k_1}\right)}{\left(\frac{1}{T_1} - \frac{1}{T_2}\right)} = \frac{8.314\frac{\text{J}}{\text{K mol}}\ln\left(\frac{2\ k_1}{k_1}\right)}{\left(\frac{1}{283.2\ \text{K}} - \frac{1}{293.2\ \text{K}}\right)} = 4.785 \times 10^4\ \frac{\text{J}}{\text{mol}} \times \frac{1\ \text{kJ}}{1000\ \text{J}} = 47.85\ \frac{\text{kJ}}{\text{mol}}$$

Check: The units (kJ/mol) are correct. The activation energy is typical for a reaction.

67. Reaction would have the faster rate because the orientation factor, *p*, would be larger for this reaction since the reactants are symmetrical.

69. Since the first reaction is the slow step, it is the rate determining step. Using this first step to determine the rate law, Rate = k_1 $[AB]^2$. Since this is the observed rate law, this mechanism is consistent with the experimental data.

71. a) The overall reaction is the sum of the steps in the mechanism:

$$Cl_2\ (g) \underset{k_2}{\overset{k_1}{\rightleftharpoons}} \cancel{2\,Cl\ (g)}$$

$$\cancel{Cl\ (g)} + CHCl_3\ (g) \xrightarrow[k_3]{} HCl\ (g) + \cancel{CCl_3\ (g)}$$

$$\cancel{Cl\ (g)} + \cancel{CCl_3\ (g)} \xrightarrow[k_4]{} CCl_4\ (g)$$

$$Cl_2\ (g) + CHCl_3\ (g) \rightarrow HCl\ (g) + CCl_4\ (g)$$

b) The intermediates are the species that are generated by one step and consumed by other steps. These are a Cl (g) and CCl_3 (g).

c) Since the second step is the rate determining step, Rate = k_3 [Cl] [$CHCl_3$]. Since Cl is an intermediate, its concentration can not appear in the rate law. Using the fast equilibrium in the first step, we see that $k_1[Cl_2] = k_2\,[Cl]^2$ or $[Cl] = \sqrt{\frac{k_1}{k_2}[Cl_2]}$. Substituting this into the first rate expression we get that Rate $= k_3\sqrt{\frac{k_1}{k_2}}[Cl_2]^{1/2}[CHCl_3]$. Simplifying this expression we see Rate = k $[Cl_2]^{1/2}$ [$CHCl_3$].

73. Heterogeneous catalysts require a large surface area because catalysis can only happen at the active sites on the surface. A greater surface area means greater opportunity for the substrate to react which results in a speedier reaction.

75. Assume Rate ratio α k ratio (since concentration terms will cancel each other) and $k = A\,e^{-E_a/RT}$.
T = 25 °C + 273.15 = 298 K $E_{a1} = 1.25 \times 10^5$ J/mol and $E_{a2} = 5.5 \times 10^4$ J/mol. Ratio of rates will be

$$\frac{k_2}{k_1} = \frac{\cancel{A}\,e^{-E_{a2}/RT}}{\cancel{A}\,e^{-E_{a1}/RT}} = \frac{e^{\frac{-5.5\times10^4\ \frac{\cancel{J}}{\cancel{mol}}}{\left(8.314\ \frac{\cancel{J}}{\cancel{K}\ \cancel{mol}}\right)298\ \cancel{K}}}}{e^{\frac{-1.25\times10^5\ \frac{\cancel{J}}{\cancel{mol}}}{\left(8.314\ \frac{\cancel{J}}{\cancel{K}\ \cancel{mol}}\right)298\ \cancel{K}}}} = \frac{e^{-22.199}}{e^{-50.453}} = 10^{12}.$$

77. **Given:** table of [CH_3CN] versus time **Find:** a) reaction order, k; b) $t_{1/2}$; and t for 90 % conversion
Conceptual plan: a and b) Look at the data and see if any common reaction orders can be eliminated. If the data does not show an equal concentration drop with time, then zero order can be eliminated. Look for changes in the half life (compare time for concentration to drop to one half of any value). If the half life is not constant, then the first order can be eliminated. If the half life is getting longer as the concentration drops, this might suggest second order. Plot the data as indicated by the appropriate rate law or if it is first order and there is an obvious half life in the date, a plot is not necessary. Determine k from the slope of the plot (or using half life equation for first order). c) Finally calculate the time to 90 % conversion using the appropriate integrated rate equation.
Solution: a and b) By the above logic, we can see that the reaction is first order. It takes 15.0 hr for the concentration to be cut in half for any concentration (1.000 M to 0.501 M; 0.794 M to 0.398 M; and 0.631 M to 0.316 M), so $t_{1/2}$ = 15.0 hr. Then use $t_{1/2} = \frac{0.693}{k}$ and rearrange to solve for k.

$$k = \frac{0.693}{t_{1/2}} = \frac{0.693}{15.0\ \text{hr}} = 0.0462\ \text{hr}^{-1}.$$

c) $[CH_3CN]_t = 0.10\ [CH_3CN]_0$. Since $ln[CH_3CN]_t = -kt + ln[CH_3CN]_0$ rearrange to solve for t

$$t = -\frac{1}{k}\ln\frac{[CH_3CN]_t}{[CH_3CN]_0} = -\frac{1}{0.0462\ \text{hr}^{-1}}\ln\frac{0.10\ \cancel{[CH_3CN]_0}}{\cancel{[CH_3CN]_0}} = 49.8\ \text{hr}.$$

Check: The units (none, hr^{-1}, hr, and hr) are correct. The rate law is a common form. The data showed a constant half life very clearly. The rate constant is consistent with the units necessary to get rate as M/s and the magnitude is reasonable since we have a first order reaction. The time to 90 % conversion is consistent with a time between three and four half lives.

79. **Given:** Rate = $k \dfrac{[A][C]^2}{[B]^{1/2}}$ = 0.0115 M/s at certain initial concentrations of A, B and C; double A and C concentration and triple B concentration **Find:** reaction rate

Conceptual plan: $[A]_1$, $[B]_1$, $[C]_1$, Rate 1, $[A]_2$, $[B]_2$, $[C]_2$ → Rate 2

$$\frac{\text{Rate 2}}{\text{Rate 1}} = \frac{k\dfrac{[A]_2[C]_2^2}{[B]_2^{1/2}}}{k\dfrac{[A]_1[C]_1^2}{[B]_1^{1/2}}}$$

Solution: $\dfrac{\text{Rate 2}}{\text{Rate 1}} = \dfrac{k\dfrac{[A]_2[C]_2^2}{[B]_2^{1/2}}}{k\dfrac{[A]_1[C]_1^2}{[B]_1^{1/2}}}$ Rearrange to solve for Rate 2. Rate 2 = $\dfrac{k\dfrac{[A]_2[C]_2^2}{[B]_2^{1/2}}}{k\dfrac{[A]_1[C]_1^2}{[B]_1^{1/2}}}$ Rate 1 $[A]_2 = 2[A]_1$, $[B]_2 = 3[B]_1$, $[C]_2 = 2[C]_1$ and Rate 1 = 0.0115 M/s so

$$\text{Rate 2} = \frac{k\dfrac{2[A]_1(2[C]_1)^2}{(3[B]_1)^{1/2}}}{k\dfrac{[A]_1[C]_1^2}{[B]_1^{1/2}}}\,0.0115\,\frac{M}{s} = \frac{2^3}{3^{1/2}}\,0.0115\,\frac{M}{s} = 0.0531\,\frac{M}{s}$$

Check: The units ($M\ s^{-1}$) are correct. The should increase because we have a factor of 8 (2^3) divided by the square root of three.

81. **Given:** N_2O_5 decomposes to NO_2 and O_2, first order in $[N_2O_5]$; $t_{1/2}$ = 2.81 hr at 25 °C; V = 1.5 L, $P°_{N2O5}$ = 745 torr **Find:** P_{O2} after 215 minutes

Conceptual plan: Write a balanced reaction. then $t_{1/2}$ → k then °C → K and torr → atm then

$N_2O_5 \rightarrow 2\ NO_2 + \frac{1}{2}\ O_2$ $\qquad t_{1/2} = \dfrac{0.693}{k}$ $\qquad K = °C + 273.15$ $\qquad \dfrac{1\ \text{atm}}{760\ \text{torr}}$

$P°_{N2O5}$, V, T → n/V then min → hr $[N_2O_5]_0$, t, k → $[N_2O_5]_t$ then $[N_2O_5]_0$, $[N_2O_5]_t$ → $[O_2]_t$ then

$PV = nRT$ $\qquad \dfrac{1\ \text{hr}}{60\ \text{min}}$ $\qquad \ln[A]_t = -kt + \ln[A]_0$ $\qquad [O_2]_t = ([N_2O_5]_0 - [N_2O_5]_t) \times \dfrac{1/2\ \text{mol } O_2}{1\ \text{mol } N_2O_5}$

$[O_2]_t$, V, T → $P°_{O2}$ then finally atm → torr

$PV = nRT$ $\qquad \dfrac{760\ \text{torr}}{1\ \text{atm}}$

Solution: $t_{1/2} = \dfrac{0.693}{k}$ and rearrange to solve for k. $k = \dfrac{0.693}{t_{1/2}} = \dfrac{0.693}{2.81\text{hr}} = 0.246619\ \text{hr}^{-1}$. then

T = 25 °C + 273.15 = 298 K. 745 torr x $\dfrac{1\ \text{atm}}{760\ \text{torr}}$ = 0.980263 atm then $PV = nRT$ Rearrange to solve for n/V. $\dfrac{n}{V} = \dfrac{P}{RT} = \dfrac{0.980263\ \text{atm}}{0.08206\ \dfrac{\text{L atm}}{\text{K mol}} \times 298\ \text{K}}$ = 0.0400862 M then 215 min x $\dfrac{1\ \text{hr}}{60\ \text{min}}$ = 3.58333 hr . Since

$\ln[N_2O_5]_t = -kt + \ln[N_2O_5]_0 = -(0.246619\ \text{hr}^{-1})(3.58333\ \text{hr}) + \ln(0.0400862\ \text{M}) = -4.10044$ →

$[N_2O_5]_t = e^{-4.10044} = 0.0165653$ M then

$[O_2]_t = ([N_2O_5]_0 - [N_2O_5]_t) \times \dfrac{1/2\ \text{mol } O_2}{1\ \text{mol } N_2O_5} = (0.0400862\ \dfrac{\text{mol } N_2O_5}{\text{L}} - 0.0165653\ \dfrac{\text{mol } N_2O_5}{\text{L}}) \times \dfrac{1/2\ \text{mol } O_2}{1\ \text{mol } N_2O_5}$ =

= 0.0117605 M O_2

then finally $PV = nRT$ rearrange to solve for P.

$$P = \frac{n}{V}RT = 0.0117605\,\frac{\text{mol}}{\text{L}} \times 0.08206\,\frac{\text{L atm}}{\text{K mol}} \times 298\ \text{K} = 0.287589\ \text{atm} \times \frac{760\ \text{torr}}{1\ \text{atm}} = 219\ \text{torr}$$

Check: The units (torr) are correct. The pressure is reasonable because it must be less than one half of the original pressure.

83. **Given:** I_2 formation from I atoms, second order in I; $k = 1.5 \times 10^{10}\ M^{-1}\,s^{-1}$, $[I]_0 = 0.0100$ M
Find: t to decrease by 95 %
Conceptual plan: $[I]_0$, $[I]_t$, k → t

$$\frac{1}{[A]_t} = kt + \frac{1}{[A]_0}$$

Solution: $[I]_t = 0.05\,[I]_0 = 0.05 \times 0.0100\ M = 0.0005$ M. Since $\frac{1}{[I]_t} = kt + \frac{1}{[I]_0}$ rearrange to solve for t

$$t = \frac{1}{k}\left(\frac{1}{[I]_t} - \frac{1}{[I]_0}\right) = \frac{1}{(1.5 \times 10^{10}\ M^{-1}\,s^{-1})}\left(\frac{1}{0.0005\ M} - \frac{1}{0.0100\ M}\right) = 1.267 \times 10^{-7}\ s = 1 \times 10^{-7}\ s.$$

Check: The units (s) are correct. We expect the time to be extremely small because the rate constant is so large.

85. a) There are two elementary steps in the reaction mechanism because there are two peaks in the reaction progress diagram.

b)

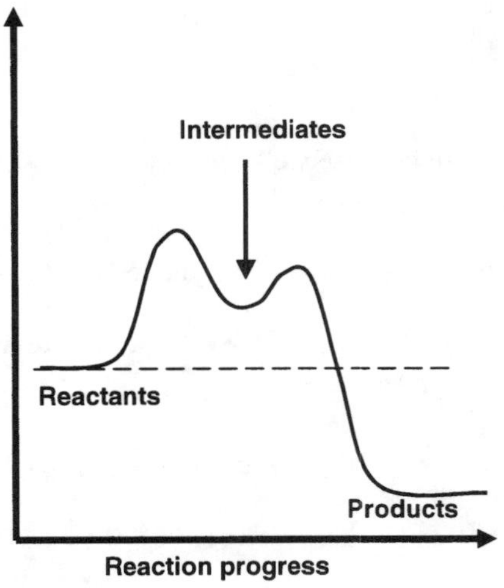

c) The first step is the rate limiting step because it has the higher activation energy.

d) The overall reaction is exothermic because the products are at a lower energy than the reactants.

87. **Given:** n-butane desorption from single crystal aluminum oxide, first order; $k = 0.128\ s^{-1}$ at 150 K; initially completely covered **Find:** a) $t_{1/2}$; b) t for 25 % and for 50 % to desorb; c) fraction remaining after 10 s and 20 s
Conceptual plan: a) k → $t_{1/2}$ b) $[C_4H_{10}]_0$, $[C_4H_{10}]_t$, k → t c) $[C_4H_{10}]_0$, t, k → $[C_4H_{10}]_t$

$$t_{1/2} = \frac{0.693}{k} \qquad \ln[A]_t = -kt + \ln[A]_0 \qquad \ln[A]_t = -kt + \ln[A]_0$$

Solution:

a) $t_{1/2} = \frac{0.693}{k} = \frac{0.693}{0.128\ s^{-1}} = 5.41\ s.$

b) $ln[C_4H_{10}]_t = -kt + ln[C_4H_{10}]_0$ rearrange to solve for t. For 25 % desorbed $[C_4H_{10}]_0 = 0.75\ [C_4H_{10}]_t$

and $t = -\frac{1}{k}\ln\frac{[C_4H_{10}]_t}{[C_4H_{10}]_0} = -\frac{1}{0.128\ s^{-1}}\ln\frac{0.75\ [C_4H_{10}]_0}{[C_4H_{10}]_0} = 2.2\ s$. For 50 % desorbed $[C_4H_{10}]_0 =$

$0.50\ [C_4H_{10}]_t$ and $t = -\frac{1}{k}\ln\frac{[C_4H_{10}]_t}{[C_4H_{10}]_0} = -\frac{1}{0.128\ s^{-1}}\ln\frac{0.50\ [C_4H_{10}]_0}{[C_4H_{10}]_0} = 5.4\ s$.

c) For 10 s $ln[C_4H_{10}]_t = -kt + ln[C_4H_{10}]_0 = -(0.128\ s^{-1})(10\ s) + \ln(1.00) = -1.2\underline{8}$ →

$[C_4H_{10}]_t = e^{-1.2\underline{8}} = 0.28$ = fraction covered and for 20 s

$ln[C_4H_{10}]_t = -kt + ln[C_4H_{10}]_0 = -(0.128\ s^{-1})(20\ s) + \ln(1.00) = -2.5\underline{6}$ →

$[C_4H_{10}]_t = e^{-2.5\underline{6}} = 0.077$ = fraction covered.

Check: The units (s, s, s, none, and none) are correct. The half life is reasonable considering the size of the rate constant. The time to 25 % desorbed is less than one half life. The time to 50 % desorbed is the half life. The fraction at 10 s is consistent with about two half lives. The fraction covered at 20 s is consistent with about four half lives.

89. a) **Given:** table of rate constant versus T **Find:** E_a, and A

Conceptual plan: First convert temperature data into Kelvins (°C + 273.15 = K). Since $\ln k = \frac{-E_a}{R}\left(\frac{1}{T}\right) + \ln A$ **a plot of ln *k* versus 1/T will have a slope = $-E_a/R$ and an intercept = ln A.**

Solution: The slope can be determined by measuring Δy/Δx on the plot or by using functions, such as "add trendline" in Excel. Since the slope = - 10$\underline{7}$59 K = $-E_a/R$ then

$E_a = -(slope)R =$

$= -(-10\underline{7}59\ K)\left(8.314\ \frac{J}{K\ mol}\right)\left(\frac{1\ kJ}{1000\ J}\right)$

$= 89.5\ \frac{kJ}{mol}$

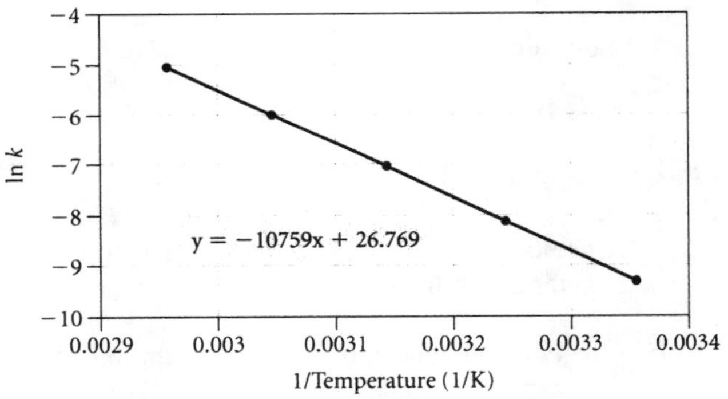

and intercept = 26.7$\underline{6}$9 = ln A then

$A = e^{intercept} = e^{26.7\underline{6}9} = 4.22 \times 10^{11}\ s^{-1}$.

Check: The units (kJ/mol and s^{-1}) are correct. The plot was extremely linear, confirming Arrhenius behavior. The activation and frequency factor are typical form many reactions.

b) **Given:** part a) results **Find:** *k* at 15 °C

Conceptual plan: °C → K then T, E_a, A → *k*

$°C + 273.15 = K$ $\ln k = \frac{-E_a}{R}\left(\frac{1}{T}\right) + \ln A$

Solution: 15 °C + 273.15 = 288 K then

$$\ln k = \frac{-E_a}{R}\left(\frac{1}{T}\right) + \ln A = \frac{-89.5\ \frac{kJ}{mol} \times \frac{1000\ J}{1\ kJ}\ E_a}{8.314\ \frac{J}{K\ mol}}\left(\frac{1}{288\ K}\right) + \ln(4.22 \times 10^{11}\ s^{-1}) = -1\underline{0}.610 \rightarrow$$

$k = e^{-1\underline{0}.610} = 2.5 \times 10^{-5}\ M^{-1}s^{-1}$.

Check: The units ($M^{-1}\ s^{-1}$) are correct. The value of the rate constant is less than the value at 25 °C.

c) **Given:** part a) results, 0.155 M C_2H_5Br and 0.250 M OH^- at 75 °C **Find:** initial reaction rate

Conceptual plan: °C → K then T, E_a, A → *k* then *k*, $[C_2H_5Br]$, $[OH^-]$ → initial reaction rate

$°C + 273.15 = K$ $\ln k = \frac{-E_a}{R}\left(\frac{1}{T}\right) + \ln A$ $Rate = k\ [C_2H_5Br]\ [OH^-]$

Solution: 75 °C + 273.15 = 348 K then

$$\ln k = \frac{-E_a}{R}\left(\frac{1}{T}\right) + \ln A = \frac{-89.5 \frac{\cancel{kJ}}{\cancel{mol}} \times \frac{1000\ \cancel{J}}{1\ \cancel{kJ}} E_a}{8.314 \frac{\cancel{J}}{\cancel{K}\ \cancel{mol}}}\left(\frac{1}{348\ \cancel{K}}\right) + \ln(4.22 \times 10^{11}\ s^{-1}) = -4.1656 \rightarrow$$

$k = e^{-4.1656} = 1.5521 \times 10^{-2}\ M^{-1}\ s^{-1}$.

Rate = $k\,[C_2H_5Br]\,[OH^-] = (1.5521 \times 10^{-2}\ \cancel{M^{-1}}\ s^{-1})\,(0.155\ \cancel{M})\,(0.250\ M) = 6.0 \times 10^{-4}\ M\ s^{-1}$

Check: The units ($M\ s^{-1}$) are correct. The value of the rate is reasonable considering the value of the rate constant (larger than in the table) and the fact that the concentrations are less than 1 M.

91. a) No, because the activation energy is zero. This means that the rate constant ($k = A\,e^{-E_a/RT}$) will be independent of temperature.

b) No bond is broken and the two radicals attract to each other.

c) Formation of diatomic gases from atomic gases.

93. **Given:** $t_{1/2}$ for radioactive decay of C-14 = 5730 years; bone has 19.5 % C-14 in living bone
Find: age of bone
Conceptual plan: radioactive decay implies first order kinetics, $t_{1/2} \rightarrow k$ then 19.5 % of $[C\text{-}14]_0$, $k \rightarrow t$

$$t_{1/2} = \frac{0.693}{k} \qquad\qquad ln[A]_t = -kt + ln[A]_0$$

Solution: $t_{1/2} = \frac{0.693}{k}$ rearrange to solve for k. $k = \frac{0.693}{t_{1/2}} = \frac{0.693}{5730\ yr} = 1.20942 \times 10^{-4}\ yr^{-1}$ then

$[C\text{-}14]_t = 0.195\ [C\text{-}14]_0$. Since $ln[C\text{-}14]_t = -kt + ln[C\text{-}14]_0$ rearrange to solve for t

$$t = -\frac{1}{k}\ln\frac{[C\text{-}14]_t}{[C\text{-}14]_0} = -\frac{1}{1.20942 \times 10^{-4}\ yr^{-1}}\ln\frac{0.195\ \cancel{[C\text{-}14]_0}}{\cancel{[C\text{-}14]_0}} = 1.35 \times 10^{4}\ yr.$$

Check: The units (yr) are correct. The time to 19.5 % decay is consistent with the time being between two and three half lives.

95. a) For each, check that all steps sum to overall reaction and that the predicted rate law is consistent with experimental data (Rate = $k\,[H_2]\,[I_2]$).
For the first mechanism, the single step is the overall reaction. The rate law is determined by the stoichiometry, so Rate = $k\,[H_2]\,[I_2]$ and the mechanism is valid.
For the second mechanism, the overall reaction is the sum of the steps in the mechanism:

$$I_2\,(g) \underset{k_2}{\overset{k_1}{\rightleftharpoons}} \cancel{2\,I\,(g)}$$

$$H_2\,(g) + \cancel{2\,I\,(g)} \xrightarrow{k_3} 2\,HI\,(g)$$ So the sum matches the overall reaction.

$$H_2\,(g) + I_2\,(g) \rightarrow 2\,HI\,(g)$$

Since the second step is the rate determining step, Rate = $k_3\,[H_2]\,[I]^2$. Since I is an intermediate, its concentration can not appear in the rate law. Using the fast equilibrium in the first step, we see that $k_1[I_2] = k_2\,[I]^2$ or $[I]^2 = \frac{k_1}{k_2}[I_2]$.

Substituting this into the first rate expression we get that Rate = $k_3\frac{k_1}{k_2}[H_2]\,[I_2]$ and the mechanism is valid.

For the third mechanism, the overall reaction is the sum of the steps in the mechanism:

$$I_2\,(g) \underset{k_2}{\overset{k_1}{\rightleftharpoons}} \cancel{2\,I\,(g)}$$

$$H_2\,(g) + \cancel{I\,(g)} \underset{k_4}{\overset{k_3}{\rightleftharpoons}} \cancel{H_2I\,(g)}$$ So the sum matches the overall reaction.

$$\cancel{H_2I\,(g)} + \cancel{I\,(g)} \xrightarrow{k_5} 2\,HI\,(g)$$

$$H_2\,(g) + I_2\,(g) \rightarrow 2\,HI\,(g)$$

Since the third step is the rate determining step, Rate = k_5 [H_2I] [I]. Since H_2I and I are intermediates, their concentrations can not appear in the rate law. Using the fast equilibrium in the first step, we see that $k_1[I_2] = k_2\,[I]^2$ or $[I] = \sqrt{\frac{k_1}{k_2}[I_2]}$. Using the fast equilibrium in the second step, we see that $k_3[H_2]$ $[I] = k_4\,[H_2I]$ or $[H_2I] = \frac{k_3}{k_4}[H_2]\,[I]$. Substituting the both of these expressions for the intermediates into the first rate expression we get that

Rate = $k_5\,[H_2I]\,[I] = k_5 \frac{k_3}{k_4}[H_2]\,[I]\,\sqrt{\frac{k_1}{k_2}[I_2]}$. Substituting again for [I] we get Rate = $k_5 \frac{k_3}{k_4}[H_2]$ $\sqrt{\frac{k_1}{k_2}[I_2]}\,\sqrt{\frac{k_1}{k_2}[I_2]}$.

Simplifying this expression we see Rate = $k_5 \frac{k_3}{k_4}\frac{k_1}{k_2}[H_2]\,[I_2]$ and the mechanism is valid.

b) To distinguish between mechanisms, you could look for the buildup of I(*g*) and/or H_2I(*g*), the intermediates.

97. a) For a zero order reaction, the rate is independent of the concentration. If the first half goes in the first 100 s, the second half will go in the second 100 s. This means that there will be none or 0 % left at 200 s.

b) For a first order reaction, the half life is independent of concentration. This means that if half of the reactant decomposes in the first 100 s, then half of this (or another 25 % of the original amount) will decompose in the second 100 s. This means that at 200 s 50 % + 25 % = 75 % has decomposed or 25 % remains.

c) For a second order reaction, $t_{1/2} = \frac{1}{k[A]_0} = 100\text{ s}$ and the integrated rate expression is $\frac{1}{[A]_t} = kt + \frac{1}{[A]_0}$.

We can rearrange the first expression to solve for k as $k = \frac{1}{100\text{ s}[A]_0}$. Substituting this and 200 s into the integrated rate expression we get $\frac{1}{[A]_t} = \frac{200\,\cancel{s}}{100\,\cancel{s}[A]_0} + \frac{1}{[A]_0}$ → $\frac{1}{[A]_t} = \frac{3}{[A]_0}$ → $\frac{[A]_t}{[A]_0} = \frac{1}{3}$ or 33 % remains.

99. Using the energy diagram show and using Hess' Law, we can see that the activation energy for the decomposition is equal to the activation energy for the formation reaction plus the heat of formation of 2 moles of HI or $E_{aformation} = E_{adecomposition} + 2\,\Delta H_f^0(\text{HI})$. So $E_{aformation} = 185\text{ kJ} + 2\text{mol}(-5.65\text{ kJ/mol}) = 174\text{ kJ}$

.

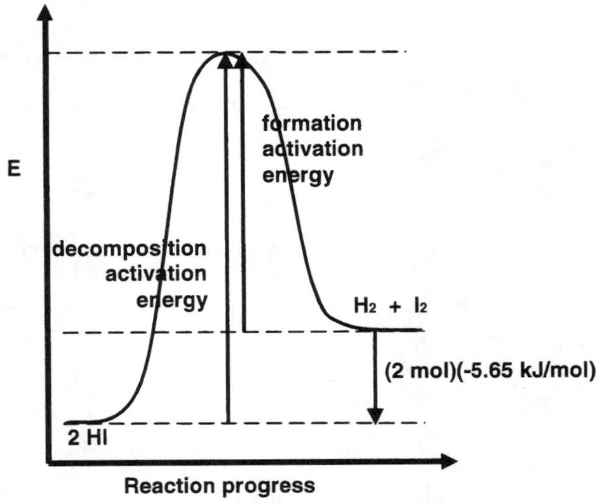

Note: energy axis is not to scale.

Check: Since the reaction is endothermic, we expect the activation energy in the reverse direction to be less in the reverse direction.

101. a) Since the rate determining step involves the collision of two molecules, the expected reaction order would be second order.

b) The proposed mechanism is

$$CH_3NC + \cancel{CH_3NC} \underset{k_2}{\overset{k_1}{\rightleftharpoons}} \cancel{CH_3NC^*} + \cancel{CH_3NC} \qquad \text{(fast)}$$

$$\cancel{CH_3NC^*} \xrightarrow{k_3} CH_3CN \qquad \text{(slow)}$$ So the sum matches the overall reaction.

$$CH_3NC \rightarrow CH_3CN$$

Since the second step is the rate determining step, Rate = $k_3[CH_3NC^*]$. Since CH_3NC^* is an intermediate, its concentration can not appear in the rate law. Using the fast equilibrium in the first step, we see that $k_1[CH_3NC]^2 = k_2[CH_3NC^*][CH_3NC]$ or $[CH_3NC^*] = \frac{k_1}{k_2}[CH_3NC]$. Substituting this into the first rate expression we get that Rate = $k_3 \frac{k_1}{k_2}[CH_3NC]$, which simplifies to Rate = $k\,[CH_3NC]$.

This matches the experimental observation of first order and the mechanism is valid.

103. a) **Given:** N_2O_5 decomposes to NO_2 and O_2, first order in $[N_2O_5]$; $k = 7.48 \times 10^{-3}\ s^{-1}$; $P°_{N2O5} = 0.100$ atm
Find: t to $P_{Total} = 0.145$ atm
Conceptual plan: Write a balanced reaction. then

$N_2O_5 \rightarrow 2\,NO_2 + ½\,O_2$

Write expression for P_{Total} in terms of amount reacted x, $P°_{N2O5}$, k → t

let x = $P_{N2O5reacted}$ $P_{Total} = P_{N2O5} + P_{NO2} + P_{O2}$ $\qquad ln[A]_t = -kt + ln[A]_0$

Solution: $P_{Total} = P_{N2O5} + P_{NO2} + P_{O2} = (0.100\,\text{atm} - x) + (2x) + (1/2\,x) = 0.100\ \text{atm} + 1.5\,x$ Set $P_{Total} = 0.145\ \text{atm} = 0.100\ \text{atm} + 1.5\,x$ and solve for x. $x = \frac{0.145\ \text{atm} - 0.100\ \text{atm}}{1.5} = 0.030\ \text{atm}$ then $P_{N2O5} = 0.100\ \text{atm} - x = 0.100\ \text{atm} - 0.030\ \text{atm} = 0.070\ \text{atm}$. Since P α n/V or M

$ln[N_2O_5]_t = -kt + ln[N_2O_5]_0$ rearrange to solve for t.

$$t = -\frac{ln\frac{[N_2O_5]_t}{[N_2O_5]_0}}{k} = -\frac{\ln\left(\frac{0.070\ \cancel{\text{atm}}}{0.100\ \cancel{\text{atm}}}\right)}{7.48 \times 10^{-3}\ s^{-1}} = 4\underline{7}.684\ s = 48\ s$$

Check: The units (s) are correct. The time is reasonable because it is less than one half life and the amount decomposing is less that 50 %.

b) **Given:** N_2O_5 decomposes to NO_2 and O_2, first order in $[N_2O_5]$; $k = 7.48 \times 10^{-3}\ s^{-1}$; $P°_{N2O5} = 0.100$ atm
Find: t to $P_{Total} = 0.200$ atm
Conceptual plan: Write a balanced reaction. then

$N_2O_5 \rightarrow 2\,NO_2 + ½\,O_2$

Write expression for P_{Total} in terms of amount reacted x, $P°_{N2O5}$, k → t

let x = $P_{N2O5reacted}$ $P_{Total} = P_{N2O5} + P_{NO2} + P_{O2}$ $\qquad ln[A]_t = -kt + ln[A]_0$

Solution: $P_{Total} = P_{N2O5} + P_{NO2} + P_{O2} = (0.100\,\text{atm} - x) + (2x) + (1/2\,x) = 0.100\ \text{atm} + 1.5\,x$ Set $P_{Total} = 0.200\ \text{atm} = 0.100\ \text{atm} + 1.5\,x$ and solve for x. $x = \frac{0.200\ \text{atm} - 0.100\ \text{atm}}{1.5} = 0.06\underline{6}667\ \text{atm}$ then $P_{N2O5} = 0.100\ \text{atm} - x = 0.100\ \text{atm} - 0.06\underline{6}667\ \text{atm} = 0.03\underline{3}333\ \text{atm}$. Since P α n/V or M

$ln[N_2O_5]_t = -kt + ln[N_2O_5]_0$ rearrange to solve for t.

$$t = -\frac{ln\frac{[N_2O_5]_t}{[N_2O_5]_0}}{k} = -\frac{\ln\left(\frac{0.03\underline{3}333\ \cancel{\text{atm}}}{0.100\ \cancel{\text{atm}}}\right)}{7.48 \times 10^{-3}\ s^{-1}} = 1\underline{4}6.873\ s = 150\ s = 1.5 \times 10^2\ s.$$

Check: The units (s) are correct. The time is reasonable because it is between one and two half lives and the amount decomposing is 67 %.

c) **Given:** N_2O_5 decomposes to NO_2 and O_2, first order in $[N_2O_5]$; $k = 7.48 \times 10^{-3}\ s^{-1}$; $P°_{N2O5} = 0.100$ atm
Find: P_{Total} after 100 s
Conceptual plan: $P°_{N2O5}$, *k*, t → P_{N2O5} then Write a balanced reaction. then

$ln[A]_t = -kt + ln[A]_0$ $\qquad N_2O_5 \rightarrow 2\,NO_2 + ½\,O_2$

$P°_{N2O5}$, P_{N2O5} → x then Write expression for P_{Total} in terms of amount reacted.

$x = P^0_{N2O5} - P_{N2O5}$ let x = $P_{N2O5reacted}$ $P_{Total} = P_{N2O5} + P_{NO2} + P_{O2}$

Solution: Since P α n/V or M

$ln[N_2O_5]_t = -kt + ln[N_2O_5]_0 = -\left(7.48\ x\ 10^{-3}\ s^{-1}\right)(100\ s) + \ln(0.100\ atm) = -3.0\underline{5}059$

$P_{N2O5} = e^{-3.0\underline{5}059} = 0.047\underline{3}312$ atm so $x = P^0_{N2O5} - P_{N2O5} = 0.100$ atm - $0.047\underline{3}312$ atm $= 0.05\underline{2}669$ atm finally

$P_{Total} = P_{N2O5} + P_{NO2} + P_{O2} = (0.100\text{atm} - x) + (2x) + (1/2x) = 0.100\text{ atm} + 1.5\,x =$

$= 0.100\text{ atm} + 1.5(0.05\underline{2}669\text{ atm}) = 0.179\text{ atm}$

Check: The units (atm) are correct. The pressure is reasonable because the time is between those for parts a and b.

105. **Given:** N_2O_3 decomposes to NO_2 and NO, first order in $[N_2O_3]$, table of $[NO_2]$ versus time, at 50,000 s = $[N_2O_3] = 0$ **Find:** *k*

Conceptual plan: Write a balanced reaction. then Write expression for P_{NO2} in terms P_{N2O3} then

$N_2O_3 \rightarrow NO_2 + NO$ $P_{N2O3} = 0.784\text{ atm} - P_{NO2}$

Plot ln P_{N2O3} versus time (this will have the same slope as ln $[N_2O_3]$ versus time). Since $ln[A]_t = -kt + ln[A]_0$, the negative of the slope will be the rate constant.

Solution: $P_{N2O3} = 0.784\text{ atm} - P_{NO2}$ Plot ln P_{N2O3} versus time (this will have the same slope as ln $[N_2O_3]$ versus time). Since $ln[A]_t = -kt + ln[A]_0$, the negative of the slope will be the rate constant. The slope can be determined by measuring Δy/Δx on the plot or by using functions, such as "add trendline" in Excel. The last point (50,000 s) can not be plotted because the concentration is 0 and the ln 0 is undefined, Thus the rate constant is 3.20 x 10^{-4} s^{-1} and the rate law is Rate = $3.20\,10^{-4}\,s^{-1}[N_2O_3]$.

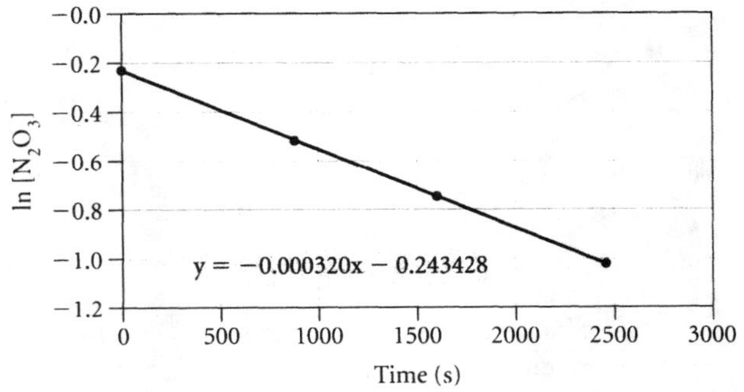

Check: The units (s^{-1}) are correct. The rate constant is typical for a reaction.

107. Reaction A is second order. A plot of 1/[A] versus time will be liner $\left(\frac{1}{[A]_t} = kt + \frac{1}{[A]_0}\right)$. Reaction B is first order. A plot of ln [A] versus time will be linear ($ln[A]_t = -kt + ln[A]_0$). Reactant concentrations drop more quickly for first order reactions than for second order reactions.

Chapter 14
Chemical Equilibrium

1. Like adult hemoglobin, fetal hemoglobin is in equilibrium with oxygen. However, the equilibrium constant for fetal hemoglobin is larger than the equilibrium constant for adult hemoglobin, meaning that the reaction tends to go farther in the direction of the product. Consequently, fetal hemoglobin loads oxygen at a lower oxygen concentration than adult hemoglobin. In the placenta, fetal blood flows in close proximity to maternal blood, without the two ever mixing. Because of the different equilibrium constants, the maternal hemoglobin unloads oxygen which the fetal hemoglobin then binds and carries into its own circulatory system. Nature has thus evolved a chemical system through which the mother's hemoglobin can in effect hand off oxygen to the hemoglobin of the fetus.

3. The general expression of the equilibrium constant is: $K = \frac{[\mathrm{C}]^c[\mathrm{D}]^d}{[\mathrm{A}]^a[\mathrm{B}]^b}$

5. If you reverse the equation, invert the equilibrium constant. $K_{\text{reverse}} = \frac{1}{K_{\text{forward}}}$

   If you multiply the coefficients in the equation by a factor, raise the equilibrium constant to the same factor. That is, if you multiply the reaction by n, raise the equilibrium constant K to the n. $K' = K^n$

7. K_c is the equilibrium constant with respect to concentration in molarity. K_p is the equilibrium constant with respect to partial pressures in atmospheres. The two can be related: $K_p = K_c(\mathrm{RT})^{\Delta n}$

9. The concentration of a solid does not change because a solid does not expand to fill its container. Its concentration, therefore, depends only on its density, which is constant as long as some solid is present. Consequently, solids are not included in the equilibrium expression. Similarly, the concentration of a pure liquid does not change, so, liquids are also excluded from the equilibrium expression.

11. When we know the initial concentrations of the reactants and products and the equilibrium concentration of one reactant or product, the other equilibrium concentrations can be deduced from the stoichiometry of the reaction. From the initial and equilibrium concentration of the one reactant, we can find the change in concentration for that reactant. Using the stoichiometry of the reaction, we can determine the equilibrium concentrations of the other reactants and products. We generally use an ICE table to keep track of the changes.

13. Standard states are defined as 1 M concentration, or 1 atm pressure. So the value of K_c and K_p when the reactants and products are in their standard states is 1.

15. To solve a problem given initial concentrations and the equilibrium constant you would: Prepare an ICE table; and then calculate Q; compare Q and K_c and predict the direction of the reaction; represent the change with x; sum the table and determine the equilibrium values; put the equilibrium values in the equilibrium expression and solve for x. Determine the reactant and product concentration.

17. According to LeChatelier's principle, when a chemical system at equilibrium is disturbed, the system shifts in a direction that minimizes the disturbance.

19. If a chemical system is at equilibrium:
    Decreasing the volume causes the reaction to shift in the direction that has the fewer moles of gas particles.
    Increasing the volume causes the reaction to shift in the direction that has the greater number of moles of gas particles.
    If a reaction has an equal number of moles of gas on both sides of the chemical equation, then a change in volume produces no effect on the equilibrium.
    Adding an inert gas to the mixture at a fixed volume has no effect on the equilibrium.

21. The equilibrium constant is defined as the concentrations of the products raised to their stoichiometric coefficients divided by the concentrations of the reactants raised to their stoichiometric coefficients.

a) $K=\dfrac{[SbCl_3][Cl_2]}{[SbCl_5]}$

b) $K=\dfrac{[NO]^2[Br_2]}{[BrNO]^2}$

c) $K=\dfrac{[CS_2][H_2]^4}{[CH_4][H_2S]^2}$

d) $K=\dfrac{[CO_2]^2}{[CO]^2[O_2]}$

23. With an equilibrium constant of 1.4 x 10^{-5}, the value of the equilibrium constant is small, therefore, the concentration of reactants will be greater than the concentration of reactants. This is independent of the initial concentration of the reactants and products.

25. (i) has 10 H_2 and 10 I_2
(ii) has 7 H_2 and 7 I_2 and 6 HI
(iii) has 5 H_2 and 5 I_2 and 10 HI
(iv) has 4 H_2 and 4 I_2 and 12 HI
(v) has 3 H_2 and 3 I_2 and 14 HI
(vi) has 3 H_2 and 3 I_2 and 14 HI
a) Concentration of (v) and (vi) are the same so the system reached equilibrium at (v).

b) If a catalyst was added to the system, the system would reach the conditions at (v) sooner, since a catalyst speeds up the reaction but does not change the equilibrium conditions.

c) The final figure (vi)would have the same amount of reactants and products since a catalyst speeds up the reaction, but does not change the equilibrium concentrations.

27. a) If you reverse the reaction, invert the equilibrium constant. So, $K' = \dfrac{1}{K_p} = \dfrac{1}{2.26x10^4} = 4.42 \text{ x } 10^{-5}$. The reactants will be favored.

b) If you multiply the coefficients in the equation by a factor, raise the equilibrium constant to the same factor.
So, $K' = (K_p)^{1/2} = (2.26 \text{ x } 10^4)^{1/2} = 1.50 \text{ x } 10^2$. The products will be favored.

c) Begin with the reverse of the reaction and invert the equilibrium constant.
$K_{reverse} = \dfrac{1}{K_p} = \dfrac{1}{2.26x10^4} = 4.42 \text{ x } 10^{-5}$
Then, multiply the reaction by 2 and raise the value of K_{revere} to the 2 power.
$K' = (K_{reverse})^2 = (4.42 \text{ x } 10^{-5})^2 = 1.96 \text{ x } 10^{-9}$. The reactants will be favored.

29. To find the equilibrium constant for reaction 3, you need to combine reactions 1 and 2 to get reaction 3. Begin by reversing reaction 2, then multiply reaction 1 by 2 and add the two new reactions. When you add reactions you multiply the values of *K*.

$N_2(g) + O_2(g) \rightleftharpoons \cancel{2NO}(g) \qquad K_1 = \frac{1}{K_p} = \frac{1}{2.1 \times 10^{30}} = 4.\underline{7}6 \times 10^{-31}$

$\cancel{2NO}(g) + Br_2(g) \rightleftharpoons 2\,NOBr(g) \qquad K_2 = (K_p)^2 = (5.3)^2 = 2\underline{8}.09$

$N_2(g) + O_2(g) + Br_2(g) \rightleftharpoons 2\,NOBr(g) \qquad K_3 = K_1K_2 = (4.\underline{7}6 \times 10^{-31})(2\underline{8}.09) = 1.3 \times 10^{-29}$

31. a) **Given:** $K_p = 6.26 \times 10^{-22}$ **Find:** K_c

Conceptual Plan: $K_p \rightarrow K_c$

$$K_p = K_c(RT)^{\Delta n}$$

Solution: Δn = mol product gas – mol reactant gas = 2 – 1 = 1

$$K_c = \frac{K_p}{(RT)^{\Delta n}} = \frac{6.26 \times 10^{-22}}{\left(0.08206 \frac{L \bullet atm}{mol \bullet \cancel{K}} \times 298\ \cancel{K}\right)^1} = 2.56 \times 10^{-23}$$

Check: Substitute into the equation and confirm that you get the original value of K_p.

$$K_p = K_c(RT)^{\Delta n} = (2.56 \times 10^{-23})\left(0.08206 \frac{L \bullet atm}{mol \bullet K} \times 298\right)^1 = 6.26 \times 10^{-22}$$

b) **Given:** $K_p = 7.7 \times 10^{24}$ **Find:** K_c

Conceptual Plan: $K_p \rightarrow K_c$

$$K_p = K_c(RT)^{\Delta n}$$

Solution: Δn = mol product gas – mol reactant gas = 4 – 2 = 2

$$K_c = \frac{K_p}{(RT)^{\Delta n}} = \frac{7.7 \times 10^{24}}{\left(0.08206 \frac{L \bullet atm}{mol \bullet \cancel{K}} \times 298\ \cancel{K}\right)^2} = 1.3 \times 10^{22}$$

Check: Substitute into the equation and confirm that you get the original value of K_p.

$$K_p = K_c(RT)^{\Delta n} = (1.3 \times 10^{22})\left(0.08206 \frac{L \bullet atm}{mol \bullet K} \times 298\right)^2 = 7.7 \times 10^{24}$$

c) **Given:** $K_p = 81.9$ **Find:** K_c

Conceptual Plan: $K_p \rightarrow K_c$

$$K_p = K_c(RT)^{\Delta n}$$

Solution: Δn = mol product gas – mol reactant gas = 2 – 2 = 0

$$K_c = \frac{K_p}{(RT)^{\Delta n}} = \frac{81.9}{\left(0.08206 \frac{L \bullet atm}{mol \bullet \cancel{K}} \times 298\ \cancel{K}\right)^0} = 81.9$$

Check: Substitute into the equation and confirm that you get the original value of K_p.

$$K_p = K_c(RT)^{\Delta n} = (81.9)\left(0.08206 \frac{L \bullet atm}{mol \bullet K} \times 298\right)^0 = 81.9$$

33. a) Since H_2O is a liquid, it is omitted from the equilibrium expression. $K_{eq} = \frac{[HCO_3^-][OH^-]}{[CO_3^{2-}]}$

b) Since $KClO_3$ and KCl are both solids they are omitted from the equilibrium expression. $K_{eq} = [O_2]^3$

c) Since H_2O is a liquid, it is omitted from the equilibrium expression. $K_{eq} = \frac{[H_3O^+][F^-]}{[HF]}$

d) Since H_2O is a liquid, it is omitted from the equilibrium expression. $K_{eq} = \frac{[NH_4^+][OH^-]}{[NH_3]}$

35. **Given:** At equilibrium: [CO] = 0.105M, $[H_2]$ = 0.114M, $[CH_3OH]$ = 0.185M **Find:** K_c
Conceptual Plan: Balanced reaction → equilibrium expression → K_c

Solution: $K_c = \frac{[CH_3OH]}{[CO][H_2]^2} = \frac{(0.185)}{(0.105)(0.114)^2} = 136$

Check: The answer is reasonable since the concentration of products is greater than the concentration of reactants and the equilibrium constant should be greater than 1.

37. At 500K: **Given:** At equilibrium: $[N_2]$ = 0.115M, $[H_2]$ = 0.105M, and $[NH_3]$ = 0.439 **Find:** K_c
Conceptual Plan: Balanced reaction → equilibrium expression → K_c

Solution: $K_c = \frac{[NH_3]^2}{[N_2][H_2]^3} = \frac{(0.439)^2}{(0.115)(0.105)^3} = 1.45 \times 10^3$

Check: The value is reasonable since the concentration of products is greater than the concentration of reactants.

At 575K: **Given:** At equilibrium: $[N_2]$ = 0.110M, $[NH_3]$ = 0.128, K_c = 9.6 **Find:** $[H_2]$
Conceptual Plan: Balanced reaction → equilibrium expression → $[H_2]$

Solution: $K_c = \frac{[NH_3]^2}{[N_2][H_2]^3}$ $9.6 = \frac{(0.128)^2}{(0.110)(x)^3}$ $x = 0.249$

Check: Plug the value for x back into the equilibrium expression, and check the value.
$9.6 = \frac{(0.128)^2}{(0.110)(0.249)^3}$

At 775K: **Given:** At equilibrium: $[N_2]$ = 0.120M, $[H_2]$ = 0.140, K_c = 0.0584 **Find:** $[NH_3]$
Conceptual Plan: Balanced reaction → equilibrium expression → $[NH_3]$

Solution: $K_c = \frac{[NH_3]^2}{[N_2][H_2]^3}$ $0.0584 = \frac{(x)^2}{(0.120)(0.140)^3}$ $x = 0.00439$

Check: Plug the value for x back into the equilibrium expression, and check the value.
$0.0584 = \frac{(0.00439)^2}{(0.120)(0.140)^3}$

39. **Given:** P(NO) = 118 torr; $P(Br_2)$ = 176 torr, K_p = 28.4 **Find:** P(NOBr)
Conceptual Plan: **torr → atm and then balanced reation → equilibrium expression → P(NOBr)**

$$\frac{1\text{ atm}}{760\text{ torr}}$$

Solution: $P_{NO} = 118\text{ torr} \times \frac{1\text{ atm}}{760\text{ torr}} = 0.1553\text{ atm}$ $P_{Br_2} = 176\text{ torr} \times \frac{1\text{ atm}}{760\text{ torr}} = 0.2316$

$K_p = \frac{P^2_{NOBr}}{P^2_{NO}P_{Br_2}}$ $28.4 = \frac{x^2}{(0.1553)^2(0.2316)}$ $x = 0.398\text{ atm} = 303\text{ torr}$

Check: Plug the value for x back into the equilibrium expression, and check the value.
$28.4 = \frac{(0.398)^2}{(0.1553)^2(0.2316)}$

41. **Given:** $[Fe^{3+}]_{initial} = 1.0 \times 10^{-3}$ M; $[SCN^-]_{initial} = 8.0 \times 10^{-4}$ M; $[FeSCN^{2+}]_{eq} = 1.7 \times 10^{-4}$ M **Find:** K_c
Conceptual Plan:
1. **Prepare ICE table**
2. **Calculate concentration change for known value**
3. **Calculate concentration changes for other reactants/products**
4. **Determine equilibrium concentration**
5. **Write the equilibrium expression and determine K_c**

Solution: $Fe^{3+}(aq) + SCN^-(aq) \leftrightarrows FeSCN^{3+}(aq)$

	$[Fe^{3+}]$	$[SCN^-]$	$[FeSCN^{3+}]$
I	1.0×10^{-3}	8.0×10^{-4}	0.00
C	-1.7×10^{-4}	-1.7×10^{-4}	$+1.7\times10^{-4}$

E	$8.3x10^{-4}$	$6.3x10^{-4}$	$1.7x10^{-4}$

$$K_c = \frac{[FeSCN^{2+}]}{[Fe^{3+}][SCN^-]} = \frac{(1.7x10^{-4})}{(8.3x10^{-4})(6.3x10^{-4})} = 3.3x10^2$$

43. **Given:** 3.67 L flask; 0.763 g H_2 initial; 96.9 g I_2 initial; 90.4 g HI equilibrium **Find:** K_c
Conceptual Plan: g → mol → M and then

$$n = \frac{g}{\text{molar mass}} \quad M = \frac{n}{V}$$

1. **Prepare ICE table**
2. **Calculate concentration change for known value**
3. **Calculate concentration changes for other reactants/products**
4. **Determine equilibrium concentration**
5. **Write the equilibrium expression and determine K_c**

Solution: $0.763 \text{ g } H_2 \times \frac{1 \text{ mol } H_2}{2.016 \text{ g } H_2} = 0.3785 \text{ mol } H_2$ $\quad \frac{0.3785 \text{ mol } H_2}{3.67 \text{ L}} = 0.103 \text{ M}$

$96.9 \text{ g } I_2 \times \frac{1 \text{ mol } I_2}{253.8 \text{ g } I_2} = 0.3818 \text{ mol } I_2$ $\quad \frac{0.3818 \text{ mol } I_2}{3.67 \text{ L}} = 0.104 \text{ M}$

$90.4 \text{ g HI} \times \frac{1 \text{ mol HI}}{127.9 \text{ g } I_2} = 0.7068 \text{ mol HI}$ $\quad \frac{0.7068 \text{ mol HI}}{3.67 \text{ L}} = 0.193 \text{ M}$

	$H_2(g)$ +	$I_2(g)$	⇆ $2HI(g)$
	$[H_2]$	$[I_2]$	$[HI]$
I	0.103	0.104	0.00
C	- 0.0965	- 0.0965	+0.193
E	0.0065	0.0075	0.193

$$K_c = \frac{[HI]^2}{[H_2][I_2]} = \frac{(0.193)^2}{(0.0065)(0.0075)} = 764$$

45. **Given:** $K_c = 8.5 \times 10^{-3}$; $[NH_3] = 0.166$; $[H_2S] = 0.166$ **Find:** Will solid form or decompose?
Conceptual Plan: Calculate Q → compare Q and K_c
Solution: $Q = [NH_3][H_2S] = (0.166)(0.166) = 0.0276$
$Q = 0.0276$ and $K_c = 8.5 \times 10^{-3}$ so $Q > K_c$ and the reaction will shift to the left, so more solid will form.

47. **Given:** 6.55 g Ag_2SO_4, 1.5 L solution, $K_c = 1.1 \times 10^{-5}$ **Find:** will more solid dissolve
Conceptual Plan:
g Ag_2SO_4 → mol Ag_2SO_4 → $[Ag_2SO_4]$ → $[Ag^+]$,$[SO_4^{-2}]$ → calculate Q and compare to K_c.

$$\frac{1 \text{ mol } Ag_2SO_4}{311.81 \text{ g}} \qquad [\,] = \frac{mol\, Ag_2SO_4}{vol\, solution} \qquad Q = [Ag^+]^2[SO_4^{2-}]$$

Solution: $6.55 \text{ g } Ag_2SO_4 \left(\frac{1 \text{ mol } Ag_2SO_4}{311.81 \text{ g } Ag_2SO_4} \right) = 0.0140 \text{ mol } Ag_2SO_4$

$$\frac{0.0210\, mol\, Ag_2SO_4}{1.5\, L\, solution} = 0.0140 M\, Ag_2SO_4$$

$[Ag^+] = 2[Ag_2SO_4] = 2(0.0140 M) = 0.0280 M \qquad [SO_4^{2-}] = [Ag_2SO_4] = 0.0140 M$

$Q = [Ag^+]^2[SO_4^{2-}] = (0.0280)^2(0.0140) = 1.1x10^{-5}$

$Q = K_c$ so the system is at equilibrium and is a saturated solution. Therefore, if more solid is added it will not dissolve.

49. a) **Given:** [A] = 1.0 M, [B] = 0.0 $K_c = 2.0$; $a = 1$, $b = 1$ **Find:** [A], [B] at equilibrium

Conceptual Plan: Prepare an ICE table; and then calculate Q; compare Q and K_c and predict the direction of the reaction; represent the change with x; sum the table and determine the equilibrium values; put the equilibrium values in the equilibrium expression and solve for x. Determine [A] and [B].

Solution:

	$A(g) \leftrightarrows$	$B(g)$
	[A]	[B]
I	1.0	0.00
C	- x	x
E	1 – x	x

$Q = \frac{[B]}{[A]} = \frac{0}{1} = 0$ $\quad$ $Q < K$ therefore, the reaction will proceed to the right by x

$K_c = \frac{[B]}{[A]} = \frac{(x)}{(1 - x)} = 2.0$ $\quad$ x = 0.67

[A] = 1 - 0.67 = 0.33 $\quad$ [B] = 0.67

Check: Plug the values into the equilibrium expression: $K_c = \frac{0.67}{0.33} = 2.0$

b) **Given:** [A] = 1.0 M, [B] = 0.0 K_c = 2.0; $a = 2$, $b = 2$ **Find:** [A], [B] at equilibrium

Conceptual Plan: Prepare an ICE table; and then calculate Q; compare Q and K_c and predict the direction of the reaction; represent the change with x; sum the table and determine the equilibrium values; put the equilibrium values in the equilibrium expression and solve for x. Determine [A] and [B].

Solution:

	$2A(g) \leftrightarrows$	$2B(g)$
	[A]	[B]
I	1.0	0.00
C	- 2x	2x
E	1 – 2x	2x

$Q = \frac{[B]^2}{[A]^2} = \frac{0}{1} = 0$ $\quad$ $Q < K$ therefore, the reaction will proceed to the right by x

$K_c = \frac{[B]^2}{[A]^2} = \frac{(2x)^2}{(1 - 2x)^2} = 2.0$ $\quad$ $x = 0.2\underline{9}3$

$[A] = 1 - 2(0.2\underline{9}3) = 0.414 = 0.41$ $\quad$ $[B] = 2(0.2\underline{9}3) = 0.586 = 0.59$

Check: Plug the values into the equilibrium expression: $K_c = \frac{(0.59)^2}{(0.41)^2} = 2.0$

c) **Given:** [A] = 1.0 M, [B] = 0.0 K_c = 2.0; $a = 1$, $b = 2$ **Find:** [A], [B] at equilibrium

Conceptual Plan: Prepare an ICE table; and then calculate Q; compare Q and K_c and predict the direction of the reaction; represent the change with x; sum the table and determine the equilibrium values; put the equilibrium values in the equilibrium expression and solve for x. Determine [A] and [B].

Solution:

	$A(g) \leftrightarrows$	$2B(g)$
	[A]	[B]
I	1.0	0.00
C	- x	2x
E	1 – x	2x

$Q = \frac{[B]^2}{[A]} = \frac{0}{1} = 0$ $\quad$ $Q < K$ therefore, the reaction will proceed to the right by x

$$K_c = \frac{[B]^2}{[A]} = \frac{(2x)^2}{(1 - x)} = 2.0 \qquad 4x^2 + 2x - 2 = 0$$

$(4x - 2)(x + 1) = 0$; $x = -1$ or $x = 0.5$, therefore, $x = 0.5$

$[A] = 1 - 0.50 = 0.50M \qquad [B] = 2x = 2(0.50) = 1.0M$

Check: Plug the values into the equilibrium expression: $K_c = \frac{(1.0)^2}{0.50} = 2.0$

51. **Given:** $[N_2O_4] = 0.0500$ M, $[NO_2] = 0.0$ $K_c = 0.513$ **Find:** $[N_2O_4]$, $[NO_2]$ at equilibrium
Conceptual Plan: Prepare an ICE table; and then calculate Q; compare Q and K_c and predict the direction of the reaction; represent the change with x; sum the table and determine the equilibrium values; put the equilibrium values in the equilibrium expression and solve for x. Determine $[N_2O_4]$ and $[NO_2]$.
Solution:

	N_2O_4 (g) ⇆	$2\ NO_2$(g)
	$[N_2O_4]$	$[NO_2]$
I	0.0500	0.00
C	- x	2x
E	0.0500 – x	2x

$$Q = \frac{[NO_2]^2}{[N_2O_4]} = \frac{0}{0.0500} = 0$$

$Q < K$ therefore, the reaction will proceed to the right by x

$$K_c = \frac{[NO_2]^2}{[N_2O_4]} = \frac{(2x)^2}{(0.0500 - x)} = 0.513 \qquad 4x^2 + 0.513x - 0.02565 = 0$$

$$\frac{-b \pm \sqrt{b^2 - 4ac}}{2a} = \frac{-0.513 \pm \sqrt{(0.513)^2 - 4(4)(-0.02565)}}{2(4)} = \frac{-0.513 \pm \sqrt{0.6735}}{2(4)}$$

$x = -0.1667$ or $x = 0.0385$, therefore, $x = 0.0385$

$[A] = 0.0500 - 0.03850 = 0.0115M \qquad [B] = 2x = 2(0.03850) = 0.0770M$

Check: Plug the values into the equilibrium expression: $K_c = \frac{(0.0770)^2}{0.0115} = 0.515$

53. **Given:** $[CO] = 0.10$ M, $[CO_2] = 0.0$ $K_c = 4.0 \times 10^3$ **Find:** $[CO_2]$ at equilibrium
Conceptual Plan: Prepare an ICE table; and then calculate Q; compare Q and K_c and predict the direction of the reaction; represent the change with x; sum the table and determine the equilibrium values; put the equilibrium values in the equilibrium expression and solve for x. Determine [CO], $[Cl_2]$. and $[COCl_2]$.
Solution:

	NiO(s) +CO (g)	⇆ Ni(s) + CO_2 (g)
	[CO]	$[CO_2]$
I	0.10	0.0
C	- x	x
E	0.10 –x	x

$$Q = \frac{[CO_2]}{[CO]} = \frac{0}{(0.10)} = 0$$

$Q < K$ therefore, the reaction will proceed to the right by x

$$K_c = \frac{[CO_2]}{[CO]} = \frac{x}{(0.10 - x)} = 4.0 \times 10^3 \qquad 4.0 \times 10^3(0.10 - x) = x$$

$x = 0.10$

$[CO_2] = 0.10$ M

Check: since the equilibrium constant is so large, the reaction goes essentially to completion, therefore, it is reasonable that the concentration of product is 0.10 M

55. **Given:** $[HC_2H_3O_2] = 0.210$M M, $[H_3O^+] = 0.0$, $[C_2H_3O_2^-] = 0.0$, $K_c = 1.8 \times 10^{-5}$
Find: $[HC_2H_3O2]$, $[H_2O^+]$, $[C_2H_3O_2^-]$ at equilibrium

Conceptual Plan: Prepare an ICE table; and then calculate Q; compare Q and K_c and predict the direction of the reaction; represent the change with x; sum the table and determine the equilibrium values; put the equilibrium values in the equilibrium expression and solve for x. Determine [CO], [Cl_2]. and [$COCl_2$].

Solution:

	$HC_2H_3O_2\,(aq)$	$+H_2O(l)$	$\leftrightarrows$	$H_3O^+\,(aq)$ +	$C_2H_3O_2^-(aq)$
	$[HC_2H_3O_2]$	$[H_2O]$		$[H_3O^+]$	$[C_2H_3O_2^-]$
I	0.210			0.0	0.0
C	- x			x	x
E	0.210–x			x	x

$$Q = \frac{[H_3O^+][C_2H_3O_2^-]}{[HC_2H_3O_2]} = \frac{0}{(0.210)} = 0$$

$Q < K$ therefore, the reaction will proceed to the right by x

$$K_c = \frac{[H_3O^+][C_2H_3O_2^-]}{[HC_2H_3O_2]} = \frac{(x)(x)}{(0.210 - x)} = 1.8 \times 10^{-5}$$

assume x is small compared to 0.210

$x^2 = 0.210(1.8 \times 10^{-5})$

$x = 0.00194$ check assumption: $\frac{0.00194}{0.210} \times 100 = 0.92\ \%$; assumption valid

$[H_3O^+] = [C_2H_3O_2^-] = 0.00194M$

$[HC_2H_3O_2] = 0.210 - 0.00194 = 0.20\underline{0}7 = 0.210M$

Check: Plug the values into the equilibrium expression: $K_c = \frac{(0.00194)(0.00194)}{(0.210)} = 1.79 \times 10^{-5}$; within 1 significant figure of true value, answers are valid.

57. **Given:** $P_{Br2} = 755$ torr, $P_{Cl2} = 735$ torr, $P_{BrCl} = 0.0$, $K_p = 1.11 \times 10^{-4}$ **Find:** P_{BrCl} at equilibrium
Conceptual Plan: Torr → atm and then:Prepare an ICE table; and then calculate Q; compare Q and K_c and predict the direction of the reaction; represent the change with x; sum the table and determine the equilibrium values; put the equilibrium values in the equilibrium expression and solve for x. Determine P_{BrCl}
Solution:

$$P_{Br_2} = 755 \text{ torr} \times \frac{1 \text{ atm}}{760 \text{ torr}} = 0.99\underline{3}4 \text{ atm} \qquad P_{Cl_2} = 735 \text{ torr} \times \frac{1 \text{ atm}}{760 \text{ torr}} = 0.96\underline{7}1 \text{ atm}$$

	$Br_2\,(g)$	$+Cl_2(g)$	$\leftrightarrows$ $2BrCl\,(g)$
	P_{Br2}	P_{Cl2}	P_{BrCl}
I	0.9934	0.9671	0.0
C	- x	-x	2x
E	0.9934–x	0.9671– x	2x

$$Q = \frac{P^2_{BrCl}}{P_{Br_2}P_{Cl_2}} = \frac{0}{(0.9934)(0.9671)} = 0$$

$Q < K$ therefore, the reaction will proceed to the right by x

$$K_c = \frac{P^2_{BrCl}}{P_{Br_2}P_{Cl_2}} = \frac{(2x)^2}{(0.9934 - x)(0.9671 - x)} = 1.11 \times 10^{-4}$$

asusme x is small compared ot 0.9934 and 0.9671

$$\frac{(2x)^2}{(0.9934)(0.9671)} = 1.11 \times 10^{-4} \qquad 4x^2 = 1.066 \times 10^{-4}$$

$x = 0.00516 = 3.92$ torr

$P_{BrCl} = 2x = 2(3.92 \text{ torr}) = 7.84$ torr

Check: Plug the values into the equilibrium expression:

$$K_c = \frac{(2(0.00516))^2}{(0.9934 - 0.00516)(0.99671 - 0.00516)} = 1.065 \text{ x } 10^{-4} = 1.12 \text{ x } 10^{-4};$$

within 1 significant figure of the true value, therefore, answers are valid.

59. a) **Given:** [A] = 1.0 M, [B] = [C] = 0.0, K_c = 1.0 **Find:** [A], [B], [C] at equilibrium

Conceptual Plan: Prepare an ICE table; and then calculate Q; compare Q and K_c and predict the direction of the reaction; represent the change with x; sum the table and determine the equilibrium values; put the equilibrium values in the equilibrium expression and solve for x. Determine [A], [B]. and [C].

Solution:

	A (g)	⇆	B (g) +	C(g)
	[A]		[B]	[C]
I	1.0		0.0	0.0
C	- x		x	x
E	1.0–x		x	x

$$Q = \frac{[B][C]}{[A]} = \frac{0}{(1.0)} = 0$$ $Q < K$ therefore, the reaction will proceed to the right by x

$$K_c = \frac{[B][C]}{[A]} = \frac{(x)(x)}{(1.0 - x)} = 1.0$$

$$x^2 = 1.0(1.0 - x)$$

$$x^2 + x - 1 = 0$$

$$\frac{-b \pm \sqrt{b^2 - 4ac}}{2a} \qquad \frac{-1 \pm \sqrt{1^2 - 4(1)(-1)}}{2(1)}$$

x = 0.61̲5 or x = - 1.615, therefore, x = 0.61̲5

[B] = [C] = x = 0.61̲5=0.62

[A] = 1.0 - 0.61̲5=0.385=0.38

Check: Plug the values into the equilibrium expression: $K_c = \frac{(0.62)(0.62)}{(0.38)} = 1.01 = 1.0$; which is the equilibrium constant so the values are correct.

b) **Given:** [A] = 1.0 M, [B] = [C] = 0.0, K_c = 0.010 **Find:** [A], [B], [C] at equilibrium

Conceptual Plan: Prepare an ICE table; and then calculate Q; compare Q and K_c and predict the direction of the reaction; represent the change with x; sum the table and determine the equilibrium values; put the equilibrium values in the equilibrium expression and solve for x. Determine [A], [B]. and [C].

Solution:

	A (g)	⇆	B (g) +	C(g)
	[A]		[B]	[C]
I	1.0		0.0	0.0
C	- x		x	x
E	1.0–x		x	x

$$Q = \frac{[B][C]}{[A]} = \frac{0}{(1.0)} = 0$$ $Q < K$ therefore, the reaction will proceed to the right by x

$$K_c = \frac{[B][C]}{[A]} = \frac{(x)(x)}{(1.0 - x)} = 0.010$$

$x^2 = 0.010(1.0 - x)$

$x^2 + 0.010x - 0.010 = 0$

$$\frac{-b \pm \sqrt{b^2 - 4ac}}{2a} \quad \frac{-(0.010) \pm \sqrt{(0.010)^2 - 4(1)(-0.010)}}{2(1)}$$

$x = 0.09\underline{5}12$ or $x = -0.1005$, therefore, $x = 0.09\underline{5}12$

$[B] = [C] = x = 0.09\underline{5}12 = 0.095$

$[A] = 1.0 - 0.09\underline{5}12 = 0.90488 = 0.90$

Check: Plug the values into the equilibrium expression:

$K_c = \frac{(0.095)(0.095)}{(0.90)} = 0.01002 = 0.010$; which is the equilibrium constant so the values are correct.

c) **Given:** $[A] = 1.0$ M, $[B] = [C] = 0.0$, $K_c = 1.0 \times 10^{-5}$ **Find:** [A], [B], [C] at equilibrium
Conceptual Plan: Prepare an ICE table; and then calculate Q; compare Q and K_c and predict the direction of the reaction; represent the change with x; sum the table and determine the equilibrium values; put the equilibrium values in the equilibrium expression and solve for x. Determine [A], [B]. and [C].
Solution:

	A (g)	⇆	B (g) +	C(g)
	[A]		[B]	[C]
I	1.0		0.0	0.0
C	- x		x	x
E	1.0–x		x	x

$Q = \frac{[B][C]}{[A]} = \frac{0}{(1.0)} = 0$ $\qquad$ $Q < K$ therefore, the reaction will proceed to the right by x

$$K_c = \frac{[B][C]}{[A]} = \frac{(x)(x)}{(1.0 - x)} = 1.0 \times 10^{-5}$$

assume x is small compared to 1.0

$x^2 = 1.0(1.0 \times 10^{-5})$

$x = 0.003\underline{1}6$ $\qquad$ check assumption: $\frac{0.003\underline{1}6}{1.0} \times 100 = 0.32\%$, assumption valid

$[B] = [C] = x = 0.003\underline{1}6 = 0.0032$

$[A] = 1.0 - 0.003\underline{1}6 = 0.9968 = 1.0$

Check: Plug the values into the equilibrium expression:

$K_c = \frac{(0.0032)(0.0032)}{(1.0)} = 1.024 \times 10^{-5} = 1.0 \times 10^{-5}$; which is the equilibrium constant so the values are correct.

61. **Given:** $CO(g) + Cl_2(g) \leftrightarrows COCl_2(g)$ at equilibrium $\qquad$ **Find:** what is the effect of each of the following
a) $COCl_2$ is added to the reaction mixture: Adding $COCl_2$ increases the concentration of $COCl_2$ and causes the reaction to shift to the left.

b) Cl_2 is added to the reaction mixture: Adding Cl_2 increases the concentration of Cl_2 and causes the reaction to shift to the right.

c) $COCl_2$ is removed from the reaction mixture: removing the $COCl_2$ decreases the concentration of $COCl_2$ and causes the reaction to shift to the right.

63. **Given:** $2KClO_3(s) \leftrightarrows 2KCl(s) + 3O_2(g)$ at equilibrium $\qquad$ **Find:** what is the effect of each of the following
a) O_2 is removed from the reaction mixture: Removing the O_2 decreases the concentration of O_2 and causes the reaction to shift to the right.

b) KCl is added to the reaction mixture: Adding KCl does not cause any change in the reaction. KCl is a solid and the concentration remains constant so the addition of more solid does not change the equilibrium concentration.

c) $KClO_3$ is added to the reaction mixture: Adding $KClO_3$ does not cause any change in the reaction. $KClO_3$ is a solid and the concentration remains constant so the addition of more solid does not change the equilibrium concentration.

d) O_2 is added to the reaction mixture: Adding O_2 increases the concentration of O_2 and causes the reaction to shift to the left.

65. a) **Given:** $I_2(g) \leftrightarrows 2I(g)$ at equilibrium **Find:** the effect of increasing the volume
The chemical equation has 2 moles of gas on the right and 1 mole of gas on the left. Increasing the volume of the reaction mixture decreases the pressure and causes the reaction to shift to the right (toward the side with more moles of gas particles).

b) **Given:** $2H_2S(g) \leftrightarrows 2H_2(g) + S_2(g)$ **Find:** the effect of decreasing the volume
The chemical equation has 3 moles of gas on the right and 2 mole of gas on the left. Decreasing the volume of the reaction mixture increases the pressure and causes the reaction to shift to the left (toward the side with fewer moles of gas particles).

c) **Given:** $I_2(g) + Cl_2(g) \leftrightarrows 2ICl(g)$ **Find:** the effect of decreasing the volume
The chemical equation has 2 moles of gas on the right and 2 mole of gas on the left. Decreasing the volume of the reaction mixture increases the pressure but causes no shift in the reaction because the moles are equal on both sides.

67. **Given:** $C(s) + CO_2(g) \leftrightarrows 2CO(g)$ is endothermic **Find:** the effect of increasing the temperature
Since the reaction is endothermic we can think of the heat as a reactant, increasing the temperature is equivalent to adding a reactant causing the reaction to shift to the right. This will cause an increase in the concentration of products and a decrease in the concentration of reactant, therefore, the value of K will increase.

Find: the effect of decreasing the temperature
Since the reaction is endothermic we can think of the heat as a reactant, decreasing the temperature is equivalent to removing a reactant causing the reaction to shift to the left. This will cause a decrease in the concentration of products and an increase in the concentration of reactants, therefore, the value of K will decrease.

69. **Given:** $C(s) + 2H_2(g) \leftrightarrows CH_4(g)$ is exothermic **Find:** determine which will favor CH_4

a) adding more C to the reaction mixture: does NOT favor CH_4. Adding C does not cause any change in the reaction. C is a solid and the concentration remains constant so the addition of more solid does not change the equilibrium concentration.

b) adding more H_2 to the reaction mixture: favors CH_4. Adding H_2 increases the concentration of H_2 causing the reaction to shift to the right.

c) raising the temperature of the reaction mixture: does NOT favor CH_4. Since the reaction is exothermic we can think of heat as a product, raising the temperature is equivalent to adding a product causing the reaction to shift to the left.

d) lowering the volume of the reaction mixture: favors CH_4. The chemical equation has 1 moles of gas on the right and 2 mole of gas on the left. Decreasing the volume of the reaction mixture increases the pressure and causes the reaction to shift to the right (toward the side with fewer moles of gas particles).

e) adding a catalyst to the reaction mixture: does NOT favor CH_4. A catalyst added to the reaction mixture only speeds up the reaction, it does not change the equilibrium concentration.

f) adding neon gas to the reaction mixture: does NOT favor CH_4. Adding an inert gas to a reaction mixture at a fixed volume has no effect on the equilibrium

71. a) To find the value of K for the new equation, combine the two given equation to yield the new equation.

Reverse equation 1, and use $1/K_1$ and then add to equation 2. To find K for equation 3 use $(1/K_1)(K_2)$.

$HBO_2(aq)$	$\leftrightarrows$	~~Hb(aq)~~ + $O_2(aq)$	$K_1 = 1/1.8$
~~Hb(aq)~~ + $CO(aq)$	$\leftrightarrows$	$HbCO(aq)$	$K_2 = 306$
$HBO_2(aq) + CO(aq)$	$\leftrightarrows$	$HbCO(aq) + O_2(aq)$	$K_3 = K_1K_2 = (1/1.8)(306) = 170$

b) **Given:** $O_2 = 20\%$, $CO = 0.10\%$ **Find:** The ratio $\frac{[HbCO]}{[HbO_2]}$

Conceptual Plan: Determine the equilibrium expression and then determine $\frac{[HbCO]}{[HbO_2]}$

Solution: $K = \frac{[HbCO][O_2]}{[HbO_2][CO]}$ $\quad 170 = \frac{[HbCO](20.0)}{[HbO_2](0.10)}$ $\quad \frac{[HbCO]}{[HbO_2]} = 170\left(\frac{0.10}{20.0}\right) = \frac{0.85}{1.0}$

Since the ratio is almost 1:1, 0.10% CO will replace about 50% of the O_2 in the blood. The CO blocks the uptake of O_2 by the blood and is therefore, highly toxic.

73. **Given:** $C_2H_4(g) + Cl_2(g) \leftrightarrows C_2H_4Cl_2(g)$ is exothermic
Find: Which of the following will maximize $C_2H_4Cl_2$

a) Increasing the reaction volume. Will not maximize $C_2H_4Cl_2$. The chemical equation has 1 mole of gas on the right and 2 moles of gas on the left. Increasing the volume of the reaction mixture decreases the pressure and causes the reaction to shift to the left (toward the side with more moles of gas particles).

b) Removing $C_2H_4Cl_2$ as it forms. Will maximize $C_2H_4Cl_2$. Removing the $C_2H_4Cl_2$ will decrease the concentration of $C_2H_4Cl_2$ and will cause the reaction to shift to the right, producing more $C_2H_4Cl_2$.

c) Lowering the reaction temperature. Will maximize $C_2H_4Cl_2$. The reaction is exothermic so we can think of heat as a product. Lowering the temperature will cause the reaction to shift to the right, producing more $C_2H_4Cl_2$.

d) Adding Cl_2. Will maximize $C_2H_4Cl_2$. Adding Cl_2 increases the concentration of Cl_2 so the reaction shifts to the right which will produce more $C_2H_4Cl_2$.

75. **Given:** Reaction 1 at equilibrium: $P_{H2} = 0.958$ atm; $P_{I2} = 0.877$ atm; $P_{HI} = 0.0200$ atm: reaction 2: $P_{H2} = P_{I2} = 0.621$ atm; $P_{HI} = 0.101$ atm **Find:** Is reaction 2 at equilibrium, if not, what is the P_{HI} at equilibrium.
Conceptual Plan: Use equilibrium Partial Pressures to determine K_p. Use K_p to determine if reaction 2 is at equilibrium. Prepare an ICE table; and then calculate Q; compare Q and K_p and predict the direction of the reaction; represent the change with x; sum the table and determine the equilibrium values; put the equilibrium values in the equilibrium expression and solve for x. Determine P_{HI}.

Solution: $H_2(g) + I_2(g) \leftrightarrows 2HI(g)$

	P_{H2}	P_{I2}	P_{HI}
Reaction 1:	0.958	0.877	0.020

$$K_p = \frac{P_{HI}^2}{P_{H_2}P_{I_2}} = \frac{(0.020)^2}{(0.958)(0.877)} = 4.7\underline{6}10 \times 10^{-4}$$

$H_2(g) + I_2(g) \leftrightarrows 2HI(g)$

Reaction 2:	P_{H2}	P_{I2}	P_{HI}
I	0.621	0.621	0.101
C	x	x	-2x
E	0.621+x	0.621+x	0.101 -2x

$$Q = \frac{P^2_{HI}}{P_{H_2}P_{I_2}} = \frac{0.101^2}{(0.621)(0.621)} = 0.0264\text{: } Q > K \text{ so the reaction shifts to the left}$$

$$K_p = \frac{P^2_{HI}}{P_{H_2}P_{I_2}} = \frac{(0.101\text{ -x})^2}{(0.621\text{+x})(0.621\text{+x})} = 4.7\underline{6}10 \text{ x } 10^{-4}$$

$$\sqrt{\frac{(0.101\text{ -2x})^2}{(0.621\text{+x})(0.621\text{+x})}} = \sqrt{4.7\underline{6}10 \text{ x } 10^{-4}}$$

$$\frac{(0.101\text{ -2x})}{(0.621\text{+x})} = 2.1\underline{8}2 \text{ x } 10^{-2}$$

x = 0.04325= 0.0433

$P_{H_2} = P_{I_2} = 0.621\text{+x} = 0.621 + 0.0433 = 0.664$ atm; $P_{HI} = 0.101 - 2x = 0.101 - 2(0.0433) = 0.0144$ atm

Check: Plug the values into the equilibrium expression:

$$K_p = \frac{(0.0144)^2}{(0.664)} = 4.703 \text{ x } 10^{-4} = 4.70 \text{ x } 10^{-4}$$; this value is close to the original equilbirium constant

77. **Given:** 200.0 L container; 1.27 kg N_2; 0.310 kg H_2; 725K; $K_p = 5.3 \text{ x } 10^{-5}$
Find: mass in g of NH_3 and % yield
Conceptual Plan:

$K_p \rightarrow K_c$ and then kg → g → mol → M and then Prepare an ICE table; represent the change

$K_p = K_c(RT)^{\Delta n}$ $\quad \frac{1000 \text{ g}}{\text{kg}} \quad \frac{\text{g}}{\text{molar mass}} \quad \frac{\text{mol}}{\text{vol}}$

with x; sum the table and determine the equilibrium values; put the equilibrium values in the equilibrium expression and solve for x. Determine [NH_3]. Then M → mol → g and then determine

M x vol mol x molar mass

theoretical yield NH_3 → % yield

determine limiting reactant $\quad \frac{\text{actual yield}}{\text{theoretical yield}}$

Solution:

$$K_p = K_c(RT)^{\Delta n} \qquad K_c = \frac{K_p}{(RT)^{\Delta n}} = \frac{5.3 \text{ x } 10^{-5}}{\left((0.0821 \frac{\text{L}\cdot\text{atm}}{\text{mol}\cdot\text{K}})(725\text{K})\right)^{-2}} = 0.1\underline{8}77$$

$$n_{N_2} = 1.27 \text{ kg } N_2 \text{ x } \frac{1000 \text{ g}}{\text{kg}} \text{ x } \frac{1 \text{ mol } N_2}{28.00 \text{ g } N_2} = 45.\underline{3}57 \text{ mol } N_2 \qquad n_{H_2} = 0.310 \text{ kg } H_2 \text{ x } \frac{1000 \text{ g}}{\text{kg}} \text{ x } \frac{1 \text{ mol } H_2}{28.00 \text{ g } H_2} = 15\underline{3}.77 \text{ mol } H_2$$

$N_2(g) + 3H_2(g) \leftrightarrows 2NH_3(g)$

Reaction 1:	$[N_2]$	$[H_2]$	$[NH_3]$
I	0.22$\underline{6}$8	0.76$\underline{8}$9	0.0
C	-x	-3x	2x
E	0.2268-x	0.7689-3x	2x

Reaction shifts to the right

$$K_c = \frac{[NH_3]^2}{[N_2][H_2]^3} = \frac{(2x)^2}{(0.2268\text{-x})(0.7689\text{-3x})^3} = 0.1877$$

Assume x is small compared to 0.2268 and 0.7689

$$\frac{(2x)^2}{(0.2268)(0.7689)^3} = 0.1877 \quad x = 0.06956 \quad \text{check assumption: } \frac{0.06956}{0.2268} \text{ x } 100 = 30.7 \text{ not valid}$$

Use method of successive substitution to solve for x.

x = 0.0461

$[NH_3] = 2x = 2(0.0461) = 0.0922M$

Check: Plug the values into the equilibrium expression:

$$K_c = \frac{(0.0922)^2}{(0.2268 - 0.0461)(0.7689 - 3(0.0461))^3} = 0.1876$$; this value is close to the original equlbirium constant

Determine grams NH_3: $0.0922 \frac{\cancel{mol\ NH_3}}{\cancel{L}} \times \frac{\cancel{200.0\ L}}{} \times \frac{17.02\ g\ NH_3}{\cancel{mol\ NH_3}} = 3\underline{1}3.8\ g = 3.1 \times 10^2\ g$

Determine the theoretical yield: Determine the limiting reactant:

$$1.27\ \cancel{kg\ N_2} \times \frac{1000\ \cancel{g}}{\cancel{kg}} \times \frac{1\ \cancel{mol\ N_2}}{28.0\ \cancel{g\ N_2}} \times \frac{2\ \cancel{mol\ NH_3}}{1\ \cancel{mol\ N_2}} \times \frac{17.02\ g\ NH_3}{\cancel{mol\ NH_3}} = 15\underline{4}4\ g\ NH_3$$

$$0.310\ \cancel{kg\ H_2} \times \frac{1000\ \cancel{g}}{\cancel{kg}} \times \frac{1\ \cancel{mol\ H_2}}{2.016\ \cancel{g\ H_2}} \times \frac{2\ \cancel{mol\ NH_3}}{3\ \cancel{mol\ H_2}} \times \frac{17.02\ g\ NH_3}{\cancel{mol\ NH_3}} = 17\underline{4}5\ g\ NH_3$$

N_2 produces the least amount of NH_3 therefore, it is the limiting reactant and the theoretical yield is 1.54×10^3 g NH_3.

$$\%\ yield = \frac{3.1 \times 10^2\ g}{1.54 \times 10^3\ g} \times 100 = 20.\%$$

79. **Given:** At equilibrium: $P_{CO} = 0.30$ atm; $P_{Cl2} = 0.10$ atm; $P_{COCl2} = 0.60$ atm, add 0.40 atm Cl_2
Find: P_{CO} when system returns to equilibrium.
Conceptual Plan: Use equilibrium Partial Pressures to determine K_p. For the new conditions: prepare an ICE table; represent the change with x; sum the table and determine the equilibrium values; put the equilibrium values in the equilibrium expression and solve for x. Determine P_{CO}.
Solution: $CO(g) + Cl_2(g) \leftrightarrows COCl_2(g)$

Condition 1:	P_{CO}	P_{Cl2}	P_{COCl2}
	0.30	0.10	0.60

$$K_p = \frac{P_{COCl_2}}{P_{CO}P_{Cl_2}} = \frac{(0.60)}{(0.30)(0.10)} = 20.$$

$CO(g) + Cl_2(g) \leftrightarrows COCl_2(g)$

Condition 2:	P_{CO}	P_{Cl2}	P_{COCl2}
I	0.30	0.10+0.40	0.60
C	-x	-x	+x
E	0.30-x	0.50-x	0.60 + x

Reaction shifts to the right because the concentration of Cl_2 was increased

$$K_p = \frac{P_{COCl_2}}{P_{CO}P_{Cl_2}} = \frac{(0.60+x)}{(0.30-x)(0.50-x)} = 20$$

$20x^2 - 17x + 2.4 = 0$

$$\frac{-b \pm \sqrt{b^2 - 4ac}}{2a} = \frac{-(-17) \pm \sqrt{(-17)^2 - 4(20)(2.4)}}{2(20)}$$

x = 0.67 or 0.18 So, x = 0.18

$P_{CO} = 0.30 - 0.18 = 0.12$ atm; $P_{Cl_2} = 0.50 - 0.18 = 0.32$; $P_{COCl_2} = 0.60 + 0.18 = 0.78$ atm

Check: Plug the values into the equilibrium expression:

$$K_p = \frac{(0.78)}{(0.12)(0.32)} = 20.3 = 20.$$; this is the same as the original equlbrium constant.

81. **Given:** $K_p = 0.76$; P_{total} at equilibrium = 1.00 **Find:** $P_{initial}$ CCl_4
Conceptual Plan: Prepare an ICE table; represent the $P(CCl_4)$ with A and the change with x; sum the table and determine the equilibrium values; use the total pressure and solve for A in terms of x. Determine partial pressure of each at equilibrium use the equilibrium expression to determine x and then determine A
Solution: $CCl_4(g) \leftrightarrows C(s) + 2Cl_2(g)$

	P_{CCl4}	P_C	P_{Cl2}
I	A	constant	0.00
C	-x		+2x
E	A-x		2x

$P_{Total} = P_{CCl_4} + P_{Cl_2}$ $1.0 = A - x + 2x$ $A = 1 - x$

$P_{CCl_4} = (A - x) = (1-x) - x = 1 - 2x$; $P_{Cl_2} = (2x)$

$$K_p = \frac{P^2_{Cl_2}}{P_{CCl_4}} = \frac{(2x)^2}{(1-2x)} = 0.76$$

$4x^2 + 1.52x - 0.76 = 0$ $x = 0.285$ or -0.665 so $x = 0.2\underline{8}5$

$A = 1 - x = 1.0 - 0.285 = 0.715 = 0.72$ atm

Check: Plug values into equilibrium expression:

$$K_p = \frac{P^2_{Cl_2}}{P_{CCl_4}} = \frac{(2x)^2}{(A-x)} = \frac{(2(0.285))^2}{(0.715-0.285)} = 0.755 = 0.76;$$ the original equilibrium constant

83. **Given:** vol = 0.654 L, T = 1000 K, $K_p = 3.9 \times 10^{-2}$ **Find:** mass CaO as equilibrium
Conceptual Plan: $K_p \rightarrow P_{CO2} \rightarrow n(CO_2) \rightarrow n(CaO) \rightarrow g$
PV = nRT stoichiometry g = n(molar mass)

Solution: Since $CaCO_3$ and CaO are solids, they are not included in the equilibrium expression.

$K_p = P_{CO_2} = 3.9 \times 10^{-2}$

$$n = \frac{PV}{RT} = \frac{(3.9 \times 10^{-2}\ \text{atm})(0.654\ \text{L})}{\left(0.0821\ \frac{\text{L}\cdot\text{atm}}{\text{mol}\cdot\text{K}}\right)(1000\ \text{K})} = 3.\underline{1}06 \times 10^{-4}\ \text{mol } CO_2$$

$$3.\underline{1}06 \times 10^{-4}\ \text{mol } CO_2 \times \frac{1\ \text{mol CaO}}{1\ \text{mol } CO_2} \times \frac{56.1\ \text{g CaO}}{1\ \text{mol CaO}} = 0.0174\ \text{g} = 0.017\ \text{g CaO}$$

Check: The small value of *K* would give a small amount of products, so we would not expect to have a large mass of CaO formed.

85. a) **Given:** NO (P = 522 torr), O_2 (P = 421 torr); at equilibrium, $P_{total} = 748$ torr **Find:** K_p
Conceptual Plan: Prepare an ICE table; represent the change with x; sum the table and determine the equilibrium values; use the total pressure and solve for x. torr → atm → K_p
Solution: $2NO(g) + O_2(g) \leftrightarrows 2NO_2(g)$

	P_{NO}	P_{O2}	P_{NO2}
I	522 torr	421 torr	0.00
C	-2x	-x	+2x
E	522-2x	421-x	2x

$P_{Total} = P_{NO} + P_{O_2} + P_{NO_2}$ $748 = 522 - 2x + (421 - x) + 2x$

$x = 195$ torr $P_{NO} = (522 - 2(195) = 132$ torr; $P_{O_2} = (421 - 195) = 226$ torr; $P_{NO_2} = 2(195) = 390$ torr

$P_{NO} = 132\ \text{torr} \times \frac{1\ \text{atm}}{760\ \text{torr}} = 0.17\underline{3}7$ atm; $P_{O_2} = 226\ \text{torr} \times \frac{1\ \text{atm}}{760\ \text{torr}} = 0.29\underline{7}4$ atm; $P_{NO_2} = 390\ \text{torr} \times \frac{1\ \text{atm}}{760\ \text{torr}} = 0.51\underline{3}2$ atm

$$K_p = \frac{P^2_{NO_2}}{P^2_{NO}P_{O_2}} = \frac{(0.51\underline{3}2)^2}{(0.17\underline{3}7)^2(0.29\underline{7}4)} = 29.34 = 29.3$$

b) **Given:** NO (P = 255 torr), O_2 (P = 185 torr), $K_p = 29.3$ **Find:** equilibrium P N_2O_4
Conceptual Plan: torr → atm and then prepare an ICE table; represent the change with x; sum the table and $\frac{\text{atm}}{760\ \text{torr}}$
determine the equilibrium values; put the equilibrium values in the equilibrium expression and solve for x. Determine P N_2O_4.

Solution: $P_{NO} = 255\ \text{torr} \times \frac{1\ \text{atm}}{760\ \text{torr}} = 0.33\underline{5}5$ atm $P_{O_2} = 185\ \text{torr} \times \frac{1\ \text{atm}}{760\ \text{torr}} = 0.24\underline{3}4$ atm

$2NO(g) + O_2(g) \leftrightarrows 2NO_2(g)$

	P_{NO}	P_{O2}	P_{NO2}
I	0.3355	0.2434	0.00
C	-2x	-x	+2x
E	0.3355-2x	0.2434-x	2x

$$K_p = \frac{P^2_{NO_2}}{P^2_{NO}P_{O_2}} = \frac{(2x)^2}{(0.3355 - 2x)^2(0.2434\text{-}x)} = 29.3$$

$x = 0.11\underline{1}25 \qquad P_{NO_2} = 2x = 2(0.11\underline{1}25) = 0.22\underline{2}5$ atm

$$0.22\underline{2}5 \text{ atm x } \frac{760 \text{ torr}}{1 \text{ atm}} = 169.1 \text{ torr} = 169 \text{ torr}$$

Check: Plug the values into the equilibrium expression:

$$K_p = \frac{(0.22\underline{2}5)^2}{(0.1130)^2(0.1322)} = 29.327 = 29.3$$; this is the same as the original equilbrium constant.

87. **Given:** P(NOCl) at equilibrium = 115 torr; $K_p = 0.27$, T = 700 K, **Find:** initial pressure NO, Cl_2

Conceptual Plan:

torr → atm and then prepare an ICE table; represent the change with x; sum the table and

$$\frac{\text{atm}}{760 \text{ torr}}$$

determine the equilibrium values; put the equilibrium values in the equilibrium expression and determine initial pressure.

Solution: $115 \text{ torr x } \frac{1 \text{ atm}}{760 \text{ torr}} = 0.15\underline{1}3 \text{ atm}$

$2NO(g) + Cl_2(g) \leftrightarrows 2NOCl(g)$

	P_{NO}	P_{Cl2}	P_{NOCl}	
				Let A = initial pressure of NO and Cl_2
I	A	A	0.00	
C	-2x	-x	+2x	
E	A-2x	A-x	0.151	2x = 0.151, so, x = 0.0756
	A-0.151	A-0.0756		

$$K_p = \frac{P^2_{NOCL}}{P^2_{NO}P_{Cl_2}} = \frac{(0.151)^2}{(A - 0.151)^2(A - 0.0756)} = 0.27$$

$$0.27A^3 - 0.1019A^2 + 0.01231A - 0.022336 = 0$$

$$A = 0.565$$

$$P_{NO} = P_{Cl_2} = A = 0.565\text{atm} = 429 \text{ torr}$$

Check: Plug the values into the equilibrium expression:

$$K_p = \frac{P^2_{NOCL}}{P^2_{NO}P_{Cl_2}} = \frac{(0.151)^2}{(0.565 - 0.151)^2(0.565 - 0.0756)} \; 0.2\underline{7}2$$;

this is within 1 significant figure of the original equilbrium constant.

89. **Given:** P= 0.750 atm, density = 0.520 g/L, T = 337°C **Find:** K_c

Conceptual Plan: Prepare an ICE table; represent the P(CCl_4) with A and the change with x; sum the table and determine the equilibrium values; use the total pressure and solve for A in terms of x. Determine partial pressure of each at equilibrium in terms of x and then use the density to determine

$$d = \frac{PM}{RT}$$

the apparent molar mass and then use the mole fraction (in terms of P) and the molar mass of each

$$\chi_A = \frac{P_A}{P_{Total}}$$ **gas, to determine x.**

Solution: $2NO_2(g) \leftrightarrows 2NO(g) + O_2(g)$

	P_{NO2}	P_{NO}	P_{O2}
I	A	0.00	0.00
C	-2x	+2x	+x
E	A-2x	2x	x

$P_{Total} = P_{NO_2} + P_{NO} + P_{O_2}$ $\quad 0.750 = A - 2x + 2x + x \quad A = 0.750 - x$

$P_{NO_2} = (A - 2x) = (0.750\text{-}x) - 2x = (0.750 - 3x);\ P_{NO} = (2x)\ ;\ P_{O_2} = x$

$$d = \frac{PM}{RT} \quad M = \frac{dRT}{P} = \frac{(0.520\frac{g}{L})(0.0821\frac{L\cdot atm}{mol\cdot K})(610\ K)}{0.750\ atm} = 34.72\ g/mol$$

$$M = \chi_{NO_2}M_{NO_2} + \chi_{NO}M_{NO} + \chi_{O_2}M_{O_2} = \frac{P_{NO_2}}{P_{total}}M_{NO_2} + \frac{P_{NO}}{P_{total}}M_{NO} + \frac{P_{O_2}}{P_{total}}M_{O_2}$$

$$P_{Total}M = P_{NO_2}M_{NO_2} + P_{NO}M_{NO} + P_{O_2}M_{O_2}$$

$(0.750)(34.7) = (0.750 - 3x)(46.0) + 2x(30.0) + x(32.0)$

$x = 0.184$

$$K_p = \frac{P^2_{NO}P_{O_2}}{P^2_{NO_2}} = \frac{(2x)^2(x)}{(0.750\text{-}3x)} = \frac{(2(0.184))^2(0.184)}{(0.750\text{-}3(0.184))^2} = 0.63\underline{5}6$$

$$K_p = K_c(RT)^{\Delta n} \qquad 0.63\underline{5}6 = K_c((0.0821\frac{L\cdot atm}{mol\cdot K})(610K))^1$$

$K_c = 1.27 \times 10^{-2}$

91. Yes, the direction will depend on the volume. If the initial moles of A and B are equal, the initial concentration of A and B are equal regardless of the volume. Since $K_c = \frac{[B]^2}{[A]} = 1$, if the volume is such that the [A] = [B] < 1.0, than $Q < K$ and the reaction goes right to reach equilibrium, but, if the volume is such that the [A] = [B] > 1.0, than $Q > K$ and the reaction goes left to reach equilibrium.

93. An examination of the data shows when $P_A = 1.0$ then $P_B = 1.0$, therefore, $K_p = \frac{P_B^b}{P_A^a} = \frac{(1.0)^b}{(1.0)^a} = 1.0$. Therefore, the value of the numerator and denominator must be equal. We see from the data that $P_B = \sqrt{P_A}$, so $P_B^2 = P_A$. Since, the stoichiometric coefficients become exponents in the equilibrium expression, $a = 1$ and $b = 2$.

Chapter 15
Acids and Bases

1. The pain of heartburn is caused by hydrochloric acid which is excreted in the stomach to kill microorganisms and activate enzymes that break down food. The hydrochloric acid can sometimes back up out of the stomach and into the esophagus, a phenomenon known as acid reflux. When hydrochloric acid comes in contact with the lining of the esophagus, the H^+ ions irritate the esophageal tissues, resulting the burning sensation. The simplest way to relieve mild heartburn is to swallow repeatedly. Saliva contains bicarbonate ion (HCO_3^-) that acts as a base and, when swallowed, neutralizes some of the acid in the esophagus. Heartburn can also be treated with antacids such as Tums, milk of magnesia, or Mylanta. These over – the – counter medications contain more base than does saliva and therefore, do a better job of neutralizing the esophageal acid.

3. A carboxylic acid contains the following group of atoms: 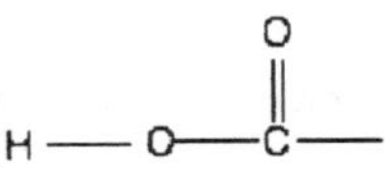 . Carboxylic acids are often found in substances derived from living organisms. Examples are citric acid, malic acid, acetic acid.

5. The hydronium ion is H_3O^+. In water, H^+ ions always associate with H_2O molecules to form hydronium ions and other associated species with the general formula $H(H_2O)_n^+$.

7. According to Huheey: " The differences between the various acid – base concepts are not concerned with which is right, but which is most convenient to use in a particular situation." There is no single correct definition, we use the definition that is best for a particular situation.

9. A conjugate acid – base pair are two substances related to each other by the transfer of a proton. In the reaction:
   $NH_3(aq) + H_2O(l) \leftrightarrows NH_4^+(aq)$ and $OH^-(aq)$, the conjugate pairs are: NH_3/NH_4^+ and H_2O/OH^-.

11. A diprotic acid is one that contains two ionizable protons. An example is H_2SO_4.
    A triprotic acid is one that contains three ionizable protons. An example is H_3PO_4.

13. The autoionization of water is: $H_2O(l) + H_2O(l) \leftrightarrows H_3O^+(aq) + OH^-(aq)$
    and has the ion product of water: $K_w = [H_3O^+][OH^-]$.
    It can also be written: $H_2O(l) \leftrightarrows H^+(aq) + OH^-(aq)$ and the ion product of water is: $K_w = [H^+][OH^-]$.
    At 25 °C, $K_w = 1.0 \times 10^{-14}$.

15. We define pH as: $pH = -\log[H_3O^+]$. An acidic solution has a pH < 7, a basic solution has a pH > 7 and a neutral solution has a pH = 7.

17. We can neglect the contribution of the autoionization of water to the H_3O^+ concentration in a solution of a strong or weak acid because it is negligible compared to the concentration from the acid itself. Even the weakest of the weak acids has an ionization constant that is four orders of magnitude larger than that of water.

19. The percent ionization is defined as: $\text{percent ionization} = \frac{\text{concentraion of ionized acid}}{\text{initial concentraion of acid}} x100$. The percent ionization of a weak acid decreases with the increasing concentration of the acid.

21. $B(aq) + H_2O(l) \leftrightarrows BH^+(aq) + OH^-(aq)$

23. At 25 °C, the product of K_a for and acid and K_b for its conjugate base is $K_w = 1.0 \times 10^{-14}$.

25. For most polyprotic acids, K_{a1} is much larger than K_{a2}. Therefore, the amount of H_3O^+ contributed by the first ionization step is much larger than that contributed by the second or third ionization step. In addition, the production of H_3O^+ by the first step inhibits additional production of H_3O^+ by the second step because of Le Chatelier's principle.

27. For an H – Y binary acid, the factors affecting the ease with which this hydrogen will be donated are the polarity of the bond and the strength of the bond.

29. A Lewis acid is an electron pair acceptor. A Lewis base is an electron pair donor.

31. The combustion of fossil fuels produces oxides of sulfur and nitrogen, which react with oxygen and water to form sulfuric and nitric acids. These acids then combine with rain to form acid rain. Acid rain is a significant problem in the northeastern United States.

33. a) acid $HNO_3(aq) \rightarrow H^+(aq) + NO_3^-(aq)$

b) acid $NH_4^+(aq) \rightarrow H^+(aq) + NH_3(aq)$

c) base $KOH(aq) \rightarrow K^+(aq) + OH^-(aq)$

d) acid $HC_2H_3O_2(aq) \rightarrow H^+(aq) + C_2H_3O_2^-(aq)$

35. a) Since H_2CO_3 donates a proton to H_2O it is the acid. After H_2CO_3 donates the proton it becomes HCO_3^-, the conjugate base. Since H_2O accepts a proton, it is the base. After H_2O accepts the proton it becomes H_3O^+ the conjugate acid.

b) Since H_2O donates a proton to NH_3 it is the acid. After H_2O donates the proton it becomes OH^-, the conjugate base. Since NH_3 accepts a proton, it is the base. After NH_3 accepts the proton it becomes NH_4^+ the conjugate acid.

c) Since HNO_3 donates a proton to H_2O it is the acid. After HNO_3 donates the proton it becomes NO_3^-, the conjugate base. Since H_2O accepts a proton, it is the base. After H_2O accepts the proton it becomes H_3O^+ the conjugate acid.

d) Since H_2O donates a proton to C_5H_5N it is the acid. After H_2O donates the proton it becomes OH^-, the conjugate base. Since C_5H_5N accepts a proton, it is the base. After C_5H_5N accepts the proton it becomes $C_5H_5NH^+$ the conjugate acid.

37. a) Cl^- $HCl(aq) + H_2O(l) \rightarrow H_3O^+(aq) + Cl^-(aq)$

b) HSO_3^- $H_2SO_3(aq) + H_2O(l) \leftrightarrows H_3O^+(aq) + HSO_3^-(aq)$

c) CHO_2^- $HCHO_2(aq) + H_2O(l) \leftrightarrows H_3O^+(aq) + CHO_2^-(aq)$

d) F^- $HF(aq) + H_2O(l) \leftrightarrows H_3O^+(aq) + F^-(aq)$

39. $H_2PO_4^- + H_2O(l) \leftrightarrows H_3O^+(aq) + HPO_4^{2-}(aq)$
$H_2PO_4^- + H_2O(l) \leftrightarrows H_3PO_4(aq) + OH^-(aq)$

41. a) HNO_3 is a strong acid

b) HCl is a strong acid

c) HBr is a strong acid

d) H_2SO_3 is a weak acid $H_2SO_3(aq) + H_2O(l) \leftrightarrows H_3O^+(aq) + HSO_3^-(aq)$

$$K_a = \frac{[H_3O^+][HSO_3^-]}{[H_2SO_3]}$$

43. a) Contains no HA, 10 H^+, and 10 A^-

b) Contains 3 HA, 3 H^+, and 7 A^-

c) Contains 9 HA, 1 H^+, and 1 A^-
So, solution a > solution b > solution c

45. a) F^- is a stronger base than Cl^-.
F^- is the conjugate base of HF(a weak acid), Cl^- is the conjugate base of HCl(a strong acid), the weaker the acid, the stronger the conjugate base.

b) NO_2^- is a stronger base than NO_3^-.
NO_2^- is the conjugate base of HNO_2(a weak acid), NO_3^- is the conjugate base of HNO_3(a strong acid), the weaker the acid, the stronger the conjugate base.

c) ClO^- is a stronger base than F^-.
F^- is the conjugate base of HF ($K_a = 3.5 \times 10^{-4}$), ClO^- is the conjugate base of HClO($K_a = 2.9 \times 10^{-8}$)
HClO is the weaker acid, the weaker the acid, the stronger the conjugate base.

47. a) **Given:** $K_w = 1.0 \times 10^{-14}$,$[H_3O^+] = 9.7 \times 10^{-9}$ M **Find:** $[OH^-]$
Conceptual Plan: $[H_3O^+] \rightarrow [OH^-]$

$$K_w = 1 \times 10^{-14} = [H_3O^+][OH^-]$$

Solution:

$K_w = 1 \times 10^{-14} = (9.7 \times 10^{-9})[OH^-]$

$[OH^-] = 1.0 \times 10^{-6}$M

$[OH^-] > [H_3O^+]$ so the solution is basic

b) **Given:** $K_w = 1.0 \times 10^{-14}$,$[H_3O^+] = 2.2 \times 10^{-6}$ M **Find:** $[OH^-]$
Conceptual Plan: $[H_3O^+] \rightarrow [OH^-]$

$$K_w = 1 \times 10^{-14} = [H_3O^+][OH^-]$$

Solution:

$K_w = 1 \times 10^{-14} = (2.2 \times 10^{-6})[OH^-]$

$[OH^-] = 4.5 \times 10^{-9}$M

$[H_3O^+] > [OH^-]$ so the solution is acidic

c) **Given:** $K_w = 1.0 \times 10^{-14}$,$[H_3O^+] = 1.2 \times 10^{-9}$ M **Find:** $[OH^-]$
Conceptual Plan: $[H_3O^+] \rightarrow [OH^-]$

$$K_w = 1 \times 10^{-14} = [H_3O^+][OH^-]$$

Solution:

$K_w = 1 \times 10^{-14} = (1.2 \times 10^{-9})[OH^-]$

$[OH^-] = 8.3 \times 10^{-6}$M

$[OH^-] > [H_3O^+]$ so the solution is basic

49. a) **Given:** $[H_3O^+] = 1.7 \times 10^{-8}$ M **Find:** pH and pOH
Conceptual Plan: $[H_3O^+] \rightarrow$ **pH** $\rightarrow$ **pOH**

$pH = -\log[H_3O^+]$ $pH + pOH = 14$

Solution: $pH = -\log(1.7 \times 10^{-8}) = 7.77$ $pOH = 14.00 - 7.77 = 6.23$

pH > 7 so the solution is basic.

b) **Given:** $[H_3O^+] = 1. \times 10^{-7}$ M **Find:** pH and pOH
Conceptual Plan: $[H_3O^+] \rightarrow$ **pH** $\rightarrow$ **pOH**

$pH = -\log[H_3O^+]$ $pH + pOH = 14$

Solution: $pH = -\log(1.0 \times 10^{-7}) = 7.00$ $pOH = 14.00 - 7.00 = 7.00$

pH = 7 so the solution is neutral.

c) **Given:** $[H_3O^+] = 2.2 \times 10^{-6}$ M **Find:** pH and pOH
Conceptual Plan: $[H_3O^+] \rightarrow$ **pH** $\rightarrow$ **pOH**

$pH = -\log[H_3O^+]$ $pH + pOH = 14$

Solution: $pH = -\log(2.2 \times 10^{-6}) = 5.66$ $pOH = 14.00 - 5.66 = 8.34$

pH < 7 so the solution is acidic.

51. pH = -log[H_3O^+] K_w = 1 x 10^{-14} = [H_3O^+][OH^-]

[H_3O^+]	[OH^-]	pH	Acidic or basic
7.1 x 10^{-4}	1.4 x 10^{-11}	**3.15**	acidic
3.7 x 10^{-9}	2.7 x 10^{-6}	8.43	basic
8 x 10^{-12}	1 x 10^{-3}	**11.1**	basic
6.2 x 10^{-4}	**1.6 x 10^{-11}**	3.20	acidic

$[H_3O^+] = 10^{-3.15} = 7.1 \times 10^{-4}$ $\quad [OH^-] = \dfrac{1 \times 10^{-14}}{7.1 \times 10^{-4}} = 1.4 \times 10^{-11}$

$[OH^-] = \dfrac{1 \times 10^{-14}}{3.7 \times 10^{-9}} = 2.7 \times 10^{-6}$ $\quad pH = -\log(3.7 \times 10^{-9}) = 8.43$

$[H_3O^+] = 10^{-11.1} = 8 \times 10^{-12}$ $\quad [OH^-] = \dfrac{1 \times 10^{-14}}{8 \times 10^{-12}} = 1 \times 10^{-3}$

$[H_3O^+] = \dfrac{1 \times 10^{-14}}{1.6 \times 10^{-11}} = 6.2 \times 10^{-4}$ $\quad pH = -\log(6.2 \times 10^{-4}) = 3.20$

53. **Given:** $K_w = 2.4 \times 10^{-14}$ at 37°C **Find:** [H_3O^+], pH

Conceptual Plan: K_w → [H_3O^+]→pH

$K_w = [H_3O^+][OH^-]$ pH = -log[H_3O^+]

Solution: $H_2O(l) + H_2O(l) \leftrightarrows H_3O^+(aq) + OH^-(aq)$

$K_w = [H_3O^+][OH^-]$

$[H_3O^+] = [OH^-] = \sqrt{K_w} = \sqrt{2.4 \times 10^{-14}} = 1.5 \times 10^{-7}$

pH = –log[H_3O^+] = –log(1.6 x 10^{-7}) = 6.82

Check: The value of K_w increased indicating more products formed, so the [H_3O^+] increases and the pH decreases from the values at 25°C

55. a) **Given:** 0.15M HCl (strong acid) **Find:** [H_3O^+],[OH^-], pH

Conceptual Plan:[HCl] → [H_3O^+] → pH and then [H_3O^+] → [OH^-]

[HCl] → [H_3O^+] pH = -log[H_3O^+] [H_3O^+][OH^-] = 1 x 10^{-14}

Solution: 0.15 M HCl = 0.15 M H_3O^+ pH = –log(0.15) = 0.82

[OH^-] = 1 x 10^{-14} / 0.15 M = 6.67 x 10^{-14}

Check: HCl is a strong acid with a relatively high concentration, so we expect the pH to be low and the [OH^-] to be small.

b) **Given:** 0.025M HNO_3 (strong acid) **Find:** [H_3O^+],[OH^-], pH

Conceptual Plan:[HNO_3] → [H_3O^+] → pH and then [H_3O^+] → [OH^-]

[HNO_3] → [H_3O^+] pH = -log[H_3O^+] [H_3O^+][OH^-] = 1 x 10^{-14}

Solution: 0.025 M HNO_3 = 0.025 M H_3O^+ pH = –log(0.025) = 1.60

[OH^-] = 1.0 x 10^{-14} / 0.025 M = 4.0 x 10^{-13}

Check: HNO_3 is a strong acid, so we expect the pH to be low and the [OH^-] to be small.

c) **Given:** 0.072M HBr and 0.015M HNO_3 (strong acids) **Find:** [H_3O^+],[OH^-], pH

Conceptual Plan:[HBr]+ [HNO_3] → [H_3O^+] → pH and then [H_3O^+] → [OH^-]

[HBr] + [HNO_3] → [H_3O^+] pH = -log[H_3O^+] [H_3O^+][OH^-] = 1 x 10^{-14}

Solution: 0.072 M HBr = 0.072 M H_3O^+ and 0.015M HNO_3 = 0.015M H_3O^+

Total H_3O^+ = 0.072M + 0.015M = 0.087M pH = –log(0.100) = 1.06

[OH^-] = 1.0 x 10^{-14} / 0.087 M = 1.1 x 10^{-13}

Check: HBr and HNO_3 are both strong acids and completely dissociate, this gives a relatively high concentration, so we expect the pH to be low and the [OH^-] to be small.

d) **Given:** HNO_3 = 0.855% by mass, $d_{solution}$ = 1.01 g/mL **Find:** [H_3O^+],[OH^-], pH

Conceptual Plan:
$$\% \text{ mass HNO}_3 \rightarrow \text{g HNO}_3 \rightarrow \text{mol HNO}_3 \text{ and then g soln} \rightarrow \text{mL soln} \rightarrow \text{L soln} \rightarrow \text{M HNO}_3$$

$$\frac{\%}{100} \qquad \frac{\text{mol HNO}_3}{63.018 \text{ g HNO}_3} \qquad\qquad \frac{1.01 \text{ g soln}}{\text{mL soln}} \quad \frac{1000 \text{ mL soln}}{\text{L soln}}$$

$$\rightarrow \text{M H}_3\text{O}^+ \rightarrow \text{pH and then } [\text{H}_3\text{O}^+] \rightarrow [\text{OH}^-]$$

$$[\text{HNO}_3] \rightarrow [\text{H}_3\text{O}^+] \qquad \text{pH} = -\log[\text{H}_3\text{O}^+] \quad [\text{H}_3\text{O}^+][\text{OH}^-] = 1 \times 10^{-14}$$

Solution: $\dfrac{0.855 \text{ g HNO}_3}{100 \text{ g soln}} \times \dfrac{1 \text{ mol HNO}_3}{63.018 \text{ g HNO}_3} \times \dfrac{1.01 \text{ g soln}}{\text{mL soln}} \times \dfrac{1000 \text{ mL soln}}{\text{L soln}} = 0.137 \text{ M HNO}_3$

$0.137 \text{ M HNO}_3 = 0.137 \text{ M H}_3\text{O}^+ \qquad \text{pH} = -\log(0.137) = 0.863$

$[\text{OH}^-] = 1.00 \times 10^{-14} / 0.137 \text{ M} = 7.30 \times 10^{-14}$

Check: HNO$_3$ is a strong acids and completely dissociates, this gives a relatively high concentration, so we expect the pH to be low and the [OH$^-$] to be small.

57. a) **Given:** pH = 1.25, 0.250 L **Find:** g HI
   Conceptual Plan: pH $\rightarrow$ [H$_3$O$^+$] $\rightarrow$ [HI] $\rightarrow$ mol HI $\rightarrow$ g HI
   $\quad$ pH = -log[H$_3$O$^+$] [H$_3$O$^+$] $\rightarrow$ [HI] mol = MV g = mol(127.9 g/mol)

   Solution:
   $[\text{H}_3\text{O}^+] = 10^{-1.25} = 0.056\text{M} = [\text{HI}]$ $\quad \dfrac{0.056 \text{ mol HI}}{\text{L}} \times 0.250 \text{ L} \times \dfrac{127.90 \text{ g HI}}{\text{mol HI}} = 1.8 \text{ g HI}$

   b) **Given:** pH = 1.75, 0.250 L **Find:** g HI
   Conceptual Plan: pH $\rightarrow$ [H$_3$O$^+$] $\rightarrow$ [HI] $\rightarrow$ mol HI $\rightarrow$ g HI
   $\quad$ pH = -log[H$_3$O$^+$] [H$_3$O$^+$] $\rightarrow$ [HI] mol = MV g = mol(127.9 g/mol)

   Solution:
   $[\text{H}_3\text{O}^+] = 10^{-1.75} = 0.0178\text{M} = [\text{HI}]$ $\quad \dfrac{0.0178 \text{ mol HI}}{\text{L}} \times 0.250 \text{ L} \times \dfrac{127.90 \text{ g HI}}{\text{mol HI}} = 0.57 \text{ g HI}$

   c) **Given:** pH = 2.85, 0.250 L **Find:** g HI
   Conceptual Plan: pH $\rightarrow$ [H$_3$O$^+$] $\rightarrow$ [HI] $\rightarrow$ mol HI $\rightarrow$ g HI
   $\quad$ pH = -log[H$_3$O$^+$] [H$_3$O$^+$] $\rightarrow$ [HI] mol = MV g = mol(127.9 g/mol)

   Solution:
   $[\text{H}_3\text{O}^+] = 10^{-2.85} = 0.0014\text{M} = [\text{HI}]$ $\quad \dfrac{0.0014 \text{ mol HI}}{\text{L}} \times 0.250 \text{ L} \times \dfrac{127.90 \text{ g HI}}{\text{mol HI}} = 0.045 \text{ g HI}$

59. **Given:** 0.100M benzoic acid. $K_a = 6.5 \times 10^{-5}$ **Find:** [H$_3$O$^+$], pH
   Conceptual Plan: Write a balanced reaction. Prepare an ICE table; represent the change with x; sum the table and determine the equilibrium values; put the equilibrium values in the equilibrium expression and solve for x. Determine [H$_3$O$^+$] and pH
   Solution: HC$_7$H$_5$O$_2$(aq) + H$_2$O(l) $\leftrightharpoons$ H$_3$O$^+$(aq) + C$_7$H$_5$O$_2$$^-$(aq)

I	0.100M	0.0	0.0
C	- x	x	x
E	0.100 – x	x	x

   $$K_a = \frac{[\text{H}_3\text{O}^+][\text{C}_7\text{H}_5\text{O}_2^-]}{[\text{HC}_7\text{H}_5\text{O}_2]} = \frac{(x)(x)}{(0.100 - x)} = 6.5 \times 10^{-5}$$

   Assume x is small compared to 0.100

   $x^2 = (6.5 \times 10^{-5})(0.100)$ $\qquad x = 2.5 \times 10^{-3} \text{ M} = [\text{H}_3\text{O}^+]$

   pH = -log(2.5 x 10^{-3}) = 2.60

61. a) **Given:** 0.500M HNO$_2$. $K_a = 4.6 \times 10^{-4}$ **Find:** [HNO$_2$], pH
   Conceptual Plan: Write a balanced reaction. Prepare an ICE table; represent the change with x; sum the table and determine the equilibrium values; put the equilibrium values in the equilibrium expression and solve for x. Determine [H$_3$O$^+$] and pH

Solution: $HNO_2(aq) + H_2O(l) \leftrightarrows H_3O^+(aq) + NO_2^-(aq)$

I	0.500M	0.0	0.0
C	- x	x	x
E	0.500 – x	x	x

$$K_a = \frac{[H_3O^+][NO_2^-]}{[HNO_2]} = \frac{(x)(x)}{(0.500 - x)} = 4.6 x 10^{-4}$$

Assume x is small compared to 0.500

$x^2 = (4.6 \text{ x } 10^{-4})(0.500)$ $\quad$ $x = 0.015M = [H_3O^+]$

Check assumption: $\frac{0.015}{0.500}$ x 100 =3.0% assumption valid

$[HNO_2] = 0.500 - 0.015 = 0.485M$ $\quad$ $pH = -\log(0.015) = 1.82$

b) **Given:** 0.100M HNO_2. $K_a = 4.6 \text{ x } 10^{-4}$ **Find:** $[HNO_2]$, pH

Conceptual Plan: Write a balanced reaction. Prepare an ICE table; represent the change with x; sum the table and determine the equilibrium values; put the equilibrium values in the equilibrium expression and solve for x. Determine $[H_3O^+]$ and pH

Solution: $HNO_2(aq) + H_2O(l) \leftrightarrows H_3O^+(aq) + NO_2^-(aq)$

I	0.100M	0.0	0.0
C	- x	x	x
E	0.100 – x	x	x

$$K_a = \frac{[H_3O^+][NO_2^-]}{[HNO_2]} = \frac{(x)(x)}{(0.100 - x)} = 4.6 x 10^{-4}$$

Assume x is small compared to 0.100

$x^2 = (4.6 \text{ x } 10^{-4})(0.100)$ $\quad$ $x = 0.0068M = [H_3O^+]$

Check assumption: $\frac{0.0068}{0.100}$ x 100 =6.8% assumption not valid

$x^2 = (4.6 \text{ x } 10^{-4})(0.100 - x)$ $\quad$ $x^2 + 4.6 \text{ x } 10^{-4}x - 4.6 \text{ x } 10^{-5} = 0$

$x = 0.00656$

$[HNO_2] = 0.100 - 0.00656 = 0.093M$

$pH = -\log(0.00656) = 2.18$

c) **Given:** 0.100M HNO_2. $K_a = 4.6 \text{ x } 10^{-4}$ **Find:** $[HNO_2]$, pH

Conceptual Plan: Write a balanced reaction. Prepare an ICE table; represent the change with x; sum the table and determine the equilibrium values; put the equilibrium values in the equilibrium expression and solve for x. Determine $[H_3O^+]$ and pH

Solution: $HNO_2(aq) + H_2O(l) \leftrightarrows H_3O^+(aq) + NO_2^-(aq)$

I	0.0100M	0.0	0.0
C	- x	x	x
E	0.0100 – x	x	x

$$K_a = \frac{[H_3O^+][NO_2^-]}{[HNO_2]} = \frac{(x)(x)}{(0.0100 - x)} = 4.6 x 10^{-4}$$

Assume x is small compared to 0.100

$x^2 = (4.6 \text{ x } 10^{-4})(0.0100)$ $\quad$ $x = 0.0021M = [H_3O^+]$

Check assumption: $\frac{0.0021}{0.0100}$ x 100 =21% assumption not valid

$x^2 = (4.6 \text{ x } 10^{-4})(0.0100 - x)$ $\quad$ $x^2 + 4.6 \text{ x } 10^{-4}x - 4.6 \text{ x } 10^{-6} = 0$

$x = 0.0019$

$pH = -\log(0.0019) = 2.72$

63. **Given:** 15.0 mL glacial acetic, d = 1.05 g/mL, dilute to 1.50 L, $K_a = 1.8 \times 10^{-5}$ **Find:** pH
Conceptual Plan:
mL acetic acid → g acetic acid → mol acetic acid → M and then write a balanced reaction.

$$\frac{1.05\text{ g}}{\text{mL}} \qquad \frac{\text{mol acetic acid}}{60.05\text{ g}} \qquad M = \frac{\text{mol}}{\text{L}}$$

Prepare an ICE table; represent the change with x; sum the table and determine the equilibrium values; put the equilibrium values in the equilibrium expression and solve for x. Determine $[H_3O^+]$ and pH

Solution: $15.0\text{ mL} \times \frac{1.05\text{ g}}{\text{mL}} \times \frac{1\text{ mol}}{60.05\text{ g}} \times \frac{1}{1.50\text{ L}} = 0.17\underline{4}8\text{ M}$

	$HC_2H_3O_2(aq) + H_2O(l) \rightleftharpoons$	$H_3O^+(aq)$	$+ C_2H_3O_2^-(aq)$
I	0.1748M	0.0	0.0
C	- x	x	x
E	0.1748 – x	x	x

$$K_a = \frac{[H_3O^+][C_2H_3O_2^-]}{[HC_2H_3O_2]} = \frac{(x)(x)}{(0.1748 - x)} = 1.8\times10^{-5}$$

Assume x is small compared to 0.1748

$x^2 = (1.8 \times 10^{-5})(0.1748)$ $\qquad x = 0.001\underline{7}9M = [H_3O^+]$

Check assumption: $\frac{0.001\underline{7}9}{0.1748} \times 100 = 1.0\%$ assumption valid

$pH = -\log(0.001\underline{7}9) = 2.75$

65. **Given:** 0.185 M HA, pH = 2.95 **Find:** K_a
Conceptual Plan: pH → $[H_3O^+]$ and then write a balanced reaction, prepare and ICE table, calculate equilibrium concentrations, and then plug into the equilibrium expression to solve for K_a.

Solution: $[H_3O^+] = 10^{-2.95} = 0.0012\text{ M} = [A^-]$

	$HA(aq) + H_2O(l) \rightleftharpoons$	$H_3O^+(aq)$	$+ A^-(aq)$
I	0.185M	0.0	0.0
C	- x	x	x
E	0.185 – 0.0012	0.0012	0.0012

$$K_a = \frac{[H_3O^+][A^-]}{[HA]} = \frac{(0.0012)(0.0012)}{(0.185 - 0.0012)} = 6.8 \times 10^{-6}$$

67. **Given:** 0.125 HCN $K_a = 4.9 \times 10^{-10}$ **Find:** % ionization
Conceptual Plan: Write a balanced reaction. Prepare an ICE table; represent the change with x; sum the table and determine the equilibrium values; put the equilibrium values in the equilibrium expression and solve for x and then x → % ionization

$$\%\text{ ionization} = \frac{x}{[HCN]_{original}} \times 100$$

Solution:

	$HCN(aq) + H_2O(l) \rightleftharpoons$	$H_3O^+(aq)$	$+ CN^-(aq)$
I	0.125M	0.0	0.0
C	- x	x	x
E	0.125 – x	x	x

$$K_a = \frac{[H_3O^+][CN^-]}{[HCN]} = \frac{(x)(x)}{(0.125 - x)} = 4.9\times10^{-10}$$

Assume x is small compared to 0.125

$x^2 = (4.9 \times 10^{-10})(0.125)$ $\qquad x = 7.\underline{8}3 \times 10^{-6}$

$$\%\text{ ionization} = \frac{7.\underline{8}3 \times 10^{-6}}{0.125} \times 100 = 0.0063\%\text{ ionized}$$

69. a) **Given:** 1.00M $HC_2H_3O_2$ $K_a = 1.8 \times 10^{-5}$ **Find:** % ionization

Conceptual Plan: Write a balanced reaction. Prepare an ICE table; represent the change with x; sum the table and determine the equilibrium values; put the equilibrium values in the equilibrium expression and solve for x and then x → % ionization

$$\% \text{ ionization} = \frac{x}{[HC_2H_3O_2]_{original}} \times 100$$

Solution: $HC_2H_3O_2$ (aq) + H_2O(l) ⇌ H_3O^+(aq) + $C_2H_3O_2^-$ (aq)

I	1.0 M	0.0	0.0
C	$-x$	x	x
E	$1.00 - x$	x	x

$$K_a = \frac{[H_3O^+][C_2H_3O_2^-]}{[HC_2H_3O_2]} = \frac{(x)(x)}{(1.00 - x)} = 1.8 \times 10^{-5}$$

Assume x is small compared to 1.00

$$x^2 = (1.8 \times 10^{-5})(1.00) \qquad x = 0.00424$$

$$\% \text{ ionization} = \frac{0.00424}{1.00} \times 100 = 0.42\% \text{ ionized}$$

b) **Given:** 0.500 M $HC_2H_3O_2$ $K_a = 1.8 \times 10^{-5}$ **Find:** % ionization

Conceptual Plan: Write a balanced reaction. Prepare an ICE table; represent the change with x; sum the table and determine the equilibrium values; put the equilibrium values in the equilibrium expression and solve for x and then x → % ionization

$$\% \text{ ionization} = \frac{x}{[HC_2H_3O_2]_{original}} \times 100$$

Solution: $HC_2H_3O_2$ (aq) + H_2O(l) ⇌ H_3O^+(aq) + $C_2H_3O_2^-$ (aq)

I	0.500 M	0.0	0.0
C	$-x$	x	x
E	$0.500 - x$	x	x

$$K_a = \frac{[H_3O^+][C_2H_3O_2^-]}{[HC_2H_3O_2]} = \frac{(x)(x)}{(0.500 - x)} = 1.8 \times 10^{-5}$$

Assume x is small compared to 0.500

$$x^2 = (1.8 \times 10^{-5})(0.500) \qquad x = 0.00300$$

$$\% \text{ ionization} = \frac{0.00300}{0.500} \times 100 = 0.60\% \text{ ionized}$$

c) **Given:** 0.100M $HC_2H_3O_2$ $K_a = 1.8 \times 10^{-5}$ **Find:** % ionization

Conceptual Plan: Write a balanced reaction. Prepare an ICE table; represent the change with x; sum the table and determine the equilibrium values; put the equilibrium values in the equilibrium expression and solve for x and then x → % ionization

$$\% \text{ ionization} = \frac{x}{[HC_2H_3O_2]_{original}} \times 100$$

Solution: $HC_2H_3O_2$ (aq) + H_2O(l) ⇌ H_3O^+(aq) + $C_2H_3O_2^-$ (aq)

I	0.100 M	0.0	0.0
C	$-x$	x	x
E	$0.100 - x$	x	x

$$K_a = \frac{[H_3O^+][C_2H_3O_2^-]}{[HC_2H_3O_2]} = \frac{(x)(x)}{(0.100 - x)} = 1.8 \times 10^{-5}$$

Assume x is small compared to 1.00

$$x^2 = (1.8 \times 10^{-5})(0.100) \qquad x = 0.00134$$

$$\% \text{ ionization} = \frac{0.00134}{0.100} \times 100 = 1.3\% \text{ ionized}$$

d) **Given:** 0.0500M $HC_2H_3O_2$ $K_a = 1.8 \times 10^{-5}$ **Find:** % ionization

Conceptual Plan: Write a balanced reaction. Prepare an ICE table; represent the change with x; sum the table and determine the equilibrium values; put the equilibrium values in the equilibrium expression and solve for x and then x → % ionization

$$\% \text{ ionization} = \frac{x}{[HC_2H_3O_2]_{original}} \times 100$$

Solution: $HC_2H_3O_2(aq) + H_2O(l) \leftrightarrows H_3O^+(aq) + C_2H_3O_2^-(aq)$

I	0.0500 M	0.0	0.0
C	- x	x	x
E	0.0500 – x	x	x

$$K_a = \frac{[H_3O^+][C_2H_3O_2^-]}{[HC_2H_3O_2]} = \frac{(x)(x)}{(0.0500 - x)} = 1.8 \times 10^{-5}$$

Assume x is small compared to 0.0500

$x^2 = (1.8 \times 10^{-5})(0.0500)$ $x = 9.\underline{4}9 \times 10^{-4}$

$$\% \text{ ionization} = \frac{9.\underline{4}9 \times 10^{-4}}{0.0500} \times 100 = 1.9\% \text{ ionized}$$

71. **Given:** 0.148 M HA 1.55% dissociation **Find:** K_a

Conceptual Plan: M → $[H_3O^+]$ → K_a and then write a balanced reaction, determine equilibrium concentration and plug into the equilibrium expression.

Solution: $(0.148 \text{ M HA})(0.0155) = 0.0022\underline{9}4$ $[H_3O^+]=[A^-]$

	$HA(aq) + H_2O(l) \leftrightarrows$	$H_3O^+(aq)$	$+ A^-(aq)$
I	0. 148M	0.0	0.0
C	- x	x	x
E	0.148 – 0.0022$\underline{9}$4	0.0022$\underline{9}$4	0.0022$\underline{9}$4

$$K_a = \frac{[H_3O^+][A^-]}{[HA]} = \frac{(0.0022\underline{9}4)(0.0022\underline{9}4)}{(0.148 - 0.0022\underline{9}4)} = 3.61 \times 10^{-5}$$

73. a) **Given:** 0.250 M HF $K_a = 3.5 \times 10^{-4}$ **Find:** pH, % dissociation

Conceptual Plan: Write a balanced reaction. Prepare an ICE table; represent the change with x; sum the table and determine the equilibrium values; put the equilibrium values in the equilibrium expression and solve for x and then x → % ionization

$$\% \text{ ionization} = \frac{x}{[HF]_{original}} \times 100$$

Solution: $HF(aq) + H_2O(l) \leftrightarrows H_3O^+(aq) + F^-(aq)$

I	0.250M	0.0	0.0
C	- x	x	x
E	0.250 – x	x	x

$$K_a = \frac{[H_3O^+][F^-]}{[HF]} = \frac{(x)(x)}{(0.250 - x)} = 3.5 \times 10^{-4}$$

Assume x is small compared to 0.250

$x^2 = (3.5 \times 10^{-4})(0.250)$ $x = 0.009\underline{3}5M = [H_3O^+]$

$$\frac{0.009\underline{3}5}{0.250} \times 100 = 3.7\%$$

$pH = -\log(0.009\underline{3}5) = 2.03$

b) **Given:** 0.100 M HF $K_a = 3.5 \times 10^{-4}$ **Find:** pH, % dissociation

Conceptual Plan: Write a balanced reaction. Prepare an ICE table; represent the change with x; sum the table and determine the equilibrium values; put the equilibrium values in the equilibrium expression and solve for x and then x → % ionization

$$\% \text{ ionization} = \frac{x}{[HF]_{original}} \times 100$$

Solution: $HF(aq) + H_2O(l) \leftrightarrows H_3O^+(aq) + F^-(aq)$

	HF	H_3O^+	F^-
I	0.1000M	0.0	0.0
C	- x	x	x
E	0.100 – x	x	x

$$K_a = \frac{[H_3O^+][F^-]}{[HF]} = \frac{(x)(x)}{(0.100 - x)} = 3.5 \times 10^{-4}$$

Assume x is small compared to 0.100

$x^2 = (3.5 \times 10^{-4})(0.100)$ $x = 0.00592M = [H_3O^+]$

$$\frac{0.00592}{0.100} \times 100 = 5.9\%$$

x > 5.0 % therefore, assumption invalid

$x^2 + 3.5 \times 10^{-4}x - 3.5 \times 10^{-5} = 0$

x = 0.00574 or -0.00609

pH = -log(0.00574) = 2.24

$$\% \text{ dissociation} = \frac{0.00574}{0.100} \times 100 = 5.7\%$$

c) **Given:** 0.050 M HF $K_a = 3.5 \times 10^{-4}$ **Find:** pH, % dissociation

Conceptual Plan: Write a balanced reaction. Prepare an ICE table; represent the change with x; sum the table and determine the equilibrium values; put the equilibrium values in the equilibrium expression and solve for x and then x → % ionization

$$\% \text{ ionization} = \frac{x}{[HF]_{original}} \times 100$$

Solution: $HF(aq) + H_2O(l) \leftrightarrows H_3O^+(aq) + F^-(aq)$

	HF	H_3O^+	F^-
I	0.050M	0.0	0.0
C	- x	x	x
E	0.050 – x	x	x

$$K_a = \frac{[H_3O^+][F^-]}{[HF]} = \frac{(x)(x)}{(0.050 - x)} = 3.5 \times 10^{-4}$$

Assume x is small compared to 0.250

$x^2 = (3.5 \times 10^{-4})(0.050)$ $x = 0.00418M = [H_3O^+]$

$$\frac{0.00418}{0.050} \times 100 = 8.4\%$$

Assumption invalid.

$x^2 + 3.5 \times 10^{-4}x - 1.75 \times 10^{-5} = 0$

x = 0.00401 or -0.00436

pH = -log(0.00401) = 2.40

$$\% \text{ dissociation} = \frac{0.00401}{0.050} \times 100 = 8.0\%$$

75. a) **Given:** 0.115 HBr (strong acid), 0.125 $HCHO_2$ (weak acid) **Find:** pH

Conceptual Plan: Since mixture is a strong acid, weak acid, the strong acid will dominate. Use the concentration of the strong acid to determine $[H_3O^+]$ and then pH

Solution: 0.115 M HBr = 0.115 M $[H_3O^+]$ pH = - log(0.115) = 0.939

b) **Given**: 0.150 M HNO_2 (weak acid), 0.085 M HNO_3 (strong acid) **Find:** pH
Conceptual Plan: Since mixture is a strong acid, weak acid, the strong acid will dominate. Use the concentration of the strong acid to determine $[H_3O^+]$ and then pH
Solution: 0.085 M HNO_3 = 0.085 M $[H_3O^+]$ pH = - log(0.085) = 1.07

c) **Given:** 0.185M $HCHO_2$, $K_a = 1.8 \times 10^{-4}$; 0.225 M $HC_2H_3O_2$, $K_a = 1.8 \times 10^{-5}$ **Find:** pH
Conceptual Plan: Since the mixture is a weak acid, weak acid and the *K* values are only 10^1 apart, you need to find $[H_3O^+]$ from each reaction. Write a balanced reaction. Prepare an ICE table; represent the change with x; sum the table and determine the equilibrium values; put the equilibrium values in the equilibrium expression and solve for x
Solution: $HCHO_2\,(aq) + H_2O(l) \leftrightarrows H_3O^+(aq) + CHO_2^-\,(aq)$

I	0.185 M	0.0	0.0
C	- x	x	x
E	0.185 – x	x	x

$$K_a = \frac{[H_3O^+][CHO_2^-]}{[HCHO_2]} = \frac{(x)(x)}{(0.185 - x)} = 1.8\text{x}10^{-4}$$

Assume x is small compared to 0.100

$x^2 = (1.8 \times 10^{-4})(0.185)$ $x = 0.005\underline{7}7$

$HC_2H_3O_2\,(aq) + H_2O(l) \leftrightarrows H_3O^+(aq) + C_2H_3O_2^-\,(aq)$

I	0.225 M	0.005$\underline{7}$7	0.0
C	- x	x	x
E	0.225 – x	x	x

$$K_a = \frac{[H_3O^+][C_2H_3O_2^-]}{[HC_2H_3O_2]} = \frac{(0.00577 + x)(x)}{(0.225 - x)} = 1.8\text{x}10^{-5}$$

Assume x is small compared to 0.0.225

$x^2 + 0.00602x - 4.05 \times 10^{-6} = 0$ $x = 0.00062$

$[H_3O^+] = 0.005\underline{7}7 + x = 0.00577 + 0.00062 = 0.006\underline{3}89$ M

pH = - log(0.006$\underline{3}$89) = 2.19

d) **Given:** 0.050 M $HC_2H_3O_2$, $K_a = 1.8 \times 10^{-5}$; 0.050 M HCN, $K_a = 4.9 \times 10^{-10}$ **Find:** pH
Conceptual Plan: Since the values of *K* are more than 10^1 apart, the acid with the larger *K* will dominate the reaction. Write a balanced reaction. Prepare an ICE table; represent the change with x; sum the table and determine the equilibrium values; put the equilibrium values in the equilibrium expression and solve for x
Solution: $HC_2H_3O_2\,(aq) + H_2O(l) \leftrightarrows H_3O^+(aq) + C_2H_3O_2^-\,(aq)$

I	0.0500 M	0.0	0.0
C	- x	x	x
E	0.0500 – x	x	x

$$K_a = \frac{[H_3O^+][C_2H_3O_2^-]}{[HC_2H_3O_2]} = \frac{(x)(x)}{(0.0500 - x)} = 1.8\text{x}10^{-5}$$

Assume x is small compared to 0.0500

$x^2 = (1.8 \times 10^{-5})(0.0500)$ $x = 9.\underline{4}9 \times 10^{-4}$

pH = - log($9.\underline{4}9 \times 10^{-4}$) = 3.02

77. a) **Given:** 0.15 M NaOH **Find:** $[OH^-]$, $[H_3O^+]$, pH, pOH
Conceptual Plan: [NaOH] → $[OH^-]$ → $[H_3O^+]$ → pH → pOH
$K_w = [H_3O^+][OH^-]$ pH = -log$[H_3O^+]$ pH + pOH =14
Solution: $[OH^-]$ = [NaOH] = 0.15M

$$[H_3O^+] = \frac{K_w}{[OH^-]} = \frac{1 \times 10^{-14}}{0.15M} = 6.7 \times 10^{-14}M$$

pH = -log(6.7×10^{-14}) = 13.18

pOH = 14.00 - 13.18 = 0.82

b) **Given:** 1.5×10^{-3} M $Ca(OH)_2$ **Find:** $[OH^-]$, $[H_3O^+]$, pH, pOH

Conceptual Plan: $[Ca(OH)_2] \rightarrow [OH^-] \rightarrow [H_3O^+] \rightarrow pH \rightarrow pOH$

$K_w = [H_3O^+][OH^-]$ $pH = -\log[H_3O^+]$ $pH + pOH = 14$

Solution: $[OH^-] = 2[Ca(OH)_2] = 2(1.5 \times 10^{-3}) = 0.0030$ M

$$[H_3O^+] = \frac{K_w}{[OH^-]} = \frac{1 \times 10^{-14}}{0.0030\ M} = 3.\underline{3}3 \times 10^{-12} M$$

$pH = -\log(3.\underline{3}3 \times 10^{-12}) = 11.48$

$pOH = 14.00 - 11.48 = 2.52$

c) **Given:** 4.8×10^{-4} M $Sr(OH)_2$ **Find:** $[OH^-]$, $[H_3O^+]$, pH, pOH

Conceptual Plan: $[Sr(OH)_2] \rightarrow [OH^-] \rightarrow [H_3O^+] \rightarrow pH \rightarrow pOH$

$K_w = [H_3O^+][OH^-]$ $pH = -\log[H_3O^+]$ $pH + pOH = 14$

Solution: $[OH^-] = [Sr(OH)_2] = 2(4.8 \times 10^{-4}) = 9.6 \times 10^{-4}$ M

$$[H_3O^+] = \frac{K_w}{[OH^-]} = \frac{1 \times 10^{-14}}{9.6 \times 10^{-4}\ M} = 1.\underline{0}4 \times 10^{-11} M$$

$pH = -\log(1.\underline{0}4 \times 10^{-11}) = 10.98$

$pOH = 14.00 - 10.98 = 3.02$

d) **Given:** 8.7×10^{-5} M KOH **Find:** $[OH^-]$, $[H_3O^+]$, pH, pOH

Conceptual Plan: $[KOH] \rightarrow [OH^-] \rightarrow [H_3O^+] \rightarrow pH \rightarrow pOH$

$K_w = [H_3O^+][OH^-]$ $pH = -\log[H_3O^+]$ $pH + pOH = 14$

Solution: $[OH^-] = [KOH] = 8.7 \times 10^{-5}$ M

$$[H_3O^+] = \frac{K_w}{[OH^-]} = \frac{1 \times 10^{-14}}{8.7 \times 10^{-5} M} = 1.\underline{1} \times 10^{-10} M$$

$pH = -\log(1.\underline{1} \times 10^{-10}) = 9.94$

$pOH = 14.00 - 9.94 = 4.06$

79. **Given:** 3.85% KOH by mass, d = 1.01 g/mL **Find:** pH

Conceptual Plan:

% mass → g KOH → mol KOH and mass soln → mL soln → L soln → M KOH → $[OH^{-1}]$

$$\frac{1\ \text{mol KOH}}{56.01\ \text{g KOH}} \qquad \frac{1.01\ \text{g soln}}{\text{mL soln}} \qquad \frac{1000\ \text{mL soln}}{\text{L soln}}$$

→ pOH → pH

$pOH = -\log[OH^-]$ $pH + pOH = 14$

Solution: $$\frac{3.85\ \text{g KOH}}{100.0\ \text{g soln}} \times \frac{1\ \text{mol KOH}}{56.01\ \text{g KOH}} \times \frac{1.01\ \text{g soln}}{\text{mL soln}} \times \frac{1000\ \text{mL soln}}{\text{L soln}} = 0.69\underline{4}2\ \text{M KOH}$$

$[OH^-] = [KOH] = 0.69\underline{4}2$ M $pOH = -\log(0.69\underline{4}2) = 0.159$ $pH = 14.000 - 0.159 = 13.841$

81. a) $NH_3(aq) + H_2O(l) \leftrightarrows NH_4^+(aq) + OH^-(aq)$ $$K_b = \frac{[NH_4^+][OH^-]}{[NH_3]}$$

b) $HCO_3^-(aq) + H_2O(l) \leftrightarrows H_2CO_3(aq) + OH^-(aq)$ $$K_b = \frac{[H_2CO_3][OH^-]}{[HCO_3^-]}$$

c) $CH_3NH_2(aq) + H_2O(l) \leftrightarrows CH_3NH_3^+(aq) + OH^-(aq)$ $$K_b = \frac{[CH_3NH_3^+][OH^-]}{[CH_3NH_2]}$$

83. Given: 0.15M NH_3 $K_b = 1.76 \times 10^{-5}$ **Find:** [OH^-],pH, pOH

Conceptual Plan: Write a balanced reaction. Prepare an ICE table; represent the change with x; sum the table and determine the equilibrium values; put the equilibrium values in the equilibrium expression and solve for x.

x = [OH^-] → pOH → pH
   pOH = -log[OH^-] pH + pOH = 14

Solution: $NH_3(aq) + H_2O(l) \leftrightharpoons NH_4^+(aq) + OH^-(aq)$

I	0.15	0.0 0.0
C	-x	x x
E	0.15 – x	x x

$$K_b = \frac{[NH_4^+][OH^-]}{[NH_3]} = \frac{(x)(x)}{(0.15 - x)} = 1.76 \times 10^{-5}$$

Assume x is small

$x^2 = (1.76 \times 10^{-5})(0.15)$ $x = [OH^-] = 0.001\underline{6}2$ M

pOH = -log(0.001\underline{6}2) = 2.79

pH = 14.00 - 2.80 = 11.21

85. Given: $pK_b = 10.4$, 455 mg/L caffeine   **Find:** pH

Conceptual Plan: pK_b → K_b and then mg/L → g/L → mol/L and then Write a balanced reaction. Prepare an ICE table; represent the change with x; sum the table and determine the equilibrium values; put the equilibrium values in the equilibrium expression and solve for x.

x = [OH^-] → pOH → pH
   pOH = -log[OH^-] pH + pOH = 14

Solution: $K_b = 10^{-10.4} = \underline{3}.98 \times 10^{-11}$

$$\frac{455 \text{ mg caffeine}}{\text{L soln}} \times \frac{\text{g caffeine}}{1000 \text{ mg caffeine}} \times \frac{1 \text{ mol caffeine}}{194.19 \text{ g}} = 0.0023\underline{4}2 \text{ M caffeine}$$

$C_8H_{10}N_4O_2(aq) + H_2O(l) \leftrightharpoons HC_8H_{10}N_4O_2^+(aq) + OH^-(aq)$

I	0.0023\underline{4}2	0.0	0.0
C	-x	x	x
E	0.0023\underline{4}2– x	x	x

$$K_b = \frac{[HC_8H_{10}N_4O_2^+][OH^-]}{[C_8H_{10}N_4O_2]} = \frac{(x)(x)}{(0.0023\underline{4}2 - x)} = 3.98 \times 10^{-11}$$

Assume x is small

$x^2 = (\underline{3}.98 \times 10^{-11})(0.0023\underline{4}2)$ $x = [OH^-] = \underline{3}.05 \times 10^{-7}$ M

pOH = -log($\underline{3}.05 \times 10^{-7}$) = 6.5

pH = 14.00 - 6.5 = 7.5

87. Given: 0.150 M morphine, pH = 10.5   **Find:** K_b

Conceptual Plan: pH → pOH → [OH^-] and then write a balanced equation, prepare and ICE table,
   pH = pOH = 14   pOH = -log[OH^-]

determine equilibrium concentrations → K_b

Solution: pOH = 14.0 – 10.5 = 3.5   [OH^-]=$10^{-3.5}$ = $\underline{3}.16 \times 10^{-4}$ = [$Hmorphine^+$]

morphine(aq) + $H_2O(l)$ $Hmorphine^+$(aq)+ OH^-(aq)

I	0.150	0.0	0.0
C	-x	x	x
E	0.150– x	3.16×10^{-4}	3.16×10^{-4}

$$K_b = \frac{[Hmorphine^+][OH^-]}{[morphine]} = \frac{(3.16 \times 10^{-4})(3.16 \times 10^{-4})}{(0.150 - 3.16 \times 10^{-4})} = \underline{6}.68 \times 10^{-7} = 7 \times 10^{-7}$$

89. a) pH neutral: Br^- is the conjugate base of a strong acid, therefore, it is pH neutral.

b) weak base: ClO^- is the conjugate base of a weak acid, therefore, it is a weak base.
    $ClO^-(aq) + H_2O(l) \leftrightharpoons HClO(aq) + OH^-(aq)$

c) weak base: CN^- is the conjugate base of a weak acid, therefore, it is a weak base.
$CN^-(aq) + H_2O(l) \leftrightarrows HCN(aq) + OH^-(aq)$

d) pH neutral: Cl^- is the conjugate base of a strong acid, therefore, it is pH neutral.

91. **Given:** $[F^-] = 0.140$ M, $K_a(HF) = 3.5 \times 10^{-4}$ **Find:** $[OH^-]$, pOH
Conceptual Plan: Determine K_b. Write a balanced reaction. Prepare an ICE table; represent the change with x;

$$K_b = \frac{K_w}{K_a}$$

sum the table and determine the equilibrium values; put the equilibrium values in the equilibrium expression and solve for x. Determine $[OH^-]$ → pOH → pH

$pOH = -\log[OH^-]$ $pH + pOH = 14$

Solution: $F^-(aq) + H_2O(l) \leftrightarrows HF(aq) + OH^-(aq)$

	F^-	HF	OH^-
I	0.140	0.0	0.0
C	-x	x	x
E	0.140– x	x	x

$$K_b = \frac{K_w}{K_a} = \frac{1 \times 10^{-14}}{3.5 \times 10^{-4}} = \frac{(x)(x)}{(0.140 - x)}$$

Assume x is small

$x = 2.0 \times 10^{-6} = [OH^-]$ $pOH = -\log(2.0 \times 10^{-6}) = 5.70$

$pH = 14.00 - 5.70 = 8.30$

93. a) Weak acid: NH_4^+ is the conjugate acid of a weak base, therefore, it is a weak acid.
$NH_4^+(aq) + H_2O(l) \leftrightarrows H_3O^+(aq) + NH_3(aq)$

b) pH neutral: Na^+ is the counterion of a strong base, therefore, it is pH neutral.

c) Weak acid: The Co^{3+} cation is a small, highly charged metal cation, therefore, it is a weak acid.
$Co(H_2O)_6^{3+}(aq) + H_2O(l) \leftrightarrows Co(H_2O)_5(OH)^{2+}(aq) + H_3O^+(aq)$

d) Weak acid: $CH_2NH_3^+$ is the conjugate acid of a weak base, therefore, it is a weak acid.
$CH_2NH_3^+(aq) + H_2O(l) \leftrightarrows H_3O^+(aq) + CH_2NH_2(aq)$

95. a) acidic: $FeCl_3$ — Fe^{3+} is a small, highly charged metal cation and is therefore, acidic. Cl^- is the conjugate base of a strong acid, therefore, it is pH neutral.

b) basic: NaF — Na^+ is the counterion of a strong base, therefore, it is pH neutral. F^- is the conjugate base of a weak acid, therefore, it is basic.

c) pH neutral: $CaBr_2$ — Ca^{2+} is the counterion of a strong base, therefore, it is pH neutral. Br^- is the conjugate base of a strong acid, therefore, it is pH neutral.

d) acidic: NH_4Br — NH_4^+ is the conjugate acid of a weak base, therefore, it is acidic. Br^- is the conjugate base of a strong acid, therefore, it is pH neutral.

e) acidic: $C_6H_5NH_3NO_2$ — $C_6H_5NH_3^+$ is the conjugate acid of a weak base, therefore, it is a weak acid. NO_2^- is the conjugate base of a weak acid, therefore, it is a weak base. To determine pH, compare K values.

$$K_a(C_6H_5NH_3^+) = \frac{1 \times 10^{-14}}{3.9 \times 10^{-10}} = 2.6 \times 10^{-5} \qquad K_b(NO_2^-) = \frac{1 \times 10^{-14}}{4.6 \times 10^{-4}} = 2.2 \times 10^{-11}$$

$K_a > K_b$ therefore, the solution is acidic.

97. Identify each species and determine acid, base, neutral.
NaCl pH neutral: Na^+ is the counterion of a strong base, therefore, it is pH neutral. Cl^- is the conjugate base of a strong acid, therefore, it is pH neutral.

NH_4Cl acidic: NH_4^+ is the conjugate acid of a weak base, therefore, it is acidic. Cl^- is the conjugate base of a strong acid, therefore, it is pH neutral.
$NaHCO_3$ basic: Na^+ is the counterion of a strong base, therefore, it is pH neutral. HCO_3^- is the conjugate base of a weak acid, therefore, it is basic.
NH_4ClO_2 acidic: $NH_4ClO_2NH_4^+$ is the conjugate acid of a weak base, therefore, it is a weak acid. ClO_2^- is the conjugate base of a weak acid, therefore, it is a weak base. $K_a(NH_4^+) = 5.6 \times 10^{-10}$ $K_b(ClO_2^-) = 9.1 \times 10^{-13}$ NaOH strong base
Increasing acidity: $NaOH < NaHCO_3 < NaCl < NH_4ClO_2 < NH_4Cl$

99. a) **Given:** 0.10 M NH_4Cl **Find:** pH
Conceptual Plan: Identify each species and determine which will contribute topH. Write a balanced reaction. Prepare an ICE table; represent the change with x; sum the table and determine the equilibrium values; put the equilibrium values in the equilibrium expression and solve for x. Determine $[H_3O^+]$ → pH
Solution: NH_4^+ is the conjugate acid of a weak base, therefore, it is acidic. Cl^- is the conjugate base of a strong acid, therefore, it is pH neutral.

	$NH_4^+(aq)$ + $H_2O(l)$ ⇆	$NH_3(aq)$ +	$H_3O^+(aq)$
I	0.10	0.0	0.0
C	-x	x	x
E	0.10– x	x	x

$$K_a = \frac{K_w}{K_b} = \frac{1 \times 10^{-14}}{1.8 \times 10^{-5}} = 5.\underline{5}6 \times 10^{-10} = \frac{(x)(x)}{(0.10 - x)}$$

Assume x is small

$x = 7.\underline{4}5 \times 10^{-6} = [H_3O^+]$ $pH = -\log(7.\underline{4}5 \times 10^{-6}) = 5.13$

b) **Given:** 0.10 M $NaC_2H_3O_2$ **Find:** pH
Conceptual Plan: Identify each species and determine which will contribute to pH. Write a balanced reaction. Prepare an ICE table; represent the change with x; sum the table and determine the equilibrium values; put the equilibrium values in the equilibrium expression and solve for x. Determine $[OH^-]$ → pOH → pH
$pOH = -\log[OH^-]$ $pH + pOH = 14$
Solution: Na^+ is the counterion of a strong base, therefore, is pH neutral. $C_2H_3O_2^-$ is the conjugate base of a weak acid, therefore, it is basic.

	$C_2H_3O_2^-(aq)$ + $H_2O(l)$ ⇆	$HC_2H_3O_2(aq)$+	$OH^-(aq)$
I	0.10	0.0	0.0
C	-x	x	x
E	0.10– x	x	x

$$K_b = \frac{K_w}{K_a} = \frac{1 \times 10^{-14}}{1.8 \times 10^{-5}} = 5.\underline{5}6 \times 10^{-10} = \frac{(x)(x)}{(0.10 - x)}$$

Assume x is small

$x = 7.\underline{4}5 \times 10^{-6} = [OH^-]$ $pOH = -\log(7.\underline{4}5 \times 10^{-6}) = 5.13$

$pH = 14.00 - 5.13 = 8.87$

c) **Given:** 0.10 M NaCl **Find:** pH
Conceptual Plan: Identify each species and determine which will contribute to pH.
Solution: Na^+ is the counterion of a strong base, therefore, it is pH neutral. Cl^- is the conjugate base of a strong acid, therefore, it is pH neutral.
pH = 7.0

101. **Given:** 0.15 M KF **Find:** concentration of all species
Conceptual Plan: Identify each species and determine which will contribute to pH. Write a balanced reaction. Prepare an ICE table; represent the change with x; sum the table and determine the equilibrium values; put the equilibrium values in the equilibrium expression and solve for x. And then, determine $[OH^-]$ → $[H_3O^+]$
$K_w = [H_3O^+][OH^-]$

Solution: K^+ is the counterion of a strong base, therefore, is pH neutral. F^- is the conjugate base of a weak acid, therefore, it is basic.

$F^-(aq)$ + $H_2O(l)$ ⇆	$HF(aq)$ +	$OH^-(aq)$
I 0.15	0.0	0.0
C -x	x	x
E 0.15 – x	x	x

$$K_b = \frac{K_w}{K_a} = \frac{1 \times 10^{-14}}{3.5 \times 10^{-4}} = \frac{(x)(x)}{(0.15 - x)}$$

Assume x is small

$x = 2.1 \times 10^{-6} = [OH^-] = [HF]$ $\quad [H_3O^+] = \frac{K_w}{[OH^-]} = \frac{1 \times 10^{-14}}{2.1 \times 10^{-6}} = 4.8 \times 10^{-9}$

$[K^+] = 0.15$ M
$[F^-] = (0.15 - 2.1 \times 10^{-6}) = 0.15$M
$[HF] = 2.1 \times 10^{-6}$
$[OH^-] = 2.1 \times 10^{-6}$
$[H_3O^+] = 4.8 \times 10^{-9}$

103. $H_3PO_4(aq) + H_2O(l) \leftrightarrows H_3O^+(aq) + H_2PO_4^-(aq)$ $\quad K_{a1} = \frac{[H_3O^+][H_2PO_4^-]}{[H_3PO_4]}$

$H_2PO_4^-(aq) + H_2O(l) \leftrightarrows H_3O^+(aq) + HPO_4^{2-}(aq)$ $\quad K_{a2} = \frac{[H_3O^+][HPO_4^{2-}]}{[H_2PO_4^-]}$

$HPO_4^{2-}(aq) + H_2O(l) \leftrightarrows H_3O^+(aq) + PO_4^{3-}(aq)$ $\quad K_{a3} = \frac{[H_3O^+][PO_4^{3-}]}{[HPO_4^{2-}]}$

105.a) **Given:** 0.350 M H_3PO_4 $\quad K_{a1} = 7.5 \times 10^{-3}$, $K_{a2} = 6.2 \times 10^{-8}$ **Find:** $[H_3O^+]$, pH
Conceptual Plan: K_{a1} much larger than K_{a2} so, use K_{a1} to calculate $[H_3O^+]$. Write a balanced reaction. Prepare an ICE table; represent the change with x; sum the table and determine the equilibrium values; put the equilibrium values in the equilibrium expression and solve for x.
Solution: $H_3PO_4(aq) + H_2O(l) \leftrightarrows H_3O^+(aq) + H_2PO_4^-(aq)$

	H_3PO_4	H_3O^+	$H_2PO_4^-$
I	0.350M	0.0	0.0
C	- x	x	x
E	0.350 – x	x	x

$$K_a = \frac{[H_3O^+][H_2PO_4^-]}{[H_3PO_4]} = \frac{(x)(x)}{(0.350 - x)} = 7.3 \times 10^{-3}$$

Assume x is small compared to 0.350

$x^2 = (7.3 \times 10^{-3})(0.350)$ $\quad x = 0.0512M = [H_3O^+]$

Check assumption: $\frac{0.0512}{0.350} \times 100 = 14.6\%$ assumption not valid

$x^2 + 7.5 \times 10^{-3}\, x - 0.002625 = 0$ $\quad x = 0.04\underline{7}62 = [H_3O^+]$

$pH = -\log(0.04\underline{7}62) = 1.32$

b) **Given:** 0.350 M $H_2C_2O_4$ $\quad K_{a1} = 6.0 \times 10^{-2}$, $K_{a2} = 6.0 \times 10^{-5}$ **Find:** $[H_3O^+]$, pH
Conceptual Plan: K_{a1} much larger than K_{a2} so, use K_{a1} to calculate $[H_3O^+]$. Write a balanced reaction. Prepare an ICE table; represent the change with x; sum the table and determine the equilibrium values; put the equilibrium values in the equilibrium expression and solve for x.

Solution: $H_2C_2O_4(aq) + H_2O(l) \leftrightarrows H_3O^+(aq) + HC_2O_4^-(aq)$

I	0.350M	0.0	0.0
C	- x	x	x
E	0.350 – x	x	x

$$K_a = \frac{[H_3O^+][HC_2O_4^-]}{[H_2C_2O_4]} = \frac{(x)(x)}{(0.350 - x)} = 6.0 \times 10^{-2}$$

$x^2 + 6.0 \times 10^{-2}\ x - 0.021 = 0$ $\quad$ $x = 0.1\underline{1}79 = 0.12\ M\ [H_3O^+]$

$pH = -\log(0.1\underline{1}79) = 0.92$

107. **Given:** 0.500 M H_2SO_3 $\quad$ $K_{a1} = 1.6 \times 10^{-2}$, $K_{a2} = 6.4 \times 10^{-8}$ $\quad$ **Find:** concentration all species

Conceptual Plan: Conceptual Plan: K_{a1} much larger than K_{a2} so, use K_{a1} to calculate $[H_3O^+]$. Write a balanced reaction. Prepare an ICE table; represent the change with x; sum the table and determine the equilibrium values; put the equilibrium values in the equilibrium expression and solve for x.

Solution: $H_2SO_3(aq) + H_2O(l) \leftrightarrows H_3O^+(aq) + HSO_3^-(aq)$

I	0.500M	0.0	0.0
C	- x	x	x
E	0.500 – x	x	x

$$K_a = \frac{[H_3O^+][HSO_3^-]}{[H_2SO_3]} = \frac{(x)(x)}{(0.500 - x)} = 1.6 \times 10^{-2}$$

$x^2 + 1.6 \times 10^{-2}\ x - 0.0080 = 0$ $\quad$ $x = 0.08\underline{1}8 = 0.082\ M\ [H_3O^+] = [HSO_3^-]$

Use the values from reaction 1 in reaction 2

$HSO_3^-(aq) + H_2O(l) \leftrightarrows H_3O^+(aq) + SO_3^{2-}(aq)$

I	0.0818M	0.0818	0.0
C	- y	y	y
E	0.0818 – y	0.0818+y	y

$$K_a = \frac{[H_3O^+][SO_3^{2-}]}{[HSO_3^-]} = \frac{(0.0818 + y)(y)}{(0.0818 - y)} = 6.4 \times 10^{-8}$$

Assume y is small $\quad$ $y = 6.4 \times 10^{-8}$

$[H_2SO_3] = 0.500 - 0.08\underline{1}8 = 0.418\ M$

$[HSO_3^-] = x = 0.08\underline{1}8 = 0.082\ M$

$[SO_3^{2-}] = y = 6.4 \times 10^{-8}\ M$

$[H_3O^+] = x + y = 0.08\underline{1}8 = 0.082\ M$

$$[OH^-] = \frac{K_w}{[H_3O^+]} = \frac{1 \times 10^{-14}}{0.0818} = 1.2 \times 10^{-13}\ M$$

109.a) **Given:** $[H_2SO_4] = 0.50\ M$ $K_{a2} = 0.012$ $\quad$ **Find:** $[H_3O^+]$,pH

Conceptual Plan: first ionization step is strong, use K_{a2} and reaction 2. Write a balanced reaction. Prepare an ICE table; represent the change with x; sum the table and determine the equilibrium values; put the equilibrium values in the equilibrium expression and solve for x.

Solution: $H_2SO_4(aq) + H_2O(l) \rightarrow H_3O^+(aq) + HSO_4^-(aq)$ strong

0.50 M

$[H_3O^+] = [HSO_4^-] = 0.50\ M$

$HSO_4^-(aq) + H_2O(l) \leftrightarrows H_3O^+(aq) + SO_4^{2-}(aq)$

I	0.500M	0.500	0.0
C	- x	x	x
E	0.500 – x	x	x

$$K_a = \frac{[H_3O^+][SO_4^{2-}]}{[HSO_4^-]} = \frac{(x)(x)}{(0.500 - x)} = 0.012$$

$x^2 + 0.512\ x - 0.006 = 0$ $\quad$ $x = 0.01\underline{1}5 = [H_3O^+]$ from second ionization step

$[H_3O^+] = 0.50 + 0.011 = 0.51M$

$pH = -\log(0.51) = 0.29$

b) **Given:** $[H_2SO_4] = 0.10$ M $K_{a2} = 0.012$ **Find:** $[H_3O^+]$,pH

Conceptual Plan: first ionization step is strong, use K_{a2} and reaction 2. Write a balanced reaction. Prepare an ICE table; represent the change with x; sum the table and determine the equilibrium values; put the equilibrium values in the equilibrium expression and solve for x.

Solution: $H_2SO_4(aq) + H_2O(l) \rightarrow H_3O^+(aq) + HSO_4^-(aq)$ strong

0.10 M

$[H_3O^+] = [HSO_4^-] = 0.10$ M

	$HSO_4^-(aq) + H_2O(l)$ ⇆	$H_3O^+(aq)$ +	$SO_4^{2-}(aq)$
I	0.10M	0.10	0.0
C	- x	x	x
E	0.10 – x	x	x

$$K_a = \frac{[H_3O^+][SO_4^{2-}]}{[HSO_4^-]} = \frac{(x)(x)}{(0.10 - x)} = 0.012$$

$x^2 + 0.112\,x - 0.0012 = 0$ $\quad x = 0.009848$

$$\frac{0.009848}{0.10} \times 100 = 9.8\%$$ contribution of second ionizations step not negligible

$[H_3O^+] = 0.10 + 0.0085 = 0.1085 = 0.11$M

pH = - log(0.11) = 0.96

c) **Given:** $[H_2SO_4] = 0.050$ M $K_{a2} = 0.012$ **Find:** $[H_3O^+]$,pH

Conceptual Plan: first ionization step is strong, use K_{a2} and reaction 2. Write a balanced reaction. Prepare an ICE table; represent the change with x; sum the table and determine the equilibrium values; put the equilibrium values in the equilibrium expression and solve for x.

Solution: $H_2SO_4(aq) + H_2O(l) \rightarrow H_3O^+(aq) + HSO_4^-(aq)$ strong

0.050 M

$[H_3O^+] = [HSO_4^-] = 0.050$ M

	$HSO_4^-(aq) + H_2O(l)$ ⇆	$H_3O^+(aq)$ +	$SO_4^{2-}(aq)$
I	0.050M	0.10	0.0
C	- x	x	x
E	0.050 – x	x	x

$$K_a = \frac{[H_3O^+][SO_4^{2-}]}{[HSO_4^-]} = \frac{(x)(x)}{(0.050 - x)} = 0.012$$

$x^2 + 0.062\,x - 0.006 = 0$ $\quad x = 0.008509$

$$\frac{0.008509}{0.05} \times 100 = 17\%$$ contribution of second ionizations step not negligible

$[H_3O^+] = 0.050 + 0.0085 = 0.0585 = 0.059$M

pH = - log(0.059) = 1.23

111. a) HCl is the stronger acid. HCl is the weaker bond, therefore, it is more acidic.

b) HF is the stronger acid. F is more electronegative than O, so the bond is more polar and more acidic.

c) H_2Se is the stronger acid. The H – Se bond is weaker, therefore, it is more acidic.

113. a) H_2SO_4 is the stronger acid because it has more oxygen atoms.

b) $HClO_2$ is the stronger acid because it has more oxygen atoms.

c) HClO is the stronger acid because Cl is more electronegative than Br.

d) CCl_3COOH is the stronger acid because Cl is more electronegative than H.

115. S^{2-} is the stronger base. Base strength is determined from the corresponding acid. The weaker the acid, the stronger the base. H_2S is the weaker acid because it has a stronger bond.

117. a) Lewis acid: Fe^{3+} has empty d orbital and can accept lone pair electrons.

b) Lewis acid: BH_3 has an empty p orbital to accept a lone pair of electrons.

c) Lewis base: NH_3 has a lone pair of electrons to donate.

d) Lewis base: F^- has lone pair electrons to donate.

119. a) Fe^{3+} accepts electron pair from H_2O, so Fe^{3+} is the Lewis acid and H_2O is the Lewis base.

b) Zn^{2+} accepts electron pair from NH_3, so Zn^{2+} is the Lewis acid and NH_3 is the Lewis base.

c) The empty p orbital on B accepts an electron pair from $(CH_3)_3N$, so BF_3 is the Lewis acid and $(CH_3)_3N$ is the Lewis base.

121. a) Weak acid. The beaker contains 10 HF molecule, $2H_3O^+$ ions and $2F^-$ ions. Since both the molecule and the ions exist in solution, the acid is a weak acid.

b) Strong acid. The beaker contains $12H_3O^+$ ions and $2OH^-$ ions. Since the molecule is completely ionized in solution, the acid is a strong acid.

c) Weak acid. The beaker contains 10 $HCHO_2$ molecules, $2H_3O^+$ ions and $2CHO_2^-$ ions. Since both the molecule and the ions exist in solution, the acid is a weak acid.

d) Strong acid. The beaker contains $12H_3O^+$ ions and $2NO_3^-$ ions. Since the molecule is completely ionized in solution, the acid is a strong acid.

123. $HbH^+(aq) + O_2(aq) \leftrightarrows HbO_2(aq) + H^+(aq)$
Using LeChatlier's Principle, if the $[H^+]$ increases, the reaction will shift left, if the $[H^+]$ decreases, the reaction will shift right. So, if the pH of blood is too acidic (low pH; $[H^+]$ increased), the reaction will shift to the left. This will cause less of the HbO_2 in the blood and decrease the oxygen – carrying capacity of the hemoglobin in the blood.

125. **Given:** 4.00×10^2 mg $Mg(OH)_2$, 2.00×10^2 mL HCl solution, pH = 1.3
Find: volume neutralized, % neutralized.
Conceptual Plan:

mg $Mg(OH)_2$ → g $Mg(OH)_2$ → mol $Mg(OH)_2$ and then pH → $[H_3O^+]$ and then mol $Mg(OH)_2$

$$\frac{\text{g Mg(OH)}_2}{1000\text{ mg}} \qquad \frac{\text{mol Mg(OH)}_2}{58.326\text{g}} \qquad \text{pH} = -\log[H_3O^+]$$

mol OH^- → mol H_3O^+ → vol H_3O^+ → % neutralized.

$$\frac{2\text{ OH}^-}{\text{Mg(OH)}_2} \quad \frac{H_3O^+}{OH^-} \qquad \frac{\text{mol }H_3O^+}{M(H_3O^+)} \qquad \frac{\text{vol HCl neutralized}}{\text{total vol HCl}} \times 100$$

Solution: $[H_3O^+] = 10^{-1.3} = 0.05011\text{ M} = 0.05\text{ M}$

$$4.00 \times 10^2\text{ mg Mg(OH)}_2 \times \frac{\text{g Mg(OH)}_2}{1000\text{ mg Mg(OH)}_2} \times \frac{1\text{ mol Mg(OH)}_2}{58.326\text{ g Mg(OH)}_2} \times \frac{2\text{ mol OH}^-}{1\text{ mol Mg(OH)}_2}$$

$$\times \frac{1\text{ mol }H_3O^+}{1\text{ mol OH}^-} \times \frac{1\text{ L}}{0.0501\text{ mol }H_3O^+} \times \frac{1000\text{ mL}}{\text{L}} = 27\underline{3}.7\text{ mL} = 274\text{ mL neutralized}$$

Stomach contains 2.00×10^2 mL HCl at pH = 1.3 and 4.00×10^2 mg will neutralize 274 mL of pH 1.3 HCl, so all of the stomach acid will be neutralized.

127. **Given:** pH of Great Lakes acid rain = 4.5, West Coast = 5.4 **Find:** $[H_3O^+]$ and ratio of Great Lakes/ West Coast
Conceptual Plan: pH → $[H_3O^+]$, and then ratio of $[H_3O^+]$ Great Lakes to West Coast.

$\text{pH} = -\log[H_3O^+]$

Solution: Great Lakes: $[H_3O^+] = 10^{-4.5} = 3.16 \times 10^{-5}M$ West Coast: $[H_3O^+] = 10^{-5.4} = 3.98 \times 10^{-6}M$

$$\frac{\text{Great Lakes}}{\text{West Coast}} = \frac{3.16 \times 10^{-5}M}{3.98 \times 10^{-6}M} = 7.94 = 8 \text{ times more acidic}$$

129. **Given:** 6.5×10^2 mg aspirin, 8 ounces water, $pK_a = 3.5$ **Find:** pH of solution

Conceptual Plan:

mg aspirin → g aspirin → mol aspirin and ounces → quart → L and then [aspirin] and pK$_a$

$$\frac{\text{g aspirin}}{1000 \text{ mg}} \quad \frac{\text{mol aspirin}}{180.16\text{g}} \qquad \frac{1 \text{ qt}}{32 \text{ ounces}} \quad \frac{1 \text{ L}}{1.0567 \text{ qt}} \quad \frac{\text{mol aspirin}}{\text{L soln}}$$

→K_a and then: Write a balanced reaction. Prepare an ICE table; represent the change with x;

$pK_a = -\log K_a$

sum the table and determine the equilibrium values; put the equilibrium values in the equilibrium expression and solve for x. Determine $[H_3O^+]$ → pH

$pH = -\log[H_3O^+]$

Solution:

$$\frac{6.5 \times 10^2 \text{ mg aspirin}}{8 \text{ ounces}} \times \frac{1 \text{ gram aspirin}}{1000 \text{ mg aspirin}} \times \frac{\text{mol aspirin}}{180.16 \text{ g}} \times \frac{32 \text{ ounces}}{\text{qt}} \times \frac{1.0567 \text{ qt}}{1 \text{ L}} = 0.0152 \text{ M}$$

$K_a = 10^{-3.5} = 3.16 \times 10^{-4}$

	aspirin(aq) + H_2O(l) ⇆	H_3O^+(aq) +	aspirin$^-$ (aq)
I	0.0152M	0.0	0.0
C	- x	x	x
E	0.0152– x	x	x

$$K_a = \frac{[H_3O^+][\text{aspirin}^-]}{[\text{aspirin}]} = \frac{(x)(x)}{(0.0152 - x)} = 3.16 \times 10^{-4}$$

$x^2 + 3.16 \times 10^{-4}x - 4.80 \times 10^{-6} = 0$ $x = 2.04 \times 10^{-3}$ M $= [H_3O^+]$

$pH = -\log(2.04 \times 10^{-3}) = 2.69 = 2.7$

131.a) **Given:** 0.0100M $HClO_4$ **Find:** pH

Conceptual Plan: $[HClO_4]$ → $[H_3O^+]$ → pH

$[HClO_4]$ → $[H_3O^+]$ $pH = -\log[H_3O^+]$

Solution: 0.0100 M $HClO_4$ = 0.0100 M H_3O^+ $pH = -\log(0.0100) = 2.000$

b) **Given:** 0.115M $HClO_2$, $K_a = 1.1 \times 10^{-2}$ **Find:** pH

Conceptual Plan: Write a balanced reaction. Prepare an ICE table; represent the change with x; sum the table and determine the equilibrium values; put the equilibrium values in the equilibrium expression and solve for x. Determine $[H_3O^+]$ and pH

Solution: $HClO_2$ (aq) + H_2O(l) ⇆ H_3O^+(aq) + ClO_2^- (aq)

	$HClO_2$	H_3O^+	ClO_2^-
I	0.115M	0.0	0.0
C	- x	x	x
E	0.115 – x	x	x

$$K_a = \frac{[H_3O^+][ClO_2^-]}{[HClO_2]} = \frac{(x)(x)}{(0.115 - x)} = 1.1 \times 10^{-2}$$

Assume x is small compared to 0.115

$x^2 = (1.1 \times 10^{-2})(0.115)$ $x = 0.0356M = [H_3O^+]$

Check assumption: $\frac{0.0356}{0.115} \times 100 = 30.9\%$ assumption not valid

$x^2 + 1.1 \times 10^{-2}x - 0.001265 = 0$

$x = 0.03049$ or -0.0415

$pH = -\log(0.03049) = 1.52$

c) **Given:** 0.045M $Sr(OH)_2$ **Find:** pH

Conceptual Plan: $[Sr(OH)_2] \rightarrow [OH^-] \rightarrow [H_3O^+] \rightarrow pH$

$K_w = [H_3O^+][OH^-]$ $pH = -\log[H_3O^+]$ $pH + pOH = 14$

Solution: $[OH^-] = [Sr(OH)_2] = 2(0.045) = 0.090\ M$

$$[H_3O^+] = \frac{K_w}{[OH^-]} = \frac{1 \times 10^{-14}}{0.090\,M} = 1.\underline{1}1 \times 10^{-13} M$$

$pH = -\log(1.\underline{1}1 \times 10^{-13}) = 12.95$

d) **Given:** 0.0852 KC_6H_5O, K_a (HC_6H_5O)= 4.9×10^{-10} **Find:** pH

Conceptual Plan: Identify each species and determine which will contribute to pH. Write a balanced reaction. Prepare an ICE table; represent the change with x; sum the table and determine the equilibrium values; put the equilibrium values in the equilibrium expression and solve for x. Determine $[OH^-] \rightarrow pOH \rightarrow pH$

$pOH = -\log[OH^-]$ $pH + pOH = 14$

Solution: K^+ is the counterion of a strong base, therefore, is pH neutral. CN^- is the conjugate base of a weak acid, therefore, it is basic.

	CN^- (aq) + H_2O(l) $\leftrightarrows$	HCN (aq)+	OH^- (aq)
I	0.0852	0.0	0.0
C	-x	x	x
E	0.0852 – x	x	x

$$K_b = \frac{K_w}{K_a} = \frac{1 \times 10^{-14}}{4.9 \times 10^{-10}} = 2.\underline{0}4 \times 10^{-5} = \frac{(x)(x)}{(0.0852 - x)}$$

Assume x is small

$x = 1.\underline{3}2 \times 10^{-3} = [OH^-]$ $pOH = -\log(1.\underline{3}2 \times 10^{-3}) = 2.88$

$pH = 14.00 - 2.88 = 11.12$

e) **Given:** 0.155 NH_4Cl, K_b (NH_3)= 1.8×10^{-5} **Find:** pH

Conceptual Plan: Identify each species and determine which will contribute to pH. Write a balanced reaction. Prepare an ICE table; represent the change with x; sum the table and determine the equilibrium values; put the equilibrium values in the equilibrium expression and solve for x. Determine $[H_3O^+] \rightarrow pH$

Solution: NH_4^+ is the conjugate acid of a weak base, therefore, it is acidic. Cl^- is the conjugate base of a strong acid, therefore, it is pH neutral.

	NH_4^+ (aq) + H_2O(l) $\leftrightarrows$	NH_3 (aq) +	H_3O^+ (aq)
I	0.155	0.0	0.0
C	-x	x	x
E	0.155– x	x	x

$$K_a = \frac{K_w}{K_b} = \frac{1 \times 10^{-14}}{1.8 \times 10^{-5}} = 5.\underline{5}6 \times 10^{-10} = \frac{(x)(x)}{(0.155 - x)}$$

Assume x is small

$x = 9.\underline{2}8 \times 10^{-6} = [H_3O^+]$ $pH = -\log(9.\underline{2}8 \times 10^{-6}) = 5.03$

133.a) **Given:** 0.0550M HI (strong acid), 0.00850M HF (weak acid) **Find:** pH

Conceptual Plan: Since mixture is a strong acid, weak acid, the strong acid will dominate. Use the concentration of the strong acid to determine $[H_3O^+]$ and then pH

Solution: 0.0550M HI = 0.0550 M $[H_3O^+]$ $pH = -\log(0.0550) = 1.260$

b) **Given:** 0.112 M NaCl (salt), 0.0953 M KF (salt) **Find:** pH

Conceptual Plan: Identify each species and determine which will contribute to pH. Write a balanced reaction. Prepare an ICE table; represent the change with x; sum the table and determine the equilibrium values; put the equilibrium values in the equilibrium expression and solve for x. Determine $[H_3O^+] \rightarrow pH$

Solution: Na^+ is the counterion of a strong base, therefore, it is pH neutral. Cl^- is the conjugate base of a strong acid, therefore, it is pH neutral. K^+ is the counterion of a strong base, therefore, it is pH neutral. F^- is the conjugate base of a weak acid. Therefore, it will produce a basic solution.

	$F^-(aq)$ + $H_2O(l)$ ⇆	$HF(aq)$ +	$OH^-(aq)$
I	0.0953	0.0	0.0
C	-x	x	x
E	0.0953– x	x	x

$$K_b = \frac{K_w}{K_a} = \frac{1 \times 10^{-14}}{3.5 \times 10^{-4}} = \frac{(x)(x)}{(0.0953 - x)}$$

Assume x is small

$x = 1.\underline{6}50 \times 10^{-6} = [OH^-]$ $pOH = -\log(1.\underline{6}50 \times 10^{-6}) = 5.78$

$pH = 14.00 - 5.78 = 8.22$

c) **Given:** 0.132 M NH_4Cl (salt), 0.150 M HNO_3 (strong acid)   **Find:** pH
Conceptual Plan: Since the mixture is a strong acid and a salt, the strong acid will dominate. Use the concentration of the strong acid to determine $[H_3O^+]$ and then pH.
Solution: 0. 150M HNO_3 = 0.150 M $[H_3O^+]$ $pH = -\log(0.150) = 0.824$

d) **Given:** 0.0887M $NaC_7H_5O_2$ (salt) 0.225 M KBr(salt)
Conceptual Plan: Identify each species and determine which will contribute to pH. Write a balanced reaction. Prepare an ICE table; represent the change with x; sum the table and determine the equilibrium values; put the equilibrium values in the equilibrium expression and solve for x. Determine $[H_3O^+]$ → pH
Solution: Na^+ is the counterion of a strong base, therefore, it is pH neutral. $C_7H_5O_2^-$ is the conjugate base of a weak acid. Therefore, it will produce a basic solution. K^+ is the counterion of a strong base, therefore, it is pH neutral. Cl^- is the conjugate base of a strong acid, therefore, it is pH neutral.

	$C_7H_5O_2^-(aq)$ + $H_2O(l)$ ⇆	$HC_7H_5O_2(aq)$ +	$OH^-(aq)$
I	0.0887	0.0	0.0
C	-x	x	x
E	0.0887– x	x	x

$$K_b = \frac{K_w}{K_a} = \frac{1 \times 10^{-14}}{6.5 \times 10^{-5}} = \frac{(x)(x)}{(0.0887 - x)}$$

Assume x is small

$x = 3.\underline{6}94 \times 10^{-6} = [OH^-]$ $pOH = -\log(3.\underline{6}94 \times 10^{-6}) = 5.43$

$pH = 14.00 - 5.43 = 8.57$

e) **Given:** 0.0450 M HCl (strong acid), 0.0225 M HNO_3 (strong acid)   **Find:** pH
Conceptual Plan: Since mixture is a strong acid and strong acid, $[H_3O^+]$ is the sum of the concentration of both acids, and then determine pH
Solution: 0.0450 M HCl = 0.020 $[H_3O^+]$, 0. 0225 M HNO_3 = 0.0225 M $[H_3O^+]$
$[H_3O^+] = 0.0450 + 0.0225 = 0.0675M$ $pH = -\log(0.0675) = 1.171$

135.a) sodium cyanide = NaCN   nitric acid = HNO_3
$H^+(aq) + CN^-(aq) \leftrightarrows HCN(aq)$

b) ammonium chloride = NH_4Cl sodium hydroxide = NaOH
$NH_4^+(aq) + OH^-(aq) \leftrightarrows NH_3(aq) + H_2O(l)$

c) sodium cyanide = NaCN   ammonium bromide = NH_4Br
$NH_4^+(aq) + CN^-(aq) \leftrightarrows NH_3(aq) + HCN(aq)$

d) potassium hydrogen sulfate = $KHSO_4$ lithium acetate = $LiC_2H_3O_2$
$HSO_4^-(aq) + C_2H_3O_2^-(aq) \leftrightarrows SO_4^{2-}(aq) + HC_2H_3O_2(aq)$

e) sodium hypochlorite = NaClO   ammonia = NH_3
No reaction, both are bases

137. **Given:** 1.0M urea, pH = 7.050 **Find:** K_a Hurea$^+$

Conceptual Plan:

pH → pOH → [OH$^-$] → K_b(urea) → K_a(Hurea$^+$). Write a balanced reaction. Prepare an ICE

pH + pOH = 14 pOH = -log[OH$^-$]

table; represent the change with x; sum the table and determine the equilibrium values; put the equilibrium values in the equilibrium expression and determine K_b.

Solution: pOH = 14.000 – 7.050 = 6.950 [OH^{-1}] = $10^{-6.950}$ = 1.122 x 10^{-7}M

	urea (aq) + H_2O(l) ⇆	Hurea$^+$ (aq)+	OH^- (aq)
I	1.0	0.0	0.0
C	-x	x	x
E	1.0– 1.122 x 10^{-7}	1.122 x 10^{-7}	1.122 x 10^{-7}

$$K_b = \frac{[\text{Hurea}^+][\text{OH}^-]}{[\text{urea}]} = \frac{(1.122 \times 10^{-7})(1.122 \times 10^{-7})}{(1.0 - 1.122 \times 10^{-7})} = 1.2589 \times 10^{-14}$$

$$K_a = \frac{K_w}{K_b} = \frac{1.00 \times 10^{-14}}{1.2589 \times 10^{-14}} = 0.7943 = 0.794$$

139. The calculation is incorrect because it neglects the contribution from the autoionization of water.
HI is a strong acid, so [H_3O^+] from HI = 1.0 x 10^{-7}

	H_2O(l) + H_2O(l) ⇆	H_3O^+(aq) +	OH^- (aq)
I		1 x 10^{-7}	0.0
C	-x	x	x
E		1 x 10^{-7} + x	x

$$K_w = [H_3O^+][OH^-] = 1.0 \times 10^{-14}$$

$$(1 \times 10^{-7} + x)(x) = 1.0 \times 10^{-14} \qquad x^2 + 1 \times 10^{-7}\,x - 1.0 \times 10^{-14} = 0$$

$$x = 6.18 \times 10^{-8}$$

$$[H_3O^+] = (1 \times 10^{-7} + x) = (1 \times 10^{-7} + 6.18 \times 10^{-8}) = 1.618 \times 10^{-7}$$

$$pH = -\log(1.618 \times 10^{-7}) = 6.79$$

141. **Given:** 0.00115M HCl, 0.01000M $HClO_2$ K_a = 1.1 x 10^{-2} **Find:** pH

Conceptual Plan: Use HCl to determine [H_3O^+], use [H_3O^+] and $HClO_2$ to determine dissociation of $HClO_2$. Write a balanced reaction, prepare and ICE table, calculate equilibrium concentrations, and then plug into the equilibrium expression.

Solution: 0.00115M HCl = 0.00115M H_3O^+

	H ClO_2 (aq) + H_2O(l) ⇆	H_3O^+(aq) +	ClO_2^- (aq)
I	0.0100M	0.00115	0.0
C	- x	x	x
E	0.01000 – x	0.00115 + x	x

$$K_a = \frac{[H_3O^+][ClO_2^-]}{[HClO_2]} = \frac{(0.00115 + x)(x)}{(0.0100 - x)} = 1.1 \times 10^{-2}$$

$$x^2 + 0.01215x - 1.1 \times 10^{-4} = 0 \qquad x = 0.006045$$

$$[H_3O^+] = 0.00115 + 0.006045 = 0.00719$$

$$pH = -\log(0.00719) = 2.14$$

143. **Given:** 1.0M HA, K_a = 1.0 x 10^{-8} K= 4.0 for reaction 2 **Find:** [H^+], [A^-], [HA_2^-]

Conceptual Plan: Combine reaction 1 and reaction 2 and determine equilibrium expression and the value of K. Prepare an ICE table, calculate equilibrium concentrations, and then plug into the equilibrium expression

Solution:

	HA(aq)	$\leftrightarrows$	$H^+(aq)$ +	$A^-(aq)$	$K = 1.0 \times 10^{-8}$
	$HA(aq) + A^-(aq)$	$\leftrightarrows$	$HA_2^+(aq)$		$K = 4.0$
	2HA(aq)	$\leftrightarrows$	$H^+(aq)$ +	$HA_2^+(aq)$	$K = 4.0 \times 10^{-8}$
I	1.0		0.0	0.0	
C	- 2x		x	x	
E	1.0 – 2x		x	x	

$$K = \frac{[H^+][HA_2^-]}{[HA]^2} = 4.0 \times 10^{-8} = \frac{(x)(x)}{(1.0 - 2x)^2}$$

$x^2 + 16 \times 10^{-8}x - 4 \times 10^{-8} = 0$ $x = 1.\underline{9}992 \times 10^{-4}$

	HA(aq)	$\leftrightarrows$	$H^+(aq)$ +	$A^-(aq)$	$K = 1.0 \times 10^{-8}$
I	1.0		1.9992×10^{-4}	0.0	
C	- 2y		y	y	
E	1.0 – 2y		1.9992×10^{-4}+y	y	

$$K = \frac{[H^+][A^-]}{[HA]} = 1.0 \times 10^{-8} = \frac{(1.9992 \times 10^{-4} + y)(y)}{(1.0 - y)}$$

$y^2 + 1.9992 \times 10^{-4}x - 1 \times 10^{-8} = 0$ $y = 4.\underline{2}9 \times 10^{-5}$

$[H^+] = x + y = 1.\underline{9}992 \times 10^{-4} + 4.\underline{2}9 \times 10^{-5} = 2.4 \times 10^{-4}$

$[A^-] = y = 4.29 \times 10^{-5}$

$[HA_2^-] = x = 2.0 \times 10^{-4}$

145. Solution b would be most acidic.

a) 0.0100 M HCl (strong acid) and 0.0100 M KOH(strong base), since the concentrations are equal , the acid and base will completely neutralize each other and the resulting solution will be pH neutral.

b) 0.0100 M HF (weak acid) and 0.0100 KBr (salt). K_a (HF) = 3.5×10^{-4}. The weak acid will produce and acidic solution. K^+ is the counterion of a strong base and is pH neutral. Br^- is the conjugate base of a strong acid and is pH neutral.

c) 0.0100 M NH_4Cl (salt) and 0.100 M CH_3NH_3Br. $K_b(NH_3) = 1.8 \times 10^{-5}$, $K_b(CH_3NH_2) = 4.4 \times 10^{-4}$. NH_4^+ is the conjugate acid of a weak base and $CH_3NH_3^+$ is the conjugate acid of a weak base. Cl^- and Br^- are the conjugate base of a strong acid and will be pH neutral. The solution will be acidic, however, K_a for the conjugate acids is much smaller K_a for HF, so the solution will be acidic, but not as acidic as HF.

d) 0.100 M NaCN (salt) and 0.100 M $CaCl_2$. Na^+ and Ca^{2+} ion are the counterion of a strong base, therefore, they are pH neutral. Cl^- is the conjugate base of a strong acid and is pH neutral. CN^- is the conjugate base of a weak acid and will produce a basic solution.

147. $CH_3COOH < CH_2ClCOOH < CHCl_2COOH < CCl_3COOH$

Since Cl is more electronegative than H, as you add Cl you increase the amount of electronegativity and pull electron density away from the O – H bond making it more acidic.

Chapter 16
Aqueous Ionic Equilibrium

1. The pH range of human blood is held between 7.36 and 7.42. This nearly constant blood pH is maintained by buffers, which is a chemical system that resists pH changes, neutralizing added acid or base. An important buffer system in blood is a mixture of carbonic acid (H_2CO_3) and bicarbonate ion (HCO_3^-).

3. The common ion effect occurs when a solution contains two substances ($HC_2H_3O_2$ and $NaC_2H_3O_2$) that share a common ion ($C_2H_3O_2^-$). The presence of the $C_2H_3O_2^-(aq)$ ion causes the acid to ionize even less than it normally would, resulting in a less acidic solution (higher pH). This effect is an example of Le Châtelier's principle shifting an equilibrium, because of the addition (or removal) of the common ion from the solution.

5. When the concentration of the conjugate acid and base components of a buffer system are equal the pH is equal to the pK_a of the weak acid of the buffer system. When more of the acid component is present the pH becomes more acidic (pH drops). When more of the base component is present the pH becomes more basic (pH rises). The pH in both cases can be calculates with the Henderson-Hasselbalch equation.

7. To find the pH of this solution, determine which component is the acid and which component is the base and substitute their concentrations into the Henderson-Hasselbalch equation to compute pH. The pK_a is the negative of the log of the equilibrium constant of the acid dissociation reaction where the weak acid component and water are the reactants, and the conjugate base and H_3O^+ are the products. At the end, confirm that the "x is small" approximation is valid by calculating the $[H_3O^+]$ from the pH. Since H_3O^+ is formed by ionization of the acid, the calculated $[H_3O^+]$ has to be less than 0.05 (or 5%) of the initial concentration of the acid in order for the "x is small" approximation to be valid.

9. The relative concentrations of acid and conjugate base should not differ by more than a factor of 10 in order for a buffer to be reasonably effective. Using the Henderson-Hasselbalch equation, this means that the pH should be with in one pH unit of the weak acid's pK_a.

11. The titration of weak acid by a strong base will always have a basic equivalence point because, at the equivalence point, all of the acid has been converted into its conjugate base, resulting in a weakly basic solution.

13. a) The initial pH of the solution is simply the pH of the strong acid. Since strong acids completely dissociate, the concentration of H_3O^+ is the concentration of the strong acid and pH = - log $[H_3O^+]$.

    b) Before the equivalence point, H_3O^+ is in excess. Calculate the $[H_3O^+]$ by subtracting the number of moles of added OH^- from the initial number of moles of H_3O^+ and dividing by the *total* volume. Then convert to pH using - log $[H_3O^+]$.

    c) At the equivalence point, neither reactant is in excess and the pH = 7.00.

    d) Beyond the equivalence point, OH^- is in excess. Calculate the $[OH^-]$ by subtracting the initial number of moles of H_3O^+ from the number of moles of added OH^- and dividing by the *total* volume. Then convert to pH using - log $[H_3O^+]$.

15. When a polyprotic acid is titrated with a strong base, and if K_{a1} and K_{a2} are sufficiently different, the pH curve will have two equivalence points because the two acidic protons will be titrated sequentially. The titration of the first acidic proton will be completed before the titration of the second acidic proton.

17. The endpoint is the point when the indicator changes color in an acid-base titration. The equivalence point is when stoichiometrically equivalent amounts of acid and base have reacted. With the correct indicator, the endpoint of the titration will occur at the equivalence point.

19. The solubility product constant (K_{sp}) is the equilibrium expression for a chemical equation representing the dissolution of an ionic compound. The expression of the solubility product constant of A_mX_n is: $K_{sp} = [A^{n+}]^m [X^{m-}]^n$.

21. In accordance with Le Châtelier's principle, the presence of a common ion in solution causes the equilibrium to shift to the left (compared to its position with pure water as the solvent), which means that less of the ionic compound dissolves. Thus, the solubility of an ionic compound is lower in a solution containing a common ion than in pure water. The exact value of the solubility can be calculated by working an equilibrium problem in which the concentration of the common ion is accounted for in the initial conditions. The molar solubility of a compound, A_mX_n, in a solution with an initial concentration of $A^{n+} = [A^{n+}]_0$ and an initial concentration of $X^{m-} = [X^{m-}]_0$ can be computed directly from K_{sp} by solving for S in the expression: $K_{sp} = ([A^{n+}]_0 + mS)^m ([X^{m-}]_0 + nS)^n$.

23. Q is the reaction quotient, the product of the concentrations of the ionic components raised to their stoichiometric coefficients, and K is the product of the concentrations of the ionic components raised to their stoichiometric coefficients at equilibrium. For a solution containing an ionic compound: If $Q < K_{sp}$, the solution is unsaturated. More of the solid ionic compound can dissolve in the solution; If $Q = K_{sp}$, the solution is saturated and the solution is holding the equilibrium amount of the dissolved ions and additional solid will not dissolve in the solution; and If $Q > K_{sp}$, the solution is supersaturated and under most circumstances, the excess solid will precipitate out of a supersaturated solution.

25. Qualitative analysis is a systematic way to determine the metal ions present in an unknown solution by the selective precipitation of the ions. The word qualitative means involving quality or kind. So qualitative analysis involves finding the kind of ions present in the solution. Quantitative analysis is concerned with quantity, or the amounts of substances in a solution or mixture.

27. The only solution that HNO_2 will ionize less in is d) 0.10 M $NaNO_2$. It is the only solution that generates a common ion NO_2^- with nitrous acid.

29. a). **Given:** 0.15 M $HCHO_2$ and 0.10 M $NaCHO_2$ **Find:** pH **Other:** K_a ($HCHO_2$) = 1.8×10^{-4}

Conceptual Plan: M $NaCHO_2$ → M CHO_2^- then M $HCHO_2$, M CHO_2^- → $[H_3O^+]$ → pH

$NaCHO_2\,(aq) \rightarrow Na^+\,(aq) + CHO_2^-\,(aq)$ ICE Chart $pH = -\log [H_3O^+]$

Solution: Since 1 CHO_2^- ion is generated for each $NaCHO_2$, $[CHO_2^-] = 0.10$ M CHO_2^-.

$$HCHO_2\,(aq) + H_2O\,(l) \rightleftharpoons H_3O^+\,(aq) + CHO_2^{2-}\,(aq)$$

	$[HCHO_2]$	$[H_3O^+]$	$[CHO_2^{2-}]$
Initial	0.15	≈ 0.00	0.10
Change	$-x$	$+x$	$+x$
Equil	$0.15 - x$	$+x$	$0.10 + x$

$$K_a = \frac{[H_3O^+][CHO_2^-]}{[HCHO_2]} = 1.8 \times 10^{-4} = \frac{x(0.10 + x)}{0.15 - x}$$

Assume x is small (x << 0.10 < 0.15) so $\frac{x(0.10 + \cancel{x})}{0.15 - \cancel{x}} = 1.8 \times 10^{-4} = \frac{x(0.10)}{0.15}$ and $x = 2.7 \times 10^{-4}$ = $[H_3O^+]$. Confirm that assumption is valid $\frac{2.7 \times 10^{-4}}{0.10} \times 100\ \% = 0.27\ \%$ so assumption is valid. Finally,

$pH = -\log [H_3O^+] = -\log (2.7 \times 10^{-4}) = 3.57$

Check: The units (none) are correct. The magnitude of the answer makes physical sense because pH should be greater than – log (0.15) = 0.82 because this is a weak acid and there is a common ion effect.

b) **Given:** 0.12 M NH_3 and 0.18 M NH_4Cl **Find:** pH **Other:** K_b (NH_3) = 1.79×10^{-5}

Conceptual Plan: M NH_4Cl → M NH_4^+ then M NH_3, M NH_4^+ → $[OH^-]$ → $[H_3O^+]$ → pH

$NH_4Cl\,(aq) \rightarrow NH_4^+\,(aq) + Cl^-\,(aq)$ ICE Chart $K_w = [H_3O^+][OH^-]$ $pH = -\log [H_3O^+]$

Solution: Since 1 NH_4^+ ion is generated for each NH_4Cl, $[NH_4^+] = 0.18$ M NH_4^+.

$$NH_3(aq) + H_2O(l) \rightleftharpoons NH_4^+(aq) + OH^-(aq)$$

	$[NH_3]$	$[NH_4^+]$	$[OH^-]$
Initial	0.12	0.18	≈ 0.00
Change	$-x$	$+x$	$+x$
Equil	$0.12 - x$	$0.18 + x$	$+x$

$$K_b = \frac{[NH_4^+][OH^-]}{[NH_3]} = 1.79 \times 10^{-5} = \frac{(0.18 + x)x}{0.12 - x}$$

Assume x is small (x << 0.12 < 0.18) so $\frac{(0.18 + \cancel{x})x}{0.12 - \cancel{x}} = 1.79 \times 10^{-5} = \frac{(0.18)x}{0.12}$ and x = $1.\underline{1}9333 \times 10^{-5}$ = $[OH^-]$. Confirm that assumption is valid $\frac{1.\underline{1}9333 \times 10^{-5}}{0.12} \times 100\ \% = 9.9 \times 10^{-3}\ \%$ so assumption is valid.

$K_w = [H_3O^+][OH^-]$ so $[H_3O^+] = \frac{K_w}{[OH^-]} = \frac{1.0 \times 10^{-14}}{1.\underline{1}9333 \times 10^{-5}} = 8.\underline{3}799 \times 10^{-10}$ M. Finally,

$pH = -\log [H_3O^+] = -\log (8.\underline{3}799 \times 10^{-10}) = 9.08$

Check: The units (none) are correct. The magnitude of the answer makes physical sense because pH should be less than 14 + log (0.12) = 13.1 because this is a weak base and there is a common ion effect.

31. **Given:** 0.15 M $HC_7H_5O_2$ in pure water and in 0.10 M $NaC_7H_5O_2$
Find: % ionization in both solutions **Other:** K_a ($HC_7H_5O_2$) = 6.5×10^{-5}
Conceptual Plan: pure water: M $HC_7H_5O_2$ → $[H_3O^+]$ → % ionization then in $NaC_7H_5O_2$ solution:

ICE Chart $\quad \%\ ionization = \frac{[H_3O^+]_{equil}}{[HC_7H_5O_2]_0} \times 100\ \% =$

M $NaC_7H_5O_2$ → M $C_7H_5O_2^-$ then M H $C_7H_5O_2$, M $C_7H_5O_2^-$ → $[H_3O^+]$ → % ionization

$NaC_7H_5O_2\ (aq) \rightarrow Na^+\ (aq) + C_7H_5O_2^-\ (aq)$ $\quad$ ICE Chart $\%\ ionization = \frac{[H_3O^+]_{equil}}{[HC_7H_5O_2]_0} \times 100\ \% =$

Solution: in pure water:

$$HC_7H_5O_2(aq) + H_2O(l) \rightleftharpoons H_3O^+(aq) + C_7H_5O_2^{2-}(aq)$$

	$[HC_7H_5O_2]$	$[H_3O^+]$	$[C_7H_5O_2^{2-}]$
Initial	0.15	≈ 0.00	0.00
Change	$-x$	$+x$	$+x$
Equil	$0.15 - x$	$+x$	$+x$

$K_a = \frac{[H_3O^+][C_7H_5O_2^-]}{[HC_7H_5O_2]} = 6.5 \times 10^{-5} = \frac{x^2}{0.15 - x}$ Assume x is small (x << 0.10) so

$\frac{x^2}{0.15 - \cancel{x}} = 6.5 \times 10^{-5} = \frac{x^2}{0.15}$ and x = $3.\underline{1}225 \times 10^{-3}$ = $[H_3O^+]$. then

$\%\ ionization = \frac{[H_3O^+]_{equil}}{[HC_7H_5O_2]_0} \times 100\ \% =$

$\frac{3.\underline{1}225 \times 10^{-3}}{0.15} \times 100\ \% = 2.1\ \%$, which also confirms that the assumption is valid (since it is less than 5 %).

In $NaC_7H_5O_2$ solution: Since 1 $C_7H_5O_2^-$ ion is generated for each $NaC_7H_5O_2$, $[C_7H_5O_2^-]$ = 0.10 M $C_7H_5O_2^-$.

$$HC_7H_5O_2(aq) + H_2O(l) \rightleftharpoons H_3O^+(aq) + C_7H_5O_2^{2-}(aq)$$

	$[HC_7H_5O_2]$	$[H_3O^+]$	$[C_7H_5O_2^{2-}]$
Initial	0.15	≈ 0.00	0.10
Change	$-x$	$+x$	$+x$
Equil	$0.15 - x$	$+x$	$0.10 + x$

$K_a = \frac{[H_3O^+][C_7H_5O_2^-]}{[HC_7H_5O_2]} = 6.5 \times 10^{-5} = \frac{x(0.10 + x)}{0.15 - x}$ Assume x is small (x << 0.10 < 0.15) so

$\frac{x(0.10 + \cancel{x})}{0.15 - \cancel{x}} = 6.5 \times 10^{-5} = \frac{x(0.10)}{0.15}$ and x = $9.\underline{7}5 \times 10^{-5}$ = $[H_3O^+]$. then

$\%\ ionization = \frac{[H_3O^+]_{equil}}{[HC_7H_5O_2]_0} \times 100\ \% = \frac{9.\underline{7}5 \times 10^{-5}}{0.15} \times 100\ \% = 0.065\ \%$, which also confirms that the assumption is valid (since it is less than 5 %). The percent ionization in the sodium benzoate solution

is less than is pure water because of the common ion effect. An increase in one of the products (benzoate ion) shifts the equilibrium to the left, so less acid dissociates.
Check: The units (%) are correct. The magnitude of the answer makes physical sense because the acid is weak and so the percent ionization is low. With a common ion present, the percent ionization decreases.

33. a) **Given:** 0.15 M HF **Find:** pH **Other:** K_a (HF) = 3.5×10^{-4}
Conceptual Plan: M HF → $[H_3O^+]$ → pH
ICE Chart $pH = -\log[H_3O^+]$

Solution:

$$HF(aq) + H_2O(l) \rightleftharpoons H_3O^+(aq) + F^-(aq)$$

	[HF]	$[H_3O^+]$	$[F^-]$
Initial	0.15	≈ 0.00	0.00
Change	$-x$	$+x$	$+x$
Equil	$0.15-x$	$+x$	$+x$

$$K_a = \frac{[H_3O^+][F^-]}{[HF]} = 3.5 \times 10^{-4} = \frac{x^2}{0.15-x}$$

Assume x is small (x << 0.15) so $\frac{x^2}{0.15 - \cancel{x}} = 3.5 \times 10^{-4} = \frac{x^2}{0.15}$ and $x = 7.2457 \times 10^{-3} = [H_3O^+]$.

Confirm that assumption is valid $\frac{7.2457 \times 10^{-3}}{0.15} \times 100\,\% = 4.8\,\% < 5\,\%$ so assumption is valid.

Finally, $pH = -\log[H_3O^+] = -\log(7.2457 \times 10^{-3}) = 2.14$.

Check: The units (none) are correct. The magnitude of the answer makes physical sense because pH should be greater than – log (0.15) = 0.82 because this is a weak acid.

b) **Given:** 0.15 M NaF **Find:** pH **Other:** K_a (HF) = 3.5×10^{-4}
Conceptual Plan: M NaF → M F^- and K_a → K_b then M F^- → $[OH^-]$ → $[H_3O^+]$ → pH
NaF (aq) → Na^+ (aq) + F^- (aq) $K_w = K_a K_b$ *ICE Chart* $K_w = [H_3O^+][OH^-]$ $pH = -\log[H_3O^+]$

Solution: Since 1 F^- ion is generated for each NaF, $[F^-]$ = 0.15 M F^-. Since $K_w = K_a K_b$, rearrange to solve for K_b. $K_b = \frac{K_w}{K_a} = \frac{1.00 \times 10^{-14}}{3.5 \times 10^{-4}} = 2.8571 \times 10^{-11}$

$$F^-(aq) + H_2O(l) \rightleftharpoons HF(aq) + OH^-(aq)$$

	$[F^-]$	[HF]	$[OH^-]$
Initial	0.15	0.00	≈ 0.00
Change	$-x$	$+x$	$+x$
Equil	$0.15-x$	$+x$	$+x$

$$K_b = \frac{[HF][OH^-]}{[F^-]} = 2.8571 \times 10^{-11} = \frac{x^2}{0.15-x}$$

Assume x is small (x << 0.15) so $\frac{x^2}{0.15 - \cancel{x}} = 2.8571 \times 10^{-11} = \frac{x^2}{0.15}$ and $x = 2.0702 \times 10^{-6} = [OH^-]$.

Confirm that assumption is valid $\frac{2.0702 \times 10^{-6}}{0.15} \times 100\,\% = 0.0014\,\% < 5\,\%$ so assumption is valid.

$K_w = [H_3O^+][OH^-]$ so $[H_3O^+] = \frac{K_w}{[OH^-]} = \frac{1.0 \times 10^{-14}}{2.0702 \times 10^{-6}} = 4.8305 \times 10^{-9}$ M. Finally,

$pH = -\log[H_3O^+] = -\log(4.8305 \times 10^{-9}) = 8.32$.

Check: The units (none) are correct. The magnitude of the answer makes physical sense because pH should be slightly basic, since the fluoride ion is a very weak base.

c) **Given:** 0.15 M HF and 0.15 M NaF **Find:** pH **Other:** K_a (HF) = 3.5×10^{-4}
Conceptual Plan: M NaF → M F^- then M HF, M F^- → $[H_3O^+]$ → pH
NaF (aq) → *Na^+* (aq) + *F^-* (aq) ICE Chart $pH = -\log[H_3O^+]$

Solution: : Since 1 F^- ion is generated for each NaF, $[F^-]$ = 0.15 M F^-.

$$HF(aq) + H_2O(l) \rightleftharpoons H_3O^+(aq) + F^-(aq)$$

	[HF]	$[H_3O^+]$	$[F^-]$
Initial	0.15	≈ 0.00	0.15
Change	$-x$	$+x$	$+x$
Equil	$0.15 - x$	$+x$	$0.15 + x$

$$K_a = \frac{[H_3O^+][F^-]}{[HF]} = 3.5 \times 10^{-4} = \frac{x(0.15 + x)}{0.15 - x}$$

Assume x is small (x << 0.15) so $\frac{x(0.15 + \cancel{x})}{0.15 - \cancel{x}} = 3.5 \times 10^{-4} = \frac{x(\cancel{0.15})}{\cancel{0.15}}$ and $x = 3.5 \times 10^{-4} = [H_3O^+]$.

Confirm that assumption is valid $\frac{3.5 \times 10^{-4}}{0.15} \times 100\ \% = 0.23\ \% < 5\ \%$ so assumption is valid.

Finally, $pH = -\log[H_3O^+] = -\log(3.5 \times 10^{-4}) = 3.46$.

Check: The units (none) are correct. The magnitude of the answer makes physical sense because pH should be greater than in part a (2.14) because of the common ion effect suppressing the dissociation of the weak acid.

35. When an acid (such as HCl) is added is will react with the conjugate base of the buffer system as follows: $HCl + NaC_2H_3O_2 \rightarrow HC_2H_3O_2 + NaCl$. When a base (such as NaOH) is added is will react with the weak acid of the buffer system as follows: $NaOH + HC_2H_3O_2 \rightarrow H_2O + NaC_2H_3O_2$. The reaction generates the other buffer system component.

37. a) **Given:** 0.15 M $HCHO_2$ and 0.10 M $NaCHO_2$ **Find:** pH **Other:** $K_a(HCHO_2) = 1.8 \times 10^{-4}$

Conceptual Plan: identify acid and base components then M $NaCHO_2$ → M CHO_2^- then

acid = $HCHO_2$ base = CHO_2^- $NaCHO_2\ (aq) \rightarrow Na^+\ (aq) + CHO_2^-\ (aq)$

K_a, M $HCHO_2$, M CHO_2^- → pH

$$pH = pK_a + \log\frac{[base]}{[acid]}$$

Solution: Acid = $HCHO_2$, so [acid] = $[HCHO_2]$ = 0.15 M. Base = CHO_2^-. Since 1 CHO_2^- ion is generated for each $NaCHO_2$, $[CHO_2^-]$ = 0.10 M CHO_2^- = [base]. then

$pH = pK_a + \log\frac{[base]}{[acid]} = -\log(1.8 \times 10^{-4}) + \log\frac{0.10\ \cancel{M}}{0.15\ \cancel{M}} = 3.57$. Note that in order to use the Henderson-Hasselbalch Equation, the assumption that x is small must be valid. This was confirmed in problem #29.

Check: The units (none) are correct. The magnitude of the answer makes physical sense because pH should be less than the pK_a of the acid because there is more acid than base. The answer agrees with problem #29.

b) **Given:** 0.12 M NH_3 and 0.18 M NH_4Cl **Find:** pH **Other:** $K_b(NH_3) = 1.79 \times 10^{-5}$

Conceptual Plan: identify acid and base components then M NH_4Cl → M NH_4^+ and

acid = NH_4^+ base = NH_3 $NH_4Cl\ (aq) \rightarrow NH_4^+\ (aq) + Cl^-\ (aq)$

K_b → pK_b → pK_a then pK_a, M NH_3, M NH_4^+ → pH

$pK_b = -\log K_b$ $14 = pK_a + pK_b$ $pH = pK_a + \log\frac{[base]}{[acid]}$

Solution: Base = NH_3, [base] = $[NH_3]$ = 0.12 M Acid = NH_4^+. Since 1 NH_4^+ ion is generated for each NH_4Cl, $[NH_4^+]$ = 0.18 M NH_4^+ = [acid]. Since $K_b(NH_3) = 1.79 \times 10^{-5}$,

$pK_b = -\log K_b = -\log(1.79 \times 10^{-5}) = 4.75$. Since $14 = pK_a + pK_b$,

$pK_a = 14 - pK_b = 14 - 4.75 = 9.25$ then $pH = pK_a + \log\frac{[base]}{[acid]} = 9.25 + \log\frac{0.12\ \cancel{M}}{0.18\ \cancel{M}} = 9.07$

Note that in order to use the Henderson-Hasselbalch Equation, the assumption that x is small must be valid. This was confirmed in problem #29.

Check: The units (none) are correct. The magnitude of the answer makes physical sense because pH should be less than the pK_a of the acid because there is more acid than base. The answer agrees with problem #29, within the error of the value.

39. a) **Given:** 0.125 M HClO and 0.150 M KClO **Find:** pH **Other:** $K_a(HClO) = 2.9 \times 10^{-8}$

Conceptual Plan: identify acid and base components then M KClO → M ClO^- then

acid = HClO base = ClO^- $KClO\ (aq) \rightarrow K^+\ (aq) + ClO^-\ (aq)$

M HClO, M ClO⁻ → pH

$$pH = pK_a + \log\frac{[base]}{[acid]}$$

Solution: Acid = HClO, so [acid] = [HClO] = 0.125 M. Base = ClO^-. Since 1 ClO^- ion is generated for each KClO, [ClO^-] = 0.150 M ClO^- = [base]. then

$$pH = pK_a + \log\frac{[base]}{[acid]} = -\log(2.9 \times 10^{-8}) + \log\frac{0.150\ \cancel{M}}{0.125\ \cancel{M}} = 7.62\,.$$

Check: The units (none) are correct. The magnitude of the answer makes physical sense because pH should be greater than the pK_a of the acid because there is more base than acid.

b) **Given:** 0.175 M $C_2H_5NH_2$ and 0.150 M $C_2H_5NH_3Br$ **Find:** pH **Other:** K_b ($C_2H_5NH_2$) = 5.6 x 10^{-4}
Conceptual Plan: identify acid and base components then M $C_2H_5NH_3Br$ → M $C_2H_5NH_3^+$ and

acid = $C_2H_5NH_3^+$ base = $C_2H_5NH_2$ $C_2H_5NH_3Br\ (aq) \rightarrow C_2H_5NH_3^+\ (aq) + Br^-\ (aq)$

K_b → pK_b → pK_a then pK_a, M $C_2H_5NH_2$, M $C_2H_5NH_3^+$ → pH

$pK_b = -\log K_b$ $14 = pK_a + pK_b$ $pH = pK_a + \log\frac{[base]}{[acid]}$

Solution: Base = $C_2H_5NH_2$, [base] = [$C_2H_5NH_2$] = 0.175 M Acid = $C_2H_5NH_3^+$. Since 1 $C_2H_5NH_3^+$ ion is generated for each $C_2H_5NH_3Br$, [$C_2H_5NH_3^+$] = 0.150 M $C_2H_5NH_3^+$ = [acid]. Since K_b ($C_2H_5NH_2$) = 5.6 x 10^{-4}, $pK_b = -\log K_b = -\log(5.6 \times 10^{-4}) = 3.25$. Since $14 = pK_a + pK_b$,

$$pK_a = 14 - pK_b = 14 - 3.25 = 10.48 \text{ then } pH = pK_a + \log\frac{[base]}{[acid]} = 10.75 + \log\frac{0.175\ \cancel{M}}{0.150\ \cancel{M}} = 10.82\,.$$

Check: The units (none) are correct. The magnitude of the answer makes physical sense because pH should be greater than the pK_a of the acid because there is more base than acid.

c) **Given:** 10.0 g $HC_2H_3O_2$ and 10.0 g $NaC_2H_3O_2$ in 150.0 mL solution **Find:** pH
Other: K_a ($HC_2H_3O_2$) = 1.8 x 10^{-5}
Conceptual Plan: identify acid and base components then mL → L and

acid = $HC_2H_3O_2$ base = $C_2H_3O_2^-$ $\frac{1\ L}{1000\ mL}$

g $HC_2H_3O_2$ → mol $HC_2H_3O_2$ then mol $HC_2H_3O_2$, L → M $HC_2H_3O_2$ and

$\frac{1\ mol\ HC_2H_3O_2}{60.05\ g\ HC_2H_3O_2}$ $M = \frac{mol}{L}$

g $NaC_2H_3O_2$ → mol $NaC_2H_3O_2$ then

$\frac{1\ mol\ NaC_2H_3O_2}{82.04\ g\ NaC_2H_3O_2}$

mol $NaC_2H_3O_2$, L → M $NaC_2H_3O_2$ → M $C_2H_3O_2^-$ then M $HC_2H_3O_2$, M $C_2H_3O_2^-$ → pH

$M = \frac{mol}{L}$ $NaC_2H_3O_2\ (aq) \rightarrow Na^+\ (aq) + C_2H_3O_2^-\ (aq)$ $pH = pK_a + \log\frac{[base]}{[acid]}$

Solution: $150.0\ \cancel{mL} \times \frac{1\ L}{1000\ \cancel{mL}} = 0.1500\ L$ and

$$10.0\ \cancel{g\ HC_2H_3O_2} \times \frac{1\ mol\ HC_2H_3O_2}{60.05\ \cancel{g\ HC_2H_3O_2}} = 0.16\underline{6}528\ mol\ HC_2H_3O_2$$

then $M = \frac{mol}{L} = \frac{0.16\underline{6}528\ mol\ HC_2H_3O_2}{0.1500\ L} = 1.1\underline{1}019\ M\ HC_2H_3O_2$ and

$$10.0\ \cancel{g\ NaC_2H_3O_2} \times \frac{1\ mol\ NaC_2H_3O_2}{82.04\ \cancel{g\ NaC_2H_3O_2}} = 0.12\underline{1}892\ mol\ NaC_2H_3O_2 \text{ then}$$

$M = \frac{mol}{L} = \frac{0.12\underline{1}892\ mol\ NaC_2H_3O_2}{0.1500\ L} = 0.81\underline{2}612\ M\ NaC_2H_3O_2$. So Acid = $HC_2H_3O_2$, so [acid] = [$HC_2H_3O_2$] =
1.1$\underline{1}$019 M and Base = $C_2H_3O_2^-$. Since 1 $C_2H_3O_2^-$ ion is generated for each $NaC_2H_3O_2$, [$C_2H_3O_2^-$] = 0.81$\underline{2}$612 M

$C_2H_3O_2^- = [base]$. then $pH = pK_a + \log\frac{[base]}{[acid]} = -\log(1.8 \times 10^{-5}) + \log\frac{0.812612\ M}{1.11019\ M} = 4.61$.

Check: The units (none) are correct. The magnitude of the answer makes physical sense because pH should be less than the pK_a of the acid because there is more acid than base.

41. a) **Given:** 50.0 mL of 0.15 M $HCHO_2$ and 75.0 mL of 0.13 M $NaCHO_2$ **Find:** pH
Other: K_a ($HCHO_2$) = 1.8×10^{-4}
Conceptual Plan: identify acid and base components then mL $HCHO_2$, mL $NaCHO_2$ → total mL
acid = $HCHO_2$ base = CHO_2^- total mL = mL $HCHO_2$ + mL $NaCHO_2$
then mL $HCHO_2$, M $HCHO_2$, total mL → buffer M $HCHO_2$ and
$M_1V_1 = M_2V_2$
mL $NaCHO_2$, M $NaCHO_2$, total mL → buffer M $NaCHO_2$ → buffer M CHO_2^- then
$M_1V_1 = M_2V_2$ $NaCHO_2\,(aq) \rightarrow Na^+\,(aq) + CHO_2^-\,(aq)$
K_a, M $HCHO_2$, M CHO_2^- → pH
$$pH = pK_a + \log\frac{[base]}{[acid]}$$
Solution: total mL = mL $HCHO_2$ + mL $NaCHO_2$ = 50.0 mL + 75.0 mL = 125.0 mL. then Since $M_1V_1 = M_2V_2$ Rearrange to solve for M_2. $M_2 = \frac{M_1V_1}{V_2} = \frac{(0.15\ M)(50.0\ mL)}{125.0\ mL} = 0.060\ M\ HCHO_2$ and

$M_2 = \frac{M_1V_1}{V_2} = \frac{(0.13\ M)(75.0\ mL)}{125.0\ mL} = 0.078\ M\ NaCHO_2$. Acid = $HCHO_2$, so [acid] = [$HCHO_2$] = 0.060 M.
Base = CHO_2^-. Since 1 CHO_2^- ion is generated for each $NaCHO_2$, [CHO_2^-] = 0.078 M CHO_2^- = [base]. then $pH = pK_a + \log\frac{[base]}{[acid]} = -\log(1.8 \times 10^{-4}) + \log\frac{0.078\ M}{0.060\ M} = 3.86$.

Check: The units (none) are correct. The magnitude of the answer makes physical sense because pH should be greater than the pK_a of the acid because there is more base than acid.

b) **Given:** 125.0 mL of 0.10 M NH_3 and 250.0 mL of 0.10 M NH_4Cl **Find:** pH
Other: K_b (NH_3) = 1.79×10^{-5}
Conceptual Plan: identify acid and base components then mL NH_3, mL NH_4Cl → total mL then
acid = NH_4^+ base = NH_3 total mL = mL NH_3 + mL NH_4Cl
mL NH_3, M NH_3, total mL → buffer M NH_3 and
$M_1V_1 = M_2V_2$
mL NH_4Cl, M NH_4Cl, total mL → buffer M NH_4Cl → buffer M NH_4^+ and K_b → pK_b → pK_a then
$M_1V_1 = M_2V_2$ $NH_4Cl\,(aq) \rightarrow NH_4^+\,(aq) + Cl^-\,(aq)$ $pK_b = -\log K_b$ $14 = pK_a + pK_b$
then pK_a, M NH_3, M NH_4^+ → pH
$$pH = pK_a + \log\frac{[base]}{[acid]}$$
Solution: total mL = mL NH_3 + mL NH_4Cl = 125.0 mL + 250.0 mL = 375.0 mL. then Since $M_1V_1 = M_2V_2$ Rearrange to solve for M_2. $M_2 = \frac{M_1V_1}{V_2} = \frac{(0.10\ M)(125.0\ mL)}{375.0\ mL} = 0.03\underline{3}333\ M\ NH_3$ and

$M_2 = \frac{M_1V_1}{V_2} = \frac{(0.10\ M)(250.0\ mL)}{375.0\ mL} = 0.06\underline{6}667\ M\ NH_4Cl$. Base = NH_3, [base] = [NH_3] = 0.03$\underline{3}$333 M
Acid = NH_4^+. Since 1 NH_4^+ ion is generated for each NH_4Cl, [NH_4^+] = 0.06$\underline{6}$6667 M NH_4^+ = [acid].
Since K_b (NH_3) = 1.79×10^{-5}, $pK_b = -\log K_b = -\log(1.79 \times 10^{-5}) = 4.75$. Since $14 = pK_a + pK_b$,

$pK_a = 14 - pK_b = 14 - 4.75 = 9.25$ then $pH = pK_a + \log\frac{[base]}{[acid]} = 9.25 + \log\frac{0.03\underline{3}333\ M}{0.06\underline{6}667\ M} = 8.95$.

Check: The units (none) are correct. The magnitude of the answer makes physical sense because pH should be less than the pK_a of the acid because there is more acid than base.

43. **Given:** NaF / HF buffer at pH = 4.00 **Find:** [NaF] / [HF] **Other:** K_a (HF) = 3.5×10^{-4}
Conceptual Plan: identify acid and base components then pH, K_a → [NaF] / [HF]
acid = HF base = F^- $pH = pK_a + \log\frac{[base]}{[acid]}$

Solution: $pH = pK_a + \log \frac{[base]}{[acid]} = -\log(3.5 \times 10^{-4}) + \log \frac{[NaF]}{[HF]} = 4.00$. Solve for [NaF] / [HF].

$\log \frac{[NaF]}{[HF]} = 4.00 - 3.46 = 0.54 \rightarrow \frac{[NaF]}{[HF]} = 10^{0.54} = 3.5$.

Check: The units (none) are correct. The magnitude of the answer makes physical sense because the pH is greater than the pK_a of the acid there needs to be more base than acid.

45. **Given:** 150.0 mL buffer of 0.15 M benzoic acid at pH = 4.25 **Find:** mass sodium benzoate
Other: K_a ($HC_7H_5O_2$) = 6.5×10^{-5}
Conceptual Plan: identify acid and base components then pH, K_a, $[HC_7H_5O_2]$ → $[NaC_7H_5O_2]$

acid = $HC_7H_5O_2$ base = $C_7H_5O_2^-$ $pH = pK_a + \log \frac{[base]}{[acid]}$

mL → L then $[NaC_7H_5O_2]$, L → mol $NaC_7H_5O_2$ → g $NaC_7H_5O_2$

$\frac{1 \text{ L}}{1000 \text{ mL}}$ $M = \frac{mol}{L}$ $\frac{144.11 \text{ g } NaC_7H_5O_2}{1 \text{ mol } NaC_7H_5O_2}$

Solution: $pH = pK_a + \log \frac{[base]}{[acid]} = -\log(6.5 \times 10^{-5}) + \log \frac{[NaC_7H_5O_2]}{0.15 \text{ M}} = 4.25$. Solve for $[NaC_7H_5O_2]$.

$\log \frac{[NaC_7H_5O_2]}{0.15 \text{ M}} = 4.25 - 4.19 = 0.0\underline{6}291 \rightarrow \frac{[NaC_7H_5O_2]}{0.15 \text{ M}} = 10^{0.0\underline{6}291} = 1.\underline{1}559 \rightarrow [NaC_7H_5O_2] = 0.1\underline{7}338 \text{ M}$.

Convert to moles using $M = \frac{mol}{L}$.

$\frac{0.1\underline{7}338 \text{ mol } NaC_7H_5O_2}{1 \text{ L}} \times 0.150 \text{ L} = 0.02\underline{6}007 \text{ mol } NaC_7H_5O_2 \times \frac{144.11 \text{ g } NaC_7H_5O_2}{1 \text{ mol } NaC_7H_5O_2} = 3.7 \text{ g } NaC_7H_5O_2$.

Check: The units (g) are correct. The magnitude of the answer makes physical sense because the volume of solution is small and the concentration is low, so much less than a mole is needed.

47. a) **Given:** 250.0 mL buffer 0.250 M $HC_2H_3O_2$ and 0.250 M $NaC_2H_3O_2$ **Find:** initial pH
Other: K_a ($HC_2H_3O_2$) = 1.8×10^{-5}
Conceptual Plan: identify acid and base components then M $NaC_2H_3O_2$ → M $C_2H_3O_2^-$ then

acid = $HC_2H_3O_2$ base = $C_2H_3O_2^-$ $NaC_2H_3O_2\ (aq) \rightarrow Na^+\ (aq) + C_2H_3O_2^-\ (aq)$

M $HC_2H_3O_2$, M $C_2H_3O_2^-$ → pH

$pH = pK_a + \log \frac{[base]}{[acid]}$

Solution: Acid = $HC_2H_3O_2$, so [acid] = $[HC_2H_3O_2]$ = 0.250 M. Base = $C_2H_3O_2^-$. Since 1 $C_2H_3O_2^-$ ion is generated for each $NaC_2H_3O_2$, $[C_2H_3O_2^-]$ = 0.250 M $C_2H_3O_2^-$ = [base]. then

$pH = pK_a + \log \frac{[base]}{[acid]} = -\log(1.8 \times 10^{-5}) + \log \frac{0.250 \text{ M}}{0.250 \text{ M}} = 4.74$.

Check: The units (none) are correct. The magnitude of the answer makes physical sense because pH is equal to the pK_a of the acid because there are equal amounts of acid and base.

b) **Given:** 250.0 mL buffer 0.250 M $HC_2H_3O_2$ and 0.250 M $NaC_2H_3O_2$, add 0.0050 mol HCl
Find: pH **Other:** K_a ($HC_2H_3O_2$) = 1.8×10^{-5}
Conceptual Plan: Part I: Stoichiometry:
mL → L then $[NaC_2H_3O_2]$, L → mol $NaC_2H_3O_2$ and $[HC_2H_3O_2]$, L → mol $HC_2H_3O_2$

$\frac{1 \text{ L}}{1000 \text{ mL}}$ $M = \frac{mol}{L}$ $M = \frac{mol}{L}$

write balanced equation then mol $NaC_2H_3O_2$, mol $HC_2H_3O_2$, mol HCl → mol $NaC_2H_3O_2$, mol $HC_2H_3O_2$ then

$HCl + NaC_2H_3O_2 \rightarrow HC_2H_3O_2 + NaCl$ set up stoichiometry table

Part II: Equilibrium:

mol $NaC_2H_3O_2$, mol $HC_2H_3O_2$, L, K_a → pH

$pH = pK_a + \log \frac{[base]}{[acid]}$

Solution: $250.0\ \cancel{mL} \times \frac{1\ L}{1000\ \cancel{mL}} = 0.2500\ L$ then $\frac{0.250\ mol\ HC_2H_3O_2}{1\ \cancel{L}} \times 0.250\ \cancel{L} = 0.0625\ mol\ HC_2H_3O_2$

and $\frac{0.250\ mol\ NaC_2H_3O_2}{1\ \cancel{L}} \times 0.250\ \cancel{L} = 0.0625\ mol\ NaC_2H_3O_2$ set up table to track changes:

	$HCl\ (aq)$ +	$NaC_2H_3O_2\ (aq)$ →	$HC_2H_3O_2\ (aq)$ +	$NaCl\ (aq)$
Before addition	≈ 0.00 mol	0.0625 mol	0.0625 mol	0.00 mol
Addition	0.0050 mol	–	–	–
After addition	≈ 0.00 mol	0.0575 mol	0.0675 mol	0.0050 mol

Since the amount of HCl is small, there are still significant amounts of both buffer components, so the Henderson-Hasselbalch Equation can be used to calculate the new pH.

$$pH = pK_a + \log\frac{[base]}{[acid]} = -\log(1.8 \times 10^{-5}) + \log\frac{\frac{0.0575\ \cancel{mol}}{\cancel{0.250\ L}}}{\frac{0.0675\ \cancel{mol}}{\cancel{0.250\ L}}} = 4.68\ .$$

Check: The units (none) are correct. The magnitude of the answer makes physical sense because the pH dropped slightly when acid was added.

c) **Given:** 250.0 mL buffer 0.250 M $HC_2H_3O_2$ and 0.250 M $NaC_2H_3O_2$, add 0.0050 mol NaOH
Find: pH **Other:** $K_a\ (HC_2H_3O_2) = 1.8 \times 10^{-5}$
Conceptual Plan: Part I: Stoichiometry:
mL → L then $[NaC_2H_3O_2]$, L → mol $NaC_2H_3O_2$ and $[HC_2H_3O_2]$, L → mol $HC_2H_3O_2$

$\frac{1\ L}{1000\ mL}$ $\qquad M = \frac{mol}{L}$ $\qquad M = \frac{mol}{L}$

write balanced equation then
$NaOH + HC_2H_3O_2 \rightarrow H_2O + NaC_2H_3O_2$
mol $NaC_2H_3O_2$, mol $HC_2H_3O_2$, mol NaOH → mol $NaC_2H_3O_2$, mol $HC_2H_3O_2$ then
set up stoichiometry table
Part II: Equilibrium:
mol $NaC_2H_3O_2$, mol $HC_2H_3O_2$, L, K_a → pH

$$pH = pK_a + \log\frac{[base]}{[acid]}$$

Solution: $250.0\ \cancel{mL} \times \frac{1\ L}{1000\ \cancel{mL}} = 0.2500\ L$ then $\frac{0.250\ mol\ HC_2H_3O_2}{1\ \cancel{L}} \times 0.2500\ \cancel{L} = 0.0625\ mol\ HC_2H_3O_2$

and $\frac{0.250\ mol\ NaC_2H_3O_2}{1\ \cancel{L}} \times 0.2500\ \cancel{L} = 0.0625\ mol\ NaC_2H_3O_2$ set up table to track changes:

	$NaOH\ (aq)$ +	$HC_2H_3O_2\ (aq)$ →	$NaC_2H_3O_2\ (aq)$ +	$H_2O\ (l)$
Before addition	≈ 0.00 mol	0.0625 mol	0.0625 mol	–
Addition	0.0050 mol	–	–	–
After addition	≈ 0.00 mol	0.0575 mol	0.0675 mol	–

Since the amount of NaOH is small, there are still significant amounts of both buffer components, so the Henderson-Hasselbalch Equation can be used to calculate the new pH.

$$pH = pK_a + \log\frac{[base]}{[acid]} = -\log(1.8 \times 10^{-5}) + \log\frac{\frac{0.0675\ \cancel{mol}}{\cancel{0.2500\ L}}}{\frac{0.0575\ \cancel{mol}}{\cancel{0.2500\ L}}} = 4.81\ .$$

Check: The units (none) are correct. The magnitude of the answer makes physical sense because the pH rose slightly when base was added.

49. a) **Given:** 500.0 mL pure water **Find:** initial pH and after adding 0.010 mol HCl
Conceptual Plan: pure water has a pH of 7.00 then mL → L then mol HCl, L → $[H_3O^+]$ → pH

$\frac{1\ L}{1000\ mL}$ $\qquad M = \frac{mol}{L} \quad pH = -\log[H_3O^+]$

Solution: pure water has a pH of 7.00 so initial pH = 7.00 then $500.0\ \cancel{mL} \times \frac{1\ L}{1000\ \cancel{mL}} = 0.5000\ L$

then $M = \frac{mol}{L} = \frac{0.010\ mol\ HCl}{0.5000\ L} = 0.020\ M\ HCl$. Since HCl is a strong acid, it dissociates completely, so $pH = -\log[H_3O^+] = -\log(0.020) = 1.70$.

Check: The units (none) are correct. The magnitudes of the answers make physical sense because the pH starts neutral and then dropped significantly when acid is added and there is no buffer present.

b) **Given:** 500.0 mL buffer 0.125 M $HC_2H_3O_2$ and 0.115 M $NaC_2H_3O_2$
Find: initial pH and after adding 0.010 mol HCl **Other:** $K_a(HC_2H_3O_2) = 1.8 \times 10^{-5}$
Conceptual Plan: initial pH: identify acid and base components then M $NaC_2H_3O_2$ → M $C_2H_3O_2^-$
acid = $HC_2H_3O_2$ base = $C_2H_3O_2^-$ $NaC_2H_3O_2\,(aq) \rightarrow Na^+\,(aq) + C_2H_3O_2^-\,(aq)$

then M $HC_2H_3O_2$, M $C_2H_3O_2^-$ → pH

$$pH = pK_a + \log\frac{[base]}{[acid]}$$

pH after HCl addition: Part I: Stoichiometry:
mL → L then [$NaC_2H_3O_2$], L → mol $NaC_2H_3O_2$ and [$HC_2H_3O_2$], L → mol $HC_2H_3O_2$

$\frac{1\ L}{1000\ mL}$ $M = \frac{mol}{L}$ $M = \frac{mol}{L}$

write balanced equation then
$HCl + NaC_2H_3O_2 \rightarrow HC_2H_3O_2 + NaCl$
mol $NaC_2H_3O_2$, mol $HC_2H_3O_2$, mol HCl → mol $NaC_2H_3O_2$, mol $HC_2H_3O_2$ then
set up stoichiometry table

Part II: Equilibrium:
mol $NaC_2H_3O_2$, mol $HC_2H_3O_2$, L, K_a → pH

$$pH = pK_a + \log\frac{[base]}{[acid]}$$

Solution: initial pH: Acid = $HC_2H_3O_2$, so [acid] = [$HC_2H_3O_2$] = 0.125 M. Base = $C_2H_3O_2^-$. Since 1 $C_2H_3O_2^-$ ion is generated for each $NaC_2H_3O_2$, [$C_2H_3O_2^-$] = 0.115 M $C_2H_3O_2^-$ = [base]. then

$$pH = pK_a + \log\frac{[base]}{[acid]} = -\log(1.8 \times 10^{-5}) + \log\frac{0.115\ \cancel{M}}{0.125\ \cancel{M}} = 4.71.$$

pH after HCl addition: $500.0\ \cancel{mL} \times \frac{1\ L}{1000\ \cancel{mL}} = 0.5000\ L$ then

$\frac{0.125\ mol\ HC_2H_3O_2}{1\ \cancel{L}} \times 0.5000\ \cancel{L} = 0.0625\ mol\ HC_2H_3O_2$ and

$\frac{0.115\ mol\ NaC_2H_3O_2}{1\ \cancel{L}} \times 0.5000\ \cancel{L} = 0.0575\ mol\ NaC_2H_3O_2$ set up table to track changes:

	$HCl\,(aq)$ +	$NaC_2H_3O_2\,(aq)$ →	$HC_2H_3O_2\,(aq)$ +	$NaCl\,(aq)$
Before addition	≈ 0.00 mol	0.0575 mol	0.0625 mol	0.00 mol
Addition	0.010 mol	–	–	–
After addition	≈ 0.00 mol	0.0475 mol	0.0725 mol	0.010 mol

Since the amount of HCl is small, there are still significant amounts of both buffer components, so the Henderson-Hasselbalch Equation can be used to calculate the new pH.

$$pH = pK_a + \log\frac{[base]}{[acid]} = -\log(1.8 \times 10^{-5}) + \log\frac{\frac{0.0475\ \cancel{mol}}{\cancel{0.5000\ L}}}{\frac{0.0725\ \cancel{mol}}{\cancel{0.5000\ L}}} = 4.56.$$

Check: The units (none) are correct. The magnitudes of the answers make physical sense because the pH started below the pK_a of the acid and it dropped slightly when acid was added.

c) **Given:** 500.0 mL buffer 0.155 M $CH_3CH_2NH_2$ and 0.145 M $CH_3CH_2NH_3Cl$
Find: initial pH and after adding 0.010 mol HCl **Other:** K_b ($CH_3CH_2NH_2$) = 5.6×10^{-4}
Conceptual Plan: initial pH: identify acid and base components then M $CH_3CH_2NH_3Cl$ → M $CH_3CH_2NH_3^+$
acid = $C_2H_5NH_3^+$ base = $C_2H_5NH_2$ $C_2H_5NH_3Cl\ (aq) \rightarrow C_2H_5NH_3^+\ (aq) + Cl^-\ (aq)$
and K_b → pK_b → pK_a then pK_a, M $CH_3CH_2NH_2$, M $CH_3CH_2NH_3^+$ → pH

$pK_b = -\log K_b \quad 14 = pK_a + pK_b$ $\quad\quad$ $pH = pK_a + \log\frac{[base]}{[acid]}$

pH after HCl addition: Part I: Stoichiometry:
mL→ L then [$CH_3CH_2NH_2$], L →mol $CH_3CH_2NH_2$ and [$CH_3CH_2NH_3Cl$], L→mol $CH_3CH_2NH_3Cl$

$\frac{1\ L}{1000\ mL}$ $\quad M = \frac{mol}{L}$ $\quad M = \frac{mol}{L}$

write balanced equation **then**
$HCl + CH_3CH_2NH_2 \rightarrow CH_3CH_2NH_3Cl$
mol $CH_3CH_2NH_2$, mol $CH_3CH_2NH_3Cl$, mol HCl → mol $CH_3CH_2NH_2$, mol $CH_3CH_2NH_3Cl$ then
set up stoichiometry table

Part II: Equilibrium:
mol $CH_3CH_2NH_2$, mol $CH_3CH_2NH_3Cl$, L, K_a → pH

$$pH = pK_a + \log\frac{[base]}{[acid]}$$

Solution: Base = $CH_3CH_2NH_2$, [base] = [$CH_3CH_2NH_2$] = 0.155 M Acid = $CH_3CH_2NH_3^+$. Since 1 $CH_3CH_2NH_3^+$ ion is generated for each $CH_3CH_2NH_3Cl$, [$CH_3CH_2NH_3^+$] = 0.145 M $CH_3CH_2NH_3^+$ = [acid]. Since K_b ($CH_3CH_2NH_2$) = 5.6×10^{-4}, $pK_b = -\log K_b = -\log(5.6 \times 10^{-4}) = 3.25$. Since $14 = pK_a + pK_b$, $pK_a = 14 - pK_b = 14 - 3.25 = 10.48$ then

$$pH = pK_a + \log\frac{[base]}{[acid]} = 10.75 + \log\frac{0.155\ \cancel{M}}{0.145\ \cancel{M}} = 10.78\ .$$

pH after HCl addition: $500.0\ \cancel{mL} \times \frac{1\ L}{1000\ \cancel{mL}} = 0.5000\ L$ then

$\frac{0.155\ mol\ CH_3CH_2NH_2}{1\ \cancel{L}} \times 0.5000\ \cancel{L} = 0.0775\ mol\ CH_3CH_2NH_2$ and

$\frac{0.145\ mol\ CH_3CH_2NH_3Cl}{1\ \cancel{L}} \times 0.5000\ \cancel{L} = 0.0725\ mol\ CH_3CH_2NH_3Cl$ set up table to track changes:

	$HCl\ (aq)$ +	$CH_3CH_2NH_2\ (aq)$ →	$CH_3CH_2NH_3Cl\ (aq)$
Before addition	≈ 0.00 mol	0.0775 mol	0.0725 mol
Addition	0.010 mol	–	–
After addition	≈ 0.00 mol	0.06$\underline{7}$5 mol	0.08$\underline{2}$5 mol

Since the amount of HCl is small, there are still significant amounts of both buffer components, so the Henderson-Hasselbalch Equation can be used to calculate the new pH. $pH = pK_a + \log\frac{[base]}{[acid]} = 10.75 + \log\frac{\frac{0.06\underline{7}5\ \cancel{mol}}{\cancel{0.5000\ L}}}{\frac{0.08\underline{2}5\ \cancel{mol}}{\cancel{0.5000\ L}}} = 10.66\ .$

Check: The units (none) are correct. The magnitudes of the answers make physical sense because the initial pH should be greater than the pK_a of the acid because there is more base than acid and the pH drops slightly when acid is added.

51. **Given:** 350.00 mL 0.150 M HF and 0.150 M NaF buffer
Find: mass NaOH to raise pH to 4.00 and mass NaOH to raise pH to 4.00 with buffer concentrations raised to 0.350 M **Other:** K_a (HF) = 3.5×10^{-4}
Conceptual Plan: identify acid and base components Since [NaF] = [HF] then initial $pH = pK_a$.
acid = HF base = F^- $\quad\quad pH = pK_a$
final pH, pK_a → [NaF]/[HF] and mL → L then [HF], L → mol HF and [NaF], L → mol NaF

$pH = pK_a + \log\frac{[base]}{[acid]}$ $\quad \frac{1\ L}{1000\ mL}$ $\quad M = \frac{mol}{L}$ $\quad M = \frac{mol}{L}$

then write balanced equation **then**
$NaOH + HF \rightarrow NaF + H_2O$

mol HF, mol NaF, [NaF]/[HF] → mol NaOH → g NaOH

set up stoichiometry table $\quad \frac{40.00 \text{ g NaOH}}{1 \text{ mol NaOH}}$

Finally, when the buffer concentrations are raised to 0.350 M, simply multiply the g NaOH by ratio of concentrations (0.350 M / 0.150 M).

Solution: initial $pH = pK_a = -\log(3.5 \times 10^{-4}) = 3.64$ then

$pH = pK_a + \log\frac{[\text{base}]}{[\text{acid}]} = -\log(3.5 \times 10^{-4}) + \log\frac{[\text{NaF}]}{[\text{HF}]} = 4.00$. Solve for [NaF] / [HF].

$\log\frac{[\text{NaF}]}{[\text{HF}]} = 4.00 - 3.46 = 0.54 \rightarrow \frac{[\text{NaF}]}{[\text{HF}]} = 10^{0.54} = 3.5$. $350.0 \text{ mL} \times \frac{1 \text{ L}}{1000 \text{ mL}} = 0.3500 \text{ L}$ then

$\frac{0.150 \text{ mol HF}}{1 \text{ L}} \times 0.3500 \text{ L} = 0.0525 \text{ mol HF}$ and $\frac{0.150 \text{ mol NaF}}{1 \text{ L}} \times 0.3500 \text{ L} = 0.0525 \text{ mol NaF}$

set up table to track changes:

	NaOH (*aq*) +	HF (*aq*)	→ NaF (*aq*) +	H_2O (*aq*)
Before addition	0.00 mol	0.0525 mol	0.0525 mol	–
Addition	*x*	–	–	–
After addition	≈ 0.00 mol	(0.0525 – *x*) mol	(0.0525 + *x*) mol	–

Since $\frac{[\text{NaF}]}{[\text{HF}]} = 3.5 = \frac{(0.0525 + x) \text{ mol}}{(0.0525 - x) \text{ mol}}$, solve for x. Note that the ratio of moles is the same as the ratio of concentrations, since the volume for both terms is the same. $3.5\,(0.0525 - x) = (0.0525 + x) \rightarrow$ $0.18375 - 3.5x = 0.0525 + x \rightarrow 0.13125 = 4.5x \rightarrow x = 0.029167 \text{ mol NaOH}$ then

$0.029167 \text{ mol NaOH} \times \frac{40.00 \text{ g NaOH}}{1 \text{ mol NaOH}} = 1.1667 \text{ g NaOH} = 1.2 \text{ g NaOH}$. Finally multiply the NaOH mass by the ratio of concentrations $1.1667 \text{ g NaOH} \times \frac{0.350 \text{ M}}{0.150 \text{ M}} = 2.7 \text{ g NaOH}$.

Check: The units (g) are correct. The magnitudes of the answers make physical sense because there is much less than a mole of each of the buffer components, so there must be much less than a mole of NaOH. The higher the buffer concentrations, the higher the buffer capacity and the mass of NaOH it can neutralize.

53. a) Yes, this will be a buffer because NH_3 is a weak base and NH_4^+ is its conjugate acid. The ratio of base to acid is 0.10/0.15 = 0.67, so the pH will be within 1 pH unit of the pK_a.

b) No, this will not be a buffer solution because HCl is a strong acid and NaOH is a strong base.

c) Yes, this will be a buffer because HF is a weak acid and the NaOH will convert 20.0/50.0 = 40 % of the acid to its conjugate base.

d) No, this will not be a buffer solution because both components are bases.

e) No, this will not be a buffer solution because both components are bases.

55. a) **Given:** blood buffer 0.024 M HCO_3^- and 0.0012 M H_2CO_3, $pK_a = 6.1$ **Find:** initial pH
Conceptual Plan: identify acid and base components then M HCO_3^-, M H_2CO_3 → pH

acid = H_2CO_3 base = HCO_3^- $\quad pH = pK_a + \log\frac{[\text{base}]}{[\text{acid}]}$

Solution: Acid = H_2CO_3, so [acid] = $[H_2CO_3]$ = 0.0012 M. Base = HCO_3^-, so [base] = $[HCO_3^-]$ = 0.024 M HCO_3^-. then $pH = pK_a + \log\frac{[\text{base}]}{[\text{acid}]} = 6.1 + \log\frac{0.024 \text{ M}}{0.0012 \text{ M}} = 7.4$.

Check: The units (none) are correct. The magnitude of the answer makes physical sense because pH is greater than the pK_a of the acid because there is more base than acid.

b) **Given:** 5.0 L of blood buffer **Find:** mass HCl to lower pH to 7.0
Conceptual Plan: final pH, pK_a → $[HCO_3^-]/[H_2CO_3]$ then $[HCO_3^-]$, L → mol HCO_3^- and

$pH = pK_a + \log\frac{[\text{base}]}{[\text{acid}]} \qquad M = \frac{\text{mol}}{L}$

$[H_2CO_3]$, L → mol H_2CO_3 then write balanced equation then

$M = \frac{mol}{L}$ $H^+ + HCO_3^- \rightarrow H_2CO_3$

mol HCO_3^-, mol H_2CO_3, $[HCO_3^-]/[H_2CO_3]$ → mol HCl → g HCl

set up stoichiometry table $\frac{36.46 \text{ g HCl}}{1 \text{ mol HCl}}$

Solution: $pH = pK_a + \log\frac{[base]}{[acid]} = 6.1 + \log\frac{[HCO_3^-]}{[H_2CO_3]} = 7.0$. Solve for $[HCO_3^-]/[H_2CO_3]$.

$\log\frac{[HCO_3^-]}{[H_2CO_3]} = 7.0 - 6.1 = 0.9 \rightarrow \frac{[HCO_3^-]}{[H_2CO_3]} = 10^{0.9} = \underline{7}.9433$. then

$\frac{0.024 \text{ mol } HCO_3^-}{1 \text{ L}} \times 5.0 \text{ L} = 0.12 \text{ mol } HCO_3^-$ and $\frac{0.0012 \text{ mol } H_2CO_3}{1 \text{ L}} \times 5.0 \text{ L} = 0.0060 \text{ mol } H_2CO_3$ Since HCl is a strong acid, [HCl] = $[H^+]$, and set up table to track changes:

	H^+ *(aq)* +	HCO_3^- *(aq)* →	H_2CO_3 *(aq)*
Before addition	≈ 0.00 mol	0.12 mol	0.0060 mol
Addition	*x*	–	–
After addition	≈ 0.00 mol	(0.12 – x) mol	(0.0060 + x) mol

Since $\frac{[HCO_3^-]}{[H_2CO_3]} = \underline{7}.9433 = \frac{(0.12 - x) \text{ mol}}{(0.0060 + x) \text{ mol}}$, solve for x. Note that the ration of moles is the same as the ratio of concentrations, since the volume for both terms is the same. $\underline{7}.9433(0.0060 + x) = (0.12 - x)$ → $0.04\underline{7}6598 + \underline{7}.9433x = 0.12 - x$ → $\underline{8}.9433x = 0.0\underline{7}234$ → $x = 0.00\underline{8}0888$ mol HCl then

$0.00\underline{8}0888 \text{ mol HCl} \times \frac{36.46 \text{ g HCl}}{1 \text{ mol HCl}} = 0.\underline{2}9492 \text{ g HCl} = 0.3 \text{ g HCl}$.

Check: The units (g) are correct. The amount of acid needed is small because the concentrations of the buffer components are very low and the buffer starts only 0.4 pH units above the final pH.

c) **Given:** 5.0 L of blood buffer **Find:** mass NaOH to raise pH to 7.8
Conceptual Plan: final pH, pK_a → $[HCO_3^-]/[H_2CO_3]$ then $[HCO_3^-]$, L → mol HCO_3^- and

$pH = pK_a + \log\frac{[base]}{[acid]}$ $M = \frac{mol}{L}$

$[H_2CO_3]$, L → mol H_2CO_3 then write balanced equation then

$M = \frac{mol}{L}$ $OH^- + H_2CO_3 \rightarrow HCO_3^- + H_2O$

mol HCO_3^-, mol H_2CO_3, $[HCO_3^-]/[H_2CO_3]$ → mol NaOH → g NaOH

set up stoichiometry table $\frac{40.00 \text{ g NaOH}}{1 \text{ mol NaOH}}$

Solution: $pH = pK_a + \log\frac{[base]}{[acid]} = 6.1 + \log\frac{[HCO_3^-]}{[H_2CO_3]} = 7.8$.

Solve for $[HCO_3^-]/[H_2CO_3]$. $\log\frac{[HCO_3^-]}{[H_2CO_3]} = 7.8 - 6.1 = 1.7 \rightarrow \frac{[HCO_3^-]}{[H_2CO_3]} = 10^{1.7} = 50.\underline{1}1872$. then

$\frac{0.024 \text{ mol } HCO_3^-}{1 \text{ L}} \times 5.0 \text{ L} = 0.12 \text{ mol } HCO_3^-$ and $\frac{0.0012 \text{ mol } H_2CO_3}{1 \text{ L}} \times 5.0 \text{ L} = 0.0060 \text{ mol } H_2CO_3$ Since NaOH is a strong base, [NaOH] = $[OH^-]$, and set up table to track changes:

	OH^- *(aq)* +	H_2CO_3 *(aq)* →	HCO_3^- *(aq)* +	H_2O *(l)*
Before addition	≈ 0.00 mol	0.0060 mol	0.12 mol	–
Addition	*x*	–	–	
After addition	≈ 0.00 mol	(0.0060 – x) mol	(0.12 + x) mol	

Since $\frac{[HCO_3^-]}{[H_2CO_3]} = 50.\underline{1}1872 = \frac{(0.12 + x) \text{ mol}}{(0.0060 - x) \text{ mol}}$, solve for x. Note that the ration of moles is the same as the ratio of concentrations, since the volume for both terms is the same.
$50.\underline{1}1872(0.0060 - x) = (0.12 + x)$ → $0.3\underline{0}071 - 50.\underline{1}1872x = 0.12 + x$ → $51.\underline{1}1872x = 0.1\underline{8}071$ →

$x = 0.0035351$ mol NaOH then $0.0035351 \text{ mol NaOH} \times \frac{40.00 \text{ g NaOH}}{1 \text{ mol NaOH}} = 0.14141 \text{ g NaOH} = 0.14 \text{ g NaOH}$.

Check: The units (g) are correct. The amount of base needed is small because the concentrations of the buffer components are very low.

57. **Given:** $HC_2H_3O_2/KC_2H_3O_2$, $HClO_2/KClO_2$, NH_3/NH_4Cl, and HClO/KClO potential buffer systems to create buffer at pH = 7.20 **Find:** best buffer system and ratio of component masses
Other: K_a ($HC_2H_3O_2$) = 1.8×10^{-5}, K_a ($HClO_2$) = 1.8×10^{-4}, K_b (NH_3) = 1.79×10^{-5}, K_a (HClO) = 2.9×10^{-8}
Conceptual Plan: calculate p K_a of all potential buffer acids; for the base $K_b \rightarrow pK_b \rightarrow pK_a$ then

$$pK_a = -\log K_a \qquad pK_b = -\log K_b \quad 14 = pK_a + pK_b$$

and choose the pK_a that is closest to 7.20. then pH, K_a → [base]/[acid] → mass base/mass acid

$$\text{pH} = \text{p}K_a + \log\frac{[\text{base}]}{[\text{acid}]} \qquad \frac{\mathfrak{M}\text{ (base)}}{\mathfrak{M}\text{ (acid)}}$$

Solution: for $HC_2H_3O_2/KC_2H_3O_2$: $pK_a = -\log K_a = -\log(1.8 \times 10^{-5}) = 4.74$; for $HClO_2/KClO_2$: $pK_a = -\log K_a = -\log(1.8 \times 10^{-4}) = 3.74$; for NH_3/NH_4Cl: $pK_b = -\log K_b = -\log(1.79 \times 10^{-5}) = 4.75$. Since $14 = pK_a + pK_b$, $pK_a = 14 - pK_b = 14 - 4.75 = 9.25$; and for HClO/KClO: $pK_a = -\log K_a = -\log(2.9 \times 10^{-8}) = 7.54$. So the HClO/KClO buffer system has the pK_a that is the closest to 7.20. So, $\text{pH} = \text{p}K_a + \log\frac{[\text{base}]}{[\text{acid}]} = 7.54 + \log\frac{[\text{KClO}]}{[\text{HClO}]} = 7.1$. Solve for [KClO]/[HClO].

$\log\frac{[\text{KClO}]}{[\text{HClO}]} = 7.20 - 7.54 = -0.34 \rightarrow \frac{[\text{KClO}]}{[\text{HClO}]} = 10^{-0.34} = 0.457088$. then convert to mass ratio using

$$\frac{\mathfrak{M}\text{ (base)}}{\mathfrak{M}\text{ (acid)}},\ 0.457088 \frac{\frac{\text{KClO mol}}{\text{L}}}{\frac{\text{HClO mol}}{\text{L}}} \times \frac{\frac{90.55 \text{ g KClO}}{\text{mol KClO}}}{\frac{52.46 \text{ g HClO}}{\text{mol HClO}}} = 0.79 \frac{\text{g KClO}}{\text{g HClO}}$$

Check: The units (none and g base/g acid)) are correct. The buffer system with the K_a closest to 10^{-7} is the best choice. The magnitude of the answer makes physical sense because the buffer needs more acid than base (and this fact is not overcome by the heavier molar mass of the base).

59. **Given:** 500.0 mL of 0.100 M HNO_2 / 0.150 M KNO_2 buffer and a) 250 mg NaOH, b) 350 mg KOH, c) 1.25 g HBr and d) 1.35 g HI **Find:** if buffer capacity is exceeded
Conceptual Plan: mL → L then $[HNO_2]$, L → mol HNO_2 and $[KNO_2]$, L → mol KNO_2 then

$$\frac{1 \text{ L}}{1000 \text{ mL}} \qquad M = \frac{\text{mol}}{\text{L}} \qquad M = \frac{\text{mol}}{\text{L}}$$

then calculate moles of acid or base to be added to the buffer mg → g → mol then

$$\frac{1 \text{ g}}{1000 \text{ mg}} \qquad \mathfrak{M}$$

compare the added amount to the buffer amount of the opposite component. Ratio of base/acid must be between 0.1 and 10 to maintain the buffer integrity.

Solution: $500.00 \text{ mL} \times \frac{1 \text{ L}}{1000 \text{ mL}} = 0.5000 \text{ L}$ then $\frac{0.100 \text{ mol } HNO_2}{1 \text{ L}} \times 0.5000 \text{ L} = 0.0500 \text{ mol } HNO_2$ and

$\frac{0.150 \text{ mol } KNO_2}{1 \text{ L}} \times 0.5000 \text{ L} = 0.0750 \text{ mol } KNO_2$

a) for NaOH: $250 \text{ mg NaOH} \times \frac{1 \text{ g NaOH}}{1000 \text{ mg NaOH}} \times \frac{1 \text{ mol NaOH}}{40.00 \text{ g NaOH}} = 0.00625 \text{ mol NaOH}$. Since the buffer contains 0.0500 mol acid, the amount of acid is reduced by 0.00625/0.0500 = 12.5 % and the ratio of base/acid is still between 0.1and 10. The buffer capacity is not exceeded.

b) for KOH: $350 \text{ mg KOH} \times \frac{1 \text{ g KOH}}{1000 \text{ mg KOH}} \times \frac{1 \text{ mol KOH}}{56.11 \text{ g KOH}} = 0.00624 \text{ mol KOH}$. Since the buffer contains 0.0500 mol acid, the amount of acid is reduced by 0.00624/0.0500 = 12.5 % and the ratio of base/acid is still between 0.1 and 10. The buffer capacity is not exceeded.

c) for HBr: $1.25 \, \cancel{g \, HBr} \times \dfrac{1 \, mol \, HBr}{80.91 \, \cancel{g \, HBr}} = 0.015\underline{4}496 \, mol \, HBr$. Since the buffer contains 0.0750 mol base, the amount of acid is reduced by 0.0154/0.0750 = 20.6 % and the ratio of base/acid is still between 0.1 and 10. The buffer capacity is not exceeded.

d) for HI: $1.35 \, \cancel{g \, HI} \times \dfrac{1 \, mol \, HI}{127.91 \, \cancel{g \, HI}} = 0.010\underline{5}545 \, mol \, HI$. Since the buffer contains 0.0750 mol base, the amount of acid is reduced by 0.0106/0.0750 = 14.1 % and the ratio of base/acid is still between 0.1 and 10. The buffer capacity is not exceeded.

61. (i) The equivalence point of a titration is where the pH rises sharply as base is added. The pH as the equivalence point is the midpoint of the sharp rise at ~ 50 mL added base. For (a) the pH = ~ 8 and for (b) the pH = ~ 7.

(ii) Graph (a) represents a weak acid and graph (b) represents a strong acid. A strong acid titration starts at a lower pH, has a flatter initial region and a sharper rise at the equivalence point than a weak acid. The pH at the equivalence point of a strong acid is neutral, while the pH at the equivalence point of a weak acid is basic.

63. **Given:** 20.0 mL 0.200 M KOH and 0.200 M CH_3NH_2 titrated with 0.100 M HI

a) **Find:** volume of base to reach equivalence point

Conceptual Plan: The answer for both titrations will be the same since the initial concentration and volumes of the bases are the same. Write balanced equation then mL → L then

$$HI + KOH \rightarrow KI + H_2O \text{ and } HI + KOH \rightarrow CH_3NH_3I \qquad \dfrac{1 \, L}{1000 \, mL}$$

[base], L → mol base   then   set mol base = mol acid   and   [HI], mol HI → L HI → mL HI

$$M = \dfrac{mol}{L} \qquad \text{balanced equation has 1:1 stoichiometry} \qquad M = \dfrac{mol}{L} \qquad \dfrac{1000 \, mL}{1 \, L}$$

Solution: $20.0 \, \cancel{mL \, base} \times \dfrac{1 \, L}{1000 \, \cancel{mL}} = 0.0200 \, L \, base$ then $\dfrac{0.200 \, mol \, base}{1 \, \cancel{L}} \times 0.0200 \, \cancel{L} = 0.00400 \, mol \, base$.

So mol base = 0.00400 mol = mol HI then

$$0.00400 \, \cancel{mol \, HI} \times \dfrac{1 \, L \, HI}{0.100 \, \cancel{mol \, HI}} = 0.0400 \, \cancel{L \, HI} \times \dfrac{1000 \, mL}{1 \, \cancel{L}} = 40.0 \, mL \, HI \text{ for both titrations.}$$

Check: The units (mL) are correct. The volume of acid is twice the volume of bases because the concentration of the base is twice that of the acid in each case. The answer for both titrations is the same because the stoichiometry is the same for both titration reactions.

b) The pH at the equivalence point will be neutral for KOH (since it is a strong base) and it will be acidic for CH_3NH_2 (since it is a weak base).

c) The initial pH will be lower for CH_3NH_2 (since it is a weak base and will only partially dissociate and not raise the pH as high as KOH (since it is a strong base and so it dissociate completely) at the same acid concentration.

d) The titration curves will look like:

KOH:

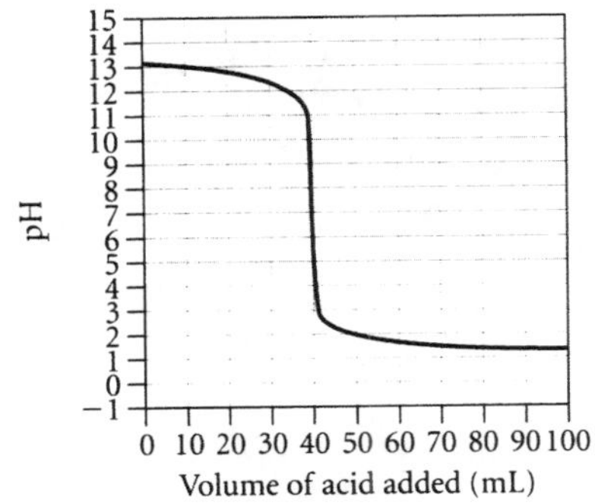

CH_3NH_2:

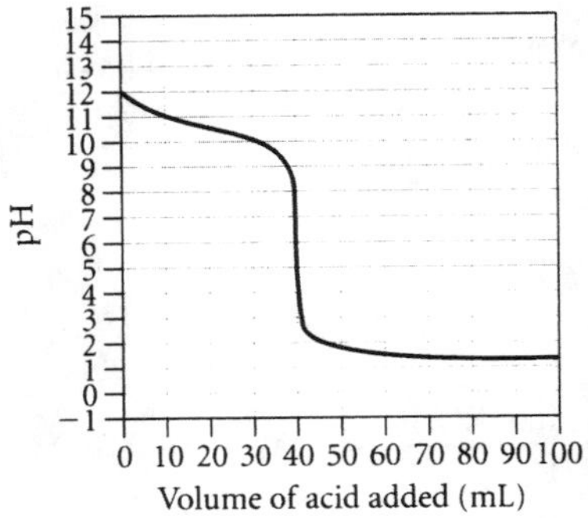

Important features to include are a high initial pH (if strong base pH is over 13 and lower for a weak base), flat initial region (very flat for strong base, not as flat for weak base where pH half-way to equivalence point is the pK_b of the base), sharp drop at equivalence point, pH at equivalence point (neutral for strong base and lower for weak base), and then flatten out at low pH.

65. a) The equivalence point of a titration is where the pH rises sharply as base is added. The volume at the equivalence point is ~ 30 mL. The pH as the equivalence point is the midpoint of the sharp rise at ~ 30 mL added base, which is a pH = ~ 9.

b) At 0 mL the pH is calculated by doing an equilibrium calculation of a weak acid in water (as done in chapter 15).

c) The pH one-half way to the equivalence point is equal to the pK_a of the acid, or ~ 15 mL.

d) The pH at the equivalence point, or ~ 30 mL, is calculated by doing an equilibrium problem with the K_b of the acid. At the equivalence point all of the acid has been converted to its conjugate base.

e) Beyond the equivalence point, or ~ 30 mL, there is excess base. All of the acid has been converted to its conjugate base and so the pH is calculated by focusing on this excess base concentration.

67. **Given:** 35.0 mL of 0.175 M HBr titrated with 0.200 M KOH

a) **Find:** initial pH

Conceptual Plan: Since HBr is a strong acid, it will dissociate completely, so initial pH = – log $[H_3O^+]$ = – log [HBr].

Solution: pH = – log [HBr] = – log 0.175 = 0.757

Check: The units (none) are correct. The pH is reasonable since the concentration is greater than 0.1 M and the acid dissociates completely, the pH is less than 1.

b) **Find:** volume of base to reach equivalence point

Conceptual Plan: Write balanced equation then mL → L then [HBr], L → mol HBr then

$HBr + KOH \rightarrow KBr + H_2O$ $\quad \frac{1\ L}{1000\ mL} \quad M = \frac{mol}{L}$

set mol acid (HBr) = mol base (KOH) and [KOH], mol KOH → L KOH → mL KOH

balanced equation has 1:1 stoichiometry $\quad M = \frac{mol}{L} \quad \frac{1000\ mL}{1\ L}$

Solution: $35.0\ \text{mL HBr} \times \frac{1\ L}{1000\ mL} = 0.0350\ \text{L HBr}$ then $\frac{0.175\ \text{mol HBr}}{1\ L} \times 0.0350\ L = 0.0061\underline{2}5\ \text{mol HBr}$.

So mol acid = mol HBr = 0.0061<u>2</u>5 mol = mol KOH then

$0.0061\underline{2}5\ \text{mol KOH} \times \frac{1\ L}{0.200\ \text{mol KOH}} = 0.030\underline{6}25\ \text{L KOH} \times \frac{1000\ mL}{1\ L} = 30.6\ \text{mL KOH}$.

Check: The units (mL) are correct. The volume of base is a little less than the volume of acid because the concentration of the base is a little greater that of the acid.

c) **Find:** pH after adding 10.0 mL of base

Conceptual Plan: Use calculations from part b. then mL → L then [KOH], L → mol KOH then

$\frac{1\ L}{1000\ mL} \quad M = \frac{mol}{L}$

mol HBr, mol KOH → mol excess HBr and L HBr, L KOH → total L then

set up stoichiometry table $\quad$ L HBr + L KOH = total L

mol excess HBr, L → [HBr] → pH

$M = \frac{mol}{L} \quad pH = -\log [HBr]$

Solution: $10.0\ \text{mL KOH} \times \frac{1\ L}{1000\ mL} = 0.0100\ \text{L KOH}$ then

$\frac{0.200\ \text{mol KOH}}{1\ L} \times 0.0100\ L = 0.00200\ \text{mol KOH}$.

Since KOH is a strong base, [KOH] = $[OH^-]$, and set up table to track changes:

	KOH (*aq*)	+ HBr (*aq*)	→ KBr (*aq*)	+ H_2O (*l*)
Before addition	0.00 mol	0.006125 mol	0.00 mol	–
Addition	0.00200 mol	–	–	–
After addition	≈ 0.00 mol	0.004125 mol	0.00200 mol	–

then 0.0350 L HBr + 0.0100 L KOH = 0.0450 L total volume. So mol excess acid = mol HBr = 0.004125 mol in 0.0450 L so

$[HBr] = \frac{0.004125 \text{ mol HBr}}{0.0450 \text{ L}} = 0.0916667 \text{ M}$ and $pH = -\log[HBr] = -\log 0.0916667 = 1.038$.

Check: The units (none) are correct. The pH is a little higher than the initial pH, which is expected since this is a strong acid.

d) **Find:** pH at equivalence point
Solution: Since this is a strong acid – strong base titration, the pH at the equivalence point is neutral or 7.

e) **Find:** pH after adding 5.0 mL of base beyond the equivalence point
Conceptual Plan: Use calculations from parts b & c. then the pH is only dependent on the amount of excess base and the total solution volumes. mL excess → L excess then

$$\frac{1 \text{ L}}{1000 \text{ mL}}$$

[KOH], L excess → mol KOH excess then

$$M = \frac{\text{mol}}{\text{L}}$$

L HBr, L KOH to equivalence point, L KOH excess → total L then

L HBr + L KOH to equivalence point + L KOH excess = total L

mol excess KOH, total L → [KOH] = [OH⁻] → [H₃O⁺] → pH

$$M = \frac{\text{mol}}{\text{L}} \qquad K_w = [H_3O^+][OH^-] \qquad pH = -\log[H_3O^+]$$

Solution: $5.0 \text{ mL KOH} \times \frac{1 \text{ L}}{1000 \text{ mL}} = 0.0050 \text{ L KOH excess}$ then

$\frac{0.200 \text{ mol KOH}}{1 \text{ L}} \times 0.0050 \text{ L} = 0.0010 \text{ mol KOH excess}$. then 0.0350 L HBr + 0.0306 L KOH + 0.0050 L KOH = 0.0706 L total volume. $[\text{KOH excess}] = \frac{0.0010 \text{ mol KOH excess}}{0.0706 \text{ L}} = 0.014164 \text{ M KOH excess}$

Since KOH is a strong base, [KOH] excess = [OH⁻]. $K_w = [H_3O^+][OH^-]$ so

$[H_3O^+] = \frac{K_w}{[OH^-]} = \frac{1.0 \times 10^{-14}}{0.014164} = 7.06 \times 10^{-13}$ M. Finally, $pH = -\log[H_3O^+] = -\log(7.06 \times 10^{-13}) = 12.15$.

Check: The units (none) are correct. The pH is rising sharply at the equivalence point, so the pH after 5 mL past the equivalence point should be quite basic.

69. **Given:** 25.0 mL of 0.115 M RbOH titrated with 0.100 M HCl

a) **Find:** initial pH
Conceptual Plan: Since RbOH is a strong base, it will dissociate completely, so [RbOH] = [OH⁻] → [H₃O⁺] → pH

$$K_w = [H_3O^+][OH^-] \qquad pH = -\log[H_3O^+]$$

Solution: Since RbOH is a strong base, [RbOH] excess = [OH⁻]. $K_w = [H_3O^+][OH^-]$ so

$[H_3O^+] = \frac{K_w}{[OH^-]} = \frac{1.0 \times 10^{-14}}{0.115} = 8.69565 \times 10^{-14}$ M and $pH = -\log[H_3O^+] = -\log(8.69565 \times 10^{-14}) = 13.06$

Check: The units (none) are correct. The pH is reasonable since the concentration is greater than 0.1 M and the base dissociates completely, the pH is greater than 13.

b) **Find:** volume of acid to reach equivalence point
Conceptual Plan: Write balanced equation then mL → L then [RbOH], L → mol RbOH then

$$HCl + RbOH \rightarrow RbCl + H_2O \qquad \frac{1 \text{ L}}{1000 \text{ mL}} \qquad M = \frac{\text{mol}}{\text{L}}$$

set mol base (RbOH) = mol acid (HCl) and [HCl], mol HCl → L HCl → mL HCl

balanced equation has 1:1 stoichiometry $\qquad M = \frac{mol}{L} \qquad \frac{1000\ mL}{1\ L}$

Solution: $25.0\ \text{mL RbOH} \times \frac{1\ L}{1000\ \text{mL}} = 0.0250\ L\ RbOH$ then

$\frac{0.115\ mol\ RbOH}{1\ L} \times 0.0250\ L = 0.0028\underline{7}5\ mol\ RbOH$. So mol base = mol RbOH = $0.0028\underline{7}5$ mol = mol HCl then $0.0028\underline{7}5\ \text{mol HCl} \times \frac{1\ L}{0.100\ \text{mol HCl}} = 0.028\underline{7}5\ \text{L HCl} \times \frac{1000\ mL}{1\ L} = 28.8\ mL\ HCl$.

Check: The units (mL) are correct. The volume of acid is a greater less than the volume of base because the concentration of the base is a little greater that of the acid.

c) **Find:** pH after adding 5.0 mL of acid

Conceptual Plan: Use calculations from part b. then mL → L then [HCl], L → mol HCl

$\frac{1\ L}{1000\ mL} \qquad M = \frac{mol}{L}$

then mol RbOH, mol HCl → mol excess RbOH and L RbOH, L HCl → total L then

set up stoichiometry table $\qquad$ L RbOH + L HCl = total L

mol excess RbOH, L → [RbOH] = [OH⁻] → [H₃O⁺] → pH

$M = \frac{mol}{L} \qquad K_w = [H_3O^+][OH^-] \quad pH = -\log [H_3O^+]$

Solution: $5.0\ \text{mL HCl} \times \frac{1\ L}{1000\ \text{mL}} = 0.0050\ L\ HCl$ then $\frac{0.100\ mol\ HCl}{1\ L} \times 0.0050\ L = 0.00050\ mol\ HCl$.

Since HCl is a strong acid, [HCl] = $[H_3O^+]$, and set up table to track changes:

	$HCl\ (aq)$	$+\ RbOH\ (aq)$	$\rightarrow\ RbCl\ (aq)$	$+\ H_2O\ (l)$
Before addition	0.00 mol	$0.0028\underline{7}5$ mol	0.00 mol	–
Addition	0.00050 mol	–	–	–
After addition	≈ 0.00 mol	$0.0023\underline{7}5$ mol	0.00050 mol	–

then 0.0250 L RbOH + 0.0050 L HCl =0.0300 L total volume. So mol excess base = mol RbOH = $0.0023\underline{7}5$ mol in 0.0300 L so $[RbOH] = \frac{0.0023\underline{7}5\ mol\ RbOH}{0.0300\ L} = 0.079\underline{1}667\ M$ Since RbOH is a strong base, [RbOH] excess = [OH⁻]. $\quad K_w = [H_3O^+][OH^-]$ so

$[H_3O^+] = \frac{K_w}{[OH^-]} = \frac{1.0 \times 10^{-14}}{0.079\underline{1}667} = 1.\underline{2}6316 \times 10^{-13}\ M$ and $pH = -\log [H_3O^+] = -\log (1.\underline{2}6316 \times 10^{-13}) = 12.90$.

Check: The units (none) are correct. The pH is a little lower than the initial pH, which is expected since this is a strong base.

d) **Find:** pH at equivalence point

Solution: Since this is a strong acid – strong base titration, the pH at the equivalence point is neutral or 7.

e) **Find:** pH after adding 5.0 mL of acid beyond the equivalence point

Conceptual Plan: Use calculations from parts b & c. then the pH is only dependent on the amount of excess acid and the total solution volumes. mL excess → L excess then

$\frac{1\ L}{1000\ mL}$

[HCl], L excess → mol HCl excess

$M = \frac{mol}{L}$

then L RbOH, L HCl to equivalence point, L HCl excess → total L then

L RbOH + L HCl to equivalence point + L HCl excess = total L

mol excess HCl, total L → [HCl] = [H₃O⁺] → pH

$M = \frac{mol}{L} \qquad pH = -\log [H_3O^+]$

Solution: $5.0\ \text{mL HCl} \times \frac{1\ \text{L}}{1000\ \text{mL}} = 0.0050\ \text{L HCl excess}$ then

$\frac{0.100\ \text{mol HCl}}{1\ \text{L}} \times 0.0050\ \text{L} = 0.00050\ \text{mol HCl excess}$. then 0.0250 L RbOH + 0.0288 L HCl + 0.0050 L HCl = 0.0588 L total volume. $[\text{HCl excess}] = \frac{0.00050\ \text{mol HCl excess}}{0.0588\ \text{L}} = 0.0085034\ \text{M HCl excess}$

Since HCl is a strong acid, [HCl] excess = $[H_3O^+]$. Finally, $pH = -\log [H_3O^+] = -\log (0.0085034) = 2.07$.

Check: The units (none) are correct. The pH is dropping sharply at the equivalence point, so the pH after 5 mL past the equivalence point should be quite acidic.

71. **Given:** 20.0 mL of 0.105 M $HC_2H_3O_2$ titrated with 0.125 M NaOH **Other:** $K_a\ (HC_2H_3O_2) = 1.8 \times 10^{-5}$

a) **Find:** initial pH

Conceptual Plan: Since $HC_2H_3O_2$ is a weak acid, set up an equilibrium problem using the initial concentration.

so M $HC_2H_3O_2$ → $[H_3O^+]$ → pH

ICE Chart $\quad pH = -\log [H_3O^+]$

Solution:

$$HC_2H_3O_2\,(aq) + H_2O(l) \rightleftharpoons H_3O^+\,(aq) + C_2H_3O_2^{2-}(aq)$$

	$[HC_2H_3O_2]$	$[H_3O^+]$	$[C_2H_3O_2^{2-}]$
Initial	0.105	≈ 0.00	0.00
Change	$-x$	$+x$	$+x$
Equil	$0.105 - x$	$+x$	$+x$

$K_a = \frac{[H_3O^+][C_2H_3O_2^-]}{[HC_2H_3O_2]} = 1.8 \times 10^{-5} = \frac{x^2}{0.105 - x}$ Assume x is small (x << 0.105) so

$\frac{x^2}{0.105 - x} = 1.8 \times 10^{-5} = \frac{x^2}{0.105}$ and $x = 1.3748 \times 10^{-3} = [H_3O^+]$. Confirm that assumption is valid

$\frac{1.3748 \times 10^{-3}}{0.105} \times 100\ \% = 1.3\ \% < 5\ \%$ so assumption is valid. Finally,

$pH = -\log [H_3O^+] = -\log (1.3748 \times 10^{-3}) = 2.86$

Check: The units (none) are correct. The magnitude of the answer makes physical sense because pH should be greater than – log (0.105) = 0.98 because this is a weak acid.

b) **Find:** volume of base to reach equivalence point

Conceptual Plan: Write balanced equation then mL → L then $[HC_2H_3O_2]$, L → mol $HC_2H_3O_2$

$HC_2H_3O_2 + NaOH \rightarrow NaC_2H_3O_2 + H_2O$ $\quad \frac{1\ \text{L}}{1000\ \text{mL}} \quad M = \frac{\text{mol}}{\text{L}}$

then set mol acid ($HC_2H_3O_2$) = mol base (NaOH) and [NaOH], mol NaOH → L NaOH

balanced equation has 1:1 stoichiometry $\quad M = \frac{\text{mol}}{\text{L}}$

then L NaOH → mL NaOH

$\frac{1000\ \text{mL}}{1\ \text{L}}$

Solution: $20.0\ \text{mL}\ HC_2H_3O_2 \times \frac{1\ \text{L}}{1000\ \text{mL}} = 0.0200\ \text{L}\ HC_2H_3O_2$ then

$\frac{0.105\ \text{mol}\ HC_2H_3O_2}{1\ \text{L}} \times 0.0200\ \text{L} = 0.00210\ \text{mol}\ HC_2H_3O_2$. So mol acid = mol $HC_2H_3O_2$ = 0.00210 mol = mol NaOH then $0.00210\ \text{mol NaOH} \times \frac{1\ \text{L}}{0.125\ \text{mol NaOH}} = 0.0168\ \text{L NaOH} \times \frac{1000\ \text{mL}}{1\ \text{L}} = 16.8\ \text{mL NaOH}$.

Check: The units (mL) are correct. The volume of base is a little less than the volume of acid because the concentration of the base is a little greater that of the acid.

c) **Find:** pH after adding 5.0 mL of base

Conceptual Plan: Use calculations from part b. then mL → L then [NaOH], L → mol NaOH

$$\frac{1\ \text{L}}{1000\ \text{mL}} \qquad M = \frac{\text{mol}}{\text{L}}$$

then mol $HC_2H_3O_2$, mol NaOH → mol excess $HC_2H_3O_2$, mol $C_2H_3O_2^-$ and

set up stoichiometry table

L $HC_2H_3O_2$, L NaOH → total L then mol excess $HC_2H_3O_2$, L → [$HC_2H_3O_2$] and

$$\text{L } HC_2H_3O_2 + \text{L NaOH} = \text{total L} \qquad M = \frac{\text{mol}}{\text{L}}$$

mol excess $C_2H_3O_2^-$, L → [$C_2H_3O_2^-$] then M $HC_2H_3O_2$, M $C_2H_3O_2^-$ → [H_3O^+] → pH

$$M = \frac{\text{mol}}{\text{L}} \qquad \text{ICE Chart} \qquad pH = -\log[H_3O^+]$$

Solution: $5.0\ \text{mL NaOH} \times \frac{1\ \text{L}}{1000\ \text{mL}} = 0.0050\ \text{L NaOH}$ then

$\frac{0.125\ \text{mol NaOH}}{1\ \text{L}} \times 0.0050\ \text{L} = 0.00062\underline{5}\ \text{mol NaOH}$. Set up table to track changes:

	NaOH (*aq*)	+ $HC_2H_3O_2$ (*aq*)	→ $NaC_2H_3O_2$ (*aq*)	+ H_2O (*l*)
Before addition	0.00 mol	0.00210 mol	0.00 mol	–
Addition	0.00062$\underline{5}$ mol	–	–	–
After addition	≈ 0.00 mol	0.0014$\underline{7}$5 mol	0.00062$\underline{5}$ mol	–

then

0.0200 L H $C_2H_3O_2$ + 0.0050 L NaOH = 0.0250 L total volume. then

$$[HC_2H_3O_2] = \frac{0.0014\underline{7}5\ \text{mol } HC_2H_3O_2}{0.0250\ \text{L}} = 0.0590\ \text{M} \text{ and } [NaC_2H_3O_2] = \frac{0.00062\underline{5}\ \text{mol } C_2H_3O_2^-}{0.0250\ \text{L}} = 0.025\ \text{M}.$$

Since 1 $C_2H_3O_2^-$ ion is generated for each $NaC_2H_3O_2$, [$C_2H_3O_2^-$] = 0.025 M $C_2H_3O_2^-$.

$$HC_2H_3O_2\,(aq) + H_2O(l) \rightleftharpoons H_3O^+\,(aq) + C_2H_3O_2^-(aq)$$

	$[HC_2H_3O_2]$	$[H_3O^+]$	$[C_2H_3O_2^-]$
Initial	0.0590	≈ 0.00	0.025
Change	$-x$	$+x$	$+x$
Equil	$0.0590 - x$	$+x$	$0.025 + x$

$$K_a = \frac{[H_3O^+][C_2H_3O_2^-]}{[HC_2H_3O_2]} = 1.8 \times 10^{-5} = \frac{x(0.025 + x)}{0.0590 - x}$$ Assume x is small (x << 0.025 < 0.0590) so

$$\frac{x(0.025 + x)}{0.0590 - x} = 1.8 \times 10^{-5} = \frac{x(0.025)}{0.0590}$$ and $x = 4.\underline{2}48 \times 10^{-5}$ = [H_3O^+]. Confirm that assumption is valid

$\frac{4.\underline{2}48 \times 10^{-5}}{0.025} \times 100\,\% = 0.17\,\% < 5\,\%$ so assumption is valid. Finally,

$pH = -\log[H_3O^+] = -\log(4.\underline{2}48 \times 10^{-5}) = 4.37$.

Check: The units (none) are correct. The pH is a little higher than the initial pH, which is expected since some of the acid has been neutralized.

d) **Find:** pH at 0ne-half of the equivalence point

Conceptual Plan: Since this is a weak acid – strong base titration, the pH at one-half the equivalence point is the pK_a of the weak acid.

Solution: $pH = pK_a = -\log K_a = -\log(1.8 \times 10^{-5}) = 4.74$.

Check: The units (none) are correct. Since this is a weak acid – strong base titration, the pH at one-half the equivalence point is the pK_a of the weak acid, so it should be a little below 5.

e) **Find:** pH at equivalence point

Conceptual Plan: Use calculations from parts b. then since all of the weak acid has been converted to its conjugate base, the pH is only dependent on the hydrolysis reaction of the conjugate base. The mol $C_2H_3O_2^-$ = initial mol $HC_2H_3O_2$ and L $HC_2H_3O_2$, L NaOH to equivalence point → total L then mol excess $C_2H_3O_2^-$, L → [$C_2H_3O_2^-$] and

$$\text{L } HC_2H_3O_2 + \text{L NaOH} = \text{total L} \qquad M = \frac{\text{mol}}{\text{L}}$$

$K_a \rightarrow K_b$ then do equilibrium calculation: $[C_2H_3O_2^-]$, K_b → $[OH^-]$ → $[H_3O^+]$ → pH

$K_w = K_a\, K_b$ set up ICE table $K_w = [H_3O^+][OH^-]$ $pH = -\log[H_3O^+]$

Solution: mol $C_2H_3O_2^-$ = initial mol $HC_2H_3O_2$ = 0.00210 mol and total volume = L $HC_2H_3O_2$ + L NaOH = 0.020 L + 0.0168 L = 0.0368 L then

$$[C_2H_3O_2^-] = \frac{0.00210 \text{ mol } C_2H_3O_2^-}{0.0368 \text{ L}} = 0.057\underline{0}652 \text{ M}$$ and $K_w = K_a\, K_b$. Rearrange to solve for K_b.

$$K_b = \frac{K_w}{K_a} = \frac{1.0 \times 10^{-14}}{1.8 \times 10^{-5}} = 5.\underline{5}556 \times 10^{-10}$$. Set up ICE table

$$C_2H_3O_2^-(aq) + H_2O(l) \rightleftharpoons HC_2H_3O_2(aq) + OH^-(aq)$$

	$[C_2H_3O_2^-]$	$[HC_2H_3O_2]$	$[OH^-]$
Initial	0.057$\underline{0}$652	≈ 0.00	≈ 0.00
Change	$-x$	$+x$	$+x$
Equil	0.057$\underline{0}$652 $- x$	$+x$	$+x$

$$K_b = \frac{[HC_2H_3O_2][OH^-]}{[C_2H_3O_2^-]} = 5.\underline{5}556 \times 10^{-10} = \frac{x^2}{0.057\underline{0}652 - x}$$ Assume x is small (x << 0.057) so

$$\frac{x^2}{0.057\underline{0}652 - x} = 5.\underline{5}556 \times 10^{-10} = \frac{x^2}{0.057\underline{0}652}$$ and $x = 5.\underline{6}305 \times 10^{-6} = [OH^-]$.

Confirm that assumption is valid $\frac{5.\underline{6}305 \times 10^{-6}}{0.057\underline{0}652} \times 100\ \% = 0.0099\ \% < 5\ \%$ so assumption is valid.

$K_w = [H_3O^+][OH^-]$ so $[H_3O^+] = \frac{K_w}{[OH^-]} = \frac{1.0 \times 10^{-14}}{5.\underline{6}305 \times 10^{-6}} = 1.\underline{7}760 \times 10^{-9}$ M . Finally,

$pH = -\log[H_3O^+] = -\log(1.\underline{7}760 \times 10^{-9}) = 8.75$.

Check: The units (none) are correct. Since this is a weak acid – strong base titration, the pH at the equivalence point is basic.

e) **Find:** pH after adding 5.0 mL of base beyond the equivalence point

Conceptual Plan: Use calculations from parts b & c. then the pH is only dependent on the amount of excess base and the total solution volumes. mL excess → L excess then

$\frac{1 \text{ L}}{1000 \text{ mL}}$

[NaOH], L excess → mol NaOH excess then

$M = \frac{\text{mol}}{\text{L}}$

L $HC_2H_3O_2$, L NaOH to equivalence point, L NaOH excess → total L then

L $HC_2H_3O_2$ + L NaOH to equivalence point + L NaOH excess = total L

mol excess NaOH, total L → [NaOH] = $[OH^-]$ → $[H_3O^+]$ → pH

$M = \frac{\text{mol}}{\text{L}}$ $K_w = [H_3O^+][OH^-]$ $pH = -\log[H_3O^+]$

Solution: $5.0 \text{ mL NaOH} \times \frac{1 \text{ L}}{1000 \text{ mL}} = 0.0050$ L NaOH excess then

$\frac{0.125 \text{ mol NaOH}}{1 \text{ L}} \times 0.0050 \text{ L} = 0.000625$ mol NaOH excess . then

0.0200 L $HC_2H_3O_2$ + 0.0168 L NaOH + 0.0050 L NaOH = 0.0418 L total volume.

$$[\text{NaOH excess}] = \frac{0.000625 \text{ mol NaOH excess}}{0.0418 \text{ L}} = 0.014\underline{9}522 \text{ M NaOH excess}$$ Since NaOH is a strong base,

[NaOH] excess = $[OH^-]$. The strong base overwhelms the weak base and is insignificant in the calculation. $K_w = [H_3O^+][OH^-]$ so $[H_3O^+] = \frac{K_w}{[OH^-]} = \frac{1.0 \times 10^{-14}}{0.014\underline{9}522} = 6.\underline{6}88 \times 10^{-13}$ M . Finally,

$pH = -\log[H_3O^+] = -\log(6.\underline{6}88 \times 10^{-13}) = 12.17$.

Check: The units (none) are correct. The pH is rising sharply at the equivalence point, so the pH after 5 mL past the equivalence point should be quite basic.

73. **Given:** 25.0 mL of 0.175 M CH_3NH_2 titrated with 0.150 M HBr **Other:** K_b (CH_3NH_2) = 4.4 x 10^{-4}

a) **Find:** initial pH

Conceptual Plan: Conceptual Plan: Since CH_3NH_2 is a weak base, set up an equilibrium problem using the initial concentration. so M CH_3NH_2 → $[OH^-]$ → $[H_3O^+]$ → pH

ICE Chart $K_w = [H_3O^+][OH^-]$ $pH = -\log[H_3O^+]$

Solution:

$$CH_3NH_2(aq) + H_2O(l) \rightleftharpoons CH_3NH_3^+(aq) + OH^-(aq)$$

	$[CH_3NH_2]$	$[CH_3NH_3^+]$	$[OH^-]$
Initial	0.175	0.00	≈ 0.00
Change	$-x$	$+x$	$+x$
Equil	$0.175 - x$	$+x$	$+x$

$$K_b = \frac{[CH_3NH_3^+][OH^-]}{[CH_3NH_2]} = 4.4 \times 10^{-4} = \frac{x^2}{0.175 - x}$$

Assume x is small (x << 0.175) so $\frac{x^2}{0.175 - \cancel{x}} = 4.4 \times 10^{-4} = \frac{x^2}{0.175}$ and x = 8.7750 x 10^{-3} = $[OH^-]$.

Confirm that assumption is valid $\frac{8.7750 \times 10^{-3}}{0.175} \times 100\ \% = 5.0\ \%$ so assumption is valid.

$K_w = [H_3O^+][OH^-]$ so $[H_3O^+] = \frac{K_w}{[OH^-]} = \frac{1.0 \times 10^{-14}}{8.7750 \times 10^{-3}} = 1.1396 \times 10^{-12}$ M . Finally,

$pH = -\log[H_3O^+] = -\log(1.1396 \times 10^{-12}) = 11.94$.

Check: The units (none) are correct. The magnitude of the answer makes physical sense because pH should be less than 14 + log (0.175) = 13.2 because this is a weak base.

b) **Find:** volume of acid to reach equivalence point

Conceptual Plan: Write balanced equation then mL → L then $[CH_3NH_2]$, L → mol CH_3NH_2

$HBr + CH_3NH_2 \rightarrow CH_3NH_3Br + H_2O$ $\frac{1\ L}{1000\ mL}$ $M = \frac{mol}{L}$

then set mol base (CH_3NH_2) = mol acid (HBr) and [HBr], mol HBr → L HBr → mL HBr

balanced equation has 1:1 stoichiometry $M = \frac{mol}{L}$ $\frac{1000\ mL}{1\ L}$

Solution: $25.0\ \cancel{mL\ CH_3NH_2} \times \frac{1\ L}{1000\ \cancel{mL}} = 0.0250\ L\ CH_3NH_2$ then

$\frac{0.175\ mol\ CH_3NH_2}{1\ \cancel{L}} \times 0.0250\ \cancel{L} = 0.004375\ mol\ CH_3NH_2$. So mol base = mol CH_3NH_2 = 0.002875 mol =

mol HBr then $0.004375\ \cancel{mol\ HBr} \times \frac{1\ L}{0.150\ \cancel{mol\ HBr}} = 0.0291667\ \cancel{L\ HBr} \times \frac{1000\ mL}{1\ \cancel{L}} = 29.2\ mL\ HBr$.

Check: The units (mL) are correct. The volume of acid is a greater less than the volume of base because the concentration of the base is a little greater that of the acid.

c) **Find:** pH after adding 5.0 mL of acid

Conceptual Plan: Use calculations from part b. then mL → L then [HBr], L → mol HBr then

$\frac{1\ L}{1000\ mL}$ $M = \frac{mol}{L}$

mol CH_3NH_2, mol HBr → mol excess CH_3NH_2 and L CH_3NH_2, L HBr → total L then

set up stoichiometry table $L\ CH_3NH_2 + L\ HBr = total\ L$

Since there are significant concentrations of both the acid and the conjugate base species, this is a buffer solution and so Henderson-Hasselbalch Equation $\left(pH = pK_a + \log\frac{[base]}{[acid]}\right)$ **can be used.**

Convert K_b to K_a using $K_w = K_a K_b$. Also note that ratio of concentrations is the same as the ration of moles, since the volume is the same for both species.

Solution: $5.0\ \cancel{mL\ HBr} \times \frac{1\ L}{1000\ \cancel{mL}} = 0.0050\ L\ HBr$ then $\frac{0.150\ mol\ HBr}{1\ \cancel{L}} \times 0.0050\ \cancel{L} = 0.00075\ mol\ HBr$.

Set up table to track changes:

	$HBr\ (aq)$	$+\ CH_3NH_2\ (aq)$	$\rightarrow\ CH_3NH_3Br\ (aq)$
Before addition	0.00 mol	0.004375 mol	0.00 mol
Addition	0.00075 mol	–	–
After addition	≈ 0.00 mol	0.003625 mol	0.000750 mol

then $K_w = K_a K_b$ so

$K_a = \frac{K_w}{K_b} = \frac{1.0 \times 10^{-14}}{4.4 \times 10^{-4}} = 2.2727 \times 10^{-11}$ M then use Henderson-Hasselbalch Equation, since the solution is a buffer. $pH = pK_a + \log\frac{[base]}{[acid]} = -\log(2.2727 \times 10^{-11}) + \log\frac{0.003625}{0.000750} = 11.33$

Check: The units (none) are correct. The pH is a little lower than the last pH, which is expected since some of the base has been neutralized.

d) **Find:** pH at one-half of the equivalence point
Conceptual Plan: Since this is a weak base – strong acid titration, the pH at one-half the equivalence point is the pK_a of the conjugate acid of weak base.
Solution: $pH = pK_a = -\log K_a = -\log(2.2727 \times 10^{-11}) = 10.64$.

Check: The units (none) are correct. Since this is a weak acid – strong base titration, the pH at one-half the equivalence point is the pK_a of the conjugate acid of the weak base, so it should be a little below 11.

e) **Find:** pH at equivalence point
Conceptual Plan: Use calculations from above. Since all of the weak base has been converted to its conjugate acid, the pH is only dependent on the hydrolysis reaction of the conjugate acid. The mol $CH_3NH_3^+$ = initial mol CH_3NH_2 and L CH_3NH_2, L HBr to equivalence point → total L

L CH_3NH_2 + L HBr = total L

then mol $CH_3NH_3^+$, L → $[CH_3NH_3^+]$

$M = \frac{mol}{L}$

then do equilibrium calculation: $[CH_3NH_3^+]$, K_a → $[H_3O^+]$ → pH

set up ICE table $\quad pH = -\log[H_3O^+]$

Solution: mol base = mol acid = mol $CH_3NH_3^+$ = 0.004375 mol.
then total volume = L CH_3NH_2 + L HBr = 0.0250 L + 0.0292 L = 0.0542 then

$[CH_3NH_3^+] = \frac{0.004375 \text{ mol mol } CH_3NH_3^+}{0.0542\text{ L}} = 0.0807196$ M. Set up ICE table

$$CH_3NH_3^+(aq) + H_2O(l) \rightleftharpoons CH_3NH_2(aq) + H_3O^+(aq)$$

	$[CH_3NH_3^+]$	$[CH_3NH_2]$	$[H_3O^+]$
Initial	0.0807196	≈ 0.00	≈ 0.00
Change	$-x$	$+x$	$+x$
Equil	$0.0807196 - x$	$+x$	$+x$

$K_a = \frac{[CH_3NH_2][H_3O^+]}{[CH_3NH_3^+]} = 2.2727 \times 10^{-11} = \frac{x^2}{0.0807196 - x}$ Assume x is small (x << 0.0807) so

$\frac{x^2}{0.0807196 - x} = 2.2727 \times 10^{-11} = \frac{x^2}{0.0807196}$ and $x = 1.3544 \times 10^{-6} = [H_3O^+]$.

Confirm that assumption is valid $\frac{1.3544 \times 10^{-6}}{0.0807106} \times 100\ \% = 0.0017\ \% < 5\ \%$ so assumption is valid.

Finally, $pH = -\log[H_3O^+] = -\log(1.3544 \times 10^{-6}) = 5.87$.

Check: The units (none) are correct. Since this is a weak base – strong acid titration, the pH at the equivalence point is acidic.

f) **Find:** pH after adding 5.0 mL of acid beyond the equivalence point
Conceptual Plan: Use calculations from parts b & c. then the pH is only dependent on the amount of excess acid and the total solution volumes. mL excess → L excess then

$\frac{1\text{ L}}{1000\text{ mL}}$

[HBr], L excess → mol HBr excess   then

$M = \frac{mol}{L}$

then L CH_3NH_2, L HBr to equivalence point, L HBr excess → total L   then

L CH_3NH_2 + L HBr to equivalence point + L HBr excess = total L

mol excess HBr, total L → [HBr] = $[H_3O^+]$ → pH

$M = \frac{mol}{L}$ $pH = -\log[H_3O^+]$

Solution: $5.0 \text{ mL HBr} \times \frac{1 \text{ L}}{1000 \text{ mL}} = 0.0050 \text{ L HBr excess}$ then

$\frac{0.150 \text{ mol HBr}}{1 \text{ L}} \times 0.0050 \text{ L} = 0.00075 \text{ mol HBr excess}$. then 0.0250 L CH_3NH_2 + 0.0292 L HBr + 0.0050 L HBr = 0.0592 L total volume. $[\text{HBr excess}] = \frac{0.00075 \text{ mol HBr excess}}{0.0592 \text{ L}} = 0.01\underline{2}669 \text{ M HBr excess}$

Since HBr is a strong acid, [HBr] excess = $[H_3O^+]$. The strong acid overwhelms the weak acid and is insignificant in the calculation. Finally, $pH = -\log[H_3O^+] = -\log(0.01\underline{2}669) = 1.90$.

Check: The units (none) are correct. The pH is dropping sharply at the equivalence point, so the pH after 5 mL past the equivalence point should be quite acidic.

75. i) acid a is more concentrated, since the equivalence point (where sharp pH rise occurs) is at a higher volume of added base.

ii) acid b has the larger K_a, since the pH at a volume of added base equal to half of the equivalence point volume is lower.

77. **Given:** 0.229 g unknown monoprotic acid titrated with 0.112 M NaOH and curve
Find: molar mass and pK_a of acid
Conceptual Plan: The equivalence point is where sharp pH rise occurs. The pK_a is the pH at a volume of added base equal to half of the equivalence point volume. then mL NaOH → L NaOH then

$\frac{1 \text{ L}}{1000 \text{ mL}}$

[NaOH], L NaOH → mol NaOH = mol acid then mol NaOH, g acid → molar mass

$M = \frac{mol}{L}$ $\frac{\text{g acid}}{\text{mol acid}}$

Solution: The equivalence point is at 25 mL NaOH. The pH at 0.5 x 25 mL = 13 mL is ~3. then

$25 \text{ mL NaOH} \times \frac{1 \text{ L}}{1000 \text{ mL}} = 0.025 \text{ L NaOH}$ then

$\frac{0.112 \text{ mol NaOH}}{1 \text{ L}} \times 0.025 \text{ L} = 0.0028 \text{ mol NaOH} = 0.0028 \text{ mol acid}$ then

$\text{Molar Mass} = \frac{0.229 \text{ g acid}}{0.0028 \text{ mol acid}} = 82 \text{ g/mol}$.

Check: The units (none and g/mol) are correct. The pK_a is consistent with a weak acid. The molar mass is reasonable for an acid (>1 g/mol).

79. **Given:** 20.0 mL of 0.125 M sulfurous acid (H_2SO_3) titrated with 0.1014 M KOH **Find:** volume of base added
Conceptual Plan: Since this is a diprotic acid, each proton is titrated sequentially.
Write balanced equations.

$H_2SO_3 + OH^- \rightarrow HSO_3^- + H_2O$ and $HSO_3^- + OH^- \rightarrow SO_3^{2-} + H_2O$

then mL → L then $[H_2SO_3]$, L → mol H_2SO_3 then set mol base (H_2SO_3) = mol acid (KOH) and

$\frac{1 \text{ L}}{1000 \text{ mL}}$ $M = \frac{mol}{L}$ balanced equation has 1:1 stoichiometry

[KOH], mol KOH → L KOH → mL KOH the volume to the second equivalence point will be double

$M = \frac{mol}{L}$ $\frac{1000 \text{ mL}}{1 \text{ L}}$

the volume to the first equivalence point.

Solution: $20.0 \text{ mL } H_2SO_3 \times \frac{1 \text{ L}}{1000 \text{ mL}} = 0.0200 \text{ L } H_2SO_3$ then

$\frac{0.115 \text{ mol } H_2SO_3}{1 \text{ L}} \times 0.0200 \text{ L} = 0.00230 \text{ mol } H_2SO_3$. So mol base = mol H_2SO_3 = 0.00230 mol = mol KOH

then $0.00230 \text{ mol KOH} \times \frac{1 \text{ L}}{0.1014 \text{ mol KOH}} = 0.02268245 \text{ L KOH} \times \frac{1000 \text{ mL}}{1 \text{ L}} = 22.7 \text{ mL KOH}$ to first equivalence point. The volume to the second equivalence point is simply twice this amount or 45.4 mL to the second equivalence point.

Check: The units (mL) are correct. The volume of base is a greater than the volume of acid because the concentration of the acid is a little greater that of the base. The volume to the second equivalence point is twice the volume to the first equivalence point.

81. The indicator will be in its acid form at an acidic pH, so the color in the HCl sample will be red. The color change will occur over the pH range from pH = $pK_a - 1.0$ to pH = $pK_a + 1.0$, so the color will start to change at pH = 5.0 – 1.0 = 4.0 and finish changing by pH = 5.0 + 1.0 = 6.0

83. Since the exact conditions of the titration are not given, a rough calculation will suffice. Looking at the pattern of earlier problems, the pH at the equivalence point of a titration of a weak acid and a strong base is the hydrolysis of the conjugate base of the weak acid that has been diluted by a factor of roughly 2 with base. If it is assumed that the initial concentration of the weak acid is ~ 0.1 M, then the conjugate base concentration will be ~ 0.05 M. From earlier calculations it can be seen that the $K_b = \frac{K_w}{K_a} = \frac{[OH^-]^2}{0.05}$ thus

$$[OH^-] = \sqrt{\frac{0.05\, K_w}{K_a}} = \sqrt{\frac{5 \times 10^{-16}}{K_a}} \text{ and the pH} = 14 + \log \sqrt{\frac{5 \times 10^{-16}}{K_a}}.$$

a) For HF, the $pK_a = 3.5 \times 10^{-4}$ and so the above equations approximates the pH at the equivalence point is ~ 8.0. Looking at Table 16.1, phenol red or *m*-nitrophenol will change at the appropriate pH range.

b) For HCl, the pH at the equivalence point is 7, since HCl is a strong acid. Looking at Table 16.1, alizarin, bromthymol blue or phenol red will change at the appropriate pH range.

c) For HCN, the $pK_a = 4.9 \times 10^{-10}$ and so the above equations approximates the pH at the equivalence point is ~ 11.0. Looking at Table 16.1, alizarin yellow R will change at the appropriate pH range.

85. For the dissolution reaction, start with the ionic compound as a solid and put it in equilibrium with the appropriate cation and anion, making sure to include the appropriate stoichiometric coefficients. The K_{sp} expression is the product of the concentrations of the cation and anion concentrations raise to their stoichiometric coefficients.

a) $BaSO_4\,(s) \rightleftharpoons Ba^{2+}\,(aq) + SO_4^{2-}\,(aq)$ and $K_{sp} = [Ba^{2+}][SO_4^{2-}]$.

b) $PbBr_2\,(s) \rightleftharpoons Pb^{2+}\,(aq) + 2\,Br^-\,(aq)$ and $K_{sp} = [Pb^{2+}][Br^-]^2$.

c) $Ag_2CrO_4\,(s) \rightleftharpoons 2\,Ag^+\,(aq) + CrO_4^{2-}\,(aq)$ and $K_{sp} = [Ag^+]^2[CrO_4^{2-}]$.

87. **Given:** ionic compound formula and Table 16.2 of K_{sp} values **Find:** molar solubility (S)

Conceptual Plan: use equation derived in problem #20 and solve for S.

for ionic compound, A_mX_n, $K_{sp} = m^m\, n^n\, S^{m+n}$.

Solution:

a) For AgBr, $K_{sp} = 5.35 \times 10^{-13}$, A = Ag^+, m = 1, X = Br^-, and n = 1 so $K_{sp} = 5.35 \times 10^{-13} = S^2$.
Rearrange to solve for S. $S = \sqrt{5.35 \times 10^{-13}} = 7.31 \times 10^{-7}$ M.

b) For $Mg(OH)_2$, $K_{sp} = 2.06 \times 10^{-13}$, A = Mg^{2+}, m = 1, X = OH^-, and n = 2 so $K_{sp} = 2.06 \times 10^{-13} = 2^2S^3$.
Rearrange to solve for S. $S = \sqrt[3]{\frac{2.06 \times 10^{-13}}{4}} = 3.72 \times 10^{-5}$ M.

c) For CaF_2, K_{sp} = 1.46 x 10^{-10}, A = Ca^{2+}, m = 1, X = F^-, and n = 2 so K_{sp} = 1.46 x 10^{-10} = 2^2S^3.

Rearrange to solve for S. $S = \sqrt[3]{\frac{1.46 \times 10^{-10}}{4}} = 3.32 \times 10^{-4}$ M .

Check: The units (M) are correct. The molar solubilities are much less than one and dependent not only on the value of the K_{sp}, but also the stoichiometry of the ionic compound. The more ions that are generated, the greater the molar solubility for the same value of the K_{sp}.

89. **Given:** ionic compound formula and molar solubility (S) **Find:** K_{sp} molar solubility (S)
Conceptual Plan: use equation derived in problem #20.
for ionic compound, A_mX_n, $K_{sp} = m^m n^n S^{m+n}$.
Solution:
a) For NiS, S = 3.27 x 10^{-11} M, A = Ni^{2+}, m = 1, X = S^{2-}, and n = 1 so $K_{sp} = S^2 = (3.27 \times 10^{-11})^2 = 1.07 \times 10^{-21}$.

b) For PbF_2, S = 5.63 x 10^{-3} M, A = Pb^{2+}, m = 1, X = F^-, and n = 2 so $K_{sp} = 2^2S^3 = 2^2 (5.63 \times 10^{-3})^3 = 7.14 \times 10^{-7}$.

c) For MgF_2, S = 2.65 x 10^{-4} M, A = Mg^{2+}, m = 1, X = F^-, and n = 2 so $K_{sp} = 2^2S^3 = 2^2 (2.65 \times 10^{-4})^3 = 7.44 \times 10^{-11}$.

Check: The units (none) are correct. The K_{sp} values are much less than one and dependent not only on the value of the solubility, but also the stoichiometry of the ionic compound. The more ions that are generated, the smaller the K_{sp} for the same value of the S.

91. **Given:** ionic compound formulas AX and AX_2 and K_{sp} = 1.5 x 10^{-5} **Find:** higher molar solubility (S)
Conceptual Plan: use equation derived in problem #20 and solve for S, then compare S values.
for ionic compound, A_mX_n, $K_{sp} = m^m n^n S^{m+n}$.
Solution: for AX, K_{sp} = 1.5 x 10^{-5}, m = 1, and n = 1 so K_{sp} = 1.5 x 10^{-5} = S^2. Rearrange to solve for S. $S = \sqrt{1.5 \times 10^{-5}} = 3.9 \times 10^{-3}$ M .
for AX_2, K_{sp} = 1.5 x 10^{-5}, m = 1, and n = 2 so K_{sp} = 1.5 x 10^{-5} = 2^2S^3. Rearrange to solve for S. $S = \sqrt[3]{\frac{1.5 \times 10^{-5}}{4}} = 1.6 \times 10^{-2}$ M .
Since 10^{-2} M > c. 10^{-3} M, AX_2 has a higher molar solubility.
Check: The units (M) are correct. The more ions that are generated, the greater the molar solubility for the same value of the K_{sp}.

93. **Given:** $Fe(OH)_2$ in 100.0 mL solution **Find:** grams of $Fe(OH)_2$ **Other:** K_{sp} = 4.87 x 10^{-17}
Conceptual Plan: use equation derived in problem #20 and solve for S, then mL → L then
for ionic compound, A_mX_n, $K_{sp} = m^m n^n S^{m+n}$. $\frac{1\ L}{1000\ mL}$

S, L → mol $Fe(OH)_2$ → g $Fe(OH)_2$
$M = \frac{mol}{L}$ $\frac{89.87\ g\ Fe(OH)_2}{1\ mol\ Fe(OH)_2}$

Solution: for $Fe(OH)_2$, K_{sp} = 4.87 x 10^{-17}, A = Fe^{2+}, m = 1, X = OH^-, and n = 2 so K_{sp} = 4.87 x 10^{-17} = 2^2S^3. Rearrange to solve for S. $S = \sqrt[3]{\frac{4.87 \times 10^{-17}}{4}} = 2.30080 \times 10^{-6}$ M . then $100.0\ \text{mL} \times \frac{1\ L}{1000\ \text{mL}} = 0.1000$ L

then
$$\frac{2.30080 \times 10^{-6}\ mol\ Fe(OH)_2}{1\ L} \times 0.1000\ L = 2.30080 \times 10^{-7}\ mol\ Fe(OH)_2 \times \frac{89.87\ g\ Fe(OH)_2}{1\ mol\ Fe(OH)_2} = 2.07 \times 10^{-5} g\ Fe(OH)_2 .$$

Check: The units (g) are correct. The solubility rules from Chapter 4 (most hydroxides are insoluble) suggest that very little $Fe(OH)_2$ will dissolve, so the magnitude of the answer is not surprising.

95. a) **Given:** BaF_2 **Find:** molar solubility (S) in pure water **Other:** K_{sp} (BaF_2) = 2.45 x 10^{-5}
Conceptual Plan: use equation derived in problem #20 and solve for S.
for ionic compound, A_mX_n, $K_{sp} = m^m n^n S^{m+n}$.
Solution: BaF_2, K_{sp} = 2.45 x 10^{-5}, A = Ba^{2+}, m = 1, X = F^-, and n = 2 so K_{sp} = 2.45 x 10^{-5} = 2^2S^3.

Rearrange to solve for S. $S = \sqrt[3]{\frac{2.45 \times 10^{-5}}{4}} = 1.83 \times 10^{-2}$ M .

b) **Given:** BaF_2 **Find:** molar solubility (S) in 0.10 M $Ba(NO_3)_2$ **Other:** K_{sp} (BaF_2) = 2.45×10^{-5}

Conceptual Plan: **M $Ba(NO_3)_2$ → M Ba^{2+} then M Ba^{2+}, K_{sp} → S**

$Ba(NO_3)_2\,(s) \rightarrow Ba^{2+}\,(aq) + 2\,NO_3^-\,(aq)$ ICE Chart

Solution: Since 1 Ba^{2+} ion is generated for each $Ba(NO_3)_2$, $[Ba^{2+}]$ = 0.10 M.

$BaF_2\,(s) \rightleftharpoons$	$Ba^{2+}\,(aq)$ +	$2\,F^-(aq)$
Initial	0.10	0.00
Change	*S*	2*S*
Equil	0.10 + *S*	2*S*

$K_{sp}(BaF_2) = [Ba^{2+}][F^-]^2 = 2.45 \times 10^{-5} = (0.10 + S)(2\,S)^2$.

Assume S << 0.10, $2.45 \times 10^{-5} = (0.10)(2\,S)^2$, and S = 7.83×10^{-3} M. Confirm that assumption is valid $\frac{7.83 \times 10^{-3}}{0.10} \times 100\,\% = 7.8\,\% > 5\,\%$ so assumption is not valid. Since expanding the expression will give a third order polynomial, that is not easily solved directly. Solve by successive approximations. Substitute S = 7.83×10^{-3} M for the S term that is part of a sum (i.e. the one in (0.10 + S)). Thus, $2.45 \times 10^{-5} = (0.10 + 7.83 \times 10^{-3})(2\,S)^2$ and S = 7.53×10^{-3} M. Substitute this new S value again. Thus, $2.45 \times 10^{-5} = (0.10 + 7.53 \times 10^{-3})(2\,S)^2$ and S = 7.55×10^{-3} M. Substitute this new S value again. Thus, $2.45 \times 10^{-5} = (0.10 + 7.55 \times 10^{-3})(2\,S)^2$ and S = 7.55×10^{-3} M. So the solution has converged and S = 7.55×10^{-3} M.

c) **Given:** BaF_2 **Find:** molar solubility (S) in 0.15 M NaF **Other:** K_{sp} (BaF_2) = 2.45×10^{-5}

Conceptual Plan: **M NaF → M F^- then M F^-, K_{sp} → S**

$NaF\,(s) \rightarrow Na^+\,(aq) + F^-\,(aq)$ ICE Chart

Solution: Since 1 F^- ion is generated for each NaF, $[F^-]$ = 0.15 M.

$BaF_2\,(s) \rightleftharpoons$	$Ba^{2+}\,(aq)$ +	$2\,F^-(aq)$
Initial	0.00	0.15
Change	*S*	2*S*
Equil	*S*	0.15 + 2*S*

$K_{sp}(BaF_2) = [Ba^{2+}][F^-]^2 = 2.45 \times 10^{-5} = (S)(0.15 + 2\,S)^2$.

Since 2 S << 0.15, $2.45 \times 10^{-5} = (S)(0.15)^2$, and S = 1.09×10^{-3} M. Confirm that assumption is valid $\frac{2\,(1.09 \times 10^{-3})}{0.10} \times 100\,\% = 2.2\,\% < 5\,\%$ so assumption is valid.

Check: The units (M) are correct. The solubility of the BaF_2 decreases in the presence of a common ion. The effect of the anion is greater because the K_{sp} expression has the anion concentration squared.

97. **Given:** $Ca(OH)_2$ **Find:** molar solubility (S) in buffers at a. pH = 4, b. pH = 7, and c. pH = 9 **Other:** $K_{sp}(Ca(OH)_2) = 4.68 \times 10^{-6}$

Conceptual Plan: **pH → $[H_3O^+]$ → $[OH^-]$ then M OH^-, K_{sp} → S**

$[H_3O^+] = 10^{-pH}$ $K_w = [H_3O^+][OH^-]$ set up ICE table

Solution:

a) pH = 4, so $[H_3O^+] = 10^{-pH} = 10^{-4} = 1 \times 10^{-4}$ M then $K_w = [H_3O^+][OH^-]$ so

$[OH^-] = \frac{K_w}{[H_3O^+]} = \frac{1.0 \times 10^{-14}}{1 \times 10^{-4}} = 1 \times 10^{-10}$ M then

$Ca(OH)_2\,(s) \rightleftharpoons$	$Ca^{2+}\,(aq)$ +	$2\,OH^-(aq)$
Initial	0.00	1×10^{-10}
Change	*S*	–
Equil	*S*	1×10^{-10}

$K_{sp}(Ca(OH)_2) = [Ca^{2+}][OH^-]^2 = 4.68 \times 10^{-6} = S\,(1 \times 10^{-10})^2$.and S = 5×10^{14} M.

b) pH = 7, so $[H_3O^+] = 10^{-pH} = 10^{-7} = 1 \times 10^{-7}$ M then $K_w = [H_3O^+][OH^-]$ so

$[OH^-] = \frac{K_w}{[H_3O^+]} = \frac{1.0 \times 10^{-14}}{1 \times 10^{-7}} = 1 \times 10^{-7}$ M then

$Ca(OH)_2\,(s) \rightleftharpoons$	$Ca^{2+}\,(aq)$ +	$2OH^-(aq)$
Initial	0.00	1×10^{-7}
Change	*S*	–
Equil	*S*	1×10^{-7}

$K_{sp}(Ca(OH)_2) = [Ca^{2+}][OH^-]^2 = 4.68 \times 10^{-6} = S\,(1 \times 10^{-7})^2$. and S = 5×10^{8} M.

c) pH = 9, so $[H_3O^+] = 10^{-pH} = 10^{-9} = 1 \times 10^{-9}$ M then $K_w = [H_3O^+][OH^-]$ so

	$Ca(OH)_2\,(s) \rightleftharpoons Ca^{2+}\,(aq) + 2\,OH^-(aq)$	
Initial	0.00	1×10^{-5}
Change	*S*	–
Equil	*S*	1×10^{-5}

$[OH^-] = \frac{K_w}{[H_3O^+]} = \frac{1.0 \times 10^{-14}}{1 \times 10^{-9}} = 1 \times 10^{-5}$ M then

$K_{sp}\,(Ca(OH)_2) = [Ca^{2+}]\,[OH^-]^2 = 4.68 \times 10^{-6} = S\,(1 \times 10^{-5})^2$. and $S = 5 \times 10^4$ M.

Check: The units (M) are correct. The solubility of the $Ca(OH)_2$ decreases as the pH increases (and the hydroxide ion concentration increases). Realize that these molar solubilities are not achievable because the density of pure $Ca(OH)_2$ is ~ 30 M. The bottom line is that as long as the hydroxide concentration can be controlled with a buffer, the $Ca(OH)_2$ will be very soluble.

99. a) $BaCO_3$ will be more soluble in acidic solutions because CO_3^{2-} is basic. In acidic solutions it can be converted to HCO_3^- and $H_2CO_3^{2-}$. These species are not CO_3^{2-} so they do not appear in the K_{sp} expression.

b) CuS will be more soluble in acidic solutions because S^{2-} is basic. In acidic solutions it can be converted to HS^- and H_2S^{2-}. These species are not S^{2-} so they do not appear in the K_{sp} expression.

c) AgCl will not be more soluble in acidic solutions because Cl^- will not react with acidic solutions, because HCl is a strong acid.

d) PbI_2 will not be more soluble in acidic solutions because I^- will not react with acidic solutions, because HI is a strong acid.

101. **Given:** 0.015 M NaF and 0.010 M $Ca(NO_3)_2$ **Find:** will a precipitate form, if so, identify it
Other: $K_{sp}\,(CaF_2) = 1.46 \times 10^{-10}$
Conceptual Plan: Look at all possible combinations and consider the solubility rules from Chapter 4. Salts of alkali metals (Na) are very soluble, so NaF and $NaNO_3$ will be very soluble. Nitrate compounds are very soluble so $NaNO_3$ will be very soluble. The only possibility for a precipitate is CaF_2. Determine if a precipitate will form by determining the concentration of the Ca^{2+} and F^- in solution. then compute the reaction quotient, Q. If Q > K_{sp} then a precipitate will form.
Solution: Since the only possible precipitate is CaF_2, calculate the concentrations of Ca^{2+} and F^-.
NaF *(s)* → Na^+ *(aq)* + F^- *(aq)*. Since 1 F^- ion is generated for each NaF, $[F^-] = 0.015$ M.
$Ca(NO_3)_2$ *(s)* → Ca^{2+} *(aq)* + 2 NO_3^- *(aq)*. Since 1 Ca^{2+} ion is generated for each $Ca(NO_3)_2$, $[Ca^{2+}] = 0.010$ M. then calculate Q (CaF_2), $A = Ca^{2+}$, m = 1, $X = F^-$, and n = 2 Since $Q = [A^{n+}]^m\,[X^{m-}]^n$, then
$Q\,(CaF_2) = [Ca^{2+}]\,[F^-]^2 = (0.010)\,(0.015)^2 = 2.3 \times 10^{-6} > 1.46 \times 10^{-10} = K_{sp}\,(CaF_2)$, so a precipitate will form.
Check: The units (none) are correct. The solubility of the CaF_2 is low, and the concentration of ions are extremely large compared to the K_{sp}, so a precipitate will form.

103. **Given:** 75.0 mL of NaOH with pH = 2.58 and 125.0 mL of 0.0018 M $MgCl_2$
Find: will a precipitate form, if so, identify it **Other:** $K_{sp}\,(Mg(OH)_2) = 2.06 \times 10^{-13}$
Conceptual Plan: Look at all possible combinations and consider the solubility rules from Chapter 4. Salts of alkali metals (Na) are very soluble, so NaOH and NaCl will be very soluble. Chloride compounds are generally very soluble so $MgCl_2$ and NaCl will be very soluble. The only possibility for a precipitate is $Mg(OH)_2$. Determine if a precipitate will form by determining the concentration of the Mg^{2+} and OH^- in solution. Since pH, not NaOH concentration, is given pOH → $[OH^-]$ then

$$[OH^-] = 10^{-pOH}$$

mix solutions and calculate diluted concentrations mL NaOH, mL $MgCl_2$ → mL total then

mL NaOH + mL $MgCl_2$ = total mL

mL, initial M → final M then compute the reaction quotient, Q.

$$M_1 V_1 = M_2 V_2$$

If Q > K_{sp} then a precipitate will form.
Solution: Since the only possible precipitate is $Mg(OH)_2$, calculate the concentrations of Mg^{2+} and OH^-.
for NaOH at pOH = 2.58, so $[OH^-] = 10^{-pOH} = 10^{-2.58} = 2.\underline{6}3027 \times 10^{-3}$ M and $MgCl_2$ (s) →
Mg^{2+} (aq) + 2 Cl^- (aq). Since 1 Mg^{2+} ion is generated for each $MgCl_2$, $[Mg^{2+}] = 0.0018$ M. then total mL = mL NaOH + mL $MgCl_2$ = 75.0 mL + 125.0 mL = 200.0 mL. then $M_1 V_1 = M_2 V_2$.

Rearrange to solve for M_2. $M_2 = M_1 \frac{V_1}{V_2} = 2.63027 \times 10^{-3}$ M OH^- x $\frac{75.0 \text{ mL}}{200.0 \text{ mL}} = 9.8635 \times 10^{-4}$ M OH^- and

$M_2 = M_1 \frac{V_1}{V_2} = 0.0018$ M Mg^+ x $\frac{125.0 \text{ mL}}{200.0 \text{ mL}} = 1.125 \times 10^{-3}$ M Mg^{2+}. Calculate Q ($Mg(OH)_2$),

A = Mg^{2+}, m = 1, X = OH^-, and n = 2 Since Q= $[A^{n+}]^m [X^{m-}]^n$, then Q ($Mg(OH)_2$) = $[Mg^{2+}] [OH^-]^2$ = $= (1.125 \times 10^{-3})(9.8635 \times 10^{-4})^2 = 1.1 \times 10^{-9} > 2.06 \times 10^{-13} = K_{sp}$ ($Mg(OH)_2$), so a precipitate will form.

Check: The units (none) are correct. The solubility of the $Mg(OH)_2$ is low, and the NaOH (a base) is high enough that the product of the concentration of ions are large compared to the K_{sp}, so a precipitate will form.

105. **Given:** KOH as precipitation agent in a) 0.015 M $CaCl_2$, b) 0.0025 M $Fe(NO_3)_2$ and c) 0.0018 M $MgBr_2$
Find: concentration of KOH necessary to form a precipitate
Other: K_{sp} ($Ca(OH)_2$) = 4.68×10^{-6}, K_{sp} ($Fe(OH)_2$) = 4.87×10^{-17}, K_{sp} ($Mg(OH)_2$) = 2.06×10^{-13}
Conceptual Plan: The solubility rules from Chapter 4 state that most hydroxides are insoluble, so all precipitate will be hydroxides. Determine the concentration of the cation in solution. Since all metals have an oxidation state of +2 and $[OH^-]$ = [KOH], all of the K_{sp} = [cation] $[KOH]^2$ and so

$[KOH] = \sqrt{\frac{K_{sp}}{[\text{cation}]}}$.

Solution:

a) $CaCl_2$ *(s)* → Ca^{2+} *(aq)* + 2 Cl^- *(aq)*. Since 1 Ca^{2+} ion is generated for each $CaCl_2$, $[Ca^{2+}]$ = 0.015 M.

then $[KOH] = \sqrt{\frac{K_{sp}}{[\text{cation}]}} = \sqrt{\frac{4.68 \times 10^{-6}}{0.015}} = 0.018$ M KOH.

b) $Fe(NO_3)_2$ *(s)* → Fe^{2+} *(aq)* + 2 NO_3^- *(aq)*. Since 1 Fe^{2+} ion is generated for each $Fe(NO_3)_2$, $[Fe^{2+}]$ = 0.0025 M. then $[KOH] = \sqrt{\frac{K_{sp}}{[\text{cation}]}} = \sqrt{\frac{4.87 \times 10^{-17}}{0.0025}} = 1.4 \times 10^{-7}$ M KOH.

c) $MgBr_2$ *(s)* → Mg^{2+} *(aq)* + 2 Br^- *(aq)*. Since 1 Mg^{2+} ion is generated for each $MgBr_2$, $[Mg^{2+}]$ = 0.0018 M. then $[KOH] = \sqrt{\frac{K_{sp}}{[\text{cation}]}} = \sqrt{\frac{2.06 \times 10^{-13}}{0.0018}} = 1.1 \times 10^{-5}$ M KOH.

Check: The units (none) are correct. Since all cations have an oxidation state of +2, it can be seen that the [KOH] needed to precipitate the hydroxide is lower the smaller the K_{sp}.

107. **Given:** solution with 0.010 M Ba^{2+} and 0.020 M Ca^{2+} add Na_2SO_4 to form precipitates
Find: a) which ion precipitates first and minimum $[Na_2SO_4]$ needed; and b) [first cation] when second cation precipitates **Other:** K_{sp} ($BaSO_4$) = 1.07×10^{-10}, K_{sp} ($CaSO_4$) = 7.10×10^{-5}
Conceptual Plan:

a) **The precipitates that will form are $BaSO_4$ and $CaSO_4$. Use equation derived in problem #19 to define K_{sp}. Substitute in concentration of cation and solve for concentration of anion to form precipitate. The cation with the lower anion concentration will precipitate first.**
for ionic compound, A_mX_n, $K_{sp} = [A^{n+}]^m [X^{m-}]^n$

b) **Substitute the higher anion concentration into the K_{sp} expression for the first cation to precipitate and calculate the amount of this first cation to remain in solution.**

Solution:

a) Derive expression for K_{sp} ($BaSO_4$), A = Ba^{2+}, m = 1, X = SO_4^{2-}, and n = 1 Since $K_{sp} = [Ba^{2+}] [SO_4^{2-}]$, then K_{sp} ($BaSO_4$) = 1.07×10^{-10} = 0.010 $[SO_4^{2-}]$. Solve for $[SO_4^{2-}]$. $[SO_4^{2-}] = 1.07 \times 10^{-8}$ M SO_4^{2-}. Since Na_2SO_4 *(s)* → 2 Na^+ *(aq)* + SO_4^{2-} *(aq)*. Since 1 SO_4^{2-} ion is generated for each Na_2SO_4, $[Na_2SO_4] = 1.1 \times 10^{-8}$ M Na_2SO_4 to precipitate $BaSO_4$. Derive expression for K_{sp} ($CaSO_4$), A = Ca^{2+}, m = 1, X = SO_4^{2-}, and n = 1 Since $K_{sp} = [Ca^{2+}] [SO_4^{2-}]$, then K_{sp} ($CaSO_4$) = 7.10×10^{-5} = 0.020 $[SO_4^{2-}]$. Solve for $[SO_4^{2-}]$. $[SO_4^{2-}]$ = 0.0036 M SO_4^{2-} = 0.0036 M Na_2SO_4 = $[Na_2SO_4]$ to precipitate $CaSO_4$. Since 1.1×10^{-8} M Na_2SO_4 << 0.0036 M Na_2SO_4, the Ba^{2+} will precipitate first.

b) Since Ca^{2+} will not precipitate until $[Na_2SO_4]$ = 0.00355 M Na_2SO_4, substitute this value into the K_{sp} expression for $BaSO_4$. So K_{sp} ($BaSO_4$) = $[Ba^{2+}] [SO_4^{2-}] = 1.07 \times 10^{-10}$ = $[Ba^{2+}]$ 0.0035. Solve for $[Ba^{2+}] = 3.0 \times 10^{-8}$ M Ba^{2+}.

Check: The units (none, M, and M) are correct. Comparing the two K_{sp} values, it can be seen that the Ba^{2+} will precipitate first since the solubility product is so much lower. Since the K_{sp} value is so low, the concentration of precipitating agent is very low. Since the $CaSO_4$ K_{sp} value is so much higher, the higher $[SO_4^{2-}]$ to precipitate Ca will force the concentration of Ba^{2+} to very low levels.

109. **Given:** solution with 1.1×10^{-3} M $Zn(NO_3)_2$ and 0.150 M NH_3 **Find:** $[Zn^{2+}]$ at equilibrium
Other: K_f $(Zn(NH_3)_4^{2+}) = 2.0 \times 10^9$
Conceptual Plan: Write balanced equation and expression for K_f. Use initial concentrations to set up ICE table. Since the K_f, assume that reaction essentially goes to completion. Solve for $[Zn^{2+}]$ at equilibrium.
Solution: $Zn(NO_3)_2$ *(s)* → Zn^{2+} *(aq)* + 2 NO_3^- *(aq)*. Since 1 Zn^{2+} ion is generated for each $Zn(NO_3)_2$, $[Zn^{2+}] = 1.1 \times 10^{-3}$ M. Balanced equation is:

$Zn^{2+}(aq) + 4\,NH_3(aq) \rightleftharpoons Zn(NH_3)_4^{2+}(aq)$ Set up ICE table with initial concentrations.

	$[Zn^{2+}]$	$[NH_3]$	$[Zn(NH_3)_4^{2+}]$	
Initial	1.1×10^{-3}	0.150	0.00	Since K_f is so large and since the initially $[NH_3]>4[Zn^{2+}]$
Change	$\approx 1.1 \times 10^{-3}$	$\approx(0.150 - 4(1.1 \times 10^{-3}))$	$\approx 1.1 \times 10^{-3}$	the reaction essentially goes to completion.
Equil	x	$0.14\underline{5}6$	1.1×10^{-3}	then write equilibrium expression and solve for x.

$K_f = \dfrac{[Zn(NH_3)_4^{2+}]}{[Zn^{2+}][NH_3]^4} = 2.0 \times 10^9 = \dfrac{1.1 \times 10^{-3}}{x(0.14\underline{5}6)^4}$ So $x = 1.2 \times 10^{-9}$ M Zn^{2+}. Since x is insignificant compared to the initial concentration, the assumption is valid.
Check: The units (M) are correct. Since K_f is so large, the reaction essentially goes to completion and $[Zn^{2+}]$ is extremely small.

111. **Given:** FeS *(s)* + 6 CN^- *(aq)* $\rightleftharpoons$ $Fe(CN)_6^{4-}$ *(aq)* + S^{2-} *(aq)* use K_{sp} and K_f values **Find:** K
Other: K_f $(Fe(CN)_6^{4-}) = 1.5 \times 10^{35}$, K_{sp} (FeS) $= 3.72 \times 10^{-19}$
Conceptual Plan: Identify the appropriate solid and complex ion. Write balanced equations for dissolving the solid and forming the complex ion. Add these two reactions to get the desired overall reaction. Using the rules from Chapter 14, multiply the individual reaction K's to get the overall K for the sum of these reactions.
Solution: Identify the solid as FeS and the complex ion as $Fe(CN)_6^{4-}$. Write the individual reactions and add them together.

$FeS(s) \rightleftharpoons \cancel{Fe^{2+}(aq)} + S^{2-}(aq)$ $K_{sp} = 1.5 \times 10^{-19}$

$\cancel{Fe^{2+}(aq)} + 6\,CN^-(aq) \rightleftharpoons Fe(CN)_6^{4-}(aq)$ $K_f = 3.72 \times 10^{35}$ Since the overall reaction is the simple

$FeS(s) + 6\,CN^-(aq) \rightleftharpoons Fe(CN)_6^{4-}(aq) + S^{2-}(aq)$

sum of the two reactions, the overall reaction $K = K_f \times K_{sp} = (1.5 \times 10^{35}) \times (3.72 \times 10^{-19}) = 5.6 \times 10^{16}$.
Check: The units (none) are correct. Since K_f is so large, it overwhelms the K_{sp} and the overall reaction is very spontaneous.

113. **Given:** 150.0 mL solution of 2.05 g sodium benzoate and 2.47 g benzoic acid **Find:** pH
Other: K_a $(HC_7H_5O_2) = 6.5 \times 10^{-5}$
Conceptual Plan: g $NaC_7H_5O_2$ → mol $NaC_7H_5O_2$ and g $HC_7H_5O_2$ → mol $HC_7H_5O_2$

$\dfrac{1 \text{ mol } NaC_7H_5O_2}{144.11 \text{ g } NaC_7H_5O_2}$ $\dfrac{1 \text{ mol } HC_7H_5O_2}{122.13 \text{ g } HC_7H_5O_2}$

Since the two components are in the same solution, the ratio of [base]/[acid] = (mol base)/(mol acid). then K_a, mol $NaC_7H_5O_2$, mol $HC_7H_5O_2$ → pH

$$pH = pK_a + \log \frac{[base]}{[acid]}$$

Solution: $2.05 \cancel{\text{ g } NaC_7H_5O_2} \times \dfrac{1 \text{ mol } NaC_7H_5O_2}{144.11 \cancel{\text{ g } NaC_7H_5O_2}} = 0.014\underline{2}252 \text{ mol } NaC_7H_5O_2$ and

$2.47 \cancel{\text{ g } HC_7H_5O_2} \times \dfrac{1 \text{ mol } HC_7H_5O_2}{122.13 \cancel{\text{ g } HC_7H_5O_2}} = 0.020\underline{2}244 \text{ mol } HC_7H_5O_2$ then

$$pH = pK_a + \log\frac{[base]}{[acid]} = pK_a + \log\frac{\text{mol base}}{\text{mol acid}} = -\log(6.5 \times 10^{-5}) + \log\frac{0.014\underline{2}252 \cancel{\text{ mol}}}{0.020\underline{2}244 \cancel{\text{ mol}}} = 4.03.$$

Check: The units (none) are correct. The magnitude of the answer makes physical sense because the pH is a little lower than the pK_a of the acid because there is more acid than base in the buffer solution.

115. **Given:** 150.0 mL of 0.25 M $HCHO_2$ and 75.0 ml of 0.20 M NaOH **Find:** pH
Other: K_a ($HCHO_2$) = 1.8×10^{-4}
Conceptual Plan: In this buffer, the base is generated by converting some of the formic acid to the formate ion. Part I: Stoichiometry:

mL → L then L, initial $HCHO_2$ M → mol $HCHO_2$ then mL → L then

$\frac{1\text{ L}}{1000\text{ mL}}$ $\quad M = \frac{\text{mol}}{\text{L}}$ $\quad \frac{1\text{ L}}{1000\text{ mL}}$

L, initial NaOH M → mol NaOH then write balanced equation then

$M = \frac{\text{mol}}{\text{L}}$ $\quad NaOH + HCHO_2 \rightarrow H_2O + NaCHO_2$

mol $HCHO_2$, mol NaOH → mol $NaCHO_2$, mol $HCHO_2$ then
set up stoichiometry table

Part II: Equilibrium:
Since the two components are in the same solution, the ratio of [base]/[acid] = (mol base)/(mol acid). then K_a, mol $NaCHO_2$, mol $HCHO_2$ → pH

$$pH = pK_a + \log\frac{[\text{base}]}{[\text{acid}]}$$

Solution: $150.0\text{ mL} \times \frac{1\text{ L}}{1000\text{ mL}} = 0.1500\text{ L}$ then $0.1500\text{ L } HCHO_2 \times \frac{0.25\text{ mol } HCHO_2}{1\text{ L } HCHO_2} = 0.03\underline{7}5\text{ mol } HCHO_2$.

then $75.0\text{ mL} \times \frac{1\text{ L}}{1000\text{ mL}} = 0.0750\text{ L}$ then $0.0750\text{ L NaOH} \times \frac{0.20\text{ mol NaOH}}{1\text{ L NaOH}} = 0.015\text{ mol NaOH}$ then

set up table to track changes:

	$NaOH\ (aq)$ +	$HCHO_2\ (aq)$ →	$NaCHO_2\ (aq)$ +	$H_2O\ (l)$
Before addition	0.00 mol	$0.03\underline{7}5$ mol	0.00 mol	–
Addition	0.015 mol	–	–	–
After addition	≈ 0.00 mol	$0.02\underline{2}5$ mol	0.015 mol	–

Since the amount of NaOH is small, there are significant amounts of both buffer components, so the Henderson-Hasselbalch Equation can be used to calculate the pH..

$$pH = pK_a + \log\frac{[\text{base}]}{[\text{acid}]} = pK_a + \log\frac{\text{mol base}}{\text{mol acid}} = -\log(1.8 \times 10^{-4}) + \log\frac{0.015\text{ mol}}{0.02\underline{2}5\text{ mol}} = 3.57\,.$$

Check: The units (none) are correct. The magnitude of the answer makes physical sense because the pH is a little lower than the pK_a of the acid because there is more acid than base in the buffer solution.

117. **Given:** 1.0 L of buffer of 0.25 mol NH_3 and 0.25 mol NH_4Cl; adjust to pH = 8.75
Find: mass NaOH or HCl **Other:** K_b (NH_3) = 1.79×10^{-5}
Conceptual Plan: To decide which reagent needs to be added to adjust pH, calculate the initial pH. Since the mol NH_3 = mol NH_4Cl, the pH = pK_a so K_b → pK_b → pK_a then

acid = NH_4^+ base = NH_3 $\quad pK_b = -\log K_b \quad 14 = pK_a + pK_b$

final pH, pK_a → $[NH_3]/[NH_4^+]$ then $[NH_3]$, L → mol NH_3 and $[NH_4^+]$, L → mol NH_4^+

$pH = pK_a + \log\frac{[\text{base}]}{[\text{acid}]}$ $\quad M = \frac{\text{mol}}{\text{L}}$ $\quad M = \frac{\text{mol}}{\text{L}}$

then write balanced equation then
$H^+ + NH_3 \rightarrow NH_4^+$

mol NH_3, mol NH_4^+, $[NH_3]/[NH_4^+]$ → mol HCl → g HCl

set up stoichiometry table $\quad \frac{36.46\text{ g HCl}}{1\text{ mol HCl}}$

Solution: Since K_b (NH_3) = 1.79×10^{-5}, $pK_b = -\log K_b = -\log(1.79 \times 10^{-5}) = 4.75$. Since $14 = pK_a + pK_b$, $pK_a = 14 - pK_b = 14 - 4.75 = 9.25$. Since the desired pH is lower (8.75) HCl (a strong acid) needs to be added. then $pH = pK_a + \log\frac{[\text{base}]}{[\text{acid}]} = 9.25 + \log\frac{[NH_3]}{[NH_4^+]} = 8.75$. Solve for $\frac{[NH_3]}{[NH_4^+]}$.

$\log\frac{[NH_3]}{[NH_4^+]} = 8.75 - 9.25 = -0.50 \rightarrow \frac{[NH_3]}{[NH_4^+]} = 10^{-0.50} = 0.3\underline{1}623$. then $\frac{0.25\text{ mol } NH_3}{1\text{ L}} \times 1.0\text{ L} = 0.25\text{ mol } NH_3$ and

$\frac{0.25 \text{ mol } NH_4Cl}{1 \text{ L}} \times 1.0 \text{ L} = 0.25 \text{ mol } NH_4Cl = 0.25 \text{ mol } NH_4^+$ Since HCl is a strong acid, [HCl] = [H^+], and set up table to track changes:

	H^+ (*aq*)	+ NH_3 (*aq*)	→ NH_4^+ (*aq*)
Before addition	≈ 0.00 mol	0.25 mol	0.25 mol
Addition	*x*	–	–
After addition	≈ 0.00 mol	(0.25 – x) mol	(0.25 + x) mol

Since $\frac{[NH_3]}{[NH_4^+]} = 0.3\underline{1}623 = \frac{(0.25 - x) \text{ mol}}{(0.25 + x) \text{ mol}}$, solve for *x*. Note that the ratio of moles is the same as the ratio of concentrations, since the volume for both terms is the same. $0.3\underline{1}623(0.25 + x) = (0.25 - x)$ → $0.07\underline{9}0575 + 0.3\underline{1}62x = 0.25 - x$ → $1.3\underline{1}623x = 0.1\underline{7}094$ → $x = 0.1\underline{2}987$ mol HCl then

$0.1\underline{2}987 \text{ mol HCl} \times \frac{36.46 \text{ g HCl}}{1 \text{ mol HCl}} = 4.7 \text{ g HCl}$.

Check: The units (g) are correct. The magnitude of the answer makes physical sense because there is much less than a mole of each of the buffer components, so there must be much less than a mole of HCl.

119. a) **Given:** potassium hydrogen phthalate = KHP = $KHC_8H_4O_4$ titration with NaOH **Find:** balanced equation

Conceptual Plan: The reaction will be a titration of the acid proton, leaving the phthalate ion in tact. The K will not be titrated since it is basic.

Solution: NaOH (*aq*) + $KHC_8H_4O_4$ (*aq*) → Na^+ (*aq*) + K^+ (*aq*) + $C_8H_4O_4^{2-}$ (*aq*) + H_2O (*l*)

Check: An acid - base reaction generates a salt (soluble here) and water. There is only on e acidic proton in KHP.

b) **Given:** 0.5527 g KHP titrated with 25.87 mL of NaOH solution **Find:** [NaOH]

Conceptual Plan: g KHP→ mol KHP→ mol NaOH and mL → L then mol NaOH and

$\frac{1 \text{ mol KHP}}{204.22 \text{ g KHP}}$ 1:1 from balance equation $\frac{1 \text{ L}}{1000 \text{ mL}}$

mL →M NaOH

$M = \frac{\text{mol}}{\text{L}}$

Solution: $0.5527 \text{ g KHP} \times \frac{1 \text{ mol KHP}}{204.22 \text{ g KHP}} = 0.00270\underline{6}395 \text{ mol KHP}$ mol KHP = mol acid = mol base = $0.00270\underline{6}395$ mol NaOH then $25.87 \text{ mL} \times \frac{1 \text{ L}}{1000 \text{ mL}} = 0.02587 \text{ L}$ then

$[NaOH] = \frac{0.00270\underline{6}395 \text{ mol NaOH}}{0.02587 \text{ L}} = 0.1046 \text{ M NaOH}$.

Check: The units (M) are correct. The magnitude of the answer makes physical sense because there is much less than a mole of acid. The magnitude of the moles of acid and base are smaller than the volume of base in liters.

121. **Given:** 0.25 mol weak acid with 10.0 mL of 3.00 M KOH diluted to 1.5000 L has pH = 3.85
Find: pK_a of acid

Conceptual Plan: mL → L then M KOH, L → mol KOH then write balanced reaction

$\frac{1 \text{ L}}{1000 \text{ mL}}$ $M = \frac{\text{mol}}{\text{L}}$ KOH + HA → NaA + H_2O

added mol KOH, initial mol acid → equil. mol KOH, equil. mol acid then

set up stoichiometry table

equil. mol KOH, equil. mol acid, pH → pK_a

$$pH = pK_a + \log\frac{[\text{base}]}{[\text{acid}]}$$

Solution: $10.00 \text{ mL} \times \frac{1 \text{ L}}{1000 \text{ mL}} = 0.01000 \text{ L}$ then $0.01000 \text{ L KOH} \times \frac{3.00 \text{ mol KOH}}{1 \text{ L KOH}} = 0.0300 \text{ mol KOH}$ then

Since KOH is a strong base, [KOH] = [OH^-], and set up table to track changes:

	$KOH\ (aq)$ +	$HA\ (aq)$ →	$KA\ (aq) + H_2O\ (l)$	
Before addition	≈ 0.00 mol	0.25 mol	0.00 mol	–
Addition	0.0300 mol	–	–	–
After addition	≈ 0.00 mol	0.23 mol	0.0300 mol	–

Since the ratio of base to acid is between 0.1 and 10, so it is a buffer solution. Note that the ratio of moles is the same as the ratio of concentrations, since the volume for both terms is the same. $pH = pK_a + \log\frac{[base]}{[acid]} = pK_a + \log\frac{0.0300\ mol}{0.23\ mol} = 3.85$.

Solve for pK_a. $pK_a = 3.85 - \log\frac{0.0300\ mol}{0.23\ mol} = 4.73$.

Check: The units (none) are correct. The magnitude of the answer makes physical sense because there is more acid than base at equilibrium, so the pK_a is higher than the pH of the solution.

123. **Given:** 0.552 g ascorbic acid dissolved in 20.00 mL and titrated with 28.42 mL of 0.1103 M KOH solution; pH = 3.72 when 10.0 mL KOH added **Find:** molar mass and K_a of acid

Conceptual Plan: mL → L then M KOH, L → mol KOH → mol acid then mol acid, g acid → $\mathfrak{M}$

$$\frac{1\ L}{1000\ mL} \qquad M = \frac{mol}{L} \qquad \text{1:1 for monoprotic acid} \qquad \mathfrak{M} = \frac{g\ acid}{mol\ acid}$$

For second part of problem: mL → L then M KOH, L → mol KOH then write balanced reaction

$$\frac{1\ L}{1000\ mL} \qquad M = \frac{mol}{L} \qquad KOH + HA \rightarrow NaA + H_2O$$

added mol KOH, initial mol acid → equil. mol KOH, equil. mol acid then

set up stoichiometry table

equil. mol KOH, equil. mol acid, pH → pK_a → K_a

$$pH = pK_a + \log\frac{[base]}{[acid]} \qquad pK_a = -\log K_a$$

Solution: $28.42\ \cancel{mL} \times \frac{1\ L}{1000\ \cancel{mL}} = 0.02842\ L$ then

$0.02842\ \cancel{L\ KOH} \times \frac{0.1103\ mol\ KOH}{1\ \cancel{L\ KOH}} = 0.003134726\ mol\ KOH$ = mol base = mol acid =

0.003134726 mol ascorbic acid then $\mathfrak{M} = \frac{g\ acid}{mol\ acid} = \frac{0.552\ g\ acid}{0.003134726\ mol\ acid} = 176\ g/mol$.

For second part of the problem, $10.00\ \cancel{mL} \times \frac{1\ L}{1000\ \cancel{mL}} = 0.01000\ L$ then

$0.01000\ \cancel{L\ KOH} \times \frac{0.1103\ mol\ KOH}{1\ \cancel{L\ KOH}} = 0.001103\ mol\ KOH$ then

Since KOH is a strong base, $[KOH] = [OH^-]$, and set up table to track changes:

	$KOH\ (aq)$ +	$HA\ (aq)$ →	$KA\ (aq) + H_2O\ (l)$	
Before addition	≈ 0.00 mol	0.003134726 mol	0.00 mol	–
Addition	0.001103 mol	–	–	–
After addition	≈ 0.00 mol	0.002031726 mol	0.001103 mol	–

Since the ratio of base to acid is between 0.1 and 10, so it is a buffer solution. Note that the ratio of moles is the same as the ratio of concentrations, since the volume for both terms is the same. $pH = pK_a + \log\frac{[base]}{[acid]} = pK_a + \log\frac{0.001103\ \cancel{mol}}{0.002031726\ \cancel{mol}} = 3.72$. Solve for pK_a.

$pK_a = 3.72 - \log\frac{0.001103}{0.002031726} = 3.98529$ and so $pK_a = -\log K_a$ or $K_a = 10^{-pK_a} = 10^{-3.98529} = 1.0 \times 10^{-4}$.

Check: The units (g/mol and none) are correct. The magnitude of the answer makes physical sense because there is much less than a mole of acid and about a half a gram of acid, so the molar mass will be high. The number is reasonable for an acid (must be > 20 g/mol – lightest acid is HF). The K_a is reasonable because the pK_a is within 1 unit of the pH when the titration solution is behaving as a buffer.

125. **Given:** saturated $CaCO_3$ solution; precipitate 1.00×10^2 mg $CaCO_3$ **Find:** volume of solution evaporated
Other: $K_{sp}(CaCO_3) = 4.96 \times 10^{-9}$
Conceptual Plan: mg $CaCO_3$ → g $CaCO_3$ → mol $CaCO_3$ then

$$\frac{1\ \text{g CaCO}_3}{1000\ \text{mg CaCO}_3} \qquad \frac{1\ \text{mol CaCO}_3}{100.09\ \text{g CaCO}_3}$$

Use equation derived in problem #20 and solve for S. then mol $CaCO_3$, S → L

for ionic compound, A_mX_n, $K_{sp} = m^m n^n S^{m+n}$. $\qquad M = \frac{\text{mol}}{\text{L}}$

Solution: $1.00 \times 10^2\ \text{mg CaCO}_3 \times \frac{1\ \text{g CaCO}_3}{1000\ \text{mg CaCO}_3} \times \frac{1\ \text{mol CaCO}_3}{100.09\ \text{g CaCO}_3} = 9.99101 \times 10^{-4}\ \text{mol CaCO}_3$ then

$K_{sp} = 4.96 \times 10^{-9}$, $A = Ca^{2+}$, m = 1, $X = CO_3^{2-}$, and n = 1 so $K_{sp} = 4.96 \times 10^{-9} = S^2$. Rearrange to solve for S.

$S = \sqrt{4.96 \times 10^{-9}} = 7.04273 \times 10^{-5}$ M. Finally, $9.99101 \times 10^{-4}\ \text{mol CaCO}_3 \times \frac{1\ \text{L}}{7.04273 \times 10^{-5}\ \text{mol CaCO}_3} = 14.2$ L.

Check: The units (L) are correct. The volume should be large since the solubility is low.

127. **Given:** $[Ca^{2+}]$ = 9.2 mg/dL and K_{sp} $(Ca_2P_2O_7) = 8.64 \times 10^{-13}$ **Find:** $[P_2O_7^{4-}]$ to form precipitate
Conceptual Plan: mg Ca^{2+} /dL → g Ca^{2+} /dL → mol Ca^{2+}/dL → mol Ca^{2+}/L then

$$\frac{1\ \text{g Ca}^{2+}}{1000\ \text{mg Ca}^{2+}} \qquad \frac{1\ \text{mol Ca}^{2+}}{40.08\ \text{g Ca}^{2+}} \qquad \frac{10\ \text{dL}}{1\ \text{L}}$$

Write balanced equation and expression for K_{sp}. then $[Ca^{2+}]$, K_{sp} → $[P_2O_7^{4-}]$

Solution: $9.2 \frac{\text{mg Ca}^{2+}}{\text{dL}} \times \frac{1\ \text{g Ca}^{2+}}{1000\ \text{mg Ca}^{2+}} \times \frac{1\ \text{mol Ca}^{2+}}{40.08\ \text{g Ca}^{2+}} \times \frac{10\ \text{dL}}{1\ \text{L}} = 2.29541 \times 10^{-3}\ \text{M Ca}^{2+}$ then write equation

$Ca_2P_2O_7$ *(s)* → 2 Ca^{2+} *(aq)* + $P_2O_7^{4-}$ *(aq)*. So $K_{sp} = [Ca^{2+}]^2 [P_2O_7^{4-}] = 8.64 \times 10^{-13} = (2.29541 \times 10^{-3})^2$ $[P_2O_7^{4-}]$. Solve for $[P_2O_7^{4-}]$. then $[P_2O_7^{4-}] = 1.6 \times 10^{-7}$ M.
Check: The units (M) are correct. Since K_{sp} is so small and the sodium concentration is fairly high, the urate concentration is driven to a very low level.

129. **Given:** CuS in 0.150 M NaCN **Find:** molar solubility (S)
Other: K_f $(Cu(CN)_4^{2-}) = 1.0 \times 10^{25}$, K_{sp} (CuS) $= 1.27 \times 10^{-36}$
Conceptual Plan: Identify the appropriate solid and complex ion. Write balanced equations for dissolving the solid and forming the complex ion. Add these two reactions to get the desired overall reaction. Using the rules from Chapter 14, multiply the individual reaction K's to get the overall K for the sum of these reactions. Then M NaCN, K → S

ICE Chart

Solution: Identify the solid as CuS and the complex ion as $Cu(CN)_4^{2-}$. Write the individual reactions and add them together.

$CuS\ (s) \rightleftharpoons Cu^{2+}(aq) + S^{2-}(aq) \qquad K_{sp} = 1.27 \times 10^{-36}$

$Cu^{2+}(aq) + 4\ CN^-(aq) \rightleftharpoons Cu(CN)_4^{2-}(aq) \qquad K_f = 1.0 \times 10^{25}$ Since the overall reaction is the simple sum

$CuS\ (s) + 4\ CN^-(aq) \rightleftharpoons Cu(CN)_4^{2-}(aq) + S^{2-}(aq)$

of the two reactions, the overall reaction $K = K_f \times K_{sp} = (1.0 \times 10^{25}) \times (1.27 \times 10^{-36}) = 1.27 \times 10^{-11}$. then.
NaCN *(s)* → Na^+ *(aq)* + CN^- *(aq)*. Since 1 CN^- ion is generated for each NaCN, $[CN^-]$ = 0.150 M.
Set up ICE table.

$CuS\ (s) + 4\ CN^-(aq) \rightleftharpoons Cu(CN)_4^{2-}(aq) + S^{2-}(aq)$

	$[CN^-]$	$[Cu(CN)_4^{2-}]$	$[S^{2-}]$
Initial	0.150	0.00	0.00
Change	$-4S$	$+S$	$+S$
Equil	$0.150 - 4S$	$+S$	$+S$

$$K = \frac{[Cu(CN)_4^{2-}][S^{2-}]}{[CN^-]^4} = 1.27 \times 10^{-11} = \frac{S^2}{(0.150 - 4S)^4}.$$

Assume S is small (4S << 0.150) so $\frac{S^2}{(0.150 - 4S)^4} = 1.27 \times 10^{-11} = \frac{S^2}{(0.150)^4}$ and $S = 8.0183 \times 10^{-8}$ =

$= 8.0 \times 10^{-8}$ M. Confirm that assumption is valid $\frac{4(8.0183 \times 10^{-8})}{0.150} \times 100\ \% = 0.00021\ \% << 5\ \%$ so assumption is valid.

Check: The units (M) are correct. Since K_f is large, the overall K is larger than the original K_{sp} and the solubility of CuS increases over that of pure water $\left(\sqrt{1.27 \times 10^{-36}} = 1.13 \times 10^{-18}\ M\right)$.

131. **Given:** 100 mL of 0.36 M NH_2OH and 50 mL of 0.26 M HCl and K_b (NH_2OH) = 1.10×10^{-8}
Find: pH
Conceptual Plan: identify acid and base components mL → L then [NH_2OH], L → mol NH_2OH then

acid = NH_3OH^+ base = NH_2OH $\qquad \frac{1\ L}{1000\ mL} \qquad M = \frac{mol}{L}$

mL → L then [HCl], L → mol HCl then write balanced equation then

$\frac{1\ L}{1000\ mL} \qquad M = \frac{mol}{L} \qquad HCl + NH_2OH \rightarrow NH_3OHCl$

mol NH_2OH, mol HCl → mol excess NH_2OH, mol NH_3OH^+

set up stoichiometry table

Since there are significant amounts of both the acid and the conjugate base species, this is a buffer solution and so Henderson-Hasselbalch Equation $\left(pH = pK_a + \log \frac{[base]}{[acid]}\right)$ **can be used. Convert K_b to K_a using $K_w = K_a K_b$. Also note that ratio of concentrations is the same as the ratio of moles, since the volume is the same for both species.**

Solution: $100\ \text{mL}\ NH_2OH \times \frac{1\ L}{1000\ \text{mL}} = 0.1\ L\ NH_2OH$ then $\frac{0.36\ mol\ NH_2OH}{1\ L} \times 0.1\ L = 0.0\underline{3}6\ mol\ NH_2OH$.

$50\ \text{mL HCl} \times \frac{1\ L}{1000\ \text{mL}} = 0.05\ L\ HCl$ then $\frac{0.26\ mol\ HCl}{1\ L} \times 0.05\ L = 0.0\underline{1}3\ mol\ HCl$.

Set up table to track changes:

	HCl (*aq*) +	NH_2OH (*aq*) →	NH_3OHCl (*aq*)
Before addition	0.00 mol	0.0$\underline{3}$6 mol	0.00 mol
Addition	0.0$\underline{1}$3 mol	–	–
After addition	≈ 0.00 mol	0.0$\underline{2}$3 mol	0.0$\underline{1}$3 mol

then $K_w = K_a K_b$ so

$K_a = \frac{K_w}{K_b} = \frac{1.0 \times 10^{-14}}{1.10 \times 10^{-8}} = 9.\underline{0}909 \times 10^{-7}\ M$ then use Henderson-Hasselbalch Equation, since the solution is a buffer. Note that the ratio of moles is the same as the ratio of concentrations, since the volume for both terms is the same. $pH = pK_a + \log \frac{[base]}{[acid]} = -\log(9.\underline{0}909 \times 10^{-7}) + \log \frac{0.0\underline{2}3\ mol}{0.0\underline{1}3\ mol} = 6.\underline{2}8918 = 6.3$.

Check: The units (none) are correct. The magnitude of the answer makes physical sense because pH should be more than the pK_a of the acid because there is more base than acid.

133. **Given:** 25.0 mL of NaOH titrated with 19.6 mL of 0.189 M HCl solution; 10.0 mL of H_3PO_4 titrated with 34.9 mL NaOH **Find:** concentration of H_3PO_4 solution
Conceptual Plan: Write 1st balanced reaction then mL → L then M HCl, L → mol HCl → mol NaOH

$HCl + NaOH \rightarrow NaCl + H_2O \qquad \frac{1\ L}{1000\ mL} \qquad M = \frac{mol}{L} \qquad 1:1$

then mL → L then mol NaOH, L → M NaOH then Write 2nd balanced reaction then mL → L

$\frac{1\ L}{1000\ mL} \qquad M = \frac{mol}{L} \qquad H_3PO_4 + 3\ NaOH \rightarrow Na_3PO_4 + 3\ H_2O \qquad \frac{1\ L}{1000\ mL}$

then M NaOH, L → mol NaOH → mol H_3PO_4 then mL → L then mol H_3PO_4, L → M H_3PO_4

$M = \frac{mol}{L} \qquad 3:1 \qquad \frac{1\ L}{1000\ mL} \qquad M = \frac{mol}{L}$

Solution: In the first titration, $19.6\ \text{mL} \times \frac{1\ L}{1000\ \text{mL}} = 0.0196\ L$ then

$0.0196\ L\ HCl \times \frac{0.189\ mol\ HCl}{1\ L\ HCl} = 0.0037\underline{0}44\ mol\ HCl \times \frac{1\ mol\ NaOH}{1\ mol\ HCl} = 0.0037\underline{0}44\ mol\ NaOH$ then

$25.0\ \text{mL} \times \frac{1\ L}{1000\ \text{mL}} = 0.0250\ L$ then $\frac{0.0037\underline{0}44\ mol\ NaOH}{0.0250\ L\ NaOH} = 0.148\underline{1}76\ M\ NaOH$. In the second titration,

$34.9\ \text{mL} \times \frac{1\ \text{L}}{1000\ \text{mL}} = 0.0349\ \text{L}$ then

$0.0349\ \text{L NaOH} \times \frac{0.148176\ \text{mol NaOH}}{1\ \text{L NaOH}} = 0.00517134\ \text{mol NaOH} \times \frac{1\ \text{mol}\ H_3PO_4}{3\ \text{mol NaOH}} = 0.00172378\ \text{mol}\ H_3PO_4$ then

$10.0\ \text{mL} \times \frac{1\ \text{L}}{1000\ \text{mL}} = 0.0100\ \text{L}$ then $\frac{0.00172378\ \text{mol}\ H_3PO_4}{0.0100\ \text{L}\ H_3PO_4} = 0.172\ \text{M}\ H_3PO_4$.

Check: The units (M) are correct. The magnitude of the answer makes physical sense because the concentration of NaOH is a little lower than the HCl (because the volume of NaOH is greater than HCl) and the concentration of H_3PO_4 is more than the NaOH (because the ratio of the volume of NaOH to volume of H_3PO_4 is just over 3 and H_3PO_4 is a triproitic acid).

135. **Given:** 10.0 L of 75 ppm $CaCO_3$ and 55 ppm $MgCO_3$ (by mass) **Find:** mass Na_2CO_3 to precipitate 90.0 % of ions **Other:** $K_{sp}\ (CaCO_3) = 4.96 \times 10^{-9}$ and $K_{sp}\ (MgCO_3) = 6.82 \times 10^{-6}$

Conceptual Plan: Assume that the density of water is 1.00 g/mL. L water → mL water → g water then

$\frac{1000\ \text{mL}}{1\ \text{L}}$ $\frac{1.00\ \text{g water}}{1\ \text{mL}}$

g water → g $CaCO_3$ → mol $CaCO_3$ → mol Ca^{2+} and g water → g $MgCO_3$ → mol $MgCO_3$ → mol Mg^{2+}

$\frac{75\ \text{g}\ CaCO_3}{10^6\ \text{g water}}$ $\frac{1\ \text{mol}\ CaCO_3}{100.09\ \text{g}\ CaCO_3}$ $\frac{1\ \text{mol}\ Ca^{2+}}{1\ \text{mol}\ CaCO_3}$ $\frac{55\ \text{g}\ MgCO_3}{10^6\ \text{g water}}$ $\frac{1\ \text{mol}\ MgCO_3}{84.32\ \text{g}\ MgCO_3}$ $\frac{1\ \text{mol}\ Mg^{2+}}{1\ \text{mol}\ MgCO_3}$

then Comparing the two K_{sp} values, essentially all of the of the Ca^{2+} will precipitate before the Mg^{2+} will begin to precipitate. Since 90.0 % of the ions are to be precipitates, there will be 10.0 % of the ions left in solution (all will be Mg^{2+}).

$(0.100)(\text{mol}\ Ca^{2+} + \text{mol}\ Mg^{2+})$

Calculate the moles of ions remaining in solution. then mol Mg^{2+}, L → M Mg^{2+} then

$M = \frac{\text{mol}}{\text{L}}$

Use equation derived in problem #19 and use to M Mg^{2+}, K_{sp} → M CO_3^{2-} then M CO_3^{2-}, L → mol CO_3^{2-}

for ionic compound, A_mX_n, $K_{sp} = [A^{n+}]^m [X^{m-}]^n$. $M = \frac{\text{mol}}{\text{L}}$

then mol CO_3^{2-} → mol Na_2CO_3 → g Na_2CO_3

$\frac{1\ \text{mol}\ CO_3^{2-}}{1\ \text{mol}\ Na_2CO_3}$ $\frac{105.99\ \text{g}\ Na_2CO_3}{1\ \text{mol}\ Na_2CO_3}$

Solution: $10.0\ \text{L} \times \frac{1000\ \text{mL}}{1\ \text{L}} \times \frac{1.00\ \text{g water}}{1\ \text{mL}} = 1.00 \times 10^4\ \text{g water}$ then

$1.00 \times 10^4\ \text{g water} \times \frac{75\ \text{g}\ CaCO_3}{10^6\ \text{g water}} \times \frac{1\ \text{mol}\ CaCO_3}{100.09\ \text{g}\ CaCO_3} \times \frac{1\ \text{mol}\ Ca^{2+}}{1\ \text{mol}\ CaCO_3} = 0.0074933\ \text{mol}\ Ca^{2+}$ and

$1.00 \times 10^4\ \text{g water} \times \frac{55\ \text{g}\ MgCO_3}{10^6\ \text{g water}} \times \frac{1\ \text{mol}\ MgCO_3}{84.32\ \text{g}\ MgCO_3} \times \frac{1\ \text{mol}\ Mg^{2+}}{1\ \text{mol}\ MgCO_3} = 0.0065228\ \text{mol}\ Mg^{2+}$ so the ions remaining in solution after 90.0 % precipitate out =

$(0.100)(\text{mol}\ Ca^{2+} + \text{mol}\ Mg^{2+}) = (0.100)(0.0074933\ \text{mol}\ Ca^{2+} + 0.0065228\ \text{mol}\ Mg^{2+}) = 0.00140607\ \text{mol ions}$.

so $\frac{0.00140607\ \text{mol}\ Ca^{2+}}{10.0\ \text{L}} = 0.000140607\ \text{M}\ Ca^{2+}$ then $K_{sp} = 6.82 \times 10^{-6}$, A = Mg^{2+}, m = 1, X = CO_3^{2-}, and n = 1

so $K_{sp} = 6.82 \times 10^{-6} = [Mg^{2+}][CO_3^{2-}] = (0.000140607)[CO_3^{2-}]$. Rearrange to solve for $[CO_3^{2-}]$.

So $[CO_3^{2-}] = 0.0485040$ M. then

$\frac{0.0485040\ \text{mol}\ CO_3^{2-}}{1\ \text{L}} \times 10.0\ \text{L} \times \frac{1\ \text{mol}\ Na_2CO_3}{1\ \text{mol}\ CO_3^{2-}} \times \frac{105.99\ \text{g}\ Na_2CO_3}{1\ \text{mol}\ Na_2CO_3} = 51.6\ \text{g}\ Na_2CO_3$.

Check: The units (g) are correct. The mass is reasonable to put in a washing machine load.

137. **Given:** excess $Mg(OH)_2$ in 1.00 L 0f 1.0 M NH_4Cl has pH = 9.00 **Find:** K_{sp} ($Mg(OH)_2$)
Other: K_b (NH_3) = 1.76×10^{-5}
Conceptual Plan: M NH_4Cl → M NH_4^+ and K_b → K_a then final pH → [H_3O^+] then

$NH_4Cl\ (aq) \rightarrow NH_4^+\ (aq) + Cl^-\ (aq)$ $\quad K_w = K_a\ K_b \quad [H_3O^+] = 10^{-pH}$

M NH_4^+, M H_3O^+, K_a → x Since x is significant compared to initial M NH_4^+ this is a buffer solution.
ICE Chart

The NH_4^+ is neutralized with $Mg(OH)_2$. Since $Mg(OH)_2\,(s) \rightleftharpoons Mg^{2+}\,(aq) + 2\ OH^-(aq)$ there are 2 moles of OH^- generated for each mole of $Mg(OH)_2$ dissolved. Thus ½ (x mol OH^-) = mol $Mg(OH)_2$ was dissolved in 1.00 L of solution. Since there is 1.00 L solution mol $Mg(OH)_2$ = [Mg^{2+}] and [H_3O^+] → [OH^-] then

$K_w = [H_3O^+][OH^-]$

Finally, write expression for K_{sp} ($Mg(OH)_2$) and substitute in values for [Mg^{2+}] and [OH^-].
Solution: Since 1 NH_4^+ ion is generated for each NH_4Cl, [NH_4^+] = 1.0 M NH_4^+. Since $K_w = K_a\ K_b$, rearrange to solve for K_a. $K_a = \dfrac{K_w}{K_b} = \dfrac{1.0 \times 10^{-14}}{1.76 \times 10^{-5}} = 5.\underline{6}818 \times 10^{-10}$ Final pH = 9.00, so

$[H_3O^+] = 10^{-pH} = 10^{-9.00} = 1.0 \times 10^{-9}\ M$ Set up ICE Chart.

$NH_4^+\,(aq) + H_2O(l) \rightleftharpoons H_3O^+\,(aq) + NH_3(aq)$

	[NH_4^+]	[H_3O^+]	[NH_3]
Initial	1.0	≈ 0.00	0.00
Change	$-x$	$+x$	$+x$
Equil	$1.0 - x$	1.0×10^{-9}	$+x$

$K_a = \dfrac{[H_3O^+]\,[NH_3]}{[NH_4^+]} = 5.\underline{6}818 \times 10^{-10} = \dfrac{(1.0 \times 10^{-9})x}{1.0 - x}$.

Solve for x. $5.\underline{6}818 \times 10^{-10}\ (1.0 - x) = (1.0 \times 10^{-9})x$ → $5.\underline{6}818 \times 10^{-10} = (1.0 \times 10^{-9} + 5.\underline{6}818 \times 10^{-10})x$ → $x = 0.3\underline{6}2318$, so this is a buffer solution. Since there is 1.00 L of solution, 0.3$\underline{6}$2318 mol of NH_4^+ is neutralized with $Mg(OH)_2$. Since $Mg(OH)_2\,(s) \rightleftharpoons Mg^{2+}\,(aq) + 2\ OH^-(aq)$ there are 2 moles of OH^- generated for each mole of $Mg(OH)_2$ dissolved. Thus ½ (0.3$\underline{6}$2318 mol OH^-) = 0.1$\underline{8}$1159 mol $Mg(OH)_2$ was dissolved in 1.00 L of solution. Thus the [Mg^{2+}] = 0.1$\underline{8}$1159 M. Since $K_w = [H_3O^+][OH^-]$ so

$[OH^-] = \dfrac{K_w}{[H_3O^+]} = \dfrac{1.0 \times 10^{-14}}{1.0 \times 10^{-9}} = 1.0 \times 10^{-5}\ M$ then K_{sp} ($Mg(OH)_2$) = [Mg^{2+}] [OH^-]^2 = (0.1$\underline{8}$1159) $(1.0 \times 10^{-5})^2$ = 1.8×10^{-11}.

Check: The units (none) are correct. The magnitude of the answer makes physical sense because the concentration of NH_4Cl is high and so it took a significant amount of $Mg(OH)_2$ to raise the pH to 9.00. Note that this number disagrees with the accepted value for the K_{sp} ($Mg(OH)_2$). This is most likely due to errors in the measurements in this experiment.

139. a) **Given:** $Au(OH)_3$ in pure water **Find:** molar solubility (S) **Other:** $K_{sp} = 5.5 \times 10^{-46}$
Conceptual Plan: use equations derived in problems #19 and #20 and solve for S. then
for ionic compound, A_mX_n, $K_{sp} = [A^{n+}]^m\,[X^{m-}]^n = m^m\,n^n\,S^{m+n}$.
Check answer for validity.
Solution: for $Au(OH)_3$, $K_{sp} = 5.5 \times 10^{-46}$, A = Au^{3+}, m = 1, X = OH^-, and n = 3 so K_{sp} = [Au^{3+}] [OH^-]^3 = $5.5 \times 10^{-46} = 3^3S^4$. Rearrange to solve for S. $S = \sqrt[4]{\dfrac{5.5 \times 10^{-46}}{27}} = 2.1 \times 10^{-12}\ M$. This answer suggests that the [OH^-] = $3(2.1 \times 10^{-12}\ M) = 6.3 \times 10^{-12}\ M$. This result is lower than found in pure water (1.0×10^{-7} M), so substitute this value for [OH^-], and solve for S = [Au^{3+}]. So K_{sp} = [Au^{3+}] [OH^-]^3 = 5.5×10^{-46} = S $(1.0 \times 10^{-7}\ M)^3$ and solving for S gives
S = 5.5×10^{-25} M.
Check: The units (M) are correct. Since K_{sp} is so small the autoionization of water must be considered and the solubility is smaller than normally anticipated.

b) **Given:** $Au(OH)_3$ in 1.0 M HNO_3 **Find:** molar solubility (S) **Other:** $K_{sp} = 5.5 \times 10^{-46}$
Conceptual Plan: Since HNO_3 is a strong acid, it will neutralize the gold (III) hydroxide (through the reaction of H^+ with OH^- to for water (the reverse of the autoionization of water equilibrium). Write balanced equations for dissolving the solid and for the neutralization reaction. Add these two reactions to get the desired overall reaction. Using the rules from Chapter 14, multiply the individual reaction K's to get the overall K for the sum of these reactions. then M HNO_3, K → S
ICE Chart

Solution: Identify the solid as $Au(OH)_3$. Write the individual reactions and add them together.

$Au(OH)_3\ (s) \rightleftharpoons Au^{3+}\ (aq) + \cancel{3\ OH^-\ (aq)}$ $\qquad K_{sp} = 5.5 \times 10^{-46}$

$3\ H^+\ (aq) + \cancel{3\ OH^-\ (aq)} \rightleftharpoons 3\ H_2O(l)$ $\qquad \left(\frac{1}{K_w}\right)^3 = \left(\frac{1}{1.0 \times 10^{-14}}\right)^3$ Since the overall reaction is the

$Au(OH)_3\ (s) + 3\ H^+\ (aq) \rightleftharpoons Au^{3+}\ (aq) + 3\ H_2O(l)$

sum of the dissolution reaction and three times the reverse of the autoionization of water reaction, the overall reaction $K = K_{sp}\left(\frac{1}{K_w}\right)^3 = \left(5.5 \times 10^{-46}\right)\left(\frac{1}{1.0 \times 10^{-14}}\right)^3 = 5.5 \times 10^{-4} = \frac{[Au^{3+}]}{[H^+]^3}$ then Since HNO_3 is a strong acid, it will completely dissociate to $\mathbf{H^+}$ and NO_3^- Set up ICE table.

$Au(OH)_3\ (s) + 3\ H^+\ (aq) \rightleftharpoons Au^{3+}\ (aq) + 3\ H_2O(l)$

	$[H^+]$	$[Au^{3+}]$
Initial	1.0	0.00
Change	$-3S$	$+S$
Equil	$1.0 - 3S$	$+S$

$K = \frac{[Au^{3+}]}{[H^+]^3} = 5.5 \times 10^{-4} = \frac{S}{(1.0 - 3S)^3}$. Assume S is small (3S << 1.0) so $\frac{S}{(1.0 - \cancel{3S})^3} = 5.5 \times 10^{-4}\ M = \frac{S}{(1.0)^3} = S$. Confirm that assumption is valid

$\frac{3(5.5 \times 10^{-4})}{1.0} \times 100\ \% = 0.017\ \% << 5\ \%$ so assumption is valid.

Check: The units (M) are correct. Since K is much larger than the original K_{sp} and so the solubility of $Au(OH)_3$ increases over that of pure water.

141. If the concentration of the acid is greater than the concentration of the base, then the pH will be less than the pK_a. If the concentration of the acid is equal to the concentration of the base, then the pH will be equal to the pK_a. If the concentration of the acid is less than the concentration of the base, then the pH will be greater than the pK_a.
   a) $pH < pK_a$
   b) $pH > pK_a$
   c) $pH = pK_a$, the OH^- will convert half of the acid to base
   d) $pH > pK_a$, the OH^- will convert more than half of the acid to base

143. Only (b) is correct. A diprotic acid will take twice as much base to fully titrate and the pH at the equivalence point decreases as the strength of the acid increases.

145. a) The solubility will be unchanged since the pH is constant and there are no common ions added.
   b) The solubility will be less because extra fluoride ions are added, suppressing the solubility of the fluoride ionic compound.
   c) The solubility will increase because some of the fluoride ion will be converted to HF, and so more of the ionic compound can be dissolved.

Chapter 17
Free Energy and Thermodynamics

1. The first law of thermodynamics states that energy is conserved in chemical processes. When we burn gasoline to run a car, for example, the amount of energy produced by the chemical reaction does not vanish, nor does any new energy appear that was not present before the combustion. Some of the energy from the combustion reaction goes toward driving the car forward (about 20%), and the rest is dissipated into the surroundings as heat (just feel the engine). However, the total energy given off by the combustion reaction exactly equals the sum of the amount of energy driving the car forward and the amount dissipated as heat—energy is conserved. In other words, when it comes to energy, you can't win; you cannot create energy that was not there to begin with.

3. A perpetual motion machine is a machine that perpetually moves without any energy input. This machine is not possible because if the machine is to be in motion, it must pay the heat tax with each cycle of its motion—over time, it will therefore run down and stop moving.

5. A spontaneous process is one that occurs without ongoing outside intervention (such as the performance of work by some external force). For example, when you drop a book in a gravitational field, the book spontaneously drops to the floor.

7. Entropy (S) is a thermodynamic function that is proportional to the number of energetically equivalent ways to arrange the components of a system to achieve a particular state. Entropy, like enthalpy, is a state function—its value depends only on the state of the system, not on how the system got to that state. Therefore, for any process, the change in entropy is just the entropy of the final state minus the entropy of the initial state or $\Delta S = S_{\text{final}} - S_{\text{intitial}}$.

9. Microstates are the number of internal arrangements. These microstates give rise to the same external arrangement, or macrostate. If I have 3 gas particles (A, B and C) and 2 containers. The fact that the first container has 2 particles and the other container has 1 particle is a macrostate. The fact that particles A and C are in the first container and particle B is in the other container is a microstate. There are at least as many microstates as there are macrostates.

11. The Second Law of Thermodynamics – For any spontaneous process, the entropy of the universe increases ($\Delta S_{\text{univ}} > 0$). The criterion for spontaneity is the entropy of the universe. Processes that increase the entropy of the universe—those that result in greater dispersal or randomization of energy— occur spontaneously. Processes that decrease the entropy of the universe do not occur spontaneously. Heat travels from a substance at higher temperature to one at lower temperature because the process disperses thermal energy. The cooler object has less thermal energy, so transferring heat from the warmer object results in greater energy randomization—the energy that was concentrated in the hot substance becomes dispersed between the two substances.

13. When water freezes at temperatures below 0 °C, the entropy of the water decreases, yet the process is spontaneous because the entropy of the universe increases ($\Delta S_{\text{univ}} > 0$). The entropy of the system can decrease ($\Delta S_{\text{sys}} < 0$) as long as the entropy of the surroundings increases by a greater amount ($\Delta S_{\text{surr}} > -\Delta S_{\text{sys}}$), so that the overall entropy of the universe undergoes a net increase. For liquid water freezing, the change in entropy for the system (ΔS_{sys}) is negative because the water becomes more ordered. For ΔS_{univ} to be positive, ΔS_{surr} must be positive and greater in absolute value (or magnitude) than ΔS_{sys}. We learned in chapter 6 that freezing is an exothermic process: it gives off heat to the surroundings. If we think of entropy as the dispersal or randomization of energy, then the release of heat energy by the system disperses that energy into the surroundings, increasing the entropy of the surroundings. The freezing of water below 0 °C increases the entropy of the universe because the heat given off to the surroundings increases the entropy of the surroundings to a sufficient degree to overcome the entropy decrease in the water. The freezing of water becomes nonspontaneous above 0 °C because the magnitude of the increase in the entropy of the surroundings due to the dispersal of energy into the surroundings is temperature dependent. The higher the temperature, the smaller the percent increase in entropy for a given rise in temeperature. Therefore, the impact of the heat released to the surroundings by the freezing of water depends on the temperature of the surroundings—the higher the temperature, the smaller the impact.

15. The change in Gibbs free energy for a process is proportional to the negative of ΔS_{univ}. Since ΔS_{univ} is a criterion for spontaneity, ΔG is also a criterion for spontaneity (although opposite in sign).

17. The third law of thermodynamics states that the entropy of a perfect crystal at absolute zero (0 K) is zero. For enthalpy we defined a standard state, so that we could define a "zero" for the scale. This is not necessary for entropy because there is an absolute zero.

19. The larger the molar mass, the greater its entropy at 25 °C. For a given state of matter, entropy generally increases with increasing molecular complexity.

21. The three ways of calculating the ΔG_{rxn} are to:
    - use tabulated values of standard enthalpies of formation to calculate ΔH_{rxn}° and use tabulated values of standard entropies to calculate ΔS_{rxn}°; then use the values of ΔH_{rxn}° and ΔS_{rxn}° calculated in these ways to calculate the standard free energy change for a reaction by using the equation: $\Delta G_{rxn}^{\circ} = \Delta H_{rxn}^{\circ} - T\,\Delta S_{rxn}^{\circ}$.
    - use tabulated values of the standard free energies of formation to calculate ΔG_{rxn}° using an equation similar to that used for standard enthalpy of a reaction: $\Delta G_{rxn}^{0} = \sum n_p \Delta G_f^0(products) - \sum n_r \Delta G_f^0(reactants)$.
    - use a reaction pathway or stepwise reaction to sum the changes in free energy for each of the steps in a manner similar to that used in chapter 6 for enthalpy of stepwise reactions.

    The method to calculate the free energy of a reaction at temperatures other than at 25°C is the first method. (The second method is only applicable at 25°C. The third method is only applicable at the temperature of the individual reactions, generally 25°C.)

23. The standard free energy change for a reaction (ΔG_{rxn}°) applies only to standard conditions. For a gas, standard conditions are those in which the pure gas is present at a partial pressure of 1 atmosphere. For nonstandard conditions, we need to calculate ΔG_{rxn} (not ΔG_{rxn}°) to predict spontaneity.

25. The free energy of reaction under nonstandard conditions (ΔG_{rxn}) can be calculated from ΔG_{rxn}° using the relationship $\Delta G_{rxn} = \Delta G_{rxn}^{\circ} + RT\ln Q$, where Q is the reaction quotient (defined in Section 14.7), T is the temperature in K, and R is the gas constant in the appropriate units (8.314 J/mol K).

27. a and c are spontaneous processes.

29. Yes, the particles may also be distributed so that one is in the 0 J level and the other is in the 20 J level. The total energy is the sum of the energies of the two particles = 0 J + 20 J = 20 J. There are two such arrangements of the particles in this fashion, so this state will have the greatest entropy and be more likely.

31. a) $\Delta S > 0$ because a gas is being generated

    b) $\Delta S < 0$ because 2 moles of gas are being converted to 1 mole of gas

    c) $\Delta S < 0$ because a gas is being converted to a solid

    d) $\Delta S < 0$ because 4 moles of gas are being converted to 2 moles of gas

33. a) $\Delta S_{sys} > 0$, because 6 moles of gas are being converted to 7 moles of gas. Since $\Delta H < 0$ $\Delta S_{surr} > 0$ and the reaction is spontaneous at all temperatures.

    b) $\Delta S_{sys} < 0$ because 2 moles of different gases are being converted to 2 moles of one gas. Since $\Delta H > 0$ $\Delta S_{surr} < 0$ and the reaction is nonspontaneous at all temperatures.

    c) $\Delta S_{sys} < 0$, because 3 moles of gas are being converted to 2 moles of gas. Since $\Delta H > 0$ $\Delta S_{surr} < 0$ the reaction is nonspontaneous at all temperatures.

    d) $\Delta S_{sys} > 0$, because 9 moles of gas are being converted to 10 moles of gas. Since $\Delta H < 0$ $\Delta S_{surr} > 0$ the reaction is spontaneous at all temperatures.

35. a) **Given:** $\Delta H^\circ_{rxn} = -287$ kJ, T = 298 K **Find:** ΔS_{surr}

Conceptual plan: **kJ → J then ΔH°_{rxn}, T → ΔS_{surr}**

$$\frac{1000\text{ J}}{1\text{ kJ}} \qquad \Delta S_{surr} = \frac{-\Delta H_{sys}}{T}$$

Solution: $-287\text{ kJ} \times \frac{1000\text{ J}}{1\text{ kJ}} = -287{,}000\text{ J}$ then $\Delta S_{surr} = \frac{-\Delta H_{sys}}{T} = \frac{-(-287{,}000\text{ J})}{298\text{ K}} = 963\frac{\text{J}}{\text{K}}$

Check: The units (J/K) are correct. The magnitude of the answer (10^3 J/K) makes sense because the kJ and the temperature started with very similar values and then a factor of 10^3 was applied.

b) **Given:** $\Delta H^\circ_{rxn} = -287$ kJ, T = 77 K **Find:** ΔS_{surr}

Conceptual plan: **kJ → J then ΔH°_{rxn}, T → ΔS_{surr}**

$$\frac{1000\text{ J}}{1\text{ kJ}} \qquad \Delta S_{surr} = \frac{-\Delta H_{sys}}{T}$$

Solution: $-287\text{ kJ} \times \frac{1000\text{ J}}{1\text{ kJ}} = -287{,}000\text{ J}$ then $\Delta S_{surr} = \frac{-\Delta H_{sys}}{T} = \frac{-(-287{,}000\text{ J})}{77\text{ K}} = 3730\frac{\text{J}}{\text{K}} = 3.73 \times 10^3\frac{\text{J}}{\text{K}}$

Check: The units (J/K) are correct. The magnitude of the answer (4 x 10^3 J/K) makes sense because the temperature is much lower than in part a, so the answer should increase.

c) **Given:** $\Delta H^\circ_{rxn} = +127$ kJ, T = 298 K **Find:** ΔS_{surr}

Conceptual plan: **kJ → J then ΔH°_{rxn}, T → ΔS_{surr}**

$$\frac{1000\text{ J}}{1\text{ kJ}} \qquad \Delta S_{surr} = \frac{-\Delta H_{sys}}{T}$$

Solution: $+127\text{ kJ} \times \frac{1000\text{ J}}{1\text{ kJ}} = +127{,}000\text{ J}$ then $\Delta S_{surr} = \frac{-\Delta H_{sys}}{T} = \frac{-127{,}000\text{ J}}{298\text{ K}} = -426\frac{\text{J}}{\text{K}}$

Check: The units (J/K) are correct. The magnitude of the answer (- 400 J/K) makes sense because the kJ are less and of the opposite sign than part a and so the answer should decrease.

d) **Given:** $\Delta H^\circ_{rxn} = +127$ kJ, T = 77 K **Find:** ΔS_{surr}

Conceptual plan: **kJ → J then ΔH°_{rxn}, T → ΔS_{surr}**

$$\frac{1000\text{ J}}{1\text{ kJ}} \qquad \Delta S_{surr} = \frac{-\Delta H_{sys}}{T}$$

Solution: $+127\text{ kJ} \times \frac{1000\text{ J}}{1\text{ kJ}} = +127{,}000\text{ J}$ then $\Delta S_{surr} = \frac{-\Delta H_{sys}}{T} = \frac{-127{,}000\text{ J}}{77\text{ K}} = -1650\frac{\text{J}}{\text{K}} = -1.65 \times 10^3\frac{\text{J}}{\text{K}}$

Check: The units (J/K) are correct. The magnitude of the answer (- 2 x 10^3 J/K) makes sense because the temperature is much lower than in part c, so the answer should increase.

37. a) **Given:** $\Delta H^\circ_{rxn} = -125$ kJ, $\Delta S_{rxn} = +253$ J/K, T = 298 K **Find:** ΔS_{univ} and spontaneity

Conceptual plan: **kJ → J then ΔH°_{rxn}, T → ΔS_{surr} then ΔS_{rxn}, ΔS_{surr} → ΔS_{univ}**

$$\frac{1000\text{ J}}{1\text{ kJ}} \qquad \Delta S_{surr} = \frac{-\Delta H_{sys}}{T} \qquad \Delta S_{univ} = \Delta S_{sys} + \Delta S_{surr}$$

Solution: $-125\text{ kJ} \times \frac{1000\text{ J}}{1\text{ kJ}} = -125{,}000\text{ J}$ then $\Delta S_{surr} = \frac{-\Delta H_{sys}}{T} = \frac{-(-125{,}000\text{ J})}{298\text{ K}} = 41\underline{9}.463\frac{\text{J}}{\text{K}}$ then

$\Delta S_{univ} = \Delta S_{sys} + \Delta S_{surr} = +253\frac{\text{J}}{\text{K}} + 41\underline{9}.463\frac{\text{J}}{\text{K}} = +672\frac{\text{J}}{\text{K}}$ so the reaction is spontaneous.

Check: The units (J/K) are correct. The magnitude of the answer (700 J/K) makes sense because both terms were positive and so the reaction is spontaneous.

b) **Given:** $\Delta H^\circ_{rxn} = +125$ kJ, $\Delta S_{rxn} = -253$ J/K, T = 298 K **Find:** ΔS_{univ} and spontaneity

Conceptual plan: **kJ → J then ΔH°_{rxn}, T → ΔS_{surr} then ΔS_{rxn}, ΔS_{surr} → ΔS_{univ}**

$$\frac{1000\text{ J}}{1\text{ kJ}} \qquad \Delta S_{surr} = \frac{-\Delta H_{sys}}{T} \qquad \Delta S_{univ} = \Delta S_{sys} + \Delta S_{surr}$$

Solution: $+125 \text{ kJ} \times \frac{1000 \text{ J}}{1 \text{ kJ}} = +125{,}000 \text{ J}$ then $\Delta S_{surr} = \frac{-\Delta H_{sys}}{T} = \frac{-125{,}000 \text{ J}}{298 \text{ K}} = -419.463 \frac{\text{J}}{\text{K}}$ then

$\Delta S_{univ} = \Delta S_{sys} + \Delta S_{surr} = -253 \frac{\text{J}}{\text{K}} - 419.463 \frac{\text{J}}{\text{K}} = -672 \frac{\text{J}}{\text{K}}$ so the reaction is nonspontaneous.

Check: The units (J/K) are correct. The magnitude of the answer (– 700 J/K) makes sense because both terms were negative and so the reaction is nonspontaneous.

c) **Given:** $\Delta H^\circ_{rxn} = -125$ kJ, $\Delta S_{rxn} = -253$ J/K, T = 298 K **Find:** ΔS_{univ} and spontaneity

Conceptual plan: kJ → J then ΔH°_{rxn}, T → ΔS_{surr} then ΔS_{rxn}, ΔS_{surr} → ΔS_{univ}

$$\frac{1000 \text{ J}}{1 \text{ kJ}} \qquad \Delta S_{surr} = \frac{-\Delta H_{sys}}{T} \qquad \Delta S_{univ} = \Delta S_{sys} + \Delta S_{surr}$$

Solution: $-125 \text{ kJ} \times \frac{1000 \text{ J}}{1 \text{ kJ}} = -125{,}000 \text{ J}$ then $\Delta S_{surr} = \frac{-\Delta H_{sys}}{T} = \frac{-(-125{,}000 \text{ J})}{298 \text{ K}} = +419.463 \frac{\text{J}}{\text{K}}$ then

$\Delta S_{univ} = \Delta S_{sys} + \Delta S_{surr} = -253 \frac{\text{J}}{\text{K}} + 419.463 \frac{\text{J}}{\text{K}} = +166 \frac{\text{J}}{\text{K}}$ so the reaction is spontaneous.

Check: The units (J/K) are correct. The magnitude of the answer (200 J/K) makes sense because the larger term was positive and so the reaction is spontaneous.

d) **Given:** $\Delta H^\circ_{rxn} = -125$ kJ, $\Delta S_{rxn} = -253$ J/K, T = 555 K **Find:** ΔS_{univ} and spontaneity

Conceptual plan: kJ → J then ΔH°_{rxn}, T → ΔS_{surr} then ΔS_{rxn}, ΔS_{surr} → ΔS_{univ}

$$\frac{1000 \text{ J}}{1 \text{ kJ}} \qquad \Delta S_{surr} = \frac{-\Delta H_{sys}}{T} \qquad \Delta S_{univ} = \Delta S_{sys} + \Delta S_{surr}$$

Solution: $-125 \text{ kJ} \times \frac{1000 \text{ J}}{1 \text{ kJ}} = -125{,}000 \text{ J}$ then $\Delta S_{surr} = \frac{-\Delta H_{sys}}{T} = \frac{-(-125{,}000 \text{ J})}{555 \text{ K}} = +225.225 \frac{\text{J}}{\text{K}}$ then

$\Delta S_{univ} = \Delta S_{sys} + \Delta S_{surr} = -253 \frac{\text{J}}{\text{K}} + 225.225 \frac{\text{J}}{\text{K}} = -28 \frac{\text{J}}{\text{K}}$ so the reaction is nonspontaneous.

Check: The units (J/K) are correct. The magnitude of the answer (– 30 J/K) makes sense because the larger term was negative and so the reaction is nonspontaneous.

39. a) **Given:** $\Delta H^\circ_{rxn} = -125$ kJ, $\Delta S_{rxn} = +253$ J/K, T = 298 K **Find:** ΔG and spontaneity

Conceptual plan: J/K → kJ/K then ΔH°_{rxn}, ΔS_{rxn}, T → ΔG

$$\frac{1 \text{ kJ}}{1000 \text{ J}} \qquad \Delta G = \Delta H_{rxn} - T\Delta S_{rxn}$$

Solution: $+253 \frac{\text{J}}{\text{K}} \times \frac{1 \text{ kJ}}{1000 \text{ J}} = +0.253 \frac{\text{kJ}}{\text{K}}$ then

$\Delta G = \Delta H_{rxn} - T\Delta S_{rxn} = -125 \text{ kJ} - (298 \text{ K})\left(0.253 \frac{\text{kJ}}{\text{K}}\right) = -200. \text{ kJ} = -2.00 \times 10^2 \text{ kJ}$ so the reaction is spontaneous.

Check: The units (kJ) are correct. The magnitude of the answer (- 200 kJ) makes sense because both terms were negative and so the reaction is spontaneous.

b) **Given:** $\Delta H^\circ_{rxn} = +125$ kJ, $\Delta S_{rxn} = -253$ J/K, T = 298 K **Find:** ΔS_{univ} and spontaneity

Conceptual plan: J/K → kJ/K then ΔH°_{rxn}, ΔS_{rxn}, T → ΔG

$$\frac{1 \text{ kJ}}{1000 \text{ J}} \qquad \Delta G = \Delta H_{rxn} - T\Delta S_{rxn}$$

Solution: $-253 \frac{\text{J}}{\text{K}} \times \frac{1 \text{ kJ}}{1000 \text{ J}} = -0.253 \frac{\text{kJ}}{\text{K}}$ then

$\Delta G = \Delta H_{rxn} - T\Delta S_{rxn} = +125 \text{ kJ} - (298 \text{ K})\left(-0.253 \frac{\text{kJ}}{\text{K}}\right) = +200. \text{ kJ} = +2.00 \times 10^2 \text{ kJ}$ so the reaction is nonspontaneous.

Check: The units (kJ) are correct. The magnitude of the answer (200 kJ) makes sense because both terms were positive and so the reaction is nonspontaneous.

c) **Given:** $\Delta H^\circ_{rxn} = -125$ kJ, $\Delta S_{rxn} = -253$ J/K, T = 298 K **Find:** ΔS_{univ} and spontaneity

Conceptual plan: **J/K → kJ/K then $\Delta H^\circ_{rxn}, \Delta S_{rxn}$, T → ΔG**

$$\frac{1\ \text{kJ}}{1000\ \text{J}} \qquad \Delta G = \Delta H_{rxn} - T\Delta S_{rxn}$$

Solution: $-253\frac{\text{J}}{\text{K}} \times \frac{1\ \text{kJ}}{1000\ \text{J}} = -0.253\frac{\text{kJ}}{\text{K}}$ then

$$\Delta G = \Delta H_{rxn} - T\Delta S_{rxn} = -125\ \text{kJ} - (298\ \text{K})\left(-0.253\frac{\text{kJ}}{\text{K}}\right) = -49.606\ \text{kJ} = -5.0 \times 10^1\ \text{kJ}$$ so the reaction is spontaneous.

Check: The units (kJ) are correct. The magnitude of the answer (-50 kJ) makes sense because the larger term was negative and so the reaction is spontaneous.

d) **Given:** $\Delta H^\circ_{rxn} = -125$ kJ, $\Delta S_{rxn} = -253$ J/K, T = 555 K **Find:** ΔS_{univ} and spontaneity

Conceptual plan: **J/K → kJ/K then $\Delta H^\circ_{rxn}, \Delta S_{rxn}$, T → ΔG**

$$\frac{1\ \text{kJ}}{1000\ \text{J}} \qquad \Delta G = \Delta H_{rxn} - T\Delta S_{rxn}$$

Solution: $-253\frac{\text{J}}{\text{K}} \times \frac{1\ \text{kJ}}{1000\ \text{J}} = -0.253\frac{\text{kJ}}{\text{K}}$ then

$$\Delta G = \Delta H_{rxn} - T\Delta S_{rxn} = -125\ \text{kJ} - (555\ \text{K})\left(-0.253\frac{\text{kJ}}{\text{K}}\right) = +15\ \text{kJ}$$ so the reaction is nonspontaneous.

Check: The units (J/K) are correct. The magnitude of the answer (+ 15 kJ) makes sense because the larger term was positive and so the reaction is nonspontaneous.

41. **Given:** $\Delta H^\circ_{rxn} = -2217$ kJ, $\Delta S_{rxn} = +101.1$ J/K, T = 25 °C **Find:** ΔS_{univ} and spontaneity

Conceptual plan: **°C → K then J/K → kJ/K then $\Delta H^\circ_{rxn}, \Delta S_{rxn}$, T → ΔG**

$$K = 273.15 + {}^\circ C \qquad \frac{1\ \text{kJ}}{1000\ \text{J}} \qquad \Delta G = \Delta H_{rxn} - T\Delta S_{rxn}$$

Solution: T = 273.15 + 25 °C = 298 K then $+101.1\frac{\text{J}}{\text{K}} \times \frac{1\ \text{kJ}}{1000\ \text{J}} = +0.1011\frac{\text{kJ}}{\text{K}}$ then

$$\Delta G = \Delta H_{rxn} - T\Delta S_{rxn} = -2217\ \text{kJ} - (298\ \text{K})\left(0.1011\frac{\text{kJ}}{\text{K}}\right) = -2247\ \text{kJ} = -2.247 \times 10^6\ \text{J}$$ so the reaction is spontaneous.

Check: The units (kJ) are correct. The magnitude of the answer (- 2250 kJ) makes sense because both terms are negative, so the reaction is spontaneous.

43.

ΔH	ΔS	ΔG	Low Temp.	High Temp.
–	+	–	Spontaneous	Spontaneous
–	–	Temp dependent	Spontaneous	Nonspontaneous
+	+	Temp dependent	Nonspontaneous	Spontaneous
+	–	+	Nonspontaneous	Nonspontaneous

45. The molar entropy of a substance increases with increasing temperatures. The kinetic energy and the molecular motion increases. The substance will have access to an increased number of energy levels.

47. a) CO_2 (*g*) because it has greater molar mass/complexity.

b) CH_3OH (*g*) because it is in the gas phase.

c) CO_2 (*g*) because it has greater molar mass/complexity.

d) SiH_4 (*g*) because it has greater molar mass.

e) $CH_3CH_2CH_3$ (*g*) because it has greater molar mass/complexity.

f) NaBr (*aq*) because a solution has more entropy than a solid crystal.

49. a) He (*g*) < Ne (*g*) < SO_2 (*g*) < NH_3 (*g*) < CH_3CH_2OH (*g*). All are in the gas phase. From He to Ne there is an increase in molar weight, beyond that, the molecules increase in complexity.

b) H_2O (*s*) < H_2O (*l*) < H_2O (*g*). Entropy increases as we go from a solid to a liquid to a gas.

c) CH_4 (*g*) < CF_4 (*g*) < CCl_4 (*g*). Entropy increases as the molar mass increases.

51. a) **Given:** C_2H_4 (*g*) + H_2 (*g*) → C_2H_6 (*g*) **Find:** $\Delta S°_{rxn}$

Conceptual plan: $\Delta S^{\circ}_{rxn} = \sum n_p S^{\circ}(products) - \sum n_r S^{\circ}(reactants)$

Solution:

Reactant/Product	S^0 (J/mol K from Appendix IIB)
C_2H_4 (g)	219.3
H_2 (*g*)	130.7
C_2H_6 (*g*)	229.2

Be sure to pull data for the correct formula and phase.

$$\Delta S^0_{rxn} = \sum n_p S^0(products) - \sum n_r S^0(reactants)$$
$$= [1(S^0(C_2H_6\ (g))] - [1(S^0(C_2H_4\ (g)) + 1(S^0(H_2\ (g))]$$
$$= [1(229.2\ \text{J/mol K})] - [1(219.3\ \text{J/mol K}) + 1(130.7\ \text{J/mol K})]$$ The moles of gas are decreasing.
$$= [229.2\ \text{J/K}] - [350.0\ \text{J/K}]$$
$$= -120.8\ \text{J/K}$$

Check: The units (J/K) are correct. The answer is negative, which is consistent with 2 moles of gas going to 1 mole of gas.

b) **Given:** C (*s*) + H_2O (*g*) → CO (*g*) + H_2 (*g*) **Find:** $\Delta S°_{rxn}$

Conceptual plan: $\Delta S^{\circ}_{rxn} = \sum n_p S^{\circ}(products) - \sum n_r S^{\circ}(reactants)$

Solution:

Reactant/Product	S^0 (J/mol K from Appendix IIB)
C (*s*)	5.7
H_2O (*g*)	188.8
CO (*g*)	197.7
H_2 (*g*)	130.7

Be sure to pull data for the correct formula and phase.

$$\Delta S^0_{rxn} = \sum n_p S^0(products) - \sum n_r S^0(reactants)$$
$$= [1(S^0(CO\ (g)) + 1(S^0(H_2\ (g))] - [1(S^0(C\ (s)) + 1(S^0(H_2O\ (g))]$$
$$= [1(197.7\ \text{J/mol K}) + 1(130.7\ \text{J/mol K})] - [1(5.7\ \text{J/mol K}) + 1(188.8\ \text{J/mol K})]$$
$$= [328.4\ \text{J/K}] - [194.5\ \text{J/K}]$$
$$= +133.9\ \text{J/K}$$

The moles of gas are increasing.
Check: The units (J/K) are correct. The answer is positive, which is consistent with 1 mole of gas going to 1 mole of gas.

c) **Given:** CO (*g*) + H_2O (*g*) → H_2 (*g*) + CO_2 (*g*) **Find:** $\Delta S°_{rxn}$

Conceptual plan: $\Delta S^{\circ}_{rxn} = \sum n_p S^{\circ}(products) - \sum n_r S^{\circ}(reactants)$

Solution:

Reactant/Product	S^0 (J/mol K from Appendix IIB)
CO (*g*)	197.7
H_2O (*g*)	188.8
H_2 (*g*)	130.7
CO_2 (*g*)	213.8

Be sure to pull data for the correct formula and phase.

$$\Delta S^0_{rxn} = \sum n_p S^0(products) - \sum n_r S^0(reactants)$$
$$= [1(S^0(H_2\,(g)) + 1(S^0(CO_2\,(g))] - [1(S^0(CO\,(g)) + 1(S^0(H_2O\,(g))]$$
$$= [1(130.7\text{ J/mol K}) + 1(213.8\text{ J/mol K})] - [1(197.7\text{ J/mol K}) + 1(188.8\text{ J/mol K})]$$
$$= [344.5\text{ J/K}] - [386.5\text{ J/K}]$$
$$= -42.0\text{ J/K}$$

The change is small because the number of moles of gas is constant.
Check: The units (J/K) are correct. The answer is small and negative, which is consistent with a constant number of moles of gas. Water molecules are bent and carbon dioxide molecules are linear, so the water has more complexity. Also, carbon monoxide is more complex than hydrogen gas.

d) **Given:** $2\ H_2S\ (g) + 3\ O_2\ (g) \rightarrow 2\ H_2O\ (l) + 2\ SO_2\ (g)$ **Find:** ΔS°_{rxn}

Conceptual plan: $\Delta S^0_{rxn} = \sum n_p S^0(products) - \sum n_r S^0(reactants)$

Solution:

Reactant/Product	S^0 (J/mol K from Appendix IIB)
$H_2S\ (g)$	205.8
$O_2\ (g)$	205.2
$H_2O\ (l)$	70.0
$SO_2\ (g)$	248.2

Be sure to pull data for the correct formula and phase.

$$\Delta S^0_{rxn} = \sum n_p S^0(products) - \sum n_r S^0(reactants)$$
$$= [2(S^0(H_2O\,(l)) + 2(S^0(SO_2\,(g))] - [2(S^0(H_2S\,(g)) + 3(S^0(O_2\,(g))]$$
$$= [2(70.0\text{J/mol K}) + 2(248.2\text{ J/mol K})] - [2(205.8\text{ J/mol K}) + 3(205.2\text{ J/mol K})]$$
$$= [636.4\text{ J/K}] - [1027.2\text{ J/K}]$$
$$= -390.8\text{ J/K}$$

The number of moles of gas is decreasing.
Check: The units (J/K) are correct. The answer is negative, which is consistent with a decrease in the number of moles of gas.

53. **Given:** $CH_2Cl_2\ (g)$ formed from elements in standard states **Find:** ΔS° and rationalize sign

Conceptual plan: Write balanced reaction. then $\Delta S^0_{rxn} = \sum n_p S^0(products) - \sum n_r S^0(reactants)$

Solution: $C\ (s) + H_2\ (g) + Cl_2\ (g) \rightarrow CH_2Cl_2\ (g)$

Reactant/Product	S^0 (J/mol K from Appendix IIB)
$C\ (s)$	5.7
$H_2\ (g)$	130.7
$Cl_2\ (g)$	223.1
$CH_2Cl_2\ (g)$	270.2

Be sure to pull data for the correct formula and phase.

$$\Delta S^0_{rxn} = \sum n_p S^0(products) - \sum n_r S^0(reactants)$$
$$= [1(S^0(CH_2Cl_2\,(g))] - [1(S^0(C\,(s)) + 1(S^0(H_2\,(g)) + 1(S^0(Cl_2\,(g))]$$
$$= [1(270.2\text{ J/mol K})] - [1(5.7\text{ J/mol K}) + 1(130.7\text{ J/mol K}) + 1(223.1\text{ J/mol K})]$$
$$= [270.2\text{ J/K}] - [359.5\text{ J/K}]$$
$$= -89.3\text{ J/K}$$

The moles of gas are decreasing.
Check: The units (J/K) are correct. The answer is negative, which is consistent with 2 moles of gas going to 1 mole of gas.

54. **Given:** $NF_4\ (g)$ formed from elements in standard states **Find:** ΔS° and rationalize sign

Conceptual plan: Write balanced reaction. then $\Delta S^0_{rxn} = \sum n_p S^0(products) - \sum n_r S^0(reactants)$

Solution: $\frac{1}{2}\ N_2\ (g) + 3/2\ F_2\ (g) \rightarrow NF_3\ (g)$

Reactant/Product	S^0 (J/mol K from Appendix IIB)
$N_2\ (g)$	191.6
$F_2\ (g)$	202.79
$NF_3\ (g)$	260.8

Be sure to pull data for the correct formula and phase.

$$\Delta S^0_{rxn} = \sum n_p S^0(products) - \sum n_r S^0(reactants)$$

$$= [1(S^0(NF_3\ (g))] - [1/2(S^0(N_2\ (g)) + 3/2(S^0(F_2\ (g))]$$

$$= [1(260.8\ \text{J/mol K})] - [1/2(191.6\ \text{J/mol K}) + 3/2(202.79\ \text{J/mol K})]$$

$$= [260.8\ \text{J/K}] - [399.\underline{9}85\ \text{J/K}]$$

$$= -139.2\ \text{J/K}$$

The moles of gas are decreasing.

Check: The units (J/K) are correct. The answer is negative, which is consistent with 2 moles of simple gases going to 1 mole of a complex gas.

55. **Given:** methanol (CH_3OH) combustion at 25 °C **Find:** ΔH°_{rxn}, ΔS°_{rxn}, ΔG°_{rxn}, and sponteneity

Conceptual plan: write balanced reaction then $\Delta H^0_{rxn} = \sum n_p \Delta H^0_f(products) - \sum n_R \Delta H^0_f(reactants)$ **then** $\Delta S^0_{rxn} = \sum n_p S^0(products) - \sum n_r S^0(reactants)$ **then °C→K then J/K→kJ/K then** $\Delta H^\circ_{rxn}, \Delta S^\circ_{rxn}, T \rightarrow \Delta G^\circ$

$$K = 273.15 + °C \qquad \frac{1\ \text{kJ}}{1000\ \text{J}} \qquad \Delta G = \Delta H_{rxn} - T\Delta S_{rxn}$$

Solution: combustion is combination with oxygen to form carbon dioxide and water

$2\ CH_3OH\ (l) + 3\ O_2\ (g) \rightarrow 2\ CO_2\ (g) + 4\ H_2O\ (g)$

Reactant/Product	ΔH°_f (kJ/mol from Appendix IIB)
$CH_3OH\ (l)$	– 238.6
$O_2\ (g)$	0.0
$CO_2\ (g)$	– 393.5
$H_2O\ (g)$	– 241.8

Be sure to pull data for the correct formula and phase.

$$\Delta H^0_{rxn} = \Sigma n_P \Delta H^\circ_f(products) - \Sigma n_R \Delta H^\circ_f(reactants)$$

$$= [2(\Delta H^\circ_f(CO_2(g))) + 4(\Delta H^\circ_f(H_2O(g)))] - [2(\Delta H^\circ_f(CH_3OH(l))) + 3(\Delta H^\circ_f(O_2(g)))]$$

$$= [2(-393.5\ \text{kJ}) + 4(-241.8\ \text{kJ})] - [2(-238.6\ \text{kJ}) + 3(0.0\ \text{kJ})]$$

$$= [-1754.2\ \text{kJ}] - [-477.2\ \text{kJ}]$$

$$= -1277\ \text{kJ}$$

then

Reactant/Product	S^0 (J/mol K from Appendix IIB)
$CH_3OH\ (l)$	126.8
$O_2\ (g)$	205.2
$CO_2\ (g)$	213.8
$H_2O\ (g)$	188.8

Be sure to pull data for the correct formula and phase.

$$\Delta S^0_{rxn} = \Sigma n_P S^\circ(products) - \Sigma n_R S^\circ(reactants)$$

$$= [2(S^\circ(CO_2(g))) + 4(S^\circ(H_2O(g)))] - [2(S^\circ(CH_3OH(l))) + 3(S^\circ(O_2(g)))]$$

$$= [2(213.8\ \text{J/mol K}) + 4(188.8\ \text{J/mol K})] - [2(126.8\ \text{J/mol K}) + 3(205.2\ \text{J/mol K})]$$ then

$$= [1182.8\ \text{J/K}] - [869.2\ \text{J/K}]$$

$$= 313.6\ \text{J/K}$$

$T = 273.15 + 25\ °C = 298\ K$ then $+313.6\frac{\text{J}}{\text{K}} \times \frac{1\ \text{kJ}}{1000\ \text{J}} = +0.3136\frac{\text{kJ}}{\text{K}}$ then

$\Delta G = \Delta H_{rxn} - T\Delta S_{rxn} = -1277\ \text{kJ} - (298\ \text{K})\left(+0.3136\frac{\text{kJ}}{\text{K}}\right) = -1370.\ \text{kJ} = -1.370 \times 10^6\ \text{J}$ so the reaction is spontaneous.

Check: The units (kJ, J/K and kJ) are correct. Combustion reactions are exothermic and we see a large negative enthalpy. We expect a large positive entropy because we have an increase in the number of moles of gas. The free energy is the sum of two negative terms so we expect a large negative free energy and the reaction is spontaneous.

57. a) **Given:** $N_2O_4\,(g) \rightarrow 2\,NO_2\,(g)$ at 25 °C
Find: $\Delta H°_{rxn}$, $\Delta S°_{rxn}$, $\Delta G°_{rxn}$, spontaneity and can temperature be changed to make it spontaneous
Conceptual plan: $\Delta H^0_{rxn} = \sum n_P \Delta H^0_f(products) - \sum n_R \Delta H^0_f(reactants)$ **then**
$\Delta S^0_{rxn} = \sum n_p S^0(products) - \sum n_r S^0(reactants)$ **then °C→K then J/K→ kJ/K then $\Delta H°_{rxn}$, ΔS_{rxn},T→ ΔG**

$K = 273.15 + °C$ $\quad \frac{1\,kJ}{1000\,J}$ $\quad \Delta G^0 = \Delta H^0_{rxn} - T\Delta S^0_{rxn}$

Solution:

Reactant/Product	ΔH^0_f(kJ/mol from Appendix IIB)
$N_2O_4\,(g)$	11.1
$NO_2\,(g)$	33.2

Be sure to pull data for the correct formula and phase.

$$\begin{aligned}\Delta H^0_{rxn} &= \sum n_P \Delta H^0_f(products) - \sum n_R \Delta H^0_f(reactants)\\ &= [2(\Delta H^0_f(NO_2\,(g))] - [1(\Delta H^0_f(N_2O_4\,(g))]\\ &= [2(33.2\text{ kJ})] - [1(11.1\text{ kJ})] \qquad \text{then}\\ &= [66.4\text{ kJ}] - [11.1\text{ kJ}]\\ &= +55.3\text{ kJ}\end{aligned}$$

Reactant/Product	S^0(J/mol K from Appendix IIB)
$N_2O_4\,(g)$	304.4
$NO_2\,(g)$	240.1

Be sure to pull data for the correct formula and phase.

$$\begin{aligned}\Delta S^0_{rxn} &= \sum n_P S^0(products) - \sum n_R S^0(reactants)\\ &= [2(S^0(NO_2\,(g))] - [1(S^0(N_2O_4\,(g))]\\ &= [2(240.1\text{ J/mol K})] - [1(304.4\text{ J/mol K})] \qquad \text{then}\\ &= [480.2\text{ J/K}] - [304.4\text{ J/K}]\\ &= +175.8\text{ J/K}\end{aligned}$$

T = 273.15 + 25 °C = 298 K then $+175.8\frac{\text{J}}{\text{K}} \times \frac{1\text{ kJ}}{1000\text{ J}} = +0.1758\frac{\text{kJ}}{\text{K}}$ then

$\Delta G^0 = \Delta H^0_{rxn} - T\Delta S^0_{rxn} = +55.3\text{ kJ} - (298\text{ K})\left(+0.1758\frac{\text{kJ}}{\text{K}}\right) = +2.9\text{ kJ} = +2.9 \times 10^3\text{ J}$ so the reaction is nonspontaneous. It can be made spontaneous by raising the temperature. .

Check: The units (kJ, J/K and kJ) are correct. The reaction requires the breaking of a bond, so we expect that this will be an endothermic reaction. We expect a positive entropy change because are increasing the number of moles of gas. Since the positive enthalpy term dominates at room temperature, the reaction is nonspontaneous. The second term can dominate if we raise the temperature high enough.

b) **Given:** $NH_4Cl\,(s) \rightarrow HCl\,(g) + NH_3\,(g)$ at 25 °C
Find: $\Delta H°_{rxn}$, $\Delta S°_{rxn}$, $\Delta G°_{rxn}$, spontaneity and can temperature be changed to make it spontaneous
Conceptual plan: $\Delta H^0_{rxn} = \sum n_P \Delta H^0_f(products) - \sum n_R \Delta H^0_f(reactants)$ **then**
$\Delta S^0_{rxn} = \sum n_p S^0(products) - \sum n_r S^0(reactants)$ **then °C→K then J/K→ kJ/K then $\Delta H°_{rxn}$, ΔS_{rxn},T →ΔG**

$K = 273.15 + °C$ $\quad \frac{1\,kJ}{1000\,J}$ $\quad \Delta G = \Delta H_{rxn} - T\Delta S_{rxn}$

Solution:

Reactant/Product	ΔH^0_f(kJ/mol from Appendix IIB)
$NH_4Cl\,(s)$	– 314.4
$HCl\,(g)$	– 92.3
$NH_3\,(g)$	– 45.9

Be sure to pull data for the correct formula and phase.

$$\Delta H^0_{rxn} = \sum n_P \Delta H^0_f(products) - \sum n_R \Delta H^0_f(reactants)$$
$$= [1(\Delta H^0_f(\text{HCl }(g)) + 1(\Delta H^0_f(\text{NH}_3\ (g))] - [1(\Delta H^0_f(\text{NH}_4\text{Cl }(g))]$$
$$= [1(-92.3\text{ kJ}) + 1(-45.9\text{ kJ})] - [1(-314.4\text{ kJ})] \quad \text{then}$$
$$= [-138.2\text{ kJ}] - [-314.4\text{ kJ}]$$
$$= +176.2\text{ kJ}$$

Reactant/Product	S^0 (J/mol K from Appendix IIB)
NH_4Cl (*s*)	94.6
HCl (*g*)	186.9
NH_3 (*g*)	192.8

Be sure to pull data for the correct formula and phase.

$$\Delta S^0_{rxn} = \sum n_P S^0(products) - \sum n_R S^0(reactants)$$
$$= [1(S^0(\text{HCl }(g)) + 1(S^0(\text{NH}_3\ (g))] - [1(S^0(\text{NH}_4\text{Cl }(g))]$$
$$= [1(186.9\text{ J/mol K}) + 1(192.8\text{ J/mol K})] - [1(94.6\text{ J/mol K})] \quad \text{then}$$
$$= [379.7\text{ J/K}] - [94.6\text{ J/K}]$$
$$= +285.1\text{ J/K}$$

$$T = 273.15 + 25\ °\text{C} = 298\text{ K} \quad \text{then} \quad +285.1\frac{\text{J}}{\text{K}} \times \frac{1\text{ kJ}}{1000\text{ J}} = +0.2851\frac{\text{kJ}}{\text{K}} \quad \text{then}$$

$$\Delta G^0 = \Delta H^0_{rxn} - T\Delta S^0_{rxn} = +176.2\text{ kJ} - (298\text{ K})\left(+0.2851\frac{\text{kJ}}{\text{K}}\right) = +91.2\text{ kJ} = +9.12 \times 10^4\text{ J}$$ so the reaction is nonspontaneous. It can be made spontaneous by raising the temperature.

Check: The units (kJ, J/K and kJ) are correct. The reaction requires the breaking of a bond, so we expect that this will be an endothermic reaction. We expect a positive entropy change because are increasing the number of moles of gas. Since the positive enthalpy term dominates at room temperature, the reaction is nonspontaneous. The second term can dominate if we raise the temperature high enough.

c) **Given:** $3\ H_2\ (g) + Fe_2O_3\ (s) \rightarrow 2\ Fe\ (s) + 3\ H_2O\ (g)$ at 25 °C

Find: $\Delta H°_{rxn}$, $\Delta S°_{rxn}$, $\Delta G°_{rxn}$, spontaneity and can temperature be changed to make it spontaneous

Conceptual plan: $\Delta H^°_{rxn} = \sum n_p \Delta H^°_f(products) - \sum n_R \Delta H^°_f(reactants)$ **then**

$\Delta S^°_{rxn} = \sum n_p S^°(products) - \sum n_r S^°(reactants)$ **then °C→ K then J/K→ kJ/K then $\Delta H°_{rxn}$, ΔS_{rxn}, T→ ΔG**

$$K = 273.15 + °C \qquad \frac{1\text{ kJ}}{1000\text{ J}} \qquad \Delta G = \Delta H_{rxn} - T\Delta S_{rxn}$$

Solution:

Reactant/Product	ΔH^0_f (kJ/mol from Appendix IIB)
H_2 (*g*)	0.0
Fe_2O_3 (*s*)	− 824.2
Fe (*s*)	0.0
H_2O (*g*)	− 241.8

Be sure to pull data for the correct formula and phase.

$$\Delta H^0_{rxn} = \sum n_P \Delta H^0_f(products) - \sum n_R \Delta H^0_f(reactants)$$
$$= [2(\Delta H^0_f(\text{Fe }(s)) + 3(\Delta H^0_f(\text{H}_2\text{O }(g))] - [3(\Delta H^0_f(\text{H}_2\ (g)) + 1(\Delta H^0_f(\text{Fe}_2\text{O}_3\ (s))]$$
$$= [2(0.0\text{ kJ}) + 3(-241.8\text{ kJ})] - [3(0.0\text{ kJ}) + 1(-824.2\text{ kJ})] \quad \text{then}$$
$$= [-725.4\text{ kJ}] - [-824.2\text{ kJ}]$$
$$= +98.8\text{ kJ}$$

Reactant/Product	S^0 (J/mol K from Appendix IIB)
H_2 (*g*)	130.7
Fe_2O_3 (*s*)	87.4
Fe (*s*)	27.3
H_2O (*g*)	188.8

Be sure to pull data for the correct formula and phase.

$$\Delta S^0_{rxn} = \sum n_P S^0(products) - \sum n_R S^0(reactants)$$

$$= [2(S^0(\text{Fe }(s)) + 3(S^0(\text{H}_2\text{O }(g))] - [3(S^0(\text{H}_2\ (g)) + 1(S^0(\text{Fe}_2\text{O}_3\ (s))]$$

$$= [2(27.3 \text{ J/mol K}) + 3(188.8 \text{ J/mol K})] - [3(130.7 \text{ J/mol K}) + 1(87.4 \text{ J/mol K})] \quad \text{then}$$

$$= [621.0 \text{ J/K}] - [479.5 \text{ J/K}]$$

$$= +141.5 \text{ J/K}$$

$T = 273.15 + 25\ °C = 298\ K$ then $+141.5 \frac{\text{J}}{\text{K}} \times \frac{1 \text{ kJ}}{1000 \text{ J}} = +0.1415 \frac{\text{kJ}}{\text{K}}$ then

$\Delta G^0 = \Delta H^0_{rxn} - T\Delta S^0_{rxn} = +98.8 \text{ kJ} - (298 \text{ K})\left(+0.1415 \frac{\text{kJ}}{\text{K}}\right) = +56.6 \text{ kJ} = +5.66 \times 10^4 \text{ J}$ so the reaction is nonspontaneous. It can be made spontaneous by raising the temperature.

Check: The units (kJ, J/K and kJ) are correct. The reaction requires the breaking of a bond, so we expect that this will be an endothermic reaction. We expect a positive entropy change because there is no change in the number of moles of gas, but the product gas is more complex. Since the positive enthalpy term dominates at room temperature, the reaction is nonspontaneous. The second term can dominate if we raise the temperature high enough. This process is the opposite of rusting, so we are not surprised that it is nonspontaneous.

d) **Given:** $N_2\ (g) + 3\ H_2\ (g) \rightarrow 2\ NH_3\ (g)$ at 25 °C

Find: $\Delta H°_{rxn}$, $\Delta S°_{rxn}$, $\Delta G°_{rxn}$, spontaneity and can temperature be changed to make it spontaneous

Conceptual plan: $\Delta H^0_{rxn} = \sum n_P \Delta H^0_f(products) - \sum n_R \Delta H^0_f(reactants)$ **then**

$\Delta S^0_{rxn} = \sum n_p S^0(products) - \sum n_r S^0(reactants)$ **then °C→K then J/K→ kJ/K then $\Delta H°_{rxn}$, ΔS_{rxn}, T → ΔG**

$$K = 273.15 + °C \qquad \frac{1 \text{ kJ}}{1000 \text{ J}} \qquad \Delta G = \Delta H_{rxn} - T\Delta S_{rxn}$$

Solution:

Reactant/Product	ΔH^0_f (kJ/mol from Appendix IIB)
$N_2\ (g)$	0.0
$H_2\ (g)$	0.0
$NH_3\ (g)$	− 45.9

Be sure to pull data for the correct formula and phase.

$$\Delta H^0_{rxn} = \sum n_P \Delta H^0_f(products) - \sum n_R \Delta H^0_f(reactants)$$

$$= [2(\Delta H^0_f(\text{NH}_3\ (g))] - [1(\Delta H^0_f(\text{N}_2\ (g)) + 3(\Delta H^0_f(\text{H}_2\ (g))]$$

$$= [2(-45.9 \text{ kJ})] - [1(0.0 \text{ kJ}) + 3(0.0 \text{ kJ})] \quad \text{then}$$

$$= [-91.8 \text{ kJ}] - [0.0 \text{ kJ}]$$

$$= -91.8 \text{ kJ}$$

Reactant/Product	S^0 (J/mol K from Appendix IIB)
$N_2\ (g)$	191.6
$H_2\ (g)$	130.7
$NH_3\ (g)$	192.8

Be sure to pull data for the correct formula and phase.

$$\Delta S^0_{rxn} = \sum n_P S^0(products) - \sum n_R S^0(reactants)$$

$$= [2(S^0(\text{NH}_3\ (g))] - [1(S^0(\text{N}_2\ (g)) + 3(S^0(\text{H}_2\ (g))]$$

$$= [2(192.8 \text{ J/mol K})] - [1(191.6 \text{ J/mol K}) + 3(130.7 \text{ J/mol K})] \quad \text{then}$$

$$= [385.6 \text{ J/K}] - [583.7 \text{ J/K}]$$

$$= -198.1 \text{ J/K}$$

$T = 273.15 + 25\ °C = 298\ K$ then $-198.1 \frac{\text{J}}{\text{K}} \times \frac{1 \text{ kJ}}{1000 \text{ J}} = -0.1981 \frac{\text{kJ}}{\text{K}}$ then

$\Delta G^0 = \Delta H^0_{rxn} - T\Delta S^0_{rxn} = -91.8 \text{ kJ} - (298 \text{ K})\left(-0.1981 \frac{\text{kJ}}{\text{K}}\right) = -32.8 \text{ kJ} = -3.28 \times 10^4 \text{ J}$ so the reaction is spontaneous.

Check: The units (kJ, J/K and kJ) are correct. The reaction requires the breaking of a bond, so we expect that this will be an endothermic reaction. We expect a positive entropy change because are increasing the number of moles of gas. Since the negative enthalpy term dominates at room temperature, the reaction is spontaneous. The second term can dominate if we raise the temperature high enough.

59. a) **Given:** $N_2O_4\ (g) \rightarrow 2\ NO_2\ (g)$ at 25 °C
Find: $\Delta G°_{rxn}$, spontaneity and compare to #57

Conceptual plan: $\Delta G^0_{rxn} = \sum n_p \Delta G^0_f(products) - \sum n_R \Delta G^0_f(reactants)$ **then compare to #57**

Solution:

Reactant/Product	ΔG^0_f(kJ/mol from Appendix IIB)
$N_2O_4\ (g)$	99.8
$NO_2\ (g)$	51.3

Be sure to pull data for the correct formula and phase.

$$\begin{aligned}\Delta G^0_{rxn} &= \sum n_P \Delta G^0_f(products) - \sum n_R \Delta G^0_f(reactants)\\ &= [2(\Delta G^0_f(NO_2\ (g))] - [1(\Delta G^0_f(N_2O_4\ (g))]\\ &= [2(51.3\text{ kJ})] - [1(99.8\text{ kJ})]\\ &= [102.6\text{ kJ}] - [99.8\text{ kJ}]\\ &= +2.8\text{ kJ}\end{aligned}$$

so the reaction is nonspontaneous.

The value is similar to in #57.

Check: The units (kJ) are correct. The free energy of the products is greater than the reactants, so the answer is positive and the reaction is nonspontaneous. The answer is the same as in #57 within the error of the calculation.

b) **Given:** $NH_4Cl\ (s) \rightarrow HCl\ (g) + NH_3\ (g)$ at 25 °C
Find: $\Delta G°_{rxn}$, spontaneity and compare to #57

Conceptual plan: $\Delta G^0_{rxn} = \sum n_p \Delta G^0_f(products) - \sum n_R \Delta G^0_f(reactants)$ **then compare to #57**

Solution:

Reactant/Product	ΔG^0_f(kJ/mol from Appendix IIB)
$NH_4Cl\ (s)$	− 202.9
$HCl\ (g)$	− 95.3
$NH_3\ (g)$	− 16.4

Be sure to pull data for the correct formula and phase.

$$\begin{aligned}\Delta G^0_{rxn} &= \sum n_P \Delta G^0_f(products) - \sum n_R \Delta G^0_f(reactants)\\ &= [1(\Delta G^0_f(HCl\ (g)) + 1(\Delta G^0_f(NH_3\ (g))] - [1(\Delta G^0_f(NH_4Cl\ (g))]\\ &= [1(-95.3\text{ kJ}) + 1(-16.4\text{ kJ})] - [1(-202.9\text{ kJ})]\\ &= [-111.7\text{ kJ}] - [-202.9\text{ kJ}]\\ &= +91.2\text{ kJ}\end{aligned}$$

so the reaction is nonspontaneous.

The result is the same as in #57.

Check: The units (kJ) are correct. The answer matches #57.

c) **Given:** $3\ H_2\ (g) + Fe_2O_3\ (s) \rightarrow 2\ Fe\ (s) + 3\ H_2O\ (g)$ at 25 °C
Find: $\Delta G°_{rxn}$, spontaneity and compare to #57

Conceptual plan: $\Delta G^0_{rxn} = \sum n_p \Delta G^0_f(products) - \sum n_R \Delta G^0_f(reactants)$ **then compare to #57**

Solution:

Reactant/Product	ΔG^0_f(kJ/mol from Appendix IIB)
$H_2\ (g)$	0.0
$Fe_2O_3\ (s)$	− 742.2
$Fe\ (s)$	0.0
$H_2O\ (g)$	− 228.6

Be sure to pull data for the correct formula and phase.

$$\Delta G^0_{rxn} = \sum n_P \Delta G^0_f(products) - \sum n_R \Delta G^0_f(reactants)$$
$$= [2(\Delta G^0_f(\text{Fe }(s)) + 3(\Delta G^0_f(\text{H}_2\text{O }(g))] - [3(\Delta G^0_f(\text{H}_2\ (g)) + 1(\Delta G^0_f(\text{Fe}_2\text{O}_3\ (s))]$$
$$= [2(0.0\text{ kJ}) + 3(-228.6\text{ kJ})] - [3(0.0\text{ kJ}) + 1(-742.2\text{ kJ})]$$
$$= [-685.8\text{ kJ}] - [-742.2\text{ kJ}]$$
$$= +56.4\text{ kJ}$$

so the reaction is nonspontaneous. The value is similar to that in #57.
Check: The units (kJ) are correct. The answer is the same as in #57 within the error of the calculation.

d) **Given:** $N_2\,(g) + 3\,H_2\,(g) \rightarrow 2\,NH_3\,(g)$ at 25 °C
Find: ΔG°_{rxn}, spontaneity and compare to #57
Conceptual plan: $\Delta G^\circ_{rxn} = \sum n_P \Delta G^\circ_f(products) - \sum n_R \Delta G^\circ_f(reactants)$ **then compare to #57**
Solution:

Reactant/Product	ΔG^0_f(kJ/mol from Appendix IIB)
$N_2\ (g)$	0.0
$H_2\ (g)$	0.0
$NH_3\ (g)$	− 16.4

Be sure to pull data for the correct formula and phase.
$$\Delta G^0_{rxn} = \sum n_P \Delta G^0_f(products) - \sum n_R \Delta G^0_f(reactants)$$
$$= [2(\Delta G^0_f(\text{NH}_3\ (g))] - [1(\Delta G^0_f(\text{N}_2\ (g)) + 3(\Delta G^0_f(\text{H}_2\ (g))]$$
$$= [2(-16.4\text{ kJ})] - [1(0.0\text{ kJ}) + 3(0.0\text{ kJ})]$$
$$= [-32.8\text{ kJ}] - [0.0\text{ kJ}]$$
$$= -32.8\text{ kJ}$$

so the reaction is spontaneous. The result is the same as in #57.
Check: The units (kJ) are correct. The answer matches #57.

Values calculated by the two methods are comparable. The method using ΔH° and ΔS° is longer, but it can be used to determine how ΔG° changes with temperature.

61. **Given:** $2\,NO\,(g) + O_2\,(g) \rightarrow 2\,NO_2\,(g)$ **Find:** ΔG°_{rxn} and spontaneity at a) 298 K, b) 715 K, and 855 K
Conceptual plan: $\Delta H^\circ_{rxn} = \sum n_P \Delta H^\circ_f(products) - \sum n_R \Delta H^\circ_f(reactants)$ then
$\Delta S^\circ_{rxn} = \sum n_P S^\circ(products) - \sum n_r S^\circ(reactants)$ **then J/K → kJ/K then ΔH°_{rxn}, ΔS_{rxn}, T → ΔG**

$$\frac{1\text{ kJ}}{1000\text{ J}} \qquad \Delta G = \Delta H_{rxn} - T\Delta S_{rxn}$$

Solution:

Reactant/Product	ΔH^0_f(kJ/mol from Appendix IIB)
NO (g)	91.3
$O_2\ (g)$	0.0
$NO_2\ (g)$	33.2

Be sure to pull data for the correct formula and phase.
$$\Delta H^0_{rxn} = \sum n_P \Delta H^0_f(products) - \sum n_R \Delta H^0_f(reactants)$$
$$= [2(\Delta H^0_f(\text{NO}_2\ (g))] - [2(\Delta H^0_f(\text{NO }(g)) + 1(\Delta H^0_f(\text{O}_2\ (g))]$$
$$= [2(33.2\text{ kJ})] - [2(91.3\text{ kJ}) + 1(0.0\text{ kJ})] \qquad \text{then}$$
$$= [66.4\text{ kJ}] - [182.6\text{ kJ}]$$
$$= -116.2\text{ kJ}$$

Reactant/Product	S^0(J/mol K from Appendix IIB)
NO (g)	210.8
$O_2\ (g)$	205.2
$NO_2\ (g)$	240.1

Be sure to pull data for the correct formula and phase.

$\Delta S^0_{rxn} = \sum n_P S^0(products) - \sum n_R S^0(reactants)$
$= [2(S^0(NO_2\ (g))] - [2(S^0(NO\ (g)) + 1(S^0(O_2\ (g))]$
$= [2(240.1\ J/mol\ K)] - [2(210.8\ J/mol\ K) + 1(205.2\ J/mol\ K)]$ then $-146.6 \frac{J}{K} \times \frac{1\ kJ}{1000\ J} = -0.1466 \frac{kJ}{K}$ then
$= [480.2\ J/K] - [626.8\ J/K]$
$= -146.6\ J/K$

a) $\Delta G^0 = \Delta H^0_{rxn} - T\Delta S^0_{rxn} = -116.2\ kJ - (298\ K)\left(-0.1466 \frac{kJ}{K}\right) = -72.5\ kJ = -7.25 \times 10^4\ J$ so the reaction is spontaneous.

b) $\Delta G^0 = \Delta H^0_{rxn} - T\Delta S^0_{rxn} = -116.2\ kJ - (715\ K)\left(-0.1466 \frac{kJ}{K}\right) = -11.4\ kJ = -1.14 \times 10^4\ J$ so the reaction is spontaneous.

b) $\Delta G^0 = \Delta H^0_{rxn} - T\Delta S^0_{rxn} = -116.2\ kJ - (855\ K)\left(-0.1466 \frac{kJ}{K}\right) = +9.1\ kJ = +9.1 \times 10^3\ J$ so the reaction is nonspontaneous.

Check: The units (kJ) are correct. The enthalpy term dominates at low temperatures making the reaction spontaneous. As the temperature increases, the decrease in entropy starts to dominate and the last case the reaction is nonspontaneous.

63. Since the first reaction has Fe_2O_3 as a product and the reaction of interest has it as a reactant, we need to reverse the first reaction. When the reaction direction is reversed ΔG changes.

$Fe_2O_3\ (s) \rightarrow 2\ Fe\ (s) + 3/2\ O_2\ (g)$ $\quad\Delta G^0 = +742.2\ kJ$

Since the second reaction has 1 mole CO as a reactant and the reaction of interest has 3 moles of CO as a reactant, we need to multiply the second reaction and the ΔG by 3.

$3\,[CO\ (g) + 1/2\ O_2\ (g) \rightarrow CO_2\ (g)]$ $\quad\Delta G^0 = 3(-257.2\ kJ) = -771.6\ kJ$

Hess' Law states the ΔG of the net reaction is the sum of the ΔG of the steps.
The rewritten reactions are:

$Fe_2O_3\ (s) \rightarrow 2\ Fe\ (s) + \cancel{3/2\ O_2\ (g)}$	$\Delta G^0 = +742.2\ kJ$
$3\ CO\ (g) + \cancel{3/2\ O_2\ (g)} \rightarrow 3\ CO_2\ (g)$	$\Delta G^0 = -771.6\ kJ$
$Fe_2O_3\ (s) + 3\ CO\ (g) \rightarrow 2\ Fe\ (s) + 3\ CO_2\ (g)$	$\Delta G^0_{rxn} = -29.4\ kJ$

65. a) **Given:** $I_2\ (s) \rightarrow I_2\ (g)$ at 25.0 °C **Find:** $\Delta G°_{rxn}$

Conceptual plan: $\Delta G^0_{rxn} = \sum n_P \Delta G^0_f(products) - \sum n_R \Delta G^0_f(reactants)$

Solution:

Reactant/Product	ΔG^0_f(kJ/mol from Appendix IIB)
$I_2\ (s)$	0.0
$I_2\ (g)$	19.3

Be sure to pull data for the correct formula and phase.

$\Delta G^0_{rxn} = \sum n_P \Delta G^0_f(products) - \sum n_R \Delta G^0_f(reactants)$
$= [1(\Delta G^0_f(I_2\ (g))] - [1(\Delta G^0_f(I_2\ (s))]$
$= [1(19.3\ kJ)] - [1(0.0\ kJ)]$
$= +19.3\ kJ$

so the reaction is nonspontaneous.

Check: The units (kJ) are correct. The answer is positive because gases have higher free energy than solids and the free energy change of the reaction is the same as free energy of formation of gaseous iodine.

b) **Given:** $I_2\ (s) \rightarrow I_2\ (g)$ at 25.0 °C (i) P_{I2} = 1.00 mm Hg; (ii) P_{I2} = 0.100 mm Hg **Find:** ΔG_{rxn}

Conceptual plan: **°C → K and mm Hg → atm then ΔG°_{rxn}, P_{I2}, T → ΔG_{rxn}**

$$K = 273.15 + °C \qquad \frac{1\ \text{atm}}{760\ \text{mm Hg}} \qquad \Delta G_{rxn} = \Delta G^0_{rxn} + RT\ln Q \quad \text{where } Q = P_{I2}$$

Solution: T = 273.15 + 25.0 °C = 298.2 K and (i) $1.00\ \text{mm Hg} \times \dfrac{1\ \text{atm}}{760\ \text{mm Hg}} = 0.00131579$ atm then

$$\Delta G_{rxn} = \Delta G^0_{rxn} + RT\ln Q = \Delta G^0_{rxn} + RT\ln P_{I2} = +19.3\ \text{kJ} + \left(8.314\frac{\text{J}}{\text{K mol}}\right)\left(\frac{1\ \text{kJ}}{1000\ \text{J}}\right)(298.2\ \text{K})\ln(0.00131579) =$$

= + 2.9 kJ so the reaction is nonspontaneous. then (ii) $0.100\ \text{mm Hg} \times \dfrac{1\ \text{atm}}{760\ \text{mm Hg}} = 0.000131579$ atm

then

$$\Delta G_{rxn} = \Delta G^0_{rxn} + RT\ln Q = \Delta G^0_{rxn} + RT\ln P_{I2} = +19.3\ \text{kJ} + \left(8.314\frac{\text{J}}{\text{K mol}}\right)\left(\frac{1\ \text{kJ}}{1000\ \text{J}}\right)(298.2\ \text{K})\ln(0.000131579) =$$

= – 2.9 kJ so the reaction is spontaneous.

Check: The units (kJ) are correct. The answer is positive at high pressures because the pressure is higher than the vapor pressure of iodine. Once the desired pressure is below the vapor pressure (0.31 mm Hg at 25.0 °C), the reaction becomes spontaneous.

c) Iodine sublimes at room temperature because there is an equilibrium between the solid and the gas phases. The vapor pressure is low (0.31 mm Hg at 25.0 °C), so a small amount of iodine can remain in the gas phase, which is consistent with the free energy values.

67. **Given:** $CH_3OH\ (g) \rightleftharpoons CO\ (g) + 2\ H_2\ (g)$ at 25 °C, P_{CH3OH} = 0.855 atm, P_{CO} = 0.125 atm, P_{H2} = 0.183 atm **Find:** ΔG

Conceptual plan: $\Delta G^\circ_{rxn} = \sum n_P \Delta G^\circ_f(products) - \sum n_R \Delta G^\circ_f(reactants)$ **then °C → K then**

$K = 273.15 + °C$

ΔG°_{rxn}, P_{CH3OH}, P_{CO}, P_{H2}, T→ΔG

$$\Delta G_{rxn} = \Delta G^\circ_{rxn} + RT\ln Q \quad \text{where } Q = \frac{P_{CO}P^2_{H2}}{P_{CH3OH}}$$

Solution:

Reactant/Product	ΔG°_f(kJ/mol from Appendix IIB)
$CH_3OH\ (g)$	– 162.3
$CO\ (g)$	– 137.2
$H_2\ (g)$	0.0

Be sure to pull data for the correct formula and phase.

$$\begin{aligned}\Delta G^0_{rxn} &= \sum n_P \Delta G^0_f(products) - \sum n_R \Delta G^0_f(reactants)\\ &= [1(\Delta G^0_f(CO\ (g)) + 2(\Delta G^0_f(H_2\ (g))] - [1(\Delta G^0_f(CH_3OH\ (g))]\\ &= [1(-137.2\ \text{kJ}) + 2(0.0\ \text{kJ})] - [1(-162.3\ \text{kJ})]\\ &= [-137.2\ \text{kJ}] - [-162.3\ \text{kJ}]\\ &= +25.1\ \text{kJ}\end{aligned}$$

T = 273.15 + 25 °C = 298 K then

$$Q = \frac{P_{CO}P^2_{H2}}{P_{CH3OH}} = \frac{(0.125)(0.183)^2}{0.855} = 0.00489605 \quad \text{then}$$

$$\Delta G_{rxn} = \Delta G^0_{rxn} + RT\ln Q = +25.1\ \text{kJ} + \left(8.314\frac{\text{J}}{\text{K mol}}\right)\left(\frac{1\ \text{kJ}}{1000\ \text{J}}\right)(298\ \text{K})\ln(0.00489605) = +11.9\ \text{kJ}$$

so the reaction is nonspontaneous.

Check: The units (kJ) are correct. The standard free energy for the reaction was positive and the fact that Q was less than one made the free energy smaller, but the reaction at these conditions is still not spontaneous.

69. a) **Given:** $2\ CO\ (g) + O_2\ (g) \rightleftharpoons 2\ CO_2\ (g)$ at 25 °C **Find:** K

Conceptual plan: $\Delta G^0_{rxn} = \sum n_P \Delta G^0_f(products) - \sum n_R \Delta G^0_f(reactants)$ **then °C→K then ΔG°$_{rxn}$, T → K**

$K = 273.15 + °C \qquad \Delta G^0_{rxn} = -RT\ln K$

Solution:

Reactant/Product	ΔG^0_f(kJ/mol from Appendix IIB)
$CO\ (g)$	− 137.2
$O_2\ (g)$	0.0
$CO_2\ (g)$	− 394.4

Be sure to pull data for the correct formula and phase.

$$\begin{aligned}\Delta G^0_{rxn} &= \sum n_P \Delta G^0_f(products) - \sum n_R \Delta G^0_f(reactants)\\ &= [2(\Delta G^0_f(CO_2\ (g))] - [2(\Delta G^0_f(CO\ (g)) + 1(\Delta G^0_f(O_2\ (g))]\\ &= [2(-394.4\ kJ)] - [2(-137.2\ kJ) + 1(0.0\ kJ)]\\ &= [-788.8\ kJ] - [-274.4\ kJ]\\ &= -514.4\ kJ\end{aligned}$$

T = 273.15 + 25 °C = 298 K then

$\Delta G^0_{rxn} = -RT\ln K$ Rearrange to solve for K.

$$K = e^{\frac{-\Delta G^0_{rxn}}{RT}} = e^{\frac{-(-514.4\ kJ)\times\frac{1000\ J}{1\ kJ}}{\left(8.314\frac{J}{K\ mol}\right)(298\ K)}} = e^{207.623} = 1.48 \times 10^{90}.$$

Check: The units (none) are correct. The standard free energy for the reaction was very negative and so we expect a very large K. The reaction is spontaneous and so mostly products are present at equilibrium.

b) **Given:** $2\ H_2S\ (g) \rightleftharpoons 2\ H_2\ (g) + S_2\ (g)$ at 25 °C **Find:** K

Conceptual plan: $\Delta G^0_{rxn} = \sum n_P \Delta G^0_f(products) - \sum n_R \Delta G^0_f(reactants)$ **then °C → K then ΔG°$_{rxn}$, T → K**

$K = 273.15 + °C \qquad \Delta G^0_{rxn} = -RT\ln K$

Solution:

Reactant/Product	ΔG^0_f(kJ/mol from Appendix IIB)
$H_2S\ (g)$	− 33.4
$H_2\ (g)$	0.0
$S_2\ (g)$	79.7

Be sure to pull data for the correct formula and phase.

$$\begin{aligned}\Delta G^0_{rxn} &= \sum n_P \Delta G^0_f(products) - \sum n_R \Delta G^0_f(reactants)\\ &= [2(\Delta G^0_f(H_2\ (g)) + 1(\Delta G^0_f(S_2\ (g))] - [2(\Delta G^0_f(H_2S\ (g))]\\ &= [2(0.0\ kJ)] + 1(79.7\ kJ)] - [2(-33.4\ kJ)]\\ &= [79.7\ kJ] - [-66.8\ kJ]\\ &= +146.5\ kJ\end{aligned}$$

T = 273.15 + 25 °C = 298 K then

$\Delta G^0_{rxn} = -RT\ln K$ Rearrange to solve for K.

$$K = e^{\frac{-\Delta G^0_{rxn}}{RT}} = e^{\frac{-146.5\ kJ\times\frac{1000\ J}{1\ kJ}}{\left(8.314\frac{J}{K\ mol}\right)(298\ K)}} = e^{-59.1305} = 2.09 \times 10^{-26}.$$

Check: The units (none) are correct. The standard free energy for the reaction was positive and so we expect a small K. The reaction is nonspontaneous and so mostly reactants are present at equilibrium.

71. a) **Given:** $CO\ (g) + 2\ H_2\ (g) \rightleftharpoons CH_3OH\ (g)$ $K_p = 2.26 \times 10^4$ at 25 °C **Find:** $\Delta G°_{rxn}$ at a) standard conditions, b) at equilibrium, and c) $P_{CH3OH} = 1.0$ atm, $P_{CO} = P_{H2} = 0.010$ atm

Conceptual plan: **°C → K then a) K, T → $\Delta G°_{rxn}$ then b) at equilibrium $\Delta G_{rxn} = 0$ then**

$K = 273.15 + °C$ $\quad \Delta G^0_{rxn} = -RT \ln K$

c) **$\Delta G°_{rxn}$, P_{CH3OH}, P_{CO}, P_{H2}, T → ΔG**

$\Delta G_{rxn} = \Delta G^0_{rxn} + RT \ln Q$ where $Q = \dfrac{P_{CH3OH}}{P_{CO}P^2_{H2}}$

Solution: T = 273.15 + 25 °C = 298 K then

a) $$\Delta G^0_{rxn} = -RT \ln K = -\left(8.314 \frac{\text{J}}{\text{K mol}}\right)\left(\frac{1 \text{ kJ}}{1000 \text{ J}}\right)(298 \text{ K}) \ln(2.26 \times 10^4) = -24.8 \text{ kJ};$$

b) at equilibrium $\Delta G_{rxn} = 0$

c) $$Q = \frac{P_{CH3OH}}{P_{CO}P^2_{H2}} = \frac{1.0}{(0.010)(0.010)^2} = 1.0 \times 10^6 \text{ then}$$

$$\Delta G_{rxn} = \Delta G^0_{rxn} + RT \ln Q = -24.8 \text{ kJ} + \left(8.314 \frac{\text{J}}{\text{K mol}}\right)\left(\frac{1 \text{ kJ}}{1000 \text{ J}}\right)(298 \text{ K}) \ln(1.0 \times 10^6) = +9.4 \text{ kJ}$$

Check: The units (kJ) are correct. The K was greater than one so we expect a negative standard free energy for the reaction. At equilibrium, by definition, the free energy change is zero. Since the conditions give a Q > K then the reactions needs to proceed in the reverse direction, which means that the reaction is spontaneous in the reverse direction.

73. a) **Given:** $2\ CO\ (g) + O_2\ (g) \rightleftharpoons 2\ CO_2\ (g)$ at 25 °C **Find:** K at 525 K

Conceptual plan: $\Delta H^0_{rxn} = \sum n_p \Delta H^0_f(products) - \sum n_R \Delta H^0_f(reactants)$ **then**

$\Delta S^0_{rxn} = \sum n_p S^0(products) - \sum n_r S^0(reactants)$ **then J/K → kJ/K then $\Delta H°_{rxn}$, ΔS_{rxn}, T → ΔG**

$\dfrac{1 \text{kJ}}{1000 \text{ J}}$ $\quad \Delta G^0 = \Delta H^0_{rxn} - T\Delta S^0_{rxn}$

then $\Delta G°_{rxn}$, T → K

$\Delta G^0_{rxn} = -RT \ln K$

Solution:

Reactant/Product	ΔH^0_f (kJ/mol from Appendix IIB)
$CO\ (g)$	−110.5
$O_2\ (g)$	0.0
$CO_2\ (g)$	−393.5

Be sure to pull data for the correct formula and phase.

$$\begin{aligned}\Delta H^0_{rxn} &= \sum n_P \Delta H^0_f(products) - \sum n_R \Delta H^0_f(reactants) \\ &= [2(\Delta H^0_f(CO_2\ (g))] - [2(\Delta H^0_f(CO\ (g)) + 1(\Delta H^0_f(O_2\ (g))] \\ &= [2(-393.5 \text{ kJ})] - [2(-110.5 \text{ kJ}) + 1(0.0 \text{ kJ})] \\ &= [-787.0 \text{ kJ}] - [-221.0 \text{ kJ}] \\ &= -566.0 \text{ kJ}\end{aligned}$$ then

Reactant/Product	S^0 (J/mol K from Appendix IIB)
$CO\ (g)$	197.7
$O_2\ (g)$	205.2
$CO_2\ (g)$	213.8

Be sure to pull data for the correct formula and phase.

$$\begin{aligned}\Delta S^0_{rxn} &= \sum n_P S^0(products) - \sum n_R S^0(reactants) \\ &= [2(S^0(CO_2\ (g))] - [2(S^0(CO\ (g)) + 1(S^0(O_2\ (g))] \\ &= [2(213.8 \text{ J/mol K})] - [2(197.7 \text{ J/mol K}) + 1(205.2 \text{ J/mol K})] \\ &= [427.6 \text{ J/K}] - [600.6 \text{ J/K}] \\ &= -173.0 \text{ J/K}\end{aligned}$$

then $-173.0 \dfrac{\text{J}}{\text{K}} \times \dfrac{1 \text{ kJ}}{1000 \text{ J}} = -0.1730 \dfrac{\text{kJ}}{\text{K}}$ then

$$\Delta G^0 = \Delta H^0_{rxn} - T\Delta S^0_{rxn} = -566.0 \text{ kJ} - (525 \text{ K})\left(-0.1730 \frac{\text{kJ}}{\text{K}}\right) = -475.2 \text{ kJ} = -4.752 \times 10^5 \text{ J} \text{ then}$$

$\Delta G^0_{rxn} = -RT \ln K$ Rearrange to solve for K.

$$K = e^{\frac{-\Delta G^0_{rxn}}{RT}} = e^{\frac{-(-4.752 \times 10^5 \text{ J})}{\left(8.314 \frac{\text{J}}{\text{K mol}}\right)(525 \text{ K})}} = e^{108.864} = 1.90 \times 10^{47}.$$

Check: The units (none) are correct. The free energy change is very negative, indicating a spontaneous reaction. This results in a very large K.

b) **Given:** $2\ H_2S\ (g) \rightleftharpoons 2\ H_2\ (g) + S_2\ (g)$ at 25 °C **Find:** K at 525 K

Conceptual plan: $\Delta H^\circ_{rxn} = \sum n_p \Delta H^\circ_f(products) - \sum n_R \Delta H^\circ_f(reactants)$ **then**

$\Delta S^\circ_{rxn} = \sum n_p S^\circ(products) - \sum n_r S^\circ(reactants)$ **then J/K → kJ/K then $\Delta H^\circ_{rxn}, \Delta S_{rxn}$, T → ΔG**

$\frac{1 \text{ kJ}}{1000 \text{ J}}$ $\Delta G^0 = \Delta H^0_{rxn} - T\Delta S^0_{rxn}$

then ΔG°_{rxn}, T → K

$\Delta G^0_{rxn} = -RT \ln K$

Solution:

Reactant/Product	ΔH^0_f (kJ/mol from Appendix IIB)
$H_2S\ (g)$	− 20.6
$H_2\ (g)$	0.0
$S_2\ (g)$	128.6

Be sure to pull data for the correct formula and phase.

$$\begin{aligned}\Delta H^0_{rxn} &= \sum n_p \Delta H^0_f(products) - \sum n_R \Delta H^0_f(reactants)\\ &= [2(\Delta H^0_f(H_2\ (g)) + 1(\Delta H^0_f(S_2\ (g))] - [2(\Delta H^0_f(H_2S\ (g))]\\ &= [2(0.0 \text{ kJ}) + 1(128.6 \text{ kJ})] - [2(-20.6 \text{ kJ})] \quad \text{then}\\ &= [128.6 \text{ kJ}] - [-41.2 \text{ kJ}]\\ &= +169.8 \text{ kJ}\end{aligned}$$

Reactant/Product	S^0 (J/mol K from Appendix IIB)
$H_2S\ (g)$	205.8
$H_2\ (g)$	130.7
$S_2\ (g)$	228.2

Be sure to pull data for the correct formula and phase.

$$\begin{aligned}\Delta S^0_{rxn} &= \sum n_p S^0(products) - \sum n_R S^0(reactants)\\ &= [2(S^0(H_2\ (g)) + 1(S^0(S_2\ (g))] - [2(S^0(H_2S\ (g))]\\ &= [2(130.7 \text{ J/mol K}) + 1(228.2 \text{ J/mol K})] - [2(205.8 \text{ J/mol K})] \text{ then } +78.0 \frac{\text{J}}{\text{K}} \times \frac{1 \text{ kJ}}{1000 \text{ J}} = +0.0780 \frac{\text{kJ}}{\text{K}} \text{ then}\\ &= [489.6 \text{ J/K}] - [411.6 \text{ J/K}]\\ &= +78.0 \text{ J/K}\end{aligned}$$

$$\Delta G^0 = \Delta H^0_{rxn} - T\Delta S^0_{rxn} = +169.8 \text{ kJ} - (525 \text{ K})\left(+0.0780 \frac{\text{kJ}}{\text{K}}\right) = +128.\underline{8}5 \text{ kJ} = +1.28\underline{8}5 \times 10^5 \text{ J} \text{ then}$$

$\Delta G^0_{rxn} = -RT \ln K$ Rearrange to solve for K.

$$K = e^{\frac{-\Delta G^0_{rxn}}{RT}} = e^{\frac{-1.28\underline{8}5 \times 10^5 \text{ J}}{\left(8.314 \frac{\text{J}}{\text{K mol}}\right)(525 \text{ K})}} = e^{-29.\underline{5}199} = 1.51 \times 10^{-13}.$$

Check: The units (none) are correct. The free energy change is positive, indicating a nonspontaneous reaction. This results in a very small K.

75. **Given:** table of K_p versus temperature **Find:** ΔH°_{rxn}, ΔS°_{rxn}

Conceptual Plan: Plot ln K verus 1/T. The slope will be $-\Delta H^\circ_{rxn}/R$ and the intercept will be $\Delta S^\circ_{rxn}/R$.

Solution: Plot ln K verus 1/T. Since $ln\, K = -\frac{\Delta H^0_{rxn}}{R}\frac{1}{T} + \frac{\Delta S^0_{rxn}}{R}$, the negative of the slope will be $-\frac{\Delta H^0_{rxn}}{R}$ and the intercept will be $\frac{\Delta S^0_{rxn}}{R}$. The slope can be determined by measuring Δy/Δx on the plot or by using functions, such as "add trendline" in Excel. Since the slope is – 6092.2 K^{-1},

$\Delta H^0_{rxn} = -\ slope\ R =$

$$= -\left(\frac{-6092.2}{K}\right)\left(8.314\frac{J}{K\ mol}\right)\left(\frac{1\ kJ}{1000\ J}\right) = 50.61\frac{kJ}{mol} = .$$

$$= 50.6\frac{kJ}{mol}$$

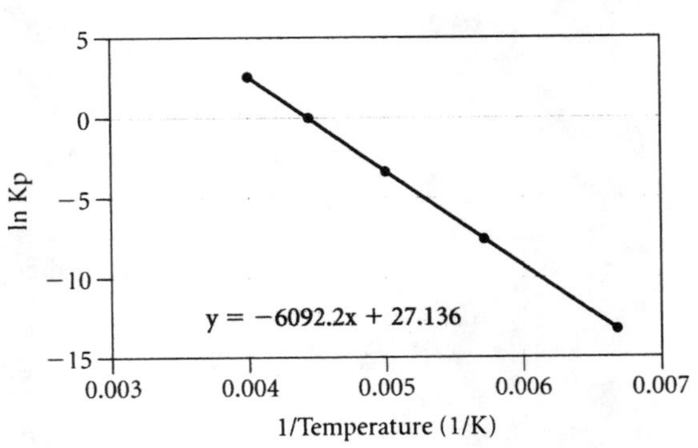

Since the intercept is 27.136, $\Delta S^0_{rxn} = intercept\ R = \left(\frac{27.136}{K}\right)\left(8.314\frac{J}{K\ mol}\right) = 225.609\frac{J}{K} = 226\frac{J}{K}$.

Check: The units are correct (kJ/mole and J/K). The plot is very linear. The numbers are typical for reactions. Since the slope is negative, the enthalpy change must be positive.

77. a) +, since vapors have higher entropy than liquids.

b) –, since solids have less entropy than liquids.

c) –, since there is only one microstate for the final macrostate and there are six microstates for the initial macrostate.

79. a) **Given:** $N_2\,(g) + O_2\,(g) \rightarrow 2\,NO\,(g)$ **Find:** $\Delta G°_{rxn}$, and K_p at 25 °C

Conceptual plan: $\Delta H^\circ_{rxn} = \sum n_p \Delta H^\circ_f(products) - \sum n_R \Delta H^\circ_f(reactants)$ **then**

$\Delta S^\circ_{rxn} = \sum n_p S^\circ(products) - \sum n_r S^\circ(reactants)$ **then °C→K then J/K→ kJ/K then $\Delta H°_{rxn}$, ΔS_{rxn}, T → ΔG**

$K = 273.15 + °C$ $\qquad \frac{1\,kJ}{1000\,J} \qquad \Delta G = \Delta H_{rxn} - T\Delta S_{rxn}$

then $\Delta G°_{rxn}$, T → K

$\Delta G^\circ_{rxn} = -RT\ln K$

Solution:

Reactant/Product	ΔH^0_f(kJ/mol from Appendix IIB)
$N_2\,(g)$	0.0
$O_2\,(g)$	0.0
NO (g)	91.3

Be sure to pull data for the correct formula and phase.

$$\Delta H^0_{rxn} = \sum n_P \Delta H^0_f(products) - \sum n_R \Delta H^0_f(reactants)$$
$$= [2(\Delta H^0_f(NO\,(g))] - [1(\Delta H^0_f(N_2\,(g)) + 1(\Delta H^0_f(O_2\,(g))]$$
$$= [2(91.3\,kJ)] - [1(0.0\,kJ) + 1(0.0\,kJ)]$$
$$= [182.6\,kJ] - [0.0\,kJ]$$
$$= +182.6\,kJ$$

then

Reactant/Product	S^0(J/mol K from Appendix IIB)
$N_2\,(g)$	191.6
$O_2\,(g)$	205.2
NO (g)	210.8

Be sure to pull data for the correct formula and phase.

$$\Delta S^0_{rxn} = \sum n_P S^0(products) - \sum n_R S^0(reactants)$$
$$= [2(S^0(\text{NO}\ (g))] - [1(S^0(\text{N}_2\ (g)) + 1(S^0(\text{O}_2\ (g))]$$
$$= [2(210.8\ \text{J/mol K})] - [1(191.6\ \text{J/mol K}) + 1(205.2\ \text{J/mol K})] \quad \text{then}$$
$$= [421.6\ \text{J/K}] - [396.8\ \text{J/K}]$$
$$= +\ 24.8\ \text{J/K}$$

$T = 273.15 + 25\ °C = 298\ K$ then $+24.8 \frac{\text{J}}{\text{K}} \times \frac{1\ \text{kJ}}{1000\ \text{J}} = +\ 0.0248 \frac{\text{kJ}}{\text{K}}$ then

$$\Delta G^0 = \Delta H^0_{rxn} - T\Delta S^0_{rxn} = +182.6\ \text{kJ} - (298\ \text{K})\left(0.0248 \frac{\text{kJ}}{\text{K}}\right) = +175.2\ \text{kJ} = +1.752 \times 10^5\ \text{J} \quad \text{then}$$

$\Delta G^0_{rxn} = -RT \ln K$ Rearrange to solve for K.

$$K = e^{\frac{-\Delta G^0_{rxn}}{RT}} = e^{\frac{-1.752 \times 10^5\ \text{J}}{\left(8.314 \frac{\text{J}}{\text{K mol}}\right)(298\ \text{K})}} = e^{-70.7144} = 1.95 \times 10^{-31}$$ so the reaction is nonspontaneous and at equilibrium mostly reactants are present.

Check: The units (kJ and none) are correct. The enthalpy is twice the enthalpy of formation of NO. We expect a very small entropy change because the number of moles of gas is unchanged. Since the positive enthalpy term dominates at room temperature, the free energy change is very positive and the reaction in the forward direction is nonspontaneous. This results in a very small K.

b) **Given:** $N_2\ (g) + O_2\ (g) \rightarrow 2\ NO\ (g)$ **Find:** $\Delta G°_{rxn}$ at 2000 K

Conceptual plan: use results from part a) $\Delta H°_{rxn}, \Delta S_{rxn}, T \rightarrow \Delta G$ **then** $\Delta G°_{rxn}, T \rightarrow K$

$$\Delta G = \Delta H_{rxn} - T\Delta S_{rxn} \qquad \Delta G^0_{rxn} = -RT \ln K$$

Solution: $\Delta G = \Delta H_{rxn} - T\Delta S_{rxn} = +182.6\ \text{kJ} - (2000\ \text{K})\left(0.0248 \frac{\text{kJ}}{\text{K}}\right) = +133.0\ \text{kJ} = +1.330 \times 10^5\ \text{J}$ then

$\Delta G^0_{rxn} = -RT \ln K$ Rearrange to solve for K.

$$K = e^{\frac{-\Delta G^0_{rxn}}{RT}} = e^{\frac{-1.330 \times 10^5\ \text{J}}{\left(8.314 \frac{\text{J}}{\text{K mol}}\right)(2000\ \text{K})}} = e^{-7.998557} = 3.36 \times 10^{-4}$$ so the forward reaction is becoming more spontaneous.

Check: The units (kJ and none) are correct. As the temperature rises, the entropy term becomes more significant. The free energy change is reduced and the K increases. The reaction is still nonspontaneous.

81. **Given:** $C_2H_4\ (g) + X_2\ (g) \rightarrow C_2H_4X_2\ (g)$ where X = Cl, Br and I
Find: $\Delta H°_{rxn}, \Delta S°_{rxn}, \Delta G°_{rxn}$ and K at 25 °C and spontaneity trends with X and temperature
Conceptual plan: $\Delta H^0_{rxn} = \sum n_P \Delta H^0_f(products) - \sum n_R \Delta H^0_f(reactants)$ **then**
$\Delta S^0_{rxn} = \sum n_P S^0(products) - \sum n_R S^0(reactants)$ **then °C→K then J/K → kJ/K then** $\Delta H°_{rxn}, \Delta S_{rxn}, T \rightarrow \Delta G$

$$K = 273.15 + °C \qquad \frac{1\ \text{kJ}}{1000\ \text{J}} \qquad \Delta G = \Delta H_{rxn} - T\Delta S_{rxn}$$

Solution:

Reactant/Product	ΔH^0_f (kJ/mol from Appendix IIB)
$C_2H_4\ (g)$	52.4
$Cl_2\ (g)$	0.0
$C_2H_4Cl_2\ (g)$	− 129.7

Be sure to pull data for the correct formula and phase.

$$\Delta H^0_{rxn} = \sum n_P \Delta H^0_f(products) - \sum n_R \Delta H^0_f(reactants)$$
$$= [1(\Delta H^0_f(\text{C}_2\text{H}_4\text{Cl}_2\ (g))] - [1(\Delta H^0_f(\text{C}_2\text{H}_4\ (g)) + 1(\Delta H^0_f(\text{Cl}_2\ (g))]$$
$$= [1(-129.7\ \text{kJ})] - [1(52.4\ \text{kJ}) + 1(0.0\ \text{kJ})] \quad \text{then}$$
$$= [-129.7\ \text{kJ}] - [52.4\ \text{kJ}]$$
$$= -182.1\ \text{kJ}$$

Reactant/Product	S^0(J/mol K from Appendix IIB)
$C_2H_4\ (g)$	219.3
$Cl_2\ (g)$	223.1
$C_2H_4Cl_2\ (g)$	308.0

Be sure to pull data for the correct formula and phase.

$$\Delta S^0_{rxn} = \sum n_P S^0(products) - \sum n_R S^0(reactants)$$
$$= [1(S^0(C_2H_4Cl_2\ (g))] - [1(S^0(C_2H_4\ (g)) + 1(S^0(Cl_2\ (g))]$$
$$= [1(308.0 \text{ J/mol K})] - [1(219.3 \text{ J/mol K}) + 1(223.1 \text{ J/mol K})]$$ then T = 273.15 + 25 °C = 298 K then
$$= [308.0 \text{ J/K}] - [442.4 \text{ J/K}]$$
$$= -134.4 \text{ J/K}$$

$$-134.4 \frac{\text{J}}{\text{K}} \times \frac{1 \text{ kJ}}{1000 \text{ J}} = -0.1344 \frac{\text{kJ}}{\text{K}}$$ then

$$\Delta G^0 = \Delta H^0_{rxn} - T\Delta S^0_{rxn} = -182.1 \text{ kJ} - (298 \text{ K})\left(-0.1344 \frac{\text{kJ}}{\text{K}}\right) = -142.0 \text{ kJ} = -1.420 \times 10^5 \text{ J}$$ then $\Delta G^0_{rxn} = -RT\ln K$

Rearrange to solve for K. $K = e^{\frac{-\Delta G^0_{rxn}}{RT}} = e^{\frac{-(-1.420 \times 10^5 \text{ J})}{\left(8.314 \frac{\text{J}}{\text{K mol}}\right)(298 \text{ K})}} = e^{57.334} = 7.94 \times 10^{24}$ so the reaction is spontaneous.

Reactant/Product	ΔH^0_f(kJ/mol from Appendix IIB)
$C_2H_4\ (g)$	52.4
$Br_2\ (g)$	30.9
$C_2H_4Br_2\ (g)$	− 38.3

Be sure to pull data for the correct formula and phase.

$$\Delta H^0_{rxn} = \sum n_P \Delta H^0_f(products) - \sum n_R \Delta H^0_f(reactants)$$
$$= [1(\Delta H^0_f(C_2H_4Br_2\ (g))] - [1(\Delta H^0_f(C_2H_4\ (g)) + 1(\Delta H^0_f(Br_2\ (g))]$$
$$= [1(-38.3 \text{ kJ})] - [1(52.4 \text{ kJ}) + 1(30.9 \text{ kJ})]$$ then
$$= [-38.3 \text{ kJ}] - [83.3 \text{ kJ}]$$
$$= -121.6 \text{ kJ}$$

Reactant/Product	S^0(J/mol K from Appendix IIB)
$C_2H_4\ (g)$	219.3
$Br_2\ (g)$	245.5
$C_2H_4Br_2\ (g)$	330.6

Be sure to pull data for the correct formula and phase.

$$\Delta S^0_{rxn} = \sum n_P S^0(products) - \sum n_R S^0(reactants)$$
$$= [1(S^0(C_2H_4Br_2\ (g))] - [1(S^0(C_2H_4\ (g)) + 1(S^0(Br_2\ (g))]$$
$$= [1(330.6 \text{ J/mol K})] - [1(219.3 \text{ J/mol K}) + 1(245.5 \text{ J/mol K})]$$ then $-134.2 \frac{\text{J}}{\text{K}} \times \frac{1 \text{ kJ}}{1000 \text{ J}} = -0.1342 \frac{\text{kJ}}{\text{K}}$ then
$$= [330.6 \text{ J/K}] - [464.8 \text{ J/K}]$$
$$= -134.2 \text{ J/K}$$

$$\Delta G^0 = \Delta H^0_{rxn} - T\Delta S^0_{rxn} = -121.6 \text{ kJ} - (298 \text{ K})\left(-0.1342 \frac{\text{kJ}}{\text{K}}\right) = -81.6 \text{ kJ} = -8.16 \times 10^4 \text{ J}$$ then $\Delta G^0_{rxn} = -RT\ln K$

Rearrange to solve for K. $K = e^{\frac{-\Delta G^0_{rxn}}{RT}} = e^{\frac{-(-8.16 \times 10^4 \text{ J})}{\left(8.314 \frac{\text{J}}{\text{K mol}}\right)(298 \text{ K})}} = e^{32.9389} = 2.02 \times 10^{14}$ so the reaction is spontaneous.

Reactant/Product	ΔH^0_f(kJ/mol from Appendix IIB)
$C_2H_4\ (g)$	52.4
$I_2\ (g)$	62.42
$C_2H_4I_2\ (g)$	66.5

Be sure to pull data for the correct formula and phase.

$$\Delta H^0_{rxn} = \sum n_P \Delta H^0_f(products) - \sum n_R \Delta H^0_f(reactants)$$
$$= [1(\Delta H^0_f(C_2H_4I_2\ (g))] - [1(\Delta H^0_f(C_2H_4\ (g)) + 1(\Delta H^0_f(I_2\ (g))]$$
$$= [1(66.5\text{ kJ})] - [1(52.4\text{ kJ}) + 1(62.42\text{ kJ})] \quad \text{then}$$
$$= [66.5\text{ kJ}] - [114.\underline{8}2\text{ kJ}]$$
$$= -48.\underline{3}2\text{ kJ}$$

Reactant/Product	S^0(J/mol K from Appendix IIB)
$C_2H_4\ (g)$	219.3
$I_2\ (g)$	260.69
$C_2H_4I_2\ (g)$	347.8

Be sure to pull data for the correct formula and phase.

$$\Delta S^0_{rxn} = \sum n_P S^0(products) - \sum n_R S^0(reactants)$$
$$= [1(S^0(C_2H_4I_2\ (g))] - [1(S^0(C_2H_4\ (g)) + 1(S^0(I_2\ (g))]$$
$$= [1(347.8\text{ J/mol K})] - [1(219.3\text{ J/mol K}) + 1(260.69\text{ J/mol K})] \quad \text{then } -132.2\frac{\text{J}}{\text{K}} \times \frac{1\text{ kJ}}{1000\text{ J}} = -0.1322\frac{\text{kJ}}{\text{K}} \text{ then}$$
$$= [347.8\text{ J/K}] - [479.\underline{9}9\text{ J/K}]$$
$$= -132.2\text{ J/K}$$

$$\Delta G^0 = \Delta H^0_{rxn} - T\Delta S^0_{rxn} = -48.\underline{3}2\text{ kJ} - (298\text{ K})\left(-0.1322\frac{\text{kJ}}{\text{K}}\right) = -8.\underline{9}244\text{ kJ} = -8.\underline{9}244 \times 10^3\text{ J then}$$

$$\Delta G^0_{rxn} = -RT\ln K \quad \text{Rearrange to solve for K. } K = e^{\frac{-\Delta G^0_{rxn}}{RT}} = e^{\frac{-(-8.\underline{9}244 \times 10^3\text{ J})}{\left(8.314\frac{\text{J}}{\text{K mol}}\right)(298\text{ K})}} = e^{3.\underline{6}021} = 37$$

and the reaction is spontaneous.

Cl_2 is the most spontaneous in the forward direction, I_2 is the least. The entropy change in the reactions is very constant. The spontaneity is determined by the standard enthalpy of formation of the dihalogenated ethane. Higher temperatures make the forward reactions less spontaneous.

Check: The units (kJ and none) are correct. The enthalpy change becomes less negative as we move to larger halogens. The enthalpy term dominates at room temperature, the free energy change is the same sign as the enthalpy change. The more negative the free energy change, the larger the K.

83. a) **Given:** $N_2O\ (g) + NO_2\ (g) \rightleftharpoons 3\ NO\ (g)$ at 298 K **Find:** $\Delta G°_{rxn}$

Conceptual plan: $\Delta G^\circ_{rxn} = \sum n_P \Delta G^\circ_f(products) - \sum n_R \Delta G^\circ_f(reactants)$

Solution:

Reactant/Product	ΔG^0_f(kJ/mol from Appendix IIB)
$N_2O\ (g)$	103.7
$NO_2\ (g)$	51.3
$NO\ (g)$	87.6

Be sure to pull data for the correct formula and phase.

$$\Delta G^0_{rxn} = \sum n_P \Delta G^0_f(products) - \sum n_R \Delta G^0_f(reactants)$$
$$= [3(\Delta G^0_f(NO\ (g))] - [1(\Delta G^0_f(N_2O\ (g)) + 1(\Delta G^0_f(NO_2\ (g))]$$
$$= [3(87.6\text{kJ})] - [1(103.7\text{ kJ}) + 1(51.3\text{ kJ})]$$
$$= [262.8\text{ kJ}] - [155.0\text{ kJ}]$$
$$= +107.8\text{ kJ}$$

The reaction is nonspontaneous.

Check: The units (kJ) are correct. The standard free energy for the reaction was positive and so the reaction is nonspontaneous.

b) **Given:** $P_{N2O} = P_{NO2} = 1.0$ atm initially **Find:** P_{N2O} when reaction ceases to be spontaneous

Conceptual Plan: Reaction will not longer be spontaneous when Q =K so then $\Delta G°_{rxn}$, T→K then

$$\Delta G^\circ_{rxn} = -RT\ln K$$

solve equilibrium problem to get gas pressures, since K << 1 the amount of NO generated will be very, very small compared to 1.0 atm, so, within experimental error, $P_{N2O} = P_{NO2} = 1.0$ atm. Simply solve of P_{NO}.

$$K = \frac{P_{NO}^3}{P_{N2O}P_{NO2}}$$

Solution: $\Delta G^0_{rxn} = -RT\ln K$ Rearrange to solve for K.

$$K = e^{\frac{-\Delta G^0_{rxn}}{RT}} = e^{\frac{-107.8\ \text{kJ} \times \frac{1000\ \text{J}}{1\ \text{kJ}}}{\left(8.314\frac{\text{J}}{\text{K mol}}\right)(298\ \text{K})}} = e^{-43.5103} = 1.27 \times 10^{-19}$$. Since $K = \frac{P_{NO}^3}{P_{N2O}P_{NO2}}$, rearrange

to solve for P_{NO}. $P_{NO}^3 = \sqrt[3]{K\,P_{N2O}P_{NO2}} = \sqrt[3]{(1.27 \times 10^{-19})(1.0)(1.0)} = 5.0 \times 10^{-7}$ atm .

Note that that assumption that P_{N2O} was very, very small was valid.

Check: The units (atm) are correct. Since the free energy change was positive, the K was very small. This leads us to expect that very little NO will be formed.

c) **Given:** $N_2O\ (g) + NO_2\ (g) \rightleftharpoons 3\ NO\ (g)$ **Find:** temperature for spontaneity

Conceptual plan: $\Delta H^0_{rxn} = \sum n_p \Delta H^0_f(products) - \sum n_R \Delta H^0_f(reactants)$ **then**

$\Delta S^0_{rxn} = \sum n_p S^0(products) - \sum n_r S^0(reactants)$ **then J/K → kJ/K then ΔH°_{rxn}, ΔS_{rxn} → T**

$\frac{1\ \text{kJ}}{1000\ \text{J}}$ $\Delta G = \Delta H_{rxn} - T\Delta S_{rxn}$

Solution:

Reactant/Product	ΔH^0_f(kJ/mol from Appendix IIB)
$N_2O\ (g)$	81.6
$NO_2\ (g)$	33.2
$NO\ (g)$	91.3

Be sure to pull data for the correct formula and phase.

$\Delta H^0_{rxn} = \sum n_P \Delta H^0_f(products) - \sum n_R \Delta H^0_f(reactants)$

$= [3(\Delta H^0_f(NO\ (g))] - [1(\Delta H^0_f(N_2O\ (g)) + 1(\Delta H^0_f(NO_2\ (g))]$

$= [3(91.3\ \text{kJ})] - [1(81.6\ \text{kJ}) + 1(33.2\ \text{kJ})]$ then

$= [273.9\ \text{kJ}] - [114.8\ \text{kJ}]$

$= +159.1\ \text{kJ}$

Reactant/Product	S^0(J/mol K from Appendix IIB)
$N_2O\ (g)$	220.0
$NO_2\ (g)$	240.1
$NO\ (g)$	210.8

Be sure to pull data for the correct formula and phase.

$\Delta S^0_{rxn} = \sum n_P S^0(products) - \sum n_R S^0(reactants)$

$= [3(S^0(NO\ (g))] - [1(S^0(N_2O\ (g)) + 1(S^0(NO_2\ (g))]$

$= [3(210.8\ \text{J/mol K})] - [1(220.0\ \text{J/mol K}) + 1(240.1\ \text{J/mol K})]$ then $+172.3\frac{\text{J}}{\text{K}} \times \frac{1\ \text{kJ}}{1000\ \text{J}} = +0.1723\frac{\text{kJ}}{\text{K}}$.

$= [632.4\ \text{J/K}] - [460.1\ \text{J/K}]$

$= +172.3\ \text{J/K}$

Since $\Delta G = \Delta H_{rxn} - T\Delta S_{rxn}$, set $\Delta G = 0$ and rearrange to solve for T.

$$T = \frac{\Delta H_{rxn}}{\Delta S_{rxn}} = \frac{+159.1\ \text{kJ}}{0.1723\frac{\text{kJ}}{\text{K}}} = +923.4\ \text{K}.$$

Check: The units (K) are correct. The reaction can be made more spontaneous by raising the temperature, because the entropy change is positive (increase in the number of moles of gas).

85. a) **Given:** ATP *(aq)* + H_2O *(l)* → ADP *(aq)* + P_i *(aq)* $\Delta G^\circ_{rxn} = -30.5$ kJ at 298 K **Find:** K

Conceptual plan: ΔG°_{rxn}, **T** → **K**

$$\Delta G^0_{rxn} = -RT\ln K$$

Solution: $\Delta G^0_{rxn} = -RT\ln K$ Rearrange to solve for K.

$$K = e^{\frac{-\Delta G^0_{rxn}}{RT}} = e^{\frac{-(-30.5\ \text{kJ})\times\frac{1000\ \text{J}}{1\ \text{kJ}}}{\left(8.314\frac{\text{J}}{\text{K mol}}\right)(298\ \text{K})}} = e^{12.3104} = 2.22 \times 10^5.$$

Check: The units (none) are correct. The free energy change is negative and the reaction is spontaneous. This results in a large K.

b) **Given:** oxidation of glucose drives reforming of ATP

Find: ΔG°_{rxn} of oxidation of glucose and moles ATP formed per mole of glucose

Conceptual plan: write balanced reaction for glucose oxidation then

$\Delta G^0_{rxn} = \sum n_P \Delta G^0_f(products) - \sum n_R \Delta G^0_f(reactants)$ **then ΔG°_{rxn}'s → moles ATP/ mole glucose**

$$\frac{\Delta G^0_{rxn}\ glucose\ oxidation}{\Delta G^0_{rxn}\ ATP\ hydrolysis}$$

Solution: $C_6H_{12}O_6$ *(s)* + 6 O_2 *(g)* → 6 CO_2 *(g)* + 6 H_2O *(l)*

Reactant/Product	ΔG^0_f(kJ/mol from Appendix IIB)
$C_6H_{12}O_6$ (*s*)	– 910.4
O_2 (*g*)	0.0
CO_2 (*g*)	– 394.4
H_2O (*l*)	– 237.1

Be sure to pull data for the correct formula and phase.

$$\Delta G^0_{rxn} = \sum n_P \Delta G^0_f(products) - \sum n_R \Delta G^0_f(reactants)$$
$$= [6(\Delta G^0_f(CO_2\ (g)) + 6(\Delta G^0_f(H_2O\ (l))] - [1(\Delta G^0_f(C_6H_{12}O_6\ (s)) + 6(\Delta G^0_f(O_2\ (g))]$$
$$= [6(-394.4\text{ kJ}) + 6(-237.1\text{ kJ})] - [1(-910.4\text{ kJ}) + 6(0.0\text{ kJ})]$$
$$= [-3789.0\text{ kJ}] - [-910.4\text{ kJ}]$$
$$= -2878.6\text{ kJ}$$

So the reaction is very spontaneous. $$\frac{2878.6\frac{\text{kJ generated}}{\text{mole glucose oxidized}}}{30.5\frac{\text{kJ needed}}{\text{mole ATP reformed}}} = 94.4\frac{\text{mole ATP reformed}}{\text{mole glucose oxidized}}$$

Check: The units (mol) are correct. The free energy change for the glucose oxidation is large compared to the ATP hydrolysis, so we expect to reform many moles of ATP.

87. a) **Given:** 2 CO *(g)* + 2 NO *(g)* → N_2 *(g)* + 2 CO_2 *(g)* **Find:** ΔG°_{rxn} and effect of increasing T on ΔG

Conceptual plan: $\Delta G^0_{rxn} = \sum n_P \Delta G^0_f(products) - \sum n_R \Delta G^0_f(reactants)$

Solution:

Reactant/Product	ΔG^0_f(kJ/mol from Appendix IIB)
CO (*g*)	– 137.2
NO (*g*)	87.6
N_2 (*g*)	0.0
CO_2 (*g*)	– 394.4

Be sure to pull data for the correct formula and phase.

$$\Delta G^0_{rxn} = \sum n_P \Delta G^0_f(products) - \sum n_R \Delta G^0_f(reactants)$$
$$= [1(\Delta G^0_f(N_2\ (g)) + 2(\Delta G^0_f(CO_2\ (g))] - [2(\Delta G^0_f(CO\ (g)) + 2(\Delta G^0_f(NO\ (g))]$$
$$= [1(0.0\text{ kJ}) + 2(-394.4\text{ kJ})] - [2(-137.2\text{ kJ}) + 2(87.6\text{ kJ})]$$
$$= [-788.8\text{ kJ}] - [-99.2\text{ kJ}]$$
$$= -689.6\text{ kJ}$$

Since the number of moles of gas is decreasing, the entropy change is negative and so ΔG will become more positive with increasing temperature.
Check: The units (kJ) are correct. The free energy change is negative since the carbon dioxide has such a low free energy of formation.

b) **Given:** $5\ H_2\ (g) + 2\ NO\ (g) \rightarrow 2\ NH_3\ (g) + 2\ H_2O\ (g)$ **Find:** $\Delta G°_{rxn}$ and effect of increasing T on ΔG

Conceptual plan: $\Delta G^0_{rxn} = \sum n_P \Delta G^0_f(products) - \sum n_R \Delta G^0_f(reactants)$

Solution: Solution:

Reactant/Product	ΔG^0_f(kJ/mol from Appendix IIB)
$H_2\ (g)$	0.0
$NO\ (g)$	87.6
$NH_3\ (g)$	– 16.4
$H_2O\ (g)$	– 228.6

Be sure to pull data for the correct formula and phase.

$$\begin{aligned}\Delta G^0_{rxn} &= \sum n_P \Delta G^0_f(products) - \sum n_R \Delta G^0_f(reactants)\\ &= [2(\Delta G^0_f(NH_3\ (g)) + 2(\Delta G^0_f(H_2O\ (g))] - [5(\Delta G^0_f(H_2\ (g)) + 2(\Delta G^0_f(NO\ (g))]\\ &= [2(-16.4\ kJ) + 2(-228.6\ kJ)] - [5(0.0\ kJ) + 2(87.6\ kJ)]\\ &= [-490.0\ kJ] - [175.2\ kJ]\\ &= -665.2\ kJ\end{aligned}$$

Since the number of moles of gas is decreasing, the entropy change is negative and so ΔG will become more positive with increasing temperature.
Check: The units (kJ) are correct. The free energy change is negative since ammonia and water have such a low free energy of formation.

c) **Given:** $2\ H_2\ (g) + 2\ NO\ (g) \rightarrow N_2\ (g) + 2\ H_2O\ (g)$ **Find:** $\Delta G°_{rxn}$ and effect of increasing T on ΔG

Conceptual plan: $\Delta G^0_{rxn} = \sum n_P \Delta G^0_f(products) - \sum n_R \Delta G^0_f(reactants)$

Solution:

Reactant/Product	ΔG^0_f(kJ/mol from Appendix IIB)
$H_2\ (g)$	0.0
$NO\ (g)$	87.6
$N_2\ (g)$	0.0
$H_2O\ (g)$	– 228.6

Be sure to pull data for the correct formula and phase.

$$\begin{aligned}\Delta G^0_{rxn} &= \sum n_P \Delta G^0_f(products) - \sum n_R \Delta G^0_f(reactants)\\ &= [1(\Delta G^0_f(N_2\ (g)) + 2(\Delta G^0_f(H_2O\ (g))] - [2(\Delta G^0_f(H_2\ (g)) + 2(\Delta G^0_f(NO\ (g))]\\ &= [1(0.0\ kJ) + 2(-228.6\ kJ)] - [2(0.0\ kJ) + 2(87.6\ kJ)]\\ &= [-457.2\ kJ] - [175.2\ kJ]\\ &= -632.4\ kJ\end{aligned}$$

Since the number of moles of gas is decreasing, the entropy change is negative and so ΔG will become more positive with increasing temperature.
Check: The units (kJ) are correct. The free energy change is negative since water has such a low free energy of formation.

d) **Given:** $2\ NH_3\ (g) + 2\ O_2\ (g) \rightarrow N_2O\ (g) + 3\ H_2O\ (g)$**Find:** $\Delta G°_{rxn}$ and effect of increasing T on ΔG

Conceptual plan: $\Delta G^0_{rxn} = \sum n_P \Delta G^0_f(products) - \sum n_R \Delta G^0_f(reactants)$

Solution:

Reactant/Product	ΔG^0_f(kJ/mol from Appendix IIB)
$NH_3\ (g)$	– 16.4
$O_2\ (g)$	0.0
$N_2O\ (g)$	103.7
$H_2O\ (g)$	– 228.6

Be sure to pull data for the correct formula and phase.

$$\Delta G^0_{rxn} = \sum n_P \Delta G^0_f(products) - \sum n_R \Delta G^0_f(reactants)$$
$$= [1(\Delta G^0_f(N_2O\ (g)) + 3(\Delta G^0_f(H_2O\ (g))] - [2(\Delta G^0_f(NH_3\ (g)) + 2(\Delta G^0_f(O_2\ (g))]$$
$$= [1(103.7\text{ kJ}) + 3(-228.6\text{ kJ})] - [2(-16.4\text{ kJ}) + 2(0.0\text{ kJ})]$$
$$= [-582.1\text{ kJ}] - [-32.8\text{ kJ}]$$
$$= -549.3\text{ kJ}$$

Since the number of moles of gas is constant the entropy change will be small and slightly negative and so the magnitude of ΔG will decrease with increasing temperature.

Check: The units (kJ) are correct. The free energy change is negative since water has such a low free energy of formation. The entropy change is negative once the S° values are reviewed ($\Delta S_{rxn} = -9.6$ J/K).

89. With one exception, the formation of any oxide of nitrogen at 298 K requires more moles of gas as reactants than are formed as products. For example 1 mole of N_2O requires 0.5 moles of O_2 and 1 mole of N_2; 1 mole of N_2O_3 requires 1 mole of N_2 and 1.5 moles of O_2 and so on. The exception is NO, where 1 mole of NO requires 0.5 moles of O_2 and 0.5 moles of N_2: ½ $N_2(g)$ + ½ $O_2(g) \rightarrow NO(g)$ This reaction has a positive ΔS because what is essentially mixing of the N and O has taken place in the product.

91. a) **Given:** glutamate *(aq)* + NH_3 *(aq)* → glutamine *(aq)* + H_2O *(l)* $\Delta G°_{rxn} = +14.2$ kJ at 298 K **Find:** K

Conceptual plan: $\Delta G°_{rxn}$, **T →** **K**

$$\Delta G^0_{rxn} = -RT\ln K$$

Solution: $\Delta G^0_{rxn} = -RT\ln K$ Rearrange to solve for K.

$$K = e^{\frac{-\Delta G^0_{rxn}}{RT}} = e^{\frac{-14.2\text{ kJ} \times \frac{1000\text{ J}}{1\text{ kJ}}}{\left(8.314\frac{\text{J}}{\text{K mol}}\right)(298\text{ K})}} = e^{-5.73142} = 3.24 \times 10^{-3}.$$

Check: The units (none) are correct. The free energy change is positive and the reaction is nonspontaneous. This results in a small K.

b) **Given:** pair ATP hydrolysis with glutamate/NH_3 reaction **Find:** show coupled reactions, $\Delta G°_{rxn}$ and K

Conceptual plan: use reaction mechanism shown, where A = NH_3 and B = glutamate ($C_5H_8O_4N^-$), then calculate $\Delta G°_{rxn}$ by adding free energies of reactions then $\Delta G°_{rxn}$, T → K

$$\Delta G^0_{rxn} = -RT\ln K$$

Solution:

$$NH_3\ (aq) + ATP\ (aq) + \cancel{H_2O\ (l)} \rightarrow \cancel{NH_3 - P_i(aq)} + ADP\ (aq) \qquad \Delta G^0_{rxn} = -30.5\text{ kJ}$$
$$\cancel{NH_3 - P_i(aq)} + C_5H_8O_4N^-\ (aq) \rightarrow C_5H_9O_3N_2^-\ (aq) + \cancel{H_2O\ (l)} + P_i(aq) \qquad \Delta G^0_{rxn} = +14.2\text{ kJ then}$$
$$NH_3\ (aq) + C_5H_8O_4N^-\ (aq) + ATP\ (aq) \rightarrow C_5H_9O_3N_2^-\ (aq) + ADP\ (aq) + P_i(aq) \quad \Delta G^0_{rxn} = -16.3\text{ kJ}$$

$\Delta G^0_{rxn} = -RT\ln K$ Rearrange to solve for K. $K = e^{\frac{-\Delta G^0_{rxn}}{RT}} = e^{\frac{-(-16.3\text{ kJ}) \times \frac{1000\text{ J}}{1\text{ kJ}}}{\left(8.314\frac{\text{J}}{\text{K mol}}\right)(298\text{ K})}} = e^{6.57902} = 7.20 \times 10^2.$

Check: The units (none) are correct. The free energy change is negative and the reaction is spontaneous. This results in a large K.

93. a) **Given:** ½ H_2 *(g)* + ½ Cl_2 *(g)* → HCl *(g)*, define standard state as 2 atm **Find:** $\Delta G°_f$

Conceptual plan: $\Delta G°_f$, P_{H2}, P_{Cl2}, P_{HCl}, **T → new** $\Delta G°_f$

$$\Delta G_{rxn} = \Delta G^0_{rxn} + RT\ln Q \quad \text{where } Q = \frac{P_{HCl}}{P_{H2}^{1/2} P_{Cl2}^{1/2}}$$

Solution: $\Delta G°_f = -95.3$ kJ/mol and $Q = \frac{P_{HCl}}{P_{H2}^{1/2} P_{Cl2}^{1/2}} = \frac{2}{2^{1/2}\, 2^{1/2}} = 1$ then

$$\Delta G_{rxn} = \Delta G^0_{rxn} + RT\ln Q = -95.3\frac{\text{kJ}}{\text{mol}} + \left(8.314\frac{\text{J}}{\text{K mol}}\right)\left(\frac{1\text{ kJ}}{1000\text{ J}}\right)(298\text{ K})\ln(1) = -95.3\frac{\text{kJ}}{\text{mol}} = -95{,}300\frac{\text{J}}{\text{mol}}.$$

Since the number of moles of reactants and products are the same, the decrease in volume affects the entropy of both equally, so there is no change in $\Delta G°_f$.

Check: The units (kJ) are correct. The Q is one so $\Delta G°_f$ is unchanged under the new standard conditions.

b) **Given:** $N_2\ (g) + \frac{1}{2}\ O_2\ (g) \rightarrow N_2O\ (g)$, define standard state as 2 atm **Find:** $\Delta G°_f$
Conceptual plan: **$\Delta G°_f$, P_{N2}, P_{O2}, P_{N2O}, T → new $\Delta G°_f$**

$$\Delta G_{rxn} = \Delta G^0_{rxn} + RT \ln Q \quad \text{where } Q = \frac{P_{N2O}}{P_{N2}\, P_{O2}^{1/2}}$$

Solution: $\Delta G°_f$ = + 103.7 kJ/mol and $Q = \frac{P_{N2O}}{P_{N2}\, P_{O2}^{1/2}} = \frac{2}{2\ 2^{1/2}} = \frac{1}{\sqrt{2}}$ then

$$\Delta G_{rxn} = \Delta G^0_{rxn} + RT \ln Q = +103.7\,\frac{\text{kJ}}{\text{mol}} + \left(8.314\,\frac{\text{J}}{\text{K mol}}\right)\left(\frac{1\ \text{kJ}}{1000\ \text{J}}\right)(298\ \text{K}) \ln\left(\frac{1}{\sqrt{2}}\right) = +102.8\,\frac{\text{kJ}}{\text{mol}} =$$

$$= +102{,}800\,\frac{\text{J}}{\text{mol}}$$

The entropy of the reactants (1.5 mol) is decreased more than the entropy of the product (1 mol). Since the product is relatively more favored at lower volume, $\Delta G°_f$ is less positive.
Check: The units (kJ) are correct. The Q is less than one so $\Delta G°_f$ is reduced under the new standard conditions.

c) **Given:** $\frac{1}{2}\ H_2\ (g) \rightarrow H\ (g)$, define standard state as 2 atm **Find:** $\Delta G°_f$
Conceptual plan: **$\Delta G°_f$, P_{H2}, P_H, T → new $\Delta G°_f$**

$$\Delta G_{rxn} = \Delta G^0_{rxn} + RT \ln Q \quad \text{where } Q = \frac{P_H}{P_{H2}^{1/2}}$$

Solution: $\Delta G°_f$ = + 203.3 kJ/mol and $Q = \frac{P_H}{P_{H2}^{1/2}} = \frac{2}{2^{1/2}} = \sqrt{2}$ then

$$\Delta G_{rxn} = \Delta G^0_{rxn} + RT \ln Q = +203.3\,\frac{\text{kJ}}{\text{mol}} + \left(8.314\,\frac{\text{J}}{\text{K mol}}\right)\left(\frac{1\ \text{kJ}}{1000\ \text{J}}\right)(298\ \text{K}) \ln\left(\sqrt{2}\right) = +204.2\,\frac{\text{kJ}}{\text{mol}} =$$

$$= +204{,}200\,\frac{\text{J}}{\text{mol}}$$

The entropy of the product (1 mol) is decreased more than the entropy of the reactant (1/2 mol). Since the product is relatively less favored, $\Delta G°_f$ is more positive.
Check: The units (kJ) are correct. The Q is greater than one so $\Delta G°_f$ is increased under the new standard conditions.

95. **Given:** $K = 3.9 \times 10^5$ at 300 K and $K = 1.2 \times 10^{-1}$ at 500 K **Find:** $\Delta H°_{rxn}$, $\Delta S°_{rxn}$
Conceptual Plan: Plot ln K verus 1/T. The slope will be $-\Delta H°_{rxn}/R$ and the intercept will be $\Delta S°_{rxn}/R$.
Solution: Plot ln K verus 1/T. Since $\ln K = -\frac{\Delta H^0_{rxn}}{R}\frac{1}{T} + \frac{\Delta S^0_{rxn}}{R}$, the negative of the slope will be $-\frac{\Delta H^0_{rxn}}{R}$ and the intercept will be $\frac{\Delta S^0_{rxn}}{R}$. The slope can be determined by measuring Δy/Δx on the plot or by using functions, such as "add trendline" in Excel. Since the slope is $+ 11246\ \text{K}^{-1}$,

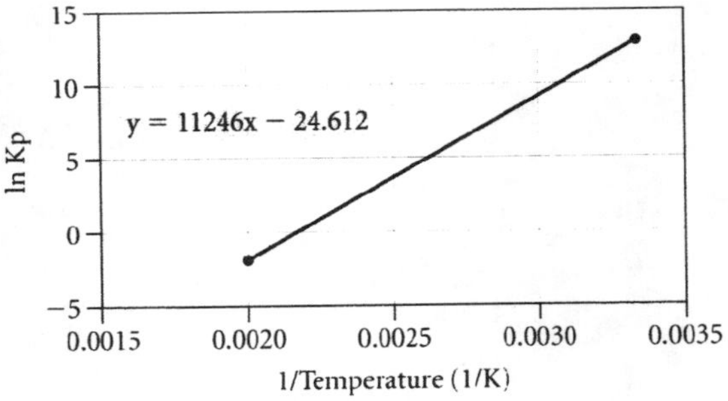

$\Delta H^0_{rxn} = -\ slope\ R =$

$$= -\left(\frac{+11246}{K}\right)\left(8.314\,\frac{\text{J}}{\text{K mol}}\right)\left(\frac{1\ \text{kJ}}{1000\ \text{J}}\right) = .$$

$$= -93.499\,\frac{\text{kJ}}{\text{mol}} = -93\,\frac{\text{kJ}}{\text{mol}}$$

Since the intercept is – 24.612, $\Delta S^0_{rxn} = intercept\ R = \left(\frac{-24.612}{K}\right)\left(8.314\,\frac{\text{J}}{\text{K mol}}\right) = -204.62\,\frac{\text{J}}{\text{K}} = -2.0 \times 10^2\,\frac{\text{J}}{\text{K}}$.

Check: The units are correct (kJ/mole and J/K mol). The numbers are typical for reactions. Since the slope is positive, the enthalpy change must be negative. The entropy change is expected to be negative since the number of moles of gas is decreasing.

97. A butane lighter is more efficient than an electric lighter since the butane lighter process is accomplished in one step. When you use an electric lighter, fuel is burned to generate electricity. This electricity is transmitted to the lighter through wires. Once it reaches the lighter, the electricity then needs to be converted to heat. Each step must pay a heat tax. Using a butane lighter has fewer steps and so it pays a much lower tax.

99. a) and c) will both increase the entropy of the surroundings because they are both exothermic reactions (adding thermal energy to the surroundings).

101. c) If the entropy of a system is increasing, the enthalpy of a reaction can be overcome (if necessary) by the entropy change as long as the temperature is high enough. If the entropy change of the system is decreasing, the reaction must be exothermic in order to be spontaneous since the entropy is working against spontaneity.

103. a) and b) are both true. Since $\Delta G_{rxn} = \Delta G^0_{rxn} + RT \ln Q$ and $\Delta G°_{rxn}$ = – 42.5 kJ, in order for ΔG_{rxn} = 0 the second term must be positive. This necessitates that Q > 1 or that we have more product than reactant. Any reaction at equilibrium has ΔG_{rxn} = 0.

Chapter 18
Electrochemistry

1. Electrochemistry uses spontaneous redox reactions to generate electrical energy.

3. Oxidation is the loss of electrons and corresponds to an increase in oxidation state. Reduction is the gain of electrons and corresponds to a decrease in oxidation state. Balancing redox reactions can be more complicated than balancing other types of reactions because both the mass (or number of each type of atom) and the charge must be balanced. Redox reactions occurring in aqueous solutions can be balanced by using a special procedure called the half-reaction method of balancing. In this procedure, the overall equation is broken down into two half-reactions: one for oxidation and one for reduction. The half-reactions are balanced individually and then added together so that the number of electrons generated in the oxidation half-reaction is the same as the number of electrons consumed in the reduction half-reaction.

5. In all electrochemical cells, the electrode where oxidation occurs is called the anode. In a voltaic cell, the anode is labeled with a negative (–) sign. The anode is negative because the oxidation reaction that occurs at the anode releases electrons. Electrons flow from the anode to the cathode (from negative to positive) through the wires connecting the electrodes.

7. A salt bridge is a pathway by which ions can flow between the half-cells, completing the electrical circuit, without the solutions in the half-cells totally mixing.

9. The standard cell potential (E^{o}_{cell}) or standard emf is the cell potential under standard conditions (1 M concentration for reactants in solution and 1 atm pressure for gaseous reactants). The cell potential is a measure of the overall tendency of the redox reaction to occur spontaneously. The more positive the cell potential, the more spontaneous the reaction. A negative cell potential indicates a nonspontaneous reaction.

11. Inert electrode, such as platinum (Pt) or graphite, are used as the anode or cathode (or both) when the reactants and products of one or both of the half-reactions are in the same phase.

13. The overall cell potential for any electrochemical cell will always be the sum of the half-cell potential for the oxidation reaction and the half-cell potential for the reduction reaction or $E^{o}_{cell} = E^{o}_{ox} + E^{o}_{red.}$

15. In general, any reduction half-reaction will be spontaneous when paired with the reverse of a half-reaction below it in the table.

17. We have seen that a positive standard cell potential (E^{0}_{cell}) corresponds to a spontaneous oxidation-reduction reaction. We also know (from Chapter 17) that the spontaneity of a reaction is determined by the sign of ΔG^{o}. Therefore, E^{o}_{cell} and ΔG^{o} must be related. We also know from Section 17.9 that ΔG^{o} for a reaction is related to the equilibrium constant (K) for the reaction. Since E^{o}_{cell} and ΔG^{o} are related, then E^{o}_{cell} and K must also be related. The equation that relates the three quantities is: $\Delta G^{0}_{rxn} = -RT\ln K = -nFE^{0}_{cell}$.

19. The Nernst equation relates concentration and cell potential as follows: $E_{cell} = E^{o}_{cell} - \dfrac{0.0592\ V}{n}\log Q$, where E_{cell} is the cell potential in V, E^{o}_{cell} is the standard cell potential in V, n is the number of moles of electrons transferred in the redox reaction, and Q is the reaction quotient. Increasing the concentration of the reactants decreases the value of Q and so it increases the cell potential. Increasing the concentration of the products increases the value of Q and so it decreases the cell potential.

21. A concentration cell is a voltaic cell in which both half reactions are the same, but in which a difference in concentration drives the current flow (since the cell potential depends, not only on the half-reactions occurring in the cell, but also on the concentrations of the reactants and products in those half-reactions). Therefore $E_{cell} = -\dfrac{0.0592\ V}{n}\log Q$.

23. Lead-acid storage batteries consist of six electrochemical cells wired in series, in which each cell produces 2 V for a total of 12 V. Each cell contains a porous lead anode where oxidation occurs according to: $Pb\ (s) + HSO_4^-\ (aq) \rightarrow PbSO_4\ (s) + H^+\ (aq) + 2\ e^-$ and a lead(IV) oxide cathode where reduction occurs according to: $PbO_2\ (s) + HSO_4^-\ (aq) + 3\ H^+\ (aq) + 2\ e^- \rightarrow PbSO_4\ (s) + 2\ H_2O\ (l)$. As electrical current

is drawn from the battery, both the anode and the cathode become coated with $PbSO_4$ *(s)* and the solution becomes depleted of HSO_4^- (aq). If the battery is run for a long time without recharging, too much $PbSO_4$ *(s)* develops on the surface of the electrodes and the battery goes dead. The lead-acid storage battery can be recharged, however, by running electrical current through it in reverse. The electrical current, which must come from an external source such as an alternator in a car, causes the preceding reaction to occur in reverse, converting the $PbSO_4$ *(s)* back to Pb *(s)* and PbO_2 *(s)*, recharging the battery.

25. Fuel cells are like batteries, but the reactants must be constantly replenished. Normal batteries lose their ability to generate voltage with use because the reactants become depleted as electrical current is drawn from the battery. In a fuel cell, the reactants—the fuel—constantly flow through the battery, generating electrical current as they undergo a redox reaction.

The most common fuel cell is the hydrogen–oxygen fuel cell. In the cell, hydrogen gas flows past the anode (a screen coated with platinum catalyst) and undergoes oxidation: $2\ H_2\ (g) + 4\ OH^-\ (aq) \rightarrow 4\ H_2O\ (l) + 4\ e^-$. Oxygen gas flows past the cathode (a similar screen) and undergoes reduction: $O_2\ (g) + 2\ H_2O\ (l) + 4\ e^- \rightarrow 4\ OH^-\ (aq)$. The half-reactions sum to the following overall reaction: $2\ H_2\ (g) + O_2\ (g) \rightarrow 2\ H_2O\ (l)$. Notice that the only product is water.

27. Applications of electrolysis include breaking water into hydrogen and oxygen; converting metal oxides to pure metals; production of sodium from molten sodium chloride; and plating metals onto other metals (for example, silver can be plated onto a less expensive metal).

29. The anion is oxidized. The cation is reduced to the metal.

31. In the electrolysis of aqueous NaCl solutions, two different reduction half-reactions are possible at the cathode, the reduction of Na^+ and the reduction of water:

$$2\ Na^+\ (l) + 2\ e^- \rightarrow 2\ Na\ (s) \qquad E^\circ_{red} = -\ 2.71\ V$$

$$2\ H_2O\ (l) + 2\ e^- \rightarrow H_2\ (g) + 2\ OH^-\ (aq) \qquad E_{red} = -\ 0.41\ V\ ([OH^-] = 10^{-7}\ M)$$

The half-reaction that occurs most easily (the one with the least negative, or most positive, half-cell potential) will be the one that actually occurs.

33. We can determine the number of moles of electrons that have flowed in a given electrolysis cell by measuring the total charge that has flowed through cell, which in turn depends on the magnitude of current and the time that the current has run. Since 1 Amphere = 1 C/s, if we multiply the amount of current (in A) flowing through the cell by the time (in s) that the current flowed, we can find the total charge that passed through the cell in that time: current (C/s) × time (s) = charge (C). The relationship between charge and the number of moles of electrons is given by Faraday's constant, which corresponds to the charge in coulombs of 1 mol of electrons, so F = 96,485 C/mole e^-. These relationships can be used to solve problems involving the stoichiometry of electrolytic cells.

35. Moisture must be present for many corrosion reactions to occur. The presence of water is necessary because water is a reactant in many oxidation reactions (such as $4\ Fe^{2+}\ (aq) + O_2\ (g) + (4 + 2n)\ H_2O\ (l) \rightarrow 2Fe_2O_3 \cdot nH_2O\ (s) + 8\ H^+\ (aq)$, and also because charge (either electrons or ions) must be free to flow between the anodic and cathodic regions.

Additional electrolytes promote more corrosion. The presence of an electrolyte (such as sodium chloride) on the surface of iron promotes rusting because it enhances current flow. This is why cars rust so quickly in cold climates where roads are salted, or in areas directly adjacent to beaches where salt water mist is present.

The presence of acids promotes corrosion. Since H^+ ions are involved in the reduction of oxygen, lower pH enhances the cathodic reaction and leads to faster corrosion.

37. **Conceptual Plan: Separate the overall reaction into two half-reactions: one for oxidation and one for reduction. → Balance each half-reaction with respect to mass in the following order: 1) balance all elements other than H and O; 2) balance O by adding H_2O; and 3) balance H by adding H^+. → Balance each half reaction with respect to charge by adding electrons. (The sum of the charges on both sides of the equation should be made equal by adding electrons as necessary.) → Make the number of electrons in both half-reactions equal by multiplying one or both half-reactions by a small whole number. → Add the two half-reactions together, canceling electrons and other species as**

necessary. → Verify that the reaction is balanced both with respect to mass and with respect to charge.
Solution:

a) Separate: $K(s) \rightarrow K^+(aq)$ and $Cr^{3+}(aq) \rightarrow Cr(s)$

Balance elements: $K(s) \rightarrow K^+(aq)$ and $Cr^{3+}(aq) \rightarrow Cr(s)$

Add electrons: $K(s) \rightarrow K^+(aq) + e^-$ and $Cr^{3+}(aq) + 3\,e^- \rightarrow Cr(s)$

Equalize electrons: $3\,K(s) \rightarrow 3\,K^+(aq) + 3\,e^-$ and $Cr^{3+}(aq) + 3\,e^- \rightarrow Cr(s)$

Add half-reactions: $3\,K(s) + Cr^{3+}(aq) + \cancel{3\,e^-} \rightarrow 3\,K^+(aq) + \cancel{3\,e^-} + Cr(s)$

Cancel electrons: $3\,K(s) + Cr^{3+}(aq) \rightarrow 3\,K^+(aq) + Cr(s)$

Check:

Reactants	Products
3 K atoms	3 K atoms
1 Cr atom	1 Cr atom
+3 charge	+3 charge

b) Separate: $Al(s) \rightarrow Al^{3+}(aq)$ and $Fe^{2+}(aq) \rightarrow Fe(s)$

Balance elements: $Al(s) \rightarrow Al^{3+}(aq)$ and $Fe^{2+}(aq) \rightarrow Fe(s)$

Add electrons: $Al(s) \rightarrow Al^{3+}(aq) + 3\,e^-$ and $Fe^{2+}(aq) + 2\,e^- \rightarrow Fe(s)$

Equalize electrons: $2\,Al(s) \rightarrow 2\,Al^{3+}(aq) + 6\,e^-$ and $3\,Fe^{2+}(aq) + 6\,e^- \rightarrow 3\,Fe(s)$

Add half-reactions: $2\,Al(s) + 3\,Fe^{2+}(aq) + \cancel{6\,e^-} \rightarrow 2\,Al^{3+}(aq) + \cancel{6\,e^-} + 3\,Fe(s)$

Cancel electrons: $2\,Al(s) + 3\,Fe^{2+}(aq) \rightarrow 2\,Al^{3+}(aq) + 3\,Fe(s)$

Check:

Reactants	Products
2 Al atoms	2 Al atoms
3 Fe atom	3 Fe atom
+6 charge	+6 charge

c) Separate: $BrO_3^-(aq) \rightarrow Br^-(aq)$ and $N_2H_4(g) \rightarrow N_2(g)$

Balance non H & O elements: $BrO_3^-(aq) \rightarrow Br^-(aq)$ and $N_2H_4(g) \rightarrow N_2(g)$

Balance O with H_2O: $BrO_3^-(aq) \rightarrow Br^-(aq) + 3\,H_2O(l)$ and $N_2H_4(g) \rightarrow N_2(g)$

Balance H with H^+: $BrO_3^-(aq) + 6\,H^+(aq) \rightarrow Br^-(aq) + 3\,H_2O(l)$ and $N_2H_4(g) \rightarrow N_2(g) + 4\,H^+(aq)$

Add electrons:

$BrO_3^-(aq) + 6\,H^+(aq) + 6\,e^- \rightarrow Br^-(aq) + 3\,H_2O(l)$ and $N_2H_4(g) \rightarrow N_2(g) + 4\,H^+(aq) + 4\,e^-$

Equalize electrons: $2\,BrO_3^-(aq) + 12\,H^+(aq) + 12\,e^- \rightarrow 2\,Br^-(aq) + 6\,H_2O(l)$ and

$3\,N_2H_4(g) \rightarrow 3\,N_2(g) + 12\,H^+(aq) + 12\,e^-$

Add half-reactions:

$2\,BrO_3^-(aq) + \cancel{12\,H^+(aq)} + 3\,N_2H_4(g) + \cancel{12\,e^-} \rightarrow 2\,Br^-(aq) + 6\,H_2O(l) + 3\,N_2(g) + \cancel{12\,H^+(aq)} + \cancel{12\,e^-}$

Cancel electrons & others: $2\,BrO_3^-(aq) + 3\,N_2H_4(g) \rightarrow 2\,Br^-(aq) + 6\,H_2O(l) + 3\,N_2(g)$

Check:

Reactants	Products
3 Br atoms	3 Br atoms
6 O atoms	6 O atoms
12 H atoms	12 H atoms
6 N atoms	6 N atoms
–2 charge	–2 charge

39. **Conceptual Plan: Separate the overall reaction into two half-reactions: one for oxidation and one for reduction. → Balance each half-reaction with respect to mass in the following order: 1) balance all elements other than H and O; 2) balance O by adding H_2O; and 3) balance H by adding H^+. → Balance each half reaction with respect to charge by adding electrons. (The sum of the charges on both sides of the equation should be made equal by adding electrons as necessary.) → Make the number of electrons in both half-reactions equal by multiplying one or both half-reactions by a small whole number. → Add the two half-reactions together, canceling electrons and other species as necessary. → Verify that the reaction is balanced both with respect to mass and with respect to charge.**

Solution:

a) Separate: $PbO_2(s) \rightarrow Pb^{2+}(aq)$ and $I^-(aq) \rightarrow I_2(s)$

Balance non H & O elements: $PbO_2(s) \rightarrow Pb^{2+}(aq)$ and $2\,I^-(aq) \rightarrow I_2(s)$

Balance O with H_2O: $PbO_2(s) \rightarrow Pb^{2+}(aq) + 2\,H_2O(l)$ and $2\,I^-(aq) \rightarrow I_2(s)$

Balance H with H^+: $PbO_2(s) + 4\,H^+(aq) \rightarrow Pb^{2+}(aq) + 2\,H_2O(l)$ and $2\,I^-(aq) \rightarrow I_2(s)$

Add electrons: $PbO_2(s) + 4\,H^+(aq) + 2\,e^- \rightarrow Pb^{2+}(aq) + 2\,H_2O(l)$ and $2\,I^-(aq) \rightarrow I_2(s) + 2\,e^-$

Equalize electrons: $PbO_2(s) + 4\,H^+(aq) + 2\,e^- \rightarrow Pb^{2+}(aq) + 2\,H_2O(l)$ and $2\,I^-(aq) \rightarrow I_2(s) + 2\,e^-$

Add half-reactions: $PbO_2\,(s) + 4\,H^+\,(aq) + \cancel{2\,e^-} + 2\,I^-\,(aq) \rightarrow Pb^{2+}\,(aq) + 2\,H_2O\,(l) + I_2\,(s) + \cancel{2\,e^-}$

Cancel electrons & others: $PbO_2\,(s) + 4\,H^+\,(aq) + 2\,I^-\,(aq) \rightarrow Pb^{2+}\,(aq) + 2\,H_2O\,(l) + I_2\,(s)$

Check:

Reactants	Products
1 Pb atom	1 Pb atom
2 O atoms	2 O atoms
4 H atoms	4 H atoms
2 I atoms	2 I atoms
+2 charge	+2 charge

b) Separate: $MnO_4^-\,(aq) \rightarrow Mn^{2+}\,(aq)$ and $SO_3^{2-}\,(aq) \rightarrow SO_4^{2-}\,(aq)$

Balance non H & O elements: $MnO_4^-\,(aq) \rightarrow Mn^{2+}\,(aq)$ and $SO_3^{2-}\,(aq) \rightarrow SO_4^{2-}\,(aq)$

Balance O with H_2O: $MnO_4^-\,(aq) \rightarrow Mn^{2+}\,(aq) + 4\,H_2O\,(l)$ and $SO_3^{2-}\,(aq) + H_2O\,(l) \rightarrow SO_4^{2-}\,(aq)$

Balance H with H^+:

$MnO_4^-\,(aq) + 8\,H^+\,(aq) \rightarrow Mn^{2+}\,(aq) + 4\,H_2O\,(l)$ and $SO_3^{2-}\,(aq) + H_2O\,(l) \rightarrow SO_4^{2-}\,(aq) + 2\,H^+\,(aq)$

Add electrons: $MnO_4^-\,(aq) + 8\,H^+\,(aq) + 5\,e^- \rightarrow Mn^{2+}\,(aq) + 4\,H_2O\,(l)$ and

$SO_3^{2-}\,(aq) + H_2O\,(l) \rightarrow SO_4^{2-}\,(aq) + 2\,H^+\,(aq) + 2\,e^-$

Equalize electrons: $2\,MnO_4^-\,(aq) + 16\,H^+\,(aq) + 10\,e^- \rightarrow 2\,Mn^{2+}\,(aq) + 8\,H_2O\,(l)$ and

$5\,SO_3^{2-}\,(aq) + 5\,H_2O\,(l) \rightarrow 5\,SO_4^{2-}\,(aq) + 10\,H^+\,(aq) + 10\,e^-$

Add half-reactions: $2\,MnO_4^-\,(aq) + 6\,\cancel{16}\,H^+\,(aq) + \cancel{10\,e^-} + 5\,SO_3^{2-}\,(aq) + \cancel{5\,H_2O(l)} \rightarrow$

$2\,Mn^{2+}\,(aq) + 3\,\cancel{8}\,H_2O(l) + 5\,SO_4^{2-}\,(aq) + \cancel{10\,H^+}\,(aq) + \cancel{10\,e^-}$

Cancel electrons: $2\,MnO_4^-\,(aq) + 6\,H^+\,(aq) + 5\,SO_3^{2-}\,(aq) \rightarrow 2\,Mn^{2+}\,(aq) + 3\,H_2O\,(l) + 5\,SO_4^{2-}\,(aq)$

Check:

Reactants	Products
2 Mn atoms	2 Mn atoms
23 O atoms	23 O atoms
6 H atoms	6 H atoms
5 S atoms	5 S atoms
–6 charge	–6 charge

c) Separate: $S_2O_3^{2-}\,(aq) \rightarrow SO_4^{2-}\,(aq)$ and $Cl_2\,(g) \rightarrow Cl^-\,(aq)$

Balance non H & O elements: $S_2O_3^{2-}\,(aq) \rightarrow 2\,SO_4^{2-}\,(aq)$ and $Cl_2\,(g) \rightarrow 2\,Cl^-\,(aq)$

Balance O with H_2O: $S_2O_3^{2-}\,(aq) + 5\,H_2O\,(l) \rightarrow 2\,SO_4^{2-}\,(aq)$ and $Cl_2\,(g) \rightarrow 2\,Cl^-\,(aq)$

Balance H with H^+: $S_2O_3^{2-}\,(aq) + 5\,H_2O\,(l) \rightarrow 2\,SO_4^{2-}\,(aq) + 10\,H^+\,(aq)$ and $Cl_2\,(g) \rightarrow 2\,Cl^-\,(aq)$

Add electrons:

$S_2O_3^{2-}\,(aq) + 5\,H_2O\,(l) \rightarrow 2\,SO_4^{2-}\,(aq) + 10\,H^+\,(aq) + 8\,e^-$ and $Cl_2\,(g) + 2\,e^- \rightarrow 2\,Cl^-\,(aq)$

Equalize electrons:

$S_2O_3^{2-}\,(aq) + 5\,H_2O\,(l) \rightarrow 2\,SO_4^{2-}\,(aq) + 10\,H^+\,(aq) + 8\,e^-$ and $4\,Cl_2\,(g) + 8\,e^- \rightarrow 8\,Cl^-\,(aq)$

Add half-reactions:

$S_2O_3^{2-}\,(aq) + 5\,H_2O\,(l) + 4\,Cl_2\,(g) + \cancel{8\,e^-} \rightarrow 2\,SO_4^{2-}\,(aq) + 10\,H^+\,(aq) + \cancel{8\,e^-} + 8\,Cl^-\,(aq)$

Cancel electrons: $S_2O_3^{2-}\,(aq) + 5\,H_2O\,(l) + 4\,Cl_2\,(g) \rightarrow 2\,SO_4^{2-}\,(aq) + 10\,H^+\,(aq) + 8\,Cl^-\,(aq)$

Check:

Reactants	Products
2 S atoms	2 S atoms
8 O atoms	8 O atoms
10 H atoms	10 H atoms
8 Cl atoms	8 Cl atoms
–2 charge	–2 charge

41. **Conceptual Plan: Separate the overall reaction into two half-reactions: one for oxidation and one for reduction. → Balance each half-reaction with respect to mass in the following order: 1) balance all elements other than H and O; 2) balance O by adding H_2O; 3) balance H by adding H^+; and 4) Neutralize H^+ by adding enough OH^- to neutralize each H^+. Add the same number of OH^- ions to each side of the equation. → Balance each half reaction with respect to charge by adding electrons. (The sum of the charges on both sides of the equation should be made equal by adding electrons as necessary.) → Make the number of electrons in both half-reactions equal by multiplying one or both half-reactions by a small whole number. → Add the two half-reactions together, canceling electrons and other species as necessary. → Verify that the reaction is balanced both with respect to mass and with respect to charge.**

Solution:

a) Separate: $ClO_2\,(aq) \rightarrow ClO_2^-\,(aq)$ and $H_2O_2\,(aq) \rightarrow O_2\,(g)$

Balance non H & O elements: $ClO_2\ (aq) \rightarrow ClO_2^-\ (aq)$ and $H_2O_2\ (aq) \rightarrow O_2\ (g)$
Balance O with H_2O: $ClO_2\ (aq) \rightarrow ClO_2^-\ (aq)$ and $H_2O_2\ (aq) \rightarrow O_2\ (g)$
Balance H with H^+: $ClO_2\ (aq) \rightarrow ClO_2^-\ (aq)$ and $H_2O_2\ (aq) \rightarrow O_2\ (g) + 2\ H^+\ (aq)$

Neutralize H^+ with OH^-:
$ClO_2\ (aq) \rightarrow ClO_2^-\ (aq)$ and $H_2O_2\ (aq) + 2\ OH^-\ (aq) \rightarrow O_2\ (g) + \underbrace{2\ H^+\ (aq) + 2\ OH^-\ (aq)}_{2\ H_2O\ (l)}$

Add electrons: $ClO_2\ (aq) + e^- \rightarrow ClO_2^-\ (aq)$ and $H_2O_2\ (aq) + 2\ OH^-\ (aq) \rightarrow O_2\ (g) + 2\ H_2O\ (l) + 2\ e^-$
Equalize electrons:
$2\ ClO_2\ (aq) + 2\ e^- \rightarrow 2\ ClO_2^-\ (aq)$ and $H_2O_2\ (aq) + 2\ OH^-\ (aq) \rightarrow O_2\ (g) + 2\ H_2O\ (l) + 2\ e^-$
Add half-reactions:
$2\ ClO_2\ (aq) + \cancel{2\ e^-} + H_2O_2\ (aq) + 2\ OH^-\ (aq) \rightarrow 2\ ClO_2^-\ (aq) + O_2\ (g) + 2\ H_2O\ (l) + \cancel{2\ e^-}$
Cancel electrons: $2\ ClO_2\ (aq) + H_2O_2\ (aq) + 2\ OH^-\ (aq) \rightarrow 2\ ClO_2^-\ (aq) + O_2\ (g) + 2\ H_2O\ (l)$

Check:

Reactants	Products
2 Cl atoms	2 Cl atoms
8 O atoms	8 O atoms
4 H atoms	4 H atoms
–2 charge	–2 charge

b) Separate: $MnO_4^-\ (aq) \rightarrow MnO_2\ (s)$ and $Al\ (s) \rightarrow Al(OH)_4^-\ (aq)$
Balance non H & O elements: $MnO_4^-\ (aq) \rightarrow MnO_2\ (s)$ and $Al\ (s) \rightarrow Al(OH)_4^-\ (aq)$
Balance O with H_2O: $MnO_4^-\ (aq) \rightarrow MnO_2\ (s) + 2\ H_2O\ (l)$ and $Al\ (s) + 4\ H_2O\ (l) \rightarrow Al(OH)_4^-\ (aq)$
Balance H with H^+:
$MnO_4^-\ (aq) + 4\ H^+\ (aq) \rightarrow MnO_2\ (s) + 2\ H_2O\ (l)$ and $Al\ (s) + 4\ H_2O\ (l) \rightarrow Al(OH)_4^-\ (aq) + 4\ H^+\ (aq)$
Neutralize H^+ with OH^-: $MnO_4^-\ (aq) + \underbrace{4\ H^+\ (aq) + 4\ OH^-\ (aq)}_{\cancel{4}2\ H_2O\ (l)} \rightarrow MnO_2\ (s) + \cancel{2\ H_2O(l)} + 4\ OH^-\ (aq)$

and $Al\ (s) + \cancel{4\ H_2O(l)} + 4\ OH^-\ (aq) \rightarrow Al(OH)_4^-\ (aq) + \underbrace{4\ H^+\ (aq) + 4\ OH^-\ (aq)}_{\cancel{4\ H_2O\ (l)}}$

Add electrons: $MnO_4^-\ (aq) + 2\ H_2O\ (l) + 3\ e^- \rightarrow MnO_2\ (s) + 4\ OH^-\ (aq)$ and
$Al\ (s) + 4\ OH^-\ (aq) \rightarrow Al(OH)_4^-\ (aq) + 3\ e^-$
Equalize electrons: $MnO_4^-\ (aq) + 2\ H_2O\ (l) + 3\ e^- \rightarrow MnO_2\ (s) + 4\ OH^-\ (aq)$ and $Al\ (s) + 4\ OH^-\ (aq)$
$\rightarrow Al(OH)_4^-\ (aq) + 3\ e^-$
Add half-reactions:
$MnO_4^-\ (aq) + 2\ H_2O\ (l) + \cancel{3\ e^-} + Al\ (s) + \cancel{4\ OH^-\ (aq)} \rightarrow MnO_2\ (s) + \cancel{4\ OH^-\ (aq)} + Al(OH)_4^-\ (aq) + \cancel{3\ e^-}$
Cancel electrons: $MnO_4^-\ (aq) + 2\ H_2O\ (l) + Al\ (s) \rightarrow MnO_2\ (s) + Al(OH)_4^-\ (aq)$

Check:

Reactants	Products
1 Mn atom	1 Mn atom
6 O atoms	6 O atoms
4 H atoms	4 H atoms
1 Al atom	1 Al atom
–1 charge	–1 charge

c) Separate: $Cl_2\ (g) \rightarrow Cl^-\ (aq)$ and $Cl_2\ (g) \rightarrow ClO^-\ (aq)$
Balance non H & O elements: $Cl_2\ (g) \rightarrow 2\ Cl^-\ (aq)$ and $Cl_2\ (g) \rightarrow 2\ ClO^-\ (aq)$
Balance O with H_2O: $Cl_2\ (g) \rightarrow 2\ Cl^-\ (aq)$ and $Cl_2\ (g) + 2\ H_2O\ (l) \rightarrow 2\ ClO^-\ (aq)$
Balance H with H^+: $Cl_2\ (g) \rightarrow 2\ Cl^-\ (aq)$ and $Cl_2\ (g) + 2\ H_2O\ (l) \rightarrow 2\ ClO^-\ (aq) + 4\ H^+\ (aq)$
Neutralize H^+ with OH^-:
$Cl_2\ (g) \rightarrow 2\ Cl^-\ (aq)$ and $Cl_2\ (g) + \cancel{2\ H_2O(l)} + 4\ OH^-\ (aq) \rightarrow 2\ ClO^-\ (aq) + \underbrace{4\ H^+\ (aq) + 4\ OH^-\ (aq)}_{\cancel{4}2\ H_2O\ (l)}$

Add electrons: $Cl_2\ (g) + 2\ e^- \rightarrow 2\ Cl^-\ (aq)$ and $Cl_2\ (g) + 4\ OH^-\ (aq) \rightarrow 2\ ClO^-\ (aq) + 2\ H_2O\ (l) + 2\ e^-$
Equalize electrons: $Cl_2\ (g) + 2\ e^- \rightarrow 2\ Cl^-\ (aq)$ and $Cl_2\ (g) + 4\ OH^-(aq) \rightarrow 2\ ClO^-(aq) + 2\ H_2O\ (l) + 2\ e^-$
Add half-reactions: $Cl_2\ (g) + \cancel{2\ e^-} + Cl_2\ (g) + 4\ OH^-\ (aq) \rightarrow 2\ Cl^-\ (aq) + 2\ ClO^-\ (aq) + 2\ H_2O\ (l) + \cancel{2\ e^-}$
Cancel electrons: $2\ Cl_2\ (g) + 4\ OH^-\ (aq) \rightarrow 2\ Cl^-\ (aq) + 2\ ClO^-\ (aq) + 2\ H_2O\ (l)$
Simplify: $Cl_2\ (g) + 2\ OH^-\ (aq) \rightarrow Cl^-\ (aq) + ClO^-\ (aq) + H_2O\ (l)$

Check:

Reactants	Products
2 Cl atoms	2 Cl atoms

2 O atoms 2 O atoms
2 H atoms 2 H atoms
–2 charge –2 charge

43. **Given:** voltaic cell overall redox reaction
Find: Sketch voltaic cell, labeling anode, cathode, all species and direction of electron flow
Conceptual plan: Separate overall reaction into 2 half-cell reactions and add electrons as needed to balance reactions. Put anode reaction on the left (oxidation = electrons as product) and cathode reaction on the right (reduction = electrons as reactant). Electrons flow from anode to cathode.
Solution:

a) $2\ Ag^+\ (aq) + Pb\ (s) \rightarrow 2\ Ag\ (s) + Pb^{2+}\ (aq)$ separates to $2\ Ag^+\ (aq) \rightarrow 2\ Ag\ (s)$ and $Pb\ (s) \rightarrow Pb^{2+}\ (aq)$ then add electrons to balance to get the cathode reaction: $2\ Ag^+\ (aq) + 2\ e^- \rightarrow 2\ Ag\ (s)$ and the anode reaction: $Pb\ (s) \rightarrow Pb^{2+}\ (aq) + 2\ e^-$.
Since we have Pb (s) as the reactant for the oxidation, it will be our anode. Since we have Ag (s) as the product for the reduction, it will be our cathode. Simplify the cathode reaction, dividing all terms by 2.

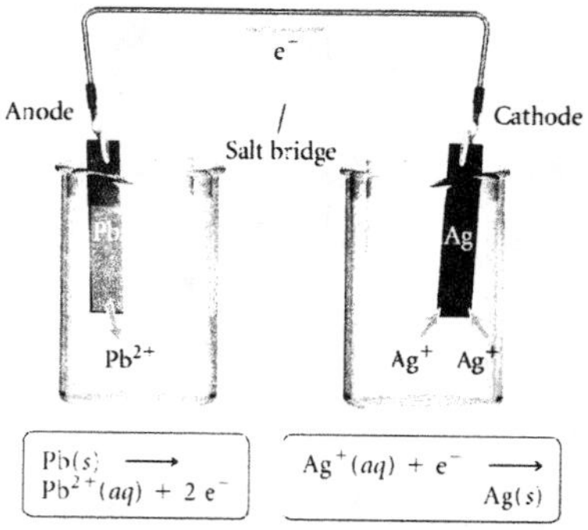

b) $2\ ClO_2\ (g) + 2\ I^-\ (aq) \rightarrow 2\ ClO_2^-\ (aq) + I_2\ (s)$ separates to $2\ ClO_2\ (g) \rightarrow 2\ ClO_2^-\ (aq)$ and $2\ I^-\ (aq) \rightarrow I_2\ (s)$ then add electrons to balance to get the cathode reaction: $2\ ClO_2\ (g) + 2\ e^- \rightarrow 2\ ClO_2^-\ (aq)$ and the anode reaction: $2\ I^-\ (aq) \rightarrow I_2\ (s) + 2\ e^-$.
Since we have $I^-\ (aq)$ as the reactant for the oxidation, we will need to use Pt as our anode. Since we have $ClO_2^-\ (aq)$ as the product for the reduction, we will need to use Pt as our cathode. Since $ClO_2\ (g)$ is our reactant for the reduction, we need to use an electrode assembly like that is used for a SHE. Simplify the cathode reaction, dividing all terms by 2.

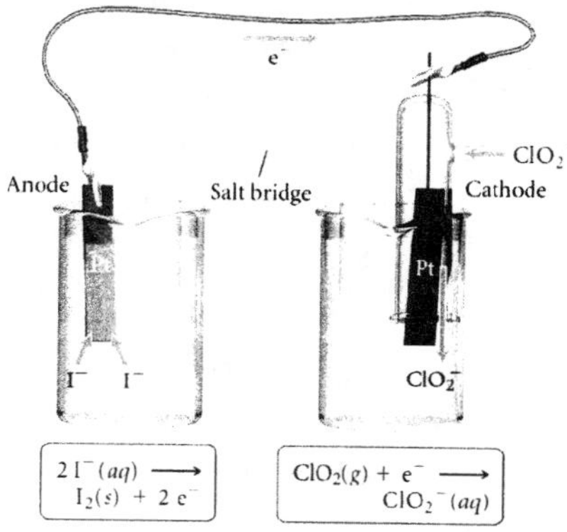

c) $O_2\ (g) + 4\ H^+\ (aq) + 2\ Zn\ (s) \rightarrow 2\ H_2O\ (l) + 2\ Zn^{2+}\ (aq)$ separates to $O_2\ (g) + 4\ H^+\ (aq) \rightarrow 2\ H_2O\ (l)$ and $2\ Zn\ (s) \rightarrow 2\ Zn^{2+}\ (aq)$ then add electrons to balance to get the cathode reaction: $O_2\ (g) + 4\ H^+\ (aq) + 4\ e^- \rightarrow 2\ H_2O\ (l)$ and the anode reaction: $2\ Zn\ (s) \rightarrow 2\ Zn^{2+}\ (aq) + 4\ e^-$.
Since we have Zn (s) as the reactant for the oxidation, it will be our anode. Since we have $H_2O\ (l)$ as the product for the reduction, we will need to use Pt as our cathode. Since $O_2\ (g)$ is our reactant for the reduction, we need to use an electrode assembly like what is used for a SHE. Simplify the anode reaction, dividing all terms by 2.

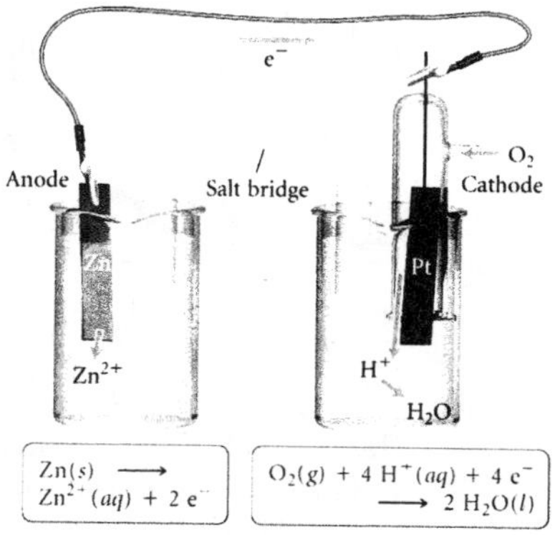

45. **Given:** overall reactions from #43 **Find:** E^o_{cell}
Conceptual plan: Look up half-reactions from solution of problem #43 in Table 18.1. In order to determine the cell potential of the oxidation reaction, reverse the sign of the standard reduction potential ($E^o_{ox} = -E^o_{red}$) then add the two half-cell potentials: $E^o_{cell} = E^o_{ox} + E^o_{red}$.
Solution:

a) $Ag^+\ (aq) + e^- \rightarrow Ag\ (s)$ $E^o_{red} = 0.80$ V and $Pb\ (s) \rightarrow Pb^{2+}\ (aq) + 2\ e^-$ $E^o_{ox} = -E^o_{red} = -(-0.13\ V) = 0.13$ V. Then $E^o_{cell} = E^o_{ox} + E^o_{red} = 0.80\ V + 0.13\ V = 0.93$ V.

b) $ClO_2\ (g) + e^- \rightarrow ClO_2^-\ (aq)$ $E^o_{red} = 0.95$ V and $2\ I^-\ (aq) \rightarrow I_2\ (s) + 2\ e^-$ $E^o_{ox} = -E^o_{red} = -0.54$ V.

Then $E^\circ_{cell} = E^\circ_{ox} + E^\circ_{red} =$ 0.95 V – 0.54 V = 0.41 V.

c) $O_2\ (g) + 4\ H^+\ (aq) + 4\ e^- \rightarrow 2\ H_2O\ (l)$ $E^\circ_{red} = 1.23$ V and $Zn\ (s) \rightarrow Zn^{2+}\ (aq) + 2\ e^-$ $E^\circ_{ox} = -E^\circ_{red} = -$ (– 0.76 V).
Then $E^\circ_{cell} = E^\circ_{ox} + E^\circ_{red} =$ 1.23 V + 0.76 V = 1.99 V.

Check: The units (V) are correct. All of the voltages are positive, which is consistent with a voltaic cell.

47. **Given:** voltaic cell drawing **Find:** a) determine electron flow direction, anode and cathode; b) write balanced overall reaction and calculate E°_{cell}; c) label electrodes as + and –; and d) directions of anions and cations from salt bridge

Conceptual plan: Look at each half-cell and write a reduction reaction, by using electrode and solution composition and adding electrons to balance. Look up half-reactions standard reduction potentials in Table 18.1. Since this is a voltaic cell, the cell potentials must be assigned to give a positive E°_{cell} (One reaction must be reversed to become the oxidation half-cell reaction and $E^\circ_{ox} = -E^\circ_{red}$.) then add the two half-cell potentials: $E^\circ_{cell} = E^\circ_{ox} + E^\circ_{red}$.

a) **Label electrode where the oxidation occurs as the anode. Label the electrode where the reduction as the cathode. Electrons flow from anode to cathode.**

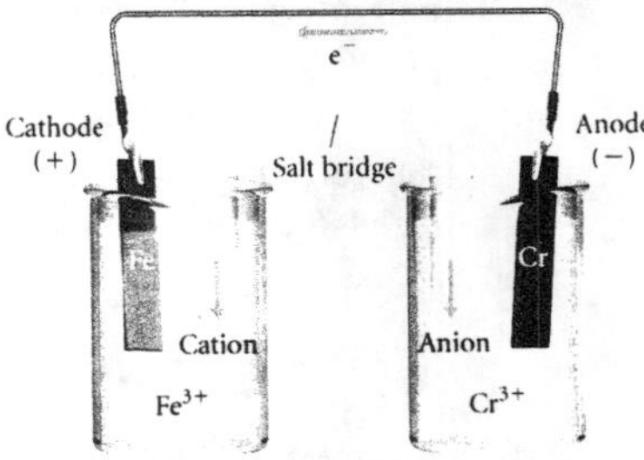

b) **Take two half-cell reactions and multiply the reactions as necessary to equalize the number of electrons transferred. Add the two half-cell reactions and cancel electrons and any other species.**

c) **Label anode as (–) and cathode as (+).**

d) **Cations will flow from the salt bridge towards the cathode and the anions will flow from salt bridge towards the anode.**

Solution: left side: $Fe^{3+}\ (aq) \rightarrow Fe\ (s)$ and right side: $Cr^{3+}\ (aq) \rightarrow Cr\ (s)$ add electrons to balance $Fe^{3+}\ (aq) + 3\ e^- \rightarrow Fe\ (s)$ and right side: $Cr^{3+}\ (aq) + 3\ e^- \rightarrow Cr\ (s)$. Look up cell standard reduction potentials: $Fe^{3+}\ (aq) + 3\ e^- \rightarrow Fe\ (s)$ $E^\circ_{red} = -$ 0.036 V and $Cr^{3+}\ (aq) + 3\ e^- \rightarrow Cr\ (s)$ $E^\circ_{red} = -$ 0.73 V. In order to get a positive cell potential, the second reaction is the oxidation reaction. $E^\circ_{cell} = E^\circ_{ox} + E^\circ_{red} = -$ (– 0.73 V) – 0.036 V = + 0.69 V.

a, c, d)

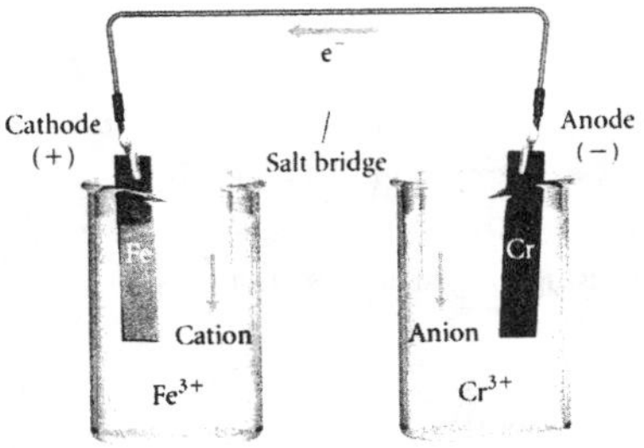

b) Add two half-reactions with the second reaction reversed.
$Fe^{3+}\ (aq) + \cancel{3e^-} + Cr\ (s) \rightarrow Fe\ (s) + Cr^{3+}\ (aq) + \cancel{3e^-}$.
Cancel electrons to get:
$Fe^{3+}\ (aq) + Cr\ (s) \rightarrow Fe\ (s) + Cr^{3+}\ (aq)$.

Check: All atoms and charge are balanced. The units (V) are correct. The cell potential is positive which is consistent with a voltaic cell.

49. **Given:** overall reactions from #43 **Find:** line notation

Conceptual plan: Use solution from problem #43. Write the oxidation half-reaction components on the left and the reduction on the right. A double vertical line (||), indicating the salt bridge, separates the two half-reactions. Substances in different phases are separated by a single vertical line (|), which represents the boundary between the phases. For some redox reactions, the reactants and products of one or both of the half-reactions may be in the same phase. In these cases, the reactants and products are simply separated from each other with a comma in the line diagram. Such cells use an inert electrode, such as platinum (Pt) or graphite, as the anode or cathode (or both).

Solution:

a) Reduction reaction: $Ag^+\ (aq) + e^- \rightarrow Ag\ (s)$ and the oxidation reaction: $Pb\ (s) \rightarrow Pb^{2+}\ (aq) + 2\ e^-$ so $Pb\ (s)|Pb^{2+}\ (aq)||Ag^+\ (aq)\ |Ag\ (s)$

b) Reduction reaction: $ClO_2\ (g) + e^- \rightarrow ClO_2^-\ (aq)$ and the oxidation reaction: $2\ I^-\ (aq) \rightarrow I_2\ (s) + 2\ e^-$ so $Pt\ (s)|I^-\ (aq)|\ I_2\ (s)||\ ClO_2\ (g)|ClO_2^-\ (aq)|Pt\ (s)$

c) Reduction reaction: $O_2\ (g) + 4\ H^+\ (aq) + 4\ e^- \rightarrow 2\ H_2O\ (l)$ and the oxidation reaction: $Zn\ (s) \rightarrow Zn^{2+}\ (aq) + 2\ e^-$ so $Zn\ (s)|Zn^{2+}\ (aq)\ ||O_2\ (g)|H^+\ (aq),\ H_2O\ (l)|Pt\ (s)$

51. **Given:** $Sn\ (s)|Sn^{2+}\ (aq)\ ||\ NO_3^-\ (aq),\ H^+\ (aq)\ ),\ H_2O\ (l\ )|NO\ (g)|Pt\ (s)$
Find: Sketch voltaic cell, labeling anode, cathode all species and direction of electron flow and E^o_{cell}
Conceptual plan: Separate overall reaction into 2 half-cell reactions knowing that the oxidation half-reaction components on the left and the reduction on the right. Add electrons as needed to balance reactions. Multiply the half-reactions by the appropriate factors to have an equal number of electrons transferred. Add the half-cell reactions and cancel electrons. Put anode reaction on the left (oxidation = electrons as product) and cathode reaction on the right (reduction = electrons as reactant). Electrons flow from anode to cathode. Look up half-reactions from solution of problem #44 in Table 18.1. In order to determine the cell potential of the oxidation reaction, reverse the sign of the standard reduction potential ($E^o_{ox} = -\ E^o_{red}$) then add the two half-cell potentials: $E^o_{cell} = E^o_{ox} + E^o_{red.}$

Solution: Oxidation reaction (anode): $Sn\ (s) \rightarrow Sn^{2+}\ (aq) + 2\ e^-$ $E^o_{ox} = -\ E^o_{red} = -\ (-\ 0.14\ V) = 0.14\ V$ and Reduction reaction (cathode): $NO_3^-\ (aq) + 4\ H^+\ (aq) + 3\ e^- \rightarrow NO\ (g) + 2\ H_2O\ (l)$ $E^o_{red} = 0.96\ V$. $E^o_{cell} = E^o_{ox} + E^o_{red} = 0.14\ V + 0.96\ V = 1.10\ V$. Multiply first reaction by 3 and the second reaction by 2 so that 6 electrons are transferred. $3\ Sn\ (s) \rightarrow 3\ Sn^{2+}\ (aq) + 6\ e^-$ and $2\ NO_3^-\ (aq) + 8\ H^+\ (aq) + 6\ e^- \rightarrow 2\ NO\ (g) + 4\ H_2O\ (l)$. Add the two half-reactions and cancel electrons $3\ Sn\ (s) + 2\ NO_3^-\ (aq) + 8\ H^+\ (aq) + \cancel{6\ e^-} \rightarrow 3\ Sn^{2+}\ (aq) + \cancel{6\ e^-} + 2\ NO\ (g) + 4\ H_2O\ (l)$. So balanced reaction is: $3\ Sn\ (s) + 2\ NO_3^-\ (aq) + 8\ H^+\ (aq) \rightarrow 3\ Sn^{2+}\ (aq) + 2\ NO\ (g) + 4\ H_2O\ (l)$.

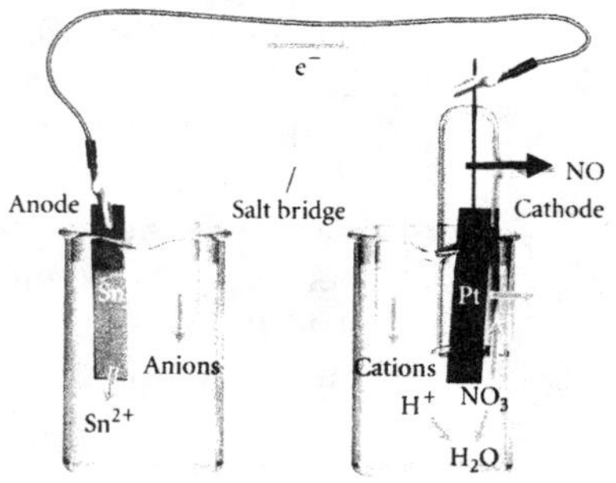

Check: All atoms and charge are balanced. The units (V) are correct. The cell potential is positive which is consistent with a voltaic cell.

53. **Given:** overall reactions **Find:** spontaneity in forward direction
Conceptual plan: Separate overall reaction into 2 half-cell reactions and add electrons as needed to balance reactions. Look up half-reactions in Table 18.1. In order to determine the cell potential of the oxidation reaction, reverse the sign of the standard reduction potential ($E^o_{ox} = -\ E^o_{red}$) then add the two half-cell potentials: $E^o_{cell} = E^o_{ox} + E^o_{red.}$ If $E^o_{cell} > 0$ the reaction is spontaneous in the forward direction.
Solution:

a) $Ni\ (s) + Zn^{2+}\ (aq) \rightarrow Ni^{2+}\ (aq) + Zn\ (s)$ separates to $Ni\ (s) \rightarrow Ni^{2+}\ (aq)$ and $Zn^{2+}\ (aq) \rightarrow Zn\ (s)$ add electrons $Ni\ (s) \rightarrow Ni^{2+}\ (aq) + 2\ e^-$ and $Zn^{2+}\ (aq) + 2\ e^- \rightarrow Zn\ (s)$. Look up cell potentials. Ni is oxidized so $E^o_{ox} = -\ E^o_{red} = -\ (-\ 0.23\ V) = 0.23\ V$. Zn^{2+} is reduced so $E^o_{red} = -\ 0.76\ V$. Then $E^o_{cell} = E^o_{ox} + E^o_{red} = 0.23\ V - 0.76\ V = -\ 0.53\ V$ and so the reaction is nonspontaneous.

b) $Ni\ (s) + Pb^{2+}\ (aq) \rightarrow Ni^{2+}\ (aq) + Pb\ (s)$ separates to $Ni\ (s) \rightarrow Ni^{2+}\ (aq)$ and $Pb^{2+}\ (aq) \rightarrow Pb\ (s)$ add electrons $Ni\ (s) \rightarrow Ni^{2+}\ (aq) + 2\ e^-$ and $Pb^{2+}\ (aq) + 2\ e^- \rightarrow Pb\ (s)$. Look up cell potentials. Ni is oxidized so $E^o_{ox} = -\ E^o_{red} = -\ (-\ 0.23\ V) = 0.23\ V$. Pb^{2+} is reduced so $E^o_{red} = -\ 0.13\ V$. Then $E^o_{cell} = E^o_{ox} + E^o_{red} = 0.23\ V - 0.13\ V = +\ 0.10\ V$ and so the reaction is spontaneous.

c) $Al\ (s) + 3\ Ag^+\ (aq) \rightarrow Al^{3+}\ (aq) + 3\ Ag\ (s)$ separates to $Al\ (s) \rightarrow Al^{3+}\ (aq)$ and $3\ Ag^+\ (aq) \rightarrow 3\ Ag\ (s)$ add electrons $Al\ (s) \rightarrow Al^{3+}\ (aq) + 3\ e^-$ and $3\ Ag^+\ (aq) + 3\ e^- \rightarrow 3\ Ag\ (s)$. Simplify the Ag reaction to: $Ag^+\ (aq) + e^- \rightarrow Ag\ (s)$. Look up cell potentials. Al is oxidized so $E^o_{ox} = -\ E^o_{red} = -\ (-\ 1.66\ V) = 1.66\ V$. Ag^+ is reduced so $E^o_{red} = 0.80\ V$. Then $E^o_{cell} = E^o_{ox} + E^o_{red} = 1.66\ V + 0.80\ V = +\ 2.46\ V$ and so the reaction is spontaneous.

d) $Pb\ (s) + Mn^{2+}\ (aq) \rightarrow Pb^{2+}\ (aq) + Mn\ (s)$ separates to $Pb\ (s) \rightarrow Pb^{2+}\ (aq)$ and $Mn^{2+}\ (aq) \rightarrow Mn\ (s)$ add electrons $Pb\ (s) \rightarrow Pb^{2+}\ (aq) + 2\ e^-$ and $Mn^{2+}\ (aq) + 2\ e^- \rightarrow Mn\ (s)$. Look up cell potentials. Pb is oxidized so $E^o_{ox} = -\ E^o_{red} = -\ (-\ 0.13\ V) = 0.13\ V$. Mn^{2+} is reduced so $E^o_{red} = -\ 1.18\ V$. Then $E^o_{cell} = E^o_{ox} + E^o_{red} = 0.13\ V - 1.18\ V = -\ 1.05\ V$ and so the reaction is nonspontaneous.

Check: The units (V) are correct. If the voltage is positive, the reaction is spontaneous.

55. In order for a metal to be able to reduce an ion, it must be above it in Table 18.1 (need positive $E^o_{cell} = E^o_{ox} + E^o_{red}$). So we need a metal that is above Mn^{2+}, but below Mg^{2+}. Aluminum is the only one in the table that meets these criteria.

57. In general, metals whose reduction half-reactions lie below the reduction of H^+ to H_2 in Table 18.1 will dissolve in acids, while metals above it will not. a) Al and c) Pb meet this criterion. To write the balanced redox reactions, pair the oxidation of the metal with the reduction of H^+ to H_2 ($2\,H^+(aq) + 2\,e^- \rightarrow H_2(g)$). For Al, $Al\,(s) \rightarrow Al^{3+}\,(aq) + 3\,e^-$. In order to balance the number of electrons transferred we need to multiply the Al reaction by 2 and the H^+ reaction by 3. So, $2\,Al\,(s) \rightarrow 2\,Al^{3+}\,(aq) + 6\,e^-$ and $6\,H^+(aq) + 6\,e^- \rightarrow 3\,H_2(g)$. Adding the two reactions: $2\,Al\,(s) + 6\,H^+(aq) + \cancel{6\,e^-} \rightarrow 2\,Al^{3+}\,(aq) + \cancel{6\,e^-} + 3\,H_2(g)$. Simplify to $2\,Al\,(s) + 6\,H^+(aq) \rightarrow 2\,Al^{3+}\,(aq) + 3\,H_2(g)$. For Pb, $Pb\,(s) \rightarrow Pb^{2+}\,(aq) + 2\,e^-$. Since each reaction involves 2 electrons we can add the two reactions. $Pb\,(s) + 2\,H^+(aq) + \cancel{2\,e^-} \rightarrow Pb^{2+}\,(aq) + \cancel{2\,e^-} + H_2(g)$. Simplify to $Pb\,(s) + 2\,H^+(aq) \rightarrow Pb^{2+}\,(aq) + H_2(g)$.

59. Nitric acid (HNO_3) oxidizes metals through the following reduction half-reaction: $NO_3^-\,(aq) + 4\,H^+\,(aq) + 3\,e^- \rightarrow NO\,(g) + 2\,H_2O\,(l)$ $E^o_{red} = 0.96$ V. Since this half-reaction is above the reduction of H^+ in Table 18.1, HNO_3 can oxidize metals (such as copper, for example) that can't be oxidizied by HCl. a) Cu will be oxidized, but b) Au (which has a reduction potential of 1.50 V will not be oxidized. To write the balanced redox reactions, pair the oxidation of the metal with the reduction of nitric acid ($NO_3^-\,(aq) + 4\,H^+\,(aq) + 3\,e^- \rightarrow NO\,(g) + 2\,H_2O\,(l)$). For Cu, $Cu\,(s) \rightarrow Cu^{2+}\,(aq) + 2\,e^-$. In order to balance the number of electrons transferred we need to multiply the Cu reaction by 3 and the nitric acid reaction by 2. So, $3\,Cu\,(s) \rightarrow 3\,Cu^{2+}\,(aq) + 6\,e^-$ and $2\,NO_3^-\,(aq) + 8\,H^+\,(aq) + 6\,e^- \rightarrow 2\,NO\,(g) + 4\,H_2O\,(l)$. Adding the two reactions: $3\,Cu\,(s) + 2\,NO_3^-\,(aq) + 8\,H^+\,(aq) + \cancel{6\,e^-} \rightarrow 3\,Cu^{2+}\,(aq) + \cancel{6\,e^-} + 2\,NO\,(g) + 4\,H_2O\,(l)$. Simplify to $3\,Cu\,(s) + 2\,NO_3^-\,(aq) + 8\,H^+\,(aq) \rightarrow 3\,Cu^{2+}\,(aq) + 2\,NO\,(g) + 4\,H_2O\,(l)$.

61. **Given:** overall reactions **Find:** E^o_{cell} and spontaneity in forward direction

Conceptual plan: Separate overall reaction into 2 half-cell reactions and add electrons as needed to balance reactions. Look up half-reactions in Table 18.1. In order to determine the cell potential of the oxidation reaction, reverse the sign of the standard reduction potential ($E^o_{ox} = -E^o_{red}$) then add the two half-cell potentials: $E^o_{cell} = E^o_{ox} + E^o_{red}$. If $E^o_{cell} > 0$ the reaction is spontaneous in the forward direction.

Solution:

a) $2\,Cu\,(s) + Mn^{2+}\,(aq) \rightarrow 2\,Cu^+\,(aq) + Mn\,(s)$ separates to $2\,Cu\,(s) \rightarrow 2\,Cu^+\,(aq)$ and $Mn^{2+}\,(aq) \rightarrow Mn\,(s)$ add electrons $2\,Cu\,(s) \rightarrow 2\,Cu^+\,(aq) + 2\,e^-$ and $Mn^{2+}\,(aq) + 2\,e^- \rightarrow Mn\,(s)$. Simplify the Cu reaction to: $Cu\,(s) \rightarrow Cu^+\,(aq) + e^-$. Look up cell potentials. Cu is oxidized so $E^o_{ox} = -E^o_{red} = -0.52$ V. Mn^{2+} is reduced so $E^o_{red} = -1.18$ V. Then $E^o_{cell} = E^o_{ox} + E^o_{red} = -0.52\text{ V} - 1.18\text{ V} = -1.70$ V and so the reaction is nonspontaneous.

b) $MnO_2\,(s) + 4\,H^+(aq) + Zn\,(s) \rightarrow Mn^{2+}\,(aq) + 2\,H_2O\,(l) + Zn^{2+}\,(aq)$ separates to $MnO_2\,(s) + 4\,H^+(aq) \rightarrow Mn^{2+}\,(aq) + 2\,H_2O\,(l)$ and $Zn\,(s) \rightarrow Zn^{2+}\,(aq)$ add electrons $MnO_2\,(s) + 4\,H^+(aq) + 2\,e^- \rightarrow Mn^{2+}\,(aq) + 2\,H_2O\,(l)$ and $Zn\,(s) \rightarrow Zn^{2+}\,(aq) + 2\,e^-$. Look up cell potentials. Zn is oxidized so $E^o_{ox} = -E^o_{red} = --0.76\text{ V} = 0.76$ V. Mn is reduced so $E^o_{red} = 1.21$ V. Then $E^o_{cell} = E^o_{ox} + E^o_{red} = 0.76\text{ V} + 1.21\text{ V} = +1.97$ V and so the reaction is spontaneous.

c) $Cl_2\,(g) + 2\,F^-\,(aq) \rightarrow 2\,Cl^-\,(aq) + F_2\,(g)$ separates to $Cl_2\,(g) \rightarrow 2\,Cl^-\,(aq)$ and $2\,F^-\,(aq) \rightarrow F_2\,(g)$ add electrons $Cl_2\,(g) + 2\,e^- \rightarrow 2\,Cl^-\,(aq)$ and $2\,F^-\,(aq) \rightarrow F_2\,(g) + 2\,e^-$. Look up cell potentials. F is oxidized so $E^o_{ox} = -E^o_{red} = -2.87$ V. Cl is reduced so $E^o_{red} = 1.36$ V. Then $E^o_{cell} = E^o_{ox} + E^o_{red} = -2.87\text{ V} + 1.36\text{ V} = -1.51$ V and so the reaction is nonspontaneous.

Check: The units (V) are correct. If the voltage is positive, the reaction is spontaneous.

63. a) Pb^{2+}. The strongest oxidizing agent is the one with the reduction reaction that is closest to the top of Table 18.1.

65. **Given:** overall reactions **Find:** ΔG^o_{rxn} and spontaneity in forward direction

Conceptual plan: Separate overall reaction into 2 half-cell reactions and add electrons as needed to balance reactions. Look up half-reactions in Table 18.1. In order to determine the cell potential of the oxidation reaction, reverse the sign of the standard reduction potential ($E^o_{ox} = -E^o_{red}$) then add the two half-cell potentials: $E^o_{cell} = E^o_{ox} + E^o_{red}$. then calculate ΔG^o_{rxn} using $\Delta G^0_{rxn} = -nFE^0_{cell}$.

Solution:

a) $Pb^{2+}\,(aq) + Mg\,(s) \rightarrow Pb\,(s) + Mg^{2+}\,(aq)$ separates to $Pb^{2+}\,(aq) \rightarrow Pb\,(s)$ and $Mg\,(s) \rightarrow Mg^{2+}\,(aq)$ add electrons $Pb^{2+}\,(aq) + 2\,e^- \rightarrow Pb\,(s)$ and $Mg\,(s) \rightarrow Mg^{2+}\,(aq) + 2\,e^-$. Look up cell potentials. Mg is oxidized so $E^\circ_{ox} = -E^\circ_{red} = -(-2.37\text{ V}) = 2.37$ V. Pb^{2+} is reduced so $E^\circ_{red} = -0.13$ V. Then $E^\circ_{cell} = E^\circ_{ox} + E^\circ_{red} = 2.37\text{ V} - 0.13\text{ V} = +2.24$ V. $n = 2$ so

$$\Delta G^0_{rxn} = -nFE^0_{cell} = -2\ \cancel{\text{mol e}^-}\ \text{x}\ \frac{96{,}485\text{ C}}{\cancel{\text{mol e}^-}}\ \text{x}\ 2.24\text{ V} = -2\ \text{x}\ 96{,}485\ \cancel{\text{C}}\ \text{x}\ 2.24\frac{\text{J}}{\cancel{\text{C}}} = -4.32\ \text{x}\ 10^5\text{ J} = -432\text{ kJ}.$$

b) $Br_2\,(l) + 2\,Cl^-\,(aq) \rightarrow 2\,Br^-\,(aq) + Cl_2\,(g)$ separates to $Br_2\,(g) \rightarrow 2\,Br^-\,(aq)$ and $2\,Cl^-\,(aq) \rightarrow Cl_2\,(g)$ add electrons $Br_2\,(g) + 2\,e^- \rightarrow 2\,Br^-\,(aq)$ and $2\,Cl^-\,(aq) \rightarrow Cl_2\,(g) + 2\,e^-$. Look up cell potentials. Cl is oxidized so $E^\circ_{ox} = -E^\circ_{red} = -1.36$ V. Br is reduced so $E^\circ_{red} = 1.09$ V. Then $E^\circ_{cell} = E^\circ_{ox} + E^\circ_{red} = -1.36\text{ V} + 1.09\text{ V} = -0.27$ V. $n = 2$ so

$$\Delta G^0_{rxn} = -nFE^0_{cell} = -2\ \cancel{\text{mol e}^-}\ \text{x}\ \frac{96{,}485\text{ C}}{\cancel{\text{mol e}^-}}\ \text{x}\ -0.27\text{ V} = -2\ \text{x}\ 96{,}485\ \cancel{\text{C}}\ \text{x} -0.27\frac{\text{J}}{\cancel{\text{C}}} = 5.2\ \text{x}\ 10^4\text{ J} = 52\text{ kJ}.$$

c) $MnO_2\,(s) + 4\,H^+(aq) + Cu\,(s) \rightarrow Mn^{2+}\,(aq) + 2\,H_2O\,(l) + Cu^{2+}\,(aq)$ separates to $MnO_2\,(s) + 4\,H^+(aq) \rightarrow Mn^{2+}\,(aq) + 2\,H_2O\,(l)$ and $Cu\,(s) \rightarrow Cu^{2+}\,(aq)$ add electrons $MnO_2\,(s) + 4\,H^+(aq) + 2\,e^- \rightarrow Mn^{2+}\,(aq) + 2\,H_2O\,(l)$ and $Cu\,(s) \rightarrow Cu^{2+}\,(aq) + 2\,e^-$. Look up cell potentials. Cu is oxidized so $E^\circ_{ox} = -E^\circ_{red} = -0.34$ V. Mn is reduced so $E^\circ_{red} = 1.21$ V. Then $E^\circ_{cell} = E^\circ_{ox} + E^\circ_{red} = -0.34\text{ V} + 1.21\text{ V} = +0.87$ V. $n = 2$ so

$$\Delta G^0_{rxn} = -nFE^0_{cell} = -2\ \cancel{\text{mol e}^-}\ \text{x}\ \frac{96{,}485\text{ C}}{\cancel{\text{mol e}^-}}\ \text{x}\ 0.87\text{ V} = -2\ \text{x}\ 96{,}485\ \cancel{\text{C}}\ \text{x}\ 0.87\frac{\text{J}}{\cancel{\text{C}}} = -1.7\ \text{x}\ 10^5\text{ J} = -1.7\ \text{x}\ 10^2\text{ kJ}.$$

Check: The units (kJ) are correct. If the voltage is positive, the reaction is spontaneous and the free energy change is negative.

67. **Given:** overall reactions from #65 **Find:** K
Conceptual plan: °C → K then ΔG°_{rxn}, T → K

$K = 273.15 + °C \qquad \Delta G^0_{rxn} = -RT\ln K$

Solution: T = 273.15 + 25 °C = 298 K then

a) $\Delta G^0_{rxn} = -RT\ln K$ Rearrange to solve for K. $K = e^{\frac{-\Delta G^0_{rxn}}{RT}} = e^{\frac{-(-432)\ \cancel{\text{kJ}}\ \text{x}\ \frac{1000\ \cancel{\text{J}}}{1\ \cancel{\text{kJ}}}}{\left(8.314\frac{\cancel{\text{J}}}{\cancel{\text{K}}\ \text{mol}}\right)(298\ \cancel{\text{K}})}} = e^{174.364} = 5.31\ \text{x}\ 10^{75}$.

b) $\Delta G^0_{rxn} = -RT\ln K$ Rearrange to solve for K. $K = e^{\frac{-\Delta G^0_{rxn}}{RT}} = e^{\frac{-52\ \cancel{\text{kJ}}\ \text{x}\ \frac{1000\ \cancel{\text{J}}}{1\ \cancel{\text{kJ}}}}{\left(8.314\frac{\cancel{\text{J}}}{\cancel{\text{K}}\ \text{mol}}\right)(298\ \cancel{\text{K}})}} = e^{-20.988} = 7.7\ \text{x}\ 10^{-10}$.

c) $\Delta G^0_{rxn} = -RT\ln K$ Rearrange to solve for K.

$$K = e^{\frac{-\Delta G^0_{rxn}}{RT}} = e^{\frac{-(-170\ \cancel{\text{kJ}})\ \text{x}\ \frac{1000\ \cancel{\text{J}}}{1\ \cancel{\text{kJ}}}}{\left(8.314\frac{\cancel{\text{J}}}{\cancel{\text{K}}\ \text{mol}}\right)(298\ \cancel{\text{K}})}} = e^{68.616} = 6.3\ \text{x}\ 10^{29}.$$

Check: The units (none) are correct. If the voltage is positive, the reaction is spontaneous and the free energy change is negative and the equilibrium constant is large.

69. **Given:** $Ni^{2+}\,(aq) + Cd\,(s) \rightarrow$ **Find:** K
Conceptual plan: Write 2 half-cell reactions and add electrons as needed to balance reactions. Look up half-reactions in Table 18.1. In order to determine the cell potential of the oxidation reaction, reverse the sign of the standard reduction potential ($E^\circ_{ox} = -E^\circ_{red}$) then add the two half-cell potentials: $E^\circ_{cell} = E^\circ_{ox} + E^\circ_{red}$. then °C → K then E°_{cell}, n, T → K

$K = 273.15 + °C \qquad \Delta G^0_{rxn} = -RT\ln K = -nFE^0_{cell}$.

Solution:
a) $Ni^{2+}\,(aq) + 2\,e^- \rightarrow Ni\,(s)$ and $Cd\,(s) \rightarrow Cd^{2+}\,(aq) + 2\,e^-$. Look up cell potentials. Cd is oxidized so $E^\circ_{ox} = -E^\circ_{red} = -(-0.40\text{ V}) = 0.40$ V. Ni^{2+} is reduced so $E^\circ_{red} = -0.23$ V. Then $E^\circ_{cell} = E^\circ_{ox} + E^\circ_{red} = 0.40\text{ V} - 0.23\text{ V} = +0.17$ V. The overall reaction is $Ni^{2+}\,(aq) + Cd\,(s) \rightarrow Ni\,(s) + Cd^{2+}\,(aq)$. $n = 2$

and T = 273.15 + 25 °C = 298 K then $\Delta G^0_{rxn} = -RT \ln K = -nFE^0_{cell}$. Rearrange to solve for K.

$$K = e^{\frac{nFE^0_{cell}}{RT}} = e^{\frac{2\ \text{mol e}^- \times \frac{96{,}485\ \text{C}}{\text{mol e}^-} \times 0.17 \frac{\text{J}}{\text{C}}}{\left(8.314 \frac{\text{J}}{\text{K mol}}\right)(298\ \text{K})}} = e^{13.241} = 5.6 \times 10^5 .$$

Check: The units (none) are correct. If the voltage is positive, the reaction is spontaneous and the equilibrium constant is large.

71. **Given:** $n = 2$ and K = 25 **Find:** $\Delta G°_{rxn}$ and $E°_{cell}$
Conceptual plan: K, T → $\Delta G°_{rxn}$ and $\Delta G°_{rxn}$, n → $E°_{cell}$

$$\Delta G^0_{rxn} = -RT \ln K \qquad \Delta G^0_{rxn} = -nFE^0_{cell}$$

Solution: $\Delta G^0_{rxn} = -RT \ln K = -\left(8.314 \frac{\text{J}}{\text{K mol}}\right)(298\ \text{K}) \ln 25 = -7.97500 \times 10^3\ \text{J} = -8.0\ \text{kJ}$ and $\Delta G^0_{rxn} = -nFE^0_{cell}$.

Rearrange to solve for $E°_{cell}$. $E^0_{cell} = \frac{\Delta G^0_{rxn}}{-nF} = \frac{-7.97500 \times 10^3\ \text{J}}{-2\ \text{mol e}^- \times \frac{96{,}485\ \text{C}}{\text{mol e}^-}} = 0.041 \frac{\text{V C}}{\text{C}} = 0.041\ \text{V}$.

Check: The units (kJ and V) are correct. If K > 1 then the voltage is positive, the free energy change is negative.

73. **Given:** Sn^{2+} *(aq)* + Mn *(s)* → Sn *(s)* + Mn^{2+} *(aq)* **Find:** a) $E°_{cell}$; b) E_{cell} when $[Sn^{2+}] = 0.0100$ M; $[Mn^{2+}] = 2.00$ M; and c) E_{cell} when $[Sn^{2+}] = 2.00$ M; $[Mn^{2+}] = 0.0100$ M
Conceptual plan: a) Separate overall reaction into 2 half-cell reactions and add electrons as needed to balance reactions. Look up half-reactions in Table 18.1. In order to determine the cell potential of the oxidation reaction, reverse the sign of the standard reduction potential ($E°_{ox} = -E°_{red}$) then add the two half-cell potentials: $E°_{cell} = E°_{ox} + E°_{red}$. b) and c) $E°_{cell}$, $[Sn^{2+}]$, $[Mn^{2+}]$, n → E_{cell}

$$E_{cell} = E°_{cell} - \frac{0.0592\ V}{n} \log Q \text{ where } Q = \frac{[Mn^{2+}]}{[Sn^{2+}]}$$

Solution:

a) separate overall reaction to: Sn^{2+} *(aq)* → Sn *(s)* and Mn *(s)* → Mn^{2+} *(aq)* add electrons Sn^{2+} *(aq)* + 2 e^- → Sn *(s)* and Mn *(s)* → Mn^{2+} *(aq)* + 2 e^-. Look up cell potentials. Mn is oxidized so $E°_{ox} = -E°_{red} = -(-1.18\ \text{V}) = 1.18$ V. Sn^{2+} is reduced so $E°_{red} = -0.14$ V. Then $E°_{cell} = E°_{ox} + E°_{red} = 1.18\ \text{V} - 0.14\ \text{V} = +1.04$ V.

b) $Q = \frac{[Mn^{2+}]}{[Sn^{2+}]} = \frac{2.00\ \text{M}}{0.0100\ \text{M}} = 200.$ and $n = 2$ the

$$E_{cell} = E^0_{cell} - \frac{0.0592\ \text{V}}{n} \log Q = 1.04\ \text{V} - \frac{0.0592\ \text{V}}{2} \log 200. = +0.97\ \text{V}$$

c) $Q = \frac{[Mn^{2+}]}{[Sn^{2+}]} = \frac{0.0100\ \text{M}}{2.00\ \text{M}} = 0.00500$ and $n = 2$ then

$$E_{cell} = E^0_{cell} - \frac{0.0592\ \text{V}}{n} \log Q = 1.04\ \text{V} - \frac{0.0592\ \text{V}}{2} \log 0.00500 = +1.11\ \text{V}$$

Check: The units (V, V and V) are correct. The Sn^{2+} reduction reaction is above the Mn^{2+} reduction reaction, so the standard cell potential will be positive. Having more products than reactants reduces the cell potential. Having more reactants than products raises the cell potential.

75. **Given:** Pb *(s)* → Pb^{2+} *(aq, 0.10 M)* + 2 e^- and MnO_4^- *(aq, 1.50 M)* + 4 H^+ *(aq, 2.0 M)* + 3 e^- → MnO_2 *(s)* + 2 H_2O *(l)* **Find:** E_{cell}
Conceptual plan: Look up half-reactions in Table 18.1. In order to determine the cell potential of the oxidation reaction, reverse the sign of the standard reduction potential ($E°_{ox} = -E°_{red}$) then add the two half-cell potentials: $E°_{cell} = E°_{ox} + E°_{red}$. Equalize the number of electrons transferred by multiplying the first reaction by 3 and the second reaction by 2. Add the two half-cell reactions and cancel the electrons. Then $E°_{cell}$, $[Pb^{2+}]$, $[MnO_4^-]$, $[H^+]$, n → E_{cell}

$$E_{cell} = E°_{cell} - \frac{0.0592\ V}{n} \log Q \text{ where } Q = \frac{[Pb^{2+}]^3}{[MnO_4^-]^2[H^+]^8}$$

Solution: Pb is oxidized so $E^{\circ}_{ox} = -E^{\circ}_{red} = --0.13$ V = 0.13 V. Mn is reduced so $E^{\circ}_{red} =$ 1.68 V. Then $E^{\circ}_{cell} = E^{\circ}_{ox} + E^{\circ}_{red} =$ 0.13 V + 1.68 V = + 1.81 V. Equalizing the electrons: 3 Pb *(s)* → 3 Pb^{2+} *(aq)* + 6 e^- and 2 MnO_4^- *(aq)* + 8 H^+ *(aq)* + 6 e^- → 2 MnO_2 *(s)* + 4 H_2O *(l)*. Adding the two reactions: 3 Pb *(s)*+ 2 MnO_4^- *(aq)* + 8 H^+ *(aq)* + ~~6 e⁻~~ → 3 Pb^{2+} *(aq)* + ~~6 e⁻~~ + 2 MnO_2 *(s)* + 4 H_2O *(l)*. Cancel the electrons: 3 Pb *(s)*+ 2 MnO_4^- *(aq)* + 8 H^+ *(aq)* → 3 Pb^{2+} *(aq)* + 2 MnO_2 *(s)* + 4 H_2O *(l)*. So $n = 6$ and

$$Q = \frac{[Pb^{2+}]^3}{[MnO_4^-]^2[H^+]^8} = \frac{(0.10)^3}{(1.50)^2(2.0)^8} = 1.\underline{7}361 \times 10^{-6} \text{ then}$$

$$E_{cell} = E^0_{cell} - \frac{0.0592 \text{ V}}{n} \log Q = 1.81 \text{ V} - \frac{0.0592 \text{ V}}{6} \log 1.\underline{7}361 \times 10^{-6} = +1.87 \text{ V} .$$

Check: The units (V) are correct. The MnO_4^- reduction reaction is above the Pb^{2+} reduction reaction, so the standard cell potential will be positive. Having more reactants than products raises the cell potential.

77. **Given:** Zn/Zn^{2+} and Ni/Ni^{2+} half-cells in voltaic cell; initially $[Ni^{2+}]$ = 1.50 M, and $[Zn^{2+}]$ = 0.100 M
Find: a) initial E_{cell}; b) E_{cell} when $[Ni^{2+}]$ = 0.500 M; and c) $[Ni^{2+}]$ and $[Zn^{2+}]$ when E_{cell} = 0.45 V
Conceptual plan:

a) **Write 2 half-cell reactions and add electrons as needed to balance reactions. Look up half-reactions in Table 18.1. In order to determine the cell potential of the oxidation reaction, reverse the sign of the standard reduction potential ($E^{\circ}_{ox} = -E^{\circ}_{red}$) then add the two half-cell potentials: $E^{\circ}_{cell} = E^{\circ}_{ox} + E^{\circ}_{red}$. Choose the direction of the half-cell reactions so that $E^{\circ}_{cell} > 0$. Add two half-cell reactions and cancel electrons to generate overall reaction. Define Q based on overall reaction. Then E°_{cell}, $[Ni^{2+}]$, $[Zn^{2+}]$, n → E_{cell}**

$$E_{cell} = E^{\circ}_{cell} - \frac{0.0592 \; V}{n} \log Q$$

b) **When $[Ni^{2+}]$ = 0.500 M, then $[Zn^{2+}]$ = 1.100 M (since the stoichiometric coefficients for Ni^{2+}: Zn^{2+} are 1:1, and the $[Ni^{2+}]$ drops by 1.00 M, the other concentration must rise by 1.00 M). Then E°_{cell}, $[Ni^{2+}]$, $[Zn^{2+}]$, n → E_{cell}**

$$E_{cell} = E^{\circ}_{cell} - \frac{0.0592 \; V}{n} \log Q$$

c) **E°_{cell}, E_{cell}, n → $[Zn^{2+}] / [Ni^{2+}]$ → $[Ni^{2+}]$, $[Zn^{2+}]$**

$$E_{cell} = E^{\circ}_{cell} - \frac{0.0592 \; V}{n} \log Q \qquad [Ni^{2+}] + [Zn^{2+}] = 1.50 \text{ M} + 0.100 \text{ M} = 1.60 \text{ M}$$

Solution:

a) Zn^{2+} *(aq)* + 2 e^- → Zn *(s)* and Ni^{2+} *(aq)* + 2 e^- → Ni *(s)*. Look up cell potentials. For Zn, $E^{\circ}_{red} = -$ 0.76 V. For Ni, $E^{\circ}_{red} = -0.23$ V. In order to get a positive E°_{cell} the sign of the Zn potential must be reversed. Zn is oxidized so $E^{\circ}_{ox} = -E^{\circ}_{red} = -(-0.76 \text{ V}) = 0.76$ V. Ni^{2+} is reduced so $E^{\circ}_{red} = -0.23$ V. Then $E^{\circ}_{cell} = E^{\circ}_{ox} + E^{\circ}_{red} =$ 0.76 V – 0.23 V = + 0.53 V. Adding the two half-cell reactions: Zn *(s)*+ Ni^{2+} *(aq)* + ~~2 e⁻~~ → Zn^{2+} *(aq)* + ~~2 e⁻~~ + Ni *(s)*. The overall reaction is: Zn *(s)*+ Ni^{2+} *(aq)* → Zn^{2+} *(aq)* + Ni *(s)*. Then $Q = \frac{[Zn^{2+}]}{[Ni^{2+}]} = \frac{0.100}{1.50} = 0.066\underline{6}667$ and $n = 2$ then

$$E_{cell} = E^0_{cell} - \frac{0.0592 \text{ V}}{n} \log Q = 0.53 \text{ V} - \frac{0.0592 \text{ V}}{2} \log 0.066\underline{6}667 = +0.56 \text{ V} .$$

b) $Q = \frac{[Zn^{2+}]}{[Ni^{2+}]} = \frac{1.100}{0.500} = 2.20$ then $E_{cell} = E^0_{cell} - \frac{0.0592 \text{ V}}{n} \log Q = 0.53 \text{ V} - \frac{0.0592 \text{ V}}{2} \log 2.20 = +0.52 \text{ V}$.

c) $E_{cell} = E^0_{cell} - \frac{0.0592 \text{ V}}{n} \log Q$ so $0.45 \text{ V} = 0.53 \text{ V} - \frac{0.0592 \text{ V}}{2} \log Q$ → $0.08 \cancel{\text{V}} = \frac{0.0592 \cancel{\text{V}}}{2} \log Q$ →

$\log Q = 2.\underline{7}0270$ → $Q = 10^{2.70270} = 5\underline{0}4.32$ then $Q = 5\underline{0}4.32 = \frac{[Zn^{2+}]}{1.60\text{M} - [Zn^{2+}]}$ solving for $[Zn^{2+}]$

$(5\underline{0}4.32)(1.60\text{M} - [Zn^{2+}]) = [Zn^{2+}]$ → $[Zn^{2+}] = \frac{8\underline{0}6.906 \text{ M}}{5\underline{0}5.32} = 1.\underline{5}9628 \text{ M} = 1.60 \text{ M}$ then

$[Ni^{2+}]$ = 1.60 M – 1.$\underline{5}$9628 M = 0.003 M.

Check: The units (V, V and V) are correct. The standard cell potential is positive and since there are more reactants than products this raises the cell potential. As the reaction proceeds, reactants are converted to products so the cell potential drops for parts b) and c).

79. **Given:** Zn/Zn^{2+} concentration cell, with $[Zn^{2+}] = 2.0$ M in one half-cell and $[Zn^{2+}] = 1.0 \times 10^{-3}$ M in other half-cell **Find:** Sketch voltaic cell, labeling anode, cathode, reactions at electrodes, all species and direction of electron flow

Conceptual plan: In a concentration cell, the half-cell with the higher concentration is always the half-cell where the reduction takes place (contains the cathode). The 2 half-cell reactions are the same, only reversed. Put anode reaction on the left (oxidation = electrons as product) and cathode reaction on the right (reduction = electrons as reactant). Electrons flow from anode to cathode.

Solution:

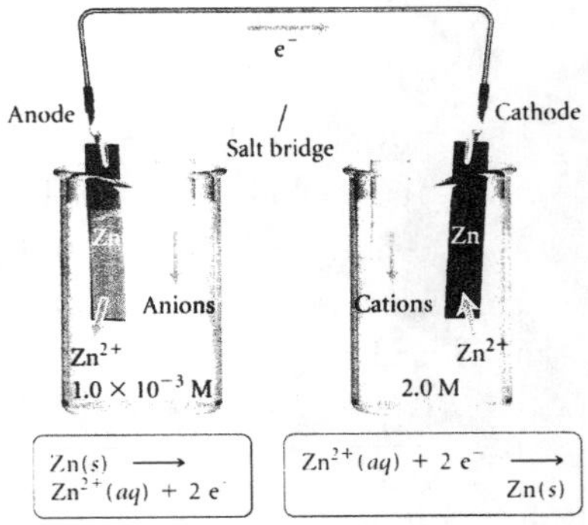

Check: The figure looks similar to the right side of Figure 18.12.

81. **Given:** Sn/Sn^{2+} concentration cell with $E_{cell} = 0.10$ V **Find:** ratio of $[Sn^{2+}]$ in two half-cells

Conceptual plan: Determine n, then E^{o}_{cell}, E_{cell}, n → Q = ratio of $[Sn^{2+}]$ in two half-cells

$$E_{cell} = E^{o}_{cell} - \frac{0.0592\ V}{n} \log Q$$

Solution: Since $Sn^{2+}\ (aq) + 2\ e^- \rightarrow Sn\ (s)$, $n = 2$. In a concentration cell, $E^{o}_{cell} = 0$ V. So

$$E_{cell} = E^{0}_{cell} - \frac{0.0592\ V}{n}\log Q \text{ so } 0.10\ V = 0.00\ V - \frac{0.0592\ V}{2}\log Q \rightarrow 0.10\ \cancel{V} = -\frac{0.0592\ \cancel{V}}{2}\log Q \rightarrow$$

$$\log Q = -3.3784 \rightarrow Q = 10^{-3.3784} = 4.2 \times 10^{-4} = \frac{[Sn^{2+}](ox)}{[Sn^{2+}](red)}.$$

Check: The units (none) are correct. Since the concentration in the reduction reaction half-cell is always greater that the concentration in the oxidation half-cell in a voltaic concentration cell, the Q or ratio of two cells is less than 1.

83. **Given:** alkaline battery **Find:** optimum mass ratio of Zn to MnO_2

Conceptual plan: Look up alkaline battery reactions. Use stoichiometry to get mole ratio. then

$Zn\ (s) + 2\ OH^-\ (aq) \rightarrow Zn(OH)_2\ (s) + 2\ e^-$ $\qquad \frac{1\ mol\ Zn}{2\ mol\ MnO_2}$

$2\ MnO_2\ (s) + 2\ H_2O\ (l) + 2\ e^- \rightarrow 2\ MnO(OH)\ (s) + 2\ OH^-\ (aq)$

mol Zn → g Zn then mol MnO_2 → g MnO_2

$$\frac{65.41\ g\ Zn}{1\ mol\ Zn} \qquad \frac{1\ mol\ MnO_2}{86.94\ g\ MnO_2}$$

Solution: $$\frac{1\ \cancel{mol\ Zn}}{2\ \cancel{mol\ MnO_2}} \times \frac{65.41\ g\ Zn}{1\ \cancel{mol\ Zn}} \times \frac{1\ \cancel{mol\ MnO_2}}{86.94\ g\ MnO_2} = 0.3762\frac{g\ Zn}{g\ MnO_2}.$$

Check: The units (mass ratio) are correct. Since more moles of $MnO_2$are needed and the molar mass is larger, the ratio is less than 1.

85. **Given:** $CH_4\ (g) + 2\ O_2\ (g) \rightarrow CO_2\ (g) + 2\ H_2O\ (g)$ **Find:** E^{o}_{cell}

Conceptual plan: $\Delta G^{0}_{rxn} = \sum n_p \Delta G^{0}_{f}(products) - \sum n_R \Delta G^{0}_{f}(reactants)$ **and determine n then ΔG^{o}_{rxn}, n → E^{o}_{cell}**

$$\Delta G^{0}_{rxn} = -nFE^{0}_{cell}$$

Solution:

Reactant/Product	ΔG_f^0(kJ/mol from Appendix IIB)
CH_4 (*g*)	− 50.5
O_2 (*g*)	0.0
CO_2 (*g*)	− 394.4
H_2O (*g*)	− 228.6

Be sure to pull data for the correct formula and phase.

$$\Delta G^0_{rxn} = \sum n_P \Delta G^0_f(products) - \sum n_R \Delta G^0_f(reactants)$$

$$= [1(\Delta G^0_f(CO_2\ (g)) + 2(\Delta G^0_f(H_2O\ (g))] - [1(\Delta G^0_f(CH_4\ (g)) + 2(\Delta G^0_f(O_2\ (g))]$$

$$= [1(-394.4 \text{ kJ}) + 2(-228.6 \text{ kJ})] - [1(-50.5 \text{ kJ}) + 2(0.0 \text{ kJ})]$$

$$= [-851.6 \text{ kJ}] - [-50.5 \text{ kJ}]$$

$$= -801.1 \text{ kJ} = -8.011 \text{ x } 10^5 \text{ J}$$

and since one C atoms goes from an oxidation state of -4 to +4 and 4 O atoms are going from 0 to -2, then n = 8 and $\Delta G^0_{rxn} = -nFE^0_{cell}$. Rearrange to solve for E^o_{cell}.

$$E^0_{cell} = \frac{\Delta G^0_{rxn}}{-nF} = \frac{-8.011 \text{ x } 10^5 \text{ J}}{-8 \text{ mol e}^- \text{ x } \frac{96{,}485 \text{ C}}{\text{mol e}^-}} = 1.038 \frac{\text{V C}}{\text{C}} = 1.038 \text{ V}.$$

Check: The units (V) are correct. The cell voltage is positive which is consistent with a spontaneous reaction.

87. In order for a metal to be able to protect iron, it must more easily oxidized than iron or be below it in Table 18.1. a) Zn and c) Mn meet this criterion.

89. **Given:** electrolytic cell sketch
Find: a) Label anode and cathode and indicate half-reactions; b) Indicate direction of electron flow; and c) Label battery terminals and calculate minimum voltage to drive reaction
Conceptual plan:
a) Write 2 half-cell reactions and add electrons as needed to balance reactions. Look up half-reactions in Table 18.1. In order to determine the cell potential of the oxidation reaction, reverse the sign of the standard reduction potential ($E^o_{ox} = -E^o_{red}$) then add the two half-cell potentials: $E^o_{cell} = E^o_{ox} + E^o_{red}$. Choose the direction of the half-cell reactions so that $E^o_{cell} < 0$.

b) Electrons flow from anode to cathode.

c) Each half cell reaction moves forward, so concentration change direction can be determined.
Solution:

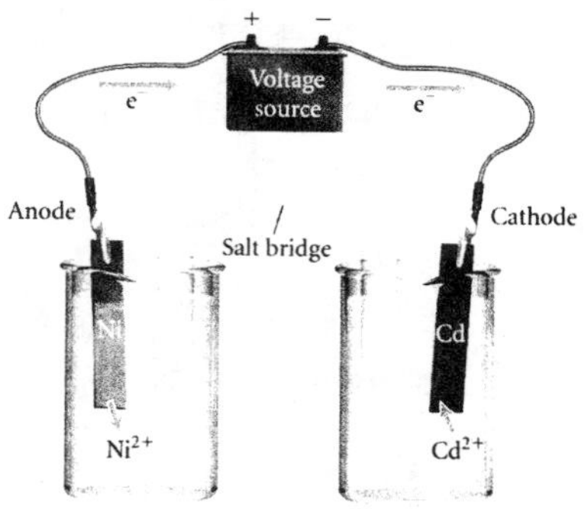

a) Ni^{2+} *(aq)* + 2 e^- → Ni *(s)* and Cd^{2+} *(aq)* + 2 e^- → Cd *(s)*. Look up cell potentials. For Ni, E^o_{red} = − 0.23 V. For Cd, E^o_{red} = − 0.40 V. In order to get a negative cell potential, Ni is oxidized so E^o_{ox} = − E^o_{red} = − (− 0.23 V) = 0.23 V. Cd^{2+} is reduced so E^o_{red} = − 0.23 V. Then $E^o_{cell} = E^o_{ox} + E^o_{red}$ = 0.23 V − 0.40 V = − 0.17 V. Since oxidation occurs at the anode, the Ni is the anode and the reaction is Ni *(s)* → Ni^{2+} *(aq)* + 2 e^-. Since reduction takes place at the cathode, Cd is the cathode and the reaction is Cd^{2+} *(aq)* + 2 e^- → Cd *(s)*.

c) Since reduction is occurring at the cathode, the battery terminal closest to the cathode is the negative terminal. Since the cell potential from part a) is = − 0.17 V, a minimum of 0.17 V must be applied by the battery.

Check: The reaction is nonspontaneous, since the reduction of Ni^{2+} is above Cd^{2+}. Electrons still flow from the anode to the cathode. The reaction can be made spontaneous with the application of electrical energy.

91. **Given:** electrolysis of molten KBr **Find:** write half-reactions
Conceptual plan: Write 2 half-cell reactions taking the cation and the anion to their elemental forms at high temperatures, adding electrons as needed to balance reactions.
Solution: KBr breaks apart to K^+ and Br^-. K^+ *(l)* + e^- → K *(l)* and 2 Br^- *(l)* → Br_2 *(g)* + 2 e^-.
Check: The cation is reduced and the anion is oxidized. Mass and charge are balanced in the reactions.

93. **Given:** electrolysis of mixture of molten KBr and molten LiBr **Find:** write half-reactions
Conceptual plan: Write 2 half-cell reactions taking the cation and the anion to their elemental forms at high temperatures, adding electrons as needed to balance reactions. Look up cation cell potentials to see which one generates the mores spontaneous reaction.
Solution: KBr breaks apart to K^+ and Br^-. K^+ *(l)* + e^- → K *(l)* and 2 Br^- *(l)* → Br_2 *(g)* + 2 e^-. LiBr breaks apart to Li^+ and Br^-. Li^+ *(l)* + e^- → Li *(l)* and 2 Br^- *(l)* → Br_2 *(g)* + 2 e^-. The anions are the same so the anode reaction is 2 Br^- *(l)* → Br_2 *(g)* + 2 e^-. Looking up the cation reduction potentials, for K^+ E°_{red} = – 2.92 V and for Li^+ E°_{red} = – 3.04 V. Since the reduction potential is more positive for K^+, the reaction at the cathode is K^+ *(l)* + e^- → K *(l)*.
Check: The cation is reduced and the anion is oxidized. Mass and charge are balanced in the reactions. The cation that is higher in Table 18.1 will be the reaction at cathode.

95. **Given:** electrolysis of aqueous solutions **Find:** write half-reactions
Conceptual plan: Write 2 half-cell reactions taking the cation and the anion to their elemental forms at standard conditions, adding electrons as needed to balance reactions. Look up cell potentials and compare to the cell potentials for the electrolysis of water to see which one generates the more spontaneous reaction.
Solution: The hydrolysis of water reactions are: 2 H_2O *(l)* → O_2 *(g)* + 4 H^+ *(aq)* + 4 e^- where E°_{ox} = – 0.82 V; and 2 H_2O *(l)* + 2 e^- → H_2 *(g)* + OH^- *(aq)* where E°_{red} = – 0.41 V.

a) NaBr breaks apart to Na^+ and Br^-. Na^+ *(aq)* + e^- → Na *(s)* and 2 Br^- *(aq)* → Br_2 *(l)* + 2 e^-. Looking up the half-cell potentials, for Na^+ E°_{red} = – 2.71 V and for Br^- E°_{ox} = – E°_{red} = – 1.09 V. Since – 0.82 V is more positive than = – 1.09 V, the oxidation reaction will be: 2 H_2O *(l)* → O_2 *(g)* + 4 H^+ *(aq)* + 4 e^- at the anode. Since – 0.41 V is more positive than = – 2.71 V, the reduction reaction will be: 2 H_2O *(l)* + 2 e^- → H_2 *(g)* + OH^- *(aq)* at the cathode.

b) PbI_2 breaks apart to Pb^{2+} and I^-. Pb^{2+} *(aq)* + 2 e^- → Pb *(s)* and 2 I^- *(aq)* → I_2 *(s)* + 2 e^-. Looking up the half-cell potentials, for Pb^{2+} E°_{red} = – 0.13 V and for I^- E°_{ox} = – E°_{red} = – 0.54 V. Since – 0.13 V is more positive than = – 0.82 V, the oxidation reaction will be: 2 I^- *(l)* → I_2 *(s)* + 2 e^- at the anode. Since – 0.13 V is more positive than = – 0.41 V, the reduction reaction will be: Pb^{2+} *(aq)* + 2 e^- → Pb *(s)* at the cathode.

c) $NaSO_4$ breaks apart to Na^+ and SO_4^{2-}. Na^+ *(aq)* + e^- → Na *(s)* and SO_4^{2-} *(aq)* + 4 H^+ *(aq)* + 2 e^- → H_2SO_3 *(aq)* + H_2O *(l)*. Notice that both of these are reductions. Since S is in such a high oxidation state (+6) it can not be oxidized. Looking up the half-cell potentials, for Na^+ E°_{red} = – 2.71 V and for SO_4^{2-} E°_{red} = 0.20 V. Since sodium sulfate solutions are neutral and not acidic, even though 0.20 V is more positive than – 0.41 V and – 2.71 V, the reduction reaction will not be: SO_4^{2-} *(aq)* + 4 H^+ *(aq)* + 2 e^- → H_2SO_3 *(aq)* + H_2O *(l)* at the cathode, which only occurs in acidic solutions. Thus, the reduction reaction will be Na^+ *(aq)* + e^- → Na *(s)*. Since the only one oxidation reaction is possible, the oxidation reaction will be: 2 H_2O *(l)* → O_2 *(g)* + 4 H^+ *(aq)* + 4 e^- at the anode.
Check: The most positive reactions are the reactions that will occur.

97. **Given:** electrolysis cell to electroplate Cu onto a metal surface **Find:** Draw cell, labeling anode and cathode; and write half-reactions
Conceptual plan: Write 2 half-cell reactions and add electrons as needed to balance reactions. The cathode reaction will be the reduction of Cu^{2+} to the metal. The anode will be the reverse reaction.
Solution:

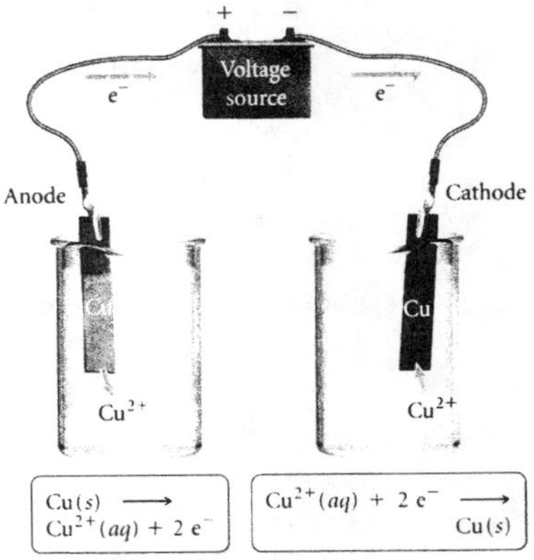

Check: The metal to be plated is the cathode, since metal ions are converted to Cu *(s)* on the surface of the metal.

99. **Given:** Cu electroplating of 225 mg Cu at a current of 7.8 A; Cu^{2+} *(aq)* + 2 e^- → Cu *(s)*
Find: time
Conceptual plan: mg Cu → g Cu → mol Cu → mol e^- → C → s

$$\frac{1\text{ g}}{1000\text{ mg}} \quad \frac{1\text{ mol Cu}}{63.55\text{ g Cu}} \quad \frac{2\text{ mol e}^-}{1\text{ mol Cu}} \quad \frac{96{,}485\text{ C}}{1\text{ mol e}^-} \quad \frac{1\text{ s}}{7.8\text{ C}}$$

Solution: $$225\text{ mg Cu} \times \frac{1\text{ g Cu}}{1000\text{ mg Cu}} \times \frac{1\text{ mol Cu}}{63.55\text{ g Cu}} \times \frac{2\text{ mol e}^-}{1\text{ mol Cu}} \times \frac{96{,}485\text{ C}}{1\text{ mol e}^-} \times \frac{1\text{ s}}{7.8\text{ C}} = 88\text{ s}.$$

Check: The units (s) are correct. Since far less than a mole of Cu is electroplated, the time is short.

101. **Given:** Na electrolysis, 1.0 kg in one hour **Find:** current
Conceptual plan: Na^+ *(l)* + e^- → Na *(l)* $\frac{\text{kg Na}}{\text{hr}} \rightarrow \frac{\text{g Na}}{\text{hr}} \rightarrow \frac{\text{mol Na}}{\text{hr}} \rightarrow \frac{\text{mol e}^-}{\text{hr}} \rightarrow \frac{\text{C}}{\text{hr}} \rightarrow \frac{\text{C}}{\text{min}} \rightarrow \frac{\text{C}}{\text{s}}$

$$\frac{1000\text{ g}}{1\text{ kg}} \quad \frac{1\text{ mol Na}}{22.99\text{ g Na}} \quad \frac{1\text{ mol e}^-}{1\text{ mol Na}} \quad \frac{96{,}485\text{ C}}{1\text{ mol e}^-} \quad \frac{1\text{ hr}}{60\text{ min}} \quad \frac{1\text{ min}}{60\text{ s}}$$

Solution:
$$\frac{1.0\text{ kg Na}}{1\text{ hr}} \times \frac{1000\text{ g Na}}{1\text{ kg Na}} \times \frac{1\text{ mol Na}}{22.99\text{ g Na}} \times \frac{1\text{ mol e}^-}{1\text{ mol Na}} \times \frac{96{,}485\text{ C}}{1\text{ mol e}^-} \times \frac{1\text{ hr}}{60\text{ min}} \times \frac{1\text{ min}}{60\text{ s}} = 1.2 \times 10^3\ \frac{\text{C}}{\text{s}} = 1.2 \times 10^3\text{ A}.$$

Check: The units (A) are correct. Since the amount per hour is so large we expect a very large current.

103. **Given:** MnO_4^- *(aq)* + Zn *(s)* → Mn^{2+} *(aq)* + Zn^{2+} *(aq)* 0.500 M $KMnO_4$ and 2.85 g Zn
Find: balance equation and volume $KMnO_4$ solution
Conceptual plan: Separate the overall reaction into two half-reactions: one for oxidation and one for reduction. → Balance each half-reaction with respect to mass in the following order: 1) balance all elements other than H and O; 2) balance O by adding H_2O; and 3) balance H by adding H^+. → Balance each half reaction with respect to charge by adding electrons. (The sum of the charges on both sides of the equation should be made equal by adding electrons as necessary.) → Make the number of electrons in both half-reactions equal by multiplying one or both half-reactions by a small whole number. → Add the two half-reactions together, canceling electrons and other species as necessary. → Verify that the reaction is balanced both with respect to mass and with respect to charge. then g Zn → mol Zn → mol MnO_4^- → L MnO_4^- → mL MnO_4^-

$$\frac{1\text{ mol Zn}}{65.41\text{ g Zn}} \quad \frac{2\text{ mol MnO}_4^-}{5\text{ mol Zn}} \quad \frac{1\text{ L MnO}_4^-}{0.500\text{ mol MnO}_4^-} \quad \frac{1000\text{ mL MnO}_4^-}{1\text{ L MnO}_4^-}$$

Solution:
Separate: MnO_4^- *(aq)* → Mn^{2+} *(aq)* and Zn *(s)* → Zn^{2+} *(aq)*
Balance non H & O elements: MnO_4^- *(aq)* → Mn^{2+} *(aq)* and Zn *(s)* → Zn^{2+} *(aq)*
Balance O with H_2O: MnO_4^- *(aq)* → Mn^{2+} *(aq)* + 4 H_2O *(l)* and Zn *(s)* → Zn^{2+} *(aq)*
Balance H with H^+: MnO_4^- *(aq)* + 8 H^+ *(aq)* → Mn^{2+} *(aq)* + 4 H_2O *(l)* and Zn *(s)* → Zn^{2+} *(aq)*
Add electrons: MnO_4^- *(aq)* + 8 H^+ *(aq)* + 5 e^- → Mn^{2+} *(aq)* + 4 H_2O *(l)* and Zn *(s)* → Zn^{2+} *(aq)* + 2 e^-
Equalize electrons:
2 MnO_4^- *(aq)* + 16 H^+ *(aq)* + 10 e^- → 2 Mn^{2+} *(aq)* + 8 H_2O *(l)* and 5 Zn *(s)* → 5 Zn^{2+} *(aq)* + 10 e^-
Add half-reactions:
2 MnO_4^- *(aq)* + 16 H^+ *(aq)* + 10 e^- + 5 Zn *(s)* → 2 Mn^{2+} *(aq)* + 8 H_2O *(l)* + 5 Zn^{2+} *(aq)* + 10 e^-
Cancel electrons: 2 MnO_4^- *(aq)* + 16 H^+ *(aq)* +5 Zn *(s)* → 2 Mn^{2+} *(aq)* + 8 H_2O *(l)* + 5 Zn^{2+} *(aq)*

$$2.85\text{ g Zn} \times \frac{1\text{ mol Zn}}{65.41\text{ g Zn}} \times \frac{2\text{ mol MnO}_4^-}{5\text{ mol Zn}} \times \frac{1\text{ L MnO}_4^-}{0.500\text{ mol MnO}_4^-} \times \frac{1000\text{ mL MnO}_4^-}{1\text{ L MnO}_4^-} = 34.9\text{ mL MnO}_4^- =$$
$= 34.9$ mL $KMnO_4$.

Check:

Reactants	Products
2 Mn atoms	2 Mn atoms
8 O atoms	8 O atoms
16 H atoms	16 H atoms
5 Zn atoms	5 Zn atoms
+14 charge	+14 charge

The units (mL) are correct. Since far less than a mole of Zinc is used, less than a mole of permanganate is consumed, so the volume is less than a liter.

105. **Given:** beaker with Al strip and Zn^{2+} ions **Find:** draw sketch after Al is submerged for a few minutes
Conceptual plan: Write 2 half-cell reactions and add electrons as needed to balance reactions. Look up half-reactions in Table 18.1. In order to determine the cell potential of the oxidation reaction, reverse the sign of the standard reduction potential ($E^o_{ox} = -E^o_{red}$) then add the two half-cell potentials: $E^o_{cell} = E^o_{ox} + E^o_{red}$. If $E^o_{cell} > 0$ the reaction is spontaneous in the forward direction and Al will dissolve and Cu will deposit.

Solution: $Al\ (s) \rightarrow Al^{3+}\ (aq)$ and $Cu^{2+}\ (aq) \rightarrow Cu\ (s)$ add electrons $Al\ (s) \rightarrow Al^{3+}\ (aq) + 3\ e^-$ and $Cu^{2+}\ (aq) + 2\ e^- \rightarrow Cu\ (s)$. Look up cell potentials. Al is oxidized so $E^o_{ox} = -E^o_{red} = --1.66\ V = 1.66\ V$. Cu^{2+} is reduced so $E^o_{red} = 0.34\ V$. Then $E^o_{cell} = E^o_{ox} + E^o_{red} = 1.66\ V + 0.34\ V = +2.00\ V$ and so the reaction is spontaneous. Al will dissolve to generate $Al^{3+}\ (aq)$ and Cu (*s*) will deposit.

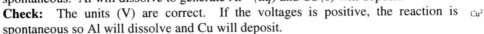

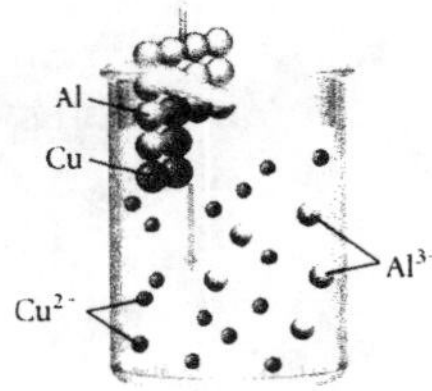

Check: The units (V) are correct. If the voltages is positive, the reaction is spontaneous so Al will dissolve and Cu will deposit.

107. **Given:** a) 2.15 g Al; b) 4.85 g Cu; and c) 2.42 g Ag in 3.5 M HI
Find: if metal dissolves, write balanced reaction and minimum amount of HI needed to dissolve metal
Conceptual Plan: In general, metals whose reduction half-reactions lie below the reduction of H^+ to H_2 in Table 18.1 will dissolve in acids, while metals above it will not. Stop here if metal does not dissolve. To write the balanced redox reactions, pair the oxidation of the metal with the reduction of H^+ to H_2 ($2\ H^+(aq) + 2\ e^- \rightarrow H_2(g)$). Balance the number of electrons transferred. Add the two reactions. Cancel electrons. then **g metal → mol metal → mol H^+ → L HI → mL HI**

$$\mathfrak{M} \qquad \frac{x\ \text{mol}\ H^+}{y\ \text{mol metal}} \qquad \frac{1\ \text{L HI}}{3.5\ \text{mol HI}} \qquad \frac{1000\ \text{mL HI}}{1\ \text{L HI}}$$

Solution: a) Al meets this criterion. For Al, $Al\ (s) \rightarrow Al^{3+}\ (aq) + 3\ e^-$. we need to multiply the Al reaction by 2 and the H^+ reaction by 3. So, $2\ Al\ (s) \rightarrow 2\ Al^{3+}\ (aq) + 6\ e^-$ and $6\ H^+(aq) + 6\ e^- \rightarrow 3\ H_2(g)$. Adding the half-reactions together: $2\ Al\ (s) + 6\ H^+(aq) + \cancel{6\ e^-} \rightarrow 2\ Al^{3+}\ (aq) + \cancel{6\ e^-} + 3\ H_2(g)$. Simplify to $2\ Al\ (s) + 6\ H^+(aq) \rightarrow 2\ Al^{3+}\ (aq) + 3\ H_2(g)$. then

$$2.15\ \cancel{g\ Al} \times \frac{1\ \cancel{mol\ Al}}{26.98\ \cancel{g\ Al}} \times \frac{6\ \cancel{mol\ H^+}}{2\ \cancel{mol\ Al}} \times \frac{1\ \cancel{L\ HI}}{3.5\ \cancel{mol\ HI}} \times \frac{1000\ mL\ HI}{1\ \cancel{L\ HI}} = 68.3\ mL\ HI\ .$$

a) Cu does not meet this criterion, so it will not dissolve in HI.

b) Ag does not meet this criterion, so it will not dissolve in HI.
Check: Only metals with negative reduction potentials will dissolve. The volume of acid needed is fairly small since the amount of metal is much less than 1 mole and the concentration of acid is high.

109. **Given:** $Pt\ (s)|H_2\ (g,\ 1\ atm)|H^+\ (aq,\ ?\ M)||Cu^{2+}\ (aq,\ 1.0\ M)|Cu\ (s)$, $E_{cell} = 355\ mV$ **Find:** pH
Conceptual plan: Write half reactions from line notation. Look up half-reactions in Table 18.1. The reaction on the left is the oxidation so reverse the sign of the standard reduction potential ($E^o_{ox} = -E^o_{red}$) then add the two half-cell potentials: $E^o_{cell} = E^o_{ox} + E^o_{red}$. Add the two half-cell reactions and cancel the electrons.
Then mV → V **E^o_{cell}, E_{cell}, P_{H2}, $[Cu^{2+}]$, n → $[H^+]$ → pH**

$$\frac{1\ V}{1000\ mV} \qquad E_{cell} = E^o_{cell} - \frac{0.0592\ V}{n}\log Q \qquad pH = -\log[H^+]$$

Solution: The half-reactions are: $H_2(g) \rightarrow 2\ H^+(aq) + 2\ e^-$ and $Cu^{2+}\ (aq) + 2\ e^- \rightarrow Cu\ (s)$. H is oxidized so $E^o_{ox} = -E^o_{red} = -(-0.00\ V) = 0.00\ V$. Cu is reduced so $E^o_{red} = 0.34\ V$. Then $E^o_{cell} = E^o_{ox} + E^o_{red} = 0.00\ V + 0.34\ V = +0.34\ V$. Adding the two reactions: $H_2(g) + Cu^{2+}\ (aq) + \cancel{2\ e^-} \rightarrow 2\ H^+(aq) + \cancel{2\ e^-} + Cu\ (s)$. Cancel the electrons: $H_2(g) + Cu^{2+}\ (aq) \rightarrow 2\ H^+(aq) + Cu\ (s)$. then $355\ \cancel{mV} \times \frac{1\ V}{1000\ \cancel{mV}} = 0.355\ V$.

So $n = 2$ and $Q = \frac{[H^+]^2}{P_{H2}[Cu^{2+}]} = \frac{(x)^2}{(1)(1.0)} = x^2$ then $E_{cell} = E^0_{cell} - \frac{0.0592\ V}{n}\log Q$ substitute in values and solve for x. $0.355\ V = 0.34\ V - \frac{0.0592\ V}{2}\log x^2 \rightarrow 0.01\underline{5}\ \cancel{V} = -\frac{0.0592\ \cancel{V}}{2}\log x^2 \rightarrow -0.\underline{5}0676 = \log x^2 \rightarrow$
$x^2 = 10^{-0.\underline{5}0676} = 0.3\underline{1}135 \rightarrow x = 0.\underline{5}5798$ then $pH = -\log[H^+] = -\log[0.\underline{5}5798] = 0.\underline{2}5338 = 0.3$.

Check: The units (none) are correct. The pH is acidic, which is consistent with dissolving a metal in acid.

111. You should be wary of the battery because the most a pairing of half-cell reactions can generate is 5 to 6 V, not 24 V.

113. **Given:** Mg oxidation and Cu^{2+} reduction; initially $[Mg^{2+}] = 1.0 \times 10^{-4}$ M and $[Cu^{2+}] = 1.5$ M in 1.0 L half-cells **Find:** a) initial E_{cell}; b) E_{cell} after 5.0 A for 8.0 hr; and c) how long can battery deliver 5. 0A
Conceptual plan:
a) Write the 2 half-cell reactions and add electrons as needed to balance reactions. Look up half-reactions in Table 18.1. In order to determine the cell potential of the oxidation reaction, reverse the sign of the standard reduction potential ($E°_{ox} = -E°_{red}$) then add the two half-cell potentials: $E°_{cell} = E°_{ox} + E°_{red}$. Add the two half-cell reactions and cancel electrons and determine *n*. then $E°_{cell}$, $[Mg^{2+}]$, $[Cu^{2+}]$, *n* → E_{cell}

$$E_{cell} = E°_{cell} - \frac{0.0592\ V}{n} \log Q$$

b) hr → min → s → C → mol e⁻ → mol Cu reduced → $[Cu^{2+}]$ and

$$\frac{60\ \text{min}}{1\ \text{hr}} \quad \frac{60\ \text{s}}{1\ \text{min}} \quad \frac{5.0\ \text{C}}{1\ \text{s}} \quad \frac{1\ \text{mol e}^-}{96{,}485\ \text{C}} \quad \frac{1\ \text{mol Cu}^{2+}}{2\ \text{mol e}^-} \quad \text{since V = 1.0L}\ [Cu^{2+}] = [Cu^{2+}] - \frac{\text{mol Cu}^{2+}\ \text{reduced}}{1.0\ \text{L}}$$

mol Cu reduced → mol Mg oxidized → $[Mg^{2+}]$

$$\frac{1\ \text{mol Mg oxidized}}{1\ \text{mol Cu}^{2+}\ \text{reduced}} \quad \text{since V = 1.0L}\ [Mg^{2+}] = [Mg^{2+}] + \frac{\text{mol Mg oxidized}}{1.0\ \text{L}}$$

c) $[Cu^{2+}]$ → mol e⁻ → C → s → min → hr

$$\frac{1\ \text{mol e}^-}{2\ \text{mol Cu}^{2+}} \quad \frac{96{,}485\ \text{C}}{1\ \text{mol e}^-} \quad \frac{1\ \text{s}}{5.0\ \text{C}} \quad \frac{1\ \text{min}}{60\ \text{s}} \quad \frac{1\ \text{hr}}{60\ \text{min}}$$

Solution:

a) Write half-reactions and add electrons Cu^{2+} *(aq)* + 2 e⁻ → Cu *(s)* and Mg *(s)* → Mg^{2+} *(aq)* + 2 e⁻. Look up cell potentials. Mg is oxidized so $E°_{ox} = -E°_{red} = -(-2.37\ V) = 2.37$ V. Cu^{2+} is reduced so $E°_{red} = 0.34$ V. Then $E°_{cell} = E°_{ox} + E°_{red} = 2.37\ V + 0.34\ V = +2.71$ V. Add the two half-cell reactions: Cu^{2+} *(aq)* + 2e⁻ + Mg *(s)* → Cu *(s)* + Mg^{2+} *(aq)* + 2e⁻. Simplify to

Cu^{2+} *(aq)* + Mg *(s)* → Cu *(s)* + Mg^{2+} *(aq)*. So $Q = \frac{[Mg^{2+}]}{[Cu^{2+}]} = \frac{1.0 \times 10^{-4}}{1.5} = 6.6667 \times 10^{-5}$ and $n = 2$ then

$$E_{cell} = E^0_{cell} - \frac{0.0592\ V}{n} \log Q = 2.71\ V - \frac{0.0592\ V}{2} \log 6.6667 \times 10^{-5} = +2.83361\ V = +2.83\ V\ .$$

b) $$8.0\ \text{hr} \times \frac{60\ \text{min}}{1\ \text{hr}} \times \frac{60\ \text{s}}{1\ \text{min}} \times \frac{5.0\ \text{C}}{1\ \text{s}} \times \frac{1\ \text{mol e}^-}{96{,}485\ \text{C}} \times \frac{1\ \text{mol Cu}^{2+}}{2\ \text{mol e}^-} = 0.74623\ \text{mol Cu}^{2+}$$ and

$$[Cu^{2+}] = [Cu^{2+}] - \frac{\text{mol Cu}^{2+}\ \text{reduced}}{1.0\ \text{L}} = 1.5\ M - \frac{0.74623\ \text{mol Cu}^{2+}}{1.0\ \text{L}} = 0.75377\ M\ Cu^{2+}$$ and

$$0.74623\ \text{mol Cu}^{2+} \times \frac{1\ \text{mol Mg oxidized}}{1\ \text{mol Cu}^{2+}\ \text{reduced}} = 0.74623\ \text{mol Mg oxidized}$$ and

$$[Mg^{2+}] = [Mg^{2+}] + \frac{\text{mol Mg oxidized}}{1.0\ \text{L}} = 1.0 \times 10^{-4}\ M + \frac{0.74623\ \text{mol Mg oxidized}}{1.0\ \text{L}} = 0.74633\ M\ Mg^{2+}$$

$$Q = \frac{[Mg^{2+}]}{[Cu^{2+}]} = \frac{0.74633}{0.75377} = 0.99013$$ and $n = 2$ then

$$E_{cell} = E^0_{cell} - \frac{0.0592\ V}{n} \log Q = 2.71\ V - \frac{0.0592\ V}{2} \log 0.99013 = +2.71013\ V = +2.71\ V\ .$$

c) In 1.0 L there are initially 1.5 moles of Cu^{2+}. So

$$1.5\ \text{mol Cu}^{2+} \times \frac{2\ \text{mol e}^-}{1\ \text{mol Cu}^{2+}} \times \frac{96{,}485\ \text{C}}{1\ \text{mol e}^-} \times \frac{1\ \text{s}}{5.0\ \text{C}} \times \frac{1\ \text{min}}{60\ \text{s}} \times \frac{1\ \text{hr}}{60\ \text{min}} = 16\ \text{hr}\ .$$

Check: The units (V, V and hr) are correct. The Cu^{2+} reduction reaction is above the Mg^{2+} reduction reaction, so the standard cell potential will be positive. Having more reactants than products increases the cell potential. As the reaction proceeds the potential drops. The concentrations drop by ½ in 8 hours (part b), so it is all consumed in 16 hours.

115. **Given:** water electrolysis at 7.8 A; $H_2(g)$: V = 25.0 L, P = 25.0 atm; T = 25 °C **Find:** time

Conceptual plan: °C → K then V, P, T → n then write half-reactions

$K = °C + 273.15$ $\qquad PV = nRT$

mol H_2 → mol e^- → C → s → min → hr

$$\frac{2\text{ mol e}^-}{1\text{ mol H}_2}\quad \frac{96{,}485\text{ C}}{1\text{ mol e}^-}\quad \frac{1\text{ s}}{7.8\text{ C}}\quad \frac{1\text{ min}}{60\text{ s}}\quad \frac{1\text{ hr}}{60\text{ min}}$$

Solution: T = 25 °C + 273.15 = 298 K, then $PV = nRT$ Rearrange to solve for n.

$$n = \frac{PV}{RT} = \frac{25.0\text{ atm x }25.0\text{ L}}{0.08206\ \frac{\text{L}\cdot\text{atm}}{\text{mol}\cdot\text{K}}\text{ x }298\text{ K}} = 25.5583\text{ mol H}_2$$ The hydrolysis of water reactions are:

$2\,H_2O\,(l) \rightarrow O_2\,(g) + 4\,H^+\,(aq) + 4\,e^-$ and $2\,H_2O\,(l) + 2\,e^- \rightarrow H_2\,(g) + OH^-\,(aq)$.

$$25.5583\text{ mol H}_2\text{ x }\frac{2\text{ mol e}^-}{1\text{ mol H}_2}\text{ x }\frac{96{,}485\text{ C}}{1\text{ mol e}^-}\text{ x }\frac{1\text{ s}}{7.8\text{ C}}\text{ x }\frac{1\text{ min}}{60\text{ s}}\text{ x }\frac{1\text{ hr}}{60\text{ min}} = 176\text{ hr}.$$

Check: The units (hr) are correct. Since we have 25 L of gas at 25 atm and 25 °C we expect ~25 moles of gas (remember 1 mole of gas at STP = 22.4 L). A very long time is expected since we have so many moles of gas to generate.

117. **Given:** Cu (s)|CuI (s)|I^- $(aq,$ 1.0 M)||Cu^+ $(aq,$ 1.0 M)|Cu (s), K_{sp} (CuI) = 1.1 x 10^{-12} **Find:** E_{cell}

Conceptual plan: Write half reactions from line notation. Since this is a concentration cell $E°_{cell}$ = 0.00 V. then K_{sp}, [I^-] → [Cu^+](ox) then $E°_{cell}$, [Cu^+](ox), [Cu^+](red), *n* → E_{cell}

$K_{sp} = [Cu^+][I^-]$ $\qquad E_{cell} = E°_{cell} - \frac{0.0592\text{ V}}{n}\log Q$

Solution: The half-reactions are: Cu (s) → $Cu^+(aq)$ + e^- and Cu^+ (aq) + e^- → Cu (s). Since this is a concentration cell $E°_{cell}$ = 0.00 V and *n* = 1. Since $K_{sp} = [Cu^+][I^-]$, rearrange to solve for [Cu^+](ox).

$$[Cu^+]\,(ox) = \frac{K_{sp}}{[I^-]} = \frac{1.1\text{ x }10^{-12}}{1.0} = 1.1\text{ x }10^{-12}\text{ M then } Q = \frac{[Cu^+](ox)}{[Cu^+](red)} = \frac{1.1\text{ x }10^{-12}}{1.0} = 1.1\text{ x }10^{-12}\text{ then}$$

$$E_{cell} = E^0_{cell} - \frac{0.0592\text{ V}}{n}\log Q = 0.00\text{ V} - \frac{0.0592\text{ V}}{1}\log(1.1\text{ x }10^{-12}) = 0.71\text{ V}.$$

Check: The units (V) are correct. Since [Cu^+](ox) is so low and [Cu^+](red) is high, the Q is very small so the voltage increase is significant.

119. **Given:** a) dispropotionation of Mn^{2+} (aq) to Mn (s) and MnO_2 (s); and b) dispropotionation of MnO_2 (s) to Mn^{2+} (aq) and MnO_4^- (s) in acidic solution **Find:** $\Delta G°_{rxn}$ and K

Conceptual plan: Separate the overall reaction into two half-reactions: one for oxidation and one for reduction. → Balance each half-reaction with respect to mass in the following order: 1) balance all elements other than H and O; 2) balance O by adding H_2O; and 3) balance H by adding H^+. → Balance each half reaction with respect to charge by adding electrons. (The sum of the charges on both sides of the equation should be made equal by adding electrons as necessary.) → Make the number of electrons in both half-reactions equal by multiplying one or both half-reactions by a small whole number. → Add the two half-reactions together, canceling electrons and other species as necessary. → Verify that the reaction is balanced both with respect to mass and with respect to charge. Look up half-reactions in Table 18.1. In order to determine the cell potential of the oxidation reaction, reverse the sign of the standard reduction potential ($E°_{ox}$ = – $E°_{red}$) then add the two half-cell potentials: $E°_{cell} = E°_{ox} + E°_{red}$. then calculate $\Delta G°_{rxn}$ using $\Delta G^0_{rxn} = -nFE^0_{cell}$. Finally

°C → K then $\Delta G°_{rxn}$, T → K

$K = 273.15 + °C$ $\qquad \Delta G^0_{rxn} = -RT\ln K$

Solution:

a) Separate: $Mn^{2+}\,(aq) \rightarrow MnO_2\,(s)$ and $Mn^{2+}\,(aq) \rightarrow Mn\,(s)$

Balance non H & O elements: $Mn^{2+}\,(aq) \rightarrow MnO_2\,(s)$ and $Mn^{2+}\,(aq) \rightarrow Mn\,(s)$

Balance O with H_2O: $Mn^{2+}\,(aq) + 2\,H_2O\,(l) \rightarrow MnO_2\,(s)$ and $Mn^{2+}\,(aq) \rightarrow Mn\,(s)$

Balance H with H^+: $Mn^{2+}\,(aq) + 2\,H_2O\,(l) \rightarrow MnO_2\,(s) + 4\,H^+\,(aq)$ and $Mn^{2+}\,(aq) \rightarrow Mn\,(s)$

Add electrons: $Mn^{2+}\,(aq) + 2\,H_2O\,(l) \rightarrow MnO_2\,(s) + 4\,H^+\,(aq) + 2\,e^-$ and $Mn^{2+}\,(aq) + 2\,e^- \rightarrow Mn\,(s)$

Equalize electrons: $Mn^{2+}\,(aq) + 2\,H_2O\,(l) \rightarrow MnO_2\,(s) + 4\,H^+\,(aq) + 2\,e^-$ and $Mn^{2+}\,(aq) + 2\,e^- \rightarrow Mn\,(s)$

Add half-reactions: Mn^{2+} *(aq)* + 2 H_2O *(l)* + Mn^{2+} *(aq)* + ~~2 e⁻~~ → MnO_2 *(s)* + 4 H^+ *(aq)* + ~~2 e⁻~~ + Mn *(s)*

Cancel electrons: 2 Mn^{2+} *(aq)* + 2 H_2O *(l)* → MnO_2 *(s)* + 4 H^+ *(aq)* + Mn *(s)*

Look up cell potentials. Mn is oxidized in the first half-cell reaction so $E^\circ_{ox} = -E^\circ_{red} = -1.21$ V. Mn is reduced in the second half-cell reaction so $E^\circ_{red} = -1.18$ V. Then $E^\circ_{cell} = E^\circ_{ox} + E^\circ_{red} = -1.21\text{ V} - 1.18\text{ V} = -2.39$ V. $n = 2$ so

$$\Delta G^0_{rxn} = -nFE^0_{cell} = -2\ \cancel{\text{mol e}^-}\ \text{x}\ \frac{96{,}485\ \text{C}}{\cancel{\text{mol e}^-}}\ \text{x} -2.39\ \text{V} = -2\ \text{x}\ 96{,}485\ \cancel{\text{C}}\ \text{x} -2.39\ \frac{\text{J}}{\cancel{\text{C}}} = 4.6\underline{1}198 \times 10^5\ \text{J} = 461\ \text{kJ}$$

T = 273.15 + 25 °C = 298 K then $\Delta G^0_{rxn} = -RT\ln K$ Rearrange to solve for K.

$$K = e^{\frac{-\Delta G^0_{rxn}}{RT}} = e^{\frac{-4.6\underline{1}198 \times 10^5\ \cancel{\text{J}}}{\left(8.314\frac{\cancel{\text{J}}}{\cancel{\text{K}}\ \text{mol}}\right)(298\ \cancel{\text{K}})}} = e^{-18\underline{6}.149} = 1.43 \times 10^{-81}.$$

Check:

Reactants	Products
2 Mn atoms	2 Mn atoms
2 O atoms	2 O atoms
4 H atoms	4 H atoms
+4 charge	+4 charge

The units (kJ and none) are correct. If the voltage is negative, the reaction is nonspontaneous and the free energy change is very positive and the equilibrium constant is extremely small.

b) Separate: MnO_2 *(s)* → Mn^{2+} *(aq)* and MnO_2 *(s)* → MnO_4^- *(aq)*

Balance non H & O elements: MnO_2 *(s)* → Mn^{2+} *(aq)* and MnO_2 *(s)* → MnO_4^- *(aq)*

Balance O with H_2O: MnO_2 *(s)* → Mn^{2+} *(aq)* + 2 H_2O *(l)* and MnO_2 *(s)* + 2 H_2O *(l)* → MnO_4^- *(aq)*

Balance H with H^+: MnO_2 *(s)* + 4 H^+ *(aq)* → Mn^{2+} *(aq)* + 2 H_2O *(l)* and
MnO_2 *(s)* + 2 H_2O *(l)* → MnO_4^- *(aq)* + 4 H^+ *(aq)*

Add electrons: MnO_2 *(s)* + 4 H^+ *(aq)* + 2 e^- → Mn^{2+} *(aq)* + 2 H_2O *(l)* and
MnO_2 *(s)* + 2 H_2O *(l)* → MnO_4^- *(aq)* + 4 H^+ *(aq)* + 3 e^-

Equalize electrons: 3 MnO_2 *(s)* + 12 H^+ *(aq)* + 6 e^- → 3 Mn^{2+} *(aq)* + 6 H_2O *(l)* and
2 MnO_2 *(s)* + 4 H_2O *(l)* → 2 MnO_4^- *(aq)* + 8 H^+ *(aq)* + 6 e^-

Add half-reactions: 3 MnO_2 *(s)* + 4 ~~12 H⁺ (aq)~~ + ~~6 e⁻~~ + 2 MnO_2 *(s)* + ~~4 H₂O (l)~~ →
3 Mn^{2+} *(aq)* + 2 ~~4 H₂O (l)~~ + 2 MnO_4^- *(aq)* + ~~8 H⁺ (aq)~~ + ~~6 e⁻~~

Cancel electrons & species: 5 MnO_2 *(s)* + 4 H^+ *(aq)* → 3 Mn^{2+} *(aq)* + 2 H_2O *(l)* + 2 MnO_4^- *(aq)*

Look up cell potentials. Mn is reduced in the first half-cell reaction so $E^\circ_{red} = 1.21$ V. Mn is oxidized in the second half-cell reaction so $E^\circ_{ox} = -E^\circ_{red} = -1.68$ V. Then $E^\circ_{cell} = E^\circ_{ox} + E^\circ_{red} = 1.21\text{ V} - 1.68\text{ V} = -0.47$ V. $n = 6$ so

$$\Delta G^0_{rxn} = -nFE^0_{cell} = -6\ \cancel{\text{mol e}^-}\ \text{x}\ \frac{96{,}485\ \text{C}}{\cancel{\text{mol e}^-}}\ \text{x} -0.47\ \text{V} = -6\ \text{x}\ 96{,}485\ \cancel{\text{C}}\ \text{x} -0.47\ \frac{\text{J}}{\cancel{\text{C}}} = 2.\underline{7}209 \times 10^5\ \text{J} = 270\ \text{kJ}$$

$= 2.7 \times 10^2$ kJ . T = 273.15 + 25 °C = 298 K then $\Delta G^0_{rxn} = -RT\ln K$ Rearrange to solve for K.

$$K = e^{\frac{-\Delta G^0_{rxn}}{RT}} = e^{\frac{-2.\underline{7}209 \times 10^5\ \cancel{\text{J}}}{\left(8.314\frac{\cancel{\text{J}}}{\cancel{\text{K}}\ \text{mol}}\right)(298\ \cancel{\text{K}})}} = e^{-1\underline{0}.982} = 2.0 \times 10^{-48}.$$

Check:

Reactants	Products
5 Mn atoms	5 Mn atoms
10 O atoms	10 O atoms
4 H atoms	4 H atoms
+4 charge	+4 charge

The units (kJ and none) are correct. If the voltage is negative, the reaction is nonspontaneous and the free energy change is very positive and the equilibrium constant is extremely small. The voltage is less than in part a) so the free energy change is not as large and the equilibrium constant is not as small.

121. **Given:** Metal, M, 50.9 g/mol, 1.20 g of metal reduced in 23.6 minutes at 6.42 A from molten chloride
Find: empirical formula of chloride
Conceptual plan:

min → s → C → mol e⁻ and g M → mol M then mol e⁻, mol M → charge → MCl_x

$$\frac{60\text{ s}}{1\text{ min}}\quad \frac{6.42\text{ C}}{1\text{ s}}\quad \frac{1\text{ mol e}^-}{96{,}485\text{ C}}\qquad \frac{1\text{ mol M}}{50.9\text{ g M}}\qquad \frac{1\text{ mol e}^-}{1\text{ mol M}}$$

Solution: $23.6\text{ min} \times \frac{60\text{ s}}{1\text{ min}} \times \frac{6.42\text{ C}}{1\text{ s}} \times \frac{1\text{ mol e}^-}{96{,}485\text{ C}} = 0.09\underline{4}2190\text{ mol e}^-$ and

$1.20\text{ g M} \times \frac{1\text{ mol M}}{50.9\text{ g M}} = 0.02\underline{3}5756\text{ mol M}$ then $\frac{0.09\underline{4}2190\text{ mol e}^-}{0.02\underline{3}57561\text{ mol M}} = 3.9\underline{9}646\ \frac{\text{e}^-}{\text{M}}$ so the empirical formula is MCl_4.

Check: The units (none) are correct. The result was an integer within the error of the measurements. The formula is typical for a metal salt. It could be vanadium, which has a +4 oxidation state.

123. **Given:** hydrogen-oxygen fuel cell; 1.2×10^3 kWh of electricity/month **Find:** V of $H_2\,(g)$ at STP/month
Conceptual plan: Write half reactions. Look up half-reactions in Table 18.1. The reaction on the left is the oxidation so reverse the sign of the standard reduction potential ($E^\circ_{ox} = -E^\circ_{red}$) then add the two half-cell potentials: $E^\circ_{cell} = E^\circ_{ox} + E^\circ_{red.}$ Add the two half-cell reactions and cancel the electrons. then kWh → J → C → mol e⁻ → mol H_2 → V

$$\frac{3.60\times10^6\text{ J}}{1\text{kWh}}\quad \frac{1\text{ C}}{0.41\text{ J}}\quad \frac{1\text{ mol e}^-}{96{,}485\text{ C}}\quad \frac{2\text{ mol H}_2}{4\text{ mol e}^-}\ \text{at STP}\ \frac{22.414\text{ L}}{1\text{ mol H}_2}$$

Solution: $2\,H_2\,(g) + 4\,OH^-\,(aq) \rightarrow 4\,H_2O\,(l) + 4\,e^-$ where $E^\circ_{ox} = -E^\circ_{red} = -(-0.83\text{ V}) = 0.83\text{ V}$; and $O_2\,(g) + 2\,H_2O\,(l) + 4\,e^- \rightarrow 4\,OH^-\,(aq)$ where $E^\circ_{red} = 0.40$ V. $E^\circ_{cell} = E^\circ_{ox} + E^\circ_{red} = 0.83\text{ V} + 0.40\text{ V} = 1.23\text{ V} = 1.23\text{ J/C}$ and $n = 4$. Net reaction is: $2\,H_2\,(g) + O_2\,(g) \rightarrow 2\,H_2O\,(l)$. then

$$1.2\times10^3\text{ kWh} \times \frac{3.60\times10^6\text{ J}}{1\text{kWh}} \times \frac{1\text{ C}}{1.23\text{ J}} \times \frac{1\text{ mole}^-}{96{,}485\text{ C}} \times \frac{2\text{ mol H}_2}{4\text{ mol e}^-} \times \frac{22.414\text{ L}}{1\text{ mol H}_2} = 4.1\times10^6\text{ L}.$$

Check: The units (L) are correct. A large volume is expected since we are trying to generate a large amount of electricity.

125. **Given:** Au^{3+}/Au electroplating; surface area = 49.8 cm², Au thickness = 1.00×10^{-3} cm, density = 19.3 g/cm³; at 3.25 A **Find:** time
Conceptual plan: Write the half-cell reaction and add electrons as needed to balance reactions. then surface area, thickness → V → g Au → mol Au → mol e⁻ → C → s

$$V = \textit{surface area} \times \textit{thickness}\quad \frac{19.3\text{ g Au}}{1\text{ cm}^3\text{ Au}}\quad \frac{1\text{ mol Au}}{196.97\text{ g Au}}\quad \frac{3\text{ mol e}^-}{1\text{ mol Au}}\quad \frac{96{,}485\text{ C}}{1\text{ mol e}^-}\quad \frac{1\text{ s}}{3.25\text{ C}}$$

Solution: write half-reaction and add electrons $Au^{3+}\,(aq) + 3\,e^- \rightarrow Au\,(s)$.
$V = \textit{surface area} \times \textit{thickness} = (49.8\text{ cm}^2)(1.00\times10^{-3}\text{ cm}) = 0.0498\text{ cm}^3$ then

$$0.0498\text{ cm}^3\text{ Au} \times \frac{19.3\text{ g Au}}{1\text{ cm}^3\text{ Au}} \times \frac{1\text{ mol Au}}{196.97\text{ g Au}} \times \frac{3\text{ mol e}^-}{1\text{ mol Au}} \times \frac{96{,}485\text{ C}}{1\text{ mol e}^-} \times \frac{1\text{ s}}{3.25\text{ C}} = 435\text{ s}.$$

Check: The units (s) are correct. Since the layer is so thin there is far less than a mole of gold, so the time is not very long. In order to be an economical process, it must be fairly quick.

127. **Given:** $C_2O_4^{2-} \rightarrow CO_2$ and $MnO_4^-\,(aq) \rightarrow Mn^{2+}\,(aq)$; 50.1 mL of MnO_4^- to titrate 0.339 g $Na_2C_2O_4$; and 4.62 g U sample titrated by 32.3 mL MnO_4^-; and $UO^{2+} \rightarrow UO_2^{2+}$
Find: percent U in sample
Conceptual plan: Separate the overall reaction into two half-reactions: one for oxidation and one for reduction. → Balance each half-reaction with respect to mass in the following order: 1) balance all elements other than H and O; 2) balance O by adding H_2O; and 3) balance H by adding H^+. → Balance each half reaction with respect to charge by adding electrons. (The sum of the charges on both sides of the equation should be made equal by adding electrons as necessary.) → Make the number of electrons in both half-reactions equal by multiplying one or both half-reactions by a small whole number. → Add the two half-reactions together, canceling electrons and other species as necessary. → Verify that the reaction is balanced both with respect to mass and with respect to charge. then mL MnO_4^- → L MnO_4^- and g $Na_2C_2O_4$ → mol $Na_2C_2O_4$ → mol MnO_4^- then

$$\frac{1\text{ L MnO}_4^-}{1000\text{ mL MnO}_4^-}\qquad \frac{1\text{ mol Na}_2\text{C}_2\text{O}_4}{134.00\text{ g Na}_2\text{C}_2\text{O}_4}\qquad \frac{2\text{ mol MnO}_4^-}{5\text{ mol Zn}}\qquad \frac{1\text{ L MnO}_4^-}{0.500\text{ mol MnO}_4^-}$$

then L MnO_4^-, mol MnO_4^- → M MnO_4^- then write U half-reactions and balance as above. →

$$M = \frac{\text{mol } MnO_4^-}{L}$$

Make the number of electrons in both half-reactions equal by multiplying one or both half-reactions by a small whole number. → Add the two half-reactions together, canceling electrons and other species as necessary. → Verify that the reaction is balanced both with respect to mass and with respect to charge. then

mL MnO_4^-, M MnO_4^- → mol MnO_4^- → mol U → g U then g U, g sample → % U

$$M = \frac{\text{mol } MnO_4^-}{L} \quad \frac{5 \text{ mol U}}{2 \text{ mol } MnO_4^-} \quad \frac{238.03 \text{ g U}}{1 \text{ mol U}} \qquad \text{percent U} = \frac{\text{g U}}{\text{g sample}} \times 100\%$$

Solution:

Separate: MnO_4^- *(aq)* → Mn^{2+} *(aq)* and $C_2O_4^{2-}$ *(aq)* → CO_2 *(g)*

Balance non H & O elements: MnO_4^- *(aq)* → Mn^{2+} *(aq)* and $C_2O_4^{2-}$ *(aq)* → 2 CO_2 *(g)*

Balance O with H_2O: MnO_4^- *(aq)* → Mn^{2+} *(aq)* + 4 H_2O *(l)* and $C_2O_4^{2-}$ *(aq)* → 2 CO_2 *(g)*

Balance H with H^+: MnO_4^- *(aq)* + 8 H^+ *(aq)* → Mn^{2+} *(aq)* + 4 H_2O *(l)* and $C_2O_4^{2-}$ *(aq)* → 2 CO_2 *(g)*

Add electrons: MnO_4^- *(aq)* + 8 H^+ *(aq)*+ 5 e^-→Mn^{2+} *(aq)* + 4 H_2O *(l)* and $C_2O_4^{2-}$ *(aq)*→ 2 CO_2 *(g)* + 2 e^-

Equalize electrons:

2 MnO_4^- *(aq)* + 16 H^+*(aq)* + 10 e^-→ 2 Mn^{2+} *(aq)* + 8 H_2O *(l)* and 5 $C_2O_4^{2-}$ *(aq)* → 10 CO_2 *(g)* + 10 e^-

Add half-reactions:

2 MnO_4^- *(aq)* + 16 H^+ *(aq)* + ~~10 e^-~~ + 5 $C_2O_4^{2-}$ *(aq)* → 2 Mn^{2+} *(aq)* + 8 H_2O *(l)* + 10 CO_2 *(g)* + ~~10 e^-~~

Cancel electrons: 2 MnO_4^- *(aq)* + 16 H^+ *(aq)* + 5 $C_2O_4^{2-}$ *(aq)* → 2 Mn^{2+} *(aq)* + 8 H_2O *(l)* + 10 CO_2 *(g)*

$$50.1 \text{ mL } MnO_4^- \times \frac{1 \text{ L } MnO_4^-}{1000 \text{ mL } MnO_4^-} = 0.0501 \text{ L } MnO_4^-$$

$$0.339 \text{ g } Na_2C_2O_4 \times \frac{1 \text{ mol } Na_2C_2O_4}{134.00 \text{ g } Na_2C_2O_4} \times \frac{2 \text{ mol } MnO_4^-}{5 \text{ mol } Na_2C_2O_4} = 0.00101194 \text{ mol } MnO_4^- \quad \text{then}$$

$$M = \frac{0.00101194 \text{ mol } MnO_4^-}{0.0501 \text{ L}} = 0.0201984 \text{ M } MnO_4^- .$$

Separate: MnO_4^- *(aq)* → Mn^{2+} *(aq)* and UO^{2+} *(aq)* → UO_2^{2+} *(aq)*

Balance non H & O elements: MnO_4^- *(aq)* → Mn^{2+} *(aq)* and UO^{2+} *(aq)* → UO_2^{2+} *(aq)*

Balance O with H_2O: MnO_4^- *(aq)* → Mn^{2+} *(aq)* + 4 H_2O *(l)* and UO^{2+} *(aq)* + H_2O *(l)* → UO_2^{2+} *(aq)*

Balance H with H^+:

MnO_4^- *(aq)* + 8 H^+ *(aq)* → Mn^{2+} *(aq)* + 4 H_2O *(l)* and UO^{2+} *(aq)* + H_2O *(l)* → UO_2^{2+} *(aq)* + 2 H^+ *(aq)*

Add electrons: MnO_4^- *(aq)* + 8 H^+ *(aq)* + 5 e^- → Mn^{2+} *(aq)* + 4 H_2O *(l)* and

UO^{2+} *(aq)* + H_2O *(l)* → UO_2^{2+} *(aq)* + 2 H^+ *(aq)* + 2 e^-

Equalize electrons: 2 MnO_4^- *(aq)* + 16 H^+ *(aq)* + 10 e^- → 2 Mn^{2+} *(aq)* + 8 H_2O *(l)* and

5 UO^{2+} *(aq)* + 5 H_2O *(l)* → 5 UO_2^{2+} *(aq)* + 10 H^+ *(aq)* + 10 e^-

Add half-reactions: 2 MnO_4^- *(aq)* + 6 ~~10 H^+ *(aq)*~~ + ~~10 e^-~~ + 5 UO^{2+} *(aq)* + ~~5 H_2O *(l)*~~ →

2 Mn^{2+} *(aq)* + 3 ~~5 H_2O *(l)*~~ + 5 UO_2^{2+} *(aq)* + ~~10 H^+ *(aq)*~~ + ~~10 e^-~~

Cancel electrons & species:

2 MnO_4^- *(aq)* + 6 H^+ *(aq)* + 5 UO^{2+} *(aq)* → 2 Mn^{2+} *(aq)* + 3 H_2O *(l)* + 5 UO_2^{2+} *(aq)*

$$32.3 \text{ mL } MnO_4^- \times \frac{0.0201984 \text{ mol } MnO_4^-}{1000 \text{ mL } MnO_4^-} \times \frac{5 \text{ mol U}}{2 \text{ mol } MnO_4^-} \times \frac{238.03 \text{ g U}}{1 \text{ mol U}} = 0.388232 \text{ g U then}$$

$$\text{percent U} = \frac{\text{g U}}{\text{g sample}} \times 100\% = \frac{0.388232 \text{ g U}}{4.63 \text{ g sample}} \times 100\% = 8.39\ \% .$$

Check: first reaction

Reactants	Products
2 Mn atoms	2 Mn atoms
28 O atoms	28 O atoms
16 H atoms	16 H atoms
10 C atoms	10 C atoms
+4 charge	+4 charge

second reaction	Reactants	Products
	2 Mn atoms	2 Mn atoms
	13 O atoms	13 O atoms
	6 H atoms	6 H atoms
	5 U atoms	5 U atoms
	+14 charge	+14 charge

The reactions are balanced. The units (%) are correct. The percentage is between 0 and 100 %.

129. a) Looking for anion reductions that are in between the reduction potentials of Cl_2 and Br_2. The only one that meets this criterion is the dichromate ion.

Chapter 19
Radioactivity and Nuclear Chemistry

1. Radioactivity is the emission of subatomic particles or high-energy electromagnetic radiation by the nuclei of certain atoms. Radioactivity was discovered in 1896 by a French scientist named Antoine-Henri Becquerel (1852-1908). Becquerel placed crystals—composed of potassium uranyl sulfate, a compound known to phosphoresce—on top of a photographic plate wrapped in black cloth. The photographic plate showed a bright exposure spot where the crystals had been. Becquerel, Marie Curie and Pierre Curie received the Nobel Prize for the discovery of radioactivity.

3. A is the mass number (number of protons + neutrons); Z is the atomic number (number of protons) and X is the chemical symbol of the element.

5. An alpha particle has the same symbol as a Helium nucleus, ${}^{4}_{2}\text{He}$. When an element emits an alpha particle, the number of protons in its nucleus decreases by 2 and the mass number decreases by 4, transforming it into a different element.

7. Gamma rays are high-energy (short-wavelength) photons and has a symbol of ${}^{0}_{0}\gamma$. A gamma ray has no charge and no mass. When a gamma-ray photon is emitted from a radioactive atom, it does not change the mass number or the atomic number of the element. Gamma rays, however, are usually emitted in conjunction with other types of radiation.

9. Electron capture occurs when a nucleus assimilates an electron from an inner orbital of its electron cloud. Like positron emission, the net effect of electron capture is the conversion of a proton into a neutron: ${}^{1}_{1}\text{p} + {}^{0}_{-1}\text{e} \rightarrow {}^{1}_{0}\text{n}$. When an atom undergoes electron capture, its atomic number decreases by one because it has one less proton, and its mass number is unchanged.

11. For the lighter elements, the N/Z ratio of stable isotopes is about one (equal numbers of neutrons and protons). However, beyond about Z = 20, the N/Z ratio of stable nuclei begins to get larger (reaching about 1.5). Above Z = 83, stable nuclei do not exist.

13. Magic numbers are certain numbers of nucleons (N or Z = 2, 8, 20, 28, 50, 82, and N= 126) that have unique stability. Nuclei containing a magic number of protons or neutrons are particularly stable.

15. The half-life is the time it takes for one-half of the parent nuclides in a radioactive sample to decay to the daughter nuclides, and is identical to the concept of half-life for chemical reactions that we covered in Chapter 13. Thus, the relationship between the half-life of a nuclide and its rate constant is given by the same expression (Equation 13.19) that we derived for a first-order reaction in Section 13.4: $t_{1/2} = \dfrac{0.693}{k}$.
Nuclides that decay quickly have short half-lives and large rate constants—they are considered very active (many decay events per unit time). Nuclides that decay slowly have long half-lives and are less active (fewer decay events per unit time).

17. The ratio of uranium-238 to lead-206 within igneous rocks (rocks of volcanic origin) measures the time that has passed since the rock solidified (at which point the "radiometric clock" was reset). Since U-238 decays into Pb-206 with a half-life of 4.5×10^9 years, the relative amounts of U-238 and Pb-206 in a uranium-containing rock reveal its age. The method uses similar half-life calculations as the carbon dating method, only the half-life is much longer, so much older materials can be dated. The oldest rocks have an age of approximately 4.0 billion years, establishing a lower bound for the age of the Earth (the Earth must be at least as old as its oldest rocks). The ages of about 70 meteorites that have struck the earth have also been extensively studied and have been found to be about 4.5 billion years old. Since the meteorites were formed at the same time as our solar system (which includes the Earth), the best estimate for the age of the Earth is therefore about 4.5 billion years.

19. The Manhattan Project was a top-secret endeavor with the main goal of building an atomic bomb before the Germans did. The project was led by physicist J.R. Oppenheimer (1904-1967) at a high-security research

facility in Los Alamos, New Mexico. Four years later, on July 16, 1945, the world's first nuclear weapon was successfully detonated at a test site in New Mexico. The first atomic bomb exploded with a force equivalent to 18,000 tons of dynamite. Ironically, the Germans—who had not made a successful nuclear bomb—had already been defeated by this time. Instead, the atomic bomb was used on Japan. One bomb was dropped on Hiroshima and a second bomb was dropped on Nagasaki. Together, the bombs killed approximately 200,000 people and forced Japan to surrender.

21. The advantages of using fission to generate electricity are: 1) a typical nuclear power plant generates enough electricity for a city of about 1 million people and uses about 50 kg of fuel per day as opposed to a coal-burning power plant using about 2,000,000 kg of fuel to generate the same amount of electricity); and 2) a nuclear power plant generates no air pollution and no greenhouse gases. (Coal-burning power plants also emit carbon dioxide, a greenhouse gas.) The disadvantages are: 1) the danger of nuclear accidents, such as overheating and the release of radiation; and 2) waste disposal, since the products of the reaction are radioactive and have long half-lives.

23. This difference in mass between the products and the reactants is known as the mass defect. The energy corresponding to the mass defect, obtained by substituting the mass defect into the equation $E = mc^2$, is known as the nuclear binding energy, the amount of energy that would be required to break apart the nucleus into its component nucleons. The nuclear binding energy per nucleon peaks at a mass number of 60. The significance of this is that the nuclides with mass numbers of about 60 are among the most stable.

25. Extremely high temperature required for fusion to occur. To date no material can withstand these temperatures. The U.S. Congress has reduced funding for these projects so that it is less likely that a viable process will be developed.

27. In a single stage linear accelerator, a charged particle such as a proton is accelerated in an evacuated tube. The accelerating force is provided by a potential difference between the ends of the tube. In multi-stage linear accelerators, such as the Stanford Linear Accelerator (SLAC) at Stanford University (Figure 19.14), a series of tubes of increasing length are connected to a source of alternating voltage, as shown in Figure 19.15. The voltage alternates in such a way that, as a positively charged particle leaves a particular tube, that tube becomes positively charged, repelling the particle to the next tube. At the same time, the tube the particle is now approaching becomes negatively charged, pulling the particle towards it. This continues throughout the linear accelerator, allowing the particle to be accelerated to velocities up to 90% of the speed of light. Linear accelerators can be used to conduct nuclear transmutations, making nuclides that don't normally exist in nature.

29. The energy associated with radioactivity can ionize molecules. When radiation ionizes important molecules in living cells, problems can develop. The ingestion of radioactive materials, especially alpha and beta emitters, is particularly dangerous because the radioactivity is then inside the body and can do even more damage. The effects of radiation can be divided into three different types: acute radiation damage, increased cancer risk, and genetic effects.

31. The biological effectiveness factor, or RBE, (for relative biological effectiveness), correct the dosage (in rads) for the type of radiation. It is a correction factor that is usually multiplied by the dose in rads to obtain the dose in a unit called the rem for roentgen equivalent man. So, dose in rads × biological effectiveness factor = dose in rem. The biological effectiveness factor for alpha radiation for example, is much higher than for gamma radiation.

33. **Conceptual Plan: Begin with the symbol for parent nuclide on the left side of the equation and the symbol for a particle on the right side (except for electron capture). → Equalize the sum of the mass numbers and the sum of the atomic numbers on both sides of the equation by writing the appropriate mass number and atomic number for the unknown daughter nuclide. → Using the periodic table, deduce the identity of the unknown daughter nuclide from the atomic number and write its symbol.**
Solution:
a) U-234 (alpha decay) ${}^{234}_{92}\text{U} \rightarrow {}^{?}_{?}? + {}^{4}_{2}\text{He}$ then ${}^{234}_{92}\text{U} \rightarrow {}^{230}_{90}? + {}^{4}_{2}\text{He}$ then ${}^{234}_{92}\text{U} \rightarrow {}^{230}_{90}\text{Th} + {}^{4}_{2}\text{He}$

b) Th-230 (alpha decay) ${}^{230}_{90}\text{Th} \rightarrow {}^{?}_{?}? + {}^{4}_{2}\text{He}$ then ${}^{230}_{90}\text{Th} \rightarrow {}^{226}_{88}? + {}^{4}_{2}\text{He}$ then ${}^{230}_{90}\text{Th} \rightarrow {}^{226}_{88}\text{Ra} + {}^{4}_{2}\text{He}$

c) Pb-214 (beta decay) ${}^{214}_{82}\text{Pb} \rightarrow {}^{?}_{?}? + {}^{0}_{-1}\text{e}$ then ${}^{214}_{82}\text{Pb} \rightarrow {}^{214}_{83}? + {}^{0}_{-1}\text{e}$ then ${}^{214}_{82}\text{Pb} \rightarrow {}^{214}_{83}\text{Bi} + {}^{0}_{-1}\text{e}$

d) N-13 (positron emission) $^{13}_{7}N \rightarrow ^{?}_{?}? + ^{0}_{+1}e$ then $^{13}_{7}N \rightarrow ^{13}_{6}? + ^{0}_{+1}e$ then $^{13}_{7}N \rightarrow ^{13}_{6}C + ^{0}_{+1}e$

e) Cr-51 (electron capture) $^{51}_{24}Cr + ^{0}_{-1}e \rightarrow ^{?}_{?}?$ then $^{51}_{24}Cr + ^{0}_{-1}e \rightarrow ^{51}_{23}?$ then $^{51}_{24}Cr + ^{0}_{-1}e \rightarrow ^{51}_{23}V$

Check: a) 234 = 230 + 4, 92 = 90 + 2, and Thorium is atomic number 90. b) 230 = 226 + 4, 90 = 88 + 2, and Radium is atomic number 88. c) 214 = 214 + 0, 82 = 83 – 1, and Bismuth is atomic number 83. d) 13 = 13 + 0, 7 = 6 + 1, and Carbon is atomic number 6. e) 51 + 0 = 51, 24 – 1 = 23, and Vanadium is atomic number 23.

35. **Given:** Th-232 decay series: α, β, β, α **Find:** balanced decay reactions

Conceptual Plan: Begin with the symbol for parent nuclide on the left side of the equation and the symbol for a particle on the right side (except for electron capture). → Equalize the sum of the mass numbers and the sum of the atomic numbers on both sides of the equation by writing the appropriate mass number and atomic number for the unknown daughter nuclide. → Using the periodic table, deduce the identity of the unknown daughter nuclide from the atomic number and write its symbol. → Use the product of this reaction to write the next reaction.

Solution:

Th-232 (alpha decay) $^{232}_{90}Th \rightarrow ^{?}_{?}? + ^{4}_{2}He$ then $^{232}_{90}Th \rightarrow ^{228}_{88}? + ^{4}_{2}He$ then $^{232}_{90}Th \rightarrow ^{228}_{88}Ra + ^{4}_{2}He$

Ra-228 (beta decay) $^{228}_{88}Ra \rightarrow ^{?}_{?}? + ^{0}_{-1}e$ then $^{228}_{88}Ra \rightarrow ^{228}_{89}? + ^{0}_{-1}e$ then $^{228}_{88}Ra \rightarrow ^{228}_{89}Ac + ^{0}_{-1}e$

Ac-228 (beta decay) $^{228}_{89}Ac \rightarrow ^{?}_{?}? + ^{0}_{-1}e$ then $^{228}_{89}Ac \rightarrow ^{228}_{90}? + ^{0}_{-1}e$ then $^{228}_{89}Ac \rightarrow ^{228}_{90}Th + ^{0}_{-1}e$

Th-228 (alpha decay) $^{228}_{90}Th \rightarrow ^{?}_{?}? + ^{4}_{2}He$ then $^{228}_{90}Th \rightarrow ^{224}_{88}? + ^{4}_{2}He$ then $^{228}_{90}Th \rightarrow ^{224}_{88}Ra + ^{4}_{2}He$

Thus the decay series is: $^{232}_{90}Th \rightarrow ^{228}_{88}Ra + ^{4}_{2}He$, $^{228}_{88}Ra \rightarrow ^{228}_{89}Ac + ^{0}_{-1}e$, $^{228}_{89}Ac \rightarrow ^{228}_{90}Th + ^{0}_{-1}e$, $^{228}_{90}Th \rightarrow ^{224}_{88}Ra + ^{4}_{2}He$.

Check: 232 = 228 + 4, 90 = 88 + 2, and Radium is atomic number 88. 228 = 228 + 0, 88 = 89 - 1, and Actinium is atomic number 89. 228 = 228 + 0, 89 = 90 – 1, and Thorium is atomic number 90. 228 = 224 + 4, 90 = 88 + 2, and Radium is atomic number 88.

37. **Conceptual Plan: Equalize the sum of the mass numbers and the sum of the atomic numbers on both sides of the equation by writing the appropriate mass number and atomic number for the unknown species. → Using the periodic table and the list of particles, deduce the identity of the unknown species from the atomic number and write its symbol.**

Solution:

a) $^{?}_{?}? \rightarrow ^{217}_{85}At + ^{4}_{2}He$ becomes $^{221}_{87}? \rightarrow ^{217}_{85}At + ^{4}_{2}He$ then $^{221}_{87}Fr \rightarrow ^{217}_{85}At + ^{4}_{2}He$

b) $^{241}_{94}Pu \rightarrow ^{241}_{95}Am + ^{?}_{?}?$ becomes $^{241}_{94}Pu \rightarrow ^{241}_{95}Am + ^{0}_{-1}?$ then $^{241}_{94}Pu \rightarrow ^{241}_{95}Am + ^{0}_{-1}e$

c) $^{19}_{11}Na \rightarrow ^{19}_{10}Ne + ^{?}_{?}?$ becomes $^{19}_{11}Na \rightarrow ^{19}_{10}Ne + ^{0}_{1}?$ then $^{19}_{11}Na \rightarrow ^{19}_{10}Ne + ^{0}_{+1}e$

d) $^{75}_{34}Se + ^{?}_{?}? \rightarrow ^{75}_{33}As$ becomes $^{75}_{34}Se + ^{0}_{-1}? \rightarrow ^{75}_{33}As$ then $^{75}_{34}Se + ^{0}_{-1}e \rightarrow ^{75}_{33}As$

Check: a) 221 = 210 + 4, 87 = 85 + 2, and Francium is atomic number 87. b) 241 = 241 + 0, 94 = 95 - 1, and the particle is a beta particle. c) 19 = 19 + 0, 11 = 80 + 1, and the particle is a positron. d) 75 = 75 + 0, 34 - 1 = 33, and the particle is an electron.

39. a) Stable, N/Z ratio is close to 1, acceptable for low Z atoms

b) Not stable, N/Z ratio much too high for low Z atom

c) Not stable, N/Z ratio is less than 1, much too low

d) Stable, N/Z ratio is acceptable for this Z

41. Sc, V, and Mn, each have odd numbers of protons. Atoms with an odd number of protons typically have fewer stable isotopes than those with an even number of protons.

43. a) Beta decay, since N/Z is too high

b) Positron emission, since N/Z is too low

c) Positron emission, since N/Z is too low

d) Positron emission, since N/Z is too low

45. a) Cs-125, since it is closer to the proper N/Z

b) Fe-62, since it is closer to the proper N/Z

47. Since the half-life of U-235 is 703 million years, 1/2 will be present after 703 million years, 1/4 will be present after 2 half-lives and 1/8 will be present after 3 half-lives or 2110 million years or 2.11 x 10^9 years.

49. **Given:** $t_{1/2}$ for isotope decay = 3.8 days; 1.55 g isotope initially **Find:** mass of isotope after 5.5 days
Conceptual plan:
radioactive decay implies first order kinetics, $t_{1/2} \rightarrow$ k then [isotope]$_0$, t, k $\rightarrow$ [isotope]$_t$

$$t_{1/2} = \frac{0.693}{k} \qquad ln\,N_t = -\,kt + ln\,N_0$$

Solution: $t_{1/2} = \frac{0.693}{k}$ rearrange to solve for k. $k = \frac{0.693}{t_{1/2}} = \frac{0.693}{3.8\text{ days}} = 0.1\underline{8}237\text{ day}^{-1}$. Since

$$ln[A]_t = -kt + \ln[A]_0 = -\,(0.1\underline{8}237\,\text{day}^{-1})\,(5.5\,\text{day}) + \ln(1.55\text{ g}) = -0.5\underline{6}478 \rightarrow$$

$$[A]_t = e^{-0.5\underline{6}478} = 0.57\text{ g}.$$

Check: The units (g) are correct. The amount is consistent with a time between one and two half-lives.

51. **Given:** F-18 initial decay rate = 1.5 x 10^5 /s, $t_{1/2}$ for F-18 = 1.83 hr **Find:** t to decay rate of 1.0 x 10^2 /s
Conceptual plan: radioactive decay implies first order kinetics, $t_{1/2} \rightarrow k$ then Rate$_0$, Rate$_t$, $k \rightarrow$ t

$$t_{1/2} = \frac{0.693}{k} \qquad ln\frac{\text{Rate}_t}{\text{Rate}_0} = -\,kt$$

Solution: $t_{1/2} = \frac{0.693}{k}$ rearrange to solve for k. $k = \frac{0.693}{t_{1/2}} = \frac{0.693}{1.83\text{ hr}} = 0.37\underline{8}689\text{ hr}^{-1}$. Since

$$ln\frac{\text{Rate}_t}{\text{Rate}_0} = -\,kt \text{ rearrange to solve for t } t = -\frac{1}{k}\ln\frac{\text{Rate}_t}{\text{Rate}_0} = -\frac{1}{0.37\underline{8}689\text{ hr}^{-1}}\ln\frac{1.0 \times 10^2\ /\text{s}}{1.5 \times 10^5\ /\text{s}} = 19.3\text{ hr}.$$

Check: The units (hr) are correct. The time is between 10 and 11 half-lives and the rate is just under $1/2^{10}$ of the original amount.

53. **Given:** boat analysis, C-14/C-12 = 72.5 % of living organism **Find:** t **Other:** $t_{1/2}$ for decay of C-14 = 5730 years
Conceptual plan: radioactive decay implies first order kinetics, $t_{1/2} \rightarrow$ k then 72.5 % of [C-14]$_0$, k $\rightarrow$ t

$$t_{1/2} = \frac{0.693}{k} \qquad ln\,N_t = -\,kt + ln\,N_0$$

Solution: $t_{1/2} = \frac{0.693}{k}$ rearrange to solve for k. $k = \frac{0.693}{t_{1/2}} = \frac{0.693}{5730\text{ yr}} = 1.2\underline{0}942 \times 10^{-4}\text{ yr}^{-1}$ then

[C-14]$_t$ = 0.725 [C-14]$_0$. Since $ln[\text{C-14}]_t = -\,kt + ln[\text{C-14}]_0$ rearrange to solve for t.

$$t = -\frac{1}{k}\ln\frac{[\text{C-14}]_t}{[\text{C-14}]_0} = -\frac{1}{1.2\underline{0}942 \times 10^{-4}\text{ yr}^{-1}}\ln\frac{0.725\,[\text{C-14}]_0}{[\text{C-14}]_0} = 2.66 \times 10^3\text{ yr}.$$

Check: The units (yr) are correct. The time to 72.5 % decay is consistent a time less than one half-life.

55. **Given:** skull analysis, C-14 decay rate = 15.3 dis/min·gC in living organisms and 0.85 dis/min·gC in skull
Find: t **Other:** $t_{1/2}$ for decay of C-14 = 5730 years
Conceptual plan: radioactive decay implies first order kinetics, $t_{1/2} \rightarrow$ k then Rate$_0$, Rate$_t$, $k \rightarrow$ t

$$t_{1/2} = \frac{0.693}{k} \qquad ln\frac{\text{Rate}_t}{\text{Rate}_0} = -\,kt$$

Solution: $t_{1/2} = \frac{0.693}{k}$ rearrange to solve for k. $k = \frac{0.693}{t_{1/2}} = \frac{0.693}{5730\text{ yr}} = 1.2\underline{0}942 \times 10^{-4}\text{ yr}^{-1}$

Since $ln\frac{\text{Rate}_t}{\text{Rate}_0} = -kt$, rearrange to solve for t.

$$t = -\frac{1}{k}\ln\frac{\text{Rate}_t}{\text{Rate}_0} = -\frac{1}{1.2\underline{0}942 \times 10^{-4}\ \text{yr}^{-1}}\ln\frac{0.85\ \cancel{\text{dis/min}\cdot\text{gC}}}{15.3\ \cancel{\text{dis/min}\cdot\text{gC}}} = 2.39 \times 10^4\ \text{yr}.$$

Check: The units (yr) are correct. The rate is 6 % of initial value and the time is consistent a time just more than four half-lives.

57. **Given:** rock analysis, 0.438 g Pb-206 to every 1.00 g U-238, no Pb-206 initially **Find:** age of rock
Other: $t_{1/2}$ for decay of U-238 to Pb-206 = 4.5 x 10^9 years
Conceptual plan: radioactive decay implies first order kinetics, $t_{1/2}$ → k then

$$t_{1/2} = \frac{0.693}{k}$$

g Pb-206 → mol Pb-206 → mol U-238 → g U-238 then $[\text{U-238}]_0$, $[\text{U-238}]_t$, k → t

$$\frac{1\ \text{mol Pb-206}}{206\ \text{g Pb-206}} \quad \frac{1\ \text{mol U-238}}{1\ \text{mol Pb-206}} \quad \frac{238\ \text{g U-238}}{1\ \text{mol U-238}} \qquad ln\,N_t = -kt + ln\,N_0$$

Solution: $t_{1/2} = \frac{0.693}{k}$ rearrange to solve for k. $k = \frac{0.693}{t_{1/2}} = \frac{0.693}{4.5 \times 10^9\ \text{yr}} = 1.\underline{5}4 \times 10^{-10}\ \text{yr}^{-1}$ then

$$0.438\ \cancel{\text{g Pb-206}} \times \frac{1\ \cancel{\text{mol Pb-206}}}{206\ \cancel{\text{g Pb-206}}} \times \frac{1\ \cancel{\text{mol U-238}}}{1\ \cancel{\text{mol Pb-206}}} \times \frac{238\ \text{g U-238}}{1\ \cancel{\text{mol U-238}}} = 0.50\underline{6}039\ \text{g U-238}.$$ Since

$ln\frac{[\text{U-238}]_t}{[\text{U-238}]_0} = -kt$, rearrange to solve for t.

$$t = -\frac{1}{k}\ln\frac{[\text{U-238}]_t}{[\text{U-238}]_0} = -\frac{1}{1.\underline{5}4 \times 10^{-10}\ \text{yr}^{-1}}\ln\frac{1.00\ \cancel{\text{g U-238}}}{(1.00 + 0.50\underline{6}039)\ \cancel{\text{g U-238}}} = 2.7 \times 10^9\ \text{yr}.$$

Check: The units (yr) are correct. The amount of Pb-206 is less than half of the initial U-238 amount and time is less than one half-life.

59. **Given:** U-235 fission induced by neutrons to Xe-144 and Sr-90 **Find:** number of neutrons produced
Conceptual Plan: Write species given on the appropriate side of the equation. →Equalize the sum of the mass numbers and the sum of the atomic numbers on both sides of the equation by writing the stoichiometric coefficient in front of the desired species.
Solution: $^{235}_{92}\text{U} + ^{1}_{0}\text{n} \rightarrow ^{144}_{54}\text{Xe} + ^{90}_{38}\text{Sr} + ?\,^{1}_{0}\text{n}$ becomes $^{235}_{92}\text{U} + ^{1}_{0}\text{n} \rightarrow ^{144}_{54}\text{Xe} + ^{90}_{38}\text{Sr} + 2\,^{1}_{0}\text{n}$ so 2 neutrons are produced.
Check: 235 + 1 = 144 + 90 + 2, 92 + 0 = 54 + 38 + 0, and no other particle is necessary to balance the equation.

61. **Given:** fusion of 2 H-2 atoms to form He-3 and 1 neutron **Find:** balanced equation
Conceptual Plan: Write species given on the appropriate side of the equation. →Equalize the sum of the mass numbers and the sum of the atomic numbers on both sides of the equation by writing the stoichiometric coefficient in front of the desired species.
Solution: $2\,^{2}_{1}\text{H} \rightarrow ^{3}_{2}\text{He} + ^{1}_{0}\text{n}$.
Check: 2(2) = 3 + 1, 2(1) = 2 + 0, and no other particle is necessary to balance the equation.

63. **Given:** U-238 bombarded by neutrons to form U-239 which undergoes 2 beta decays to form Pu-239
Find: balanced equations
Conceptual Plan: Write species given on the appropriate side of the equation. →Equalize the sum of the mass numbers and the sum of the atomic numbers on both sides of the equation by writing the stoichiometric coefficient in front of the desired species. → Use the product of this reaction to write the next reaction until process is complete.
Solution: $^{238}_{92}\text{U} + ?\,^{1}_{0}\text{n} \rightarrow ^{239}_{92}\text{U}$ becomes $^{238}_{92}\text{U} + ^{1}_{0}\text{n} \rightarrow ^{239}_{92}\text{U}$. then
beta decay $^{239}_{92}\text{U} \rightarrow ^{?}_{?}? + ^{0}_{-1}\text{e}$ becomes $^{239}_{92}\text{U} \rightarrow ^{239}_{93}? + ^{0}_{-1}\text{e}$ then $^{239}_{92}\text{U} \rightarrow ^{239}_{93}\text{Np} + ^{0}_{-1}\text{e}$. then
beta decay $^{239}_{93}\text{Np} \rightarrow ^{?}_{?}? + ^{0}_{-1}\text{e}$ becomes $^{239}_{93}\text{Np} \rightarrow ^{239}_{94}? + ^{0}_{-1}\text{e}$ then $^{239}_{93}\text{Np} \rightarrow ^{239}_{94}\text{Pu} + ^{0}_{-1}\text{e}$.
The entire process is $^{238}_{92}\text{U} + ^{1}_{0}\text{n} \rightarrow ^{239}_{92}\text{U}$, $^{239}_{92}\text{U} \rightarrow ^{239}_{93}\text{Np} + ^{0}_{-1}\text{e}$, $^{239}_{93}\text{Np} \rightarrow ^{239}_{94}\text{Pu} + ^{0}_{-1}\text{e}$.

Check: 238 + 1 = 239, 92 + 0 = 92, and no other particle is necessary to balance the equation. 239 0, 92 = 93 – 1, and Neptunium is atomic number 93. 239 = 239 + 0, 93 = 94 – 1, and Plutonium number 94.

65. **Given:** Rf-257 synthesized by bombarding Cf-249 with C-12 **Find:** balanced equation
Conceptual Plan: Write species given on the appropriate side of the equation. → Equalize the sum of the mass numbers and the sum of the atomic numbers on both sides of the equation by writing the stoichiometric coefficient in front of the desired species. → Use the product of this reaction to write the next reaction until process is complete.
Solution: ${}^{249}_{98}\text{Cf} + {}^{12}_{6}\text{C} \rightarrow {}^{257}_{?}\text{Rf} + ?\,{}^{?}_{?}?$ becomes ${}^{249}_{98}\text{Cf} + {}^{12}_{6}\text{C} \rightarrow {}^{257}_{104}\text{Rf} + 4\,{}^{1}_{0}\text{n}$.
Check: 249 + 12 = 257 = 4, 98 + 6 = 104 + 0, and Rutherfordium is atomic number 104 and 4 neutrons are needed (they have no protons) to balance the equation.

67. **Given:** 1.0 g of matter converted to energy **Find:** energy
Conceptual plan: g → kg → E

$$\frac{1\text{kg}}{1000\text{ g}} \qquad E = m\,c^2$$

Solution: $1.0\,\text{g} \times \frac{1\text{kg}}{1000\,\text{g}} = 0.0010\text{ kg}$ then $E = m\,c^2 = (0.0010\text{ kg})\left(2.9979 \times 10^8\ \frac{\text{m}}{\text{s}}\right)^2 = 9.0 \times 10^{13}\text{ J}$.

Check: The units (J) are correct. The magnitude of the answer makes physical sense because we are converting a highe quantity of amu's to energy.

69. **Given:** a) O-16 = 15.9949145 amu; b) Ni-58 = 57.935346 amu; and c) Xe-129 = 128.904780 amu
Find: mass defect and nuclear binding energy per nucleon
Conceptual plan: ${}^{A}_{Z}\text{X}$**, isotope mass → mass defect → nuclear binding energy per nucleon**

$$\text{mass defect} = Z(\text{mass } {}^{1}_{1}\text{H}) + (A - Z)(\text{mass } {}^{1}_{0}\text{n}) - \text{mass of isotope} \qquad \frac{931.5\text{ MeV}}{(1\text{ amu})(A\text{ nucleons})}$$

Solution: $\text{mass defect} = Z(\text{mass } {}^{1}_{1}\text{H}) + (A - Z)(\text{mass } {}^{1}_{0}\text{n}) - \text{mass of isotope}$.

a) O-16 mass defect = 8(1.00783 amu) + (16 - 8)(1.00866 amu) - 15.9949145 amu = 0.1370055 amu = = 0.13701 amu

and $0.1370055\text{ amu} \times \frac{931.5\text{ MeV}}{(1\text{ amu})(16\text{ nucleons})} = 7.976\ \frac{\text{MeV}}{\text{nucleons}}$.

b) Ni-58 mass defect = 28(1.00783 amu) + (58 - 28)(1.00866 amu) - 57.935346 amu = 0.543694 amu = = 0.54369 amu

and $0.543694\text{ amu} \times \frac{931.5\text{ MeV}}{(1\text{ amu})(58\text{ nucleons})} = 8.732\ \frac{\text{MeV}}{\text{nucleons}}$.

c) Xe-129 mass defect = 54(1.00783 amu) + (129 - 54)(1.00866 amu) - 128.904780 amu = 1.16754 amu

and $1.16754\text{ amu} \times \frac{931.5\text{ MeV}}{(1\text{ amu})(129\text{ nucleons})} = 8.431\ \frac{\text{MeV}}{\text{nucleons}}$.

Check: The units (amu and MeV/nucleon) are correct. The mass defect increases with an increasing number of nucleons, but the MeV/nucleon does not change by as much (on a relative basis).

71. **Given:** ${}^{235}_{92}\text{U} + {}^{1}_{0}\text{n} \rightarrow {}^{144}_{54}\text{Xe} + {}^{90}_{38}\text{Sr} + 2\,{}^{1}_{0}\text{n}$, U-235 = 235.043922 amu, Xe-144 = 143.9385 amu, and Sr-90 = 89.907738 amu **Find:** energy per g of U-235
Conceptual plan: mass of products & reactants → mass defect → mass defect / g of U-235 then

$$\text{mass defect} = \sum \text{mass of reactants} - \sum \text{mass of products} \qquad \frac{\text{mass defect}}{235.043922\text{ g U-235}}$$

g → kg → E

$$\frac{1\text{kg}}{1000\text{ g}} \qquad E = m\,c^2$$

Solution: mass defect = $\sum$ mass of reactants - $\sum$ mass of products notice that we can cancel a neutron from each side to get: ${}^{235}_{92}U \rightarrow {}^{144}_{54}Xe + {}^{90}_{38}Sr + {}^{1}_{0}n$ and

mass defect = 235.043922 g − (143.9385 g +89.907738 g + 1.00866 g) = 0.189024 g

$$\text{then } \frac{0.189024 \text{ g}}{235.043922 \text{ g U-235}} \times \frac{1 \text{ kg}}{1000 \text{ g}} = 8.04207 \times 10^{-7} \frac{\text{kg}}{\text{g U-235}} \text{ then}$$

$$E = m c^2 = \left(8.04207 \times 10^{-7} \frac{\text{kg}}{\text{g U-235}}\right)\left(2.9979 \times 10^8 \frac{\text{m}}{\text{s}}\right)^2 = 7.228 \times 10^{10} \frac{\text{J}}{\text{g U-235}}.$$

Check: The units (J) are correct. A large amount of energy is expected per gram of fuel in a nuclear reactor.

73. **Given:** $2\,{}^{2}_{1}H \rightarrow {}^{3}_{2}He + {}^{1}_{0}n$, H-2 = 2.014102 amu, and He-3 = 3.016029 amu
Find: energy per g reactant
Conceptual plan: mass of products & reactants → mass defect → mass defect / g of H-2 then

$$\text{mass defect} = \sum \text{mass of reactants} - \sum \text{mass of products} \qquad \frac{\text{mass defect}}{2(2.014102 \text{ g H-2})}$$

g → kg → E

$$\frac{1 \text{ kg}}{1000 \text{ g}} \qquad E = m c^2$$

Solution: mass defect = $\sum$ mass of reactants - $\sum$ mass of products and

mass defect = 2(2.014102 g) − (3.016029 g + 1.00866 g) = 0.003515 g

$$\text{then } \frac{0.003515 \text{ g}}{2(2.014102 \text{ g H-2})} \times \frac{1 \text{ kg}}{1000 \text{ g}} = 8.72597 \times 10^{-7} \frac{\text{kg}}{\text{g H-2}} \text{ then}$$

$$E = m c^2 = \left(8.72597 \times 10^{-7} \frac{\text{kg}}{\text{g H-2}}\right)\left(2.9979 \times 10^8 \frac{\text{m}}{\text{s}}\right)^2 = 7.84 \times 10^{10} \frac{\text{J}}{\text{g H-2}}.$$

Check: The units (J) are correct. A large amount of energy is expected per gram of fuel in a nuclear reactor.

75. **Given:** 75 kg human exposed to 32.8 rad and falling from chair **Find:** energy absorbed in each case
Conceptual plan: rad, kg → J and assume d = 0.50 m chair height then mass, d → J

$$\frac{0.01 \text{ J}}{1 \text{ kg body tissue}} \qquad E = F \cdot d = m\, g\, d$$

Solution: $32.8 \text{ rad} = 32.8 \frac{0.01 \text{ J}}{1 \text{ kg body tissue}} \times 75 \text{ kg} = 25 \text{ J}$ and

$$E = F \cdot d = m\, g\, d = 75 \text{ kg} \times 9.8 \frac{\text{m}}{\text{s}^2} \times 0.50 \text{ m} = 370 \text{ kg} \frac{\text{m}^2}{\text{s}^2} = 370 \text{ J}.$$

Check: The units (J and J) are correct. Allowable radiation exposures are low, since the radiation is very ionizing and, thus, damaging to tissue. Falling may have more energy, but it is not ionizing.

77. **Given:** $t_{1/2}$ for F-18 = 1.83 hr, 65 % of F-18 makes it to the hospital traveling at 60.0 miles/hour
Find: distance between hospital and cyclotron
Conceptual plan: $t_{1/2}$ **→** k **then [F-18]$_0$, [F-18]$_t$,** k **→ t then hr → mi**

$$t_{1/2} = \frac{0.693}{k} \qquad \ln\frac{[\text{F-18}]_t}{[\text{F-18}]_0} = -k\,t \qquad \frac{60.0 \text{ mi}}{1 \text{ hr}}$$

Solution: $t_{1/2} = \frac{0.693}{k}$ rearrange to solve for k. $k = \frac{0.693}{t_{1/2}} = \frac{0.693}{1.83 \text{ hr}} = 0.378689 \text{ hr}^{-1}$. Since

$\ln\frac{[\text{F-18}]_t}{[\text{F-18}]_0} = -k\,t$ rearrange to solve for t

$$t = -\frac{1}{k}\ln\frac{[\text{F-18}]_t}{[\text{F-18}]_0} = -\frac{1}{0.378689 \text{ hr}^{-1}}\ln\frac{0.65\,[\text{F-18}]_t}{[\text{F-18}]_0} = 1.1376 \text{ hr}. \text{ then } 1.1376 \text{ hr} \times \frac{60.0 \text{ mi}}{1 \text{ hr}} = 68 \text{ mi}.$$

Check: The units (mi) are correct. The time less than one half-life, so the distance is less than 1.83 times the speed of travel.

79. **Given:** Incomplete reactions **Find:** balanced reaction and energy (in J/mol reactant)
Conceptual Plan: Equalize the sum of the mass numbers and the sum of the atomic numbers on both sides of the equation by writing the appropriate mass number and atomic number for the unknown species. → Using the periodic table and the list of particles, deduce the identity of the unknown species from the atomic number and write its symbol. then
mass of products & reactants → mass defect in g → mass defect in kg → E

$$\text{mass defect} = \sum \text{mass of reactants} - \sum \text{mass of products} \quad \frac{1\,\text{kg}}{1000\ \text{g}} \qquad E = m\,c^2$$

Solution:

a) ${}^{?}_{?}? + {}^{9}_{4}\text{Be} \rightarrow {}^{6}_{3}\text{Li} + {}^{4}_{2}\text{He}$ becomes ${}^{1}_{1}? + {}^{9}_{4}\text{Be} \rightarrow {}^{6}_{3}\text{Li} + {}^{4}_{2}\text{He}$ then ${}^{1}_{1}\text{H} + {}^{9}_{4}\text{Be} \rightarrow {}^{6}_{3}\text{Li} + {}^{4}_{2}\text{He}$

mass defect = $\sum$ mass of reactants - $\sum$ mass of products and

mass defect = (1.00783 g + 9.012182 g) − (6.015122 g + 4.002603 g) = 0.002287 g

$$\text{then}\ \frac{0.002287\ \cancel{\text{g}}}{2\ \text{mol reactants}} \times \frac{1\,\text{kg}}{1000\ \cancel{\text{g}}} = 1.1435 \times 10^{-6}\ \frac{\text{kg}}{\text{mol reactants}}\ \text{then}$$

$$E = m\,c^2 = \left(1.1435 \times 10^{-6}\ \frac{\text{kg}}{\text{mol reactants}}\right)\left(2.9979 \times 10^{8}\ \frac{\text{m}}{\text{s}}\right)^2 = 1.03 \times 10^{11}\ \frac{\text{J}}{\text{mol reactants}}.$$

Check: 1 + 9 = 6 + 4, 1 + 4 = 3 + 2. The units (J) are correct.

b) ${}^{209}_{83}\text{Bi} + {}^{64}_{28}\text{Ni} \rightarrow {}^{272}_{111}\text{Rg} + {}^{?}_{?}?$ becomes ${}^{209}_{83}\text{Bi} + {}^{64}_{28}\text{Ni} \rightarrow {}^{272}_{111}\text{Rg} + {}^{1}_{0}?$ then ${}^{209}_{83}\text{Bi} + {}^{64}_{28}\text{Ni} \rightarrow {}^{272}_{111}\text{Rg} + {}^{1}_{0}\text{n}$

mass defect = (208.980384 g + 63.927969 g) − (272.1535 g + 1.00866 g) = −0.253807 g (Note: Since this is negative energy must be put in.)

$$\text{then}\ \frac{0.253807\ \cancel{\text{g}}}{2\ \text{mol reactants}} \times \frac{1\,\text{kg}}{1000\ \cancel{\text{g}}} = 1.269035 \times 10^{-4}\ \frac{\text{kg}}{\text{mol reactants}}\ \text{then}$$

$$E = m\,c^2 = \left(1.269035 \times 10^{-4}\ \frac{\text{kg}}{\text{mol reactants}}\right)\left(2.9979 \times 10^{8}\ \frac{\text{m}}{\text{s}}\right)^2 = 1.141 \times 10^{13}\ \frac{\text{J}}{\text{mol reactants}}$$

Check: 209 + 64 = 272 + 1, 83 + 28 = 111 + 0. The units (J) are correct.

c) ${}^{179}_{74}\text{W} + {}^{?}_{?}? \rightarrow {}^{179}_{73}\text{Ta}$ becomes ${}^{179}_{74}\text{W} + {}^{0}_{-1}? \rightarrow {}^{179}_{73}\text{Ta}$ then ${}^{179}_{74}\text{W} + {}^{0}_{-1}\text{e} \rightarrow {}^{179}_{73}\text{Ta}$

mass defect = (178.94707 g + 0.00055 g) − 178.94593 g = 0.00169 g then

$$\frac{0.00169\ \cancel{\text{g}}}{2\ \text{mol reactants}} \times \frac{1\,\text{kg}}{1000\ \cancel{\text{g}}} = 8.45 \times 10^{-7}\ \frac{\text{kg}}{\text{mol reactants}}\ \text{then}$$

$$E = m\,c^2 = \left(8.45 \times 10^{-7}\ \frac{\text{kg}}{\text{mol reactants}}\right)\left(2.9979 \times 10^{8}\ \frac{\text{m}}{\text{s}}\right)^2 = 7.59 \times 10^{10}\ \frac{\text{J}}{\text{mol reactants}}$$

Check: 179 + 0 = 179, 74 - 1 = 73. The units (J) are correct.

81. **Given:** a) Ru-114, b) Ra-216, c) Zn-58, and d) Ne-31 **Find:** write nuclear equation for most likely decay
Conceptual Plan: Decide on most likely decay mode depending on N/Z (too large = beta decay, too low = positron emission) → Write the symbol for parent nuclide on the left side of the equation and the symbol for a particle on the right side. → Equalize the sum of the mass numbers and the sum of the atomic numbers on both sides of the equation by writing the appropriate mass number and atomic number for the unknown daughter nuclide. → Using the periodic table, deduce the identity of the unknown daughter nuclide from the atomic number and write its symbol.
Solution:

a) Ru-114 will undergo beta decay ${}^{114}_{44}\text{Ru} \rightarrow {}^{?}_{?}? + {}^{0}_{-1}\text{e}$ then ${}^{114}_{44}\text{Ru} \rightarrow {}^{114}_{45}? + {}^{0}_{-1}\text{e}$ then ${}^{114}_{44}\text{Ru} \rightarrow {}^{114}_{45}\text{Rh} + {}^{0}_{-1}\text{e}$

b) Ra-216 will undergo positron emission ${}^{216}_{88}\text{Ra} \rightarrow {}^{?}_{?}? + {}^{0}_{+1}\text{e}$ then ${}^{216}_{88}\text{Ra} \rightarrow {}^{216}_{87}? + {}^{0}_{+1}\text{e}$ then ${}^{216}_{88}\text{Ra} \rightarrow {}^{216}_{87}\text{Fr} + {}^{0}_{+1}\text{e}$

c) Zn-58 will undergo positron emission $^{58}_{30}Zn \rightarrow ^{?}_{?}? + ^{0}_{+1}e$ then $^{58}_{30}Zn \rightarrow ^{58}_{29}? + ^{0}_{+1}e$ then $^{58}_{30}Zn \rightarrow ^{58}_{29}Cu + ^{0}_{+1}e$

d) Ne-31 will undergo beta decay $^{31}_{10}Ne \rightarrow ^{?}_{?}? + ^{0}_{-1}e$ then $^{31}_{10}Ne \rightarrow ^{31}_{11}? + ^{0}_{-1}e$ then $^{31}_{10}Ne \rightarrow ^{31}_{11}Na + ^{0}_{-1}e$

Check: a) 114 = 114 + 0, 44 = 45 - 1, and Rhodium is atomic number 45. b) 216 = 216 + 0, 88 = 87 + 1, and Francium is atomic number 87. c) 58 = 58 + 0, 30 = 29 + 1, and Copper is atomic number 29. d) 31 = 31 + 0, 10 = 11 - 1, and Sodium is atomic number 11.

83. **Given:** Bi-210, $t_{1/2}$ = 5.0 days, 1.2 g Bi-210, 209.984105 amu, 5.5 % absorbed
Find: beta emissions in 15.5 days and dose (in Ci)
Conceptual plan: $t_{1/2} \rightarrow k$ **then** $[\text{Bi-210}]_0, t, k \rightarrow [\text{Bi-210}]_t$ **then**

$$t_{1/2} = \frac{0.693}{k} \qquad \ln N_t = -kt + \ln N_0$$

$g_0, g_t \rightarrow$ **g decayed → mol decayed → beta decays then day → hr → min → s then**

$$g_0 - g_t = \text{g decayed} \quad \frac{1 \text{ mol Bi-210}}{209.984105 \text{ g Bi-210}} \quad \frac{6.022 \times 10^{23} \text{ beta decays}}{1 \text{ mol Bi-210}} \quad \frac{1 \text{ day}}{24 \text{ hr}} \quad \frac{60 \text{ min}}{1 \text{ hr}} \quad \frac{60 \text{ s}}{1 \text{ min}}$$

beta decays, s → beta decays / s → Ci available → Ci absorbed

$$\text{take ratio} \quad \frac{1 \text{ Ci}}{\frac{3.7 \times 10^{10} \text{ decays}}{\text{s}}} \quad \frac{5.5 \text{ Ci absorbed}}{100 \text{ Ci emitted}}$$

Solution: $t_{1/2} = \frac{0.693}{k}$ rearrange to solve for k. $k = \frac{0.693}{t_{1/2}} = \frac{0.693}{5.0 \text{ days}} = 0.1\underline{3}86 \text{ day}^{-1}$. Since

$\ln[\text{Bi-210}]_t = -kt + \ln[\text{Bi-210}]_0 = -(0.1\underline{3}86 \text{ day}^{-1})(13.5 \text{ day}) + \ln(1.2 \text{ g}) = -1.\underline{6}888 \rightarrow$

$[\text{Bi-210}]_t = e^{-1.\underline{6}888} = 0.1\underline{8}475$ g . then $g_0 - g_t$ = g decayed = 1.2 g - 0.1$\underline{8}$475 g = 1.$\underline{0}$153 g Bi-210 then

$$1.\underline{0}153 \text{ g Bi-210} \times \frac{1 \text{ mol Bi-210}}{209.984105 \text{ g Bi-210}} \times \frac{6.022 \times 10^{23} \text{ beta decays}}{1 \text{ mol Bi-210}} = 2.\underline{9}116 \times 10^{21} \text{ beta decays} =$$

$= 2.9 \times 10^{21}$ beta decays

then $13.5 \text{ day} \times \frac{24 \text{ hr}}{1 \text{ day}} \times \frac{60 \text{ min}}{1 \text{ hr}} \times \frac{60 \text{ s}}{1 \text{ min}} = 1.1\underline{6}64 \times 10^6 \text{ s}$ then

$$\frac{2.\underline{9}116 \times 10^{21} \text{ beta decays}}{1.1\underline{6}64 \times 10^6 \text{ s}} \times \frac{1 \text{ Ci}}{\frac{3.7 \times 10^{10} \text{ decays}}{\text{s}}} = 6.\underline{7}466 \times 10^4 \text{ Ci emitted} \times \frac{5.5 \text{ Ci absorbed}}{100 \text{ Ci emitted}} = 3700 \text{ Ci}$$

Check: The units (decays and Ci) are correct. The amount that decays is large since the time is over 3 half-lives and we have a relatively large amount of the isotope. Since the decay is large, the dosage is large.

85. **Given:** Ra-226 (226.05402 amu) decays to Rn-224, $t_{1/2} = 1.6 \times 10^3$ yr, 25.0 g Ra-226, T = 25.0 °C, P =1.0 atm **Find:** V of Rn-224 gas produced in 5.0 day
Conceptual plan: day → yr then $t_{1/2} \rightarrow k$ **then** $[\text{Ra-226}]_0, t, k \rightarrow [\text{Ra-226}]_t$

$$\frac{1 \text{ yr}}{365.24 \text{ day}} \qquad t_{1/2} = \frac{0.693}{k} \qquad \ln N_t = -kt + \ln N_0$$

then g_0, g_t **→ g decayed → mol decayed → mol Rn-224 formed then °C → K then**

$$g_0 - g_t = \text{g decayed} \quad \frac{1 \text{ mol Ra-226}}{226.05402 \text{ g Ra-226}} \quad \frac{1 \text{ mol Rn-224}}{1 \text{ mol Ra-226}} \qquad K = °C + 273.15$$

then P, n, T → V

$$PV = nRT$$

Solution: $5.0 \text{ day} \times \frac{1 \text{ yr}}{365.24 \text{ day}} = 0.01\underline{3}690 \text{ yr}$ then $t_{1/2} = \frac{0.693}{k}$ rearrange to solve for k.

$k = \frac{0.693}{t_{1/2}} = \frac{0.693}{1.6 \times 10^3 \text{ yr}} = 4.\underline{3}3125 \times 10^{-4} \text{ yr}^{-1}$. Since

$ln[\text{Ra-226}]_t = -kt + \ln[\text{Ra-226}]_0 = -(4.33125 \times 10^{-4}\ \text{yr}^{-1})(0.013690\ \text{yr}) + \ln(25.0\ \text{g}) = 3.21887$ →

$[\text{Ra-226}]_t = e^{3.21887} = 24.9999\ \text{g}$. then

$mg_0 - mg_t$ = mg decayed = 25.0 g - 24.9999 g = 0.000148 g Ra-226 then

$$0.000148\ \text{g Ra-226} \times \frac{1\ \text{mol Ra-226}}{226.05402\ \text{g Ra-226}} \times \frac{1\ \text{mol Rn-224}}{1\ \text{mol Ra-226}} = 6.5576 \times 10^{-7}\ \text{mol Rn-224}$$ then

and T = 25.0 °C + 273.15 = 298.2 K, then $PV = nRT$ Rearrange to solve for V.

$$V = \frac{nRT}{P} = \frac{6.5576 \times 10^{-7}\ \text{mol} \times 0.08206\ \frac{\text{L} \cdot \text{atm}}{\text{mol} \cdot \text{K}} \times 298.2\ \text{K}}{1.0\ \text{atm}} = 1.6047 \times 10^{-5}\ \text{L} = 1.6 \times 10^{-5}\ \text{L}.$$

Two significant figures are reported as requested in problem.
Check: The units (L) are correct. The amount that gas is small since the time is so small compared to the half-life.

87. **Given:** ${}^{0}_{+1}e + {}^{0}_{+1}e \rightarrow 2\ {}^{0}_{0}\gamma$ **Find:** a) energy (in kJ/mol) and b) wavelength of gamma ray photons
Conceptual plan:
a) mass of products & reactants → mass defect (g) → kg → kg/mol → E (J/mol) → E (kJ/mol)

mass defect = $\sum$ mass of reactants - $\sum$ mass of products; $\frac{1\ \text{kg}}{1000\ \text{g}}$; ÷ 2 mol; $E = mc^2$; $\frac{1\ \text{kJ}}{1000\ \text{J}}$

b) Answer in part a) is for 2 moles of γ, so E (J/2 mol γ) → E (J/ γ photon) → λ

$\frac{1\ \text{mol photons}}{6.022 \times 10^{23}\ \text{photons}}$; $E = \frac{hc}{\lambda}$

Solution:
a) mass defect = $\sum$ mass of reactants - $\sum$ mass of products = (0.00055 g + 0.00055 g) − 0 g = 0.00110 g

then $\frac{0.00110\ \text{g}}{2\ \text{mol}} \times \frac{1\ \text{kg}}{1000\ \text{g}} = 5.50 \times 10^{-7}\ \frac{\text{kg}}{\text{mol}}$ then

$$E = mc^2 = \left(5.50 \times 10^{-7}\ \frac{\text{kg}}{\text{mol}}\right)\left(2.9979 \times 10^{8}\ \frac{\text{m}}{\text{s}}\right)^2 = 4.94307 \times 10^{10}\ \frac{\text{J}}{\text{mol}} \times \frac{1\ \text{kJ}}{1000\ \text{J}} = 4.94 \times 10^{7}\ \frac{\text{kJ}}{\text{mol}}$$

b) $E = 4.94307 \times 10^{10}\ \frac{\text{J}}{\text{mol}\ \gamma} \times \frac{1\ \text{mol}\ \gamma}{6.022 \times 10^{23}\ \gamma\text{photons}} = 8.20835 \times 10^{-14}\ \frac{\text{J}}{\gamma\text{photons}}$. Then $E = \frac{hc}{\lambda}$.

Rearrange to solve for λ. $$\lambda = \frac{hc}{E} = \frac{(6.626 \times 10^{-34}\ \text{J} \cdot \text{s})\left(2.9979 \times 10^{8}\ \frac{\text{m}}{\text{s}}\right)}{8.20835 \times 10^{-14}\ \frac{\text{J}}{\gamma\text{photons}}} = 2.42 \times 10^{-12}\ \text{m} = 2.42\ \text{pm}.$$

Check: The units (kJ/mol and pm) are correct. A large amount of energy is expected per mole of mass lost. The photon is in the gamma ray region of the electromagnetic spectrum.

89. **Given:** ^{3}He = 3.016030 amu **Find:** nuclear binding energy per nucleon
Conceptual plan: ${}^{A}_{Z}$**X, isotope mass → mass defect → nuclear binding energy per nucleon**

mass defect = Z(mass ${}^{1}_{1}$H) + (A - Z)(mass ${}^{1}_{0}$n) - mass of isotope; $\frac{931.5\ \text{MeV}}{(1\ \text{amu})(A\ \text{nucleons})}$

Solution: mass defect = Z(mass ${}^{1}_{1}$H) + (A - Z)(mass ${}^{1}_{0}$n) - mass of isotope.
He-3 mass defect = 2(1.00783 amu) + (3 - 2)(1.00866 amu) - 3.016030 amu = 0.00829 amu

and $0.00829\ \text{amu} \times \frac{931.5\ \text{MeV}}{1\ \text{amu}} = 7.72\ \text{MeV}$.

Check: The units (MeV) are correct. The number of nucleons is small, so the MeV is not that large.

91. **Given:** ^{247}Es and 5 neutrons made by bombarding ^{238}U **Find:** identity of bombarding particle

Conceptual Plan: Begin with the symbols for nuclides given. → Equalize the sum of the mass numbers and the sum of the atomic numbers on both sides of the equation by writing the appropriate mass number and atomic number for the unknown daughter nuclide. → Using the periodic table, deduce the identity of the unknown nuclide from the atomic number and write its symbol.

Solution:

$^{238}_{92}U + ^{?}_{?}? \rightarrow ^{247}_{99}Es + 5\,^{1}_{0}n$ then $^{238}_{92}U + ^{14}_{7}? \rightarrow ^{247}_{99}Es + 5\,^{1}_{0}n$ then $^{238}_{92}U + ^{14}_{7}N \rightarrow ^{247}_{99}Es + 5\,^{1}_{0}n$

Check: 238 + 14 = 247 + 5(1), 92 + 7 = 99 + 5(0), and Nitrogen is atomic number 7.

93. **Given:** $t_{1/2}$ for decay of $^{238}U = 4.5 \times 10^9$ years, 1.6 g rock, 29 dis/s all radioactivity from U-238
Find: percent by mass ^{238}U in rock
Conceptual plan: $t_{1/2}$ **→** k **and s → min → hr → day → yr then Rate,** k **→ N → mol** ^{238}U **→ g** ^{238}U

$$t_{1/2} = \frac{0.693}{k} \qquad \frac{1\text{ min}}{60\text{ s}} \quad \frac{1\text{ hr}}{60\text{ min}} \quad \frac{1\text{ day}}{24\text{ hr}} \quad \frac{1\text{ yr}}{365.24\text{ day}} \qquad \text{Rate} = k\,N \qquad \frac{1\text{ mol dis}}{6.022 \times 10^{23}\text{ dis}} \quad \frac{238\text{ g }^{238}U}{1\text{ mol }^{238}U}$$

then g ^{238}U**, g rock → percent by mass** ^{238}U

$$\text{percent by mass }^{238}U = \frac{\text{g }^{238}U}{\text{g rock}} \times 100\ \%$$

Solution: $t_{1/2} = \frac{0.693}{k}$ rearrange to solve for k. $k = \frac{0.693}{t_{1/2}} = \frac{0.693}{4.5 \times 10^9\text{ yr}} = 1.\underline{5}4 \times 10^{-10}\text{ yr}^{-1}$ and

$$1\text{ s} \times \frac{1\text{ min}}{60\text{ s}} \times \frac{1\text{ hr}}{60\text{ min}} \times \frac{1\text{ day}}{24\text{ hr}} \times \frac{1\text{ yr}}{365.24\text{ day}} = 3.168\underline{8}9554 \times 10^{-8}\text{ yr}.$$

Rate = $k\,N$ Rearrange to solve for N.

$$N = \frac{\text{Rate}}{k} = \frac{29\,\frac{\text{dis}}{3.16889554 \times 10^{-8}\text{ yr}}}{1.\underline{5}4 \times 10^{-10}\text{ yr}^{-1}} = 5.\underline{9}425 \times 10^{18}\text{ dis} \quad \text{then}$$

$$5.\underline{9}425 \times 10^{18}\text{ dis} \times \frac{1\text{ mol dis}}{6.022 \times 10^{23}\text{ dis}} \times \frac{238\text{ g }^{238}U}{1\text{ mol }^{238}U} = 2.\underline{3}486 \times 10^{-3}\text{ g }^{238}U \quad \text{then}$$

$$\text{percent by mass }^{238}U = \frac{\text{g }^{238}U}{\text{g rock}} \times 100\ \% = \frac{2.\underline{3}486 \times 10^{-3}\text{ g }^{238}U}{1.6\text{ g rock}} \times 100\ \% = 0.15\ \%.$$

Check: The units (%) are correct. The mass percent is low because the dis/s is low.

95. a) **Given:** 72,500 kg Al *(s)* and 10 Al *(s)* + 6 NH_4ClO_4 *(s)* → 4 Al_2O_3 *(s)* + 2 $AlCl_3$ *(s)* + 12 H_2O *(g)* + 3 N_2 *(g)* and 608,000 kg O_2 *(g)* that reacts with hydrogen to form gaseous water
Find: energy generated (ΔH°_{rxn})
Conceptual plan: write balanced reaction for O_2 *(g)* **then**

$$\Delta H^0_{rxn} = \sum n_P \Delta H^0_f(\text{products}) - \sum n_R \Delta H^0_f(\text{reactants}) \quad \textbf{then}$$

kg → g → mol → energy then add the results from the two reactions

$$\frac{1000\text{ g}}{1\text{ kg}} \qquad \mathfrak{M} \qquad \Delta H^\circ_{rxn}$$

Solution:

Reactant/Product	ΔH^0_f (kJ/mol from Appendix IIB)
Al *(s)*	0.0
NH_4ClO_4 *(s)*	−295
Al_2O_3 *(s)*	−1675.7
$AlCl_3$ *(s)*	−704.2
H_2O *(g)*	− 241.8
N_2 *(g)*	0.0

Be sure to pull data for the correct formula and phase.

$$\Delta H^0_{rxn} = \sum n_P \Delta H^0_f(products) - \sum n_R \Delta H^0_f(reactants)$$
$$= [4(\Delta H^0_f(Al_2O_3\ (s)) + 2(\Delta H^0_f(AlCl_3\ (s)) + 12(\Delta H^0_f(H_2O\ (g)) + 3(\Delta H^0_f(N_2\ (g))] + - [10(\Delta H^0_f(Al\ (s)) + 6(\Delta H^0_f(NH_4ClO_4\ (s))]$$
$$= [4(-1675.7\text{ kJ}) + 2(-704.2\text{ kJ}) + 12(-241.8\text{ kJ}) + 3(0.0\text{ kJ})] - [10(0.0\text{ kJ}) + 6(-295\text{ kJ})]$$
$$= [-11012.8\text{ kJ}] - [-1770.\text{ kJ}]$$
$$= -924\underline{2}.8\text{ kJ}$$

then $72{,}500\ \cancel{\text{kg Al}} \times \frac{1000\ \cancel{\text{g Al}}}{1\ \cancel{\text{kg Al}}} \times \frac{1\ \cancel{\text{mol Al}}}{26.98\ \cancel{\text{g Al}}} \times \frac{924\underline{2}.8\text{ kJ}}{10\ \cancel{\text{mol Al}}} = 2.4\underline{8}3703 \times 10^9\text{ kJ}$.

Balanced reaction: $H_2\,(g) + ½\ O_2\,(g) \rightarrow H_2O\,(g)$ $\Delta H^\circ_{rxn} = \Delta H^\circ_f(H_2O\,(g)) = -241.8$ kJ/mol then

$608{,}000\ \cancel{\text{kg }Q_2} \times \frac{1000\ \cancel{\text{g }Q_2}}{1\ \cancel{\text{kg }Q_2}} \times \frac{1\ \cancel{\text{mol }Q_2}}{32.00\ \cancel{\text{g }Q_2}} \times \frac{241.8\text{ kJ}}{0.5\ \cancel{\text{mol }Q_2}} = 9.1\underline{8}84 \times 10^9\text{ kJ}$. So the total is

$2.4\underline{8}3703 \times 10^9\text{ kJ} + 9.1\underline{8}84 \times 10^9\text{ kJ} = 1.16\underline{7}2103 \times 10^{10}\text{ kJ} = 1.167 \times 10^{10}\text{ kJ}$.

Check: The units (kJ) are correct. The answer very large because the reactions are very exothermic and the weight of reactants is so large.

b) **Given:** ${}^{1}_{1}H + {}^{-1}_{-1}p + {}^{0}_{+1}e \rightarrow {}^{0}_{0}\gamma$ **Find:** mass of antimatter to give same energy as part a)

Conceptual plan: since the reaction is an annihilation reaction, no matter will be left, so the mass of antimatter is the same as the mass of the hydrogen. so kJ → J → kg → g

$\frac{1000\text{ J}}{1\text{ kJ}}$ $E = m\,c^2$ $\frac{1000\text{ g}}{1\text{ kg}}$

Solution: $1.16\underline{7}2103 \times 10^{10}\ \cancel{\text{kJ}} \times \frac{1000\text{ J}}{1\ \cancel{\text{kJ}}} = 1.16\underline{7}2103 \times 10^{13}\text{ J}$. Since $E = m\,c^2$, rearrange to solve

for m. $m = \frac{E}{c^2} = \frac{1.16\underline{7}2103 \times 10^{13}\ \text{kg}\frac{\cancel{m^2}}{\cancel{s^2}}}{\left(2.9979 \times 10^8\ \frac{\cancel{m}}{\cancel{s}}\right)^2} = 1.299 \times 10^{-4}\ \cancel{\text{kg}} \times \frac{1000\text{ g}}{1\ \cancel{\text{kg}}} = 0.1299\text{ g}$.

Check: The units (g) are correct. A small mass is expected since nuclear reactions to generate a large amount of energy.

97. **Given:** ${}^{235}_{92}U \rightarrow {}^{206}_{82}Pb$ and ${}^{232}_{90}Th \rightarrow {}^{206}_{82}Pb$ **Find:** decay series

Conceptual Plan: Write species given on the appropriate side of the equation. →Equalize the sum of the mass numbers and the sum of the atomic numbers on both sides of the equation by writing the stoichiometric coefficient in front of the desired species.

Solution: ${}^{235}_{92}U \rightarrow {}^{?}_{82}Pb + ?{}^{4}_{2}He + ?{}^{0}_{-1}e$ becomes ${}^{235}_{92}U \rightarrow {}^{207}_{82}Pb + 7{}^{4}_{2}He + 4{}^{0}_{-1}e$.

${}^{232}_{90}Th \rightarrow {}^{?}_{82}Pb + ?{}^{4}_{2}He + ?{}^{0}_{-1}e$ becomes ${}^{232}_{90}Th \rightarrow {}^{208}_{82}Pb + 6{}^{4}_{2}He + 4{}^{0}_{-1}e$.

U-235 forms Pb-207 in 7 α-decays and 4 β-decays and Th-232 forms Pb-208 in 6 α-decays and 4 β-decays.

Check: 235 = 207 + 7(4) + 4(0), and 92 = 82 + 7(2) + 4(-1). 232 = 208 + 6(4) + 4(0), and 90 = 82 + 6(2) + 4(-1). The mass of the Pb can be determined because alpha particles are large and need to be included as integer values.

99. 7. Since $1/2^6 = 1.6\%$ and $1/2^7 = 0.8\ \%$.

101. The gamma emitter is a greater threat while you sleep because it can penetrate more tissue. The alpha emitter is a greater threat if you ingest it since it is more ionizing.

Chapter 20
Organic Chemistry

1. Most common smells are caused by organic molecules, molecules containing carbon combined with several other elements including hydrogen, nitrogen, oxygen, and sulfur.

3. Carbon is unique in the vast number of compounds that it can form. Life needs diversity to exist and carbon has the ability to form more compounds than any other element.

5. Silicon can form chains with itself. However, silicon's affinity for oxygen—the Si-O bond is 142 kJ/mol stronger than the Si-Si bond—coupled with the prevalence of oxygen in our atmosphere means that silicon-silicon chains are readily oxidized to form silicates (the silicon-oxygen compounds that compose a significant proportion of minerals). By contrast, the C-C bond (347 kJ/mol) and the C-O bond (359 kJ/mole) are nearly the same strength, allowing carbon chains to exist relatively peacefully in an oxygen rich environment. In other words, silicon's affinity for oxygen robs it of the rich diversity that catenation provides to carbon.

7. Hydrocarbons—compounds that contain only carbon and hydrogen—are the simplest organic compounds. However, because of the uniqueness of carbon (just discussed), many different kinds of hydrocarbons exist. Hydrocarbons are commonly used as fuels. Candle wax, oil, gasoline, liquid propane (LP) gas, and natural gas are all composed of hydrocarbons. Hydrocarbons are also the starting materials in the synthesis of many different consumer products including fabrics, soaps, dyes, cosmetics, drugs, plastic, and rubber.

9. A structural formula shows not only the numbers of each kind of atoms, but also how the atoms are bonded together. The condensed structural formula groups the hydrogen atoms together with the carbon atom to which they are bonded. Condensed structural formulas may show some of the bonds or none at all. The carbon skeleton formula shows the carbon-carbon bonds only as lines. Each end or bend of a line represents a carbon atom bonded to as many hydrogen atoms as necessary to form a total of four bonds. Space-filling or ball-and-stick models are three-dimensional representations that show how atoms are bonded together. The space-filling models show the relative size of the atoms that are bonded together.

11. Optical isomers are two molecules that are nonsuperimposable mirror images of one another. Optical isomers contain a carbon atom with four different substituent groups. Most properties of optical isomers are the same. The differences appear when they interact with polarized light and when they are placed in environments that can interact with one isomer and not the other.

13. Alkanes are often called saturated hydrocarbons because they are saturated (loaded to capacity) with hydrogen. Unsaturated hydrocarbons contain multiple bonds and, therefore, contain fewer hydrogen atoms.

15. Since double bonds are composed of a sigma bond and a pi bond. Sigma bonds allow for free rotation about the bond. The pi bond restricts the rotational motion about the bond, resulting in isomers that are referred to as geometric isomers (same connectivity, but different geometric positions). The two isomers are designated as cis (meaning "same side") and trans (meaning "opposite sides"). Cis-trans isomerism is common in alkenes and results in different boiling points and melting points and in different abilities to interact with rigid molecules.

17. The most common types of reactions of alkanes are:
    - Hydrocarbon combustion in the presence of oxygen to form carbon dioxide and water. An example is $C_8H_{18}\ (l) + 25/2\ O_2\ (g) \rightarrow 8\ CO_2\ (g) + 9\ H_2O\ (g)$; and
    - Halogen substitution in the presence of a halogen gas to form a halogenated alkane and a hydrohalic acid. An example is $CH_4\ (g) + Cl_2\ (g) \rightarrow CH_3Cl\ (g) + HCl\ (g)$.

19. The most common types of reactions of alkenes are:
    - Hydrocarbon combustion in the presence of oxygen to form carbon dioxide and water. An example is $C_2H_4\ (g) + 3\ O_2\ (g) \rightarrow 2\ CO_2\ (g) + 2\ H_2O\ (g)$;
    - Hydrogenation in the presence of hydrogen to form an alkane. An example is $C_2H_4\ (g) + H_2\ (g) \rightarrow C_2H_6\ (g)$; and
    - Halogen addition in the presence of a halogen gas or hydrohalic acid to form an halogenated alkane. An example is $CH_2CHCH_2CH_3\ (g) + HCl\ (g) \rightarrow CH_3CClHCH_2CH_3\ (g)$.

21. The structure of benzene, C_6H_6, is a six member ring where three pi bonds are delocalized on all six of the C-C bonds that form the ring. Benzene rings are represented as one or both of the Kekulé structures :

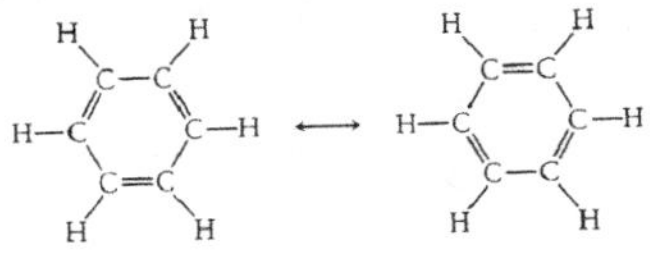

, in a shorthand notation: 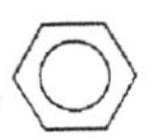 , or as a ball and stick diagram with molecular orbitals:  .

23. A functional group is a characteristic atom or group of atoms that is inserted into a hydrocarbon. Examples of functional groups are alcohols (-OH), halogens (-X, where X = F, Cl, Br and I), and carboxylic acid (-COOH).

25. In organic chemistry, we think of oxidation and reduction from the point of view of the carbon atoms in the organic molecule. Thus, oxidation is the gaining of oxygen or the losing of hydrogen by a carbon atom. Reduction is then the loss of oxygen or the gaining of hydrogen by a carbon atom.

27. Aldehydes and ketones have the following general structural formulas of RCHO and RCOR, respectively, where R represents a hydrocarbon group. Both aldehydes and ketones contain a carbonyl group, an oxygen double bonded to a carbon atom $\left(\overset{\displaystyle O}{\overset{\|}{-C-}}\right)$. Ketones have an R group attached to both sides of the carbonyl, while aldehydes have one R group and a hydrogen atom. An exception is formaldehyde, which is an aldehyde with two H atoms attached to the carbonyl group. Acetaldehyde is $CH_3-\overset{\displaystyle O}{\overset{\|}{C}}-H$ and propanone is $CH_3-\overset{\displaystyle O}{\overset{\|}{C}}-CH_3$.

29. Carboxylic acids and esters have the following general structural formulas of RCOOH and RCOOR, respectively, where R represents a hydrocarbon group. Both carboxylic acids and esters contain a carbonyl group, an oxygen double bonded to a carbon atom $\left(\overset{\displaystyle O}{\overset{\|}{-C-}}\right)$. In carboxylic acids the molecule ends in $-\overset{\displaystyle O}{\overset{\|}{C}}-O-H$ and esters have $-\overset{\displaystyle O}{\overset{\|}{C}}-O-$ inserted into a hydrocarbon chain. Acetic acid is $CH_3-\overset{\displaystyle O}{\overset{\|}{C}}-O-H$ and ethylethanoate is $CH_3-\overset{\displaystyle O}{\overset{\|}{C}}-O-CH_2-CH_3$.

31. Ethers have the following general structural formula of ROR', where R and R' represents hydrocarbon groups. The two hydrocarbon groups are linked though an oxygen atom. Diethyl ether is $CH_3-CH_2-O-CH_2-CH_3$ and ethyl propyl ether is $CH_3-CH_2-O-CH_2-CH_2-CH_3$.

33. Polymers are long chainlike molecules composed of repeating units called monomers. Copolymers are long chainlike molecules composed of two different monomers.

35. a) C_5H_{12} is an alkane, since it follows the general formula C_nH_{2n+2}, where n = 5.

    b) C_3H_6 is an alkene, since it follows the general formula C_nH_{2n}, where n = 3.

    c) C_7H_{12} is an alkyne, since it follows the general formula C_nH_{2n-2}, where n = 7.

d) $C_{11}H_{22}$ is an alkene, since it follows the general formula C_nH_{2n}, where $n = 11$.

37. $CH_3-CH_2-CH_2-CH_2-CH_2-CH_2-CH_3$, $CH_3-CH-CH_2-CH_2-CH_2-CH_3$
                                                                       $|$
                                                                     CH_3 ,

$CH_3-CH_2-CH-CH_2-CH_2-CH_3$,
                     $|$
                  CH_3

           CH_3
           $|$
$CH_3-C-CH_2-CH_2-CH_3$,
           $|$
           CH_3

$CH_3-CH-CH-CH_2-CH_3$,
           $|$ $|$
           CH_3 CH_3

           CH_3
           $|$
$CH_3-CH_2-C-CH_2-CH_3$,
           $|$
           CH_3

$CH_3-CH-CH_2-CH-CH_3$
           $|$ $|$
           CH_3 CH_3 ,

$CH_3-CH_2-CH-CH_2-CH_3$
                  $|$
                  CH_2-CH_3 , and

           CH_3
           $|$
$CH_3-C-CH-CH_3$.
           $|$ $|$
           CH_3 CH_3

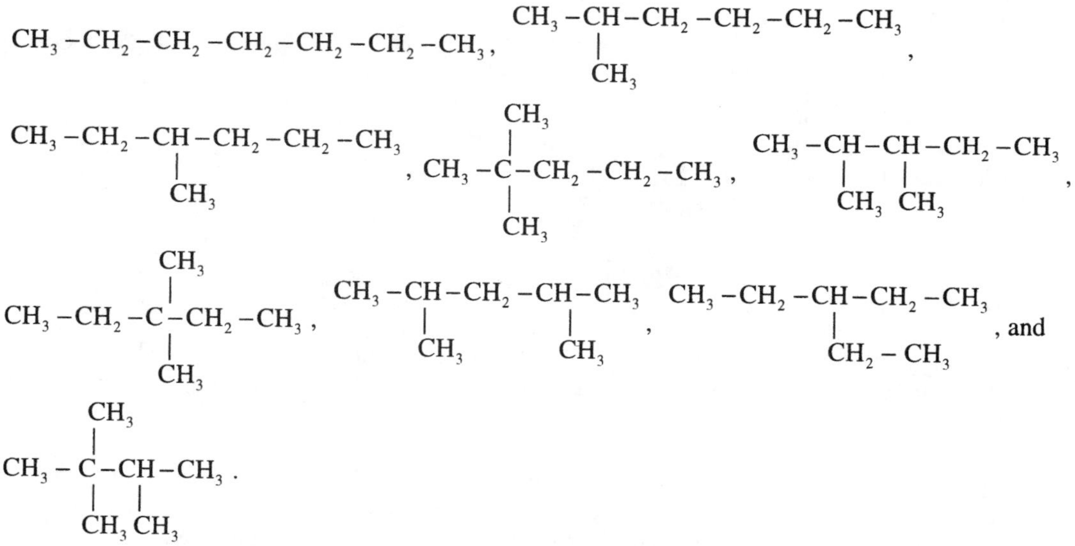

39. a) No, this molecule will not because all four of the substituents are Cl atoms.

b) Yes, this molecule will because the third carbon has four different substituents groups.

c) Yes, this molecule will because the second carbon has four different substituents groups.

d) No, each carbon has at most three different substituents groups.

41. a) They are enantiomers, because they are mirror images of each other.

b) They are the same, because you can get the second molecule by rotating the first molecule counterclockwise about the C-H bond.

c) They are enantiomers, because they are mirror images of each other.

43. **Given:** alkane structures   **Find:** name
Conceptual plan: Count the number of carbon atoms in the longest continuous carbon chain to determine the base name of the compound. Find the prefix corresponding to this number of atoms in Table 20. 5 and add the ending –ane to form the base name. → Consider every branch from the base chain to be a substituent. Name each substituent according to Table 20.6. →Beginning with the end closest to the branching, number the base chain and assign a number to each substituent. (If two substituents occur at equal distances from each end, go to the next substituent to determine from which end to start numbering.) →Write the name of the compound in the following format: (subst. #)-(subst. name)(base name). → If there are two or more substituents, give each one a number and list them alphabetically with hyphens between words and numbers. →If a compound has two or more identical substituents, designate the number of identical substituents with the prefix di- (2), tri- (3), or tetra- (4) before the substituent's name. Separate the numbers indicating the positions of the substituents relative to each other with a comma. The prefixes are not taken into account when alphabetizing.
Solution:
a) $CH_3-CH_2-CH_2-CH_2-CH_3$ has 5 carbons as the longest continuous chain. The prefix for 5 is penta-. There are no substituent groups on any of the carbons, so the name is pentane.

b) $\boxed{CH_3-CH_2-CH-CH_3}$
                  $|$ has 4 carbons as the longest continuous chain. The prefix for 4 is but- and the
                  CH_3

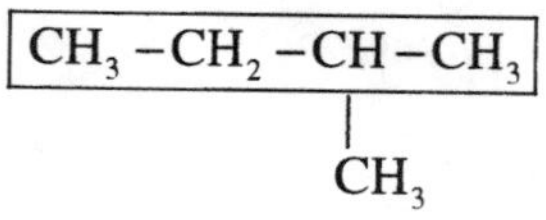

base name is butane. The only substituent group is a methyl group. 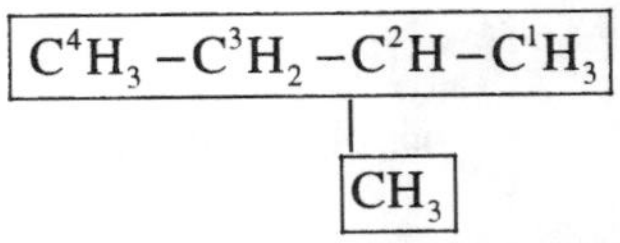 If we start numbering the chain at the end closest to the methyl group, the methyl substituent is assigned the number 2. The name of the compound is 2-methylbutane.

c) 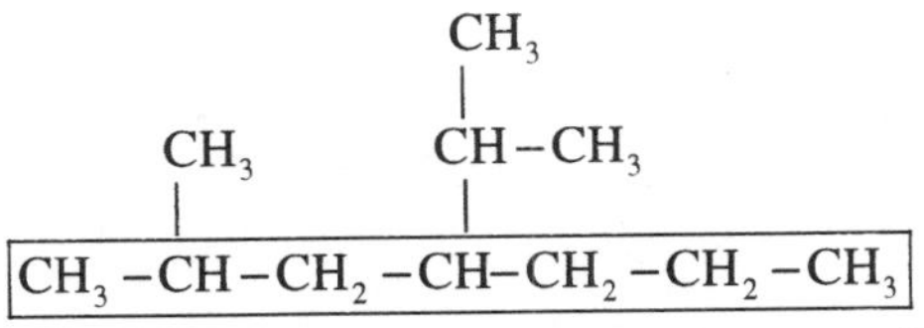 has 7 carbons as the longest continuous chain. The prefix for 7 is hept- and the base name is heptane. The substituent groups are methyl and isopropyl groups.

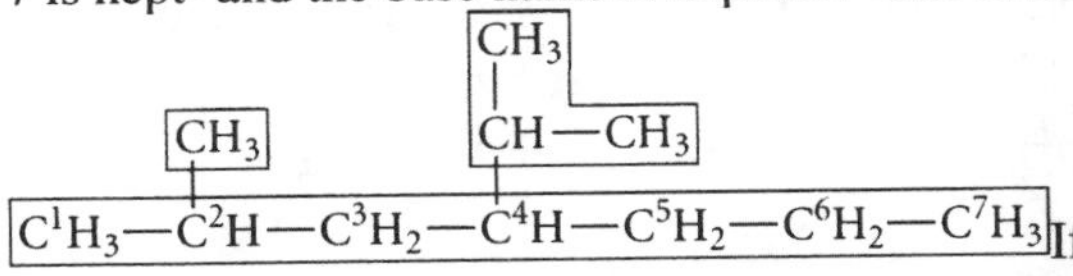 If we s 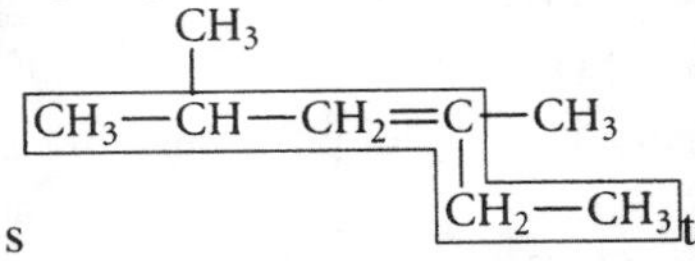 tart numbering the chain at the end closest to the methyl group, the methyl substituent is assigned the number 2 and the isopropyl group is assigned the number 4. Since i comes before m, the name of the compound is 4-isopropyl-2-methylheptane.

d) 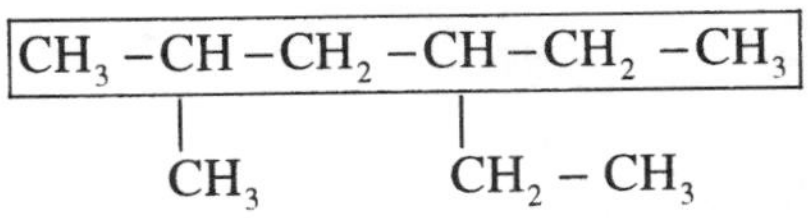 has 6 carbons as the longest continuous chain. The prefix for 6 is hex- and the base name is hexane. The only substituent groups are methyl and ethyl groups.

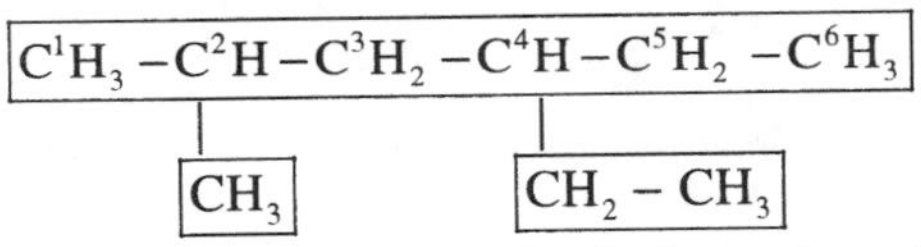 If we start numbering the chain at the end closest to the methyl group, the methyl substituent is assigned the number 2 and the ethyl group is assigned the number 4. Since e comes before m, the name of the compound is 4-ethyl-2-methylhexane.

45. **Given:** alkane names **Find:** structure

Conceptual plan: Find the number of carbon atoms corresponding to prefix of the base name in Table 20. 5. → Draw the base chain and number the carbons from left to right. → Using Table 20.6 and the prefix di- (2), tri- (3), or tetra- (4) before the substituent's name, determine each substituent. → Add the substituent to the proper carbon position in the chain. → Add hydrogen atoms to the base chain so that each carbon has 4 bonds.

Solution:

a) 3-ethylhexane. The base name hexane designates that there are 6 carbon atoms in the base chain. $C^1-C^2-C^3-C^4-C^5-C^6$. 3-ethyl designates that a $-CH_2CH_3$ group in the 3rd position.

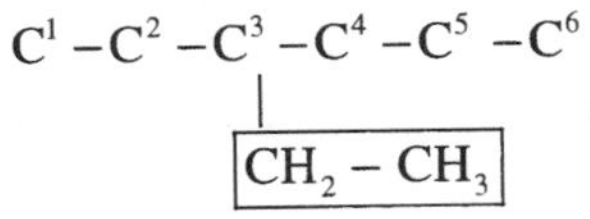 . Add hydrogens to the base chain to get the final molecule

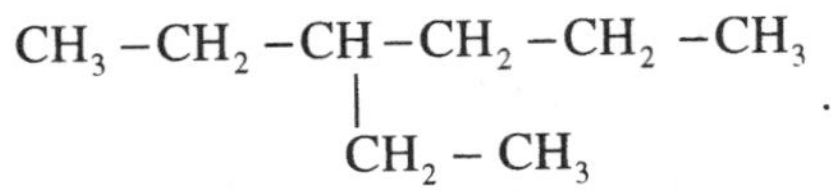 .

b) 3-ethyl-3-methylpentane. The base name pentane designates that there are 5 carbon atoms in the base chain. $C^1-C^2-C^3-C^4-C^5$. 3-ethyl designates that a $-CH_2CH_3$ group in the 3rd position; and 3 methyl designates that a $-CH_3$ group in the 3rd position . 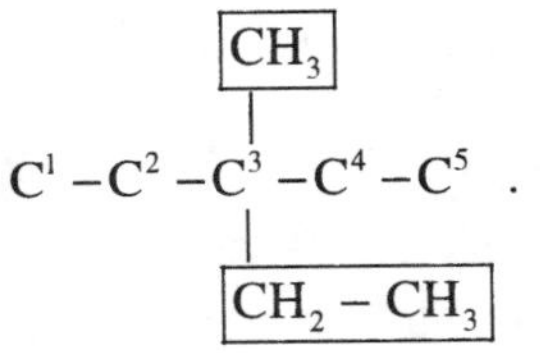 . Add hydrogens to the base chain to get the final molecule

$$\begin{array}{c} \quad\quad\quad\quad\quad\quad CH_3 \\ \quad\quad\quad\quad\quad\quad | \\ CH_3-CH_2-CH-CH_2-CH_3 \\ \quad\quad\quad\quad\quad\quad | \\ \quad\quad\quad\quad\quad\quad\quad CH_2-CH_3 \end{array}.$$

c) 2,3-dimethylbutane. The base name butane designates that there are 4 carbon atoms in the base chain. $C^1-C^2-C^3-C^4$. 2,3-dimethyl designates that are $-CH_3$ groups in the 2nd and 3rd positions.

$$\begin{array}{l} C^1-C^2-C^3-C^4 \\ \quad\;\;\; | \quad\;\;\; | \\ \quad\; \boxed{CH_3}\boxed{CH_3} \end{array}$$

. Add hydrogens to the base chain to get the final molecule

$$\begin{array}{l} CH_3-CH-CH-CH_3 \\ \quad\quad\;\;\; | \quad\;\;\; | \\ \quad\quad\; CH_3 \; CH_3 \end{array}.$$

d) 4,7-diethyl-2,2-dimethylnonane. The base name nonane designates that there are 9 carbon atoms in the base chain. $C^1-C^2-C^3-C^4-C^5-C^6-C^7-C^8-C^9$. 4,7-diethyl designates that are $-CH_2CH_3$ groups in the 4th and 7th positions; and 2,2-dimethyl designates that are two $-CH_3$ groups in the 2nd position .

$$\begin{array}{l} \quad\;\; \boxed{CH_3} \\ \quad\quad | \\ C^1-C^2-C^3-C^4-C^5-C^6-C^7-C^8-C^9 \\ \quad\quad | \quad\quad\quad\quad | \quad\quad\quad\quad\quad | \\ \quad\; \boxed{CH_3} \quad \boxed{CH_2-CH_3} \; \boxed{CH_2-CH_3} \end{array}$$

. Add hydrogens to the base chain to get the final molecule

$$\begin{array}{l} \quad\quad\; CH_3 \\ \quad\quad\quad | \\ CH_3-C-CH_2-CH-CH_2-CH_2-CH-CH_2-CH_3 \\ \quad\quad\quad | \quad\quad\quad\quad | \quad\quad\quad\quad\quad\quad\quad\quad | \\ \quad\quad\; CH_3 \quad\quad CH_2-CH_3 \quad\quad\quad CH_2-CH_3 \end{array}.$$

47. Hydrocarbon combustion in the presence of oxygen forms carbon dioxide and water. Balance the reaction.

a) $CH_3CH_2CH_3\,(g) + 5\,O_2\,(g) \rightarrow 3\,CO_2\,(g) + 4\,H_2O\,(g)$.

b) $CH_3CH_2CH{=}CH_2\,(g) + 6\,O_2\,(g) \rightarrow 4\,CO_2\,(g) + 4\,H_2O\,(g)$.

c) $2\,CH{\equiv}CH\,(g) + 5\,O_2\,(g) \rightarrow 4\,CO_2\,(g) + 2\,H_2O\,(g)$.

49. Halogen substitution reactions remove a hydrogen atom from the alkane and replace it with a halogen atom and generate a hydrohalic acid. Assume one substitution on the hydrocarbon.

a) $CH_3CH_3 + Br_2 \rightarrow CH_3CH_2Br + HBr$. Only one carbon-containing product is possible since the C-C bond freely rotates.

b) $CH_3CH_2CH_3 + Cl_2 \rightarrow [CH_3CH_2CH_2Cl$ and $CH_3CHClCH_3] + HCl$. Two carbon-containing products are possible, either on the end carbon or the middle carbon, since the C-C bond freely rotates and the end carbons are equivalent before reaction.

c) $CH_2Cl_2 + Br_2 \rightarrow CHBrCl_2 + HBr$. Only one carbon-containing product is possible since halogen substitution reactions only remove hydrogen atoms.

d) $$\begin{array}{l} CH_3-CH-CH_3 \\ \quad\quad\;\; | \\ \quad\quad CH_3 \end{array} + Cl_2 \rightarrow \left[\begin{array}{l} \quad\quad\;\; H \\ \quad\quad\;\; | \\ CH_3-C-CH_2Cl \\ \quad\quad\;\; | \\ \quad\quad CH_3 \end{array} \text{ and } \begin{array}{l} \quad\quad\;\; Cl \\ \quad\quad\;\; | \\ CH_3-C-CH_3 \\ \quad\quad\;\; | \\ \quad\quad CH_3 \end{array} \right] + HCl.$$ Two carbon-containing products are possible, either on the end carbon or the middle carbon, since the C-C bond freely rotates and the end carbons are all equivalent before reaction.

51. $CH_2=CH-CH_2-CH_2-CH_2-CH_3$, $CH_3-CH=CH-CH_2-CH_2-CH_3$ and $CH_3-CH_2-CH=CH-CH_2-CH_3$ are the only structural isomers. Remember that cis-trans isomerism generates geometric isomers, not structural isomers.

53. **Given:** alkene structures **Find:** name

Conceptual plan: **Count the number of carbon atoms in the longest continuous carbon chain that contains the multiple bond to determine the base name of the compound. Find the prefix corresponding to this number of atoms in Table 20. 5 and add the ending –ene to form the base name.** → **Consider every branch from the base chain to be a substituent. Name each substituent according to Table 20.6.** → **Beginning with the end closest to the multiple bond, number the base chain and assign a number to each substituent.** → **Write the name of the compound in the following format: (subst. #)-(subst. name)(base name).** → **If there are two or more substituents, give each one a number and list them alphabetically with hyphens between words and numbers.** → **If a compound has two or more identical substituents, designate the number of identical substituents with the prefix di- (2), tri- (3), or tetra- (4) before the substituent's name. Separate the numbers indicating the positions of the substituents relative to each other with a comma. The prefixes are not taken into account when alphabetizing.**

Solution:

a) $\boxed{CH_2 = CH-CH_2-CH_3}$ has 4 carbons as the longest continuous chain. The prefix for 4 is but- and the base name is butene. There are no substituent groups. $\boxed{C^1H_2 = C^2H-C^3H_2-C^4H_3}$ Start numbering on the left since it is closer to the double bond. Since the double bond is between position number 1 and 2, the name of the compound is 1-butene.

b)
$$\overset{CH_3 \quad CH_3}{\underset{\boxed{CH_3-CH-C=CH-CH_3}}{|\qquad |}}$$
has 5 carbons as the longest continuous chain. The prefix for 5 is pent- and the base name is pentene. The substituent groups are two methyl groups.

$$\overset{\boxed{CH_3} \quad \boxed{CH_3}}{\underset{\boxed{C^5H_3-C^4H-C^3=C^2H-C^1H_3}}{|\qquad |}}$$
If we start numbering the chain at the end closest to the double bond, the double bond is between position number 2 and 3, and the methyl substituents are assigned the numbers 3 and 4. The name of the compound is 3, 4-dimethyl-2-pentene.

c)
$$\boxed{CH_2 = HC-CH-CH_2-CH_2-CH_3}$$
$$\underset{\underset{CH_3}{|}}{CH_3-CH}$$
has 6 carbons as the longest continuous chain. The prefix for 6 is hex- and the base name is hexene. The only substituent group is an isopropyl group.

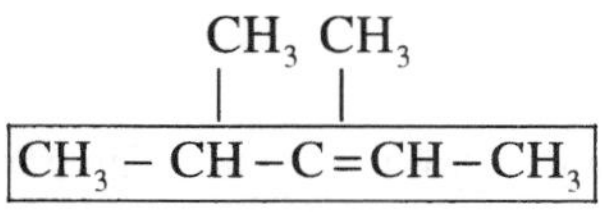

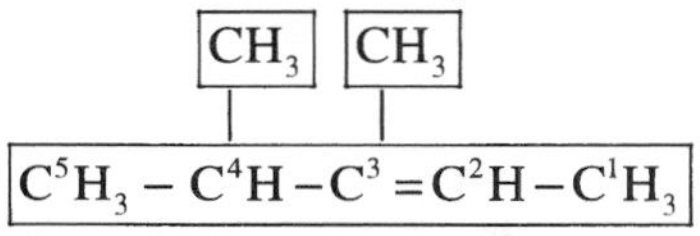

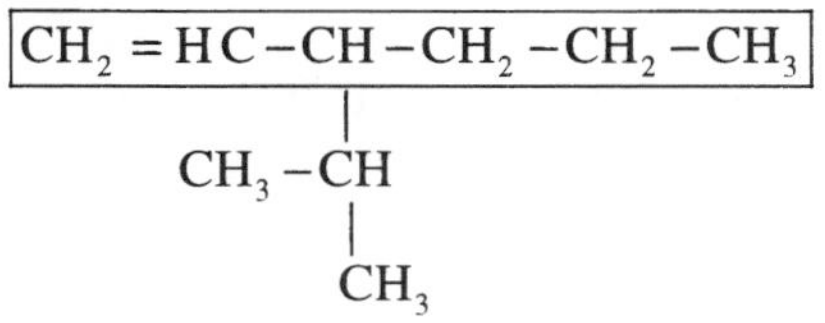

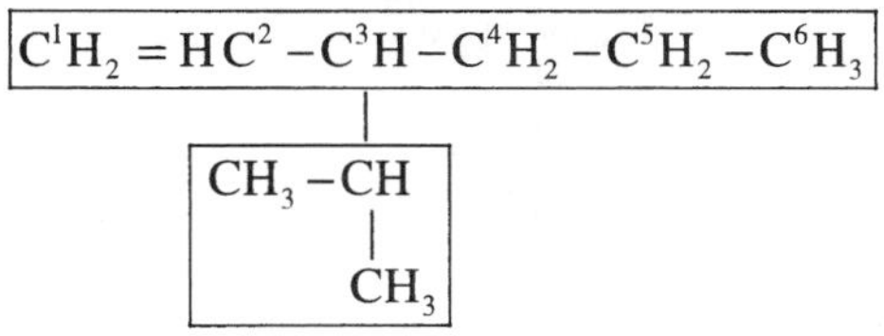

$$\boxed{C^1H_2 = HC^2-C^3H-C^4H_2-C^5H_2-C^6H_3}$$
$$\underset{\underset{\boxed{CH_3}}{|}}{\boxed{CH_3-CH}}$$
Start numbering on the left since it is closer to the double bond. Since the double bond is between position number 1 and 2, and the isopropyl group is at position 3, the name of the compound is 3-isopropyl-1-hexene.

d)
$$\overset{CH_3}{\underset{\boxed{CH_3-CH-CH_2=C-CH_3}}{|}}$$
$$\boxed{CH_2-CH_3}$$
has 6 carbons as the longest continuous chain. The prefix for 6 is hex- and the base name is hexene. The only substituent groups are two methyl groups.

$$\overset{\boxed{CH_3}}{\underset{\boxed{C^1H_3-C^2H-C^3H_2=C^4-CH_3}}{|}}$$
$$\boxed{C^5H_2-C^6H_3}$$
Since the double bond is in the middle of the chain, it will be at position 3 in both numbering schemes. If we start numbering the chain at the end closest to the left methyl group (closest to an end), the methyl substituents are assigned the numbers 2 and 4. The name of the compound is 2,4-dimethyl-3-hexene.

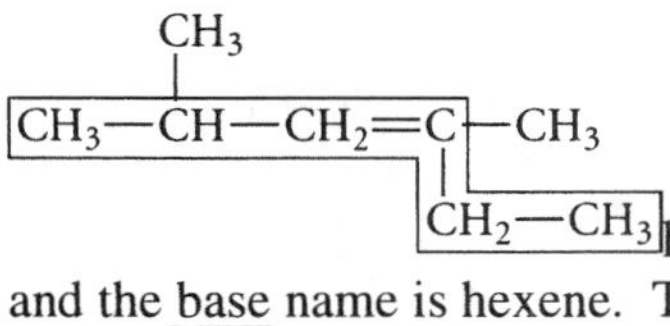

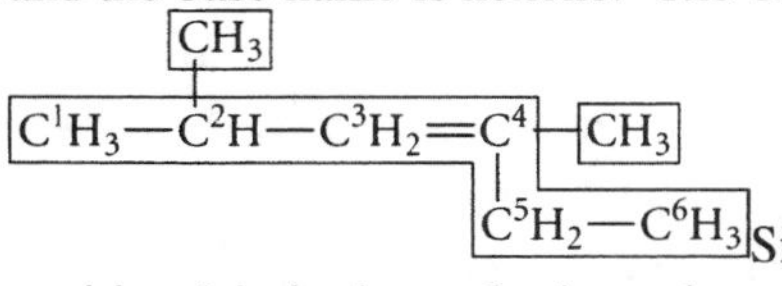

55. **Given:** alkyne structures   **Find:** name

Conceptual plan: **Count the number of carbon atoms in the longest continuous carbon chain that contains the multiple bond to determine the base name of the compound. Find the prefix corresponding to this number of atoms in Table 20. 5 and add the ending –yne to form the base name. → Consider every branch from the base chain to be a substituent. Name each substituent according to Table 20.6. →Beginning with the end closest to the multiple bond, number the base chain and assign a number to each substituent. →Write the name of the compound in the following format: (subst. #)-(subst. name)(base name). → If there are two or more substituents, give each one a number and list them alphabetically with hyphens between words and numbers. →If a compound has two or more identical substituents, designate the number of identical substituents with the prefix di- (2), tri- (3), or tetra- (4) before the substituent's name. Separate the numbers indicating the positions of the substituents relative to each other with a comma. The prefixes are not taken into account when alphabetizing.**

Solution:

a) $\boxed{CH_2 - C \equiv C - CH_3}$ has 4 carbons as the longest continuous chain. The prefix for 4 is but- and the base name is butyne. There are no substituent groups. $\boxed{C^1H_2 - C^2 \equiv C^3 - C^4H_3}$ Since the triple bond is in the middle of the molecule it does not matter which end the numbering is started. Since the triple bond is between position number 2 and 3, the name of the compound is 2-butyne.

b)
$$\overset{\displaystyle CH_3}{\underset{\displaystyle CH_3}{\boxed{CH_3 - C \equiv C - C - CH_2 - CH_3}}}$$
has 6 carbons as the longest continuous chain. The prefix for 6 is hex- and the base name is hexyne. The only substituent groups are two methyl groups.

$$\overset{\displaystyle \boxed{CH_3}}{\underset{\displaystyle \boxed{CH_3}}{\boxed{C^1H_3 - C^2 \equiv C^3 - C^4 - C^5H_2 - C^6H_3}}}$$
If we start numbering the chain at the end closest to the triple bond, the triple bond is between position number 2 and 3, and the methyl substituents are assigned the numbers 4 and 4. The name of the compound is 4,4-dimethyl-2-hexyne.

c)
$$\underset{\displaystyle \underset{\displaystyle CH_3}{\overset{\displaystyle |}{CH - CH_3}}}{\overset{\displaystyle |}{\boxed{CH \equiv C - CH - CH_2 - CH_2 - CH_3}}}$$
has 6 carbons as the longest continuous chain. The prefix for 6 is hex- and the base name is hexyne. The only substituent group is an isopropyl group.

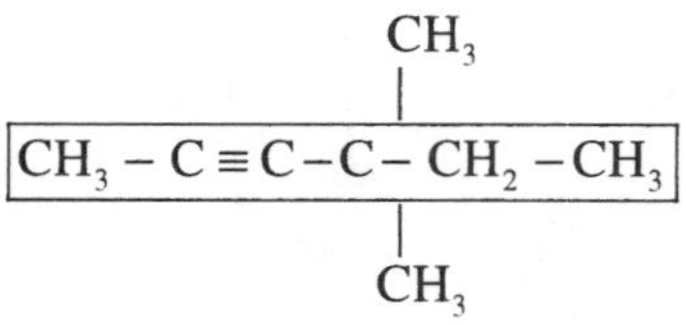

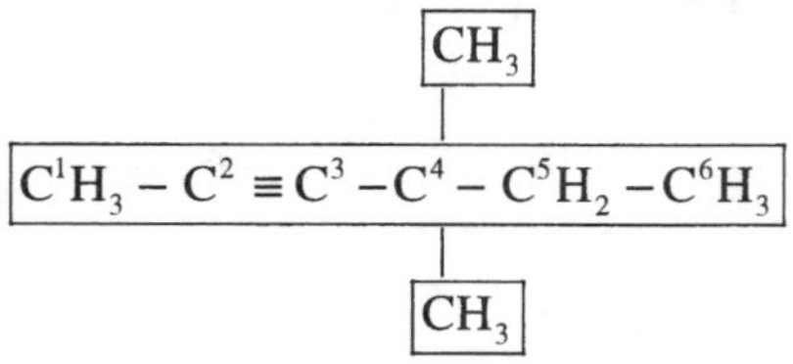

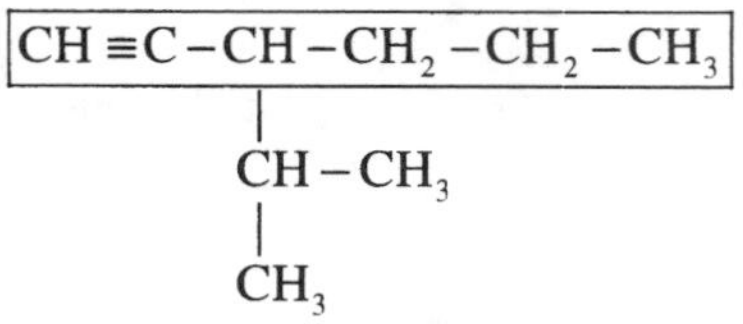

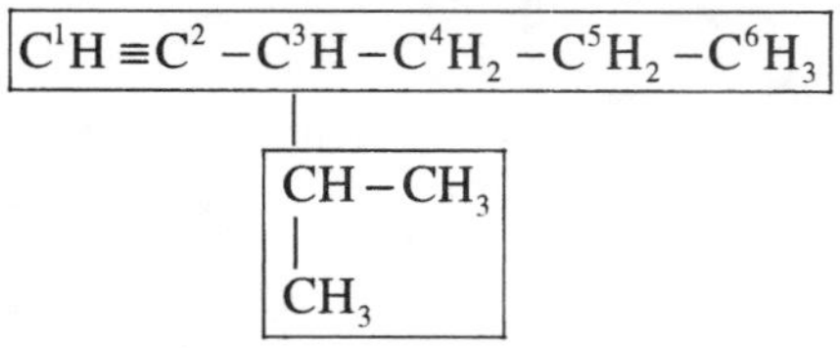

$$\underset{\displaystyle \boxed{\underset{\displaystyle CH_3}{\overset{\displaystyle |}{CH - CH_3}}}}{\overset{\displaystyle |}{\boxed{C^1H \equiv C^2 - C^3H - C^4H_2 - C^5H_2 - C^6H_3}}}$$
Start numbering at the end closest to the triple bond. The isopropyl substituent is assigned number 3. The name of the compound is 3-isopropyl-1-hexyne.

d)
$$\underset{\displaystyle CH_3 - \boxed{\underset{\displaystyle \underset{CH_3}{\overset{|}{CH_2}}}{\overset{|}{CH}} - C \equiv C - \underset{\displaystyle \underset{CH_3}{\overset{|}{CH_2}}}{\overset{\displaystyle \overset{CH_3}{\overset{|}{CH}}}{CH}} - CH_2}}{}$$
has 9 carbons as the longest continuous chain. The prefix for 9 is non- and the base name is nonyne. The substituent groups are two methyl groups.

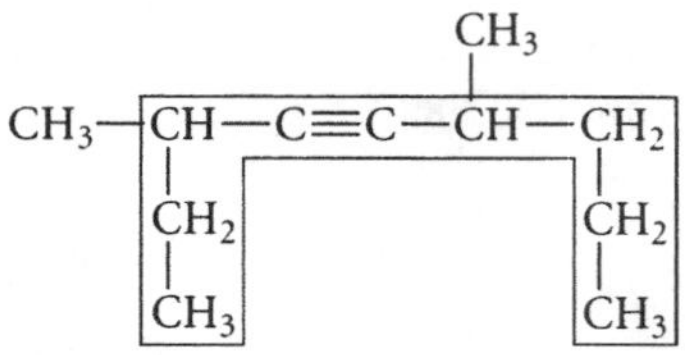

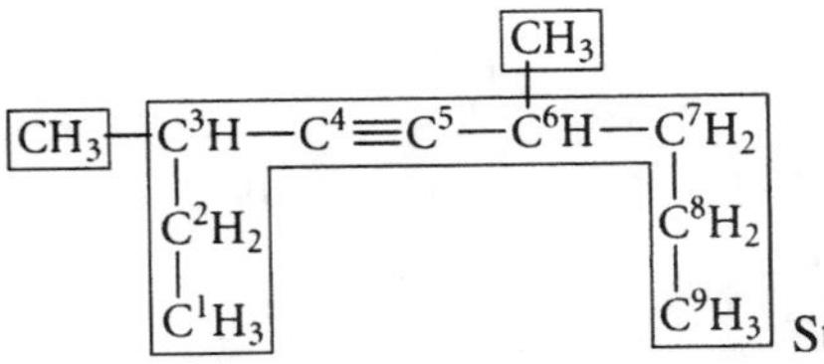

Start numbering at the bottom left and count clockwise along the chain. The triple bond is between positions 4 and 5 and the methyl substituents are assigned the numbers 3 and 6. The name of the compound is 3,6-dimethyl-4-nonyne.

57. **Given:** hydrocarbon names **Find:** structure

Conceptual plan: Find the number of carbon atoms corresponding to prefix of the base name in Table 20. 5. → Draw the base chain and number the carbons from left to right. → Determine the multiply bond type from the ending of the base name (-ene = double bond, and –yne = triple bond) and place it in the appropriate position in the chain. → Using Table 20.6 and the prefix di- (2), tri- (3), or tetra- (4) before the substituent's name, to determine each substituent. → Add the substituent to the proper carbon position in the chain. → Add hydrogen atoms to the base chain so that each carbon has 4 bonds.

Solution:

a) 4-octyne. The base name octyne designates that there are 8 carbon atoms in the base chain. $C^1-C^2-C^3-C^4-C^5-C^6-C^7-C^8$. The -yne ending designates that there is a triple bond and the 4 prefix designates that it is between the 4th and 5th positions. $C^1-C^2-C^3-C^4\equiv C^5-C^6-C^7-C^8$. Add hydrogens to the base chain to get the final molecule $CH_3-CH_2-CH_2-C\equiv C-CH_2-CH_2-CH_3$.

b) 3-nonene. The base name nonene designates that there are 9 carbon atoms in the base chain. $C^1-C^2-C^3-C^4-C^5-C^6-C^7-C^8-C^9$. The -ene ending designates that there is a double bond and the 3 prefix designates that it is between the 3rd and 4th positions. $C^1-C^2-C^3=C^4-C^5-C^6-C^7-C^8-C^9$. Add hydrogens to the base chain to get the final molecule $CH_3-CH_2-CH{=}CH-CH_2-CH_2-CH_2-CH_2-CH_3$.

c) 3,3-dimethyl-1-pentyne. The base name pentyne designates that there are 5 carbon atoms in the base chain. $C^1-C^2-C^3-C^4-C^5$. The -yne ending designates that there is a triple bond and the 1 prefix designates that it is between the 1st and 2nd positions. $C^1\equiv C^2-C^3-C^4-C^5$ 3,3-dimethyl designates that are two -CH_3 groups in the 3rd position . $\begin{array}{c} \boxed{CH_3} \\ | \\ C^1\equiv C^2-C^3-C^4-C^5 \\ | \\ \boxed{CH_3} \end{array}$. Add hydrogens to the base chain to get the final molecule $\begin{array}{c} \qquad\qquad CH_3 \\ \qquad\qquad | \\ CH\equiv C-C-CH_2-CH_3 \\ \qquad\qquad | \\ \qquad\qquad CH_3 \end{array}$.

d) 5-ethyl-3,6-dimethyl-2-heptene. The base name heptene designates that there are 7 carbon atoms in the base chain. $C^1-C^2-C^3-C^4-C^5-C^6-C^7$. The -ene ending designates that there is a double bond and the 2 prefix designates that it is between the 2nd and 3rd positions. $C^1-C^2=C^3-C^4-C^5-C^6-C^7$. 5-ethyl designates that is a -CH_2CH_3 group in the 5th position; and 3,6-dimethyl designates that are -

CH_3 groups in the 3rd and 6th positions. 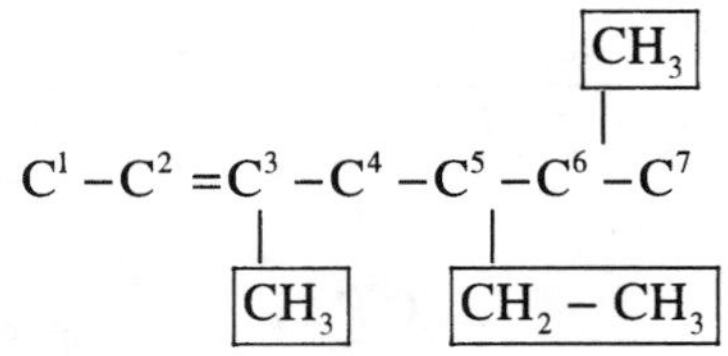 . Add hydrogens to the base chain to get the final molecule 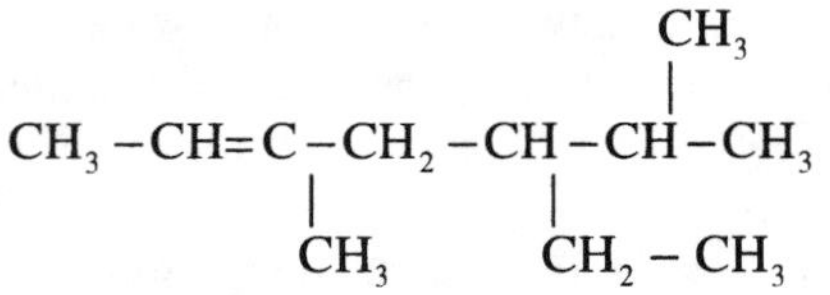 .

59. Alkene addition reactions convert a double bond to a single bond and places the two halves of the other reactant on the two carbons that were in the double bond. If the other reactant is not a homonuclear diatomic, place the hydrogen atom on the carbon that has the greater number of hydrogens (Markovnikov's rule).

a) 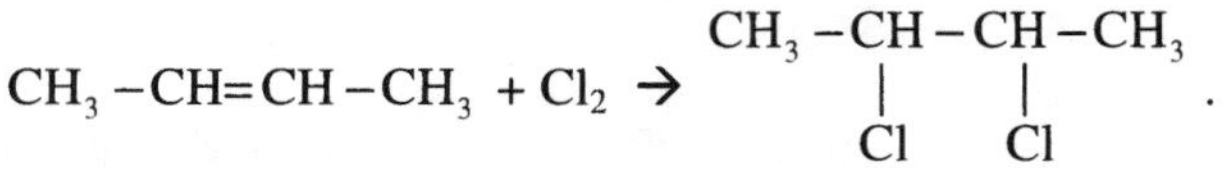 $CH_3-CH{=}CH-CH_3 + Cl_2 \rightarrow$ 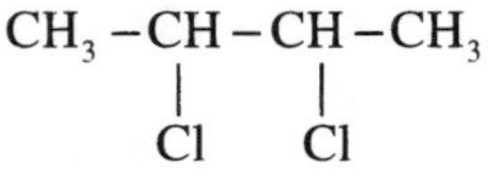 .

b) 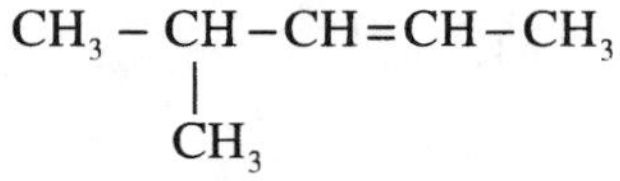 $+ HBr \rightarrow$ 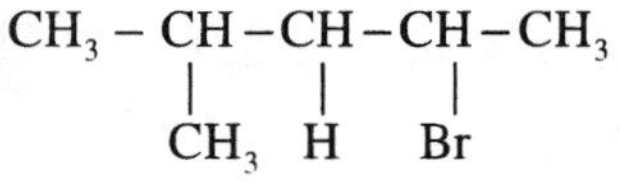 and 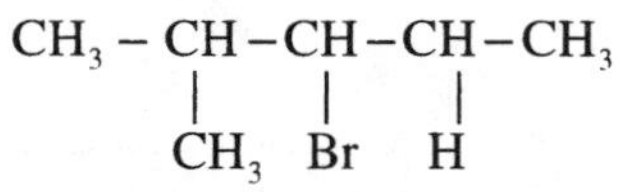 . Two carbon-containing products are possible, since both carbons on the double bond had one hydrogen each.

c) 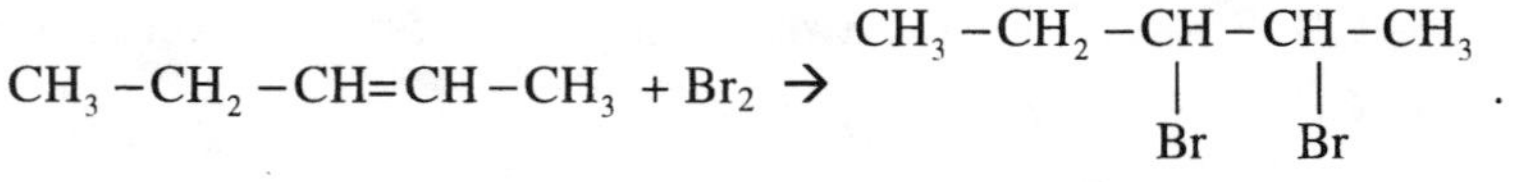 $CH_3-CH_2-CH{=}CH-CH_3 + Br_2 \rightarrow$ 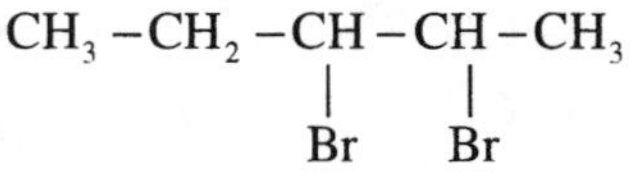 .

d) 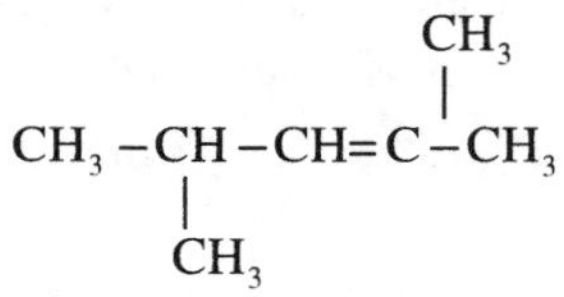 $+ HCl \rightarrow$ 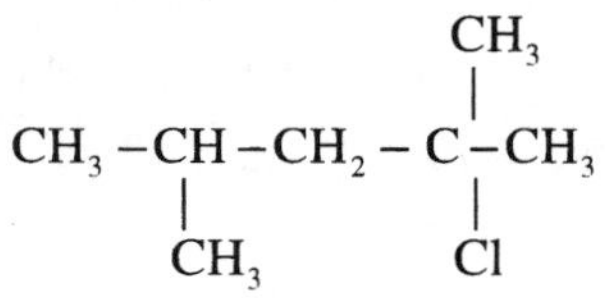 . Only one product is formed according to Markovnikov's rule .

61. Hydrogenation reactions convert a double bond to a single bond and places the hydrogen atoms on each of the two carbons that were in the double bond.

a) $CH_2{=}CH-CH_3 + H_2 \rightarrow CH_3-CH_2-CH_3$.

b) 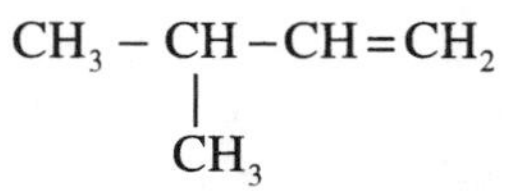 $+ H_2 \rightarrow$ 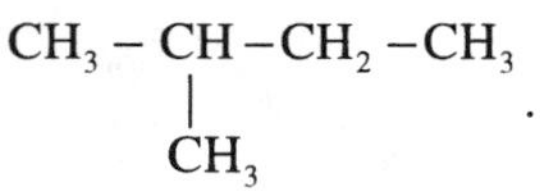 .

c) 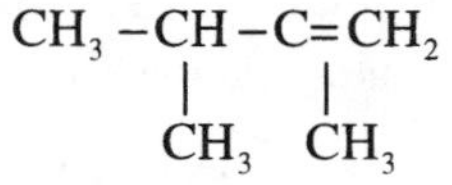 $+ H_2 \rightarrow$ 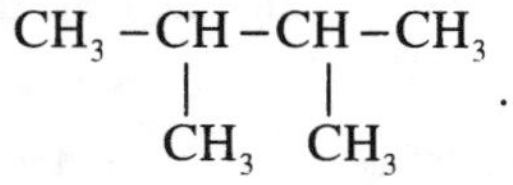 .

63. **Given:** monosubstituted benzene structures **Find:** name

Conceptual plan: Consider the branch from the benzene ring to be a substituent. Name the substituent according to Table 20.6 or with the base name of a halogen with a "o" added at the end. → Write the name of the compound in the following format: (name of substituent)benzene.

Solution:

a) $-CH_3$ or methyl is the substituent group, so the name is methylbenzene.

b) –Br or bromo is the substituent, so the name is bromobenzene.

c) –Cl or chloro is the substituent, so the name is chlorbenzene.

65. **Given:** hydrocarbon structures **Find:** name

Conceptual plan: Count the number of carbon atoms in the longest continuous carbon chain (if a multiple bond is present, make sure that this chain contains the multiple bond) to determine the base name of the compound. Find the prefix corresponding to this number of atoms in Table 20. 5 and add the ending –ane for an alkane, -ene for an alkene and –yne for an alkyne to form the base name. → Consider every branch from the base chain to be a substituent. Name each substituent according to Table 20.6. If a benzene ring is a substituent, use the phenyl group name. →Beginning with the end closest to the multiple bond, number the base chain and assign a number to each substituent. →Write the name of the compound in the following format: (subst. #)-(subst. name)(base name). → If there are two or more substituents, give each one a number and list them alphabetically with hyphens between words and numbers. →If a compound has two or more identical substituents, designate the number of identical substituents with the prefix di- (2), tri- (3), or tetra- (4) before the substituent's name. Separate the numbers indicating the positions of the substituents relative to each other with a comma. The prefixes are not taken into account when alphabetizing.

Solution: Note: Ø = 

a) 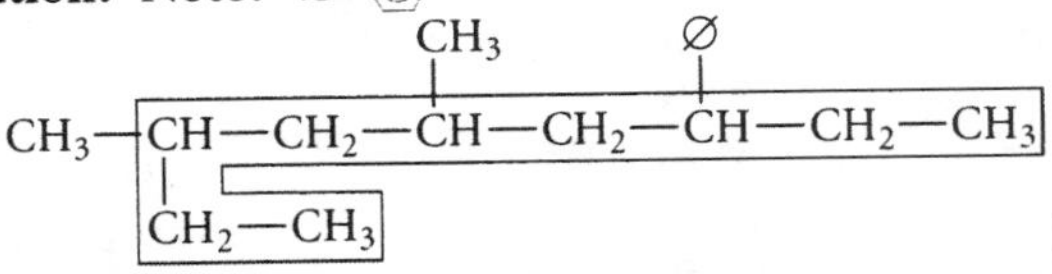 has 9 carbons as the longest continuous chain. The prefix for 9 is non- and the base name is nonane. The substituent groups are two methyl groups and a phenyl group. 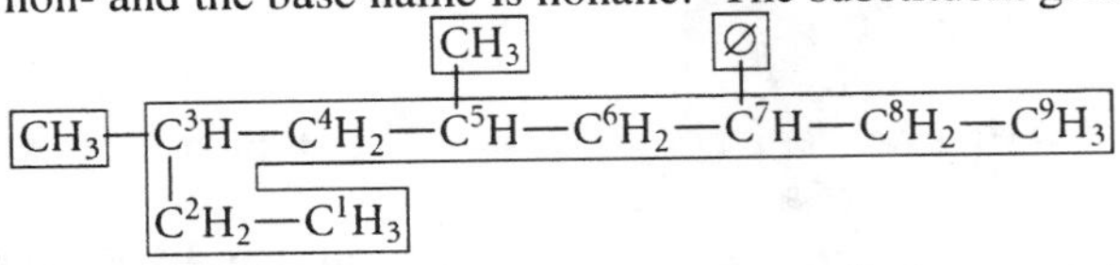 Start numbering at the end closest to the methyl group. The substituent groups are two methyl groups at positions 3 and 5; and a phenyl group at position 7. Since m comes before p, the name of the compound is 3, 5-dimethyl-7-phenylnonane. If the numbering is in the opposite direction, the name will be 5,7-dimethyl-3-phenylnonane.

b) 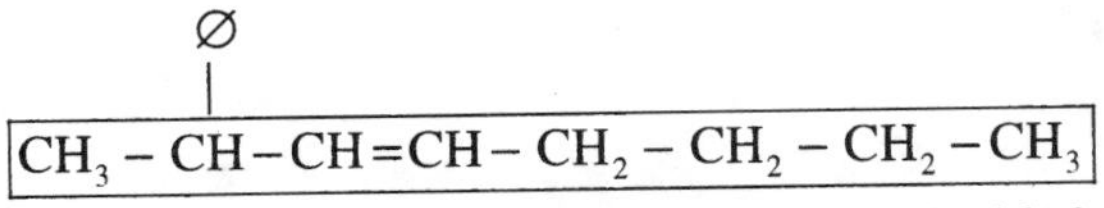 has 8 carbons as the longest continuous chain. The prefix for 8 is oct- and since there is a double bond, the base name is octene. The only substituent group is a phenyl group. 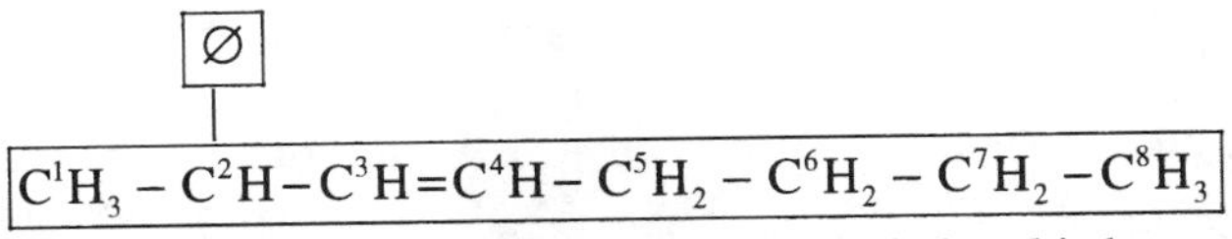 If we start numbering the chain at the end closest to the double bond, the double bond is between position number 3 and 4, and the phenyl substituent is assigned the numbers 2. The name of the compound is 2-phenyl-3-octene.

c) 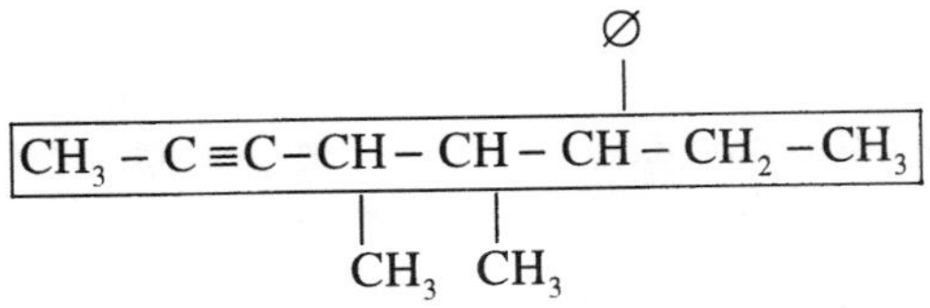 has 8 carbons as the longest continuous chain. The prefix for 8 is oct-and since there is a triple bond, the base name is octyne. The substituent groups are two methyl groups and a phenyl group. 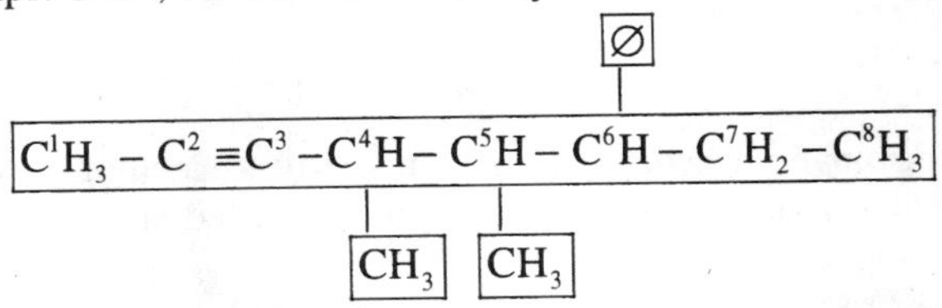 Start numbering at the end closest to the triple bond. The triple bond is between position numbers 2 and 3, the methyl groups are assigned numbers 4 and 5, and the phenyl group is assigned number 6. Since m is before p, the name of the compound is 4,5-dimethyl-6-phenyl-2-octyne.

67. **Given:** disubstituted benzene structures **Find:** name

Conceptual plan: Consider the branches from the benzene ring to be substituents. Name the substituent according to Table 20.6 or with the base name of a halogen with a "o" added at the end. → Number the benzene ring starting with the substituent that is first alphabetically and count in the direction that gets to the second substituent with the lower position number. → Write the name of the compound in the following format: (name of substituent)benzene, listing them alphabetically with hyphens between words and numbers. →If a compound has two identical substituents, designate the number of identical substituents with the prefix di- (2) before the substituent's name. Separate the numbers indicating the positions of the substituents relative to each other with a comma. → Alternate names are 1,2 = *ortho* or *o*; 1,3 = *meta* or *m*; and 1,4 = *para* or *p* replacing the numbers.

Solution:

a) Both of the substituent groups are -Br or bromo groups. Start by giving one bromo an assignment of 1 and count in either direction to give the second bromo group an assignment of 4. The name is 1,4-dibromobenzene or *p*- dibromobenzene.

b) Both of the substituent groups are -CH_2CH_3 or ethyl is the substituent groups. Start by giving the top ethyl an assignment of 1 and count in the clockwise direction to give the second ethyl group an assignment of 3. The name is 1,3-diethylbenzene or *m*-diethylbenzene.

c) One substituent group is –Cl or chloro and the other is –F or fluoro. Start by giving the chloro group an assignment of 1 and count in the counterclockwise direction to give the fluoro group an assignment of 2. The name is 1-chloro-2-fluorobenzene or *o*-chlorofluorobenzene.

69. **Given:** substituted benzene names **Find:** structures

Conceptual plan: Start with a benzene ring. Identify the structure of the substituent(s) according to Table 20.6 or with the base name of a halogen with a "o" added at the end. → If only one substituent is present, simply attach it to the benzene ring. →If a compound has two identical substituents, they are designated with the prefix di- (2) before the substituent's name. → If there are two substituents, place the first one on the benzene ring and count clockwise around the ring to determine the location to attach the second substituent. Note, 1,2 = *ortho* or *o*; 1,3 = *meta* or *m*; and 1,4 = *para* or *p* replacing the numbers.

Solution:

a) isopropylbenzene. The isopropyl group is
$$-CH-CH_3$$
$$|$$
$$CH_3$$
. Attach this to a benzene ring to make

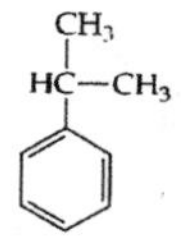

b) *meta*-dibromobenzene. There are two bromo or –Br groups. The positions are 1 and 3, since the designation is *meta*. Attach the –Br's at the 1st and 3rd positions to make

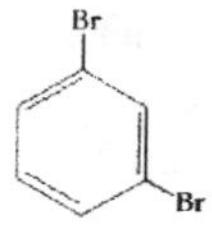

c) 1-chloro-4-methylbenzene. One substituent group is –Cl or chloro and the other is –CH_3 or methyl. Start by giving the chloro group an assignment of 1 and count in the clockwise direction to give the methyl group an assignment of 4. Attach the two groups to make

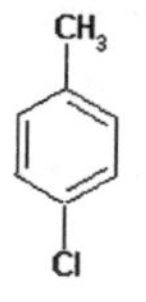

71. Aromatic substitution reaction substitute a halogen or alkyl group for a hydrogen (with a preference for the alkyl group) to generate a monosubstituted benzene and a hydrohalic acid.

a)

$$+ Br_2 \xrightarrow{FeBr_3} \quad -Br \; + \; HBr$$

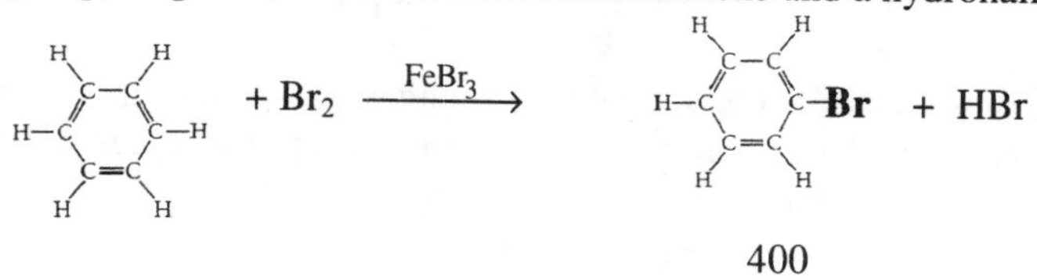

b) 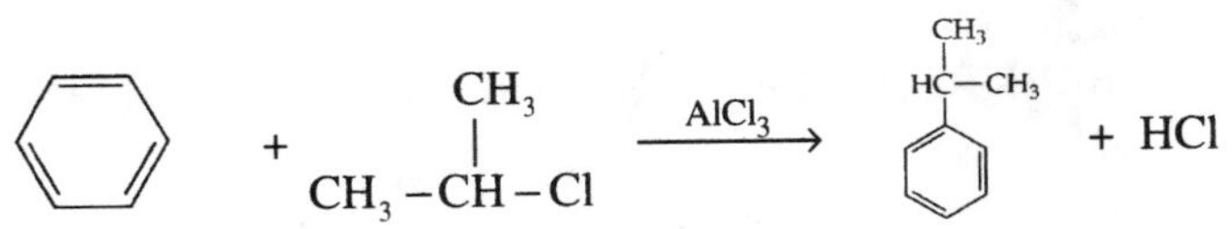

73. **Given:** alcohol structures **Find:** name
Conceptual plan: Alcohols are named like alkanes with the following differences: (1) The base chain is the longest continuous carbon chain that contains the -OH functional group; (2) The base name has the ending –ol; (3) The base chain is numbered to give the -OH group the lowest possible number; and (3) A number indicating the position of the -OH group is inserted just before the base name.
Solution:

a) $\boxed{CH_3-CH_2-CH_2}-OH$ The longest carbon chain has 3 carbons. The prefix for 3 is prop-, so the base name is propanol. The alcohol group is on the end, or 1st carbon, so the name is 1-propanol.

b) 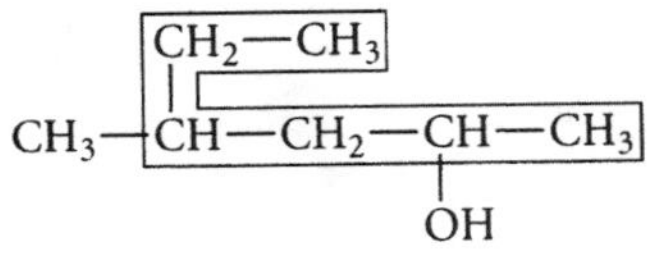

The longest carbon chain has 6 carbons. The prefix for 6 is hex-, so the base name is hexanol. There is a methyl substituent group.

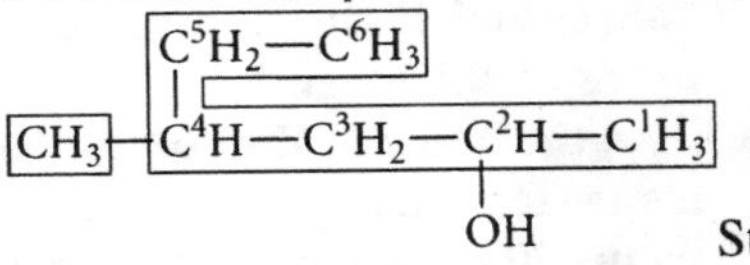

Start numbering at the end closest to the alcohol group. The alcohol group is assigned number 2 and the methyl group is assigned number 4. The name is 4-methyl-2-hexanol.

c) 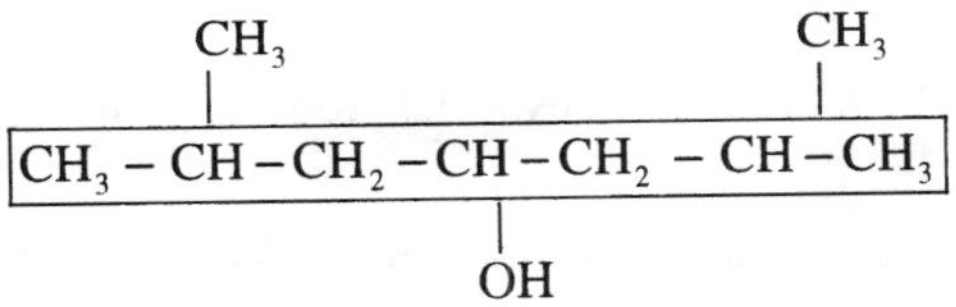

The longest carbon chain has 7 carbons. The prefix for 7 is hept-, so the base name is heptanol. There are two methyl substituent groups.

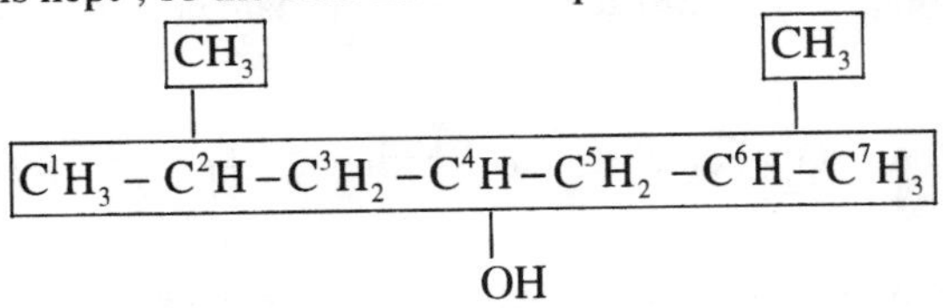

It does not matter which end the numbering is started, since substitutions are symmetrically attached. The alcohol group is assigned number 4 and the methyl groups are assigned numbers 2 and 6. The name is 2,6-dimethyl-4-heptanol.

d) 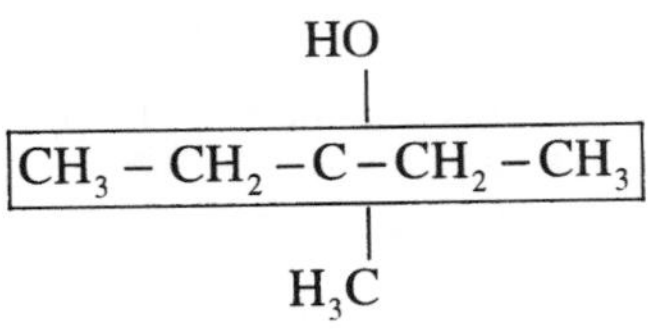

The longest carbon chain has 5 carbons. The prefix for 5 is pent-, so the base name is pentanol. There is a methyl substituent group.

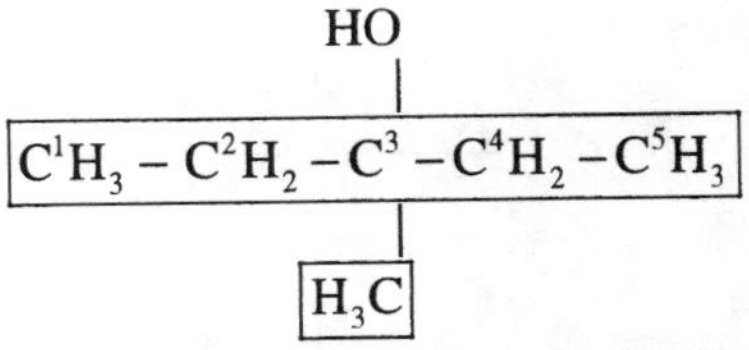

It does not matter which end the numbering is started, since substitutions are symmetrically attached. The alcohol group is assigned number 3 and the methyl group is assigned number 3. The name is 3-methyl-3-pentanol.

75. In a substitution reaction an alcohol reacts with an acid, such as HBr, to form halogenated hydrocarbons and water. In an elimination (or dehydration) reaction, concentrated acids, such as H_2SO_4, react with alcohols to eliminate water, forming an alkene. In an oxidation reaction carbon atoms gain oxygen atoms and/or lose

hydrogen atoms. In these reactions, the alcohol becomes a carboxylic acid. Alcohols react with active metals, such as alkali metals, forming an organic base and hydrogen gas.

a) This is a substitution reaction, so $CH_3-CH_2-CH_2-OH + HBr \rightarrow CH_3-CH_2-CH_2-Br + H_2O$.

b) This is an elimination reaction, so

$$CH_3-\underset{\underset{CH_3}{|}}{CH}-CH_2-OH \xrightarrow{H_2SO_4} CH_3-\underset{\underset{CH_3}{|}}{C}=CH_2 + H_2O .$$

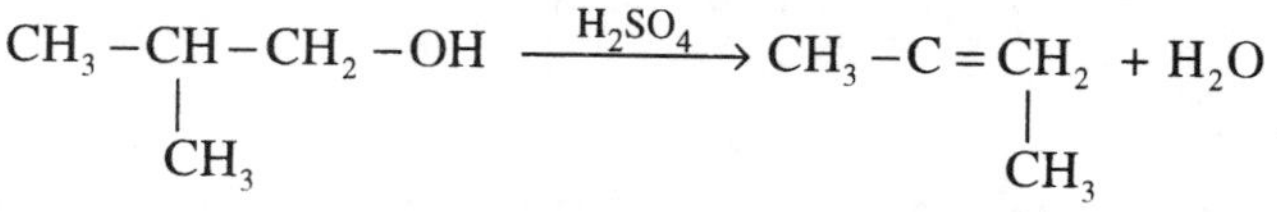

c) Sodium is an active metal, so $CH_3-CH_2-OH + Na \rightarrow CH_3-CH_2-ONa + 1/2\ H_2$.

d) This is an oxidation reaction, so

$$CH_3-\underset{\underset{CH_3}{|}}{\overset{\overset{CH_3}{|}}{C}}-CH_2-CH_2-OH \xrightarrow[H_2SO_4]{Na_2Cr_2O_7} CH_3-\underset{\underset{CH_3}{|}}{\overset{\overset{CH_3}{|}}{C}}-CH_2-\overset{\overset{O}{\|}}{C}-OH .$$

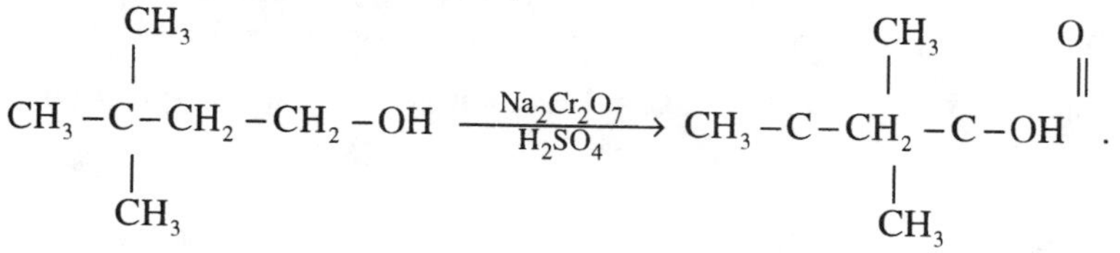

77. **Given:** aldehyde or ketone structures   **Find:** name
Conceptual Plan: Simple aldehydes are systematically named according to the number of carbon atoms in the longest continuous carbon chain that contains the carbonyl group. Form the base name from the name of the corresponding alkane by dropping the -e and add the ending -al. Simple ketones are systematically named according to the longest continuous carbon chain containing the carbonyl group. Form the base name from the name of the corresponding alkane by dropping the letter -e and adding the ending -one. For ketones, number the chain to give the carbonyl group the lowest possible number.
Solution:

a) $\boxed{CH_3-\overset{\overset{O}{\|}}{C}-CH_2-CH_3}$ The longest carbon chain has 4 carbons. The prefix for 4 is but-, since this is a ketone the base name is butanone. The position of the carbonyl carbon does not need to be specified, since there is only one place for it to be in the molecule (if it were on the end carbon it would be an aldehyde, not a ketone). Since there are no other substituent groups, the name is butanone.

b) $\boxed{CH_3-CH_2-CH_2-CH_2-\overset{\overset{O}{\|}}{CH}}$ The longest carbon chain has 5 carbons. The prefix for 5 is pent-, since this is an aldehyde the base name is pentanal. Since there are no other substituent groups, the name is pentanal.

c) $\boxed{CH_3-\underset{\underset{CH_3}{|}}{\overset{\overset{CH_3}{|}}{C}}-CH_2-\underset{\underset{}{}}{\overset{\overset{CH_3}{|}}{CH}}-CH_2-\overset{\overset{O}{\|}}{C}}-H$ The longest carbon chain has 6 carbons. The prefix for 6 is hex-, since this is an aldehyde the base name is hexanal. There are 3 methyl substituent groups.

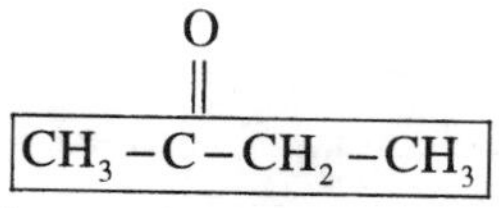

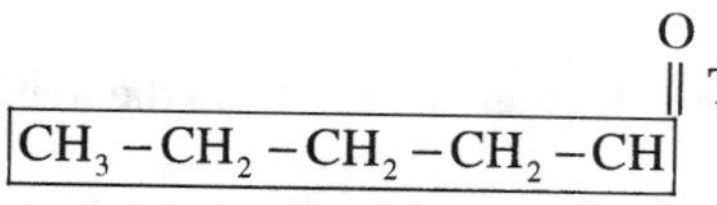

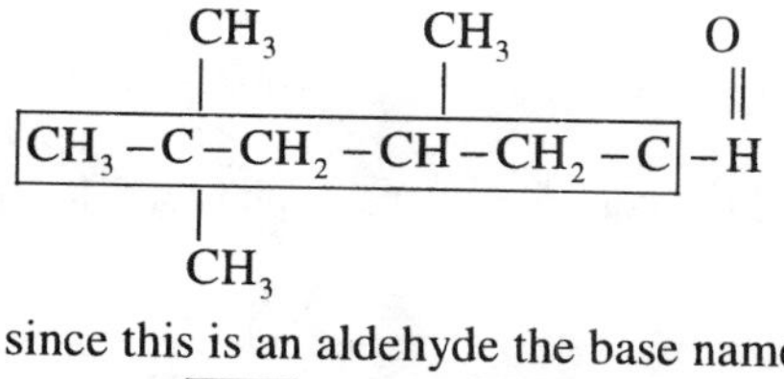

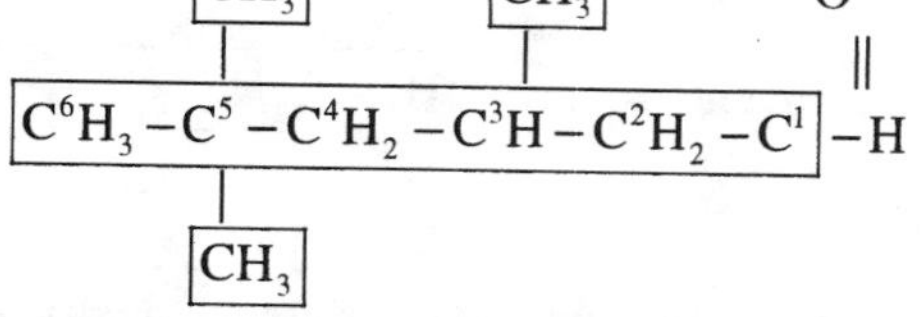

$\boxed{C^6H_3-\underset{\underset{\boxed{CH_3}}{|}}{\overset{\overset{\boxed{CH_3}}{|}}{C^5}}-C^4H_2-\overset{\overset{\boxed{CH_3}}{|}}{C^3H}-C^2H_2-C^1}-H$ Start numbering at the carbonyl carbon. The methyl groups

are at positions 3, 5, and 5. The name of the molecule is 3,5,5-trimethylhexanal.

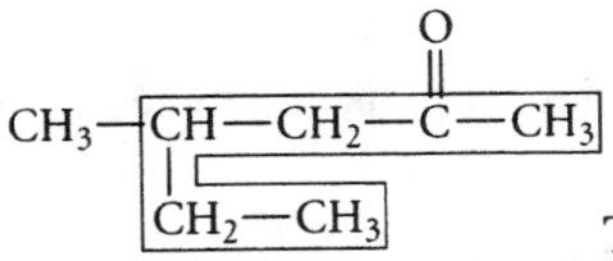

d) The longest carbon chain has 6 carbons. The prefix for 6 is hex-, since this is a ketone the base name is hexanone. There is one methyl substituent group.

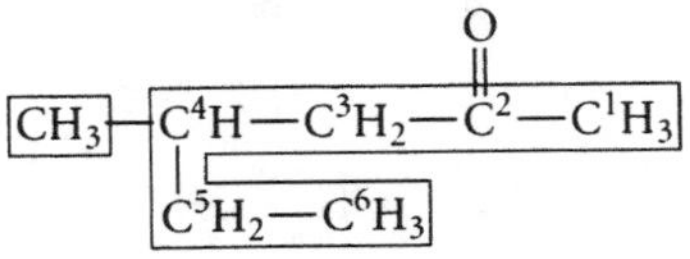

Start numbering at the end closest to the carbonyl carbon. The carbonyl carbon is in position 2 and the methyl group is at positions 4. The name of the molecule is 4-methyl-2-hexanal.

79. This is an addition reaction,

$$CH_3-CH_2-CH_2-\overset{\overset{O}{\|}}{C}-H + H-C\equiv N \xrightarrow{NaCN} CH_3-CH_2-CH_2-\overset{\overset{OH}{|}}{\underset{\underset{H}{|}}{C}}-C\equiv N \ .$$

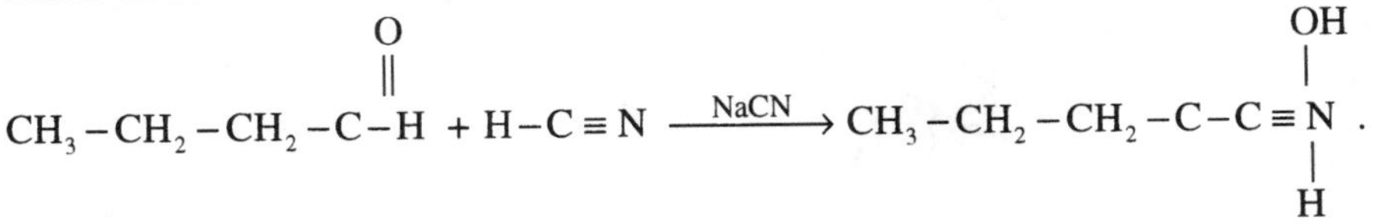

81. **Given:** carboxylic acid or ester structures   **Find:** name

Conceptual Plan: Carboxylic acids are systematically named according to the number of carbon atoms in the longest chain containing the -COOH functional group. Form the base name by dropping the -e from the name of the corresponding alkane, and adding the ending –oic acid. Esters are systematically named as if they were derived from a carboxylic acid by replacing the H on the OH with an alkyl group. The R group from the parent acid forms the base name of the compound. Change the –ic on the name of the corresponding carboxylic acid to –ate, and drop "acid". The R group that replaced the H on the carboxylic acid is named as an alkyl group with the ending –yl.

Solution:

a) $\boxed{CH_3-CH_2-CH_2-\overset{\overset{O}{\|}}{C}}-O-\boxed{CH_3}$ The carbon chain that has the carbonyl carbon has 4 carbons. The prefix for 4 is but-, so the base name is butanoate. The R group that replaces the H of the carboxylic acid is a methyl group. The name is methylbutanoate.

b) $\boxed{CH_3-CH_2-\overset{\overset{O}{\|}}{C}}-OH$ The carbon chain has 3 carbons. The prefix for 3 is prop-, since this is a carboxylic acid, and there are no other substituents, the name is propanoic acid.

c) $\boxed{CH_3-\underset{\underset{CH_3}{|}}{CH}-CH_2-CH_2-CH_2-\overset{\overset{O}{\|}}{C}}-OH$ The carbon chain has 6 carbons. The prefix for 6 is hex-, since this is a carboxylic acid, the base name is hexanoic acid. There is one methyl substituent group.

$\boxed{C^6H_3-\underset{\underset{CH_3}{|}}{C^5H}-C^4H_2-C^3H_2-C^2H_2-C^1}-OH$ Number the chain starting with the carbonyl carbon.

The methyl is in the 5th position. The name of the molecule is 5-methylhexanoic acid.

d) $\boxed{CH_3-CH_2-CH_2-CH_2-\overset{\overset{O}{\|}}{C}}-O-\boxed{CH_2-CH_3}$ The carbon chain that has the carbonyl carbon has 5 carbons. The prefix for 5 is pent-, so the base name is pentanoate. The R group that replaces the H of the carboxylic acid is an ethyl group. The name is ethylpentanoate.

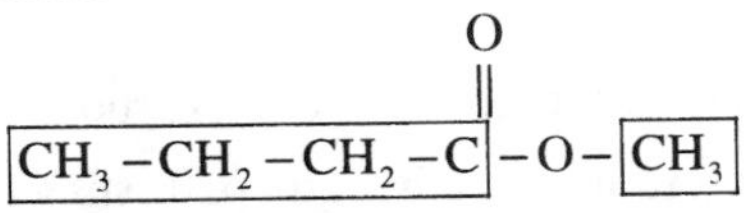

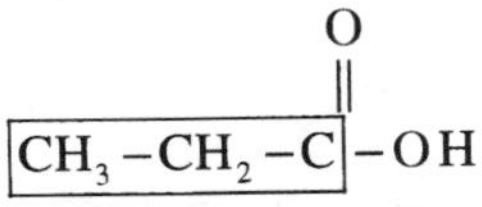

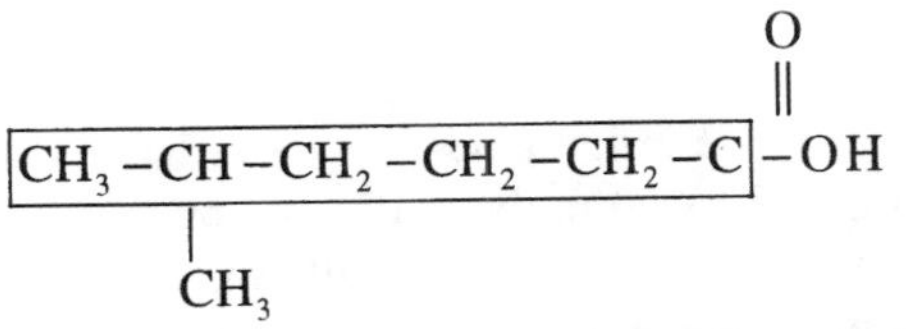

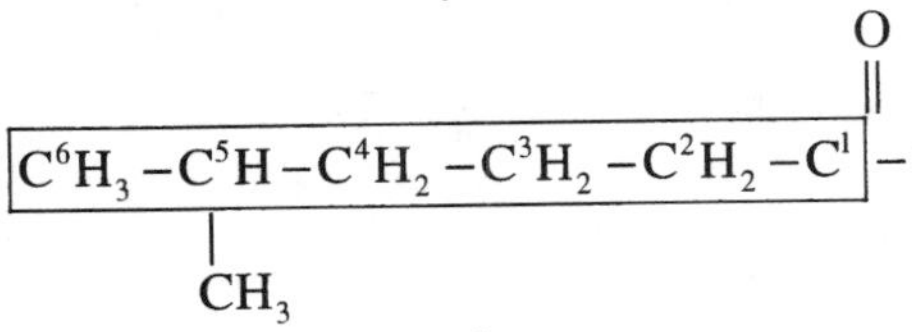

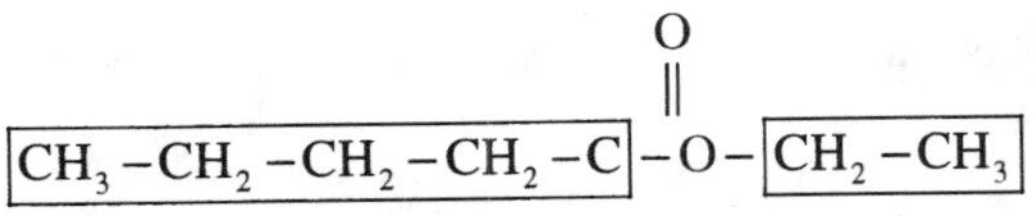

83. The reactions are condensation reactions that link the two molecules (or parts of a molecule) where the non-carbonyl oxygens are in the reactants and generate water.

a)
$$CH_3-CH_2-CH_2-CH_2-\overset{\overset{\displaystyle O}{\|}}{C}-OH + CH_3-CH_2-OH \xrightarrow{H_2SO_4}$$

$$CH_3-CH_2-CH_2-CH_2-\overset{\overset{\displaystyle O}{\|}}{C}-O-CH_2-CH_3 + H_2O.$$

b)

$$
\begin{array}{c}
\overset{O}{\|} \\
C \\
/ \;\backslash \\
CH_2 \quad OH \\
| \\
CH_2 \quad C-OH \\
\backslash \;/ \;\| \\
CH_2O
\end{array}
\xrightarrow{\text{Heat}}
\begin{array}{c}
\overset{O}{\|} \\
C \\
/ \;\backslash \\
CH_2 \quad O \\
| \quad\quad | \\
CH_2 \quad C=O \\
\backslash \;/ \\
CH_2
\end{array}
+ H_2O.
$$

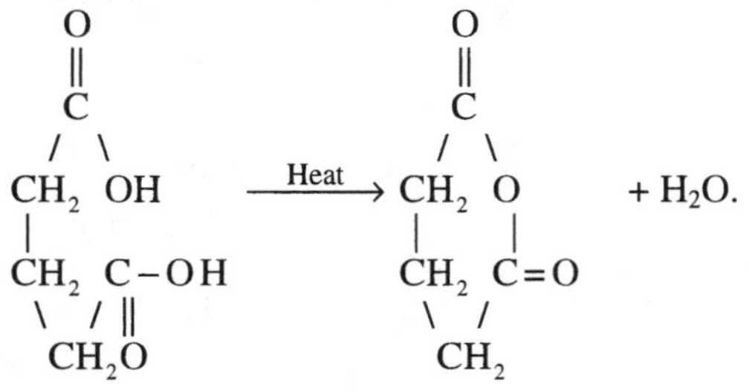

85. **Given:** ether structures   **Find:** name
Conceptual Plan: Common names for ethers have the following format: (R group 1)(R group 2) ether, where the alkyl groups are those found in Table 20.5. List the names alphabetically. If the two R groups are different, use each of their names. If the two R groups are the same, use the prefix di-.
Solution:

a) $\boxed{CH_3-CH_2-CH_2}-O-\boxed{CH_2-CH_3}$ The left carbon chain has 3 carbons and so it is a propyl group. The right carbon chain has 2 carbons and so it is an ethyl group. Listing them alphabetically the name is ethyl propyl ether.

b) $\boxed{CH_3-CH_2-CH_2-CH_2-CH_2}-O-\boxed{CH_2-CH_3}$ The left carbon chain has 5 carbons and so it is a pentyl group. The right carbon chain has 2 carbons and so it is an ethyl group. Listing them alphabetically the name is ethyl pentyl ether.

c) $\boxed{CH_3-CH_2-CH_2}-O-\boxed{CH_2-CH_2-CH_3}$ Both carbon chains have 3 carbons and so they are propyl groups. The name is dipropyl ether.

d) $\boxed{CH_3-CH_2}-O-\boxed{CH_2-CH_2-CH_2-CH_3}$ The left carbon chain has 2 carbons and so it is an ethyl group. The right carbon chain has 4 carbons and so it is an butyl group. Listing them alphabetically the name is butyl ethyl ether.

87. **Given:** amine structures   **Find:** name
Conceptual Plan: Amines are systematically named by alphabetically listing the alkyl groups that area attached to the nitrogen atom and adding an –amine ending. The alkyl groups are those found in Table 20.5. Use a prefix of di- or tri- if the alkyl groups are the same.
Solution:

a) $\boxed{CH_3-CH_2}-\underset{H}{N}-\boxed{CH_2-CH_3}$ Both carbon chains have 2 carbons and so they are ethyl groups. The name is diethylamine.

b) $\boxed{CH_3-CH_2-CH_2}-\underset{H}{N}-\boxed{CH_3}$ The left carbon chain has 3 carbons and so it is a propyl group. The right carbon chain has 1 carbon and so it is a methyl group. Listing them alphabetically the name is methylpropylamine.

c)

```
                          CH3
                           |
[CH3 - CH2 - CH2] - N - [CH2 - CH2 - CH2 - CH3]
```

The left carbon chain has 3 carbons and so it is a propyl group. The top carbon chain has 1 carbon and so it is a methyl group. The right carbon chain has 4 carbons and so it is a butyl group. Listing them alphabetically the name is butylmethylpropylamine.

89. a) This reaction is an acid-base reaction, CH_3NHCH_3 *(aq)* + HCl *(aq)* → $(CH_3)_2NH_2^+$ *(aq)* + Cl^- *(aq)*.

b) Since the acid is a carboxylic acid, this reaction is a condensation reaction, $CH_3CH_2NH_2$ *(aq)* + CH_3CH_2COOH *(aq)* → $CH_3CH_2CONHCH_2CH_3$ *(aq)* + H_2O*(l)*.

c) This reaction is an acid-base reaction, CH_3NH_2 *(aq)* + H_2SO_4 *(aq)* → $CH_3NH_3^+$ *(aq)* + HSO_4^- *(aq)*.

91. Since the monomer is

```
F     F
 \   /
  C=C
 /   \
F     F
```

, as the polymer forms the double bond breaks to link with another monomer. The structure is

```
 F  FF   FF  FF   FF
  \ / \  / \ / \  / \
   C   C   C   C    C
  / \ / \ / \ / \  /
 C   C   C   C    C
  \ / \ / \ / \  / \
  FF  FF  FF  FF   FF
```

93. In a condensation polymer atoms are eliminated. Here water is eliminated. The structure is

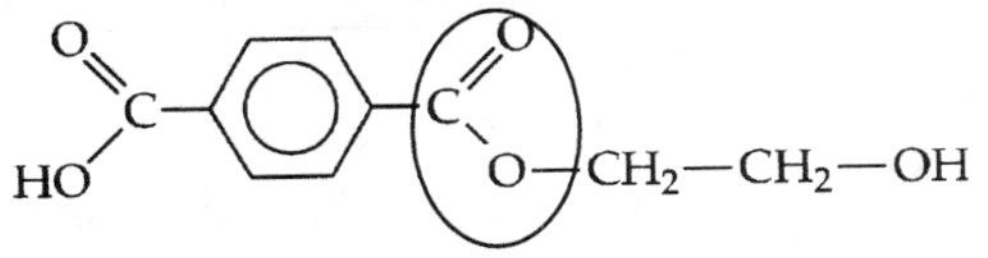

95. **Given:** structures **Find:** identify molecule type and name

Conceptual Plan: Look for functional groups to identify molecule type. Review appropriate set of naming rules to name molecule.

Solution:

a)

```
                    O
                    ||
CH3 - CH2 - CH2 - C - O - CH3
       |
      CH3
```

The compound has the formula of RCOOR', so it is an ester.

```
                    O
                    ||
[CH3 - CH2 - CH2 - C] - O - [CH3]
        |
       CH3
```

The left carbon chain has 4 carbons, which translates to but- and a base name of butanoate. There is a methyl substituent on this chain. The right carbon chain has 2 carbons, so it is a methyl group.

```
                            O
                            ||
[C4H3 - C3H2 - C2H2 - C1] - O - [CH3]
          |
        [CH3]
```

Number the left chain starting with the carbonyl carbon. The methyl group is at position 3. The name of the compound is methyl-3-methylbutanoate.

b) $$\boxed{CH_3-CH_2-\underset{\displaystyle |}{\overset{\displaystyle CH_3}{CH}}-CH_2}-O-\boxed{CH_2-CH_3}$$

The compound has the formula of ROR', so it is an ether.

The left carbon chain has 4 carbons and so it is a butyl group. There is a methyl substituent on this chain. The right carbon chain has 2 carbons and so it is an ethyl group.

$$\boxed{C^4H_3-C^3H_2-\overset{\boxed{CH_3}}{C^2H}-C^1H_2}-O-\boxed{CH_2-CH_3}$$

Number the left chain starting with the carbon next to the oxygen. The methyl substituent is in position 2, so the alkyl group is 2-methylbutyl. Listing the alkyl group alphabetically the name is ethyl 2-methylbutyl ether.

c) 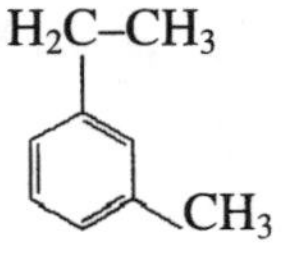

The molecule is a disubstituted benzene, so it is an aromatic hydrocarbon. The substituents are an ethyl group and a methyl group. If the ethyl position is assigned the 1st position, then the methyl group is in the 3rd position. Listing the functional groups alphabetically the name is 1-ethyl-3-methylbenzene or m-ethylmethylbenzene.

d) $$\boxed{CH_3-C\equiv C-\underset{CH_3}{CH}-\overset{CH_2-CH_3}{CH}-CH_2-CH_3}$$

The molecule contains a triple bond, so it is an alkyne. The longest carbon chain is 7. The prefix for 7 is hept- and the base name is heptyne. The substituent groups are a methyl and an ethyl group.

$$\boxed{C^1H_3-C^2\equiv C^3-\underset{\boxed{CH_3}}{C^4H}-\overset{\boxed{CH_2-CH_3}}{C^5H}-C^6H_2-C^7H_3}$$

If we start numbering the chain at the end closest to the triple bond, the triple bond is between position number 2 and 3, and the methyl substituent is assigned the number 4 and the ethyl substituent is assigned number 5. Since e is before m, the name of the compound is 5-ethyl-4-methyl-2-heptyne.

e) $$\boxed{CH_3-CH_2-CH_2-CH_2-\overset{\overset{\displaystyle O}{\|}}{CH}}$$

The molecule has the formula RCHO, is it is an aldehyde. The longest carbon chain has 4 carbons. The prefix for 4 is but-, since this is an aldehyde the base name is butanal. Since there are no other substituent groups, the name is butanal.

f) $$\boxed{CH_3-\underset{H_3C}{CH}-\overset{OH}{CH_2}}$$

The molecule has the formula ROH, so is it is an alcohol. The longest carbon chain has 3 carbons. The prefix for 3 is prop-, so the base name is propanol. There is a methyl substituent group.

$$\boxed{C^3H_3-\underset{\boxed{H_3C}}{C^2H}-\overset{OH}{C^1H_2}}$$

Start numbering with the carbon with the –OH group. The alcohol group is assigned number 1 and the methyl group is assigned number 2. The name is 2-methyl-1-propanol.

97. **Given:** structures **Find:** name

Conceptual Plan: Look for functional groups to identify molecule type. Review appropriate set of naming rules to name molecule.

Solution:

a) 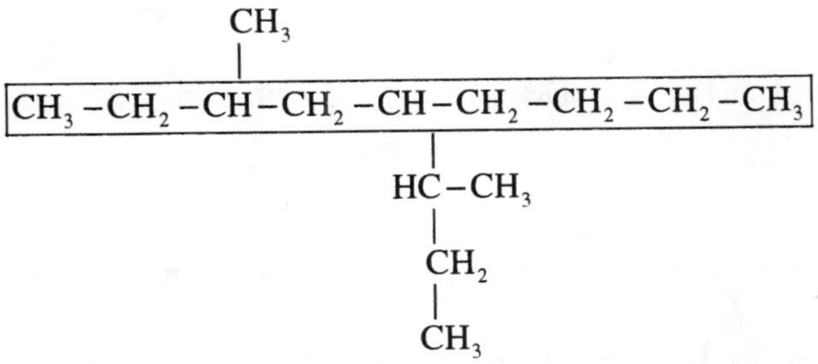

The molecule contains only single bonds, so it is an alkane. The longest continuous chain has 9 carbons. The prefix for 9 is non- and the base name is nonane. The substituent groups are a methyl group and a isobutyl group.

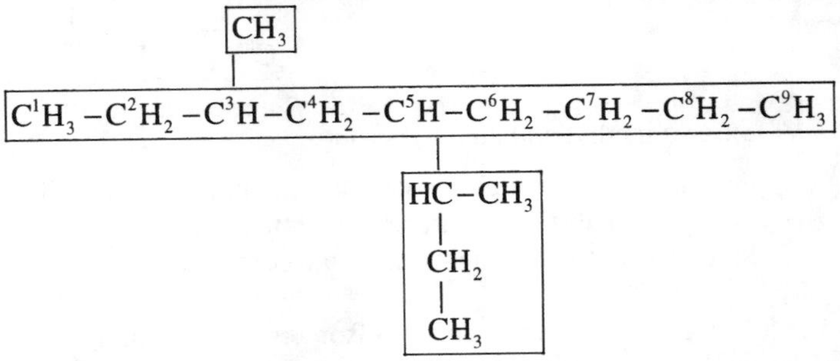

If we start numbering the chain at the end closest to the methyl group, the methyl substituent is assigned the numbers 3 and the isobutyl group is assigned the number 5. List the groups alphabetically to get the name of the compound is 5-isobutyl-3-methylnonane.

b) 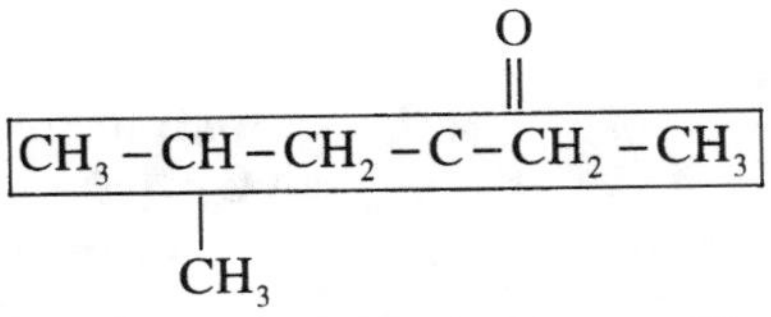

The molecule contains a carbonyl carbon in the interior of the molecule and so it is a ketone. The longest carbon chain has 6 carbons. The prefix for 6 is hex-, since this is a ketone the base name is hexanone. There is one methyl substituent group.

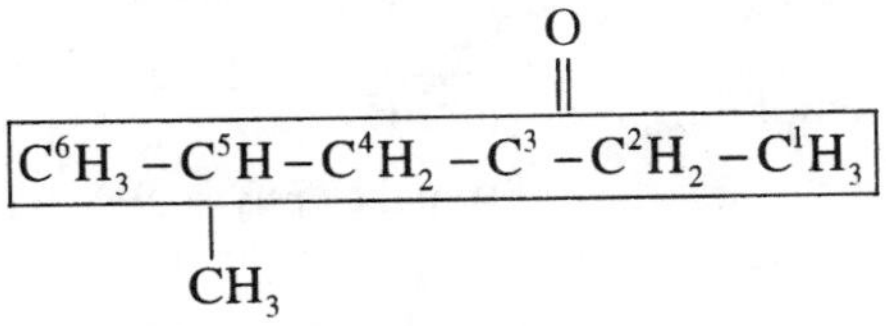

Start numbering at the right side of the molecule. The carbonyl carbon is at position number 3. The methyl group is at position 5. The name of the molecule is 5-methyl-3-hexanone.

c) 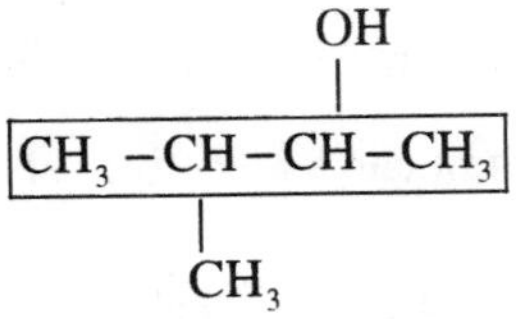

The molecule has the formula ROH, so is it is an alcohol. The longest carbon chain has 4 carbons. The prefix for 4 is but-, so the base name is butanol. The only substituent group is a methyl group.

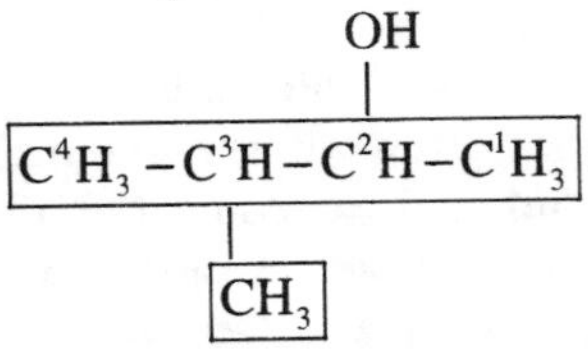

Start numbering with the end with the –OH group. The alcohol group is assigned number 2 and the methyl group is assigned number 3. The name is 3-methyl-2-butanol.

d)

$$CH_3$$
$$|$$
$$CH_2 \quad CH_3$$
$$| \qquad |$$
$$\boxed{CH_3 - CH - CH - CH - C \equiv C - H}$$
$$|$$
$$CH_3$$

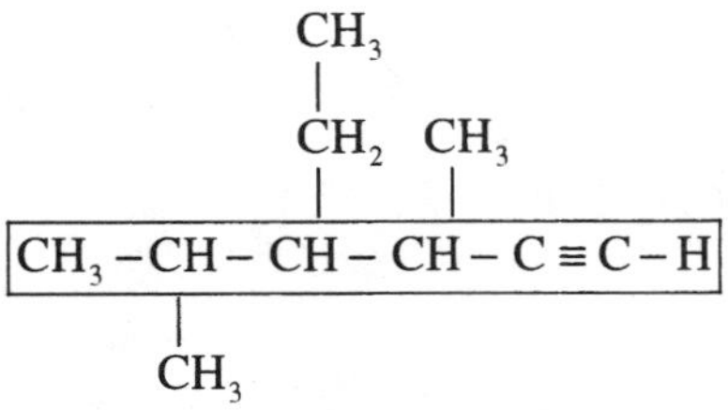

The molecule contains a triple bond, so it is an alkyne. The longest carbon chain is 6. The prefix for 6 is hex- and the base name is hexyne. The substituent groups

are two methyl and an ethyl groups. If we start numbering the chain at the end closest to the triple bond, the triple bond is between position number 1 and 2, and the methyl substituents are assigned the numbers 3 and 5 and the ethyl substituent is assigned number 4. Since e is before m, the name of the compound is 4-ethyl-3,5-dimethyl-1-hexyne.

99. a) These structures are structural isomers. The methyl and ethyl groups have been swapped.

b) These structures are isomers. The iodo second group has been moved.

c) These structures are the same molecule. The first methyl group is simply drawn in a different orientation.

101. **Given:** 15.5 kg 2-butene hydrogenation **Find:** minimum g of H_2 gas
Conceptual Plan: Write a balanced equation. kg C_4H_8 → g C_4H_8 → mol C_4H_8 → mol H_2 → g H_2

$$\frac{1000\ g}{1\ kg} \qquad \frac{1\ mol\ C_4H_8}{56.10\ g\ C_4H_8} \qquad \frac{1\ mol\ H_2}{1\ mol\ C_4H_8} \qquad \frac{2.02\ g\ H_2}{1\ mol\ H_2}$$

Solution: $C_4H_8\ (g) + H_2\ (g) \rightarrow C_4H_{10}\ (g)$ then

$$15.5\ \cancel{kg\ C_4H_8} \times \frac{1000\ \cancel{g\ C_4H_8}}{1\ \cancel{kg\ C_4H_8}} \times \frac{1\ \cancel{mol\ C_4H_8}}{56.10\ \cancel{g\ C_4H_8}} \times \frac{1\ \cancel{mol\ H_2}}{1\ \cancel{mol\ C_4H_8}} \times \frac{2.02\ g\ H_2}{1\ \cancel{mol\ H_2}} = 558\ g\ H_2$$

Check: The units (g) are correct. The magnitude of the answer (600) makes physical sense because we are reacting 15,500 g of butane. Hydrogen is lighter and the molar ratio is 1:1 so the amount of hydrogen will be significant, but less than the mass of butene.

103. a) Combustion reaction – reaction with oxygen to form CO_2 and H_2O.

b) Alkane (halogen) substitution – reaction with halogen gas to substitute a hydrogen with a halogen.

c) Alcohol elimination – ROH reaction with concentrated acid to eliminate water from the alcohol.

d) Aromatic (halogen) substitution – benzene reaction with halogen gas to substitute a hydrogen with a halogen.

105. **Given:** alkene names **Find:** structure
Conceptual plan: Find the number of carbon atoms corresponding to prefix of the base name in Table 20. 5. → Draw the base chain and number the carbons from left to right. → Place the double bond in the appropriate position in the chain. → Using Table 20.6 and the prefix di- (2), tri- (3), or tetra- (4) before the substituent's name, determine each substituent. → Add the substituent to the proper carbon position in the chain. → Add hydrogen atoms to the base chain so that each carbon has 4 bonds. → Look for substitution around double bond to determine if cis-trans isomerism is applicable and look for carbons with four different alkyl groups attached.
Solution:
a) 3-methyl-1-pentene. The base name pentene designates that there are 5 carbon atoms in the base chain.
$C^1 - C^2 - C^3 - C^4 - C^5$. The -ene ending designates that there is a double bond and the 1 prefix

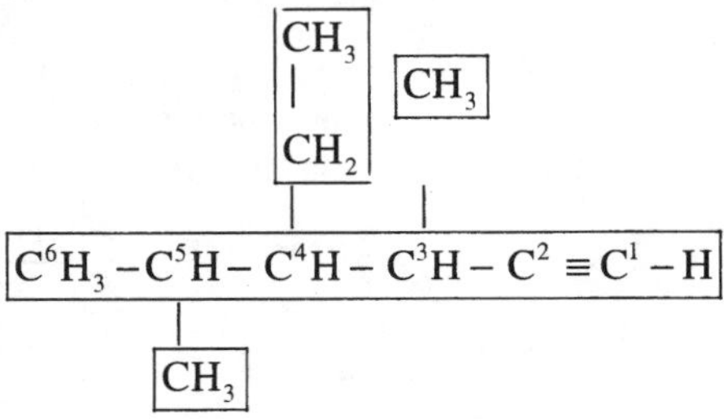

designates that it is between the 1st and 2nd positions. $C^1=C^2-C^3-C^4-C^5$ 3-methyl designates that is a $-CH_3$ group in the 3rd position. $\begin{array}{c}\boxed{CH_3}\\ | \\ C^1=C^2-C^3-C^4-C^5\end{array}$. Add hydrogens to the base chain to get the final molecule $\begin{array}{c}CH_3\\ | \\ CH_2=CH-CH-CH_2-CH_3\end{array}$. There are stereoisomerism since the 3rd carbon has four different groups attached.

b) 3,5-dimethyl-2-hexene. The base name hexene designates that there are 6 carbon atoms in the base chain. $C^1-C^2-C^3-C^4-C^5-C^6$. The -ene ending designates that there is a double bond and the 2 prefix designates that it is between the 2nd and 3rd positions. $C^1-C^2=C^3-C^4-C^5-C^6$. 3,5-dimethyl designates that are $-CH_3$ groups in the 3rd and 5th positions. $\begin{array}{c}\boxed{CH_3}\quad\quad\boxed{CH_3}\\ |\quad\quad\quad\quad | \\ C^1-C^2=C^3-C^4-C^5-C^6\end{array}$. Add hydrogens to the base chain to get the final molecule $\begin{array}{c}CH_3\quad\quad CH_3\\ |\quad\quad\quad | \\ CH_3-CH=C-CH_2-CH-CH_3\end{array}$. Cis-trans isomerism is possible around the double bond since all of the groups surrounding the double bond are different.

c) 3-propyl-2-hexene. The base name hexene designates that there are 6 carbon atoms in the base chain. $C^1-C^2-C^3-C^4-C^5-C^6$. The -ene ending designates that there is a double bond and the 2 prefix designates that it is between the 2nd and 3rd positions. $C^1-C^2=C^3-C^4-C^5-C^6$. 3-propyl designates that is a $-CH_2CH_2CH_3$ group in the 3rd position. $\begin{array}{c}\boxed{CH_2-CH_2-CH_3}\\ | \\ C^1-C^2=C^3-C^4-C^5-C^6\end{array}$. Add hydrogens to the base chain to get the final molecule $\begin{array}{c}CH_2-CH_2-CH_3\\ | \\ CH_3-CH=C-CH_2-CH-CH_3\end{array}$. Cis-trans isomerism is not possible around the double bond since the two groups on the 3rd carbon are the same.

107. The 11 isomers are $\begin{array}{c}O\\ \| \\ HC-CH_2-CH_2-CH_3\end{array}$ = aldehyde, $\begin{array}{c}O\\ \| \\ CH_3-C-CH_2-CH_3\end{array}$ = ketone, $CH_2=C-CH_2-CH_2-OH$ = alcohol & alkene, $CH_3-CH=CH-CH_2-OH$ = alcohol & alkene, $CH_3-CH_2-CH=CH-OH$ = alcohol & alkene, $CH_3-CH_2-O-CH=CH_2$ = ether & alkene, $CH_3-O-CH_2-CH=CH_2$ = ether & alkene, $CH_3-O-CH=CH-CH_3$ = ether & alkene, $\begin{array}{c}OH\\ | \\ CH_3-CH_2-C=CH_2\end{array}$ = alcohol & alkene, $\begin{array}{c}OH\\ | \\ CH_3-CH-CH=CH_2\end{array}$ = alcohol & alkene, and $\begin{array}{c}OH\\ | \\ CH_3-C=CH-CH_3\end{array}$ = alcohol & alkene. There is also a 12th isomer that is cyclic: $\begin{array}{c}H_2C-CH_2\\ |\quad\quad | \\ H_2C\quad CH_2\\ \backslash\ / \\ O\end{array}$ = ether.

109. a) This is a internal condensation reaction. It only requires heat to cause it to happen.

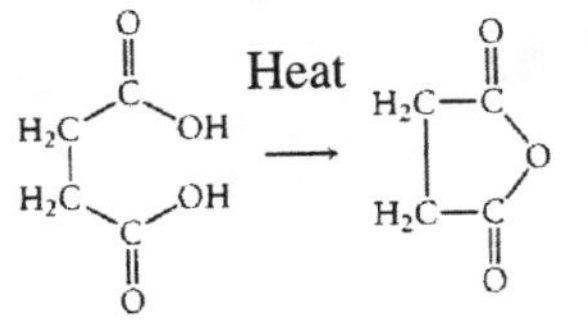

+ H_2O.

b) This is a dehydration reaction,

$$CH_3-\underset{\underset{\displaystyle CH_2-CH_3}{|}}{CH}-CH_2-\underset{\underset{\displaystyle CH_2-OH}{|}}{CH}-CH_3 \xrightarrow{H_2SO_4} CH_3-\underset{\underset{\displaystyle CH_2-CH_3}{|}}{CH}-CH_2-\underset{\overset{\displaystyle CH_2}{||}}{C}-CH_3 + H_2O.$$

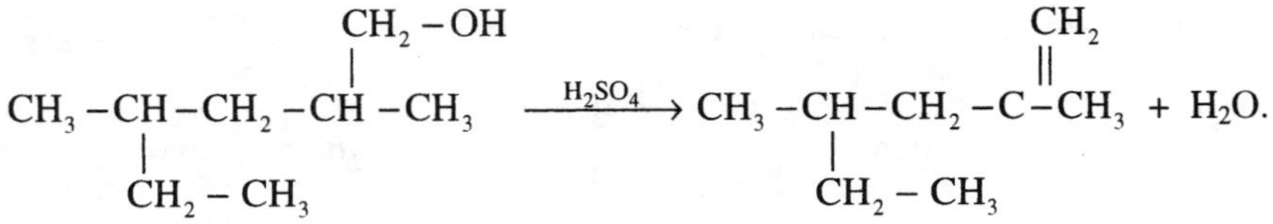

c) This is an addition reaction that follows Markovnikov's rule, $CH_3-CH_2-\underset{\underset{\displaystyle CH_3}{|}}{C}=CH_2$ + HBr →

$$CH_3-CH_2-\underset{\underset{\displaystyle CH_3}{|}}{\overset{\overset{\displaystyle Br}{|}}{C}}-CH_3 .$$

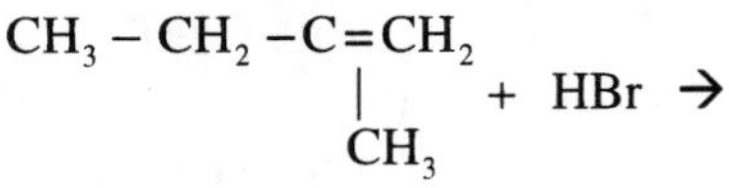

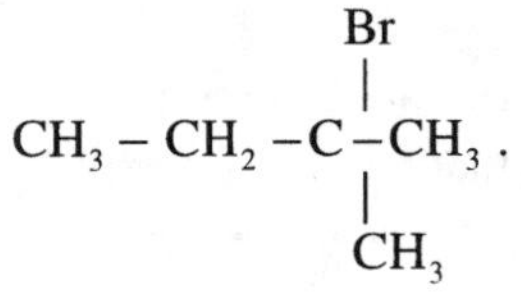

111. a) Since there are 6 primary hydrogen atoms and 2 secondary hydrogen atoms, if they are equally reactive we expect a ration of 6:2 or 3:1.

   b) Assume that we generate 100 product molecules. The yield = (# hydrogen atoms)(reactivity). So $1° = 45 = (3)$(reactivity $1°$) and $2° = 55 = (1)$(reactivity $2°$). Taking the ratio, $2°: 1° = 55: (45/3) = 55 : 15 = 11: 3$. The $2°$ hydrogens are much more reactive.

113. The chiral products have four different groups on a carbon are: $CH_2Cl-CHCl-CH_2-CH_3$ (2nd carbon), $CH_2Cl-CH_2-CHCl-CH_3$ (3rd carbon), and $CH_3-CHCl-CHCl-CH_3$ (2nd and 3rd carbons).

115. In order for the structure to have only one product after a single bromination, it must have all of the hydrogens be equivalent. The structure is $CH_3-\underset{\underset{\displaystyle CH_3}{|}}{\overset{\overset{\displaystyle CH_3}{|}}{C}}-\underset{\underset{\displaystyle CH_3}{|}}{\overset{\overset{\displaystyle CH_3}{|}}{C}}-CH_3$. To name it, find the longest carbon chain:

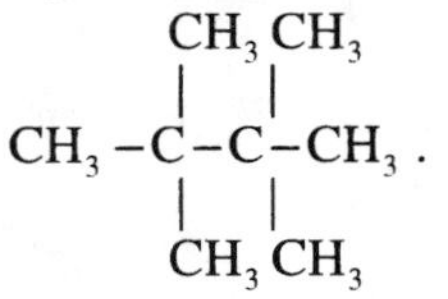

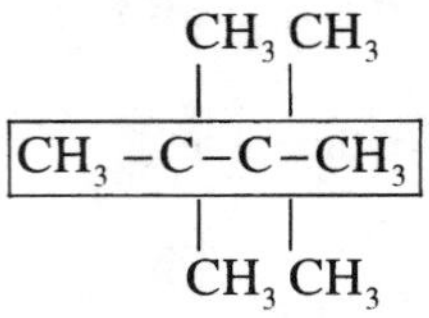

$\boxed{CH_3-\underset{\underset{\displaystyle CH_3}{|}}{\overset{\overset{\displaystyle CH_3}{|}}{C}}-\underset{\underset{\displaystyle CH_3}{|}}{\overset{\overset{\displaystyle CH_3}{|}}{C}}-CH_3}$. Since there are 4 carbons in this chain and it is an alkane, the base name is butane.

There are 4 methyl substituent groups. $\boxed{C^1H_3-\underset{\underset{\displaystyle \boxed{CH_3}}{|}}{\overset{\overset{\displaystyle \boxed{CH_3}}{|}}{C^2}} - \underset{\underset{\displaystyle \boxed{CH_3}}{|}}{\overset{\overset{\displaystyle \boxed{CH_3}}{|}}{C^3}} - C^4H_3}$ It does not matter how the chain is numbered since it is symmetrically substituted. The methyl groups are assigned the numbers 2, 2, 3, and 3. The name is 2,2,3,3-tetramethylbutane.

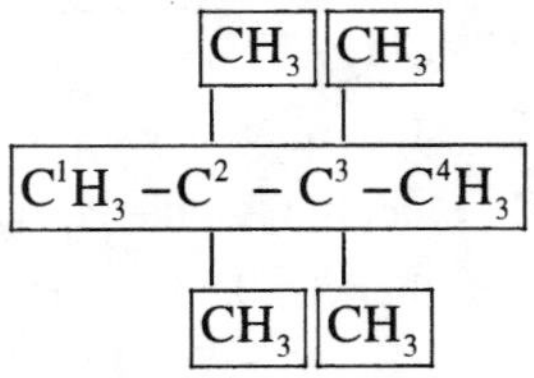

Chapter 21
Biochemistry

1. Biochemistry is that area at the interface between chemistry and biology that strives to understand living organisms at the molecular level. It is the study of the chemistry occurring in living organisms. To help diabetics, Frederick Sanger discovered the detailed chemical structure of human insulin. This allowed the synthesis of insulin. This was further enhanced by biotechnology advances, particularly by Genentech, that allowed researchers to insert the human gene for insulin into the DNA of bacterial cells. This new way to manufacture human insulin revolutionized the treatment of diabetes since the human insulin is now readily available.

3. A fatty acid is a carboxylic acid with a long hydrocarbon tail. The general structure for a fatty acid is:

$$R-\overset{\overset{\textstyle O}{\|}}{C}-OH$$

    where R represents a hydrocarbon chain containing 3 to 19 carbon atoms.

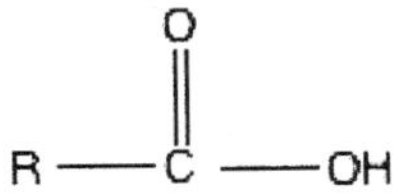

5. Fats and oils or triglycerides, are triesters composed of glycerol with three fatty acids attached. The general

$$\begin{array}{l} H_2C-O-\overset{\overset{\textstyle O}{\|}}{C}-R \\[4pt] R-\overset{\overset{\textstyle O}{\|}}{C}-O-CH \\[4pt] H_2C-O-\overset{\overset{\textstyle O}{\|}}{C}-R \end{array}$$

    structure of a triglyceride is:

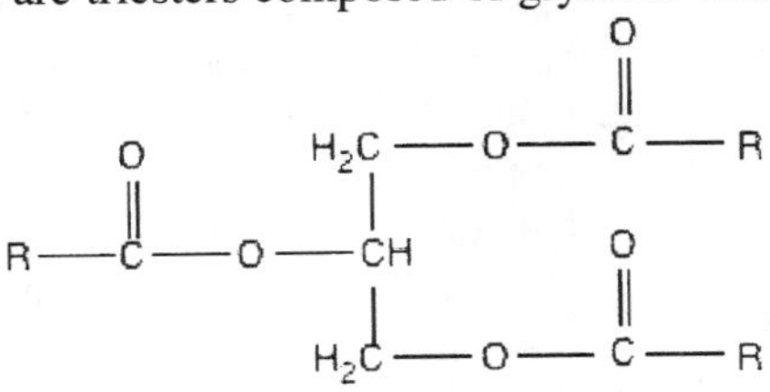

7. Phospholipids have the same basic structure as triglycerides, except that one of the fatty acid groups is replaced with a phosphate group. Unlike a fatty acid, which is nonpolar, the phosphate group is polar. The phospholipid molecule has a polar region and a nonpolar region. The polar part of the molecule is hydrophilic while the nonpolar part is hydrophobic.
    Glycolipids have similar structures and properties. The nonpolar section of a glycolipid is composed of a fatty acid chain and a hydrocarbon chain. However, the polar section is a sugar molecule.
    The structure of phospholipids and glycolipids makes them ideal as components of cell membranes, where the polar parts can interact with the aqueous environments inside and outside the cell and the nonpolar parts interact with each other, forming a double – layered structure called a lipid bilayer. Lipid bilayers encapsulate cells and many cellular structures.

9. Carbohydrates have the general formula $(CH_2O)_n$. Structurally, carbohydrates are polyhydroxy aldehydes or ketones. Carbohydrates are responsible for short – term storage of energy in living organisms, and they make up the main structural components of plants.

11. Simple carbohydrates are the simple sugars, monosaccharides and disaccharides. Polysaccharides, which are long, chainlike molecules composed of many monosaccharide units bonded together, are complex carbohydrates.

13. Proteins are the workhorse molecules in living organisms and are involved in virtually all facets of cell structure and function. Most of the chemical reactions that occur in living organisms are enabled by enzymes, proteins that act as catalysts in biochemical reactions. Proteins are also the structural elements of muscle, skin, and cartilage. Proteins transport oxygen in the blood, act as antibodies to fight disease, and function as hormones to regulate metabolic processes.

15. The R groups, or side chains, differ chemically from one amino acid to another. When amino acids are strung together to make a protein, the chemical properties of the R group determine the structure and properties of the protein.

$$\text{H}_2\text{N} - \overset{\overset{\displaystyle \text{H}}{|}}{\underset{\underset{\displaystyle \text{CH}_3}{|}}{\text{C}}} - \overset{\overset{\displaystyle \text{O}}{\|}}{\text{C}} - \text{OH} \qquad \overset{+}{\text{H}_3\text{N}} - \overset{\overset{\displaystyle \text{H}}{|}}{\underset{\underset{\displaystyle \text{CH}_3}{|}}{\text{C}}} - \overset{\overset{\displaystyle \text{O}}{\|}}{\text{C}} - \text{O}^-$$

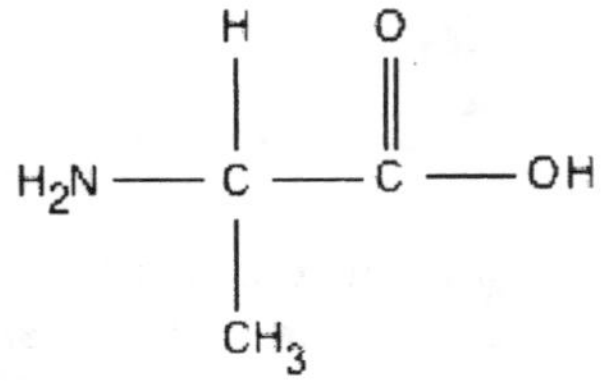

Alanine (Ala)

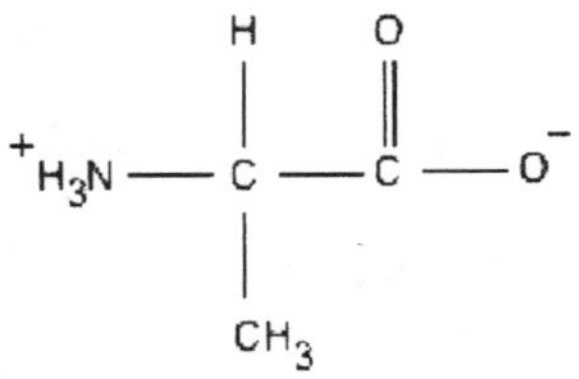

19. The 10 essential amino acids are phenylalanine, tryptophan, isoleucine, histidine, lysine, valine, threonine, methionine, arginine, and leucine. These amino acids are called essential amino acids because they can not be synthesized in the body. They must come from food and in the proportions that match the needs of protein synthesis.

21. The primary structure is simply the amino acid sequence.
The secondary structure refers to the repeating patterns in the arrangement of protein chains.
The tertiary structure refers to the large – scale twists and folds of globular proteins.
The quarternary structure refers to the arrangement of subunits in proteins that have more than one polypeptide chain.

23. In the α – helix structure, the amino acid chain is wrapped into a tight coil in which the side chains extend outward from the coil. The structure is maintained by hydrogen bonding interaction between NH and CO groups along the peptide backbone of the protein.
In the β – pleated sheet the chain is extended and forms a zigzag pattern. The peptide backbones of neighboring chains interact with one another through hydrogen bonding to form zigzag shaped sheets.

25. Nucleic acids are polymers. The individual units composing nucleic acids are called nucleotides. Each nucleotide has three parts: a sugar, a base, and a phosphate groups, which serves as a link between sugars.

27. A codon is a sequence of three bases that codes for one amino acid.
A gene is a sequence of codons within a DNA molecule that codes for a single protein.
Genes are contained in structures called chromosomes. There are 46 chromosomes in the nuclei of human cells.

29. When a cell is about to divide, the DNA unwinds and the hydrogen bonds joining the complementary bases break, forming two daughter strands. With the help of the enzyme DNA polymerase, a complement to each daughter strand with the correct complementary bases in the correct sequence is formed. The hydrogen bonds between the strands then re – form, resulting in two complete copies of the original DNA, one for each daughter cell.

31. a) No. It is an alkane chain.
    b) No. It is an amino acid.
    c) Yes. It is a saturated fatty acid.
    d) Yes. It is the steroid, cholesterol.

33. a) Yes. It is a saturated fatty acid.
    b) No. It is an alkane.
    c) No. It is an ether.
    d) Yes. It is a monounsaturated fatty acid.

35.

$$\begin{array}{l} \text{H}_2\text{C} - \text{OH} \\ | \\ \text{HO} - \text{CH} \qquad + \quad 3\left[\text{HOOC}(\text{CH}_2)_6 \Big(\text{CH}_2\text{CH}=\text{CH} \Big)_2 (\text{CH}_2)_4 \text{CH}_3 \right] \\ | \\ \text{H}_2\text{C} - \text{OH} \end{array}$$

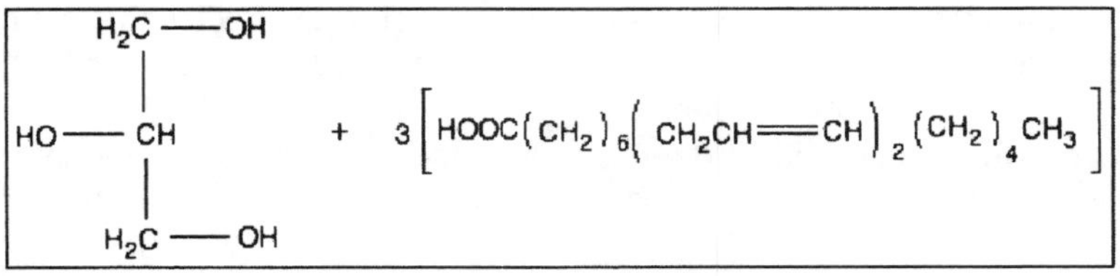

$$\text{H}_2\text{C} - \text{O} - \overset{\overset{\displaystyle O}{\|}}{\text{C}} - (\text{CH}_2)_6 \Big(\text{CH}_2\text{CH}=\text{CH} \Big)_2 (\text{CH}_2)_4 \text{CH}_3$$

$$\longrightarrow \quad \text{HC} - \text{O} - \overset{\overset{\displaystyle O}{\|}}{\text{C}} - (\text{CH}_2)_6 \Big(\text{CH}_2\text{CH}=\text{CH} \Big)_2 (\text{CH}_2)_4 \text{CH}_3$$

$$\text{H}_2\text{C} - \text{O} - \overset{\overset{\displaystyle O}{\|}}{\text{C}} - (\text{CH}_2)_6 \Big(\text{CH}_2\text{CH}=\text{CH} \Big)_2 (\text{CH}_2)_4 \text{CH}_3$$

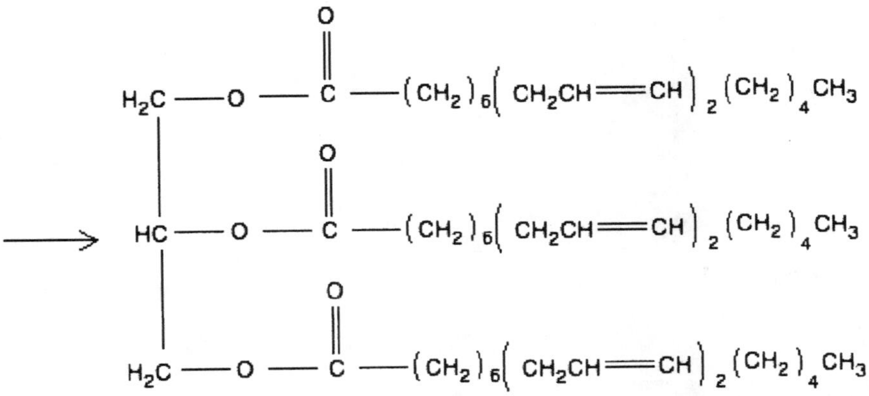

The double bonds in the linoleic acid would most likely make this triglyceride form an oil.

37. a) Yes, it is a carbohydrate, it is a monosaccharide.
    b) No, it is not a carbohydrate, it is a steroid: β – estridiol.
    c) Yes, it is a carbohydrate, it is a disaccharide.
    d) No, it is not a carbohydrate, it is a peptide.

39. a) The C – O double bond is on a terminal carbon, so it is an aldose, there are six carbon atoms, so it is a hexose.
    b) The C – O double bond is on a terminal carbon, so it is an aldose, there are five carbon atoms, so it is a pentose.
    c) The C – O double bond is NOT on a terminal carbon, so it is a ketose, there are four carbon atoms, so it is a tetrose.
    d) The C – O double bond is on a terminal carbon, so it is an aldose, there are four carbon atoms, so it is a tetrose.

41. a) Five of the six carbon atoms have 4 different substituents attached; therefore, there are five chiral centers.
    b) Three of the five carbon atoms have 4 different substituents attached; therefore, there are three chiral centers.
    c) Only one of the four carbon atoms has 4 different substituents attached; therefore, there is only one chiral center.
    d) Three of the four carbon atoms have 4 different substituents attached; therefore, there are four chiral centers.

43. Glucose

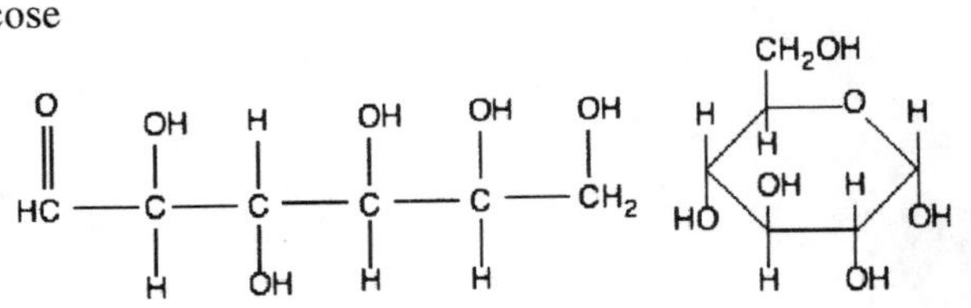

45. Hydrolysis reacts with H_2O to split the glycosidic linkage and makes monosaccharides.

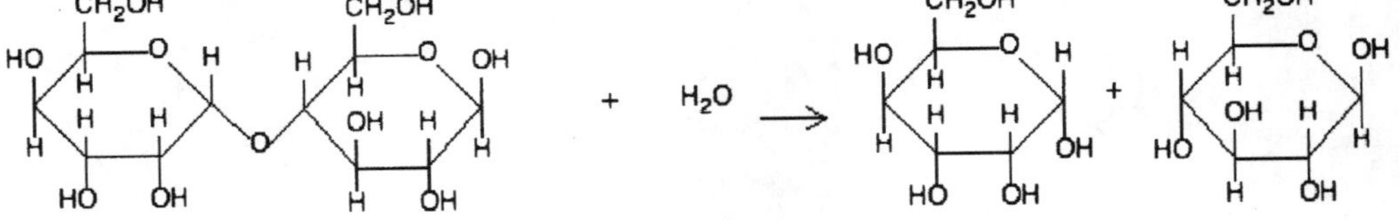

47.

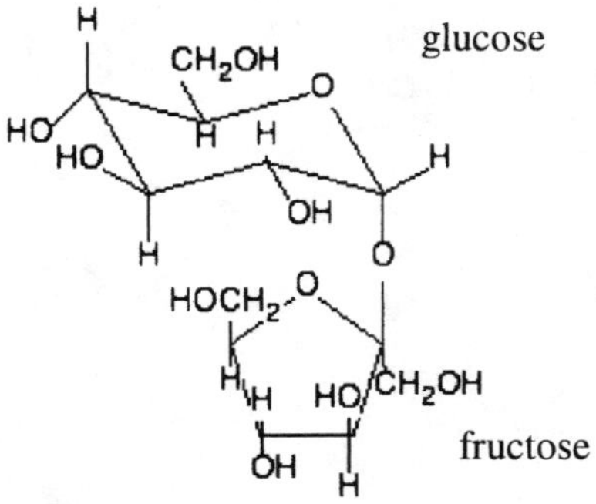

49. a) Thr = Threonine

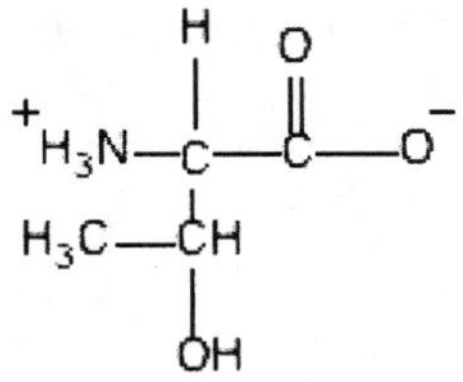

b) Ala = Alanine

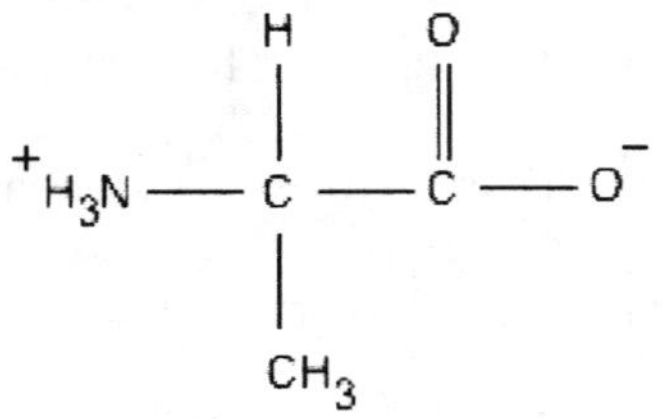

c) Leu = Leucine

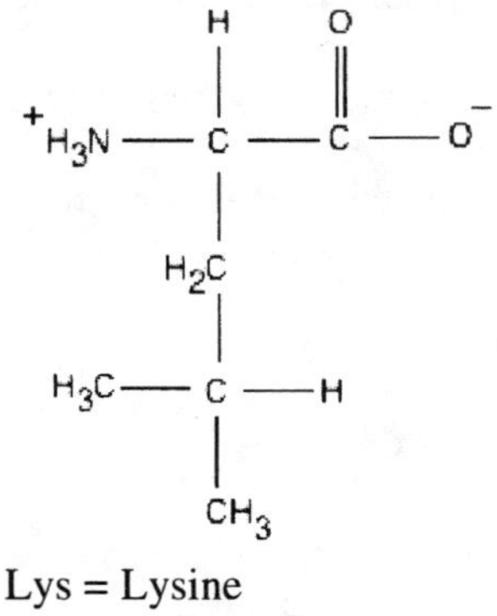

d) Lys = Lysine

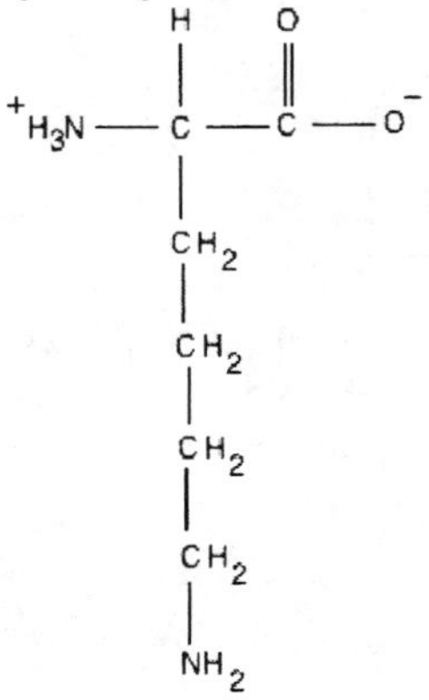

51. L – Alanine

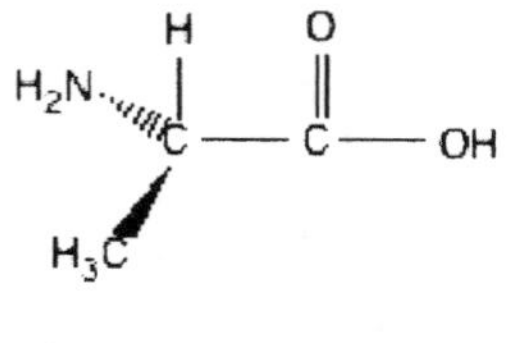

D – Alanine

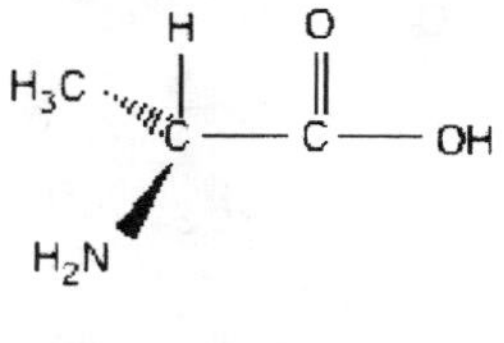

53. Because the N – terminal end and the C – terminal end are specific in a peptide, six tripeptides can be made from the three amino acids: serine, glycine, and cysteine.
Ser – Gly – Cys   Ser – Cys – Gly   Gly – Cys – Ser   Gly – Ser – Cys
Cys – Ser – Gly   Cys – Gly – Ser

55. Peptide bonds are formed when the carboxylic end of one amino acid reacts with the amine end of another amino acid.

Serine   Tyrosine

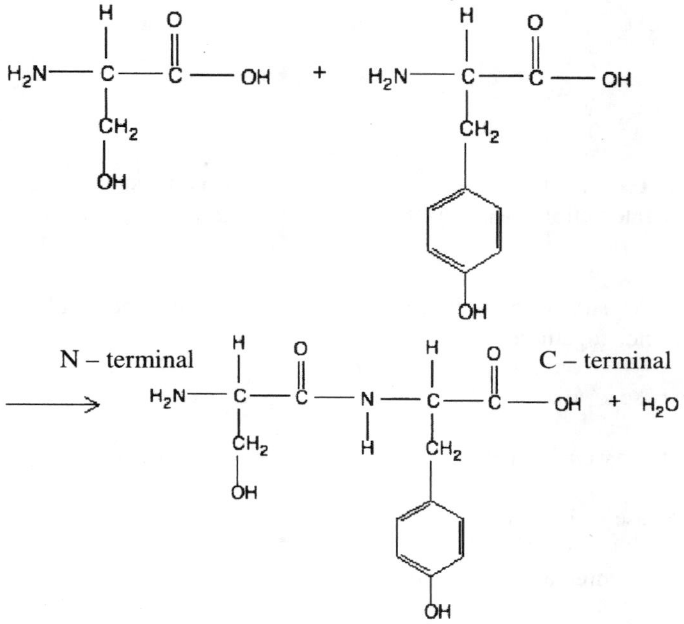

57. Peptide bonds are formed when the carboxylic end of one amino acid reacts with the amine end of another amino acid.

a) Gln – Met – Cys

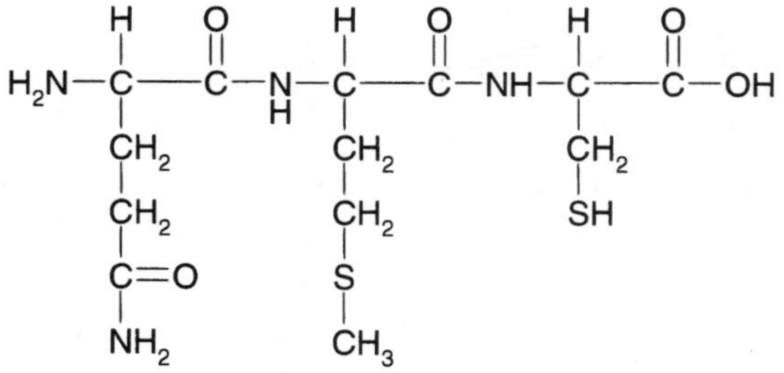

b) Ser – Leu – Cys

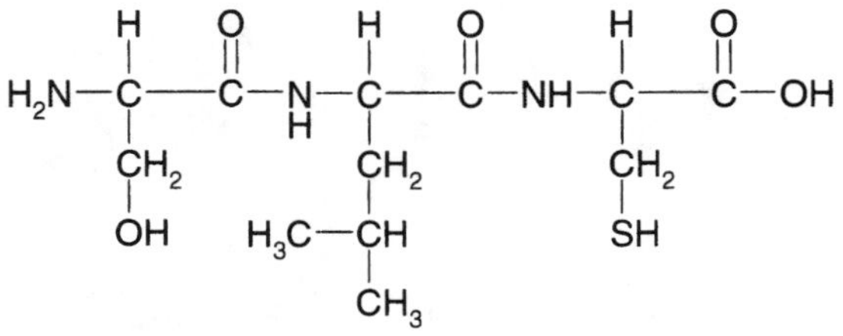

c) Cys – Leu – Ser

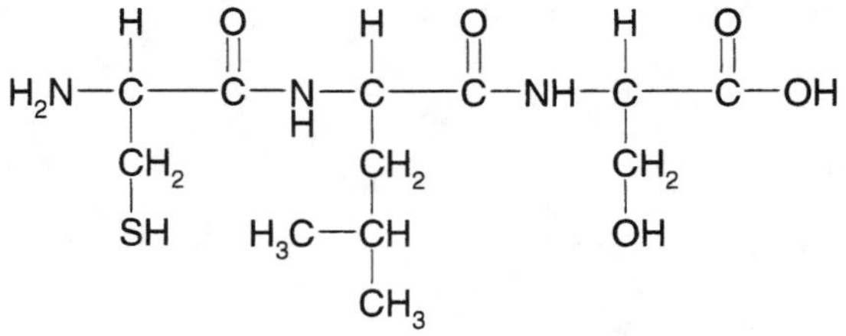

59. This would be a tertiary structure. Tertiary structure refers to the large – scale twists and folds of globular proteins. These are maintained by interactions between the R groups of amino acids that are separated by long distances in the chain sequence.

61. This would be a primary structure. Primary structure is simply the amino acid sequence. It is maintained by the peptide bonds that hold amino acids together.

63. a) Yes, it is a nucleotide and the base is adenine.

b) No, it is not a nucleotide, it is the base adenine.

c) Yes, it is a nucleotide, and the base is thymine.

d) No, it is not a nucleotide, it is the amino acid methionine.

65. The two purine bases are:

Adenine

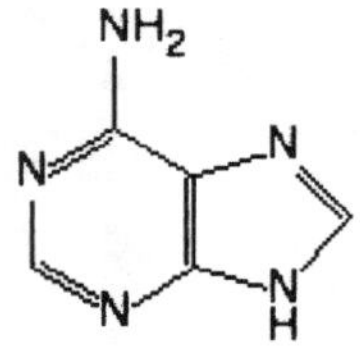

Guanine

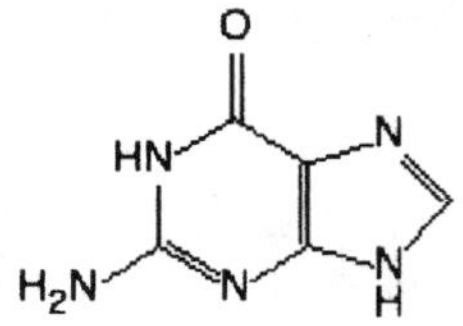

67. The bases pair in the specific arrangement: adenine always pairs with thymine and guanine always pairs with cytosine.

The sequence:

TGT ACGC

Would have the complementary sequence:

ACATG CG

69. Every amino acid has a specific codon that is comprised of three nucleotides. So, 154 amino acids, would need 154 codons and 462 nucleotides.

71. a) Peptide bonds link amino acids to form proteins.
    b) A glycosidic linkage is the connection between sugar molecules to form carbohydrates.
    c) An ester linkage forms between the glycol molecule and a fatty acid to form lipids.

73. A codon is a sequence of three nucleotides with their associated bases. A codon codes for one amino acid. A nucleotide is a combination of a sugar group, a phosphate group, and a base. While a gene is a sequence of codons that codes for a specific protein.

75. Write the Lewis structure:

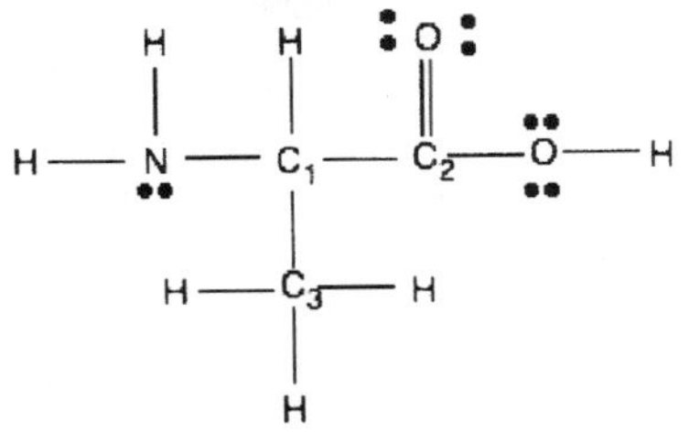

Determine the VSEPR geometry:

N has three bonding pair and one lone pair which is trigonal pyramidal molecular geometry.
C_1 has four bonding pair of electrons which gives tetrahedral molecular geometry.
C_2 has three bonding pair of electrons which gives trigonal planar molecular geometry.
C_3 has four bonding pair of electrons which gives tetrahedral molecular geometry.
O has two bonding pair of electrons and two lone pair which gives bent molecular geometry.

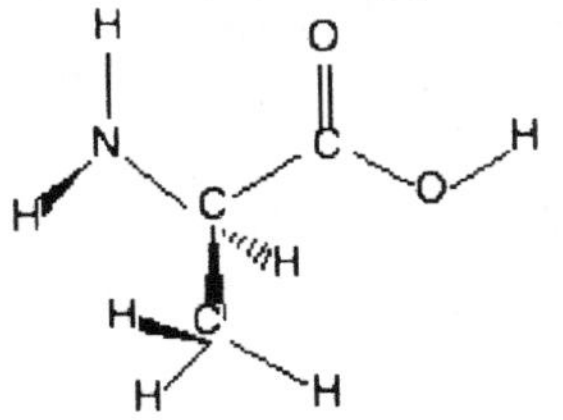

77. Amino acids which are most likely to be involved in hydrophobic interactions would be ones that have a non polar R group. These would be: valine, leucine, isolucine, phenylalanine.

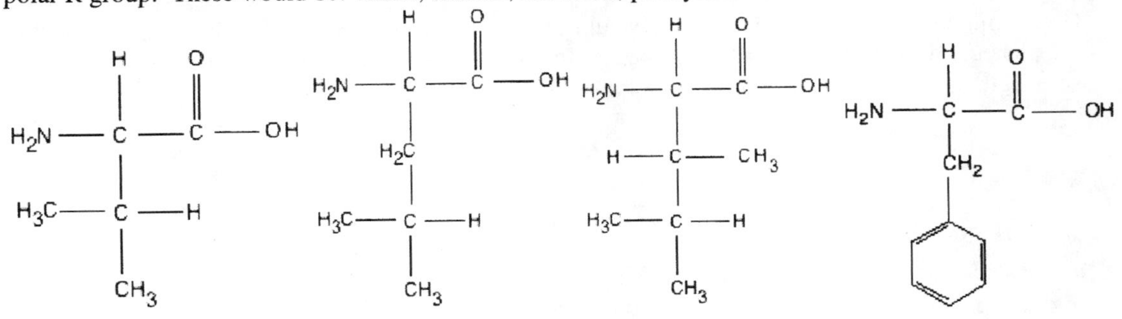

79. The reagents added break up the polypeptide but do not rearrange it. Therefore, by comparing the fragments from the two reagents we can deduce the structure of the polypeptide.

Reagent 1:	Ala – Leu – Phe – Gly – Asn – Lys	Trp – Glu – Cys	Gly – Arg
Reagent 2:	Gly – Arg – Ala – Leu – Phe	Gly – Asn – Lys – Trp	Glu – Cys
So: the sequence is:	Gly – Arg – Ala – Leu – Phe – Gly – Asn – Lys – Trp – Glu – Cys		

81. The DNA chain forms with a phosphate linkage between the C_5 carbon on one sugar and the C_3 carbon on the next sugar. When the fake thymine is introduced into the cell, DNA replication cannot occur because the N chain is attached at the C_3 position instead of the – OH group and this will keep future phosphate linkages from forming between the sugars and stop the replication.

83. First rearrange the Michaelis – Menton equation to the form of a straight line: y = mx + b

$$V_o = \frac{V_{max}[\text{glucose}]}{K_f + [\text{glucose}]} \qquad \frac{1}{V_o} = \frac{K_f + [\text{glucose}]}{V_{max}[\text{glucose}]} \qquad \frac{1}{V_o} = \frac{K_f}{V_{max}[\text{glucose}]} + \frac{\cancel{[\text{glucose}]}}{V_{max}\cancel{[\text{glucose}]}}$$

$$\frac{1}{V_o} = \frac{K_f}{V_{max}[\text{glucose}]} + \frac{1}{V_{max}} \qquad \text{So: } y = \frac{1}{V_o};\ x = \frac{1}{[\text{glucose}]};\ m = \frac{K_f}{V_{max}};\ b = \frac{1}{V_{max}}$$

[glucose]	Vo	1/[glucose]	1/Vo
0.5	12	2.0	0.083
1.0	19	1.0	0.053
2.0	27	0.50	0.037
3.0	32	0.33	0.031
4.0	35	0.25	0.029
		0	

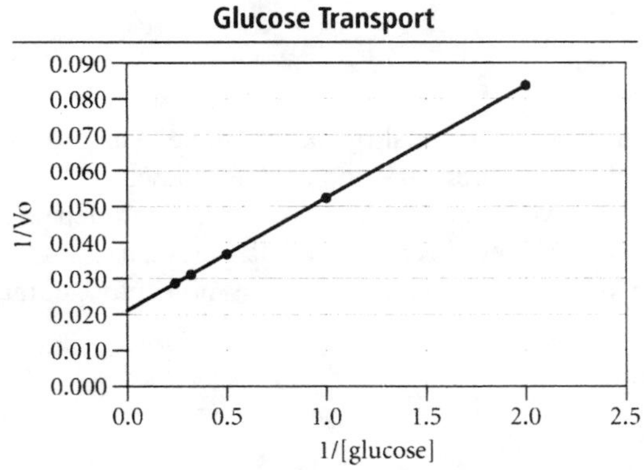

$$y = 0.021 = \frac{1}{V_{max}};\ m = 0.0353 = \frac{K_f}{V_{max}};\ \text{So } V_{max} = 47.6 \text{ and } K_f = 1.68$$

85. Four different amino acids could produce 4 factorial different tetrapeptides which is 24 different tetrapeptides.

87. A completely different genetic code would mean that there is no commonality between the life on another planet and the life on Earth. A genetic code that is similar would mean that there is commonality and the origin of the different lifeforms is the same.

Chapter 22
Chemistry of the Nonmetals

1. Boron nitride contains BN units that are isoelectronic with carbon in the sense that each BN unit contains 8 valence electrons, or 4 per atom (just like carbon). Also the size and electronegativity of the carbon atom is almost equal to the average of a boron atom and nitrogen atom. Therefore, BN forms a number of structures that are similar to those formed by carbon, including nanotubes.

3. The metallic character of the main-group elements increases going down each column due to the increasing size of the atoms.

5. Silica melts when heated above 1500 °C. After melting, if cooled quickly, silica does not crystallize back into the quartz structure. Instead, the Si atoms and O atoms form a randomly ordered or amorphous structure called a glass. A slow cooling will result in the crystalline quartz structure.

7. a) Orthosilicates (or nesosilicates) are minerals, where silicon bonds to oxygen in tetrahedral SiO_4^{4-} polyatomic anions (four electrons are gained, satisfying the octet rule for the oxygen atoms) that are isolated in the structure (not bonded to other tetrahedrons). The minerals require cations that have a total charge of 4+ to neutralize the negative charge.

   b) Amphiboles are minerals with double silicate chains and the repeating unit in the crystal is $Si_4O_{11}^{6-}$. Half of the tetrahedrons are bonded by two of the four corner O atoms, and half of the tetrahedrons are bonded by three of the four corners, bonding the two chains together. The bonding within the double chains is very strong, but the bonding between the double chains is not as strong, often resulting in fibrous-type minerals such as asbestos.

   c) Pyroxenes (or inosilicates) are minerals in which many of the silica tetrahedrons are bonded together forming chains. The formula unit for these chains is the SiO_3^{2-} unit, and the repeating unit in the structure is two formula units ($Si_2O_6^{4-}$). Two of the oxygen atoms are bonded to two silicon atoms (and thus to two other tetrahedrons) on two of the four corners of each tetrahedron. The silicate chains are held together by ionic bonding to metal cations that lie between the chains.

   d) Pyrosilicates (or sorosilicates) are minerals in which the silicate tetrahedrons can also form structures in which two tetrahedrons share one corner, forming the disilicate ion, which has the formula $Si_2O_7^{6-}$. This group requires cations that balance the 6- charge on $Si_2O_7^{6-}$.

   e) Feldspar is a common aluminosilicate where SiO_2 units becomes AlO_2^- upon substitution of aluminum. The negative charge is balanced by a positive counter ion. This leads to a covalent network of $AlSi_3O_8^-$ or $Al_2Si_2O_8^{2-}$ units.

9. Boron tends to form electron-deficient compounds (compounds in which boron lacks an octet) because it has only three valence electrons. Examples of these compounds are boron halides, such as BF_3 and BCl_3.

11. Solid carbon dioxide is referred to as dry ice because at 1 atmosphere it goes directly from a solid to a gas, without going through a liquid or wet phase.

13. About 21% of Earth's atmosphere is composed of O_2.

15. Earth's early atmosphere was reducing (instead of oxidizing) and contained hydrogen, methane, ammonia, and carbon dioxide. About 2.7 billion years ago, cyanobacteria (blue-green algae) began to convert the carbon dioxide and water to oxygen by photosynthesis.

17. a) +4. In SiO_2, (Si ox state) + 2(O ox state) = 0. Since O ox state = -2, Si ox state = - 2 (-2) = +4.

   b) +4. In SiO_4^{4-}, (Si ox state) + 4(O ox state) = - 4. Since O ox state = -2, Si ox state = - 4 - 4 (-2) = +4.

   c) +4. In $Si_2O_7^{6-}$, 2(Si ox state) + 7(O ox state) = - 6. Since O ox state = -2, Si ox state = ½ (- 6 - 7 (-2)) = +4.

19. Three SiO_4^{4-} units have a total charge of – 12. In trying to combine Ca^{2+} and Al^{3+} to get a total charge of + 12, the only possible combination is to use 3 Ca^{2+} and 2 Al^{3+} ions. The formula unit is $Ca_3Al_2(SiO_4)_3$.

21. Since OH has a -1 charge, Al has a +3 charge and Si_2O_5 has a -2 charge, 2(Al ox state) + 1(Si_2O_5 charge) + x (OH^- charge) = 0 so 2(+3) + 1(-2) + x(-1) = 0 and x = 4.

23. In Table 22.2, SiO_4 has a - 4 charge and it belongs to the class of orthosilicates, where the silicate tetrahedra stand alone, not linked to each other. The – 4 charge balances with a common oxidation state of Zr of +4.

25. Ca has a +2 charge, Mg has a +2 charge, Fe commonly has a charge of +2 or +3, and OH has a charge of – 1. So far the charge balance is 2(+2) + 4(+2) + 1(+2 or +3) + (Si_7AlO_{22} charge) + 2(-1) = 0 or Si_7AlO_{22} charge = - 12 or – 13. Looking at Table 22.2, $Si_4O_{11}^{6-}$ units are the best match. If we double the unit and replaced 1/8th of the Si's by Al, the new unit will be Si_7AlO_{22} and the charge will be 2(- 6) -1 = -13. The charge on Fe will be +3 and the class is amphibole, with a double chain structure.

27. **Given:** $Na_2[B_4O_5(OH)_4] \cdot 3\ H_2O$ = kernite, 1.0 x 10^3 kg of kernite-bearing ore, 0.98 % by mass kernite, 65 % B recovery **Find:** g boron recovered

Conceptual plan: kg ore → g ore → g kernite → g B possible → g B recovered

$$\frac{1000\ g}{1\ kg} \quad \frac{0.98\ g\ kernite}{100\ g\ ore} \quad \frac{4(10.81)\ g\ B}{291.23\ g\ kernite} \quad \frac{65\ g\ B\ recovered}{100\ g\ B}$$

Solution:

$$1.0 \times 10^3\ \cancel{kg\ ore} \times \frac{1000\ \cancel{g\ ore}}{1\ \cancel{kg\ ore}} \times \frac{0.98\ \cancel{g\ kernite}}{100\ \cancel{g\ ore}} \times \frac{4(10.81)\ \cancel{g\ B}}{291.23\ \cancel{g\ kernite}} \times \frac{65\ g\ B\ recovered}{100\ \cancel{g\ B}} = 950\ g\ B\ recovered$$

Check: The units (g) are correct. The magnitude of the answer makes physical sense because there is very little mineral in the ore, a low recovery and little B in the ore.

29. The bond angles are different in BCl_3 versus NCl_3 because there B has 3 valence electrons and N has 5 valence electrons. The boron compound does not obey the octet rule and B has sp^2 hybridization and a trigonal planar geometry. The nitrogen compound does obey the octet rule, has a lone pair of electrons and N has sp^3 hybridization and a trigonal pyramidal geometry (based on a tetrahedral geometry).

31. a) Looking at Figure 22.11, $B_6H_6^{2-}$ has 6 vertices and 8 faces, since it has an octahedral shape.

    b) $B_{12}H_{12}^{2-}$ has the formula of a *closo*-borane ($B_nH_n^{2-}$) has 12 vertices and 20 faces, since it has an icosohedral shape.

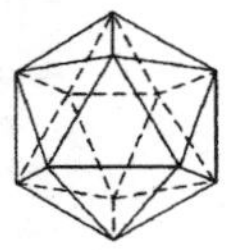

33. *Closo*-boranes have the formula $B_nH_n^{2-}$ and form fully closed polyhedra with triangular sides. Each of the vertices in the polyhedra is a boron atom with an attached hydrogen atom. *Nido*-boranes, named from the Latin word for *net*, have the formula B_nH_{n+4}. They consist of a cage of boron atoms missing one corner. *Arachno*-boranes, named from the Greek word for *web*, have the formula B_nH_{n+6}. They consist of a cage of boron atoms that is missing two or three corners.

35. Graphite is a good lubricant, because the structure has strong covalent bonds within the sheets of the structure, but weak interactions between the sheets. The diamond structure consists of carbon atoms connected to four other carbon atoms at the corners of a tetrahedron. This bonding extends throughout three dimensions, making giant molecules described as network covalent solids. There are no weak interactions in the diamond structure.

37. Regular charcoal is in large chunks that still resemble the wood from which it is made. Activated charcoal is treated with steam to break up the chunks into a finely divided powder with a high surface area.

39. Ionic carbides are composed of carbon and low-electronegativity metals such as the alkali metals and alkaline earth metals. Most of the ionic carbides contain the dicarbide ion, C_2^{2-}, commonly called the acetylide ion. Covalent carbides are composed of carbon and low-electronegativity nonmetals or metalloids. The most important covalent carbide is silicon carbide (SiC) a very hard material.

41. a) As the pressure is reduced, the solid will be converted directly to a gas.

b) As the temperature is reduced, the gas will be converted to a liquid and then to a solid.

c) As the temperature is increased, the solid will be converted directly to a gas.

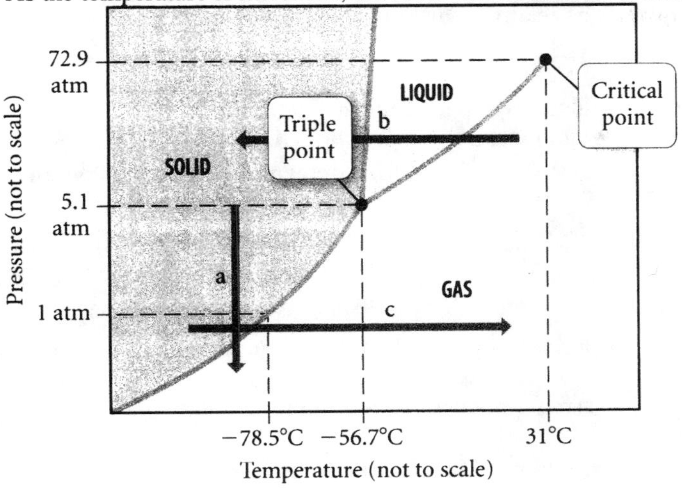

43. a) The reaction will be similar to that of iron oxide and carbon monoxide: $CO\,(g) + CuO\,(s) \xrightarrow{\text{Heat}} CO_2\,(g) + Cu\,(s)$. Note that the oxidation state of carbon increases by 2 and the oxidation state of Cu decreases by 2.

b) This reaction is discussed in section 22.5, $SiO_2\,(s) + 3\,C\,(s) \xrightarrow{\text{Heat}} SiC\,(s) + 2\,CO\,(g)$.

c) This reaction is discussed in section 22.5, $CO\,(g) + S\,(s) \xrightarrow{\text{Heat}} COS\,(g)$.

45. a) +2. In CO, (C ox state) + (O ox state) = 0. Since O ox state = -2, C ox state = +2.

b) +4. In CO_2, (C ox state) + 2(O ox state) = 0. Since O ox state = -2, C ox state = - 2 (-2) = +4.

c) +4/3. In C_3O_2, 3(C ox state) + 2(O ox state) = 0. Since O ox state = -2, C ox state = 1/3 (- 2 (-2)) = +4/3.

47. Even though there is a tremendous amount of N_2 gas in the atmosphere, the strong triple bond in elemental nitrogen renders it unusable by plants. In order to be used as fertilizer, elemental nitrogen has to be fixed, which means that it has to be converted into a nitrogen-containing compound such as NH_3.

49. White phosphorus consists of P_4 molecules in a tetrahedral shape, with the phosphorus atoms at the corners of the tetrahedron as shown in Figure 22.19. The bond angles between the three P atoms on any one face of the tetrahedron is small (60°) and strained, making the P_4 molecule unstable and reactive. When heated to about 300 °C in the absence of air, white phosphorus slightly changes its structure to a different allotrope called red phosphorus, which is amorphous. The general structure of red phosphorus is similar to that of white phosphorus, except that one of the bonds between two phosphorus atoms in the tetrahedron is broken, as shown in Figure 22.20. These two phosphorus atoms then link to other phosphorus atoms making chains that can vary in structure. The red phosphorous structure is more stable because there is less strain when the linkage is changed.

51. **Given:** saltpeter (KNO_3) and Chile saltpeter ($NaNO_3$) **Find:** mass percent N in each mineral
Conceptual plan: mineral formula → mass percent nitrogen

$$\text{mass percent nitrogen} = \frac{14.01\text{ g}}{\text{formula weight of mineral}} \times 100\ \%$$

Solution:
for KNO_3:

$$\text{mass percent nitrogen} = \frac{14.01\text{ g}}{\text{formula weight of mineral}} \times 100\ \% = \frac{14.01\ \cancel{\text{g}}}{101.11\ \cancel{\text{g}}} \times 100\ \% = 13.86\ \%$$

and for $NaNO_3$:

$$\text{mass percent nitrogen} = \frac{14.01 \text{ g}}{\text{formula weight of mineral}} \times 100\ \% = \frac{14.01 \text{ g}}{85.00 \text{ g}} \times 100\ \% = 16.48\ \%.$$

Check: The units (%) are correct. The relative magnitudes of the answers make physical sense because K is heavier than Na.

53. **Given:** decomposition of hydrogen azide (HN_3) to its elements
Find: stability at room temperature, stability at any temperature **Other:** Appendix IIB
Conceptual plan: Write reaction. Evaluate $\Delta G°_{rxn}$ then determine if $\Delta G_{rxn} > 0$ at any T.

$$\Delta G = \Delta H_{rxn} - T\Delta S_{rxn}$$

Solution: $HN_3\ (g) \rightarrow 3/2\ N_2\ (g) + 1/2\ H_2\ (g)$, at 25 °C $\Delta G°_{rxn} = -\Delta G°_f\ (HN_3\ (g)) = -1422$ kJ/mol so the decomposition is spontaneous at room temperature. $\Delta H°_{rxn} = -\Delta H°_f\ (HN_3\ (g)) = -1447$ kJ/mol, and $\Delta S°_{rxn} > 0$ since the decomposition generates more moles of gas. Since $\Delta G = \Delta H_{rxn} - T\Delta S_{rxn}$ we see that both the enthalpy and the entropy term favor decomposition and so ΔG_{rxn} is always negative and hydrogen azide is unstable at all temperature.
Check: Azides are used as explosives, so it is not surprising that they are unstable.

55. a) This reaction is a decomposition reaction, $NH_4NO_3\ (aq)$ + heat $\rightarrow$ $N_2O\ (g) + 2\ H_2O\ (l)$.

b) This reaction is discussed in section 22.6, $3\ NO_2\ (g) + H_2O\ (l) \rightarrow 2\ HNO_3\ (aq) + NO\ (g)$.

c) This reaction is discussed in section 22.6, $2\ PCl_3\ (g) + O_2\ (g) \rightarrow 2\ POCl_3\ (l)$.

57. For N_3^-: 3(N ox state) = - 1, so N ox state = - 1/3.
For $N_2H_5^+$: 2(N ox state) + 5(H ox state) = +1. Since H ox state = +1, N ox state = ½ (1 - 5(+1)) = - 2.
For NO_3^-: 1(N ox state) + 3(O ox state) = -1. Since O ox state = - 2, N ox state = - 1 - 3(-2) = +5.
For NH_4^+: 1(N ox state) + 4(H ox state) = +1. Since H ox state = +1, N ox state = +1 - 4(+1) = - 3.
For NO_2^-: 1(N ox state) + 2(O ox state) = -1. Since O ox state = - 2, N ox state = - 1 - 2(-2) = +3.
So $NO_3^- > NO_2^- > N_3^- > N_2H_5^+ > NH_4^+$.

59. P has 5 valence electrons and Cl has 7 valence electrons and so PCl_3 has 5 + 3(7) = 26 valence electrons and PCl_5 has 5 + 5(7) = 40 valence electrons. Since the electronegativity of P is less than the electronegativity of Cl, the P is in the middle. So the Lewis Dot structures are :

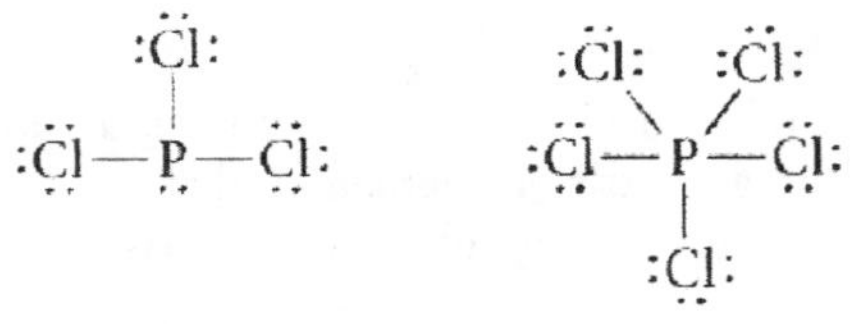

Notice that the PCl_5 violates the octet rule, which is allowed since P is in the 3rd row of the periodic table and Cl is a fairly small atom.
Since the PCl_3 has four electron groups surrounding the P atom, the orbital geometry is tetrahedral. Since there is one lone pair the VSEPR geometry is trigonal pyramidal.
Since the PCl_5 has five electron groups surrounding the P atom, the VSEPR geometry is trigonal bipyramidal.

61. **Given:** urea ($CO(NH_2)_2$) react with water to produce 23 g ammonium carbonate **Find:** g of urea needed
Conceptual Plan:
Write a balanced equation. g $(NH_4)_2CO_3$ → mol $(NH_4)_2CO_3$ → mol $CO(NH_2)_2$ → g $CO(NH_2)_2$

$$\frac{1 \text{ mol } (NH_4)_2CO_3}{96.04 \text{ g } (NH_4)_2CO_3} \qquad \frac{1 \text{ mol } CO(NH_2)_2}{1 \text{ mol } (NH_4)_2CO_3} \qquad \frac{60.03 \text{ g } CO(NH_2)_2}{1 \text{ mol } CO(NH_2)_2}$$

Solution: $CO(NH_2)_2\ (aq) + 2\ H_2O\ (l) \rightarrow (NH_4)_2CO_3\ (aq)$ then

$$23 \text{ g } (NH_4)_2CO_3 \times \frac{1 \text{ mol } (NH_4)_2CO_3}{96.04 \text{ g } (NH_4)_2CO_3} \times \frac{1 \text{ mol } CO(NH_2)_2}{1 \text{ mol } (NH_4)_2CO_3} \times \frac{60.03 \text{ g } CO(NH_2)_2}{1 \text{ mol } CO(NH_2)_2} = 14 \text{ g } CO(NH_2)_2$$

Check: The units (g) are correct. The magnitude of the answer (14) makes physical sense because we are generating 23 g of $(NH_4)_2CO_3$. Since we are adding water to the urea, the amount of urea must be less than 23 g.

63. Tetraphosphorus hexaoxide, $P_4O_6\ (s)$, forms when the amount of oxygen is limited, and tetraphosphorus decaoxide, $P_4O_{10}\ (s)$, is formed when greater amounts of oxygen are available.

65. The major commercial production method of elemental oxygen is the fractionation of air. Air is cooled until its components liquefy. Then the air is warmed, and components such as N_2 and Ar are separated, leaving oxygen behind.

67. In oxides the oxidation state of O is -2, in peroxides it is -1 and in superoxides it is – ½. Alkali metals and alkaline earth metals have only two possible oxidation states 0 and +1 for the alkali metals and +2 for alkaline earth metals.
   a) LiO_2 is a superoxide, since 1(Li ox state) + 2(O ox state) = 0 and O ox state = - ½(1) = - 1/2.

   b) CaO is an oxide, since 1(Ca ox state) + 2(O ox state) = 0 and O ox state = - 1(+2) = - 2.

   c) K_2O_2 is a peroxide, since 2(K ox state) + 2(O ox state) = 0 and O ox state = - ½(2(+1)) = - 1.

69. Initially, liquid sulfur becomes less viscous when heated because the S_8 rings have greater thermal energy which overcomes intermolecular forces. Above 150 °C the rings break and the broken rings entangle one another, causing greater viscosity. Above 180 °C the intermolecular forces weaken and the solution becomes less viscous.

71. **Given:** 1.0 L of 5.00×10^{-5} M Na_2S **Find:** maximum g of metal sulfide that can dissolve
Other: K_{sp} (PbS) = 9.04×10^{-29}, K_{sp} (PbS) = 2×10^{-25}
Conceptual Plan: M Na_2S → M S^{2-} → M metal sulfide → mol metal sulfide → g metal sulfide

$Na_2S\ (s) \rightarrow 2\ Na^+(aq) + S^{2-}\ (aq)$ ICE Chart $M = \frac{mol}{L}$ $\mathfrak{M}$

Solution: Since 1 S^{2-} ion is generated for each Na_2S, $[S^{2-}] = 5.00 \times 10^{-5}$ M Na_2S.

a)

	$PbS(s) \rightleftharpoons$	$Pb^{2+}\ (aq)$ +	$S^{2-}\ (aq)$
Initial		0.00	5.00×10^{-5}
Change		*S*	*S*
Equil		*S*	$5.00 \times 10^{-5} + S$

$K_{sp}\ (PbS) = [Pb^{2+}]\ [S^{2-}] = 9.04 \times 10^{-29} = S\ (5.00 \times 10^{-5} + S)$.

Since S << 5.00×10^{-5}, $9.04 \times 10^{-29} = S\ (5.00 \times 10^{-5})$, and $S = 1.81 \times 10^{-24}$ M. (Note that assumption that S was very small was good.) Since there is 1.0 L of solution we can dissolve at most 1.8×10^{-24} moles of PbS then $1.8 \times 10^{-24}\ \text{mol PbS} \times \frac{239.3\ \text{g PbS}}{1\ \text{mol PbS}} = 4.3 \times 10^{-22}$ g PbS.

b)

	$ZnS(s) \rightleftharpoons$	$Zn^{2+}\ (aq)$ +	$S^{2-}\ (aq)$
Initial		0.00	5.00×10^{-5}
Change		*S*	*S*
Equil		*S*	$5.00 \times 10^{-5} + S$

$K_{sp}\ (ZnS) = [Zn^{2+}]\ [S^{2-}] = 2 \times 10^{-25} = S\ (5.00 \times 10^{-5} + S)$.

Since S << 5.00×10^{-5}, $2 \times 10^{-25} = S\ (5.00 \times 10^{-5})$, and $S = 4 \times 10^{-21}$ M. (Note that assumption that S was very small was good.) Since there is 1.0 L of solution we can dissolve at most 4×10^{-21} moles of ZnS then $4 \times 10^{-21}\ \text{mol ZnS} \times \frac{97.48\ \text{g ZnS}}{1\ \text{mol ZnS}} = 4 \times 10^{-19}$ g ZnS.

Check: The units (g) are correct. The magnitude of the answer makes physical sense because the solubility constants are so small. The amount of PBS that can dissolve is less than the amount of ZnS because the solubility constant is much smaller.

73. **Given:** iron pyrite (FeS_2) roasted generating $S_2\ (g)$, 5.5 kg of iron pyrite
Find: balanced reaction and V S_2 at STP
Conceptual plan: Write reaction then kg FeS_2 → g FeS_2 → mol FeS_2 → mol S_2 → V S_2

$\frac{1000\ g}{1\ kg}$ $\frac{1\ mol\ FeS_2}{119.99\ g\ FeS_2}$ $\frac{1\ mol\ S_2}{2\ mol\ FeS_2}$ $\frac{22.414\ L\ S_2}{1\ mol\ S_2}$

Solution: $2\ FeS_2\ (s) \xrightarrow{\text{Heat}} 2\ FeS\ (s) + S_2\ (g)$, then

$$5.5\ \text{kg FeS}_2 \times \frac{1000\ \text{g FeS}_2}{1\ \text{kg FeS}_2} \times \frac{1\ \text{mol FeS}_2}{119.99\ \text{g FeS}_2} \times \frac{1\ \text{mol S}_2}{2\ \text{mol FeS}_2} \times \frac{22.414\ \text{L S}_2}{1\ \text{mol S}_2} = 510\ \text{L S}_2 .$$

Check: The units (L) are correct. The magnitude of the answer makes physical sense because there is much more than a mole of iron pyrite, so the volume of gas is many time 22 L.

75. a) + 2. In XeF_2, 1(Xe ox state) + 2(F ox state) = 0. Since F ox state = - 1, Xe ox state = - 2(- 1) = + 2. Xe has 8 valence electrons and F has 7 valence electrons and so XeF_2 has 8 + 2(7) = 22. Since the electronegativity of Xe is less than the electronegativity of F, the Xe is in the middle. So the Lewis Dot structure is : 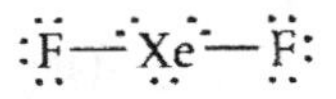

Notice that the XeF_2 violates the octet rule, which is allowed since Xe is in the 5th row of the periodic table and F is a fairly small atom. Since there are five electron groups surrounding the Xe atom, the orbital geometry is trigonal bipyramidal. Since there are three lone pairs the VSEPR geometry is linear.

b) +6. In XeF_6, 1(Xe ox state) + 6(F ox state) = 0. Since F ox state = - 1, Xe ox state = - 6(- 1) = + 6. Xe has 8 valence electrons and F has 7 valence electrons and so XeF_6 has 8 + 6(7) = 50. Since the electronegativity of Xe is less than the electronegativity of F, the Xe is in the middle. So the Lewis Dot structure is shown to the right. Notice that the XeF_6 violates the octet rule, which is allowed since Xe is in the 5th row of the periodic table and F is a fairly small atom. Since there are six electron groups surrounding the Xe atom, the VSEPR geometry is octahedral.

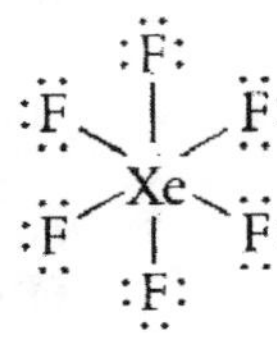

c) +6. In $XeOF_4$, 1(Xe ox state) + 1(O ox state) + 4(F ox state) = 0. Since F ox state = - 1, and O ox state = -2, Xe ox state = - (1(-2) + 4(- 1) = + 6. Xe has 8 valence electrons, O has 6 valence electrons, and F has 7 valence electrons and so $XeOF_4$ has 8 + 1(6) + 4(7) = 42. Since the electronegativity of Xe is less than the electronegativity of F, the Xe is in the middle. Oxygen needs a double bond. So the Lewis Dot structure is shown to the right. Notice that the $XeOF_4$ violates the octet rule, which is allowed since Xe is in the 5th row of the periodic table and O and F are fairly small atoms. Since there are six electron groups surrounding the Xe atom, the orbital geometry is octahedral. Since there is one lone pair the VSEPR geometry is square pyramidal.

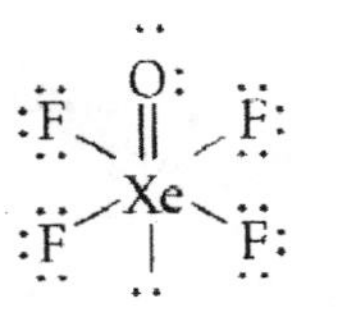

77. The red color is from molecular bromine (Br_2), so $Cl_2\ (g) + 2\ Br^-\ (aq) \rightarrow 2\ Cl^-\ (aq) + Br_2\ (l)$. Cl is being reduced and Br is being oxidized, so is the Cl_2 oxidizing agent and Br^- is the reducing agent.

79. **Given:** 55 g SiO_2 and 111 L of 0.032 M HF **Find:** limiting reagent and amount of excess reagent left
Conceptual plan: Write reaction then M HF, L HF → mol HF → mol SiO_2 → g SiO_2 then

$$M = \frac{\text{mol}}{\text{L}} \qquad \frac{1\ \text{mol SiO}_2}{4\ \text{mol HF}} \qquad \frac{60.09\ \text{g SiO}_2}{1\ \text{mol SiO}_2}$$

determine limiting reagent then determine amount of excess reagent left

amount of excess reacent = inital amount - amount reacted

Solution: $4\ HF\ (g) + SiO_2\ (s) \rightarrow SiF_4\ (g) + 2\ H_2O\ (l)$ then

$$\frac{0.032\ \text{mol HF}}{1\ \text{L}} \times 111\ \text{L HF} = 3.552\ \text{mol HF then}$$

$$3.552\ \text{mol HF} \times \frac{1\ \text{mol SiO}_2}{4\ \text{mol HF}} \times \frac{60.09\ \text{g SiO}_2}{1\ \text{mol SiO}_2} = 53.3599\ \text{g SiO}_2 .$$ Since 55 g of SiO_2 is available, HF is the limiting reagent. The amount of excess SiO_2 is $55\ g - 53.3599\ g = 2\ g\ \text{SiO}_2$.

Check: The units (g) are correct. The magnitude of the answer makes physical sense because the amount required to react all of the HF is just under the amount of glass available.

81. **Given:** lignite coal 1 mol % S and bituminous coal 5 mol, % S, 1.00 x 10^2 kg of coal
Find: g H_2SO_4 produced

Conceptual plan: Table 22.3 Composition → $\mathfrak{M}$ then

$$\mathfrak{M} = \frac{(\text{mol \% C})(12.01\text{ g/mol}) + (\text{mol \% H})(1.001\text{ g/mol}) + (\text{mol \% O})(16.00\text{ g/mol}) + (\text{mol \% S})(32.07\text{g/mol})}{100\ \%}$$

kg coal → g coal → mol coal → mol S → mol H_2SO_4 → g H_2SO_4

$$\frac{1000\text{ g}}{1\text{ kg}} \qquad \mathfrak{M} \qquad \frac{1\text{ or }5\text{ mol S}}{100\text{ mol coal}} \qquad \frac{1\text{ mol } H_2SO_4}{1\text{ mol S}} \qquad \frac{98.07\text{ g } H_2SO_4}{1\text{ mol } H_2SO_4}$$

Solution: lignite:

$$\mathfrak{M} = \frac{(\text{mol \% C})(12.01\text{ g/mol}) + (\text{mol \% H})(1.001\text{ g/mol}) + (\text{mol \% O})(16.00\text{ g/mol}) + (\text{mol \% S})(32.07\text{g/mol})}{100\ \%} =$$

$$= \frac{(71\text{ mol \% C})(12.01\text{ g/mol}) + (4\text{ mol \% H})(1.001\text{ g/mol}) + (23\text{ mol \% O})(16.00\text{ g/mol}) + (1\text{ mol \% S})(32.07\text{g/mol})}{100\ \%} =$$

$= 1\underline{2}.5678$ g/mol

$$1.00 \times 10^2 \cancel{\text{kg coal}} \times \frac{1000\ \cancel{\text{g coal}}}{1\ \cancel{\text{kg coal}}} \times \frac{1\ \cancel{\text{mol coal}}}{1\underline{2}.5678\ \cancel{\text{g coal}}} \times \frac{1\ \cancel{\text{mol S}}}{100\ \cancel{\text{mol coal}}} \times \frac{1\ \cancel{\text{mol } H_2SO_4}}{1\ \cancel{\text{mol S}}} \times \frac{98.07\text{ g } H_2SO_4}{1\ \cancel{\text{mol } H_2SO_4}} = .$$

$= 7.803 \times 10^3$ g H_2SO_4 = 8 kg H_2SO_4

bitumenous:

$$\mathfrak{M} = \frac{(\text{mol \% C})(12.01\text{ g/mol}) + (\text{mol \% H})(1.001\text{ g/mol}) + (\text{mol \% O})(16.00\text{ g/mol}) + (\text{mol \% S})(32.07\text{g/mol})}{100\ \%} =$$

$$= \frac{(80\text{ mol \% C})(12.01\text{ g/mol}) + (6\text{ mol \% H})(1.001\text{ g/mol}) + (8\text{ mol \% O})(16.00\text{ g/mol}) + (5\text{ mol \% S})(32.07\text{g/mol})}{100\ \%} =$$

$= 1\underline{2}.55156$ g/mol

$$1.00 \times 10^2 \cancel{\text{kg coal}} \times \frac{1000\ \cancel{\text{g coal}}}{1\ \cancel{\text{kg coal}}} \times \frac{1\ \cancel{\text{mol coal}}}{1\underline{2}.55156\ \cancel{\text{g coal}}} \times \frac{5\ \cancel{\text{mol S}}}{100\ \cancel{\text{mol coal}}} \times \frac{1\ \cancel{\text{mol } H_2SO_4}}{1\ \cancel{\text{mol S}}} \times \frac{98.07\text{ g } H_2SO_4}{1\ \cancel{\text{mol } H_2SO_4}} = .$$

$= 3.9067 \times 10^4$ g H_2SO_4 = 40 kg H_2SO_4

Check: The units (kg and kg) are correct. The magnitude of the answer makes physical sense because sulfuric acid is heavier than sulfur and it more than balances out the sulfur content in the coal.

83. The strength of an acid is determined by the strength of the bond of the H to the rest of the molecule – the weaker the bond, the stronger the acid. Since Cl is more electronegative than I, it pulls density away from the O more than I does and so the O can not interact with the H as strongly.

85. **Given:** HCl versus X = a) Cl_2, b) HF, and c) HI **Find:** ratio of effusion rates HCl / X

Conceptual plan: $\mathfrak{M}$ **(HCl) ,** $\mathfrak{M}$ **(X) → Rate (HCl) / Rate (X)**

$$\frac{Rate(HCl)}{Rate(X)} = \sqrt{\frac{\mathfrak{M}(X)}{\mathfrak{M}(HCl)}}$$

Solution: HCl: $\mathfrak{M} = \frac{36.45\ \cancel{\text{g}}}{1\text{ mol}} \times \frac{1\text{ kg}}{1000\ \cancel{\text{g}}} = 0.03645$ kg/mol,

a) Cl_2: $\mathfrak{M} = \frac{70.90\ \cancel{\text{g}}}{1\text{ mol}} \times \frac{1\text{ kg}}{1000\ \cancel{\text{g}}} = 0.07090$ kg/mol,

$$\frac{Rate(HCl)}{Rate(Cl_2)} = \sqrt{\frac{\mathfrak{M}(Cl_2)}{\mathfrak{M}(HCl)}} = \sqrt{\frac{0.07090\ \cancel{\text{kg / mol}}}{0.03645\ \cancel{\text{kg / mol}}}} = 1.395.$$

b) HF: $\mathfrak{M} = \frac{20.00\ \cancel{\text{g}}}{1\text{ mol}} \times \frac{1\text{ kg}}{1000\ \cancel{\text{g}}} = 0.02000$ kg/mol,

$$\frac{Rate(HCl)}{Rate(HF)} = \sqrt{\frac{\mathfrak{M}(HF)}{\mathfrak{M}(HCl)}} = \sqrt{\frac{0.02000\ \cancel{\text{kg / mol}}}{0.03645\ \cancel{\text{kg / mol}}}} = 0.7407.$$

b) HI: $\mathfrak{M} = \frac{127.90 \text{ g}}{1 \text{ mol}} \times \frac{1 \text{ kg}}{1000 \text{ g}} = 0.012790 \text{ kg/mol}$,

$$\frac{Rate(HCl)}{Rate(HI)} = \sqrt{\frac{\mathfrak{M}(HI)}{\mathfrak{M}(HCl)}} = \sqrt{\frac{0.012790 \text{ kg/mol}}{0.03645 \text{ kg/mol}}} = 0.5924.$$

Check: The units (none) are correct. The magnitudes of the answers make sense because; the heavier molecules have the lower effusion rate (because they move slower).

87. Sodium peroxide is Na_2O_2, so the reaction is 4 Na_2O_2 + 3 Fe → 4 Na_2O + Fe_3O_4. Note that the oxidation state of O is decreasing from -1 to -2, needing 8 electrons and the oxidation state of Fe is increasing from 0 to + 8/3 generating 8 electrons.

89. **Given:** O_2^+, O_2, O_2^-, O_2^{2-} and MO Theory **Find:** Explain bond lengths and state which are diamagnetic
Conceptual Plan: Use diagram from chapter 10 and fill with appropriate number of electrons to determine bond order and if diamagnetic

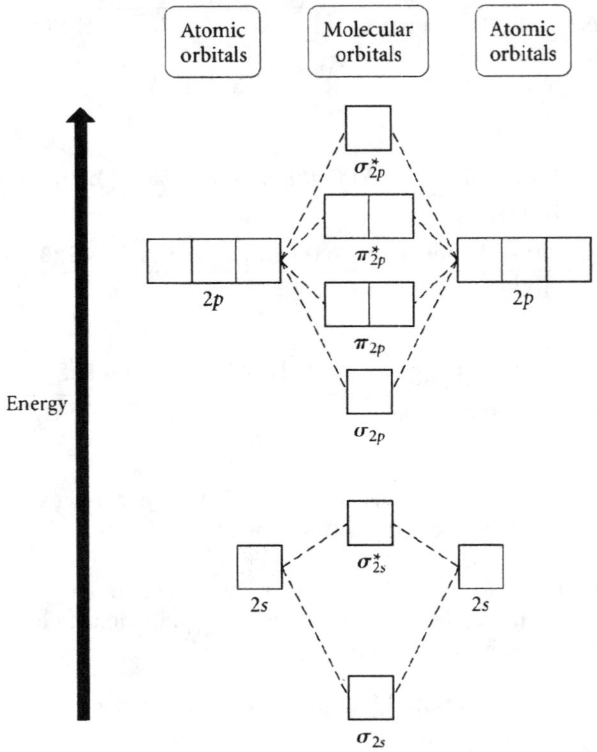

O_2, F_2, Ne_2

Solution: O_2^+ has 11 electrons in n = 2 shell and so its MO configuration σ_{2s}^2, σ_{2s}^{*2}, σ_{2p}^2, π_{2p}^4, π_{2p}^{*1}, the bond order =

$$\frac{\text{bonding electrons - antibonding electrons}}{2} = \frac{8-3}{2} = 2.5$$

and it is paramagnetic (unpaired electrons).
O_2 has 12 electrons in n = 2 shell and so its MO configuration σ_{2s}^2, σ_{2s}^{*2}, σ_{2p}^2, π_{2p}^4, π_{2p}^{*2}, the bond order =

$$\frac{\text{bonding electrons - antibonding electrons}}{2} = \frac{8-4}{2} = 2$$

and it is paramagnetic (unpaired electrons).
O_2^- has 13 electrons in n = 2 shell and so its MO configuration σ_{2s}^2, σ_{2s}^{*2}, σ_{2p}^2, π_{2p}^4, π_{2p}^{*3}, the bond order =

$$\frac{\text{bonding electrons - antibonding electrons}}{2} = \frac{8-5}{2} = 1.5$$

and it is paramagnetic (unpaired electrons).
O_2^{2-} has 14 electrons in n = 2 shell and so its MO configuration σ_{2s}^2, σ_{2s}^{*2}, σ_{2p}^2, π_{2p}^4, π_{2p}^{*4}, the bond order =

$$\frac{\text{bonding electrons - antibonding electrons}}{2} = \frac{8-6}{2} = 1$$

and it is diamagnetic (all paired electrons).
The bond order is decreasing because the electrons are being added to an antibonding orbital. As expected, the higher the bond order, the shorter the bond.
Check: The higher the bond order, the more electron orbital overlap and the shorter and stronger the bond. Molecular oxygen is paramagnetic and because of the number of π_{2p} and π_{2p}^* orbitals, many configurations are paramagnetic.

91. **Given:** iron oxides + CO *(g)* → Fe *(s)* + CO_2 *(g)*, where a) Fe_3O_4, b) FeO, and c) Fe_2O_3 **Find:** ΔH°_{rxn}
Conceptual plan: Write reaction. then $\Delta H^\circ_{rxn} = \sum n_P \Delta H^\circ_f(products) - \sum n_R \Delta H^\circ_f(reactants)$
Solution: a) Fe_3O_4 *(s)* + 4 CO *(g)* → 3 Fe *(s)* + 4 CO_2 *(g)*

Reactant/Product	ΔH^0_f (kJ/mol from Appendix IIB)
Fe_3O_4 *(s)*	− 1118.4
CO *(g)*	−110.5
Fe *(s)*	0.0
CO_2 *(g)*	− 393.5

Be sure to pull data for the correct formula and phase.

$$\Delta H^0_{rxn} = \sum n_P \Delta H^0_f(products) - \sum n_R \Delta H^0_f(reactants)$$

$$= [3(\Delta H^0_f(\text{Fe }(s)) + 4(\Delta H^0_f(CO_2 (g))] - [1(\Delta H^0_f(Fe_3O_4 (s)) + 4(\Delta H^0_f(CO (g))]$$

$$= [3(0.0 \text{ kJ}) + 4(-393.5 \text{ kJ})] - [1(-1118.4 \text{ kJ}) + 4(-110.5 \text{ kJ})]$$

$$= [-1574.0 \text{ kJ}] - [-1560.4 \text{ kJ}]$$

$$= -13.6 \text{ kJ}$$

b) $FeO (s) + CO (g) \rightarrow Fe (s) + CO_2 (g)$

Reactant/Product	ΔH^0_f (kJ/mol from Appendix IIB)
FeO (s)	-272.0
CO (g)	-110.5
Fe (s)	0.0
CO$_2$ (g)	-393.5

Be sure to pull data for the correct formula and phase.

$$\Delta H^0_{rxn} = \sum n_P \Delta H^0_f(products) - \sum n_R \Delta H^0_f(reactants)$$

$$= [1(\Delta H^0_f(\text{Fe }(s)) + 1(\Delta H^0_f(CO_2 (g))] - [1(\Delta H^0_f(FeO (s)) + 1(\Delta H^0_f(CO (g))]$$

$$= [1(0.0 \text{ kJ}) + 1(-393.5 \text{ kJ})] - [1(-272.0 \text{ kJ}) + 1(-110.5 \text{ kJ})]$$

$$= [-393.5 \text{ kJ}] - [-382.5 \text{ kJ}]$$

$$= -11.0 \text{ kJ}$$

c) $Fe_2O_3 (s) + 3 CO (g) \rightarrow 2 Fe (s) + 3 CO_2 (g)$

Reactant/Product	ΔH^0_f (kJ/mol from Appendix IIB)
Fe$_2$O$_3$ (s)	-824.2
CO (g)	-110.5
Fe (s)	0.0
CO$_2$ (g)	-393.5

Be sure to pull data for the correct formula and phase.

$$\Delta H^0_{rxn} = \sum n_P \Delta H^0_f(products) - \sum n_R \Delta H^0_f(reactants)$$

$$= [2(\Delta H^0_f(\text{Fe }(s)) + 3(\Delta H^0_f(CO_2 (g))] - [1(\Delta H^0_f(Fe_2O_3 (s)) + 3(\Delta H^0_f(CO (g))]$$

$$= [2(0.0 \text{ kJ}) + 3(-393.5 \text{ kJ})] - [1(-824.2 \text{ kJ}) + 3(-110.5 \text{ kJ})]$$

$$= [-1180.5 \text{ kJ}] - [-1155.7 \text{ kJ}]$$

$$= -24.8 \text{ kJ}$$

Fe$_2$O$_3$ is the most exothermic because it has the highest oxidation state and therefore oxidizes the most CO per mole of Fe.

Check: The units (kJ) are correct. The answers are negative, which means that the reactions are exothermic.

93. a) This is a linear molecule. Each C has 4 valence electrons and each O has 6 valence electrons, so C_3O_2 has $3(4) + 2(6) = 24$ valence electrons. All of the bonds are double bonds so the Lewis structure is:
$\ddot{O} = C = C = C = \ddot{O}$.

b) Since each carbon has two electron groups and needs to make 2 pi bonds, the hybridization is sp.

c) $\ddot{O} = C = C = C = \ddot{O} + 2\ H - \ddot{O} - H \rightarrow$ H$-$O$-$C$-$C$-$C$-$O$-$H The reaction involves breaking 2 C=C

bonds   and 2 O $-$ H bonds and making 2 C $-$ H bonds, 2 C $-$ C bonds and 2 C $-$ O bonds. Thus,

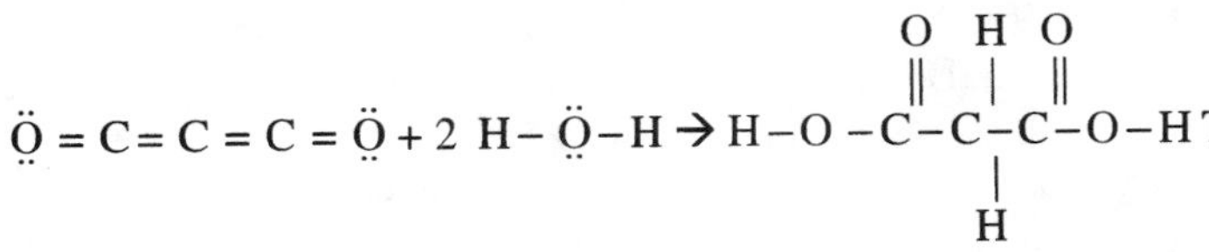

$$\Delta H^0_{rxn} = \sum \Delta H's(bonds\,broken) - \sum \Delta H's(bonds\,formed)$$

$$= [2(611\text{ kJ}) + 2(464\text{ kJ})] + [2(-414\text{ kJ}) + 2(-347\text{ kJ}) + 2(-360\text{ kJ})] = [2150\text{ kJ}] + [-2242\text{ kJ}] =$$

$$= -92\text{ kJ}$$

95. a) K_b for the overall reaction will be the product of the two individual K_b's, so $K_b = K_{b1} \cdot K_{b2} = (8.5 \times 10^{-7})(8.9 \times 10^{-16}) = 7.6 \times 10^{-22}$.

b) $K_{a1} = K_w / K_{b1} = (1.0 \times 10^{-14})/(8.5 \times 10^{-7}) = 1.2 \times 10^{-8}$.

c) **Given:** pH = 8.5, 0.012 mol N_2H_4 in 1.0 L **Find:** $[N_2H_4]$, $[N_2H_5^+]$, and $[N_2H_6^+]$
Other: $K_{b1} = 8.5 \times 10^{-7}$, $K_{b2} = 8.9 \times 10^{-16}$
Conceptual Plan:
pH → pOH → [OH⁻] and mol N_2H_4, L → M N_2H_4 then [OH⁻], $[N_2H_4]_0$ → $[N_2H_4]$, $[N_2H_5^+]$

$pH + pOH = 14 \quad pOH = -\log[OH^-]$ $\qquad M = \frac{mol}{L}$ $\qquad$ ICE *Chart*

[OH⁻], $[N_2H_4]_0$ → $[N_2H_5^+]$, $[N_2H_6^{2+}]$
ICE Chart

Solution: $pH + pOH = 14$ so pOH = 14 – pH = 14.0 – 8.5 = 5.5 then
$[OH^-] = 10^{-[OH^-]} = 10^{-5.5} = 3.16 \times 10^{-6}$.
There are 0.012 mol N_2H_4 in 1.0 L so $[N_2H_4]_0 = 0.012$ M. Set up first ICE Chart.

	$N_2H_4(aq)$ + $H_2O(l)$ ⇌	$N_2H_5^+(aq)$ +	$OH^-(aq)$
Initial	0.012	0.00	3.16×10^{-6}
Change	$-x$	x	
Equil	$0.012 - x$	x	3.16×10^{-6}

Since the solution is buffered, [OH⁻] will be unchanged.

$$K_{b1} = \frac{[N_2H_5^+][OH^-]}{[N_2H_4]} = \frac{(x)(3.16 \times 10^{-6})}{0.012 - x} = 8.5 \times 10^{-7} \text{ Solve for x.} \quad (x)(3.16 \times 10^{-6}) = 8.5 \times 10^{-7}(0.012 - x) \rightarrow$$

$x(3.16 \times 10^{-6} + 8.5 \times 10^{-7}) = 1.02 \times 10^{-8} \rightarrow x = 0.0025$ so $[N_2H_4] = 0.012\text{ M} - 0.0025\text{ M} = 0.009$ M and $[N_2H_5^+] = 0.0025$ M. then Set up second ICE Chart.

	$N_2H_5^+(aq)$ + $H_2O(l)$ ⇌	$N_2H_6^{2+}(aq)$ +	$OH^-(aq)$
Initial	0.0025	0.00	3.16×10^{-6}
Change	$-x$	x	
Equil	$0.0025 - x$	x	3.16×10^{-6}

Since the solution is buffered, [OH⁻] will be unchanged.

$$K_{b2} = \frac{[N_2H_6^{2+}][OH^-]}{[N_2H_5^+]} = \frac{(x)(3.16 \times 10^{-6})}{0.0025 - x} = 8.9 \times 10^{-16} \text{ Since } x \ll 0.0025 \text{ then } \frac{(x)(3.16 \times 10^{-6})}{0.0025} = 8.9 \times 10^{-16}$$

and
$x = 7.0 \times 10^{-13}$ M= $[N_2H_6^{2+}]$ (Note that assumption that x was very small was good.)
Check: The units (M) are correct. The magnitude of the answer makes physical sense because the only a portion of the hydrazine dissociates and the product concentrations get smaller and smaller.

97. Fine particles of activated charcoal have a much greater total surface area than does a charcoal briquette, allowing it to interact with more impurities.

99. Nitrogen can either gain three electrons or lose five electrons to obtain an octet, giving it the possibility of oxidation states between – 3 and + 5.

101. H_2S has a different bond angle than H_2O because S is larger than O and since S is less electronegative than O. H_2S is more reactive because the O – H bond is much shorter and stronger than the S – H bond.

103. Fluorine has a greater electronegativity than oxygen and is more reactive.

105. Cl is small enough to fit around Br, but Br is too large to fit around Cl and form stable bonds.

Chapter 23
Metals and Metallurgy

1. The source of vanadium in some crude oil is the very animals from which the oil was formed. It appears that some extinct animals used vanadium for oxygen transport.

3. Even though over 2 % of the Earth's mass is nickel, most of this is in the Earth's core, and because the core is so far from the surface, it is not accessible.

5. The mineral is the phase that contains the desired element(s) and the gangue is the undesired material. These are generally separated by physical methods.

7. Sodium cyanide has been traditionally used to leach gold by forming a soluble gold complex ($Au(CN)_2^-$). The problem with this process is that cyanide is very toxic and has often resulted in the contamination of streams and rivers with cyanide. New alternatives, using the thiosulfate ion ($S_2O_3^-$), are being investigated to replace it.

9. The body-centered cubic unit cell consists of a cube with one atom at each corner and one atom in the very center of the cube. In the body-centered unit cell, the atoms do not touch along each edge of the cube, but rather touch along the diagonal line that runs from one corner, through the middle of the cube, to the opposite corner. The body-centered unit cell contains two atoms per unit cell because the center atom is not shared with any other neighboring cells. The coordination number of the body-centered cubic unit cell is 8, and the packing efficiency is 68%. The face-centered cubic unit cell is characterized by a cube with one atom at each corner and one atom in the center of each cube face. In the face-centered unit cell (like the body-centered unit cell), the atoms do not touch along each edge of the cube. Instead, the atoms touch along the face diagonal. The face-centered unit cell contains four atoms per unit cell because the center atoms on each of the six faces are shared between two unit cells. So there are ½ x 6 = 3 face-centered atoms plus 1/8 x 8 = 1 corner atoms, for a total of four atoms per unit cell. The coordination number of the face-centered cubic unit cell is 12 and its packing efficiency is 74%. In the face-centered structure, any one atom strongly interacts with more atoms than in either the simple cubic unit cell or the body-centered cubic unit cell.

11. Copper was one of the first metals used because it can be found in nature in its elemental form.

13. Bronze is an alloy of copper and tin. Brass is an alloy of copper and zinc.

15. Metals are typically opaque, are good conductors of heat and electricity, and are ductile and malleable, meaning that they can be drawn into wires and flattened into sheets.

17. Aluminum, iron, calcium, magnesium, sodium and potassium are all over 1 % of the Earth's crust.

19. Hematite (Fe_2O_3) and magnetite (Fe_3O_4) are important mineral sources of iron. Cinnabar (HgS) is an important mineral source of mercury. Vanadite [$Pb_5(VO_4)Cl$], carnotite [$K_2(UO_2)_2(VO_4)_2 \cdot 3H_2O$] are important mineral sources of vanadium. Columbite [$Fe(NbO_3)_2$] is an important mineral source of niobium.

21. $MgCO_3\ (s) \xrightarrow{\text{Heat}} MgO\ (s) + CO_2\ (g)$ and $Mg(OH)_2\ (s) \xrightarrow{\text{Heat}} MgO\ (s) + H_2O\ (g)$.

23. A flux is a material that will react with the gangue to form a substance with a low melting point. MgO is the flux.

25. Hydrometallurgy is used to separate metals from ores by selectively dissolving the metal in a solution, filtering out impurities, and then reducing the metal to its elemental form.

27. The Bayer process is a hydrometallurgical process by which Al_2O_3 is selectively dissolved, leaving other oxides as solids. The soluble form of aluminum is $Al(OH)_4^-$.

29. Sponge powdered iron contains many small holes in the iron particles due to escaping of oxygen when the iron is reduced. Water atomized powdered iron has much more smooth and dense particles as the powder is formed from molten iron.

31. a) When one-half of the V atoms are replaced by Cr atoms, the composition will be 50 % Cr by moles and 50 % V by moles. To get percent by mass assume 100 moles of alloy, so

$$\text{percent by mass Cr} = \frac{(\text{mol \% Cr})(52.00 \text{ g/mol})}{(\text{mol \% Cr})(52.00 \text{ g/mol}) + (\text{mol \% V})(50.94 \text{ g/mol})} \times 100\ \% =$$

$$= \frac{(50 \cancel{\text{mol}})(52.00 \cancel{\text{g}}/\cancel{\text{mol}})}{(50 \cancel{\text{mol}})(52.00 \cancel{\text{g}}/\cancel{\text{mol}}) + (50 \cancel{\text{mol}})(50.94 \cancel{\text{g}}/\cancel{\text{mol}})} \times 100\ \% = 50.5\ \% \text{ Cr by mass}$$ and percent by mass V = 100.0 % - 50.5 % = 49.5 % V.

b) When one-fourth of the V atoms are replaced by Fe atoms, the composition will be 25 % Fe by moles and 75 % V by moles. To get percent by mass assume 100 moles of alloy, so

$$\text{percent by mass Fe} = \frac{(\text{mol \% Fe})(55.85 \text{ g/mol})}{(\text{mol \% Fe})(55.85 \text{ g/mol}) + (\text{mol \% V})(50.94 \text{ g/mol})} \times 100\ \% =$$

$$= \frac{(25 \cancel{\text{mol}})(55.85 \cancel{\text{g}}/\cancel{\text{mol}})}{(25 \cancel{\text{mol}})(55.85 \cancel{\text{g}}/\cancel{\text{mol}}) + (75 \cancel{\text{mol}})(50.94 \cancel{\text{g}}/\cancel{\text{mol}})} \times 100\ \% = 26.8\ \% \text{ Fe by mass}$$ and percent by mass V = 100.0 % - 26.8 % = 73.2 % V.

c) When one-fourth of the V atoms are replaced by Cr atoms and one-fourth of the V atoms are replaced by Fe atoms, the composition will be 25 % Cr by moles, 25 % Fe by moles and 50 % V by moles. To get percent by mass assume 100 moles of alloy, so

$$\text{percent by mass Cr} = \frac{(\text{mol \% Cr})(52.00 \text{ g/mol})}{(\text{mol \% Cr})(52.00 \text{ g/mol}) + (\text{mol \% Fe})(55.85 \text{ g/mol}) + (\text{mol \% V})(50.94 \text{ g/mol})} \times 100\ \% =$$

$$= \frac{(25 \cancel{\text{mol}})(52.00 \cancel{\text{g}}/\cancel{\text{mol}})}{(25 \cancel{\text{mol}})(52.00 \cancel{\text{g}}/\cancel{\text{mol}}) + (25 \cancel{\text{mol}})(55.85 \cancel{\text{g}}/\cancel{\text{mol}}) + (50 \cancel{\text{mol}})(50.94 \cancel{\text{g}}/\cancel{\text{mol}})} \times 100\ \% = 24.8\ \% \text{ Cr by mass}$$

$$\text{percent by mass Fe} = \frac{(\text{mol \% Fe})(55.85 \text{ g/mol})}{(\text{mol \% Cr})(52.00 \text{ g/mol}) + (\text{mol \% Fe})(55.85 \text{ g/mol}) + (\text{mol \% V})(50.94 \text{ g/mol})} \times 100\ \% =$$

$$= \frac{(25 \cancel{\text{mol}})(55.85 \cancel{\text{g}}/\cancel{\text{mol}})}{(25 \cancel{\text{mol}})(52.00 \cancel{\text{g}}/\cancel{\text{mol}}) + (25 \cancel{\text{mol}})(55.85 \cancel{\text{g}}/\cancel{\text{mol}}) + (50 \cancel{\text{mol}})(50.94 \cancel{\text{g}}/\cancel{\text{mol}})} \times 100\ \% = 26.6\ \% \text{ Fe by mass}$$

and percent by mass V = 100.0 % - (24.8 % + 26.6 %) = 48.6 % V.

33. Cr and Fe are very close together in the periodic table (atomic numbers 24 and 26, respectively), so there respective atomic radii are probably close enough to form an alloy. Also, both form body-centered cubic structures.

35. This phase diagram indicates that the solid and liquid phases are completely miscible. The composition is determined by determining the x-axis value at the point.

A: solid with 20 % Cr and 100 % - 20 % = 80 % Fe.

B: liquid with 50 % Cr and 100 % - 50 % = 50 % Fe.

37. This phase diagram indicates that the solid and liquid phases are not completely miscible. Single phases only exist between the pure component and the red line. The composition will be a mixture of the two structures. The composition of the two phases is determined by moving to the left and right until reaching the red lines. The x-axis value is the composition of that structure. According to the lever rule, the phase that is closer to the point is the dominant phase.

A: solid with 20 % Co and 100 % – 20 % = 80 % Cu overall composition. One phase will be the copper structure with ~ 4 % Co; and the other phase will be the Co structure with ~ 7 % Cu. According to the lever rule there will be more of the Cu structure, since point A is closer to the Cu structure phase boundary line.

B: solid single phase with Co structure and composition of 90 % Co and 100 % - 90 % = 10 % Cu.

39. Since C (77 pm) is much smaller than Fe (126 pm), it will fill interstitial holes. Since Mn (130 pm) and Si (118 pm) are close to the same size as Fe, they will substitute for Fe in the lattice.

41. a) Since there are the same number of octahedral holes as there are metal atoms in a closest packed structure and half are filled with N, the formula is Mo_2N.

b) Since there twice as many tetrahedral holes as there are metal atoms, and all of the tetrahedral holes are occupied, the formula is CrH_2.

43. a) Zn, since Sphalerite is ZnS.

b) Cu, since malachite is $Cu_2(OH)_2CO_3$.

c) Mn, since hausmannite is Mn_3O_4.

45. **Given:** Calcination of rhodochrosite ($MnCO_3$)
Find: heat of reaction
Conceptual Plan: Write reaction. then $\Delta H^0_{rxn} = \sum n_P \Delta H^0_f(products) - \sum n_R \Delta H^0_f(reactants)$

Solution: $MnCO_3\,(s) + \frac{1}{2}\,O_2\,(g) \xrightarrow{\text{Heat}} MnO_2\,(s) + CO_2\,(g)$

Reactant/Product	ΔH^0_f (kJ/mol from Appendix IIB)
$MnCO_3\ (s)$	– 894.1
$O_2\ (g)$	0.0
$MnO_2\ (s)$	– 520.0
$CO_2\ (g)$	– 393.5

Be sure to pull data for the correct formula and phase.

$$\Delta H^0_{rxn} = \sum n_P \Delta H^0_f(products) - \sum n_R \Delta H^0_f(reactants)$$
$$= [1\Delta H^0_f(MnO_2\ (s)) + 1\Delta H^0_f(CO_2\ (g))] - [1\Delta H^0_f(MnCO_3\ (s)) + 1/2\Delta H^0_f(O_2\ (g))]$$
$$= [1(-520.0\text{ kJ}) + 1(-393.5\text{ kJ})] - [1(-894.1\text{ kJ}) + 1/2(0.0\text{ kJ})]$$
$$= [-913.5\text{ kJ}] - [-894.1\text{ kJ}]$$
$$= -19.4\text{ kJ}$$

Check: The units (kJ) are correct. The answer is negative, which means that the reactions are exothermic.

47. When Cr is added to steel it reacts with oxygen in steel to prevent it from rusting. A Cr steel alloy would be used in any situation where a steel might be easily oxidized, such as when it comes in contact with water.

49. rutile = TiO_2: The composition will be 33.3 % Ti by moles and 66.7 % O by moles. To get percent by mass assume 100 moles of atoms, so

$$\text{percent by mass Ti} = \frac{(\text{mol \% Ti})(47.88\text{ g/mol})}{(\text{mol \% Ti})(47.88\text{ g/mol}) + (\text{mol \% O})(16.00\text{ g/mol})} \times 100\ \% =$$

$$= \frac{(33.3\ \cancel{\text{mol}})(47.88\ \text{g}/\cancel{\text{mol}})}{(33.3\ \cancel{\text{mol}})(47.88\ \text{g}/\cancel{\text{mol}}) + (66.7\ \cancel{\text{mol}})(16.00\ \text{g}/\cancel{\text{mol}})} \times 100\ \% = 59.9\ \%\ \text{Ti by mass}.$$

ilmenite = $FeTiO_3$: The composition will be 20.0 % Ti by moles, 20.0 % Fe by moles, and 60.0 % O by moles. To get percent by mass assume 100 moles of atoms, so

$$\text{percent by mass Ti} = \frac{(\text{mol \% Ti})(47.88\text{ g/mol})}{(\text{mol \% Ti})(47.88\text{ g/mol}) + (\text{mol \% Fe})(55.85\text{ g/mol}) + (\text{mol \% O})(16.00\text{ g/mol})} \times 100\ \% =$$

$$= \frac{(20.0 \text{ mol})(47.88 \text{ g/mol})}{(20.0 \text{ mol})(47.88 \text{ g/mol}) + (20.0 \text{ mol})(55.85 \text{ g/mol}) + (60.0 \text{ mol})(16.00 \text{ g/mol})} \times 100\ \% = 31.6\ \%\ \text{Ti by m}$$

51. Titanium must be arc-melted in an inert atmosphere because the high temperature and flow of electrons would cause the metal to oxidize in a normal atmosphere.

53. TiO_2 is the most important industrial product of titanium and is often used as a pigment in white paint.

55. The Bayer process is a hydrometallurgical process used to separate the bauxite ($Al_2O_3 \cdot n\ H_2O$) from the iron and silicon oxide with which it is usually found. In this process, the bauxite is digested with a hot concentrated aqueous NaOH solution under high pressure. The basic aluminum solution is separated from the oxide solids, and then the aluminum oxide is precipitated out of solution by neutralizing it. Calcination of the precipitate at temperatures greater than 1000 °C yields anhydrous alumina (Al_2O_3).

57. Carbaloy steel contains cobalt and tungsten.

59. **Given:** ilmenite, 2.0 x 10^4 kg of ore, 0.051 % by mass ilmenite, 87 % Fe recovery and 63 % Ti recovery
Find: g Fe and Ti recovered
Conceptual plan: ilmenite = $FeTiO_3$ kg ore → g ore → g ilmenite → g Ti possible → g Ti recovered

$$\frac{1000 \text{ g}}{1 \text{ kg}} \quad \frac{0.051 \text{ g ilmenite}}{100 \text{ g ore}} \quad \frac{1(47.88) \text{ g Ti}}{151.73 \text{ g ilmenite}} \quad \frac{63 \text{ g Ti recovered}}{100 \text{ g Ti}}$$

and g ilmenite → g Fe possible → g Fe recovered

$$\frac{1(55.85) \text{ g Fe}}{151.73 \text{ g ilmenite}} \quad \frac{87 \text{ g Fe recovered}}{100 \text{ g Fe}}$$

Solution: $2.0 \times 10^4 \text{ kg ore} \times \frac{1000 \text{ g ore}}{1 \text{ kg ore}} \times \frac{0.051 \text{ g ilmenite}}{100 \text{ g ore}} = 1.0\underline{2} \times 10^4 \text{ g ilmenite}$ then

$$1.0\underline{2} \times 10^4 \text{ g ilmenite} \times \frac{1(47.88) \text{ g Ti}}{151.73 \text{ g ilmenite}} \times \frac{63 \text{ g Ti recovered}}{100 \text{ g Ti}} = 2.0 \times 10^3 \text{ g Ti recovered} = 2.0 \text{ kg Ti recovered}$$

and

$$1.0\underline{2} \times 10^4 \text{ g ilmenite} \times \frac{1(55.85) \text{ g Fe}}{151.73 \text{ g ilmenite}} \times \frac{87 \text{ g Fe recovered}}{100 \text{ g Ti}} = 3.3 \times 10^3 \text{ g Fe recovered} = 3.3 \text{ kg Fe recovered}.$$

Check: The units (g) are correct. The magnitudes of the answers make physical sense because there is very little mineral in the ore. The amount of Fe recovered is higher than Ti recovered because the atomic weight of Fe and the percent recovery of Fe are both higher than Ti.

61. Four atoms surround a tetrahedral hole and six atoms surround an octahedral hole. The octahedral hole is larger because it is surrounded by a greater number of atoms.

63. Manganese has one more electron orbital available for bonding than chromium.

65. Ferromagnetic atoms, like paramagnetic atoms, have unpaired electrons. However, in ferromagnetic atoms, these electrons align with their spin oriented in the same direction, resulting in a permanent magnetic field.

67. **Given:** cylinder, h = 5.62 cm after pressing, r = 4.00 cm; d (before pressing) = 2.41 g/mL, d (after pressing) = 6.85 g/mL, d (solid iron) = 7.78 g/mL
Find: a) original height, b) theoretical height if d = d(solid iron), and c) % voids in component
Conceptual plan:

a) **Since $d = \frac{m}{V}$, $V = \pi r^2 h$, and the mass of iron is constant, this means that $d \ \alpha \ \frac{1}{h}$. So d (before pressing), d (after pressing), h(after pressing) → h (before pressing)**

$$\frac{d_1}{d_2} = \frac{h_2}{h_1}$$

b) d (solid iron), d (after pressing), h(after pressing) → h (solid iron)

$$\frac{d_1}{d_2} = \frac{h_2}{h_1}$$

c) Assume d (air) = 0, so d (solid iron), d (after pressing) → % voids

$$\%\ voids = \frac{d_{\text{after pressing}} - d_{\text{solid iron}}}{d_{\text{after pressing}}} \text{ x } 100\ \%$$

Solution: a) $\frac{d_1}{d_2} = \frac{h_2}{h_1}$ Rearrange to solve for h_2. $h_2 = \frac{d_1 h_1}{d_2} = \frac{6.85 \frac{\cancel{g}}{\cancel{mL}} \text{ x } 5.62 \text{ cm}}{2.41 \frac{\cancel{g}}{\cancel{mL}}} = 16.0 \text{ cm}$.

b) $\frac{d_1}{d_2} = \frac{h_2}{h_1}$ Rearrange to solve for h_2. $h_2 = \frac{d_1 h_1}{d_2} = \frac{6.85 \frac{\cancel{g}}{\cancel{mL}} \text{ x } 5.62 \text{ cm}}{7.78 \frac{\cancel{g}}{\cancel{mL}}} = 4.95 \text{ cm}$.

c) $\%\ voids = \frac{d_{\text{solid iron}} - d_{\text{after pressing}}}{d_{\text{after pressing}}} \text{ x } 100\ \% = \frac{7.78 \frac{\cancel{g}}{\cancel{mL}} - 6.85 \frac{\cancel{g}}{\cancel{mL}}}{6.85 \frac{\cancel{g}}{\cancel{mL}}} \text{ x } 100\ \% = 14\ \%$

Check: The units (cm, cm and %) are correct. The higher the density, the smaller the height, so the answer for a) is greater than 5.62 cm and the answer for b) is smaller than 5.62 cm. Since the density of the pressed component is close to the density of solid iron, the volume of voids is low.

69. Since there are the same number of octahedral holes as there are metal atoms in a closest packed structure and twice as many tetrahedral holes as there are metal atoms there are a total of three holes for each metal atom and the formula would be LaH_3. Since the formula is $LaH_{2.76}$, the percentage of holes filled is (2.76/3) x 100 % = 92.0 % filled.

71. Gold and silver are found in their elemental forms because of their low reactivity. Sodium and calcium are Group 1A and Group 2A metals, respectively, and are highly reactive as they readily lose their valence electrons to obtain octets.

Chapter 24
Transition Metals and Coordination Compounds

1. A transition metal atom forms an ion by losing the ns^2 (valence shell) electrons first.

3. The +2 oxidation state is common because most of the transition metals have two electrons occupying the ns orbitals. These electrons are lost first by the metal.

5. The electronegativity of the transition elements generally increases across a row, following the main group trend. However, in contrast to main – group trend, electronegativity increases from the first transition row to the second. There is little electronegativity difference between the second and third transition row. There is a slight increase from silver to gold. Therefore, Au is the most electronegative of the transition metals.

7. A ligand can be considered a Lewis base because it donates a pair of electrons. The transition metal ion would be a Lewis acid because it accepts the pair of electrons.

9. Cis – trans isomerism occurs in square planar complexes of the general formula MA_2B_2 or octahedral complexes of the general formula MA_4B_2.

11. Because of the spatial arrangement of the ligands, the normally degenerate d orbitals are split in energy. The difference between these split d orbitals is the crystal field splitting energy. The magnitude of the splitting depends on the particular complex. In strong – field complexes, the splitting is large, and in weak – field complexes, the splitting is small. The magnitude of the crystal field splitting depends in large part on the ligands attached to the central metal ion.

13. Zn^{2+} has a filled d subshell, while Cu^{2+} has 9 d electrons. Since Zn^{2+} has this filled d subshell, the color of the compounds will be white. Cu^{2+} has an incomplete d subshell, so different ligands will cause a different crystal field splitting and the compounds will have color.

15. Almost all tetrahedral complexes are high – spin because of reduced ligand – metal interactions. The *d* orbitals in a tetrahedral complex are interacting with only four ligands, as opposed to six in the octahedral complex, so the value of Δ is generally smaller.

17. Identify the noble gas that precedes the element and put it in square brackets.
Determine the outer principal quantum level for the *s* orbital. Subtract one to obtain the quantum level for the *d* orbital. If the element is in the third or fourth transition series, include $(n-2)f$ electrons in the configuration.
Count across the row to see how many electrons are in the neutral atom.
For an ion, remove the required number of electrons, first from the *s* and then from the *d* orbitals.

a) Ni; Ni^{2+}
The noble gas that precedes Ni is Ar, Ni is in the fourth period so the orbitals we use are 4s and 3d and Ni has 10 more electrons than Ar.
Ni $[Ar]4s^2 3d^8$
Ni will lose electrons from the 4s and then from the 3d
Ni^{2+} $[Ar]4s^0 3d^8$

b) Mn; Mn^{4+}
The noble gas that precedes Mn is Ar, Mn is in the fourth period so the orbitals we use are 4s and 3d and Mn has 7 more electrons than Ar.
Mn $[Ar]4s^2 3d^5$
Mn will lose electrons from the 4s and then from the 3d
Mn^{4+} $[Ar]4s^0 3d^3$

c) Y; Y^+
The noble gas that precedes Y is Kr, Y is in the fifth period so the orbitals we use are 5s and 4d and Y has 3 more electrons than Kr.
Y $[Kr]5s^2 4d^1$
Y will lose electrons from the 5s and then from the 4d
Y^+ $[Kr]5s^1 4d^1$

d) Ta; Ta^{2+}

The noble gas that precedes Ta is Xe, Ta is in the sixth period so the orbitals we use are 6s, 5d and 4f and Ta has 19 more electrons than Xe.

Ta $[Xe]6s^2 4f^{14} 5d^3$

Ta will lose electrons from the 6s and then from the 5d

Ta^{2+} $[Xe]6s^0 4f^{14} 5d^3$

19. a) V Highest oxidation state = +5. V = $[Ar]4s^2 3d^3$. Since V is to the left of Mn, it can lose all of the 3s and 3d electrons so the highest oxidation state is +5

b) Re Highest oxidation state = +7. Re = $[Xe]6s^2 4f^{14} 5d^5$. Re can lose all of the 6s and 5d electrons, so the highest oxidation state is +7.

c) Pd Highest oxidation state = +4. Pd = $[Kr]5s^2 4d^8$. Pd can lose the 5s and then 2 electrons from the 4d orbitals, so the highest oxidation state is +4.

21. a) $[Cr(H_2O)_6]^{3+}$ H_2O is neutral, so Cr has an oxidation state of +3, there are 6 H_2O molecules attached to each Cr, so the coordination number is 6

b) $[Co(NH_3)_3Cl_3]^-$ NH_3 is neutral and Cl has charge of – 1. The sum of the oxidation state of Co and the charge of chloride ion = 1–. x+(3(–1)) = –1, x = +2, therefore the oxidation state of Co is +2. The 3 NH_3 molecules and the 3 Cl^- ions are bound directly to the Co atom, therefore the coordination number is 6.

c) $[Cu(CN)_4]^{2-}$ CN has a charge of –1. The sum of the oxidation state of Cu and the charge of the cyanide ion = 2–. x+(4(–1)) = 2–, x = +2, therefore, the oxidation state of Cu is +2. The 4 cyanide ions are directly bound to the Cu atom, therefore, the coordination number is 4.

d) $[Ag(NH_3)_2]^+$ NH_3 is neutral, so Ag has an oxidation number of +1, there are 2 NH_3 molecules attached to each Ag atom so the coordination number is 2.

23. a) $[Cr(H_2O)_6]^{3+}$ is hexaaquachromium(III) ion.

$[Cr(H_2O)_6]^{3+}$ is a complex cation

Name the ligand: H_2O is aqua

Name the metal ion: Cr^{3+} is chromium(III)

Name the complex ion by adding the prefixes to indicate the number of each ligand, followed by the name of each ligand, followed by the name of the metal ion: hexaquachromium(III) ion.

b) $[Cu(CN)_4]^{2-}$ is tetracyanocuprate(II) ion.

$[Cu(CN)_4]^{2-}$ is a complex anion

Name the ligand: CN^- is cyano

Name the metal ion: Cu^{2+} is cuprate(II) because the complex is an anion

Name the complex ion by adding the prefix to indicate the number of each ligand, followed by the name of each ligand, followed by the name of the metal ion: tetracyanocuprate(II) ion.

c) $[Fe(NH_3)_5Br]SO_4$ is pentaamminebromoiron(III) sulfate

$[Fe(NH_3)_5Br]^{2+}$ is a complex cation, SO_4^{2-} is sulfate.

Name the ligands in alphabetical order: NH_3 is ammine, Br^{1-} is bromo

Name the metal cation: Fe^{3+} is iron(III)

Name the complex ion by adding prefixes to indicate the number of each ligand, followed by the name of the ligand, followed by the name of the metal ion: pentaamminebromoiron(III)

Name the compound by writing the name of the cation before the anion. The only space is between the ion names: pentaamminebromoiron(III) sulfate

d) $[Co(H_2O)_4(NH_3)(OH)]Cl_2$ is amminetetraaquahydroxycobalt(III) chloride

$[Co(H_2O)_4(NH_3)(OH)]^{2+}$ is a complex cation, Cl^{1-} is chloride.

Name the ligands in alphabetical order: NH_3 is ammine, H_2O is aqua, OH^- is hydroxo

Name the metal cation: Co^{3+} is cobalt(III)

Name the complex ion by adding prefixes to indicate the number of each ligand, followed by the name of the ligand, followed by the metal ion: amminetetraaquahydroxocobalt(III)

Name the compound by writing the name of the cation before the anion. The only space is between the ion names: amminetetraaquahydroxycobalt(III) chloride

25. a) hexamminechromium(III) is a complex ion with Cr^{3+} metal ion and 6 NH_3 ligands. $[Cr(NH_3)_6]^{3+}$

b) potassium hexacyanoferrate(III) is a compound with a $3K^+$ cations and a complex anion with Fe^{3+} metal ion and 6 CN^- ligands. $K_3[Fe(CN)_6]$

c) ethylenediaminedithiocyanatocopper(II) is a compound with a Cu^{2+} metal ion, a ethylenediamine ligand and two SCN^- ligands. $[Cu(en)(SCN)_2]$

d) tetraaquaplatinum(II) hexachloroplatinate(IV) is a complex compound with a complex cation that contains a Pt^{2+} metal ion, and 4 H_2O ligands and a complex anion that contains a Pt^{4+} metal ion and 6 Cl^- ligands. $[Pt(H_2O)_4][PtCl_6]$

27. a) $[Co(NH_3)_3(CN)_3]$ is triamminetricyanocobalt(III)

b) Since ethylenediamine is a bidentate ligand, you need three to have a coordination number of six. $[Cr(en)_3]^{3+}$ is tris(ethylenediamine)chromium(III) ion

29. In linkage isomers the ligand coordinates to the metal in different ways.

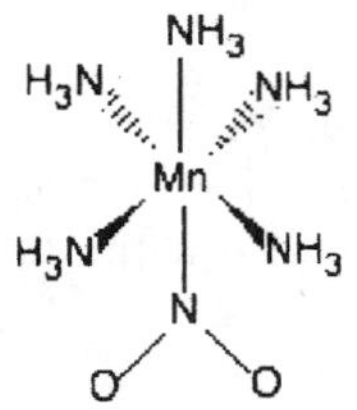

pentaamminenitromanganese(III) ion

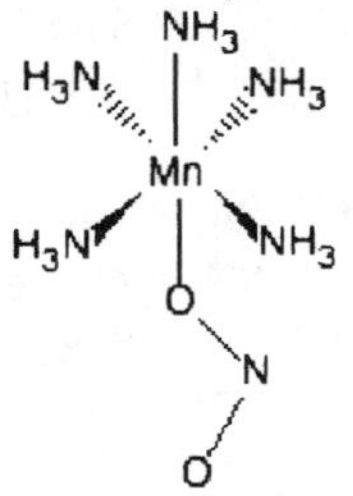

pentaamminenitritomanganese(III) ion

31. Coordination isomers occur when a coordinated ligand exchanges places with the uncoordinated counterion.
$[Fe(H_2O)_5Cl]Cl \cdot H_2O$ pentaaquachloroiron(II) chloride monohydrate

$[Fe(H_2O)_4Cl_2] \cdot 2H_2O$ tetraaquadichloroiron(II) dihydrate

33. Geometric isomers result when the ligands bonded to the metal have a different spatial arrangement.
a) No, an octahedral complex has to have at least two different ligands to have geometric isomers.

b) Yes, there will be cis – trans isomers.

c) Yes, there will be fac – mer isomers.

d) No, a square planar complex has to have at least two different ligands to have geometric isomers.

e) Yes, there will be cis – trans isomers.

35. a) Square – planar $[NiWXYZ]^{2+}$ would have 3 geometric isomers.

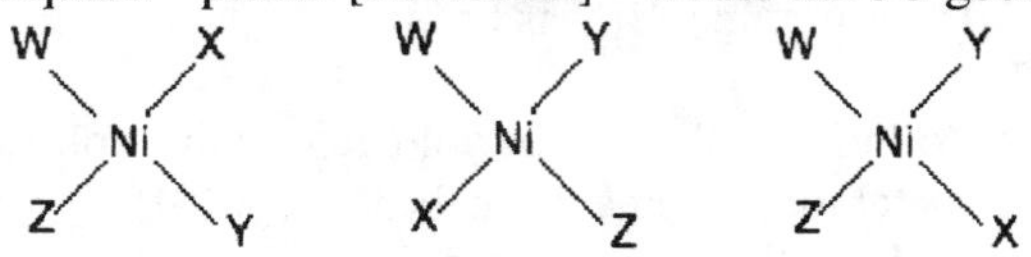

b) Tetrahedral $[ZnWXYZ]^{2+}$ would have 2 geometric isomers.

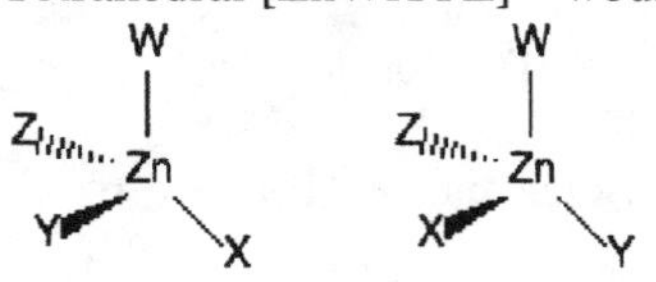

37. a) $[Cr(CO)_3(NH_3)_3]^{3+}$ has a coordination number of 6, and is octahedral. There will be fac and mer isomers and no optical isomers because rotation of the mirror images are superimposable upon each other.

fac

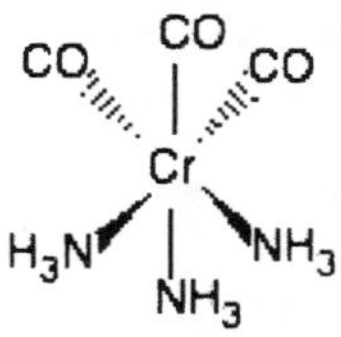

mer

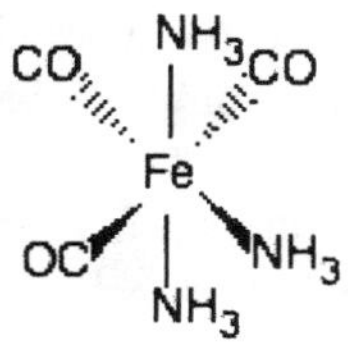

b) $[Pd(CO)_2(H_2O)Cl]^+$ has a coordination number of 4, and is d^8 complex so it is square – planar. There will be cis and trans isomers. There will be no optical isomers because rotation of the mirror images are superimposable upon each other.

cis

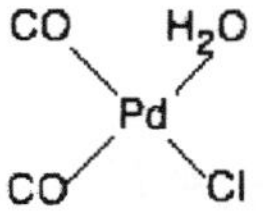

trans

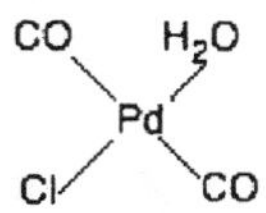

39. $[Cr(CO)_2(ox)_2]^-$ has a coordination number of 6 and is octahedral. There will be cis and trans isomers. The cis isomer has a mirror image that is nonsuperimposable.

trans

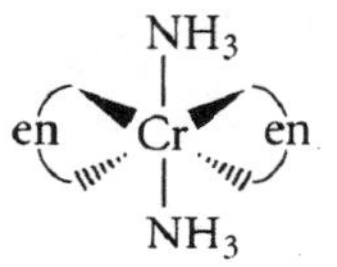

cis

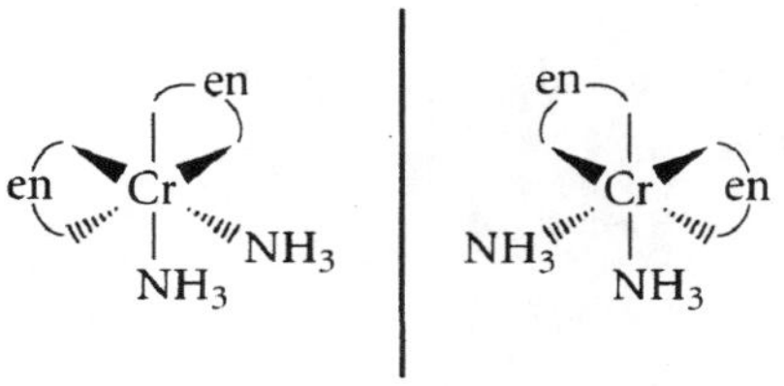

41. a) Zn^{2+} d^{10}

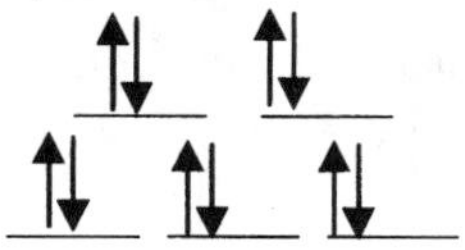

b) Fe^{3+} d^5 high spin

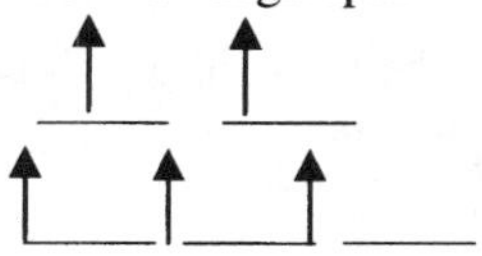

low spin

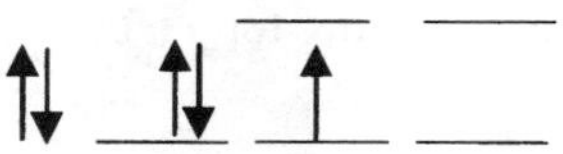

c) V^{3+} d^2

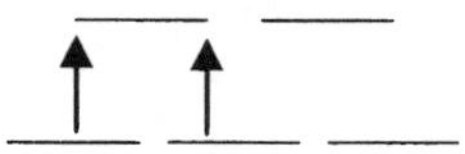

d) Co^{2+} d^7 high spin

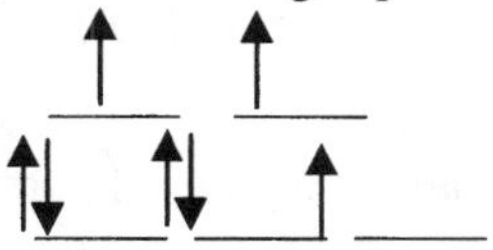

43. **Given:** $[CrCl_6]^{3-}$; λ = 735 absorbance maximum **Find:** crystal field energy

Conceptual Plan: $\boldsymbol{\lambda \rightarrow \Delta\ (J) \rightarrow}$ **kJ/ion** $\boldsymbol{\rightarrow}$ **kJ/mol**

$$\Delta = \frac{hc}{\lambda} \quad \frac{kJ}{1000\ J} \quad \frac{6.02 \times 10^{23}\ \text{ions}}{\text{mol}}$$

Solution:
$$\frac{(6.626 \times 10^{-34}\ J \cdot s)(3.00 \times 10^{8}\ \frac{m}{s})}{(735\ nm)\left(\frac{1 \times 10^{-9}\ m}{nm}\right)} = 2.7\underline{0}4 \times 10^{-19}\ \text{J/ion}$$

$$2.7\underline{0}4 \times 10^{-19}\ \frac{J}{\text{ion}} \times \frac{6.02 \times 10^{23}\ \text{ions}}{\text{mol}} \times \frac{kJ}{1000\ J} = 163\ \text{kJ/mol}$$

Check: Cl is a weak field ligand and would be expected to have a relatively small Δ which is consistent with a value of 163 kJ.

45. The crystal field ligand strength would be $CN^- > NH_3 > F^-$. The smaller the wavelength, the larger the energy, the greater the crystal field splitting that would be observed. So $[Co(CN)_6]^{3-}$ would have a smaller wavelength than $[Co(NH_3)_6]^{3+}$ which would have a smaller wavelength than $[CoF_6]^{3-}$.
So: $[Co(CN)_6]^{3-}$ = 290 nm which absorbs in the UV, and would have a colorless solution.
$[Co(NH_3)_6]^{3+}$ = 440 nm which absorbs in the blue and would have a yellow solution.
$[CoF_6]^{3-}$ = 770 nm which absorbs in the red and would have a green solution.

47. Mn^{2+} is d^5 and there are 5 unpaired electrons so the crystal field splitting energy, Δ, is small compared to the energy to pair the electrons. Therefore, NH_3 induces a weak field with Mn^{2+}.

49. a) $[RhCl_6]^{3-}$ Rh^{3+} d^6 Cl^- is a weak field ligand, so the value of Δ will be small, so there are 4 unpaired electrons.

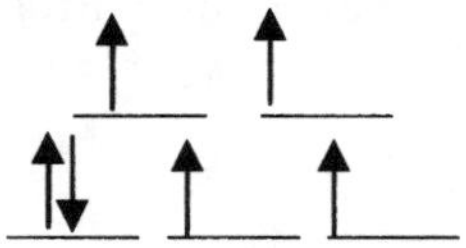

b) $[Co(OH)_6]^{4-}$ Co^{2+} d^7 OH^- is a weak field ligand, so the value of Δ will be small, so there are 3 unpaired electrons.

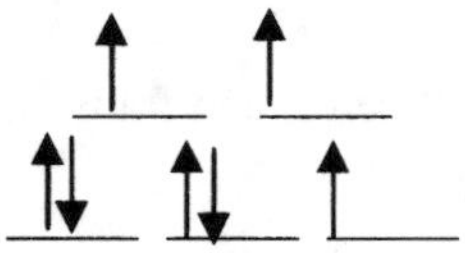

c) cis - $[Fe(en)(NO_2)_2]^+$ Fe^{3+} d^5 en and NO_2^- are strong field ligands, so the value of Δ will be large, so there is 1 unpaired electrons.

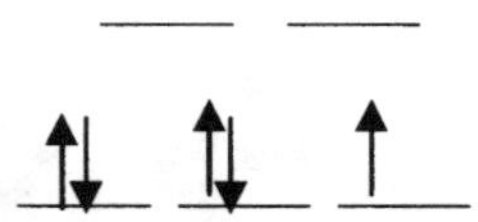

51. $[CoCl_4]^{2-}$ Co^{2+} is d^7, Cl^- is a weak field ligand, so the value of Δ will be small and there will be 3 unpaired electrons. The crystal field splitting for a tetrahedral structure is:

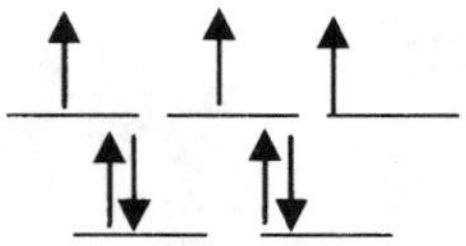

53. Hemoglobin, cytochrome c, and chlorlphyll all contain a porphyrin ligand.

55. Oxyhemoglobin is low spin and a red color. Since it is low spin, the crystal field splitting energy must be large. The red color means that the complex absorbs in the green region (~500 nm) which also indicates a large crystal field splitting energy, therefore, O_2 must be a strong field ligand.
Deoxyhemoglobin is high spin and a blue color. Since it is high spin, the crystal field splitting energy must be small. The blue color means that the complex absorbs in the orange region (~600 nm) which also indicates a small crystal field splitting energy. Both of these are consistent with H_2O as a weak field ligand.

57. a)

Cr	$[Ar]4s^13d^5$
Cr^+	$[Ar]4s^03d^5$
Cr^{2+}	$[Ar]4s^03d^4$
Cr^{3+}	$[Ar]4s^03d^3$

b)

Cu	$[Ar]4s^13d^{10}$
Cu^+	$[Ar]4s^03d^{10}$
Cu^{2+}	$[Ar]4s^03d^9$

$4d^8$

59. a)

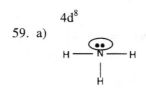

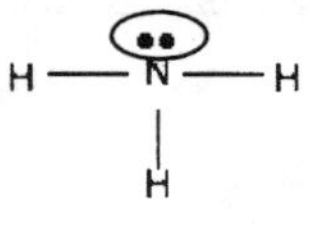

b)

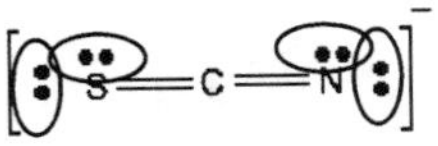 ligand can bond from either end.

c)

61. An octahedral complex has 6 ligands.
$MA_2B_2C_2$ will have cis/trans isomers: all cis; A trans, B,C cis; B trans, A,C cis; C trans, A,B cis; all trans
MAB_2C_3 will have fac/mer isomers
MA_2B_3C will have fac/mer isomers
MAB_3C_2 will have fac/mer isomers
MA_3B_2C will have fac/mer isomers
MA_2BC_3 will have fac.mer isomers
MA_3BC_2 will have fac/mer isomers
$MABC_4$ will have AB cis and trans
MAB_4C will AC have cis and trans
MA_4BC will BC have cis and trans

63. $[Fe(ox)_3]^{3-}$ Fe^{3+} d^5 coordination number = 6, octahedral, structure has a nonsuperimposable mirror image.

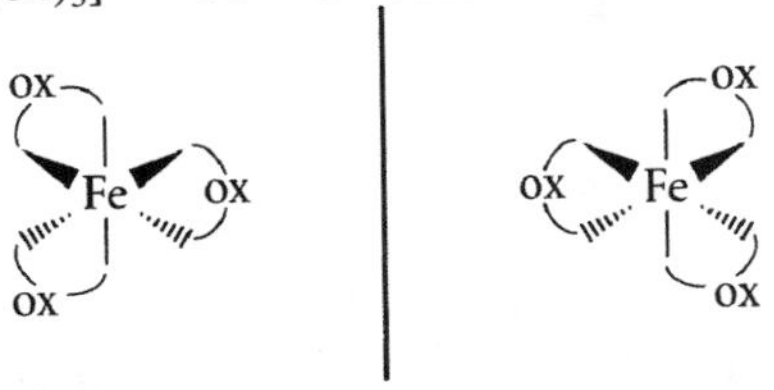

65. $[Mn(CN)_6]^{3-}$ Mn^{3+} d^4 has a coordination number of 6 and is octahedral. CN^- is a strong field ligand and will cause a large crystal field splitting energy so the ion is low spin.

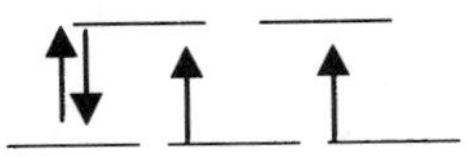 The ion will be paramagnetic with 2 unpaired electrons.

67. **Given:** 46.7% Pt, 17.0% Cl, 14.8% P, 17.2% C, 4.34% H; **Find:** formula, structures, and names for both compounds.
Conceptual Plan: % compostion → pseudoformula → formula

$$n = \frac{g}{\text{molar mass}} \quad \text{divide by smallest}$$

Solution: $46.7 \text{ g Pt} \times \frac{1 \text{ mol Pt}}{195.1 \text{ g}} = 0.23936 \text{ mol Pt}$ $\quad 17.0 \text{ g Cl} \times \frac{1 \text{ mol Cl}}{35.45 \text{ g Cl}} = 0.47955 \text{ mol Cl}$

$14.8 \text{ g P} \times \frac{1 \text{ mol P}}{30.97 \text{ g}} = 0.47788 \text{ mol P}$ $\quad 17.2 \text{ g C} \times \frac{1 \text{ mol C}}{12.01 \text{ g C}} = 1.4321 \text{ mol C}$

$4.34 \text{ g H} \times \frac{1 \text{ mol H}}{1.008 \text{ g}} = 4.3056 \text{ mol H}$

$Pt_{0.23936}Cl_{0.47955}P_{0.47788}C_{1.4321}H_{4.3056}$

$Pt_{\frac{0.23936}{0.23936}}Cl_{\frac{0.47955}{0.23936}}P_{\frac{0.47788}{0.23936}}C_{\frac{1.4321}{0.23936}}H_{\frac{4.3056}{0.23936}}$

$PtCl_2P_2C_6H_{18}$

$[Pt(P(CH_3)_3)_2Cl_2]$

cis – dichlorobis(trimethylphosphine)platinum(II) trans – dichlorobis(trimethylphosphine)platinum(II)

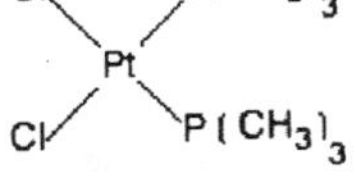

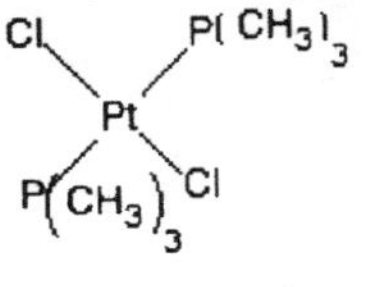

69.

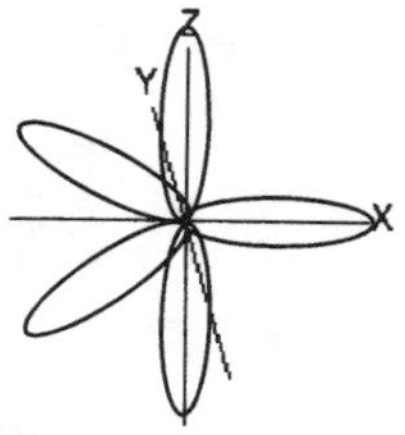

The trigonal bipyramidal complex ion has lobes along the x axis and between the x and y axes and along the z axis.. So, the ligands will interact most strongly with z^2 orbital and then with the $x^2 - y^2$ and the xy orbitals and will not interact with the xz , or the yz orbitals. So the crystal field splitting would look like:

_______ z^2

_______ $x^2 - y^2$ _______ xy

_______ xz _______ yz

71\. a) **Given:** $K_{sp}(NiS) = 3 \times 10^{-16}$ **Find:** Solubility in water

Conceptual plan: write the reaction, prepare an ICE table, substitute into the equilibrium expression and solve for S(molar solubility).

Solution: $NiS(s) \rightleftharpoons Ni^{2+}(aq) + S^{2-}(aq)$

	Ni^{2+}	S^{2-}
I	0.0	0.0
C	S	S
E	S	S

$K_{sp} = [Ni^{2+}][S^{2-}]$ $\quad 3 \times 10^{-16} = S^2$

$S = 1.7 \times 10^{-8} = 2 \times 10^{-8}$

b) **Given:** $K_{sp}(NiS) = 3 \times 10^{-16}$ $K_f\,[Ni(NH_3)_6]^{2+} = 2.0 \times 10^8$ **Find:** Solubility in 3.0 M NH_3

Conceptual Plan: Sum the reaction, prepare an ICE table, substitute into the equilibrium expression and solve for S(molar solubility).

Solution:

Reaction 1: $NiS(s) \rightleftharpoons Ni^{2+}(aq) + S^{2-}(aq)$ $\quad K_{sp}(NiS) = 3 \times 10^{-16}$

Reaction 2: $Ni^{2+}(aq) + 6NH_3 \rightleftharpoons Ni(NH_3)_6^{2+}(aq)$ $\quad K_f = 2.0 \times 10^8$

Reaction 3: $NiS(s) + 6NH_3 \rightleftharpoons Ni(NH_3)_6^{2+} + S^{2-}(aq)$ $\quad K = K_{sp}\,K_f = 6.0 \times 10^{-8}$

	NH_3	$Ni(NH_3)_6^{2+}$	S^{2-}
I	3.0 M	0.0	0.0
C	– 6S	S	S
E	3.0 – 6S	S	S

$$K = \frac{[Ni(NH_3)_6]^{2+}[S^{2-}]}{[NH_3]^6} \qquad 6.0 \times 10^{-8} = \frac{(S)(S)}{(3.0 - S)^6}$$

Assume S is small compared to 3.0

$S^2 = 4.37 \times 10^{-5}$

$S = 6.6 \times 10^{-3}$

c) NiS is more soluble in ammonia because the formation of the $[Ni(NH_3)_6]^{2+}$ complex ion is highly favorable. The formation removes Ni^{2+} ion from the solution, causing more of the NiS to dissolve.

73\. Au would have the higher ionization energy. Ionization energy increases as you go down a group in the transition metals. Because there is a large increase in the number of protons and not a large increase in size, the ionization energy increases.